Air Quality Management

Eric Taylor • Ann McMillan
Editors

Air Quality Management

Canadian Perspectives on a Global Issue

Springer

Editors
Eric Taylor
Environmental Standards Branch
British Columbia Ministry of Environment
Victoria
British Columbia
Canada

Ann McMillan
Storm Consulting
Stittsville
Ontario
Canada

ISBN 978-94-007-7556-5 ISBN 978-94-007-7557-2 (eBook)
DOI 10.1007/978-94-007-7557-2
Springer Dordrecht Heidelberg New York London

Library of Congress Control Number: 2013947600

Printed on acid-free paper

Springer is part of Springer Science+Business Media (www.springer.com)

This book is dedicated to two Canadian air quality champions, Ken Maybee and David Pengelly, who each in their own unique way worked tirelessly to raise awareness of the health risks of air pollution and to improve air quality in Canada through the application of the latest science to air quality management.

Preface

A World Clean Air Congress of the International Union of Air Pollution Prevention and Environmental Protection Associations (IUAPPA) was held in Vancouver, British Columbia in September 2010, and contained a number of sessions on air quality management. This book originated in a session that Eric Taylor organized on Canadian Air Quality Management. Seven people presented information at this session on topics ranging from a history of air quality management in Canada to the fledgling Air Quality Health Index. Dr. Alan Gertler, technical chair of the World Clean Air Congress, attended this session and recommended that its seven presentations be brought together into a book. This was the spark that ignited this project.

The scope and depth of the original idea has since mushroomed, with the book now having a much expanded scope. Of the original seven presenters, only three ended up contributing a chapter to this book. As we discussed the project, we soon realized that Canadian approaches to air quality management are unique. This offered an opportunity for a team of authors to document the Canadian approach with all its quirks and promise. As a result, the final tally of chapters has blossomed to twenty and the number of authors to forty-three. We wholeheartedly thank these authors for their enthusiasm and generosity in contributing to this book.

One of the key principles that Alan Gertler stressed in sending us on this long and unfamiliar journey was that the book should be written in a way that would help developing countries in planning an air quality program. It was this approach that persuaded us to take on this project. It has shaped the style and content of the book, and required us to make it readable by a wide audience. Hopefully the result of this long project has gone some distance in achieving Alan's vision.

May 2013

Eric Taylor
Ann McMillan

Contents

Contributors

Tyler Abel Levelton Consultants Ltd., Richmond, Canada

Randolph P. Angle R. Angle Consulting, Edmonton, Alberta, Canada

Wendy Bailey Transport Canada, Ottawa, Canada

D. Laurie Bates-Frymel Metro Vancouver, Canada

S. Bittman Pacific Agrifood Research Centre, Agassiz, British Columbia, Canada

Albert Bleeker Energy Research Centre of the Netherlands, Heerhugowaard, Netherlands

Jeffrey R. Brook Environment Canada, Toronto, Canada

T. W. Bruulsema Northeast Region, North American Program, International Plant Nutrition Institute, Ontario, Canada

Brian Bukoski Environment Canada, Winnipeg, Canada

Jean Pierre Charland Environment Canada, Ottawa, Canada

M. Cournoyer Groupe Conseil UDA Inc., St. Jean, Quebec, Canada

Tom F. Dann RS Environmental, Ottawa, Canada

Ashu Dastoor Environment Canada, Montreal, Canada

Peter Eggleton TELLIGENCE Group Enr., St-Lambert, Quebec, Canada

G. Foley US EPA, Washington DC, USA

Rebecca Freedman BC Ministry of Environment, Victoria BC, Canada

Elisabeth Galarneau Environment Canada, Toronto, Canada

Monique Gilbert Ville de Montréal, Canada

Andrew Green Lynnwood, Washington, U.S.A.

Dennis Herod Environment Canada, Ottawa, Canada

Bob Humphries Levelton Consultants Ltd., Richmond, Canada

Deniz Karman Carleton University, Ottawa, Canada

Judi Krzyzanowski Krzyzanowski Consulting, Stirling, Canada

Ling Liu Environmental Health Science and Research Bureau, Health Canada, Ottawa, ON, Canada

D. I. Massé Dairy and Swine Research and Development Centre, Quebec, Canada

Warren McCormick BC Ministry of Environment, Victoria, BC, Canada

Ann McMillan Storm Consulting, Ottawa, Canada

Sonia Melancon Ville de Montréal, Canada

Jean O. Melious Western Washington University, Bellingham, United States

Michael D. Moran Environment Canada, Gatineau, Canada

Gilles Morneau Environment Canada, Gatineau, Canada

A. Narjoux Odotech Inc., Montreal, Canada

E. Pattey Eastern Cereal and Oilseed Research Centre, Ottawa, Canada

Ed Piché Environmental Business Consultant, Toronto, Canada

G. Qiu Ontario Ministry of Environment, Ontario, Canada

Greg Rideout Environment Canada, Ottawa, Canada

Steven Sakiyama BC Ministry of Environment Victoria, British Columbia, Canada

S. C. Sheppard ECOMatters, Inc., Pinawa, Manitoba, Canada

Sharon Stevens Air Shift Group, Kamloops, Canada

David M. Stieb Environmental Health Science and Research Bureau, Health Canada, Burnaby, BC, Canada

Natalie Suzuki BC Ministry of Environment, Victoria BC, Canada

Eric Taylor British Columbia Ministry of Environment, Victoria, BC, Canada

James Young SENES Consultants, Richmond Hill, Canada

A. Van der Zaag Eastern Cereal and Oilseed Research Centre, Ottawa, Canada

Norm Zirnhelt Cariboo Environmental Quality Consulting Ltd., Williams Lake, Canada

Introduction

1

Steven Sakiyama and Randolph P. Angle

Abstract

Air quality management in Canada is comprised of several inter-related processes and tools that are introduced in this chapter, thus providing the foundation for a more detailed treatment of these aspects in the following chapters. The Canadian approach involves an adaptive process that starts with problem recognition and then proceeds through goal setting, plan design, plan implementation, tracking of progress and if necessary, making adjustments to achieve the goals of the program. Stakeholders are involved throughout the process using a variety of informational, consultative and collaborative techniques. The entire management process is supported by scientific and technical knowledge gained from air quality monitoring, emission inventories, air quality modelling, socio-economic analysis, air quality research and technological development. Implementation of air quality management plans relies on prescriptive regulations, social marketing and outcome regulations that may employ economic instruments. The mix of tools depends on the nature of the problem, its spatial scale, and jurisdictional preferences.

Keywords

Air quality management · Air quality scales · Air quality goals · Air quality management tools · Air quality monitoring · Air quality modelling · Emission reductions · Stakeholder engagement · Air quality plan implementation · Air quality research and development

1.1 Managing Air Quality

Air quality management in Canada has rapidly evolved from the control of a few pollutants emitted from industrial stacks to a complicated web of management approaches that address a host of new, inter-related air quality issues. These issues can be local to global in nature, have time scales that range from seconds to decades, and involve multiple pollutants emitted by a wide variety of sources. Some of these pollutants can transform into other pollutants, posing challenges in identifying and controlling the emissions from an array of contributing sources.

The development of air quality management has been driven by the accumulation of scientific evidence on the negative effects of air pollution on the natural environment and human health. For example, hundreds of peer-reviewed studies from around the world over the last several decades show that exposure to outdoor air pollutants not only poses a risk to human health, but the very young, the very old and those with chronic cardio-pulmonary disease are more vulnerable to its harmful effects. In addition, there is emergening evidence that these risks occur at lower air pollutant concentrations than previously thought. This awareness is pushing Canadian regulatory agencies to develop more comprehensive management approaches in order to acheive more stringent air quality goals.

This book is built on the collective wisdom and experiences of Canadian and American professionals who have

S. Sakiyama (✉)
BC Ministry of Environment Victoria, British Columbia, Canada
e-mail: steve.sakiyama@gov.bc.ca

R. P. Angle
R. Angle Consulting, Edmonton, Alberta, Canada
e-mail: rangle2009@gmail.com

E. Taylor, A. McMillan (eds.), *Air Quality Management,* DOI 10.1007/978-94-007-7557-2_1,

worked for many years in their respective fields to improve air quality in Canada. It describes our present understanding of the impacts of air pollutants on health and visibility, the state of knowledge of pollutant concentrations across Canada and how they move, disperse and transform in the atmosphere. It also assesses how Canada has developed management systems to reduce the environmental and public health risk of air pollutants in a wide variety of sectors and details how emissions from important sectors are managed. Although there is considerable diversity in management approaches across Canada, the book will feature many common elements that are employed across different jurisdictional, geographic and source sectors.

This Introductory chapter provides an overview of the process and tools used for air quality management in Canada and provides some Canadian examples. As such, it provides the ground work for the following chapters that cover each of the topics introduced here in-depth and provides the framework to link the pieces to the larger process that is set to acheive goals.

Air Quality Management is defined as the organization and coordination of human activities for the achievement of air quality goals. Air quality goals are end results for the relative amounts of substances in the atmosphere or the conditions of valued biological and physical resources affected by those substances. Goals are usually set for some undetermined time in the future, with a number of observable and measureable intermediate objectives to be achieved within fixed timeframes.

In general the goals of Air Quality Management in Canada are twofold (Mennell and Bhattacharrya 2002):

- Protect human health and the environment
- Clean "dirty areas" and keep "clean areas clean"

Under these general goals, air quality programs and air quality management systems within various jurisdictions often have more specific goals:

- Minimize, reduce or prevent emissions
- Meet ambient air quality standards
- Maintain good visibility
- Reduce acid deposition to reverse ecosystem damage (e.g. Canada-Wide Acid Rain Strategy)
- Reduce the risk of adverse health effects (e.g. Canada-Wide Standards for Particulate Matter and Ozone)

These goals are typically translated into standards for ambient air quality (or a related parameter) and standards for emissions from industrial processes, common equipment or consumer products. These will be discussed in a later section. Air quality goals may also be part of larger management plans for energy, sustainability, climate change, land use, or community involvement.

Although there are differences in geographic extent, pollutant focus and control emphasis, the most successful air management programs would be described as:

- Multi-pollutant, considering several pollutants together
- Optimized, ensuring decisions fair, just and cost-effective
- Co-ordinated, seeking local, regional and global benefits at the same time
- Science-informed, making use of the best available scientific knowledge
- Polluter-targeted, costs being borne by those responsible for the emissions
- Flexible, allowing for uncertainties and mid-course revisions
- Transparent and understandable, providing clarity to stakeholders
- Supported by stakeholders, reducing resistance to change
- Integrated, being part of other planning processes such as regional strategies, community plans, and energy plans.

1.1.1 Spatial Scales

Air quality management occurs over a wide range of geographic scales: international, national, provincial, regional, and local.

International scale management deals with issues that are global in scope, such as climate change, stratospheric ozone depletion and inter-continental transport. An early example of a management approach on this scale is the UNECE Convention on Long Range Transport of Air Pollution. This convention, adopted in 1979 and currently has 49 countries (including Canada) that have signed the convention, set the broad framework for the United Nations Economic Commission for Europe region to work cooperatively on the transport of pollutants through the atmosphere and over borders, oceans, and continents. From this convention, 8 different protocols have been produced—including the Aarhus Protocol on Persistent Organic Compounds which entered into force in 2003 and bans the production and use of some substances, while scheduling others for elimination or severe restriction at a later stage.

On a national scale, the Canada-Wide Acid Rain Strategy for Post 2000 (FPT Ministers 1998) had a primary long-term goal of meeting the environmental threshold of critical loads for acid deposition across Canada. The Strategy built on the success of the 1985 Eastern Canada Acid Rain Program, which achieved a 50% reduction in SO_2 (sulphur dioxide) emissions in eastern Canada since 1980. With the national program, Canada met all its international commitments on SO_2 emissions, both under the UNECE Convention and under the Canada-United States Air Quality Agreement. Chapter 16 discusses this Agreement and its implications on air quality management in Canada.

An example of a Provincial scale air quality management program is the framework developed by the Alberta Clean Air Strategic Alliance (CASA 2003) in order to achieve the

Canada Wide Standards for $PM_{2.5}$ (Particulate Matter below 2.5 μm in diameter) and O_3 (ozone) by 2010. Also included in the plan are the CWS provisions for "Keeping Clean Areas Clean and Continuous Improvement" (KCAC/CI) that apply at ambient concentrations below the numeric CWS, as well as provisions on monitoring and reporting of progress and activities. The plan is comprehensive in that it considers monitoring, stakeholder involvement and science studies.

On a regional scale, there is the Metro Vancouver Air Quality Management Plan (MV 2005), and the Ville de Montreal air quality management program (Ville de Montreal 2010) which is part of their Sustainable Development Plan. Such plans focus on regional scale issues such as urban smog. Included in the plan are specific emission control actions that are needed to meet the CWS for $PM_{2.5}$ as well as further studies to understand the sources of PM (spatial distribution and amounts).

There are also examples of air quality management on a local scale (called "Local Airshed Management"), where the geographic scope is focused on stakeholders in a specific area given that International, National, or Provincial efforts may not be comprehensive enough to meet their needs. One example is the airshed management efforts in Prince George, British Columbia (PGAQIC 2006). This program contains many of the science and policy elements common to programs on a broader geographic scale. Previous versions of this program include several recommendations to improve air quality, as well as a number of scientific and monitoring studies to further the local understanding for the underlying causes of poor air quality. Another example is the Action Plan for the SW Greater Toronto Area Oakville-Clarkson Airshed (Bastille, 2010), where a Task Force made 30 recommendations regarding a new Airshed Management System that includes strategies, emission targets, timelines and the assignment of roles and responsibilities. These recommendations reflect the local concerns that current National and Provincial programs will not be sufficient to manage and improve air quality in this region. Chapter 17 discusses further aspects of the process of airshed planning on a community level.

1.2 The Air Quality Management Process

Figure 1.1 is an illustration of a comprehensive air quality management program (after NAS 2004) and is essentially the application of the adaptive management process to the issue of air quality. The system is represented as a roundtable with the top showing the process of air quality management supported by a base of Science and Technology. The system is interconnected with multiple dependences as air issues are often inextricably influenced by one another, and thus need to be addressed in an integrated manner through inter-agency approaches and multi-disciplinary teams within agencies (Yap, 2011). For example, reductions in emissions of combustion related sources often have co-benefits for other air issues such as acid rain, air toxics and climate change (see Brook et al. 2009 for a discussion on the need for an integrated, multi-pollutant approach). The following sections describe the components of these two main parts of the comprehensive program.

1.2.1 Emergence of an Air Quality Issue

Before a management process begins, a problem must be recognized by the government. Problems may be identified through various indicators, but often come to the attention of governments through environmental group, the media, or influential citizens (after Pal 2006). Recognition that something needs attention is the first step in obtaining a commitment of resources for better definition of the problem (characterization of the air quality issue). Problem definition can be a long and circuitous process that attempts to answer three questions (Pal (2006): (1) Are the indicators valid? (2) What are the causes of the observed effects? (3) What actions can be taken if the problem is "real"? Structuring the problem relies on the accumulated evidence of scientific studies that analyze measurements of concentrations or depositions of pollutants and related human and ecosystem health effects. Models (air quality, human and ecosystem, economic) may be applied to determine the spatial and temporal extent, the severity (or magnitude) of the effects, or the cost of various solutions. As a particular air quality issue emerges, government agencies conduct science reviews or assessments in order to answer the three questions and inform policy makers on potential management actions.

This front-end process is illustrated in Fig. 1.2.

The acid rain issue is a perfect example of the emergence process. Early monitoring and scientific studies by Gorham (1957) and Gorham and Gordon (1960) reported acidification of lakes occurring in eastern Canada and around Sudbury. Beamish and Harvey (1972) documented the decrease in fish in northern Ontario lakes. In 1976 the Toronto Star ran a series of articles on acid rain which served to awaken public interest in the issue (Schmandt et al. 1989). In 1981, the Canadian Coalition on Acid Rain (CCAR) was formed by 12 environmental groups. Growing ultimately to 58 member groups representing over 2 million Canadians, Canada's largest environmental group played a central role in raising awareness of the acid rain issue through advocacy and educational programs as well as by lobbying the governments of both Canada and the United States for legislation restricting acid rain-causing emissions. Subsequently there were further studies by federal and provincial governments, universities and industries to characterize and predict the extent of acid rain and its effects on Canadian surface waters. From this growing pool of scientific literature, a sequence of review or

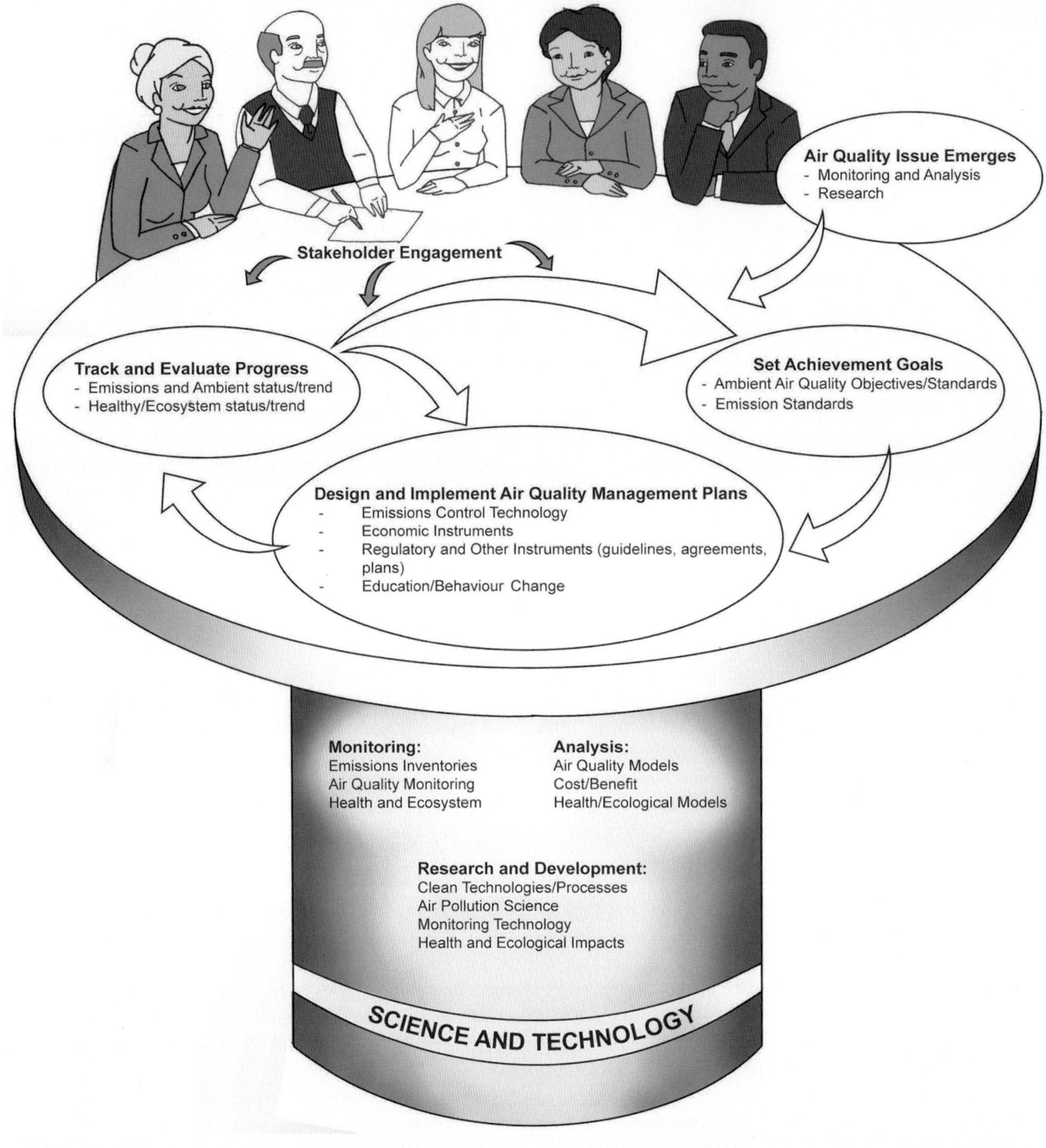

Fig. 1.1 The process and supporting tools of air quality management. Adapted and reprinted with permission from Air Quality Management in the United States, by the National Academy of Sciences, Courtesy of the National Academies Press, Washington, D.C. (Committee 2004)

assessment reports were undertaken such as by Harvey et al. (1981), RMCC (1986) and Jeffries et al. (2001).

In response to the scientific evidence, policy makers responded with various management plans for SO_2 and NO_x emissions controls. For example, in 1970 Canada signed the United Nations Economic Commission for Europe (UN ECE) Convention on Long-Range Transboundary Air Pollution (LRTAP) to reduce and prevent this phenomenon. In 1980 the Parliamentary Standing Committee on Fisheries and Forestry established a subcommittee on acid rain with members

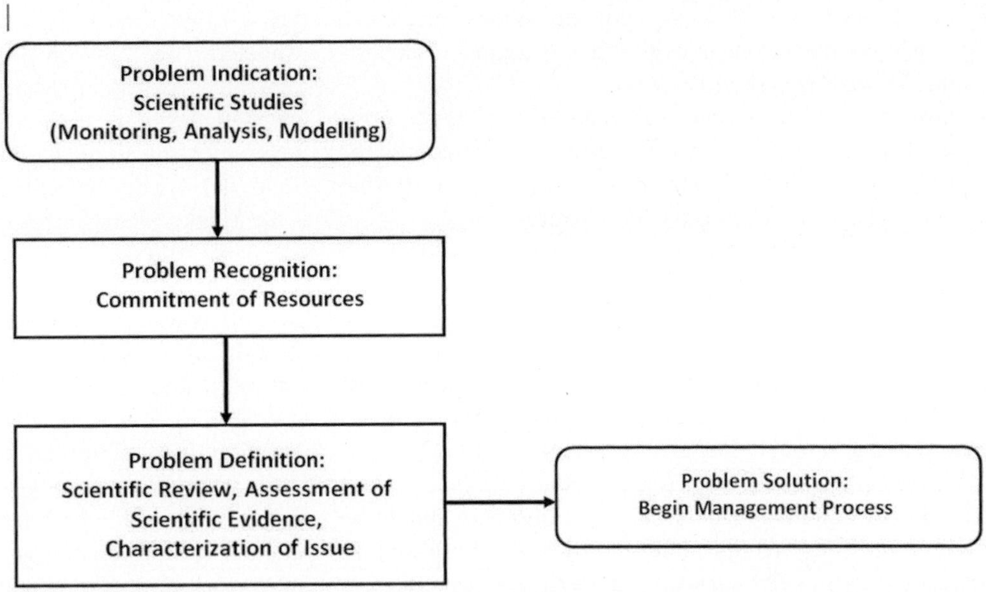

Fig. 1.2 Emergence of an air quality issue

from all three federal parties in the House of Commons. The subcommittee received evidence at four public hearings, visited a number of industrial installations, and travelled to Europe to observe acid rain damage. Their report "Still Waters: The Chilling Reality of Acid Rain" recommended 38 actions (Subcommittee 1981). When the subcommittee observed that little process was being made, they published another report "Time Lost: A Demand for Action on Acid Rain" (Subcommittee 1984) with an additional 16 recommendations. The earliest domestic program was the Eastern Canadian Acid Rain Program which represented the first coordinated effort of the federal government and the seven easternmost provinces to address this issue. This program, launched in 1985, achieved its goal of reducing SO_2 emissions to 50% of 1980 levels by 1994. Several other programs followed; with more recent examples is the Canada-Wide Acid Rain Strategy for Post 2000 (FPT Ministers 1998).

Another example of this emergence process is the growth of scientific studies that point to the adverse effects of air pollution on human health, and the realization that these effects occur at lower and lower concentrations that previously thought (Chapter 7 provides a detailed look at this issue). This evidence helped to inform the development of the Canada Wide Standard for Ozone and Particulate Matter which were established in June, 2000 (CCME 2000) after an extensive process that involved ancillary studies and consultation. Chapter 14 describes the process involved in the setting of these ambient standards.

Once an air quality issue has been identified and defined, then it can be "put on the table" so the management process can begin to solve it. The next step is to establish the goals of the management program.

1.2.2 Setting Air Quality Achievement Goals

An air quality management program is designed to achieve specific goals which reflect the overarching goals described earlier. These include goals for the ambient environment (which may be specified by ambient air quality objectives for different pollutants) and goals for emissions (as specified by emission standards for different types of sources).

Ambient Air Quality Objectives are established in terms of concentration or deposition magnitudes that provide a level of protection for human health and the environment. They are used to assess the current state of air quality, and to determine the success of the Plan. One example is the National Ambient Air Quality Objective for SO_2. In Canada ambient air quality objectives are sometimes referred to as standards (as in the Canada-Wide Standards for $PM_{2.5}$). Chapter 14 elaborates on the terminology and the process involved in setting ambient goals in Canada.

Ambient objectives are typically established by government agencies through a stakeholder consultation process, and are based on scientific evidence on air pollution effects on human health and ecosystems. An example would be the process involved in the establishment of the Canada Wide Standards for Ozone and $PM_{2.5}$. For some pollutants such as $PM_{2.5}$, the scientific evidence indicates there are adverse effects at all non-zero levels; in other words, there is no threshold effect. Therefore setting a Canada-Wide Standard represented a balance between complete protection and reasonable achievement.

Ambient air quality standards/objectives are comprised of up to 4 parts:

- Indicator: the pollutant of interest
- Averaging Time: the period over which pollutant exposure is defined (1 hr, 1 day, 1 week, 1 year)

- Level: the concentration as a volume/volume ratio (such as ppb) or mass/volume units (such as $\mu g/m^3$) or deposition in mass/area (kg/ha)
- Form or Metric: the statistical expression of the objective used to assess performance. For example the metric of the $PM_{2.5}$ Canada Wide Standard (CWS) is the annual 98th percentile value, averaged over 3 years.

In addition to providing the broad achievement goals of an air quality management plan, ambient air quality objectives may be used to: determine whether a regulated emission source needs to reduce emissions, indicate when specific preventative actions may be required (no open burning, woodstove bans), or indicate when specific protective actions may be required (e.g. limit outdoor activity).

Emission standards are based on the premise that the amounts of pollutants emitted to the atmosphere should be minimized. They are sometimes called source performance standards, technology standards, or equipment design standards. They may be either quantitative or qualitative. Although there is no guarantee that ambient air quality will not be compromised even if these emissions standards are met, they provide an achievement goal that can be readily imposed on different emission producing activities. Quantitative standards specify numerical maximum values for properties such as intensity (mass per unit production), concentration in a fuel or effluent (for example in mg/kg), or opacity (darkness of the emitted plume). Qualitative standards specify the type of control devices that must be installed, or the type of the fuel used.

For example, Canada-Wide Standards for Mercury emissions from coal-fired power plants have been established by CCME (2006) in two parts: (1) a cap on mercury emissions for each province, to be achieved by 2010, and (2) the use of best available technology for new plants. Another example is the emission standards for wood stoves, inserts and small fireplaces developed by the Canadian Standards Association (CSA B415). This standard recommends that emission levels of wood stoves manufactured in Canada meet requirements equivalent to the 1990 standards of the US Environmental Protection Agency.

Emission standards reflect advances in processes or control technology but are balanced by practicality and demonstrated effectiveness.

1.2.3 Design and Implementation of Air Quality Management Plans (AQMP)

An *Air Quality Management Plan* (**AQMP**) details what needs to be done, the timelines, how it will be achieved and who will do it. It requires coordinated actions by government, business, industry, and the public. Examples of this on a National scale are the NO_x/VOC Management Plan (CCME 1990), the Acid Deposition Post-2000 Strategy (FPT Ministers 1998) and on a more regional scale—the Metro Vancouver Air Quality Management Plan (MV 2005). These are examples of the relationships between different levels of governments in air quality management (discussed further in Chap. 16). Finally an example of a Plan with a specific focus is the Smoke Management Framework for British Columbia (BC MoE 2011).

The AQMP comprises various methods (discussed in the following sections) that drive emissions reductions using regulatory instruments, economic incentives, and social marketing. These methods can apply to the reduction of emissions from a variety of source sectors such as transportation, agriculture and industry (each sector is covered in more detail in Chap. 9–12). Details about the reductions such as the specific pollutants targeted and their respective amounts are informed by science and technology: emissions inventories, air quality models and ambient monitoring.

Emission Reduction Methods

Emissions can be limited or reduced through process changes, energy efficiency improvements, or the application of control technology. Examples of end of tailpipe treatments include baghouses to reduce particulate matter from sawmill operations, and Selective Catalytic Reduction (SCR) for NO_x reduction from thermal power generation plants. Improved combustion can be achieved through fluidized beds for coal fired power plants, natural gas combined cycle turbines for thermal power generation, and the switching from high sulphur to low sulphur fuel. As an example, Canadian regulations impose maximum limits for sulphur in on-road, off-road, rail (locomotive) and marine (vessel) diesel fuels to facilitate the effective operation of advanced emission control technologies installed on vehicles and engines:

- After June 1, 2006, 15 mg Sulphur/kg limit for *on-road* diesel fuel (Ultra Low Sulphur Diesel) (was 500 mg/kg)
- 500 mg/kg for *off-road,* rail and marine diesel fuels in June 2007;
- 15 mg/kg for off-road diesel fuel in June 2010; and
- 15 mg/kg for rail and marine diesel fuels in June 2012.

Sometimes emissions reductions can be achieved through the application of best management practices. For example a small percentage of cars can contribute the majority of emissions from the vehicle fleet if they are poorly operated and maintained. Vehicle inspection and maintenance programs such as the AirCare program in the Lower Mainland of British Columbia (Aircare 2003) and the Drive Clean program in Ontario (ERG 2005) test vehicles for compliance with the manufacturer specifications for emissions. Vehicles that do not meet the respective specifications must be repaired. Refer to Chap. 10 for more detail regarding the transportation sector emissions management.

Implementation Methods

Regulatory instruments are the legal means by which governments implement air quality management plans. These consist of: (a) legislation (by elected representatives), (b) statutory regulations (under legislation) and (c) quasi-regulations such as standards, guidelines permits, approvals, by-laws, and mandatory codes of practice issued by government departments under authority of an Act or Regulation. All regulatory instruments fall somewhere along a spectrum in terms of how much detail is specified and what is left to interpretation or dependent upon external factors. *Prescriptive regulation,* sometimes called "command and control", defines how activities are to be undertaken (eg, what techniques or equipment to use, what qualifications must be held by those doing the work, where and when an activity may be performed). *Outcome regulation,* also known as performance-based regulation, specifies a desired end-state and does not constrain how compliance is to be achieved. An example of this is the regulation for new marine spark-ignition engines which specifies performance standards, allowing manufacturer's flexibility in their achievement (Government of Canada 2011). While outcome regulation is more consistent with innovation and efficiency, prescriptive regulation may deliver more certain results.

Canada's federal environmental regulatory system is enabled by the Canadian Environmental Protection Act (CEPA 1999) which:

- made pollution prevention the cornerstone of national efforts to reduce toxic substances in the environment;
- set out processes to assess the risks to the environment and human health posed by substances in commerce;
- imposed timeframes for managing toxic substances;
- provided a wide range of tools to manage toxic substances, other pollution and wastes;
- ensured the most harmful substances are phased out or not released into the environment in any measurable quantity;
- included provisions to regulate vehicle, engine and equipment emissions;
- encouraged citizen input into decision-making; and
- provided for cooperation and partnership with other governments and Aboriginal peoples.

In addition, each province and territory has its own regulatory system and Chap. 14 covers their relationship with the federal system.

Economic instruments raise costs for high emitters or provide rewards for low emitters. *Emission charges* impose a cost upon the emitter as economic rent for the use of the atmosphere as a repository for waste products. *Tradable emission permits* define a property right that has economic value. Such "rights" can be sold, traded, auctioned, banked, or retired. Emitters with low cost control options can make reductions and sell their permits to generate extra revenue. *Fines* impose economic loss for emitters who violate standards. *Tax breaks* reward actions to lower emissions by reducing the amount of taxes payable. *Subsidies* provide grants or low interest loans to those who invest in reducing emissions.

In *emission trading,* as introduced for example by Ontario (OME 2005), the government first sets a goal for emissions reduction called an emissions target for a given area or region and specific caps on emissions for key industrial sectors. The total emissions allowed under the cap are divided into allowances, with each allowance equal to one tonne of emissions. Each capped emitter is allocated a specific number of allowances by the government, which equals the amount of pollutants that the emitter is allowed to release into the atmosphere in a given year. Each capped emitter is required to monitor its actual emissions throughout the year and report annual or seasonal emissions amounts. At the end of the year, if the actual emissions are equal to the number of allowances, the capped emitter has achieved compliance and does not have to take further action. If the actual emissions are greater than the number of allowances, the capped emitter can "buy" allowances from another capped emitter that has excess allowances. If the actual emissions are less than the number of allowances, the capped emitter can "sell" its allowances to another capped emitter which exceeded its allowances, or "bank" them to meet its own reduction commitments in future years. This allows a company with high environmental performance to gain financially, while a company with lower environmental performance must pay for its higher emissions. With total emissions capped in a region, emitters with low cost control options make reductions and sell their surplus permits to another emitter who has high cost control options. Such a program offers greater flexibility and economic efficiency than would a command and control approach.

Emissions fees may be regarded as "green taxes" levied on emissions of pollutants that lead to environmental problems. Since the fees depend on the amount of pollutants emitted, in theory they provide an incentive for regulated sources to reduce emissions, and thus reduce the amount of fees paid. Examples of this are the Metro Vancouver Air Quality Management Fees Regulation Bylaw (MV 2008) which specifies fees that depend on the pollutant and the amount emitted for those permitted sources in their jurisdiction. British Columbia has a similar program.

Incentive programs reduce emissions indirectly through the development or cleaner energy systems. Examples include the Federal Government ecoEnergy Innovation Initiative (NRCan 2011) that will cover up to 75 % of the development costs for new technology that produces cleaner and more efficient energy, and the New Brunswick Existing Commercial Buildings Retrofit Program (Efficiency New Brunswick 2011) that offers up to $ 50,000 for energy saving retrofitting costs.

Social Marketing uses education and public outreach to bring about changes in behaviour. For example, vehicle emission standards can be imposed to limit vehicle emissions, however the gains can be offset by citizens who drive when they could take mass transit, idle their vehicles unnecessarily, travel alone when they could carpool, or duplicate trips through poor personal planning. For communication to be effective, that is, to facilitate an intended societal response or desired social change, it must accomplish two things: (1) sufficiently elevate and maintain the motivation to change a practice or policy, and (2) contribute to lowering barriers and resistance to change. One example is the woodstove exchange program by BC Ministry of Environment for the Bulkley Valley—Lakes district which encouraged residents to change out their older, smoky wood stoves for new, clean burning appliances by applying principles of Community Based Social Marketing (Schmidt 2009). Chapter 19 further examines the aspects of social marketing and its role air quality management.

1.2.4 Tracking and Evaluating Progress

As the AQMP is implemented there is a need to monitor and summarize various types of related changes, specially, the total emissions, ambient air quality and the condition of selected receptors (health and ecosystem). Reviewing the trends will determine whether the AQMP is achieving the anticipated improvements in the specified timeframes. If there are significant deviations, then course corrections may need to be taken. If there is little progress toward the air quality goals, then the AQMP must be revisited.

1.2.5 Stakeholder Engagement

Canada has experimented with different forms of public consultation for many years. Canada's Green Plan (Govt. of Canada 1990) included opportunities for public review and comment at a number of public meetings and workshops across the country. One of the principal outcomes was a recommendation that "the public must be involved in decision-making, beginning with the conceptual stage and continuing throughout the decision-making process." The NO_x/VOC Management Plan (CCME 1990) included extensive public consultation. The Canadian Council of Ministers of the Environment has long been committed (CCME 1993) to provide opportunities for participation in its decision making processes, believing that meaningful input by Canadians "will result in better and more informed decisions and better working relationships with its stakeholders". CCME identified three types of stakeholder involvement representing various levels of engagement as indicated in Table 1.1. This characterization is similar to that used by OECD (2001) but simpler than the five types used the International Association for Public Participation (IAP2 2007).

There is no single recipe that guarantees successful stakeholder engagement. Primarily, citizens, the public and stakeholders want to know that their advice is valued and useful. CCME (1993) bases its consultations on ten principles:

1. Relevance: focused on the issue at hand
2. Effectiveness: clear scope and objectives
3. Full and fair access: opportunity to influence and participate
4. Respect for diverse interests: range of value, knowledge & perspectives
5. Efficiency: realistic deadlines and costs
6. Participant funding: support for participation
7. Flexibility: responsive to changing needs
8. Access to Information: equal and timely
9. Accountability: to constituencies and process
10. Implementation and feedback: commitment to act and report back.

Small problems may be handled on the left hand side of the spectrum in Table 1.1, larger ones more toward the middle. The development of the Canada-Wide Standards (CCME 2000) involved extensive stakeholder consultation. As noted above, the Strategic Options Process operated by Environment Canada and Health Canada to manage CEPA-toxics is a multi-stakeholder consultation. On the right hand side of the spectrum, Alberta conducts its strategic air quality planning through an independent multi-stakeholder organization, the Clean Air Strategic Alliance (CASA 2010).

Engaging all sectors (those that emit, and those that receive) in the development of an air quality management plan helps to ensure support for the resulting plan. Implementation can then proceed smoothly without the opposition that might otherwise delay or derail the planned actions. The stakeholders involved in air quality planning will depend on the problem being solved. Source sector stakeholders could include: transportation (marine, rail, air, commercial trucks, public); agriculture (greenhouses, different types of farms, equipment); government institutions (federal, provincial, municipal); fuel distribution (gas stations, distribution centers); industry (e.g. aluminum and alumina, asphalt plants, base metal smelting, breweries, cement, chemicals, electricity, iron ore pellets, iron and steel, mining, motor vehicle manufacturing, oil sands, petroleum refineries, pipelines, potash, pulp and paper, and upstream oil and gas, wood products). Typical receptor sector stakeholders include: sensitive populations (cancer, lung disease); environmental groups; municipal councils; public/community interest groups; medical community. Most sectors have umbrella organizations which will supply representatives to the involvement process; sometimes it is necessary to invite specific individuals to participate on behalf of a sector.

Table 1.1 Three types of stakeholder involvement representative of a broader spectrum of possibilities

	Inform	Consult	Collaborate
Description	One-way flow delivering information from government to stakeholders	Two-way communication between governments and stakeholders	Interactive process in which stakeholders actively participate with government
Purpose	Provide information to assist stakeholders in understanding the problems and potential solutions	Obtain feedback from stakeholders on analysis, alternatives or decisions	Partner with stakeholders in defining the problem, generating alternative solutions, and choosing the preferred course of action
Commitment	We will keep you informed	We will listen and acknowledge your concerns and show how they influenced the final decision	We look to you for advice and innovation and will use it to the maximum extent possible
Example techniques	Open house, factsheets, newsletters, websites	Workshops, advisory committees, focus groups, public meetings	Joint working groups, steering committee, consensus decision-making

1.3 Scientific and Technical Support

All of the foregoing steps are supported by a base of science and technology, tools or methods that help inform decisions made at each step in the management process. These tools are described below.

1.3.1 Ambient Air Quality Monitoring

This involves the measurement the atmospheric concentrations or depositions (the rate at which pollutants are deposited to the earth's surface) of pollutants at certain locations. Ideally, monitoring would occur over several years so that a trend can be established. The purposes of such a monitoring program can include the following:

- Establish current conditions
- Identify and characterize the air quality issue
- Assess whether an AQMP is required to improve and protect
- Determine whether any actions implemented under the AQMP are making a difference
- Determine regulatory compliance—operation in accordance with the terms of its air emissions permit
- Make short-term predictions (extrapolate trends for public advisories, possibly implement emission control measures to avoid a problem)
- Study the level of exposure and human/ecosystem health (to establish dose response relationships)
- Describe transport and chemical transformation of airborne substances
- Evaluate air quality model performance
- Identify source contributions in multi-source airsheds
- Understand trans-boundary air pollution
- Assess quality of life issues (visibility, odours)
- Inform the public about ambient air quality

There are many monitoring methods and the choice is determined by the purposes of the monitoring and available resources. Canadian agencies that operate air quality monitoring typically have a network of fixed monitoring stations, where monitoring equipment is housed in specially designed enclosures with specific operating requirements in terms of power, temperature, supplies, data storage, data transfer, and routine maintenance. Supporting meteorological measurements may also be made.

In these networks the *criteria* (or *common*) *air pollutants* are measured at fixed locations using continuously operating instruments that provide minute-by-minute concentrations from which can be calculated 1-h, 3-h, 8-h, 24-h averages. Such information may be used to produce an overall summary of the air quality through an index such as the Air Quality Health Index (see Chap. 18 for more detail on the AQHI). Integrated 24-hour average sampling is also conducted for $PM_{2.5}$, VOCs and Air Toxics. The samples are sent to a certified laboratory for chemical and/or gravimetric (weight) analysis.

The National Air Pollutant Surveillance Program (NAPS) is operated by Environment Canada through cooperative agreements with the provinces and territories. The purpose of this network is to monitor and assess the quality of the air in Canadian urban areas continuously. Currently there are 286 sites in 203 communities located in every province and territory. In addition to continuously measure criteria pollutants (those that contribute to smog, acid rain and/or poor air quality) such as SO_2, NO_2, CO, O_3, $PM_{2.5}$, other types of monitoring and analysis techniques are employed to obtain concentrations for more than 340 types of chemicals at typical urban NAPS sites. For example, air collected in canisters is analyzed for more than 167 volatile organic compounds that contribute to smog formation (Environment Canada 2011a).

In addition, the Provinces, NW Territories, and the Regions of Metro Vancouver, Hamilton, and Ville de Montreal operate their own monitoring networks, where real-time measurements of different criteria pollutants are available on-line. Some of the stations in these networks are part of the NAPs program. In addition to these provincial and regional networks, air quality monitoring can also be conducted

Fig. 1.3 Ontario Ministry of Environment TAGA Mobile Monitoring Unit (with permission from the Ontario Ministry of Environment)

by industrial permit holders (for example the Sudbury SO_2 monitoring network (Clean Air Sudbury 2011)) or in local airsheds. The data obtained through these systems may or may not be linked into the provincial system. Chapter 3 has more information on these Canadian monitoring networks and the national patterns of air quality that are derived from the collected data.

Some agencies have a mobile monitoring unit for specific investigations such as Metro Vancouver's Mobile Air Monitoring Unit (MV 2010) and the Ontario Trace Atmospheric Gas Analyzer (TAGA) unit shown in Fig. 1.3. In these units instruments (both for continuous and non-continuous monitoring) are installed in a specially designed vehicle (truck or bus). These mobile platforms provide flexibility for special studies or environmental emergencies where they can be temporarily set up at a location of interest, or used to track pollutant plumes (delineate impact areas) and identify sources.

Finally, there are other air quality monitoring programs (Environment Canada 2011b) that are designed for specific issues such as acid deposition and the global transport of persistent organic pollutants, or for a specific interest area (the Great Lakes). The Canadian Air and Precipitation Monitoring Network (CAPMoN), operated by Environment Canada, is designed to study the regional patterns and trends of atmospheric pollutants such as acid rain, smog, particulate matter and mercury, in both air and precipitation. As of 2010 there are 33 CAPMoN sites across Canada. The sites are predominantly located in central and eastern Canada but new sites are being developed in the west. The Global Atmospheric Passive Sampling (GAPS Network) is a program for producing comparable global-scale data for persistent organic pollutants (POPs). This program was initiated in December 2004 as a two-year pilot study before evolving into a network, and consists of more than 50 sites on seven continents. Twelve sites are located in Canada. The Canadian Aerosol Baseline Measurement (CABM) Program monitors changes in the physical and chemical properties of aerosols which influence Canadian air quality and climate. The network includes four monitoring sites managed by Environment Canada in operation since 2005. The Integrated Atmospheric Deposition Network determines the atmospheric loadings and trends (both spatial and temporal) of priority toxic chemicals to the Great Lakes and its basin (17 sites in total).

1.3.2 Emission Inventory

An emissions inventory (EI) is a database that lists, by source, estimates of the quantity of specified air pollutants discharged into the atmosphere from within a defined area during a given time period (typically a specified year). Although simple in concept, an emissions inventory is a powerful tool to assist in air quality management decision making. For example, it can be used to:

- identify the highest individual or sectoral emitters to set priorities for source management
- establish emission trends to assess the effectiveness of source management efforts

- forecast emissions (based on economic and growth projections) to identify emerging trends and growth sectors where action should be taken now to avoid problems in the future.
- provide inputs to dispersion models which estimate the impacts of emitted pollutants on ambient air quality.

The characteristics of the inventory depend on its purpose. For example the pollutants included in an ozone depleting substance inventory can be different than a emissions inventory for air quality management purposes which focuses on the criteria pollutants (CO, PM_{10}, $PM_{2.5}$, SO_2, NO_x, VOC's, NH_3).

For EI's that support air quality management, the emissions are often grouped into broad categories such as Point, Area, Mobile and Natural Sources. Point sources are those from stationary industrial facilities (i.e. facilities with stacks) that operate under an authorization (i.e. they have a permit, or operate under a code of practice). Area sources are stationary emission sources that occur over an area. For example, a large forest fire with smoke emanating over the burned area or the vapours from waste water settling pond are considered area sources. In addition, area sources could be comprised of several small point sources that become significant when considered collectively such as residential wood stoves in a municipality or a series of piles of wood burned in a cleared area. Mobile Sources are emission sources that move. Some examples are motor vehicles (e.g., passenger cars, trucks and motorcycles), aircraft, marine vessels, trains, off-road vehicles, and small off-road engines (e.g., agricultural, lawn/garden, construction and recreational equipment). Natural Sources are those emissions which occur in nature without the influence of humans, such as wildfires, volcanoes, marine aerosols and emissions from vegetation.

Often the sources within the broad categories are further broken down to subcategories. For example in the mobile source category, vehicle emissions from on-road vehicles can be broken down into several types such as light duty cars, vans, SUV's, light and heavy duty trucks, etc.

The federal government (Environment Canada) collects and compiles emissions information and produces National, Provincial and Territorial Emissions Inventories for criteria air pollutants, certain heavy metals and persistent organic pollutants together with historical national emission trends data. The emissions from industrial facilities are based on the Environment Canada's National Pollutant Release Inventory (NPRI) program. As required by the Canadian Environmental Protection Act, for the latest inventory year (2009), over 8400 facilities reported to the NPRI on more than 300 listed substances (Environment Canada 2011c). This information is compiled along with air pollutant emissions data from all non-industrial sources (including both human and natural emissions) in collaboration with provincial, territorial and regional agencies using the latest emission estimation methodologies and statistics available. The resulting NPRI summaries are a publicly accessible inventory of pollutants from facilities across Canada as well as estimates for other sources such as motor vehicles, residential heating, forest fires and agriculture.

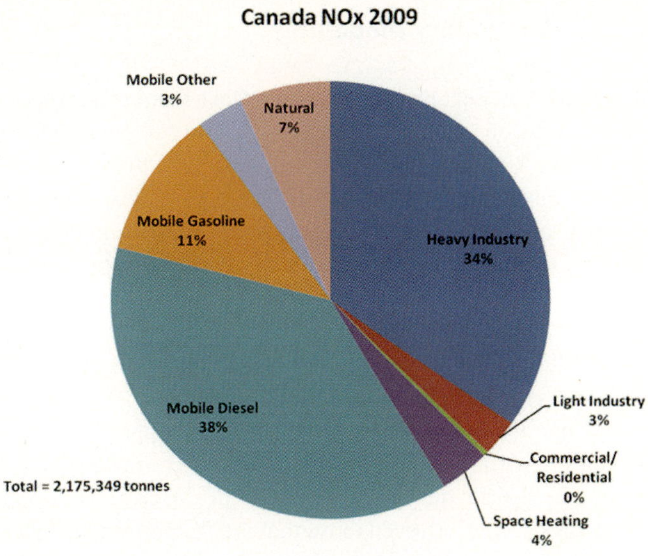

Fig. 1.4 Summary of 2009 NO_x Emissions and Sources in Canada

Some provinces have developed their own emissions inventory where this is an interest for more refined estimate techniques (i.e. less uncertainty) for a specific pollutant or source sector than used in the NPRI, or to provide better detail on historical and future trends, such as the Alberta Emissions Inventory (AENV 2008) or additional information on the spatial distribution of the emission sources. The 2008 Quebec (MDDEP 2011) and 2005 British Columbia Emissions Inventories (BC MoE 2009) are examples of inventories developed by the provinces.

There are many examples of emission inventories for smaller geographic areas. Examples of emissions inventories on a city-wide scale is the City of Hamilton, Ontario GHG and Air Pollutant Inventory (Stantec 2009) or the comprehensive Metro Vancouver EI that includes both forecasts and backcast estimates of emissions within their boundaries (MV 2007).

The EI information is typically summarized in a series of tables that are specific to each pollutant, showing the different contributing source sectors and their respective amounts for the total area defined by the inventory, and may be further broken into sub-areas (defined by political or geographical boundaries). Finally pie charts, corresponding to each criteria pollutant are often used to illustrate the EI summary (Fig. 1.4). Chap. 13 treats the topic of Emissions Inventories in Canada in detail.

1.3.3 Air Quality Modelling

Air quality models provide a cause-effect link between pollutant emissions and the resulting air quality impacts. There are different types of air quality models used to support air quality management such as dispersion models, receptor models, physical models and statistical models. Although dispersion models are most frequently used, the other model types offer certain abilities that are needed to understand the increasingly complex picture of air quality management. Due to their popularity for these purposes, the first two types will be reviewed here.

Air quality models are powerful tools to support air quality management decision-making. They can be used to:

- Assess the air quality implications of different emission control scenarios, even those that are "experimental" (for example, what would be the air quality if all vehicles were electric powered?).
- Assist in the design of stacks (height, exit conditions, location) in order to minimize air quality impacts.
- Determine emission limits in permits
- Identify sources that are contributing to the air quality problem
- Understand why poor air quality episodes occur
- Determine the importance of geophysical conditions (topography, surface conditions) on air quality
- Aid in forecasting air quality (see Chap. 6)

Dispersion models use science-based formulations that describe the behaviour of a pollutant in the atmosphere. The complexity of dispersion models varies greatly, so the specific dispersion model used is determined largely by factors such as: the geographic extent and the geophysical characteristics of the area considered, the pollutant(s) and characteristics of the source(s) of concern, and the output and the corresponding level of confidence required from the model. Despite the variety in the different types of dispersion models, they all require the following basic inputs:

- Emissions and source information
- Meteorology
- Geophysical characteristics

Regional scale transformation and transport models have been applied to support air quality management decisions across Canada. The Atmospheric Deposition Oxidant Model (ADOM) (Venkatram et al. 1992) was used in the development of the Canada-Wide Acid Rain Strategy for Post 2000. The Regional Lagrangian Acid Deposition Model (RELAD) is used by Alberta Environment (Cheng 1994) in its acid deposition management framework. Environment Canada has developed (Gong et al. 2006) and applied a Unified Regional Air-quality Modelling System (AURAMS) to forecast concentrations and depositions for a variety of uses such as critical load exceedances (Moran et al. 2008). AURAMS has also been applied to the support the development of low-sulphur marine fuel regulations associated with the establishment of the Emission Control Area (ECA) near the coast of North America (MEPC, 2009). Finally, the Community Multi-Scale Air Quality modeling system (CMAQ) has been used in a variety of applications, such as investigating transboundary influences on O_3 and $PM_{2.5}$ over Ontario (Chtcherbakov et al. 2007) and air quality forecasting for the Lower Fraser Valley in British Columbia (Della Monache et al. 2007). Chap. 4 provides details on global and regional scale air quality modeling as it relates to air quality management in Canada.

Regulatory agencies often rely on local scale dispersion models as part of their approval process for industrial projects. For example, if a new incinerator was proposed at a specific location, a local scale dispersion model would be used to predict the pollutant concentrations that would occur within a 10 km downwind distance (although depending on the model, this could be 50–100 km). Provincial regulatory agencies rely on the US Environmental Protection Agency dispersion models (AERMOD, CALPUFF, SCREEN3) and many provinces have produced a guideline on model selection and application for their respective province, for example, Newfoundland (NDEC 2006), Ontario (OME 2009) and British Columbia (BC MoE 2008). Chapter 5 provides more detail on the topic of regulatory dispersion models.

Receptor models are statistical procedures for identifying and quantifying the sources that contribute to measured air quality at a specific location. Since they do not use emissions, meteorology or geophysical information to calculate concentrations, they offer an advantage over dispersion models for situations where the sources of a pollutant (such as PM) are difficult or impossible to characterize (location, emission rates). Instead, receptor models use the chemical and physical characteristics of gases and particles measured at the monitoring site. Examples of three such models are: Chemical Mass Balance (CMB), which uses source profiles and speciated ambient data to quantify source contributions; UNMIX, which uses factor analysis to estimate the number of sources, the source compositions, and source contributions to each sample; and Positive Matrix Factorization (PMF), which uses a form factor analysis to describe the underlying co-variability of many variables with a smaller set of factors. CMB has found application for some time (e.g. Lowenthal et al. 1994; Cheng et al. 1998) while PMF has been used more recently. For example PMF was applied to chemically speciated $PM_{2.5}$ measurements to identify the sources that contribute to $PM_{2.5}$ in the town of Golden, BC (Willis 2006).

1.3.4 Socio-Economic Analysis (SEA)

The purpose of a socio-economic analysis (SEA) is to evaluate what costs and benefits an action will create for society by comparing what will happen if this action is implemented

Table 1.2 Types of socio-economic analysis (SEA) tools

Name of SEA Tool	Description of the Tool
Simple screening & choice methods	Includes checklists of mandatory criteria, minimum and maximum values, ranking using numerical or categorical scales, pairwise comparison, trend indications
Abatement cost function analysis	Maps the increase in per unit cost of emissions or risk reduction associated with tighter levels of control
Financial analysis	Determines the impact a proposed action on industry and the associated impacts on overall competitiveness
Cost-benefit analysis (CBA)	Converts all costs and benefits into money values for comparison
Risk-benefit analysis	CBA that involves only partial monetary valuation
Cost-effectiveness analysis (CEA)	Compares the costs of alternative means of achieving pre-set goals
Input-output models (I-O)	Quantify the transactions within sectors and the linkages between various sectors in an economy
General equilibrium models (GE)	Look at the economic system as a whole and observes all changes in prices and quantities simultaneously
Multi-criteria analysis (MCA)	Relies on scoring and weighting techniques to evaluate a decision problems characterized by a large number of diverse attributes

as compared to the situation where the action is not implemented. The analysis typically attempts to also include the effects that are indirect or incompletely reflected by market transactions. The analysis can be used to better understand how the various costs and benefits are distributed over the various affected parties in society. A SEA ideally covers all relevant effects related to the introduction of such an action: impacts on health, impacts on the environment; and impacts on the economy, (e.g. the costs to different actors in the supply chain, changes in consumer satisfaction, plant closures, competitiveness, inflation), and social effects (e.g. unemployment, labour quality). A wide range of tools is available (OECD 2000) as illustrated in Table 1.2.

The Government of Canada has integrated SEA into the development of all legislation with a requirement that "the benefits (of a proposed regulation) outweigh the costs to Canadians, their governments and businesses" (Govt. of Canada 1999). In administering the Canadian Environmental Protection Act, Environment Canada and Health Canada have developed "a time limited process to establish environmental and health goals and targets, to identify and evaluate a range of management options for meeting the goals and targets, and to make recommendations to the accountable federal ministers on the most cost effective and efficient options to implement. The Strategic Options Process is a multi-stakeholder consultation process; stakeholders are invited to determine their level of participation in the process. The principles for SOPs are: open, transparent, inclusive, timely/disciplined, cost-effective, defensible/predictable, flexible, harmonization and cross-sectoral equity" (Env Canada 1994).

Ontario uses economic analysis in the implementation of its air quality standards (OME 2009). The Canadian Council of Ministers of the Environment uses SEA in establishing standards (e.g. CCME 2003) and at one time had a framework for SEA (CCME 1998). A cost-benefit analysis for the Canada-Wide Standard on Particulate Matter and Ozone was the subject of a Royal Society of Canada review (Royal Society 2001) which recommended a number of improvements to SEA practices in Canada. Cost-benefit analysis has also been used in support of strategic choices (DSS Management 2005) and emission regulation (Env Canada 2007).

Metro Vancouver (Crane Management 2005) assessed potential air quality management initiatives using a variety of SEA tools in four categories:

1. Economic Efficiency: benefit-cost, benefits of avoided health damage, benefits of avoided visibility degradation, benefits of avoided agricultural crop losses, co-benefits such as reduced traffic congestion, financial costs of implementation, price and substitution costs, tax effect costs, and innovation costs
2. Economic Development: changes in employment
3. Competitiveness: changes in profits, exports, imports and share of national job estimates
4. Social quality of life: public perceptions about air quality and air quality management from opinion surveys, emissions levels by subregions (differential geographical impact), changes in health outcomes (morbidity and mortality levels).

1.3.5 Research and Development

The Science and Technology support base includes research to further understand the physical processes involved in the whole life cycle of air pollutants: their generation, transport and dispersion, chemical transformation, their ultimate fate and effects on receptors. This understanding is fundamental to the development of scientific and technical tools used to support air quality management.

In Canada, research and development related to the broad topic of "air quality" is largely undertaken by three groups: the Federal Government, Industry and Academia. This section provides just a few examples of the Canadian efforts in order to illustrate the range and types of studies as well as the different groups active in this area.

Federal Government The Emissions Research and Measurement Division of Environment Canada conducts laboratory and field studies related to air pollution that originates from mobile and stationary sources. An example of the research by the Division are projects that help inform decision making and regulations regarding the impacts and effects from vehicles and engines utilizing renewable fuels, different control technologies and propulsion systems under representative Canadian conditions.

The Atmospheric Science and Research Directorate of Environment Canada conducts research in measurement techniques (e.g. remote sensing), new data analysis techniques to facilitate the assessment of risk associated with air pollution, and the development of models that describe the behaviour of air emissions in the atmosphere. Two examples of the wide range of projects undertaken in the Directorate include the development of the Canadian photochemical model called AURAMS (Gong et al. 2006), and the measurement, and characterization of $PM_{2.5}$ in Canada (Brook et al. 1997).

Health Canada has lead several studies to further understanding on the effects of air quality on human health within the Canadian context. For example, research conducted by Burnett et al (1997, 2000) on air quality and health for Canadian specific communities helped inform the science basis for the establishment of the Canada Wide Standards for $PM_{2.5}$. In addition, these studies combined with other studies on air quality and health, have helped in the development of the Air Quality Benefits Assessment Tool (AQBAT)—a computer simulation tool designed to estimate the human health and welfare benefits or risks associated with changes in Canada's ambient air quality (Judek and Stieb 2006).

Industry Research and development by this group is largely conducted under the banner of Industry Associations. For example, under the auspices of The Canadian Electrical Association, eight coal-fired power generation companies in Canada jointly developed and implemented a Mercury Program designed to improve the information base around the measurement and control of mercury emissions from coal-fired generation. Member companies are also working together under the Canadian Clean Power Coalition (CCPC), created in 2000, in association with American electricity producers to advance the technologies needed to build cleaner, more efficient, more economical coal-fueled power plants. The CCPC and its members have spent more than $ 50 million furthering this objective.

The Canadian Association of Petroleum Producers (CAPP), representing 150 companies that produce more than 95 % of Canada's natural gas and crude oil, started the Environmental Research Advisory Council (ERAC) in the mid- 1970's to initiate research and technology development on environmental issues relating to oil and natural gas production in western Canada. ERAC, with the help of other associations (notably the Petroleum Technology Alliance Canada (PTAC)) and individual companies, have conducted peer-reviewed air quality research studies aimed at methods to reduce flaring and venting emissions. PTAC also keeps members up to date on new air quality regulatory programs, and the status of industry-sponsored research on issues such as soot formation during flaring and leak detection monitoring.

Academia Several Universities in Canada are involved in research in atmospheric science, but specific to the field of air pollution, there are three notable examples. One is the University of British Columbia (UBC) where research on photochemical smog (Ainslie and Steyn 2007) and trans-Pacific transport of air pollution (McKendry et al. 2008) has been the subject of several studies. UBC is also internationally recognized leader in the field of health effects of air pollution such as those related to wood smoke (Brauer et al. 2011) and traffic pollution (Gehring et al. 2010).

The University of Toronto is involved in research into remote sensing techniques that involve satellite measurements of atmospheric trace gases together with global three-dimensional modelling of tropospheric chemistry and transport (He et al. 2011).

In the field of atmospheric chemistry, the York University Centre for Atmospheric Chemistry (CAC) is involved in both laboratory and field studies on various pollutants such as $PM_{2.5}$, VOC's and O_3 as well as different methods for detection and analysis of pollutants (Rudolph et al 2002). They have been key collaborators with Environment Canada in large field campaigns such as BAQS-Met 2007 (Bottenheim et al. 2010), and Pacific 2001 (Vingarzan and Li 2006).

Collaboration Finally, collaborative approaches are common as research and development projects require expertise and resources from a broad range of disciplines and thus often conducted in partnership with other groups described earlier. For example the Emissions Research and Measurement Division of Environment Canada collaborates with other government agencies (e.g. National Research Council, Health Canada, Natural Resources Canada), several academic institutions (Carleton University, University of British Columbia, universities of Windsor, Alberta and Toronto), industries (such as the vehicle manufacturers), other levels of government (including municipalities), and international agencies (e.g. US EPA).

Another example of collaborative research is the Border Air Quality Study conducted in South Western Ontario. This involved the federal and Ontario provincial governments, as well as several Ontario universities. The study was designed to improve understanding of the airshed in the US-Canada border region of South Western Ontario as well as to gain insight into the sources of air pollution, the impacts of the transboundary flow of pollutants, and lake effects during smog episodes in this region (Bottenheim et al. 2010).

References

AENV (2008) Alberta air emissions, trends and projections. Alberta Environment (AENV) Information Centre, Edmonton

AirCare (2003) Public report—highlights from the 2001–2002 technical review of AirCare. Air Care, Burnaby

Ainslie B, Steyn DG (2007) Spatiotemporal trends in episodic ozone pollution in the lower Fraser Valley, B.C. in relation to Mesoscale atmospheric circulation patterns and emissions. J App Met 46(10):1631–1644

Bastille D (2010) Action plan. Report of the air quality task force to the Honorable John Gerretson, Ontario Minister of Environment. South West Greater Toronto Area, Oakville-Clarkson Airshed. http://www.ene.gov.on.ca/environment/en/resources/STDPROD_080704.html. Accessed 31 May 2012

BC MoE (2008) Guidelines for air quality dispersion modelling in British Columbia. British Columbia Ministry of environment (BC MoE), Victoria

BC MoE (2009) BC emissions inventory of criteria air contaminants. BC Ministry of Environment (BC MoE), Victoria

BC MoE (2011) A smoke management framework for British Columbia: A cross-government approach to reduce human exposure to smoke from biomass burning. BC Ministry of Environment (BC MoE), Victoria

Beamish RJ, Harvey HH (1972) Acidification of the La Cloche Mountain lakes, Ontario and resulting fish mortalities. J Fish Res Board Canada 29:1131–1143

Bottenheim JD, Hastie D, Abbatt J, McLaren R (2010) The Border Air Quality and Meteorology Study (BAQS-Met). Special issue atmos. Chem Phys Discuss, 10. http://www.atmos-chem-phys-discuss.net/special_issue120.html. Accessed 17 Oct 2011

Brauer M, Karr C, MacIntyre E, Clark N, Su J, Baribeau A-M, Tamburic L, Lencar C, Demers P, Buzzelli M, Larson T (2011) Woodsmoke and children's health: findings from the border air quality study. J Epidemiology 22(1):186

Brook JR, Dann TF, Burnett RT (1997) The relationship among TSP, PM_{10}, $PM_{2.5}$ and inorganic constituents of atmospheric particulate matter at multiple Canadian locations. J Air Waste Manag Assoc 47:2–19

Brook JR, Demerjian KL, Hidy G, Molina LT, Pennell WT, Sheffe W (2009) New directions: results-oriented Multi-pollutant approach to air quality management. Atmos Env 43(12):2091–2093

Burnett RT, Brook JR, Yung WT, Dales RE, Krewski D (1997) Association between ozone and hospitalization for respiratory diseases in 16 Canadian cities. Environ Res 72:24–31

Burnett RT, Brook J, Dann T, Delocla C, Philips O (2000). Association between particulate and gas-phase components of urban air pollution and daily mortality in eight Canadian cities. Inhal Toxicology 12(Suppl. 1):15–39

CEPA/FPAC (1999) National ambient air quality objectives for particulate matter part1: science assessment document

CCME (1990) Management plan for oxides of nitrogen and volatile organic compounds. Phase I. Canadian council of ministers of environment, Winnipeg

CCME (1993) Guidelines for consultations and partnerships involving stakeholders in CCME. Canadian council of ministers of the environment, Winnipeg

CCME (1998) Framework for the application of socio-economic analyses in setting environmental standards. Canadian council of ministers of the environment, Winnipeg

CCME (2000) Canada-Wide standards for particulate matter (PM) and ozone. Canadian council of ministers of the environment, Winnipeg

CCME (2003) Economic analysis: update in support of the Canada-Wide standards for particulate matter and ozone. Canadian council of ministers of the environment, Winnipeg

CCME (2006) Canada wide standards for mercury emissions from coal-fired power plants. canadian council of ministers of the environment, Winnipeg

CASA (2003) Particulate matter and ozone management framework. Clean Air Strategic Alliance, Edmonton

CASA (2010) The CASA Way: a better way to manage air issues in Alberta. Clean Air Strategic Alliance, Edmonton

Chtcherbakov A, Bloxam R, Yap D, Fraser D, Reid N, Wong S (2007) Models-3/CMAQ simulations to estimate transboundary influences on ozone and particulate matter concentrations over ontario in spring-summer, 1998 Air pollution modelling and Its application XVII 2007 Editors: C. Borrego, A-L Norman. Springer Publishing ISBN: 978-0-387-28255-8

Cheng L (1994) Concentration and deposition of anthropogenic pollutants in Alberta. Alberta department of environmental protection, Edmonton

Cheng L, Sandhu HS, Angle RP, Myrick RH (1998) Characteristics of inhalable particulate matter in Alberta Cities. Atmos Environ 32(22):3835–3844

Clean Air Sudbury (2011) Sudbury's Air quality. http://www.cleanair-sudbury.ca/index_files/Page608.html. Accessed 10 Oct 2011.

Committee on Air Quality Management in the United States (2004) Air quality management in the United States. National research council, national academies press. ISBN: 0-309-51142-8

Crane Management (2005) Socio-economic considerations of cleaning greater Vancouver's Air. Crane management consultants Ltd, Vancouver

Della Monache L, Deng X, Zhou Y, Modzelewski H, Hicks G, Cannon T, Stull RB, di Cenzo C (2007) Air quality ensemble forecast over the Lower Fraser Valley, British Columbia. Air pollution modelling and Its application XVII 2007 Editors: C. Borrego, A-L Norman. Springer Publishing ISBN: 978-0-387-28255-8

DDEP (2011) Inventaire des émissions des principaux contaminants atmosphériques au Québec en 2008 et évolution depuis 1990. Development durable Environment et Parcs, Quebec

DSS Management Consultants Inc, RWDI Air (2005) Cost benefit analysis: replacing Ontario's coal-fired electricity generation. Ontario Ministry of Energy, Toronto

Efficiency New Brunswick (2011) Energy smart program guide: efficiency NB commercial buildings energy efficiency retrofit program. Efficiency New Brunswick, Saint John

ERG, Eastern Research Group (2005) Evaluation of Ontario drive clean program. Ontario Ministry of Environment, Toronto

Environment Canada (1994) Guidance document on the options evaluation process. Environment Canada, Ottawa

Environment Canada (2007) Backgrounder benefits and costs of the regulatory framework for Air emissions. Environment Canada, Ottawa

Environment Canada (2011a) National Air Pollution Surveillance Program. http://www.ec.gc.ca/rnspa-naps/default.asp?lang=En&n=5C0D33CF-1. Accessed 10 Oct 2011

Environment Canada (2011b) Air Quality Monitoring Networks. http://www.ec.gc.ca/rs-mn/default.asp?lang=En. Accessed 10 Oct 2011

Environment Canada (2011c) About the National Pollutant Release Inventory (NPRI). http://www.ec.gc.ca/inrp-npri/default.asp?lang=En&n=CA8D2224-1. Accessed 10 Oct 2011

Federal/Provincial/Territorial Ministers of Environment and Energy (1998) Canada-Wide Acid Rain Strategy-Post 2000: strategy and supporting document. Minister of Public Works and Government Services Canada, Ottawa

Gehring U, Wijga A, Brauer M, Fischer P, Jongste JC de, Kerkhof M, Oldenwening M, Brunekreef B (2010) Traffic-related air pollution and the development of asthma and allergies during the first 8 years of life—The PIAMA study. Am J Resp Critical Care Med 181(6):596–603

Gong W, Dastoor AP, Bouchet VS, Gong S, Makar PA, Moran MD, Pabla B, Ménard S, Crevier LP, Cousineau S, Venkatesh S (2006) Cloud processing of gases and aerosols in a regional air quality model (AURAMS). Atmos Res 82:248–275

Gorham E (1957) The chemical composition of lake waters in Halifax County, Nova Scotia. Limnol Oceanogr 2:12

Government of Canada (1990) A report on the Green Plan Consultations. Minister of Supply and Services Canada, Ottawa

Government of Canada (1999) Government of Canada Regulatory Policy. Privy Council Office, Ottawa

Government of Canada (2011) Marine Spark-Ignition Engine, Vessel and Off-Road Recreational Vehicle Emission Regulations. http://canadagazette.gc.ca/rp-pr/p2/2011/2011-02-16/html/sor-dors10-eng.html. Accessed 18 Oct 2011

Gorham E, Gordon AG (1960) The influence of smelter fumes upon the chemical composition of lake waters near Sudbury, Ontario and upon the surrounding vegetation. Can J Bot 38:477

Harvey HH, Pierce RC, Dillon PJ, Kramer JR, Whelpdale DM (1981) Acidification in the Canadian aquatic environment: scientific criteria for assessing the effects of acidic deposition on aquatic ecosystems. National Research Council Canada, Ottawa

He H, Tarasick DW, Hocking WK, Carey-Smith TK, Rochon Y, Zhang J, Makar PA, Osman M, Brook J, Moran MD, Jones DBA, Mihele C, Wei JC, Osterman G, Argall PS, McConnell J, Bourqui MS (2011) Transport analysis of ozone enhancement in Southern Ontario during BAQS-Met. Atmos Chem Phys 11:2569–2583

IAP2 (2007) IAP2 spectrum of public participation. International Association for Public Particpation. Thornton, Colorado

Jeffries DS, Brydges TG, Dillon PJ, Dupont, J, Gunn JM, Harvey HH, Keller W, Lam DCL (2001) Chapter 7 aquatic acidification. In: threats to sources of drinking water and aquatic ecosystem health in Canada. Environment Canada, Burlington

Judek, S. Stieb D (2006). Introduction to the air quality benefits assessment tool. Healthy Environments and Consumers Safety Branch (HECSB), Health Canada. October, 2006.

Lowenthal D, Wittorff D, Gertler A (1994). CMB source apportionment during REVEAL (REgional Visibility Experiment in the Lower Fraser Valley). British Columbia, Ministry of Environment, Victoria.

McKendry IG, Macdonald AM, Leaitch LR, van Donkelaar A, Zhang Q, Duck T, Martin RV (2008) Trans-Pacific dust events observed at Whistler, British Columbia during INTEX-B. Atmos Chem Phys 8:10275–10300

MEPC (2009) Proposal to designate an emission control area for nitrogen oxides, sulphur oxides and particulate matter. Marine Environment Protection Committee (MEPC 59/6/5). www.epa.gov/nonroad/marine/ci/mepc-59-eca-proposal.pdf. Accessed 17 Oct 2011

Mennell M, Bhattacharrya K (2002) Air quality management. Chapter 9 in a citizen's guide to air pollution (Edited by Bates D and Caton R). David Suzuki Foundation, Vancouver

MV (2005) Air quality management plan for Greater Vancouver. Metro Vancouver (MV), Burnaby

MV (2007) 2005 Air emissions inventory for the Lower Fraser Valley and forecast and backcast. Metro Vancouver (MV), Burnaby

MV (2008) Greater Vancouver Regional District Air quality management fees regulation Bylaw No. 1083. Metro Vancouver (MV), Burnaby

MV (2010) 2009 Lower Fraser Valley air quality report. Metro Vancouver (MV), Burnaby

Moran MD, Zheng Q, Pavlovic R, Cousineau S, Bouchet VS, Sassi M, Makar PA, Gong W, Stroud C (2008). Predicted acid deposition critical-load exceedances across Canada from a one-year simulation with a regional particulate-matter model. Proc. 15th Joint AMS/A & WMA Conference on Applications of Air Pollution Meteorology, Jan. 21–24, New Orleans

Newfoundland Dept of Environment and Conservation (2006) Guidance document: guidelines for plume dispersion modelling. Government of Newfoundland and Labrador, St. John's

NRCan (2011) The ecoENERGY Innovation initiative. Natural Resources Canada (NRCan). www.nrcan.gc.ca/eneene/science/ecii-eng.php. Accessed 10 Oct 2011

OECD (2000) Framework for integrating Socio-economic analysis in chemical risk management decision making. Organization for Economic Cooperation and Development (OECD), Paris

OECD (2001) Engaging citizens in policy-making: information, consultation and Public participation. Organization for Economic Cooperation and Development (OECD), Paris

OME (2005) Emissions trading fact sheet. Ontario Ministry of Environment, Toronto

OME (2009) Air dispersion modelling guideline for Ontario. Ontario Ministry of Environment, Toronto

OME (2009) Guideline for the implementation of air standards in Ontario. Ontario Ministry of the Environment, Toronto

Pals LA (2006) Beyond policy analysis: public issue management in Turbulent Times, 3rd ed. Thomson Nelson, Toronto

PGAIC (2006) Prince George air quality management plan—phase two. Prince George Air Quality Implementation Committee (PGAIC), Prince George

RMCC, Federal-Provincial Research and Monitoring Coordinating Committee (1986) Assessment of the state of knowledge on the long-range transport of air pollutants and acid deposition: Part 3–aquatic effects. Environment Canada, Ottawa

Royal Society of Canada (2001) Report of an expert panel to review the Socio-economic models and related components supporting the development of Canada-Wide standards for particulate matter and ozone. Royal Society of Canada, Ottawa

Rudolph J, Czuba E, Norman A-L, Huang L, Ernst D (2002) Stable carbon composition of nonmethane hydrocarbons in emissions from transportation related sources and atmospheric observations in an urban atmosphere. Atmos Env 36:1173–1181

Schmandt J, Clarkson J, Roderich H (Eds) (1989) Acid rain and friendly neighbours. Duke University Press, Durham

Schmidt E (2009) Highway 16 wood stove exchange—positioning and marketing recommendations. The FRAUSE Group. www.bcairquality.ca/reports/pdfs/highway16_woodstove_exchange.pdf. Accessed Oct 17 2011

Stantec (2009) Final report—city of Hamilton, GHG and air pollution emissions inventory project. City of Hamilton, Hamilton

Subcommittee on Acid Rain (1981) Still Waters: the chilling reality of acid rain. House of Commons, Ottawa

Subcommittee on Acid Rain (1984) Time lost: a demand for action on acid rain. House of Commons, Ottawa

Venkatram A, Karamchandani P, Kuntasal G, Misra PK, Davies DL (1992) The development of the acid deposition and oxidant model (ADOM). Environ Pollut 75(2):189–198

Ville de Montreal (2010) Montreal community sustainable action plan 2010–2015: together for a sustainable metropolis. Ville de Montreal, Montreal.

Vingarzan R, Li S-M (2006) The pacific 2001 air quality study—synthesis of findings and policy implications. Atmos Env 40(15):2637–2649

Willis P (2006) Particulate matter source apportionment in Golden, British Columbia. British Columbia Ministry of Environment, Cranbrook

Yap D (2011) Personal communication

Part I

Air Pollution Science

A History of Air Quality Management

2

Ann McMillan and G. Foley

Abstract

A brief history of air pollution management in Canada is provided in this chapter. Whether the issue be acid rain, ground level ozone, particulate matter, air toxics, or emergency releases of air pollutants, Canada has traditionally built a management approach on: strong science; a desire to move in a similar direction to our southern friends, the U.S.A.; connections to the rest of the world; all delivered by a cadre of world class scientists feeding their results to decision makers.

Beyond the research and modelling that has explained some of the complexities of chemical behavior in the atmosphere, it is clear that Canada's management systems would not be as effective as they have been without a strong federal/provincial commitment to monitoring emissions and air quality as well as a balanced approach among governments, non-governmental organizations, private industry and academia to model, analyze and make sense of the observations. Canada's unique approach has been led by a number of exceptional people over the years and their contributions have kept Canada in the forefront of the scientific understanding and bridging the gap into regulation needed for effective air quality management.

Keywords

Trail smelter · Sudbury · Canada wide standards · Critical loads · Acid rain · Ground level ozone · Particulate matter · CEPA · Persistent organic pollutants · Heavy metals

2.1 Introduction

Building on the Introduction in Chap. 1, this chapter will provide a summary of the history of air pollution management in Canada from several perspectives. In Canada, air quality management has been driven by a rather small number of key issues which have required some decisions to be made and actions taken. In general, these key issues have been driven by the presence of a "smoking gun" which is scientifically based evidence of harm or effect caused by air pollution. There have been a number of key people who have contributed their expertise and their ingenuity at the right time to ensure that science drove air quality management forward. As a result, air pollution management in Canada tends to be fact-based and well-grounded in the scientific method. This leads to a different process for each issue and a need for consolidation at the management level. Adding to the complexity of this approach is the fact that Canada is a unique blend of jurisdictions all having an interest in, and some responsibility for, air quality management. This chapter focuses on the federal role while Chap. 18 provides some details about the specific management approaches of the provinces.

Our long peaceful border with the U.S.A. has been important in the management of air quality on both sides of the border. Both countries have been actively engaged

A. McMillan (✉)
Storm Consulting, Ottawa, Canada
e-mail: mcmillan@storm.ca

G. Foley
US EPA, Washington DC, USA
e-mail: foley.gary@epa.gov

E. Taylor, A. McMillan (eds.), *Air Quality Management,* DOI 10.1007/978-94-007-7557-2_2,

Fig. 2.1 Rough timeline of historical AQ management

1900	First Trail Smelter was built.
	Acid rain recognized as an issue.
	Pollution incident in Glasgow.
	100 m stack built in Trail.
	Tetraethyl lead added to gasoline
	Muese incident.
	Special tribunal report on Trail awarding damages to U.S.
	Detroit/Winsor Reference to IJC.
	Deterioration in Great Lakes observed.
1950	ELA established.
	Donora incident.
	Great Smog of London.
	Beamish work on the La Cloche Lakes.
	Establishment of NAPs.
	First U.S. and Canadian Clean Air Acts.
	US Acid Rain program.
	Canadian Department of the Environment formed.
	Superstack constructed in Sudbury.
1975	Canadian Council of Ministers of the Environment (CCME).
	Great Lakes Water Quality Agreement.
	UN ECE LRTAP Convention.
	U.S. Clean Air Act. U.S. standard for PM_{10}.
	CARE established.
	Canadian Acid Rain Program.
	First NAPAP Assessment
	Task Force on Cleaner Vehicles and Fuels.
	Annex 15 added to GLWQA.
1990	Canada/US Air Quality Agreement.
	CEPA passed.
	CAMNet established.
	Canadian NOx/VOC Assessment.
	Scrubbers added to the Superstack.
	Northern Contaminants Program Established.
	CWS for particulate matter and ozone.
	Canada-Wide Acid Rain Strategy for Post 2000.
	Comprehensive Nuclear Test Ban Treaty
2000	U.S. standard for $PM_{2.5}$.
	UN ECE Heavy Metals Protocol.
	2004 Canadian Acid Deposition Science Assessment and Summary of Key Results.
	UNEP POPs Convention.
	Canada/U.S. Border Air Quality Strategy.
2007	Guidance Document for determination of achievement of CWS

internationally in the management of global air pollution and both have maintained leadership positions in international Air Quality Management circles. There has been an ongoing focus on being good neighbours.

The span of the chapter is large, it could be the topic of a whole book on its own, so the authors have chosen "representative" examples in many cases rather than attempting an exhaustive treatment in order to best tell a bit of a story. If readers find that their favorite situation or person is absent from the discussion, their forgiveness is requested, it is acknowledged that this is only a start and perhaps future writings can pick up on the wealth of material that is not found in this chapter.

Detailed descriptions of emission management for the various sectors are not given here but in more detailed chapters to come, for example, a thorough description of "Industrial Emissions Management" is given in Chap. 10.

2.2 Key Issues that have Driven Air Quality Management in Canada

Throughout this book there are many air quality issues mentioned and discussed to illustrate a number of the concepts and processes that make up Air Quality Management in Canada. There are, however, a few high profile issues which seem to have galvanized attention and action across the country and thus served to drive the process of Air Quality Management forward. The following issues and events are in this category:

- The Trail Smelter issue
- Acid rain – Sudbury
- Smog
  - Major incidents
  - Ground level ozone
  - Particulate matter
  - Sulphur in gas

- Air toxics
  - Great Lakes Water Quality Agreement
  - The Northern Contaminants Program
  - Canadian Environmental Protection Act (CEPA)
- Emergencies

In the following it is notable how there are similarities and differences in the way these situations were identified, described, recognized as important, characterized, studied, and finally resolved. The results of this resolution make up Canada's air quality regime, how we got here often provides insight as to why air pollution is managed the way it is today. A "history" is often linear in time but this one is not. Fig. 2.1 illustrates the timeline for some key air quality events mentioned in this chapter.

2.2.1 The Trail Smelter Case

The Trail smelter emissions ensured that air pollution became a recognized issue in Canada as early as 1896 when the first smelter was built. As the facility was expanded over the years, locals started to protest the smoke plume, and in the early 1920s a stack over 100 m tall was built. This served to disperse the fumes further down the Columbia River Valley and caused vegetation damage on farms on the American side of the border, over 20 km away.

Eventually, protests got so insistent that President Franklin D. Roosevelt wrote directly to the then Prime Minister of Canada, Richard Bennett, a very unusual step. A special tribunal was set up in response to the letter to decide whether "the Trail smelter should be required to refrain from causing damage in the State of Washington in the future and, if so, to what extent" (Allum 2006) and to consider the question of responsibility for damages.

Morris Katz (1901–1987) had graduated from McGill in 1929 with a PhD in chemistry and biochemistry. At the time the tribunal was set up, he was a chemist for the National Research Council of Canada. He was requested to study the Trail situation from a scientific perspective. His work on the effects of smelter fumes on vegetation (Katz et al. 1939), is a classic, and was generalized in his further work over the next decade (Katz 1949). Plant health was the "smoking gun" in this case although there were plenty of other effects observed.

The special tribunal reported its findings in 1941 in a landmark decision stating "No state has the right to use or permit the use of its territory in such a manner as to cause injury by fumes in or to the territory of another or the properties of the persons therein, when the case is of serious consequences and the injury is established by clear and convincing evidence" (Kaijser 2011).

Canada and the U.S.A. accepted the findings of the tribunal and $428,000 was paid to affected farmers. Through sulphur recovery and an air quality management plan based on meteorology, the problem was partly solved. Dr. Katz writes in 1963 "A large new industry was created to convert the waste sulfur oxides to sulfuric acid, ammonium sulfate, ammonium nitrate, and phosphate fertilizer. Today the Trail Smelter recovers about 91 % of the sulphur dioxide, formerly wasted, by conversion into these valuable by-products" (Katz 1963).

While this was not the end of the situation and several rounds have taken place since, the first Trail Smelter case set many precedents for air quality management not only in Canada, but worldwide. It established that pollutants from an industrial source could have impacts very far away from their emission. It also set the scene for the great care that the U.S.A. and Canada have taken since in managing the air pollution that crosses their lengthy border.

Morris Katz went on to be chairman, Canadian Section, Technical Advisory Board on Air Pollution, Defense Research Council Chemical Laboratories, International Joint Commission (IJC), where he continued his pioneering work on transboundary air pollution by looking at public health in the Detroit/Windsor area, (Katz 1955). This work laid the foundation for the creation of the International Air Quality Advisory Program (IAQAP) and the Michigan-Ontario Board created under the IJC in 1966.

In his later years, Dr. Katz became a Professor of Chemistry at Syracuse University before returning to Canada to work at York University where he remained as Professor, and Professor Emeritus of Chemistry until his death in 1987. Always ahead of his time, his work at York was on polycyclic aromatic hydrocarbons and their mutagenic properties.

2.2.2 Acid Rain

Some of the pollutants emitted from the Trail Smelter were sulphur and nitrogen oxides from combustion. "Acid rain" was identified as far back as the 19th century (Smith 1872) as a potential ecosystem stressor. Scandinavian scientists (e.g., Oden 1968) brought the issue to worldwide attention. In Europe, decades of work were done to show that ecosystem impacts were experienced in Scandanavian countries as a result of industrial activity in Germany and other European countries. In 1972 the First Stockholm Convention on Environment recognized acid rain as an issue.

In Canada, there was concern that the levels of air pollutants were not known, leading to the establishment of the National Air Pollution Surveillance Program (NAPS) in 1969. It was established under an agreement between the federal and eight provincial governments to monitor sulphur dioxide and particulate matter. In 1972 the first NAPS report was issued, based on the results of measurements at 36 stations (NAPS 2013). Chap. 3 illustrates the essential role NAPS plays today in Canada's air pollution management.

Technically, "acid rain" is quite a narrow issue, focussing on rainwater which has had its acidity increased due to contamination by chemicals such as sulphur and nitrogen oxides. The sulphur and nitrogen oxides combine with water in the atmosphere to form acids, most commonly sulphuric and nitric acid. These can be washed out with rain, snow and fog, sometimes a hundreds of kilometers from where they were emitted, even in another country, giving rise to the term "long-range transboundary air pollution" or LRTAP. The chemicals can also dry deposit on surfaces and this aspect is also considered part of the acid rain issue.

We now know that fish reproduction is affected about pH 5 and by pH 3 adult fish cannot survive, with some species being more sensitive than this. Impacts on biota are complex since impacts on bacteria and other aspects of the ecosystem can have effects much higher up the food chain (Harvey 1982). In addition to ecosystem impacts, health impacts as well as impacts on physical structures have been studied. It has been the ecosystem effects that have driven this issue, however.

Canada and the U.S. were engaged in the early discussions on acid rain in Europe, and in fact, Canada signed the United Nations Economic Commission for Europe (UN ECE) Convention on Long-Range Transboundary Air Pollution (LRTAP) in 1979 (UN ECE 2013b). This convention was based on one of the first "assessments" of long range transport of air pollutants (LRTAP) for Western Europe. A whole new methodology was developed and agreed to by scientists across Europe (as well as Canada and the U.S.A) which was based on using "transfer matrices" to represent the complex science describing the transfer of emissions from some countries to acid deposition in others.

In Canada, early papers by Gorham reported acid lakes near Halifax, Nova Scotia and as a result of the smelter in Sudbury, Ontario (Gorham and Gordon 1963). However it wasn't until a 1972 paper written by and Richard Beamish and Harold Harvey on fish mortality in the La Cloche Lakes near Sudbury provided a "smoking gun" that "acid rain" became the poster child for air pollution management (Beamish and Harvey 1972).

Richard Beamish first observed the effect of acid rain on fish in Ontario lakes while pursuing his PhD at the University of Toronto in the late 1960s, and the work on the La Cloche Lakes was part of his thesis work. He began his career with the Department of Fisheries and Oceans in Winnipeg after doing further studies at Woods Hole. After this he went to the Pacific Biological Station in Nanaimo where he headed the Groundfish Section, served as Station Director from 1980 to 1993. He was Commissioner and President of the Pacific Halibut Commission, a delegate to the North Pacific Marine Science Organization as well as the North Pacific Anadromous Fish Commission, and an affiliate professor of marine fisheries and aquaculture at Malaspina University College. He is an example of a long term government scientist who has held a leadership position in interdisciplinary environmental science in Canada for decades.

The regional nature of the issue was soon reflected in the work of Likens (Likens 1972; Likens et al. 1972), Kramer (Kramer 1973), Peter Dillon (Dillon et al. 1978), and many others, and what had been recognized as an international issue came close to home for Canadians.

During the 1960's, a group of small Ontario lakes south of the TransCanada highway and east of Lake of the Woods was set aside by the Ontario Department of Lands and Forests as the Experimental Lakes Area (ELA) to investigate the eutrophication process which was plaguing the Great Lakes. David Schindler, who was at that time at Trent University, applied for a position at the ELA where he first identified detergent phosphates as being the culprit killing lakes. Subsequently, using the controlled environment of the ELA, he demonstrated that some fish are extremely sensitive to pH and that acidification can cause dramatic shifts in food chains by killing sensitive species (Schindler et al 1985).

Beyond the details of the acid rain issue itself, methodologies began to emerge by which scientists could assess each component of the issue, and beyond this to communicate their results to policy makers and the general public. In Canada, for example, emissions of sulphur and nitrogen were estimated (Environment Canada 1973) and extensive work was done to measure the emissions using in-stack techniques as well as various means of measuring the plumes emitted. It was determined that major sources were non-ferrous metal smelters, coal-fired generators as well as more distributed sources such as transportation (Summers and Whelpdale 1975). Work was done on the chemistry of the atmosphere to figure out how the acidification mechanism worked.

The physical behaviour of the pollutants was studied as they moved from emission to deposition point. The mechanisms of deposition were studied. All this information was assembled into models to describe the overall phenomenon (Clarke et al. 1989; Voldner et al 1981). And, finally, the effects on ecosystems and health were studied (Puckett 1979, Schindler 1988). It took scientists who had different specialities to be authoritative on each of these key pieces of the issue and hence it marked one of the very first times that an integrated scientific response was provided to such a complex and challenging issue.

This early work led to a deluge of science from both sides of the Canada/U.S.A. border working toward defining the extent and magnitude of the biological and chemical effects on surface waters of eastern North America. In 1975, the United States Environmental Protection Agency (US EPA) received $134 million (USD) per year for research on environmental issues associated with current energy production and use as well as the energy technologies under development for use in the future. Of course, air quality research received a major portion of this budget. However, it was also significant that acid rain research was specifically identified in this budget to receive approximately $12 million. This drove work in Canada as well. Canada and the US convened a group under the International Joint Commission to summarize what was known about the issue. Lester Machta and Howard Ferguson were two of the leaders of the work and assembled a set of four reports. These reports drove the signing, on August 5, 1980, of a Memorandum of Intent (MOI) between Canada and the USA. This work led in turn to the preparation of a series of "assessments" of the issue (e.g. Harvey et al. 1981; US/Canada Work Groups 3A 1981, 2 and 3B 1982, 1 1983; Schindler 1988; Atmospheric Environment Service 1998). This work proceeded in the midst of complaints that the costs of dealing with acid rain would lead to uncompetitive industry and huge overhead costs.

Nonetheless, in the 1970s and 1980s the emerging consensus on aquatic effects justified action on sulphur dioxide and nitrogen oxides emission controls. In 1985, a domestic "Acid Rain Control Program" was established. In the seven eastern provinces, a 40% sulphur dioxide emission reduction from 1980 levels by 1994 was required.

In Canada, one of the scientific breakthroughs in communicating the serious nature of the acid rain issue was the concept of "critical loads" which is the amount of acid deposition that a particular region can receive without being adversely affected. It was found that deposition of sulphates in precipitation in excess of 20 kg/ha per year damaged moderately sensitive aquatic ecosystems (UN ECE 2013a). The United States never did accept the concept of critical loads.

In 1981 U.S. President Ronald Reagan eliminated energy and environment research, which started at the US EPA and other agencies in 1975, but retained the $12 million (USD) per year for acid rain research. Work on the Canada/U.S. MOI was stopped until more scientific evidence could be presented on LRTAP and effects.

Meanwhile, in Europe, in 1981 the Organization for Economic Co-operation and Development (OECD) Air Management Group completed the first cost-benefit analysis of sulphur dioxide, sulphate aerosol and acid rain for Western Europe (OECD 1981), showing that benefits were at least comparable to costs and might considerably exceed them in monetary terms. In 1982, the Second Stockholm Convention focussed on acid rain and the progress made since the 1972 Convention. Canadian and American scientists and air quality managers participated in expert meetings and high level discussions.

In addition, the UN Economic Commission for Europe (ECE) began to focus on acid rain and the issue moved from OECD to ECE. The European Monitoring and Evaluation Program (EMEP) was established to coordinate and perform European acid rain monitoring and modelling. Canada and the USA contributed to the work of EMEP which established a Working Group on Effects to study air quality and deposition effects, and especially to define critical levels. In 1983, the UN ECE Protocol on sulphur dioxide was approved at a high level meeting in Munich and opened to be signed by countries. Canada signed this Protocol although the U.S.A. declined to do so (UN ECE 2013b).

In 1984–85, then US EPA Administrator Ruckelshaus proposed to do a lake survey in the United States to determine the extent of acid rain effects. A large supplemental appropriation (about $80 M USD/year) was obtained, with $10 M USD/year going to the United States Forestry Service (USFS) for surveys of forests. This support led to increased independent effort to build regional acid rain deposition models for North America. Increased effort was put into modelling photochemical oxidants on a regional basis (Clarke et al. 1989).

In 1990 the final National Acid Precipitation Assessment Program (NAPAP) Assessment of Acid Deposition was released as a multi-volume, comprehensive compilation of the state of science and technology for this issue. Around this time also, the US EPA, Environment Canada, the Ontario Ministry of the Environment, and the Electric Power Research Institute (EPRI) conducted a monitoring study for 2 years in order to produce an integrated air quality and atmospheric deposition data set for sulphur compounds which could be used to evaluate the performance of U.S. and Canadian regional atmospheric deposition models. Model improvements resulted from this work. Meanwhile, science showed that the 20 kg per hectare number used as a target under the Eastern Canada Acid Rain program was only sufficient to protect lakes that were "moderately sensitive".

The Canada/US Air Quality Agreement, discussed in Chap. 16, was signed on March 31, 1991, establishing national and eastern emission caps and requiring a substantial sulphur dioxide emission reduction in the United States as well.

In 1993, the Canadian Council of Ministers of the Environment (CCME) met with the Council of Energy Ministers in the first Joint Ministers Meeting (JMM). They created a Comprehensive Air Quality Management Framework for Canada along with a National Air Issues Steering Committee (NAISC) and a National Air Issues Coordinating Committee (NAICC) to implement it. In 1994, an Acid Rain Task Group was established under this mechanism to revisit the acid rain issue. In 1998, all 26 of the federal-provincial-territorial governments signed a "Canada-Wide Acid Rain Strategy for Post-2000" (CCME 2011) which has the long term goal of meeting critical loads through reduction of emissions. An important foundation of this strategy was its design as a framework for addressing the problem by protecting lakes and forests. The second phase of sulphur dioxide reduction was designed to bring wet sulphate deposition throughout eastern Canada to below "critical load" levels. In some areas these "critical loads" are as low as 8 kg/ha/year.

According to the 2012 Progress Report under the 1991 Canada/US Air Quality Agreement (Canada/United States 2012),

"Canada's total emissions of sulphur dioxide have decreased by 57% from 1990 levels while the U.S. has reduced total sulphur dioxide emissions from covered sources by 67% from their 1990 emission levels. Between 2000 and 2010, Canada reduced total emissions of nitrogen oxides by 40% in the transboundary ozone region while U.S. total nitrogen oxide emissions decreased by 42% in the region."

The "acid rain" issue is often thought of as a "poster child" issue from a science/policy perspective. While publicly it is touted as a "solved" issue, and Canada and the US have been very successful in meeting the requirements of the Canada/US Air Quality Agreement, scientists have continued to raise concerns that critical loads are still being exceeded in many sensitive lakes, and that impacts are occurring on forests as well. Environment Canada's website as of October, 2012, states that "Between 21 and 75% of eastern Canada, continues to receive levels of acid deposition in excess of critical loads." Lakes and their ecosystems have not recovered to pre-acid rain status and sensitive ecosystems are gone forever from many lakes.

With the realization that damage can occur at levels substantially below those that the emissions restrictions were designed to achieve, and with the continued focus federally being on eastern Canada, the Prairie Acid Rain Coalition was formed with Martha Kostuch as chair. In 2006, she argued that with the huge increases in energy production, Western Canada could see impacts from acid rain, especially forest health and growth, and demanded that the federal government do more monitoring in Saskatchewan and Manitoba where the problems could manifest. Unfortunately, Martha passed away in 2008. Efforts to address acid rain in the west refocused specifically on the Tar Sands development in Alberta.

In the east, resources for acid rain research were refocused on other emerging air quality issues and with the success of the Air Quality Agreement, there is little impetus to take a deeper look at the legacy of acid rain. It is known that acidification of lake systems cannot be "undone" and can have effects such as the release of heavy metals from soil and rock into the environment. Sect. 2.2.4.3 discusses heavy metals.

Beyond the research aspects of the acid rain issue, it served as a catalyst in Canada for the formation of a small, cohesive group of scientists who were recognized worldwide as not only being leaders in their fields but also able to communicate their messages effectively (Brydges 2004). Many of this group spent their whole careers working in scientific areas to support air quality management but have retired over a period from 2000 through the present. Perhaps one of their biggest successes was communicating the importance of air pollution to Canadians.

The Sudbury Smelter

Prior to the construction of the Superstack, the waste gases from the smelter in Sudbury contributed to severe local ecological damage. The use of open coke beds in the early to mid-20th century as well as logging for fuel resulted in a near-total loss of native vegetation. Exposed rocky outcrops were stained black, first by the pollution from the roasting yards, then by the acid rain. In some places there was a blackened and acidified layer that penetrated up to three inches into the once pink-gray granite.

The Superstack was built in 1972 at an estimated cost of 25 million dollars to disperse sulphur gases and other byproducts of the smelting process away from the city of Sudbury. After the stack was built, these gases, from the largest nickel smelting operation in the world at the Copper Cliff processing facility could be detected in the atmosphere around Greater Sudbury in a radius of 240 km. The stack is 380 m (1,250 ft) tall. This makes it the second tallest free-standing structure (behind the CN Tower) in Canada at the same height as the Empire State Building.

Construction of the Superstack was followed by an environmental reclamation project which included rehabilitation of existing landscapes and selected water bodies such as Lake Ramsey. Over three million new trees were planted within the Greater Sudbury area in an ambitious greening program. In 1992, Inco and the city were given an award by the United Nations in honour of these environmental rehabilitation programs.

While the Superstack lowered the ground-level pollution in the city, it dispersed sulfur dioxide, and nitrogen dioxide gases over a much larger area. The heavily industrialized Ohio Valley contributed to the ecological problem of lakes as far north as northern Ontario. Research from data gleaned up to the late 1980s demonstrated acid rain to have affected the biology of some 7,000 lakes.

Prior to Vale's purchase of Inco, a major construction effort by Inco in the early 1990s added scrubbers to cleanse waste gases before pumping them up the Superstack. These upgrades were completed in 1994 and emissions from then on have been much reduced. Despite the 90% reduction in the sulfur dioxide and other gases, carbon dioxide and water vapour remain the most visible components and continue to contribute to the Sudbury Superstack's image as a pollution source.

SO_2 reductions have reached the point where the natural draught from the heat of the plume is no longer sufficient to provide enough buoyancy and natural gas burners and fans are now needed to move the SO_2 up the stack.

In contrast to the reduction of SO_2 emissions, Inco's Superstack still stands out in North America in its arsenic, nickel and lead emissions to the atmosphere. Using data compiled by the Commission for Environmental Cooperation (Taking Stock 1997), Inco alone accounts for 20% of all of the arsenic emitted in North America, 13% of the lead and 30% of the nickel. Although it is not strictly fair to compare a nickel-copper smelter to a lead smelter, by so doing one can get an idea of how poor the containment of lead is at Copper Cliff. In 1998, Inco emitted 146.7 metric tons of lead at Copper Cliff with a smelter production of 238,500 metric tons of nickel-copper matte. The EPA regulations in the United States require a primary lead smelter to limit emissions of lead to 3.0 gm per MT of product. With this emission factor, Copper Cliff would be required to limit emissions of lead to approximately 1.0 MT per year, demonstrating that the actual emission is about 150 times greater than allowed by US regulations for a lead smelter. Steps are underway to reduce emissions. As a result of the lead emissions from the Inco Superstack, the surrounding community of Copper Cliff was found to have levels of lead in soil tests at a level sufficient to cause harm to young children (Pollution Probe 2003).

This illustrates that although air quality issues are generally considered and managed as independent entities, there is a large overlap in them with some sources contributing to a range of pollution issues. Air toxics are discussed further in Sect. 2.2.4.

2.2.3 Smog (Ground Level Ozone and PM)

"Smog" is a sort-of generic term which bows to the importance of smoke and fog, but does not really address the components in a manner that is conducive to understanding the air quality implications. Smog has a number of "active ingredients" including ground level ozone and particulate matter which, while they often appear together in the air, present rather different air quality management issues.

Major Incidents

The global history of air pollution management goes as far back as the discovery of fire. Since the industrial revolution there has been recorded public outcry, without which, it seems, no action is taken. In England, for example, under Henry V (1413–1422) steps were taken to regulate the movement of coal in London and taxation was employed to restrict its use. Over 1,000 smog-related deaths occurred in 1909 in Edinburgh and Glasgow, Scotland (Encyclopedia Britannica 2013). The word "smoke-fog" was first used by Mr. De Voeux in his report to the Manchester Conference of the Smoke Abatement League on this event.

While severe pollution incidents have occurred many times throughout history, perhaps the three most famous "air pollution episodes" occurred in the Meuse Valley in 1930, in Donora in 1948 and in London in 1952 (Phalen and Phalen 2012).

In the 25 km from Huy to Liege along the Meuse River in the 1920s there were 4 coke ovens, 3 steel mills, 4 glass factories and 3 zinc smelters. On December 1, 1930 a heavy fog descended. At first people became ill, but then on the third day, 60 people died. It wasn't until December 5 that the deadly fumes dissipated. As well as people, many cattle died and effects were reported in wild animals as well. The severity of the incident did not lead to anti-pollution measures and in September, 1972, an industrial accident coupled with a severe fog produced another incident heavy in sulfur dioxide. Fortunately, there were no deaths from that one.

Donora was a heavily industrialized town of almost 13,000 residents situated in inner bend of the high, narrow river valley of the Monongahela River, about 32 km south of Pittsburgh. Various United Steel facilities including a steel and wire works, zinc works and coal mining emitted chloride, fluoride, hydrogen sulfide, sulfur dioxide, cadmium oxide, as well as soot and ash into the air. On Tuesday, October 26, 1948, an inversion formed in the valley and persisted for 5 days until it rained on Sunday, October 31, trapping the pollutants in the USA's worst air pollution incident. A medical doctor at the scene, Dr. R.W. Koehler, wrote as he observed a passing train: "they were firing up for the grade and the smoke was belching out, but it did not rise…It just spilled out over the lip of the stack like a black liquid, like ink or oil, and rolled down to the ground and lay there." Twenty people died due to respiratory tract infections from hypoxia and due to obstruction in their air passages by pus from infected lungs. Most of the 5 women and 15 men who died were elderly and succumbed on the third day of the episode. 6,000 people were affected. Although US Steel settled out of court, this incident provided some of the impetus behind the passage of the U.S. Federal Clean Air Act in 1955 (Shenk 1970).

But the episode that galvanized the world's attention on air pollution as an issue was the "great smog of London" which started on Friday, December 5, 1952. The meteorological situation was such that an "inversion" limited the dispersion of pollutants in the vertical and the circulation pattern held them together in a vortex over the city. The situation persisted for five days with visibility so poor that people were afraid to leave their homes in case they would get lost.

Hospitals were crowded and eventually the numerical tally was made. There had been about 4,000 more deaths than normal for a five-day period, with many of those who died having pre-existing heart or lung disease. Deaths due to chronic respiratory disease increased tenfold and hospital admissions for respiratory illness increased threefold. Claims to the national health insurance system were 108% above normal.

The British didn't immediately recognize the horrifying nature of the incident but as details emerged, public thinking about air pollution and how it should be managed was changed forever, not only in Britain but in Canada and other countries as well. One person who experienced the London Smog first hand and brought his interest in impacts of pollution on human health to Canada, was David Bates.

In Canada, this issue was first recognized in the late 1950s when damage was noticed to crops of tobacco grown along the north shore of Lake Erie. The "weather fleck" damage was shown to occur following higher than normal concentrations of ozone (50–150 ppb). Air quality management in this period was driven by voluntary actions. For example, the Sarnia region, heavily industrialized already, initiated the Sarnia-Lambton Environmental Association (SLEA) in 1952. It still monitors ambient environmental conditions today, and shares information with the community as well as government agencies.

At that time, air quality objectives were set in a two stage process, with NRC providing a synthesis of the science of the pollutant, and a Federal-Provincial Committee establishing objectives by factoring in other dimensions.

Ground Level Ozone

Ozone was discovered by Schonbein in Basel, Switzerland in the middle of the 19th century. He met Michael Faraday when he was in England teaching German at a boarding school and noted that an electric spark is often followed by a memorable smell. Later, when he became a professor of chemistry back in Basel, he was able to show that the smell was caused by ozone, which he went on to show could be intensely irritating (Ruben 2001).

It was noted in early aircraft such as the deHavilland Comet that oxygen masks used at altitudes over 10,000 m deteriorated rapidly. When Dr. David Bates joined the Royal Victoria Hospital in Montreal in 1956, he was interested in this problem in the context of how ozone might react with peoples' lungs. He did an experiment with the first DC-8 by putting rubber bands in places onboard the airplane and running a control experiment with rubber bands in a box on the ground, measuring ozone simultaneously in both places. This experiment was made possible because ozone measurement methods had just been developed. From this work it was estimated that ozone levels in the cockpit would average about 50 ppb. Subsequently, the compressors controlling cabin pressure in aircraft were modified to minimize ozone releases reducing the symptoms that aircrew had complained about (Young et al. 1962).

These early results encouraged Dr. Bates to focus on the small airways of the lung and over the next few decades he and other researchers developed a good understanding of the effects of ozone exposure (Bates 2006). Inflammation in the lung was found to be an early effect of ozone exposure, persisting up to 24 h after the exposure ceased. It was later found that although the lung responds to ozone exposure by thickening its coating of mucus to recover functionally, this underlying inflammation may persist and damage the pulmonary structure.

It was recognized that ozone and other photochemical oxidants in lesser amounts could be formed in the atmosphere by the photochemical oxidation of hydrocarbons and other organic pollutants in the presence of nitrogen dioxide and other pollutants such as aldehydes, ketones, alcohols, and organic acids. In the Los Angeles area, peak values of 500–750 parts per billion (ppb) in air by volume were measured, and other American cities began to record elevated levels. Mexico City and other big urban centers worldwide were beginning to take a much more serious interest in air pollution. While it was still not absolutely clear what the specific contributions of ozone and particulate matter were, it was now very clear that air pollution causes significant human health effects.

There are several key differences between the air pollution caused by ground level ozone and that caused by acid rain. Ozone is not "released" from sources as a pollutant, rather it forms in the atmosphere under certain conditions (such as the presence of sunlight) from "precursors" which are emitted. In fact nitrogen oxides, that are also part of the acid rain issue, play an important role in the atmospheric chemistry causing the formation of ozone. The other chemical family that is important in the process are the volatile organic compounds (VOCs). In addition, it was later discovered that the oxidants that were formed with ozone played a role in the formation of acid rain and other acidic deposition, thus relationships between air pollution and health can be complex (Bates and Sizto 1987).

Over the next decades, in addition to Dr. Bates, other Canadian scientists became active in studying the health effects of air pollution and this link between health scientists and air quality became an important feature of Canadian air quality management.

Dr. Rick Burnett, of Health Canada, pioneered use of data available through our health care system to do statistical analyses of large data bases on hospital admissions among others. Dr. Burnett worked with colleagues in Canada at Harvard and worldwide, who followed up with work substantiating the health consequences of air pollution. He established links with all the foremost authorities in the area, using powerful statistical analyses to reveal relationships in huge data bases. Studies done in the 1990s established the links between air pollution and health in many useful ways (Burnett et al. 1994a, 1994b, 1995, 1997, 1998, 1999).

Dr. Jeff Brook pioneered making the complex measurements needed to support studying the linkages between air quality and health in the field. Once associations are found from data, specific studies linking pollutant exposure to effects directly are important, as are characterizations of pollutants, (Brook and Johnson 2000, Brook et al. 2002). More about this in the next section on particulates.

The results of these studies have been important in designing the most effective air quality management approaches. The ozone issue revolutionized the engagement of the health community in air quality issues. With early analysis based on Canadian hospital admission data, it was shown that health effects were significant, piquing the interest and engagement of a wider health expert community. A Canadian NOx/VOC Assessment was published by Environment Canada in 1996 through the Canadian Council of Ministers of the Environment (CCME), which had issued a Management Plan for Nitrogen Oxides and Volatile Organic Compounds (Phase 1) in 1990. The Assessment included several volumes and presented a compelling, integrated overview of what was known about ground level ozone at the time, including the health dimension. As with the assessment work done on acid rain, governments began to take action.

The CCME, previously mentioned, had coordinated work on National Ambient Air Quality Objectives (NAAQOs) since the mid-eighties. They also worked towards "harmonization" of federal and provincial environmental management policies and programs as described in Chap. 15. While it wasn't until the Canada-Wide Accord on Environmental Harmonization was signed in 1998 that this was formalized, this kind of thinking drove much of the air quality management of the 1990s. The Accord was principle based including consideration of: polluter pays, precaution, pollution prevention, science-based, transparent, consensus-based and inclusive (role for Aboriginals) and flexible implementation.

The Federal-Provincial Working Group on Air Quality Objectives and Guidelines (WGAQOG) had developed objectives for carbon monoxide, total suspended particulate and sulphur dioxide in previous years. They now worked to determine a "reference level" for ozone (a level above which there are no demonstrated effects on human health and/or the environment). This technical group took into account all the latest work on effects and concluded that there was no such level for ground level ozone. They then took a pragmatic approach to protecting the health of Canadians (Federal-Provincial Working Group on Air Quality Objectives and Guidelines 1996, 1998, 1999). "The Canada-wide Standards for PM and Ozone… represent a balance between achieving the best health and environmental protection possible and the feasibility and costs of reducing the pollutant emissions that contribute to PM and ground-level ozone in ambient air" (CCME 2013).

The CCME then coordinated action to lower smog levels in problem areas of the country. In June 2000, the federal, provincial and territorial governments (except Quebec) signed the Canada-wide Standards (CWS) for Particulate Matter (PM) and Ozone which committed governments to reduce PM and ground-level ozone by 2010. The CWS and related provisions for ozone are: A CWS of 65 ppb, 8-hour averaging time, by 2010; achievement to be based on the fourth highest measurement annually, averaged over 3 consecutive years.

The CWS reflected driving principles such as: continuous improvement and keeping clean areas clean. Since Canada has an enviably clean atmospheric environment, these principles were to stress that it is not acceptable to pollute up to the level of the CWS, these are levels to be attained in areas of higher pollution. Hence special measures are needed in these areas as they develop to prevent pollution.

There is still further research to be done to clarify the time scales on which health effects depend and to sort out the finer details of the volatile organic compounds' role in the chemistry of ozone. If there are health outcomes due to chronic low level exposure that differ from outcomes of acute exposure, the form of the standard may need revision to protect health (Jerrett et al. 2009).

This is a very different approach to that taken in the USA to protect their Class 1 areas. It is also notable that the CWS does not explicitly consider visibility as an issue the way it has been considered in the United States. See Chap. 8 and 16 for more details.

A number of "joint initial actions" by federal and provincial governments were agreed to, including: providing more thorough and timely air quality information to governments, industry and the public by linking jurisdictional databases of ambient air quality data and facilitating access to existing public information; reducing emissions from the transportation sector and from residential wood burning; and developing national multi-pollutant emission reduction strategies for: Pulp and Paper; Lumber and Allied Wood Products; Electric Power; Iron and Steel; Base Metals Smelting; and Concrete Batch Mix and Asphalt Mix Plants.

The complex atmospheric chemistry that takes precursors and converts them over hours and hundreds of kilometers into ozone presents a challenge not only to the chemist, but also to the air quality manager. Recognizing that there is a significant transboundary component to the ground level ozone issue, in December, 2000, the Canada/U.S. Air Quality Agreement was modified by the addition of an annex on transboundary management of the precursors of ozone including specific objectives for volatile organic compounds and nitrogen oxides, to reduce transboundary flows of tropospheric ozone and their precursors.

Specific provisions related to transboundary flow were written into the CWS as well. Canada and the U.S.A. each declared a Pollution Emission Management Area (PEMA) for the application of the Annex. For Canada it included 301,330 km^2 south of the 48th parallel from the Ottawa River to east of Lake Superior. For the United States, the area was comprised of the states of Connecticut, Delaware, Illinois, Indiana, Kentucky, Maine, Maryland, Massachusetts, Michigan, New Hampshire, New York, New Jersey, Ohio, Pennsylvania, Rhode Island, Vermont, West Virginia and Wisconsin as well as the District of Columbia. For Canada, the response was focused less on NOx from stationary sources and more on VOCs from a number of vehicular and off-road engine applications.

In 2003, an update "Atmospheric Science of Ground-level Ozone: Update in Support of the Canada-Wide Standards for Particulate Matter and Ozone" was released (CCME 2003). It concluded that the overall science had not changed dramatically since the 1996 assessment but that action should continue on the basis of those recommendations.

In 2007 a detailed "Guidance Document on Achievement Determination" for CWS was developed under the CCME by a Working Group on Monitoring and Reporting having largely similar membership to the Working Group on Air Quality Objectives and Guidelines (CCME 2007).

In the 2012 Progress Report under the Canada/US Air Quality Agreement, there is a table that shows preliminary 2010 emissions from the United States and Canadian PEMA. There are NOx reductions mainly from on-road mobile sources and electric power generation. VOC reductions are primarily form on-road and nonroad mobile sources and solvent utilization. In general, over the last decade, on-road vehicle emission reduction has driven the overall decline in emissions.

Particulate Matter

As has already been discussed, advances in the management of air quality in Canada have been linked closely to advances in associated sciences, especially monitoring. It is similarly true with particulate matter. Of course the Trail Smelter and other large industries had been known to produce large amounts of "smoke" and the "smog" incidents associated with ozone exposure had also been associated with visibility impairments, but it wasn't until the mid-twentieth century that the solid dust which was transported by the atmosphere began to be characterized as "particulate matter" and studied and eventually managed as a separate issue. The smog" events discussed in the section on ground level ozone were, of course, particulate events as well and the two issues tended to be managed as one until the science developed sufficiently to provide relatively clear characterizations of the pieces requiring different approaches to management.

In Canada and the United States, the highly industrialized Detroit River area was the subject of complaints from people in the 1940s. This area contains the cities of Detroit and Windsor as well as a number of smaller municipalities. Governments gave a reference to the International Joint Commission in January, 1949, to protect public health and welfare on either side of the international border from industrial emissions, including those from vessel traffic on the river.

Under the leadership of Morris Katz, a sampling study was designed using accordion-pleated filters in high volume samplers which filtered the air at a measured rate of about 50 cubic feet per minute. First, a six week study was conducted at 32 sampling sites in Detroit, which was followed by a similar study at 25 sites in and around Windsor. The total weight of particulates was measured and samples were analyzed for silicon, calcium, aluminum, iron, magnesium, lead, manganese, copper, zinc, titanium, tin, molybdenum, barium, nickel, vanadium, chromium, cadmium, beryllium, antimony, and cobalt (in decreasing order of results by median weight). These measurements were used to select high and low pollution areas in Detroit and Windsor which contained well defined population groups. This work was followed up by a health pretest in 1953 in which field trials of questionnaires were carried out. The Detroit Health Department and the City of Windsor participated by studying general health and medical care records.

Several interesting outcomes resulted from this work:

- Vessel emissions of black smoke, fly ash, and gaseous combustion products were found to be particularly objectionable because of their nearness to residential, recreational and civic land uses. A voluntary control program was initiated in 1954 sponsored by the IJC with the cooperation of the Lake Carriers and Dominion Maritime Associations as well as conversion of vessels for more efficient fuel-burning.
- In the organic fraction, more than 20 metallic elements were identified as well as chlorides, sulfates, nitrates, fluorides, and carbonates. The distribution of lead was correlated with the density of vehicular traffic.
- Organic constituents began to receive increased attention as they were recognized as potent carcinogens. The increasing incidence of lung cancer was associated with exposure to polycyclic aromatic compounds in particular the benzopyrenes in the atmosphere.

This work and its results indicate that it was recognized early on that particulate matter is exceptionally inhomogeneous in the atmosphere and that it varied enormously in composition depending on its source and the atmospheric processes to which it was exposed.

In the 1950s, dustfall was measured in many Canadian cities. Suburban levels were found to be rather constant with large variations in the industrial and commercial areas of a city. The air in downtown and industrial sections contains about twice to more than three times the air-borne dust found in suburban zones. It was recognized early that smoke particles of submicron size contribute very little to the weight of air-borne dust but influence soiling and visibility characteristics. Filtering air through paper tape and measuring the optical density of the stain deposited led to the Coefficient of Haze (CoH) per 1000 ft of air. The highest levels were found during the heating season winter months and minimum levels in the summer. Weekly and daily cycles were also noted with levels lowest on weekends and highest in the mornings, about 8–9 a.m.

Robert Edward Munn was born in 1919 in Winnipeg, Manitoba and graduated from McMaster University in 1941. He joined the Meteorological Division of the Canadian Department of Transport, forerunner of the Atmospheric Environment Service (AES) after graduating, and trained as a meteorologist in Short Course and Advanced Course #3. After initial postings, he completed an M.A. degree in 1946, was posted to Gander, Newfoundland, and then moved to the Public Weather Office in Halifax after the war ended. There he began to write and publish technical papers such as "A Survey of the Persistence of Precipitation at Halifax".

In 1956, Ted went to Windsor, Ontario, as an air pollution meteorologist. He published "Suspended particulate concentrations: Spatial correlations in the Detroit-Windsor area" in Tellus in 1975 based on the data set collected under the auspices of the IJC on total suspended particulate (Munn 1975). The data were collected every third day from September, 1967 to November, 1968. This paper linked the air quality data firmly to meteorological conditions and, by noting the patterns in the urban areas were less homogeneous than in the rural areas, gave early hints that particulates could be both a short range and a longer range air pollution issue.

After retirement from the public service after 35 years, Ted Munn started a new career as an Associate Professor at the Institute for Environmental Studies at the University of Toronto in 1977. He led the environment program at the International Institute for Applied Systems Analysis (IIASA) and was on the editorial boards of over 15 journals. He edited "Boundary Layer Meteorology" for 25 years and was recognized in the 25th Anniversary Volume in 1995 (Taylor et al. 1995). In 2003, the Encyclopedia of Global Environmental Change, consisting of five theme-oriented volumes each edited by a distinguished expert was published by Wiley. Ted Munn, as the overall editor, endeavoured to have the 3400 page story consistently in one voice for ease of reading.

When the National Air Pollution Surveillance Program (NAPs) was set up in 1969, in addition to sulphur dioxide and nitrogen dioxide, it measured Total Suspended Particulates (TSP). The changes in understanding of the health impacts of particulate matter over the years have translated into major monitoring transitions which make the derivation of long term trends difficult (Sirois 1998).

Air pollution scientists over the next decades focussed on the management of the issues most amenable to management, acid rain and ground level ozone and stuck with a simple model of particulate matter. In 1971, the United States Environmental Protection Agency (US EPA) set standards for Total Suspended Particulate: an annual geometric mean of 75 $\mu g/m^3$ (primary health based) and a 24 h average of 260 $\mu g/m^3$ (primary) and 150 $\mu g/m^3$ (secondary environmental) not to be exceeded more than once per year.

In the meantime, a wide variety of research was done looking at health effects of air pollution (Samet et al. 2000). These studies, while often exploratory, laid the foundation for an explosion of studies on air pollution and health starting in the 1980s, which focussed on particulates. For example, the aforementioned OECD report of 1981 drew from these studies and focussed on the inhalation of sulphate aerosols as the key acidic atmospheric component in acid deposition causing the morbidity health effects that were estimated in the study. The role of nitrogen oxides was also explored (Brook et al. 2007a). The role of the transportation sector in urban settings was also studied (Brook et al. 2007b).

Unfortunately, with the advent of such a focus on health effects as a driver for management decisions, there are still great uncertainties as to the role that the constituents of particles play in their responsibility for observed health effects. Size of particles has been shown to be significantly associated with health effects, with coarse particles being responsible for less health effects than fine. Typically particulate matter with diameters between 10 and 2.5 μm is considered 'coarse', particulate matter of 2.5 μm or less in diameter, $PM_{2.5}$, is classified as 'fine' (PM_{10} is the combination of coarse and fine) and particulate matter with diameter less than 0.1 μm or less being referred to as ultrafine. Hence, efforts to regulate particles focus on standards and guidelines for ambient levels of particles of certain sizes. Scientists also determined that sulphate and nitrate aerosols accounted for a very large portion of the effect of $PM_{2.5}$. This connected $PM_{2.5}$ to acid rain and ozone and supported the concept of "one atmosphere". Air Quality Managers could no longer address each pollutant separately and atmospheric models that were used in assessing air quality control strategies embraced this concept.

The emissions of particulates are also complex with both primary and secondary routes being important. Coarse particles come mainly from crushing and abrasion processes (e.g. road dust, airborne soil, pollen, plant and animal fragments). Both primary (directly emitted) and secondary (formed in the atmosphere) fine particulate come from combustion processes such as transportation and power generation and can include a wide array of constituents from heavy metals to elemental and organic carbon as well as sulfates and nitrates. One comprehensive summary of the state of knowledge about emissions was done under NARSTO (Hidy et al. 2003). NARSTO was a tripartite organization (Canada/US/Mexico) which carried out several large assessments including one on particulate matter.

Particulate matter can remain suspended in the atmosphere for timeframes up to weeks, depending on size and other properties. It is removed from the atmosphere by wet and dry deposition. Levels can be quite variable spatially. Some idea of this with respect to $PM_{2.5}$ has been established from monitoring, but measurements for PM_{10} and ultrafines are not dense enough to define the distributions and their variability.

The United States revised its approach to managing particulate matter when in 1987 it brought in a standard for PM_{10}. A 24 h average of 150 $\mu g/m^3$ was not to be exceeded more than once per year on average over a three-year period and an annual arithmetic mean, averaged over 3 years was not to exceed 50 $\mu g/m^3$.

Particulate matter was recognized as much more complex than earlier air quality issues. From a scientific perspective was it the size of the particle that was the major concern, the chemical components of the particle, or its chemical reactivity, or perhaps a mixture of all of these factors? The sheer volume of scientific work done in the past several decades is formidable, and the bottom line is that in spite of active air quality management programs in place, the final word is not in on a number of important scientific issues. While this leaves challenges for air quality management, it is clear that reduction of air pollution can have significant health benefits, see Chap. 7.

In order to get a firm grip on the effects of air particles on human health and hence design management approaches to reduce these effects, it would be helpful to understand that the observed effects are biologically plausible and thus that the air pollution is causing the health detriment. It turns out that this is very difficult to establish. One of the key tools which would help air quality managers would be a dose-response curve relating exposure to the severity of health effects.

Inhaled particulate from smoking is much more concentrated than that experienced from most ambient air. The dangers of smoking were established over this period and a profound cultural change has taken place to see far fewer smokers than in the past. If dose response was linear, we would expect that these changes would constitute a powerful air quality management program which would significantly reduce health impacts of particulates. It is emerging, however, that an empirically estimated dose-response function is relatively steep at very low levels of exposure, has no "threshold" for effects, and flattens out at higher levels of exposure.

To support air quality management, it is important not only to show that health effects are caused by an issue, but also that reductions in exposure result in risk reductions. There remains work to be done to clarify why combustion-related fine particles appear to cause the bulk of health effects and what the synergistic effects of co-pollutants might be among other things.

In the presence of arguably more science than any other air quality issue, but in the absence of a clear concise and well agreed 'story' integrating all the pieces, air quality management of particulate matter in Canada is still evolving as will be seen in Chap. 7. However, the CCME included particulate matter in the development of CWS with ground level ozone in 2000. The CWS and related provisions for PM are: A CWS for $PM_{2.5}$ of 30 μg/m^3, 24 h averaging time, by year 2010; achievement to be based on the 98th percentile ambient measurement annually, averaged over 3 consecutive years.

In 2004, a joint scientific assessment of particulate matter was done under the auspices of the Canada/US Air Quality Agreement and found that atmospheric particulate is a binational issue with transboundary significance. It was agreed to discuss the issue further with a view toward eventually developing an Annex to the agreement similar to the Ozone Annex. There have been further discussions since this time and it will be interesting to see whether, and in what forum this issue is pursued.

The US EPA revised standards in 2006 to include a $PM_{2.5}$ 24 h standard of 35 μg/m^3 as 98th percentile, averaged over 3 years and an annual arithmetic mean, averaged over 3 years of less than 15 μg/m^3 and a PM_{10} 24 h standard of 150 μg/m^3 not to be exceeded more than once per year on average over a 3 year period. The review cycle of the PM_{25} standard will be completed in late 2016 and will include yet another level of science.

In Canada, a number of management issues were recognized early on. While particle size is proxy for health effects to a degree, with smaller particles being more harmful, until recently there has been no national monitoring program in place, based on standardized methodologies and having appropriate levels of data management and analysis to develop the basic data about particulate levels across Canada. As already pointed out, historically it has not been clear what would be best to measure to support air quality management. Chap. 3 gives an overview of the monitoring ongoing at present in Canada. Jeff Brook has also conducted a number of specific local studies to establish particulate characteristics (Brook et al. 2004) and pioneered sophisticated mobile techniques to do so (Levy et al. 2012).

In contrast to acid rain, which tends to have a few major sources which dominate (smelters, electrical generating stations), the sources of particulate matter are extremely varied. The National Pollutant Release Inventory (NPRI) was designed to capture the emissions of large sources of pollutants; small sources are not required to report.

Beyond the issue of particle size, there is the issue of particle composition. There are indications that some particles are much more "active" in causing health responses than others. Currently there are no measurements of ambient constituents of particulate matter being carried out nationally according to agreed protocols. Thus it will be some time (decades at least) before we see a fully evidence-based approach to management of particulates. While the health science is moving towards certainty about health impacts of particulate matter, the physical science side is still not in a good position to respond to the health driver with crisp approaches to management of the issue. See Chap. 20 for some future directions.

In 2008, a Canada-United States Border Air Quality Strategy was announced having three joint projects:

- The Great Lakes Basin Airshed Management Framework;
- The Georgia Basin/Puget Sound International Airshed Strategy; and
- A study on the feasibility of Emissions Trading for nitrogen oxides and sulphur dioxide.

How these initiatives would intersect with the proposed way forward on air quality management laid out in a Notice of

Intent published in the October, 2006, Canada Gazette by the Ministers of Environment and Health "to develop and implement regulations and other measures to reduce air emissions", mainly under CEPA and Canada's Clean Air Act, is not very clear. The departments link air quality to climate change in the Gazette and propose a variety of measures to address both, with an early focus on the transportation sector. This sector has seen the development of a number of new regulations. Key elements include: emission targets and timelines, compliance options, and compliance assessment, monitoring and reporting, mostly bringing Canada in line with American regulations. A key part of this approach was asking the National Round Table on Environment and Energy (NRTEE 2008) for advice on national objectives for ambient air for particulate matter and ozone for the future and national emission reduction targets for a variety of chemicals for 2050 as well as actions on greenhouse gas reductions. With the dissolution of the NRTEE at the end of March 2013, it will fall to the CCME to continue to manage the direction. Some possible future directions are outlined in Chap. 20.

Sulphur in Gasoline

One piece of a uniquely Canadian approach to dealing with these complex air quality issues has been to use the Canadian Environmental Protection Act (see Sect. 2.2.4.4) to regulate the level of sulphur in gasoline. The Regulations now limit sulphur in gasoline produced, imported or sold to an average level of 30 mg/kg with a never-to-be exceeded maximum of 80 mg/kg. While a thorough discussion of the current state of "Transportation Emissions; Sources and Regulations" is given in Chap. 10, the historical approach to regulating this area is a Canadian success story.

This initiative started well before the Notice of Intent when the CCME initiated a Task Force on Cleaner Vehicles and Fuels in 1994 to develop options for "a national approach to new vehicle emission and efficiency standards and fuel formulations for Canada, recognizing regional/urban realities". The Task Force undertook the first explicit cost/benefit studies nationally in Canada with respect to air quality management "Environmental and Health Benefits of Cleaner Vehicles and Fuels (Hagler Bailly Study)". The results were based on the latest science of the time, and, although not perfect, followed study protocols that set the bar for work to follow. They also convincingly made the case for benefits from regulations in the tens of millions of dollars range.

In their report (Environment Canada 1995), the Task Force recommended, that:

- Canada adopt tighter new vehicle emission standards in harmony with the U.S. (including that Transport Canada should update regulations under the Motor Vehicle Safety Act to harmonize with standards currently in regulations under the U.S. 1990 Clean Air Act).
- The federal government in concert with the provinces and stakeholders through the National Air Issues Coordinating Committee move toward making advanced technology vehicles available for sale in a timely manner.
- The provincial governments consider a menu of actions including development of inspection and maintenance programs, early retirement of high-emitting vehicles, transportation demand management, remote sensing for high emitting vehicles and vapour recovery at gasoline service stations.
- Ministers endorse fuel efficiency improvement through changes in driver behaviour, improved on-road efficiency, purchase decisions of more efficient vehicles, and improved vehicle fuel efficiency technology.
- Environment Canada lead the development and implementation of a regulated national standard to ensure provision of 100 % on-road, low-sulphur diesel.
- Environment Canada, in consultation with provinces and stakeholders, lead in the development and implementation of a regulated minimum national standard for gasoline.
- Environment Canada, with direct involvement of the provinces and other stakeholders, establish an effective, continuing process to ensure a coordinated approach on vehicle and fuel emission control programs; and Environment Canada report to CCME Ministers on actions taken with respect to implementation of these recommendations.

It is interesting to note that these recommendations have been followed up and resulted in a remarkably coherent and effective regulatory approach to on-road transportation in Canada. The 2006 Notice of Intent and all that has followed show remarkable consistency over a lengthy period of time in federal-provincial discussion.

2.2.4 Air Toxics

Management of Air Toxics has taken very different routes depending on a variety of factors. There are air toxics, such as heavy metals and persistent organic pollutants (POPS) that have been seen to have environmental and health effects, and hence, like the issues already discussed, have had actions to address them due to the smoking gun. There are many other toxic chemicals that may or may not have an air pathway or effects related to an air pathway but which are subject to screening and the possibility of management. These families of toxics have had rather different management pathways.

Toxic chemicals in the atmosphere are even more complex than particulate matter because of their huge range of physical and chemical properties and the ways they can interact with other chemicals, the environment and human health. Because of this complexity, a comprehensive presentation of the history of air toxics management is beyond the scope of this chapter. Rather than attempt to cover the whole, a few key elements of the history of air toxics management will be presented in the context of the chemicals that are managed. There remains a lot to be done in this area, not only in Canada, but globally as well. That being said, the science has ensured that the most dangerous chemicals are being addressed.

The issues that have already been discussed have been isolated from atmospheric components at large because of their specific impacts on health or the environment. Part of the challenge with air toxics is that the impacts are extremely diverse and, in some cases, subtle. In addition, while the atmosphere may be a transport mechanism for some of these toxics, the atmospheric pathway may not be the source of the exposure of concern, which might be through consumption of contaminated fish, for example.

The Great Lakes Water Quality Agreement (GLWQA)

Growing public concern about the deterioration of environmental quality in the Great Lakes stimulated research on the inputs and behaviour of pollutants in the Lakes in the 1960s. At that time governments moved toward regulating discharges work on source, fate and effects of pollutants in the Lakes. The first Great Lakes Water Quality Agreement (GLWQA) was signed between Canada and the United States in 1972. The focus of the first work was to limit the input of phosphorus to the Lakes, which soon had success.

In the 1970s it was observed that the nests of the common terns around Hamilton Harbour were filled with eggs that had failed to hatch. Some of the eggs contained dead embryos while some of the chicks that hatched had severe deformities such as crossed bills. Meanwhile, on Lake Superior, double-crested cormorants were having reproductive problems, with their eggs having much thinner shells than usual. In addition, more and more fish throughout the Lakes were found with tumours, particularly of the liver.

A number of studies were done which linked such chemicals as dicholorodiphnyl dicholorethylene (DDE), a form of the insecticide dichlorodiphenyl trichloroethane (DDT), and polychlorinated biphenyls (PCBs) to the observed effects. Levels of these sorts of chemicals in the environment had previously been in the parts per billion, but autopsies of wild birds showed levels as high as 150 ppb of DDE and 300 ppb of PCBs. It was recognized that the birds were bioaccumulating toxic chemicals through the food chain. Micro-organisms such as plankton absorbed toxic chemicals and were consumed by fish. The fish stored the toxics in their fatty tissue. The birds, being at the top of the food chain, ate the fish and accumulated levels of toxics many times higher than in the environment.

In 1977, at Isle Royale, in Lake Superior, a study of fish from Lake Siskiwit (a landlocked lake on the island) revealed the presence of toxic chemicals including the pesticide toxaphene, which had never been used on the island. In addition, toxaphene was detected in the rain.

Scientists at the then Atmospheric Environment Service of Environment Canada began a measurement study at the Centre for Atmospheric Research (CARE) at Egbert, Ontario, a rural site away from obvious urban influences. They found that highest values occurred in the summer, and that these high values were episodic in nature. They used back trajectory modelling techniques to show that sources of these chemicals were often the southern USA. Many of the pesticides that were found were known to be extremely persistent. While chemicals such as toxaphene and DDT had been banned in the US and Canada, they were still used in Mexico, Central America and other parts of the world. Thus, for the Great Lakes, atmospheric transport accounts for a surprisingly high amount of toxic chemical deposition into the Lakes (Environment Canada 1994).

This work triggered an interest in a group of chemicals which are persistent organic pollutants or POPs. POPs are very stable chemical compounds and consequently can last in the environment for years or decades. They are also bioaccumulative, meaning they can concentrate in living organisms and accumulate up the food chain through fish, predatory birds, mammals and humans (Hoff et al. 1994).

The use and production of polychlorinated biphenyls (PCBs) peaked in the 1960s and are now declining. The Lakes effectively "breathe" such POPS with concentrations in the open lake waters (around 100–300 pg/L) being in equilibrium with the air. The atmosphere, especially in urban industrial areas, can be the major source of PCBs to the Lakes (Strachan and Eisenreich 1988).

When the Great Lakes Water Quality Agreement (GLWQA) was signed in 1978, little was known about this issue. However, by 1987, when the Agreement was revised, Annex 15 was added and was, perhaps, the most prescriptive agreement in place between two countries anywhere in the world. Under this Agreement, Canada and the United States were to operate an Integrated Atmospheric Deposition Network (IADN) which was to monitor atmospheric levels of a slate of toxic chemicals to be developed by the countries and reviewed by them. The US Congress helped get IADN established by giving EPA a 1 year deadline in 1990 to get this network established. Thus, the Canada-US partnership received a boost. Operation of the Network was governed by a formal Binational Implementation Plan approved by Canada and the United States in June, 1990. The stations were set up as "background" stations to measure pollutants at a regional scale.

In the United States, the Great Lakes Toxic Substances Control Agreement was agreed to by state governors in 1986 followed by a $100 million Great Lakes Protection Fund in 1989. On the Canadian side, in the 1980s the Ontario Ministry of the Environment spent $280 million on a program to remediate Areas of Concern and local hot spots (MacDowell 2012). In April, 1997, Canada and the United

States adopted the Great Lakes Binational Toxics Strategy (GLBTS) for virtual elimination of persistent toxic substances in the Great Lakes.

Regular peer reviews of the IADN were done over the years and the most recent, in 2010, included the suggestion that IADN should be integrated with other monitoring activities in the U.S.A. and Canada.

The new GLWQA, which has been revised in September, 2012 has removed the air quality aspects. Sadly, there is no other mechanism under which such monitoring of air toxics will be done on a routine basis.

The Northern Contaminants Program (NCP)

When chemicals were measured in the Arctic that were not used or produced there, it was postulated that these were deposited from atmospheric transport due to the "grasshopper" effect. Warm temperatures lift or volatilize toxic chemicals out of the soil to be carried by the wind. The chemical can condense onto particulates if the temperature drops. The chemical can deposit directly on a surface in dry deposition or can be washed and wet deposited by rain or snow. This process can happen many times over days or decades like the hopping of a grasshopper. On the global scale, toxics volatilize from tropical and temperate soils and migrate to colder regions. Hence, a few years after banned pesticides were found in the Great Lakes, they were found in the breast milk of Northern peoples.

POPs can enter the human system through traditional foods such as beluga muktuk (skin) and seal blubber. Aboriginal peoples, who rely heavily on such country foods, are particularly affected. Some POPs can be passed on from mother to child across the placenta, or through breast milk.

The Northern Contaminants Program (NCP), co-ordinated Canada's action on northern contaminants, including POPs, both nationally and internationally. A multi-disciplinary initiative, funded by the Government of Canada, the NCP addressed health, science and communications issues related to contaminants in Canada's Arctic. It was established in 1991 through the Government of Canada's Green Plan and Arctic Environmental Strategy (AES). Data immediately began to be collected and analyzed on air pollution issues in the North (Fellin et al. 1996) building on a solid base of existing Canadian research (Barrie 1967).

The NCP aimed at reducing, and where possible, eliminating contaminants in country foods harvested in the North, while providing information that assists decision-making by individuals and communities in their food use. Managed by a committee chaired by then named Indian and Northern Affairs Canada (INAC), comprised of four federal government departments (Indian and Northern Affairs Canada, Health Canada, Environment Canada, Fisheries and Oceans Canada), the territorial governments (Nunavut, Northwest Territories, the Yukon) and representatives of Northern Aboriginal organizations including Inuit Tapirisat of Canada (ITC), Inuit Circumpolar Conference (ICC), Dene Nation and the Council of Yukon First Nations, the NCP set new standards for transparency and applicability of atmospheric science.

Phase I of the Northern Contaminants Program (1991–1997) was followed by Phase II (1998–2003), which emphasized quantifying risks and focussing research on the effects of contaminants on human health. A Canadian Arctic Contaminants Assessment Report II (NCP 2003) focussed on health issues relevant to all northern consumers of country foods, including Quebec and Labrador Inuit. Canadian scientists stepped forward as world leaders in understanding the complexities of mercury in the Arctic, the linkages between the coming of Arctic spring to complex atmospheric chemistry, and particulate matter in the North among many other topics.

POPs can travel great distances around the globe through the atmosphere and hence require international action to manage. Most POPs substances of concern have been banned or severely restricted in Canada for years, but they are still produced, used and stored as waste in a number of other countries. The vast majority of POPs entering Canada's environment, as a result of transport through the atmosphere, come from foreign sources, in particular: the United States, Mexico and Central America, certain eastern European countries, including Russia, and certain southern and southeastern Asian countries. As a result, reductions of international releases of POPs are required to ensure continued environmental progress in Canada generally and in our North in particular.

Efforts to reduce POPs and also heavy metals such as mercury and cadmium (see Sect. 2.2.4.3) began in many international fora. The Canada-US-Mexico Commission for Economic Cooperation (CEC) established a North American program on Sound Management of Chemicals which set up working groups on DDT, mercury and other POPs. The LRTAP program of the UNECE set up working groups on POPs and Heavy Metals which developed protocols for the member Countries to sign. It was also realized that the "transfer matrix" concept developed for acid rain could be applied to some of the POPs and heavy metals.

While these regional programs were successful at regional scales, the most significant negotiations to reduce or eliminate emissions of POPs on a global scale began under the auspices of the United Nations Environment Programme (UNEP) in Montreal in June of 1998. The intent of the UNEP POPs Convention is to bring all countries under the umbrella of a single global agreement. Twelve POPs were targeted by the draft agreement reached in Johannesburg, South Africa, December 10, 2000. These fall into three broad categories:

- Pesticides—DDT, chlordane, toxaphene, mirex, aldrin, dieldrin, endrin, heptachlor.
- Industrial chemicals—PCBs, hexachlorobenzene.
- By-products and contaminants—dioxins and furans.

In March of 2000, Canada became the first country to make a specific funding commitment, $20 million, for POPs capacity building in developing countries and countries with economies in transition. This funding helped those countries find alternatives to the use of POPs, such as DDT. This commitment was well received by the developing world, and helped the final negotiating session to reach agreement to provide new and additional funding and technical assistance to developing countries and countries in transition to meet their obligations to minimize and eliminate POPs.

Canada is a leader in the science of identifying and assessing past and current sources of POPs, and in predicting global movement through the atmosphere. Canadian scientists have improved the ability to detect POPs in rain and snow and have contributed to tracking the accumulation of these chemicals up the food chain and into humans. These developments are the basis for policy decisions and action both in Canada and on the international scene. In parallel with some of this work, the Arctic Council was formed and initiated an Arctic Monitoring and Assessment Program (AMAP). Canadian scientists were active in the production of the first AMAP assessment (AMAP 1998) and in fact Canadian Russel Shearer of Aboriginal Affairs and Northern Development Canada (AANDC) currently chairs AMAP internationally.

As introduced above, within the North American region, Canada has also developed regional action plans with Mexico and the United States on chlordane, DDT and PCBs under the North American Free Trade Agreement's Commission on Environmental Cooperation. Regional action plans are being considered for lindane as well as for dioxins, furans and hexachlorobenzene.

While the United States is interested in northern contaminants, especially as they impact on the State of Alaska, there has not been the same level of collaboration on air issues along the Northern border between the US and Canada as in the Great Lakes Region. The International Air Quality Board of the IJC presented recommendations to the IJC Commission regarding air pollution along the northern border (IJC March 2010, August 2010, April 2012), but there is no mechanism such as the Canada/US Air Quality Agreement by which to take them forward.

In 2008, Canada participated in the International Polar Year (IPY). Under that program some of the work led by Haley Hung was supported to look at the movement of POPs across the Pacific and into Canada's North, in a natural extension of some of her earlier NCP work (Hung et al. 2005). Canadian scientists led many significant projects on a wide range of topics which was culminated by a major international conference held in Montreal in 2012. Following this major effort, support for Arctic air quality research has dwindled except in direct connection to initiatives such as the Beaufort Regional Environmental Assessment (BREA).

Heavy Metals

In Sect. 2.2.2.1 the Sudbury Superstack was discussed in the context of acid rain, but identified as a source of heavy metals as well. No air pollutant has a longer history than heavy metals, because these have occurred naturally in the environment since the dawn of time. There are many anthropogenic sources of heavy metals as well. Some major sources are smelting, iron and steel production, fossil-fired generating stations, industrial boilers, cement kilns, vehicles and engines, waste incineration and some products such as fluorescent lights.

The range of effects of heavy metals is wide. Some, like lead and mercury are toxic, while others such as zinc and selenium are essential micronutrients for good health. In this chapter a brief summary of the history of lead and mercury management in Canada is given as an introduction to the topic.

One of the biggest success stories in the management of air pollution is lead. Being very soft and pliable and highly resistant to corrosion, it was ideal for use in plumbing as well as for the manufacture of pewter. Ancient Romans also used it as a cosmetic. More recently, it was also used in paints and to solder food cans and water pipes. In the early 1900s, lead was already recognized as toxic, with the main observed cause being leaded paint. If children ingested this paint it was known to cause seizure, coma, and possibly learning disabilities or death.

In 1922 tetraethyl lead was added to gasoline to improve engine performance. While lead exposure had been largely due to ingestion, this step made the atmospheric pathway for lead more important. Five workers in an American Ethyl plant died of exposure in 1924 after going insane.

Industries refused at first to take any responsibility for these health impacts, claiming that children who ate paint were subnormal to begin with and that plant workers were working too hard. They also pointed to the fact that lead is an element that is present in soils and air naturally.

It wasn't until Clair Patterson (an American geologist) analyzed ice cores from Greenland in 1965 and showed that levels of lead in the recent years were much higher than in the past, that industry took note. In the face of data showing that people in the U.S. had blood concentrations of lead 100 times higher than the "natural" level, action was quick to come.

United States President Nixon signed the Clean Air Act (CAA) of 1970 into law on December 31st, and the Environmental Protection Agency, formed on December 2, was given the responsibility of lowering emissions of hydrocarbons, carbon monoxide and nitrogen oxides as well as the emissions of lead. Eventually the Consumer Product Safety Commission followed and lead in paint was banned in 1976. It wasn't until the 1990 amendments to the CAA that lead in gasoline was to be phased out by 1995. In 2008, the EPA tightened standards again requiring industries to reduce levels to.15 μg/m^3. The new standard is 10 times more restrictive than previous requirements.

In Canada, the first Clean Air Act was passed in 1970 and the Canadian Department of the Environment was formed in 1971. One of the first items on the agenda was lead (the others were asbestos, mercury and vinyl chloride). Lead was one of the first substances to be considered under the Environmental Contaminants Act and was then added to the List of Toxic Substances (Schedule 1) of the original Canadian Environmental Protection Act (CEPA).

These measures have worked and levels of lead in the atmosphere have plummeted. While continued efforts to reduce levels are necessary given the serious nature of potential health impacts from lead and the fact that people today have several hundred times more lead in their blood than people did a century ago, it remains one of the best success stories in environmental management to date.

Meanwhile, one of the ongoing challenges is mercury in the environment.

Throughout the ages mercury has found uses with the compound cinnabar (HgS) providing the red colour used in pre-historic cave paintings. Metallic mercury was known in ancient Greece where it was used to lighten the skin. In medicine mercury was used as a cure for syphilis, mercury compounds have also been used as diuretics (calomel (Hg_2Cl_2)), and mercury amalgam is still used for filling teeth in many countries.

Mercury is one of the few heavy metals that can be in the gaseous form in the air. It has a long residence time and hence can be transported thousands of kilometers in the atmosphere. In addition to being released from industrial sources as outlined above, mercury is found in a number of products including fluorescent lamps, thermometers, batteries and dental amalgam to name but a few.

In the environment, inorganic mercury is converted to organic compounds, such as methyl mercury, which is very stable and accumulates in the food chain. Until the 1970s, methyl mercury was commonly used for control of fungi on seed grain.

Health effects of inorganic mercury are rare but include neurological damage that is reversible after exposure is stopped. Mercury is an allergen and may cause contact eczema. The effects of mercury amalgams used in dentistry are still controversial with some people claiming that they have symptoms associated with their fillings.

Health effects of organic mercury include nervous system damage. The Minamata catastrophe in Japan in the 1950s was caused by methyl mercury poisoning from fish contaminated by mercury discharges to the surrounding sea. Another mercury tragedy occurred in Iraq in 1971, when 6,500 people were hospitalized and more than 400 died after eating grain treated with a methyl mercury fungicide. The grain was intended for planting, but the residents mistook it as edible and ground it into flour, unaware that the bread they were making was deadly poisonous. Most of what we know today about the effects of mercury poisoning comes from studies of the people who were poisoned in Japan and Iraq.

Several serious incidents of mercury contamination occurred in Canada soon as well. In 1969, a pulp and paper mill contaminated the English-Wabigoon River system near Dryden in northwestern Ontario. The mill's chlor-alkali plant used mercury to manufacture chlorine that, in turn, was used to bleach paper. Eventually the mercury was discharged to the local waterway, polluting the fish in the English-Wabigoon River system, making them unfit to eat and threatening the health and disrupting the livelihood of the local population that depended on the fish. The White Dog and Grassy Narrows First Nations people experienced high levels of mercury in their blood and hair. Since the closure of the chlor-alkali plant, mercury levels in local fish species have dropped, but remain, for the most part, elevated.

The route of exposure of the most concern from an air pollution management perspective, then, is the atmospheric route by which watersheds become contaminated with methyl mercury which then accumulates in the fish.

One of the biggest challenges in managing mercury in Canada was the lack of a national monitoring program to provide a baseline of mercury levels in the environment. At Environment Canada around this time, Keith Puckett and Bill Schroeder decided to initiate a mercury measurement program to get this necessary base line data. In 1994, the Canadian Atmospheric Mercury Measurement Network (CAMNet) was established to measure gaseous mercury. Today there are 11 sites in operation, according to the website, and it has been rolled into NAPS (CAMNet 2013). Mercury was also found in the Northern atmosphere and went through depletion events in the spring (Schroeder et al. 1998).

In June 2000, the CCME announced Canada Wide Standards for mercury emissions (CCME 2000) stating that "reduced deposition will contribute, in time to reduced impacts". Two major source categories were targeted: products and major point source emissions.

In 2006, a Risk Management Strategy for mercury-containing products was drafted and released for comment by Environment Canada. This was followed in 2007 by "Proposed Risk Management Instruments for Mercury-containing Products". Point source emissions of mercury have been substantially reduced.

As introduced in the previous section with POPs, there is a lot of international work on heavy metals.

The 1978, the Canada/US Great Lakes Water Quality Agreement set water limits for arsenic, cadmium, chromium, copper, iron, lead, mercury, nickel, selenium and zinc. Annex 15, which dealt with air toxics only, listed mercury (Muir et al. 2009).

Under the North American Free Trade Agreement's Commission on Environmental Cooperation, Canada helped Mexico establish mercury deposition monitoring stations and become part of the Canada-US Mercury Deposition Monitoring Network.

The UN ECE Convention on LRTAP developed the Aarhus Heavy Metals Protocol in 1998 which targets three particularly harmful metals: cadmium, lead and mercury. This Protocol amended in 2012 to adopt more stringent controls of heavy metals emissions and to introduce ways to include new Parties, notably countries in Eastern Europe, the Caucasus and Central Asia, covers emissions from industry, products and underscores the removal of lead from gasoline.

In 2011, North American scientists studying mercury in the Pacific Ocean determined that this global heavy metal was not only transported eastward through the atmosphere but also was transported in the ocean water across the Pacific from Asia to North America.

The Canadian Environmental Protection Act (CEPA)

In 1969, amidst acid rain concerns, the federal Department of National Health and Welfare established an Air Pollution Control Division, and in 1971 the Department of the Environment was formed assuming those duties. As has already been mentioned, the Canada Clean Air Act was passed in 1971. One of its main objectives was to promote a uniform approach to managing air pollution across Canada, since it was recognized that the provinces would have an active role in regulating industry. As already noted, this Act was used to regulate asbestos from mining, lead from smelters, mercury from chlor-alkali pulp and paper plants and vinyl chloride.

The Clean Air Act was followed, in 1975, by the Environmental Contaminants Act which was jointly administered by the Departments of Health and the Environment and which responded to international initiatives such as the OECD work on Polychlorinated biphenyls (PCBs). Unfortunately, control measures were difficult to impose and none went forward.

The Environmental Contaminants Act (ECA) was reviewed in 1985 resulting in the decision that a more comprehensive approach to managing toxic substances was needed. Bill C-74 was drafted, the Canadian Environmental Protection Act (CEPA), and eventually proclaimed in 1988. CEPA was based on a cradle to grave management philosophy for toxic materials and initiated a National Advisory Committee (CEPA NAC) to provide advice. Environment Canada or Health Canada chair the group which, in addition to representatives of all the provinces and territories, includes up to six Aboriginal government representatives.

The Government of Canada encourages emission reductions on the domestic front. Under the Toxic Substances Management Policy (TSMP) of 1995, toxic substances that are determined to be persistent, bioaccumulative and resulting primarily from human activity are known as Track 1 substances, and targeted for virtual elimination from the environment. The key pieces of federal legislation used to implement the objectives outlined in the TSMP include the Canadian Environmental Protection Act (CEPA), the Pest Control Products Act, the Fisheries Act and the Hazardous Products Act (HPA). The Canada Consumer Product Safety Act of 2010 (Canadian Consumer Product Safety Act 2013) eventually replaced Part One of HPA.

The Canadian Council of Ministers of the Environment (CCME) identified the management and reduction of toxic substances in the environment as a national priority through the CCME Policy for the Management of Toxic Substances. The CCME Policy supports the coordination of government actions on the management of toxic substances, ensuring that the approach is complementary to the TSMP and other activities nation-wide.

The Canada-Wide Standards process is a framework for the CCME to work together in addressing key environmental protection and health risk reduction issues that require common environmental standards across the country. In June of 2000, standards were approved in principle by the Ministers for two priority sectors emitting dioxins and furans: incineration and coastal boilers burning salt laden wood.

The conventional chemicals that cause air pollution (sulphur dioxide, nitrogen oxides, VOCs, particulate matter) have been declared toxic under the Canadian Environmental Protection Act and hence are now linked in to management of toxic substances more generally, countering some of the discomfort initially felt by the air community when the air pollution management focus was watered down by inclusion in CEPA.

In 1999 the revised CEPA was passed after a significant review. The "principles" basis has been expanded to include: Sustainable Development; Pollution Prevention "the use of processes, practices, materials, products, substances or energy that avoid or minimize the creation of pollutants and waste and reduce the overall risk to the environment or human health"; Virtual Elimination, the reduction of releases to the environment of a substance to a level below which its release cannot be accurately measured; Ecosystem Approach; Precautionary Principle "where there are threats of serious or irreversible damage, lack of full scientific certainty shall not be used as a reason for postponing cost-effective measures to prevent environmental degradation"; Intergovernmental Cooperation; National Standards; Polluter Pays Principle; and Science-based Decision-Making emphasizes the integral role of science and traditional aboriginal knowledge (where available) in decision-making and that social, economic and technical issues are to be considered in the risk management process.

CEPA 1999 is based on the assessment of existing or new substances as toxics. Existing substances are those on the

Domestic Substances List (DSL). The DSL includes substances that were, between January 1, 1984, and December 31, 1986, in commercial use in Canada, or were used for commercial manufacturing purposes, or were manufactured in or imported into Canada in a quantity of 100 kg or more in any one calendar year. The list is regularly amended to include additional substances that have been assessed under the Act and allowed into Canada. The DSL currently contains approximately 23 000 substances from the original list along with an additional 1954 substances that have been added to the list following assessments of new substances.

Determining a substance to be toxic under CEPA 1999 is a function of its release or possible release into the environment, the resulting concentrations in environmental media and its inherent toxicity. Section 64 of CEPA 1999 defines a substance as toxic "if it is entering or may enter the environment in a quantity or concentration or under conditions that:

- have or may have an immediate or long-term harmful effect on the environment or its biological diversity;
- constitute or may constitute a danger to the environment on which life depends; or
- constitute or may constitute a danger in Canada to human life or health."

Substances that meet the definition of toxic are put on Schedule 1, the List of Toxic Substances for potential further action. The current list is at http://www.ec.gc.ca/lcpe-cepa/default.asp?lang=En&n=0DA2924D-1&wsdoc=4ABEFFC8-5BEC-B57A-F4BF-11069545E434.

CEPA is based on a risk management approach and gives a wide range of authorities to seek further information as a basis for regulation.

CEPA has a time clock associated with it, a proposed regulation or instrument establishing "preventive or control actions" for managing the substance must be developed with 24 months. The proposal is published in the Canada Gazette, Part I, for a 60-day comment period. Once proposed, the Ministers have a further 18 months to finalize the regulation or instrument.

Previously the comment has been made that emissions are an area of great uncertainty in the management of air quality, particularly when chemicals can change in the atmosphere due to physical or chemical reactions. This is even more true of toxic chemicals, and there are so many of them that it is a real challenge to assemble data on which industries emit which toxic chemicals and how much. While the big emitters such as refineries and power plants usually have sufficient knowledge of their processes to be able to provide data through the National Pollutant Release Inventory (NPRI) in Canada and the National Emissions Inventory (NEI) in the USA, the thousands of small businesses such as dry cleaners, paint shops, and small manufacturing may not have the expertise to be able to do this, and yet may be significant emitters of certain toxics.

In the 1960s, in the United States, the situation was similar to that in Canada. The EPA debated which pollutants required regulation and how stringent regulations should be. The EPA listed only eight HAPS and established national emission standards for only seven of them (asbestos, benzene, beryllium, inorganic arsenic, mercury, radionuclides, and vinyl chloride) from 1970 to 1990. During that period, the Clean Air Act (CAA) directed EPA to regulate toxic air pollutants based on the risks of hazardous air pollutants to human health, and then within a year to promulgate standards to reduce emissions of the HAPs to levels that provided a margin of safety protection to the public. Setting a national emission standard for a hazardous air pollutant (NESHAP) was fraught with difficulty since there was little available data on air toxics and there were problems in defining the "ample margin of safety protection" required. While EPA and the scientific community gained valuable knowledge about risk assessment methods through this work, the chemical-by-chemical regulatory approach based solely on risk proved difficult.

Under Sect. 112 of the 1990 Clean Air Act Amendments (CAAA), Congress changed the approach to hazardous air pollutants to a technology-based regulation, so a hazardous air pollutant is defined as any air pollutant listed in Sect. 112 (b).|| In this section, 189 substances were listed as hazardous air pollutants to which EPA may add or delete chemicals. The CAAA also required EPA to pass technology-based emission standards (referred to as maximum achievable control technology (MACT) standards) for all major source categories. This required the source categories have MACT standards promulgated and, by 2004, EPA had passed all (96) MACTs standards.

The 1990 CAAA air toxic program had two phases (DeRose 2009). In the first phase, EPA developed MACT standards requiring sources to meet specific emissions limits based on emission levels already being achieved by similar sources in the country. In the second phase, EPA must establish additional requirements to control any "residual risk" that exists 8 years after promulgation of MACT. The EPA completed development of its strategy for addressing residual risks from air toxics in March of 1999 in its Residual Risk Report to Congress.

Also included in the 1990 CAAA are provisions that EPA study several specific topics (including the Mercury, Great Waters, and Utilities) and continues to study these and others. Additionally, EPA's Urban Air Toxics Strategy released in August 1998 proposes to address the problems of cumulative exposures to air toxics in urban areas through an integrated approach that considers stationary and mobile sources of urban air toxics. These programs, in combination with the residual risk program, will provide a coordinated federal approach to address air toxics.

To avoid controversy over which pollutants would be listed as HAPs, Congress included an initial list of 189 HAPs into the 1990 CAAA. EPA (or an individual) may add or delete (delist) pollutants from this list. Also, an individual can petition the EPA to add or delist a substance from the HAP list. To add it must be shown that emissions, ambient concentrations, bioaccumulation, or deposition of the air pollutant are known to cause, or may reasonably be anticipated to cause, adverse effects to human health or adverse environment effects and vice versa to delist. Since passage of these Amendments, EPA and has delisted only two HAPs: caprolactam and methyl ethyl ketone putting the current list of HAPs at 187. EPA has a current list of all regulated HAPs on its Air Toxics Web page: http://www.epa.gov/ttn/atw/orig189.html.

In both Canada and the United States, there is some overlap between HAPs and criteria pollutants. This is important because many programs aimed at the reduction of particulate matter (PM) and volatile organic compounds (VOCs) will have a beneficial effect on air toxics (the reverse also being true). For example, ozone, which is formed by the interaction of NOX, VOC, and sunlight, will be reduced when a HAP VOC is reduced. Another example would be a program that reduces PM. Since PM is comprised of many chemicals, some which may contain various HAPs, a reduction in PM may also reduce the specific HAP that is in its makeup.

2.2.5 Air Pollution Emergencies

Most air quality management focuses on the long term management of airsheds in a manner that protects the health and safety of people and their environment. However, we've seen that attention to air pollution management was captured through some short term incidents that caused the public to focus their attention on air quality. There are another class of air quality issues that require a different style of management because large amounts of substances are released over a short period which can cause serious local effects as well as impacts around the globe. To make matters more complicated, these incidents are not all manmade, the eruptions of volcanoes, for example, can cause major releases of pollutants into the atmosphere. Volcanic dust is a particular hazard for air traffic.

In Canada, in contrast to some other countries, air pollution management at the federal level has evolved in conjunction with the weather service. Hence the Atmospheric Environment Service, now known as the Meteorological Service of Canada, pioneered studies of "chemical weather" in which air pollution models were introduced into a variant of the operational weather prediction model instead of using a meteorological scenario generator as is done in most other countries, including the United States. This opens the door for near real time simulations of these sudden air quality issues, as well as a variety of analysis and hindcast options as described in Chap. 4.

For example, by chance, when the Chernobyl accident occurred in April, 1986, the World Meteorological Organization was already doing research related to long-range transport of airborne pollutants through the use of atmospheric transport models. Work was done, in near real time in Dorval, by Janusz Pudykiewicz to demonstrate that the radionuclides from Chernobyl were circling the northern globe and passing over Canada. This work informed the world as to where the plume would go and how it would disperse over the following days and became the first of a number of models to play this role. While this was not an accepted approach at the time and hence results were not used to guide practical aspects of the emergency management close to the disaster area, the results did comfort Canadians. While radiation was being measured in Canada from Chernobyl, the levels were very low and not of major concern either in terms of exposure or contamination of foodstuffs (Pudykiewicz 1988).

Following the cold war there was heighted interest in banning nuclear explosions and the Comprehensive Nuclear Test Ban Treaty (CTBT) was opened for signature in September 1996. As part of this Treaty, an International monitoring system (IMS) of 337 monitoring facilities is being established. This monitoring system was designed, using a variety of monitoring techniques, to detect and identify the nuclear origin of a nuclear explosion unambiguously. The design was based on many scenarios run with an atmospheric transport model. If something is detected by the monitoring system, the model can be used to determine the place of origin of a radionuclide after detection. Of course, the models can also be used for Treaty verification purposes.

Canada has played a leadership role throughout this process from the initial science done by Dr. Pudykiewicz to leadership in the development of the CTBT. For example, the CTBT Preparatory Commission's technical working group established a Sub-Task Group on cooperation with the World Meteorological Organization (WMO) with Michel Jean, now Director General of the Weather Service, as chair. A joint CTBT-WMO backtracking modeling system is now in place and is regularly exercised to test and refine procedures. The first test was held in May 2000 and after a series of workshops and improvements the work was summarized in a joint paper in 2007 (Becker et al. 2007).

Meanwhile, due to their interest in possible radionuclides in the atmosphere, the International Civil Aviation Organization (ICAO) participated in the CTBT program. They developed the idea of monitoring explosive volcanic eruptions in order to warn aviation of potential volcanic ash hazards. In 1995, a Rapporteur on CTBT matters was established, a role that was filled by another Canadian, Peter Chen. ICAO recognized the hazard that volcanic ash presented to the aviation industry and nine Volcanic Ash Advisory Centers (VAACS) were created, one of which is located in Dorval. Each VAAC provides volcanic ash advisories when a volcano erupts in their area of responsibility. The Canadian VAAC cooperates closely with the two American VAACs, one in Anchorage and one in Washington DC (Chen 2009).

The Canadian Meteorological Center (CMC) has been involved in a number of other simulations which couple atmospheric pollutants with the weather. During the forest fire season of 2002, an extremely hot and dry summer created conditions for more than 85 forest wildfires in Quebec. A version of the model GEM-AQ was run allowing the emissions to be tracked. Smoke was observed as far away as Washington DC and MODIS satellite imagery was used to verify the model results. (Lavoue et al. 2007).

The major area of uncertainty when predicting forest fire impacts is details about the emission source such as how large and hot it is and what chemicals are being produced. While there have been a number of simulations of forest fires done over the years, this system is not yet operational. In western Canada the BlueSky system also predicts smoke from forest fires.

In the US a different approach has been taken with the development of the RAINS system for forest fire prediction. In this system, a meteorological preprocessor is used instead of the "real" weather forecasting program. This system is operational in the States but is a separate entity from the Weather Service.

2.2.6 Summary

The issues described here served to highlight to Canadians a number of principles about air pollution that have underlain the approach to air quality management in Canada. The Trail case could be seen to be one of the earliest acknowledgements of precaution in addition to an early acknowledgement of the responsibilities of good neighbours to each other. This was picked up in the acid rain era when the U.S.A. accepted stringent emission limitations in order to reduce acidification of lakes that were both American and Canadian.

The management of the Sudbury Smelter once again illustrates that with ingenuity air pollution can be reduced in the context of maintaining competitive industry. The great smog and other pollution events of that type really brought home the point that it is not just ecosystem damage that is caused by air pollution. Affected vegetation and fish do not resonate in people's minds in the same way as do the deaths and illnesses of people.

Hence, from these historical cases, it is clear that Canada has been responsive to the impacts of air pollution and, when it is clear that there is an issue arising, action has been taken. In fact, these historical issues have driven science programs that have laid the basis for modern air quality management as we know it today.

One overriding principle appears to drive Canadian air quality management. For a nation characterized by a stoic "eh?" we sure do seem to talk about it a lot. Every successful management advance has involved scientists talking nationally and internationally across borders amongst themselves to get the facts straight, and then talking to air quality managers locally, regionally, nationally and internationally, followed by the air quality managers consulting widely with stakeholders before settling on an approach to move forward.

Unfortunately, actions tend to be taken only in response to a perceived "smoking gun", and once the immediate issue of interest is addressed, actions are cut back or abandoned without necessarily solving the problem as well as possible. Thus, acid rain, smog including ground level ozone and particulate, and air toxics continue to damage ecosystems and to injure and kill Canadians. While it is important to continually revisit science and monitoring programs to increase their efficiency and minimize costs, making changes without assessing the impacts of the changes on program effectiveness and output is short sighted indeed.

2.3 The People, Beyond Jurisdictional Role and Key Issues, that have made Canada a Leader in Air Quality Management

In all cases, there have been an important cadre of people who have been in the front lines and have recognized the strength of the science coming forward and the need to provide the science in a way useful for decision makers to take action on whichever issue was at the forefront of concern.

Some of the scientists who played a key role in the historical issues discussed so far have been highlighted, whether they have been academics, government employees or both. There are a couple of other groups of individuals, not government employees, who have played key historical roles in the management of air quality in Canada. They have unfortunately been given short shrift simply because of space limitations.

Receptive individuals in industry, like Charlie Ferguson of Inco in Sudbury, who was a leader in figuring out the possibilities to modify the plant to both be profitable and cleaner for the community, must be recognized as important figures in air quality management. Gordon Lloyd, who led the development of the Responsible Care program for the Chemical industry and Sid Barton of ORTECH who worked to develop the Sarnia Lambton Environmental Society which put the responsibility on industries to manage their emissions also made huge contributions to air quality management in Canada.

Finally, nongovernmental organizations have played a strong role in air quality management in Canada. Bruce Walker of STOP has kept a series of players honest in meeting the Acid Rain commitments signed onto decades ago. In Montreal he is famous as a watchdog for their Bylaw 90. Martha Kostuch who was the face of the Friends of the Old Man River, repeatedly called the Alberta and Canadian governments on a number of key issues. No one will forget colourful Ken Maybee, the leader of the New Brunswick Lung Association, who for years was the lone health representative advocating for clean air and who passed away in October 2012. He co-chaired the national group leading the development and implementation of the new Air Quality Health Index, was instrumental in having Canada proclaim Clean Air Day, was an official negotiator for the Ozone Annex to the Canada/US Air Quality Agreement and was to be recognized by presentation of the Order of Canada.

One whole group of outstanding contributors to air quality management that has not been included here are the group of provincial contributors. Their work appears in other chapters.

2.4 Conclusion

Canada is a large and diverse country with vast pristine vistas as well as large, competitive industrial sites and rapidly growing urban areas. The lessons that have been learned over the 100+ years that air pollution has been an issue in parts of Canada are lessons that form the basis of our air quality management systems of today. As our perceptions of air quality management broaden from concern with smoke to such issues as noise, smell, visibility and specific emissions from sectors removed from the traditional fossil fuel burners to agricultural practices, for example, our approach may need to continue to change (see chapters in this book addressing some of these issues).

Nonetheless, Canadian air quality managers have learned some lessons from history. These lessons include:

- A credible process with a strong scientific basis including air quality monitoring supported by modelling provides a basis for designing air quality management strategies that work (acid rain).
- Meteorology is an important driver of air quality, not to be ignored (Detroit-Windsor).
- Stakeholders can play a big role in the decision making and organizations that move forward without them do so at their peril (Notice of Intent 2006).
- Air pollution has impacts not only on plants and animals but also on people. These impacts remain serious, shortening the lives of thousands of Canadians every year (Burnett et al. 1998).
- Management of air pollution issues leads industry to be more ingenious and competitive; the costs of air pollution management have historically been recovered in greater productivity and new products (INCO).

While, as is shown in several chapters in this book, air quality management in Canada is more ad hoc than that in many other countries, most notably the United States, in general the management actions target the major issues and are effective. While CEPA gives a general approach for dealing with pollution, our history shows that the interventions of superb Canadian scientists have led to early recognition of the most important air pollution issues as well as early action to address them. With less focus on doing Canadian monitoring and science and more on a 'process' approach to air quality management, will Canadians enjoy a clean atmosphere in the face of unprecedented development both nationally and globally? Only time will tell.

References

Allum JR (2006) An outcrop of hell: history, environment and politics of the Trail Smelter Dispute. In: Bratspies R, Miller R (eds) Transboundary harm in international law: lessons from the trail smelter arbitration. Cambridge University Press, Cambridge, pp 13–26

AMAP (1998) AMAP Assessment Report: arctic pollution issues. Arctic monitoring and assessment programme (AMAP). Oslo, pp xii + 859

Atmospheric Environment Service (1998) 1997 Canadian acid rain assessment ISBN: 0662631188 ISBN13: 9780662631187 DDC: 363.738 Edition: (v. 1)

Barrie LA (1967) Arctic air pollution: an overview of current knowledge. Atmos Environ 20(4):643–663

Bates DV, Sizto R (1987) Air pollution and hospital admissions in Southern Ontario: the acid summer haze effect. Environ Res 43:317–331

Bates DV (2006) Ozone—42 Years later. Presentation at a continuing medical education course at St Paul's Hospital

Beamish R, Harvey H (1972) Acidification of the La Cloche Mountain lakes, Ontario, and resulting fish mortalities. J Fish Res Board Can 29:1131–1143

Becker A, Wotawa G, De Geer L-E, Seibert P, Draxler RR, Sloan C, D'Amours R, Hort M, Glaab H, Heinrich P, Grillon V, Shershakov V, Katayama K, Zhang Y, Stewart P, Hirtl M, Jean M, Chen P (2007) Global backtracking of anthropogenic radionuclides by means of a receptor oriented ensemble dispersion modelling system in support of nuclear-test-ban treaty verification. Atmos Environ 41:4520–4534

Brook JR, Brook RD, Urch B et al (2002) Inhalation of fine particulate air pollution and ozone causes acute arterial vasoconstriction in healthy adults. Circulation 105:1534–1536

Brook JR, Johnson DH (2000) Identification of representative warm season periods for regional air quality (ozone) model simulations. Atmos Environ 34:1591–1599

Brook JR, Strawbridge K, Snyder BJ et al (2004) Towards an understanding of the fine particle variations in the LFV: integration of chemical, physical and meteorological observations. Atmos Envir 38:5775–5788

Brook JR, Burnett RT, Dann TF, Cakmak S, Goldberg MS, Fan X, Wheeler AJ (2007) Further interpretation of the acute effect of nitrogen dioxide observed in Canadian time-series studies. J Expo Sci and Environ Epidemiol 17:S36–S44

Brook JR, Graham L, Charland JP et al (2007) Investigation of the motor vehicle exhaust contribution to primary fine particle organic carbon in urban air. Atmos Environ 41:119–135

Brydges T (2004) Acid rain in story and song. Thomas and Marilyn Brydges, Brampton

Canadian Consumer Product Safety Act (CCPSA) (2013) http://www.hc-sc.gc.ca/cps-spc/legislation/acts-lois/ccpsa-lcspc/index-eng.php. Accessed 5 May 2013

Burnett RT, Bartlett S, Krewski D, Roberts G, Raad-Young M (1994a) Air pollution effects on hospital admissions: a statistical analysis of parallel time series. Environ Ecol Stat 1:325–332

Burnett RT, Dales RE, Raizenne ME, Krewski D, Summers PW, Roberts GR, Raad-Young M, Dann T, Brook J (1994b) Effects of low ambient levels of ozone and sulfates on the frequency of respiratory admissions to Ontario hospitals. Environ Res 65:172–194

Burnett RT, Dales RE, Krewski D, Vincent R, Dann T, Brook J (1995) Associations between ambient particulate sulfate and admissions to Ontario hospitals for cardiac and respiratory diseases. Am J Epidemiol 142:15–22

Burnett RT, Brook JR, Yung WT, Dales RE, Krewski D (1997) Association between ozone and hospitalization for respiratory diseases in 16 Canadian cities. Environl Res 72:24–31

Burnett RT, Cakmak S, Brook J (1998) The effect of the urban ambient air pollution mix on daily mortality rates in 11 Canadian cities. Can J Public Health 89:152–156

Burnett RT, Smith-Doiron M, Stieb D, Cakmak S, Brook J (1999) Effects of particulate and gaseous air pollution on cardiorespiratory hospitalizations. Arch Environ Health 54:130–139

Canadian Atmospheric Mercury Measurement Network (CAMNet) http://www.ec.gc.ca/natchem/default.asp?lang=en&n=4285446C-1. Accessed 5 May 2013

Canadian Council of Ministers of the Environment (2003) Atmospheric science of ground-level ozone: update in support of the Canada-wide standards for particulate matter and ozone. http://www.ccme.ca/assets/pdf/scrvw_oz_atmsphrc_sc_e.pdf. Accessed 23 April 2013

Canadian Council of Ministers of the Environment (2000) Canada-wide standards for mercury emissions endorsed. Quebec City, 5–6 June 2000

Canadian Council of Ministers of the Environment (2007) Guideline document on achievement determination Canada-wide standards for particulate matter and .zone. PN 1391 978-1-896997-74-2 PDF. Accessed 5 May 2013

Canadian Council of Ministers of the Environment (2011) 2008-2009 Progress report on the Canada-wide acid rain strategy for post-2000. PN1458, ISSN 1911–1541 PDF

Canadian Council of Ministers of the Environment (2013) www.ccme.ca. Accessed 30 Apr 2013

Canada/United States (2012) Air quality agreement progress report. ISBN: 1910–5223, pp 81.

Chen P (2009) Contributing to a safer and more secure world: WMO support to the comprehensive nuclear-test ban treaty. MeteoWorld http://www.wmo.int/pages/publications/meteoworld/archive/aug09/ctbto_en.html. Accessed 28 Feb 2013

Clark TL, Voldner EC, Dennis RL, Seilkop SK, Alvo M, Olson MP (1989) The evaluation of long-term sulfur deposition models. Atmos Environ 23:2267–2288

DeRose L (2009) Introduction to air toxics student manual. Lake Michigan Air Directors Consortium, APTI 400

Dillon PJ et al (1978) Acidic precipitation in southcentral Ontario, recent observations. J Fish Res Board Can 35:809

Encyclopaedia Britannica (2013) www.britannica.com/EBchecked/topic/550008/smog. Accessed 5 May 2013

Environment Canada (1973) A nationwide inventory of air pollutant emissions—1970, Report EPS 3-AP-73-2/ Environmental Protection Service, Ottawa, pp 123

Environment Canada (1994) Mystery on the Great Lakes. ISBN 0-662-20831-5, pp 16

Environment Canada (1995) Report to the Canadian Council of Ministers of the Environment by the Task Force on Cleaner Vehicles and Fuel, Environmental and Health Benefits of Cleaner Vehicles and Fuels (Hagler Bailly Study). http://www.ec.gc.ca/Publications/default.asp?lang=En&xml=05B86327-5AC7-46D4-A4B2-6EAB-916C5B8C. Accessed 30 Apr 2013

Environment Canada (2004) Canadian acid deposition science assessment. Environment Canada, Ottawa

Federal-Provincial Working Group on Air Quality Objectives and Guidelines (1996) A protocol for the development of national ambient objectives, Part 1. Science assessment document and derivation of the reference level(s). Environment Canada and Health Canada, Toronto and Ottawa

Federal-Provincial Working Group on Air Quality Objectives and Guidelines (1998) National ambient air quality objectives for particulate matter. Executive summary. Public Works and Government Services, Ottawa

Federal-Provincial Working Group on Air Quality Objectives and Guidelines (1999) National ambient air quality objectives for ground-level ozone: science assessment document. Health Canada and Environment Canada, Ottawa

Fellin P, Dougherty D, Barrie LA, Toom D, Muir D, Grift N, Lockhart L, Billeck B (1996) Air monitoring in the arctic: results for selected persistent organic pollutants for 1992. Environ Toxicol Chem 15(3):253–261

Gorham E, Gordon AG (1963) Some effects of smelter pollution upon aquatic vegetation near Sudbury, Ontario. Can J Bot 38:477–487

Harvey HH, Pierce RC, Dillon PJ, Kramer JP, Whelpdale DM (1981) Acidification in the Canadian environment. NRCC Report No 18475:369

Harvey HH (1982) Population responses of fish in acidified waters. In: Johnson RE (ed) Proceedings of an international symposium on acid precipitation, Cornell 1981. Cornell, pp 227–242

Hidy G, Niem Di, Pace T (2003) Chapter 4: emission characterization. In: McMurray P, Shepherd M, Vickery J (eds) Particulate matter science for policy makers: a NARSTO assessment. Cambridge University Press, Cambridge

Hoff RM, Strachan WMJ, Sweet CW, Chan CH, Shackleton M, Bidleman TF, Brice KA, Burniston DA, Cussion S, Gatz DF, Harlin K, Schroeder WH (1994) Atmospheric deposition of toxic chemicals to the Great Lakes: a review of data through 1994. Proceedings of IADN Workshop

Hung H, Blanchard P, Halsal CJl, Bidleman TF, Stern GA, Fellin P, Muir DCG, Barrie LA, Jantunen LM, Helm PA, Ma JKonoplev A (2005) Temporal and spatial variabilities of atmospheric polychlorinated biphenyls (PCBs), organochlorine (OC) pesticides and polycyclic aromatic hydrocarbons (PAHs) in the Canadian arctic: results from a decade of monitoring. Sci Tot Environ 342:119–144

International Joint Commission, International Air Quality Advisory Board (March 2010), Expert consultation meeting-air quality issues related to the northern boundary region between the United States and Canada. http://www.ijc.org/php/publications/pdf/ID1636.pdf. Accessed 30 Apr 2013

International Joint Commission, International Air Quality Advisory Board (August 2010) Second expert consultation meeting—Air quality issues related to the northern boundary region between the United States and Canada. http://ijc.org/php/publications/pdf/IAQAB_-_Whitehorse_Report_april-11.pdf. Accessed 30 Apr 2013

International Joint Commission, International Air Quality Advisory Board (April 2012) Report on air quality issues related to the northern boundary region between Canada and the United States. http://www.ijc.org/php/publications/pdf/IAQAB-Report-to-IJC-Northern-Air-Quality-April-2012.pdf. Accessed 30 Apr 2013

Jerrett M, Burnett RT, Pope III CA, Ito K, Thurston G, Krewski D, Shi Y, Calle E, Thun M (2009) Long-term ozone exposure and mortality. N Engl J Med 360:1085–1095. doi: 10.1056/NEJMoa0803894

Kaijser A (2011) The trail from trail: new challenges for historians of technolgy. Technol Cult 52(1)

Katz M et al. (1939) Effect of sulfur dioxide on vegetation. National Research Council, Ottawa

Katz M (1949) Sulfur dioxide in the atmosphere and its relation to plant life. Ind Eng Chem 2450–2465

Katz M (1955) Atmospheric pollution: a growing problem in public health. Am J Public Health 298–305
Katz M (1963) Air pollution in Canada—current status report. AJPM 539(2):173–184
Kramer JR (1973) Fate of atmospheric sulphur dioxide and related substances as indicated by the chemistry of precipitation, a report to the air management branch. Ministry of Environment, Ontario, pp 143
Lavoue D, Gong S, Stocks BJ (2007) Modelling emissions from Canadian wildfires: a case study of the 2002 Quebec fires. Int J Wildland Fire 16:649–663
Levy I, Mihele C, Lu G, Narayan J, Hilker N, Brook JR (2012) Elucidating multipollutant exposure aross a complex metropolitan area by systematic deployment of a mobile lab. Atmos Chem Phys Discuss 12:31585–31627
Likens GE (1972) The chemistry of precipitation in the central Finger Lakes Region. Cornell University Water Resources and Marine Science Centre Technical Report No. 50, pp 47
Likens GE, Bormann FH, Johnson NM (1972) Acid rain: a serious regional environmental problem. Science 184:1176–1179
MacDowell LS (2012) An environmental history of Canada. UBC Press, ISBN 978-0-7748-2103–2
Muir DCG, Wang X, Yang F, Nguyen N, Jackson TA, Evans MS, Douglas M, Köck G, Lamoureux S, Pienitz R, Smol J, Vincent WF, Dastoor AP (2009) Spatial trends and historical deposition of mercury in eastern and northern Canada inferred from lake sediment cores. Environ Sci Technol 43:4802–4809
Munn RE (1975) Suspended particulate concentrations: spacial correlations in the Detroit-Windsor area. Tellus 27(4):397–405
National Air Pollution Surveillance Program (NAPS) (2013) www.ec.gc.ca/natchem/default.asp?lang=en&n=EEOE2169-1. Accessed 12 Feb 2013
National Round Table on the Environment and the Economy (2008) Developing ambient air quality objectives for Canada: advice to the Minister of Environment. ISBN 978-1-894737-15–9
Northern Contaminants Program (2003) Canadian arctic contaminants assessment Report II. Available in hard copy and at http://www.ainc-inac.gc.ca/NCP/index_e.html. Accessed 20 Apr 2013
Oden S (1968) The acidification of air and precipitation and its consequences in the natural environment. Ecol Community Bulletin 1
Organization for Economic Co-operation and Development (OECD) (1981) The costs and benefits of sulphur dioxide control: a methodological study. Paris
Phalen RF, Phalen RN (2012) Introduction to air pollution science. Burlington, Jones & Bartlett, pp 333
Pollution Probe (2003) Sulphur dioxide and toxic metal emissions from smelters in Ontario, pp 25
Puckett KJ (1979) The effects of acid precipitation on lichens. Presentation at workshop on the effects of acid rain on wildlife. Ottawa, Canadian Wildlife Service
Pudykiewicz J (1988) Numerical simulation of the transport of radioactive cloud from the Chernobyl nuclear accident. Tellus B40(4):391–412
Ruben Mordecai B (2001) The history of ozone: the Shonbein period, 1839–1868. Bull Hist Chem 26(1):40–56
Samet JM, Dominici F, Curriero FC, Coursac I, Zeger SD (2000) Fine particulate air pollution and mortality in 20 U.S. cities, 1987–1994. N Engl J Med 343:1742–1749
Schindler DW, Mills KH, Malley DF, Findlay DL, Shearer JA, Davies IJ, Turner MA, Undsey GA, Cruikshank DR (1985) Long-term ecosystem stress: the effects of experimental acidification on a small lake. Science 228:1395–1398
Schindler DW (1988) Effects of acid rain on freshwater ecosystems. Science 239:150 p
Schroeder WH, Anlauf WK, Barrie LA, Lu JY, Steffen A, Schneeberger DR, Berg T (1998) Arctic springtime depletion of mercury. Nature 394:331–332
Shrenk HH (1970) Air pollution in Donora: an analysis of the extreme effects of smog. Elmsford, Maxwell Reprint Company, pp 173
Sirois A (1998) A brief and biased overview of time-series analysis or how to find that evasive trend. In: Proceedings of the workshop on advanced statistical methods and their application to air quality data sets, Report WMO TD No 956, Word Meteorological Organization, Geneva, Switzerland
Smith RA (1872) Air and rain. The beginnings of a chemical climatology. Longmans, Green and Co, London
Strachan WMJ, Eisenreich SJ (1988) Mass balancing of toxic chemicals in the Great Lakes: the role of atmospheric deposition. International Joint Commission, Windsor Ontario
Summers PW, Whelpdale DM (1975) Acid precipitation in Canada. Atmos Environ Service pp 78
Taking Stock (1997) North American pollutant releases and transfers. Commission for Environmental Cooperation, 2000
Taylor P, Thomas M, Truhlar E, Whelpdale D (1995) R.E. (Ted) Munn—founding editor; a mini-biography. Bound-Lay Meteorol 78:3–8
UN ECE (2013a) ICP modelling and mapping critical loads and levels approach. http://www.unece.org/env/lrtap/WorkingGroups/wge/definitions.html. Accessed 29 Apr 2013
UN ECE (2013b) Convention on long-range transboundary air pollution. http://www.unece.org/env/lrtap/. Accessed 30 Apr 2013
US/Canada Work Group 3A (1981) Strategies, development and implementation. Interim report, United States-Canada, memorandum of intent on transboundary air pollution, pp 56 (with 5 appendices)
US/Canada Work Group 3B (1982) Emissions, costs and engineering assessment. Final Report, Ibid, June 1982
US/Canada Work Group 2 (1982) Atmospheric sciences and analysis. Final Report, Ibid
US/Canada Work Group 1 (1983) Impact assessment. Final Report, Ibid
Voldner EC, Olson M (1981) Comparison between measured and computed concentrates of sulphur compounds in eastern North America. J Geophys Res 86:5339–5346
Young WA, Shaw DB, Bates DV (1962) Presence of ozone in aircraft flying at 35,000 ft. Aerospace Med 33:311

The State of Air Quality in Canada: National Patterns

3

Jeffrey R. Brook, Tom F. Dann, Elisabeth Galarneau, Dennis Herod and Jean Pierre Charland

Abstract

This chapter provides a description of what air pollutants are routinely monitored in Canada and then focuses on ambient concentrations, geographic differences, temporal patterns and recent trends. In the context of the past 30–40 years, Canadian air quality has been an environmental management success story. Levels throughout the populated regions of the country are considerably lower than in the 1970's and 1980's. The 1990's also brought reductions and this chapter shows that even up through the 2000's improvements in air quality have been realized. However, these recent measurements indicate that there are areas where the current Canadian standards or provincial guidelines are still exceeded for ozone, $PM_{2.5}$ and toxics. Ozone has been the most problematic although there has been improvement in recent years. Similarly, on a national scale $PM_{2.5}$ has decreased. However, pollutants such as $PM_{2.5}$ can have effects on the population at low levels. Thus, air quality remains an important public health issue to track in Canada through routine monitoring; nationally, regionally and at the local scale, where areas of concentrated emissions due to industrial activity or population behaviour (e.g., traffic, wood burning) can lead to higher population exposures. Monitoring is critical in regions of projected industrial or population growth in order to maintain current levels, to identify options for improvement and to inform adaptive management approaches.

Keywords

Particulate matter · Ground-level ozone · Nitrogen dioxide · Volatile organic compounds · Air pollutants · Canadian trends · Air quality monitoring · Hazardous air pollutants

3.1 Introduction

Ambient air quality measurements provide crucial information to the public, policymakers and scientists. This chapter provides a description of what air pollutants are routinely monitored in Canada and then focuses on ambient concentrations, geographic differences, temporal patterns and recent trends.

Ground-level ozone (O_3) and fine particulate matter ($PM_{2.5}$—particulate matter below 2.5 μm in diameter) have traditionally been of greatest interest for monitoring and policy because they are considered to be the main contributors to photochemical smog events. One of the reasons for this is that they tend to build-up in the atmosphere and levels can increase over relatively large areas, from hundreds to thousands of square kilometers. Their slow rate of removal

J. R. Brook (✉) · E. Galarneau
Environment Canada, Toronto, Canada
e-mail: Jeff.Brook@ec.gc.ca

T. F. Dann
RS Environmental, Ottawa, Canada
e-mail: tom-dann@rogers.com

E. Galarneau
e-mail: elisabeth.galarneau@ec.gc.ca

D. Herod · J. P. Charland
Environment Canada, Ottawa, Canada
e-mail: dennis.herod@ec.gc.ca

J. P. Charland
e-mail: jean-pierre.charland@ec.gc.ca

E. Taylor, A. McMillan (eds.), *Air Quality Management*, DOI 10.1007/978-94-007-7557-2_3,

from the atmosphere and their formation processes, which involve conversion of other air pollutants, such as sulphur dioxide, nitrogen oxides and volatile organic compounds, in the presence of sunshine, are the main reasons for this build-up.

Another important reason for focus on O_3 and $PM_{2.5}$ is their human health effects. Numerous studies have shown that O_3 has significant respiratory effects, contributing to increases in emergency department visits (e.g., Stieb et al. 1996) and hospital admissions (e.g. Burnett et al. 1997). Ozone has also been linked to an increased rate of mortality for respiratory illnesses (e.g., Jerrett et al. 2009). Long-term exposure to $PM_{2.5}$ is considered to present the greatest mortality risk among air pollutants due to its significant cardiovascular effects (e.g., Brook et al. 2010; Pope et al. 2004). Recently, Crouse and co-workers (2012) found that long term outdoor $PM_{2.5}$ exposures across Canada during the 1991–2001 period (i.e., outdoor concentrations), which, compared with many countries in the world, were relatively low, are having a detectable impact on life expectancy. Furthermore, the magnitude of this effect is similar to what has been found in other countries. These results, as well as the large body of recent health effects research (Pope and Dockery 2006), suggest that public health benefits can be achieved by further reducing average $PM_{2.5}$ in Canada.

Several other air pollutants, known collectively as air toxics or hazardous air pollutants (HAPs) have also been identified as having the potential to induce adverse health effects. This category of pollutants generally excludes criteria air contaminants (i.e., those mentioned above) whose concentrations tend to be higher than those of most HAPs.

National scale patterns of air pollutant concentrations, temporal patterns and $PM_{2.5}$ and volatile organic compound (VOC) composition during the decade of the 2000's are presented in this chapter. First, the main national monitoring networks are described. Ozone and $PM_{2.5}$ are then the main focus starting with a discussion of background levels and finishing with information on recent trends in annual concentration. The latter sections present data and discuss spatial and temporal patterns for nitrogen dioxide (NO_2), nitric oxide (NO), VOCs, sulphur dioxide (SO_2) and hazardous air pollutants.

Policy Context Canada-wide Standards (CWS) In June 2000, the Canadian Council of Ministers of the Environment (CCME), except Québec*, signed the Canada-wide Standards for Particulate Matter and Ozone, which comprise ambient targets to be achieved by 2010. Each jurisdiction is committed to actions aimed at meeting the Standards for PM and Ozone by reporting on compliance once the target date is reached in 2010. This includes comprehensive reports every five years on all provisions of the CWS beginning in 2006 and annual reports on compliance and maintenance of the CWS beginning in 2011.

The CWS for $PM_{2.5}$ of 30 $\mu g\ m^{-3}$ is based on a three year average of the 98th percentile of 24-hr measurements and requires daily sampling. The CWS for O_3 of 65 ppbv is based on a three-year average of the 4th highest daily 8 h average.

Management of HAPS The compounds considered to be HAPs differ by jurisdiction. The United States (U.S.) has developed a formal list of HAPs through its Clean Air Act (US EPA 2010) and the U.S. Environmental Protection Agency has released several National Air Toxics Assessments in recent years (US EPA 2013). In Canada, no such formality exists, though many atmospheric contaminants found toxic under the Canadian Environmental Protection Act (CEPA) fit the common definition of HAPs. Compounds declared CEPA-toxic have been managed individually under the Toxic Substances Management Policy (TSMP) (Canada 1995). CEPA-toxic compounds that are persistent, bioaccumulative, and predominantly anthropogenic are assigned to TSMP's Track 1 requiring virtual elimination from the environment. Compounds not meeting all TSMP Track 1 criteria are assigned to Track 2 with a goal of preventing or minimizing releases to the environment.

3.2 Routine Monitoring in Canada

Ambient air quality monitoring across Canada is conducted by government, industry, and in some provinces, non-government organizations (NGOs). Many of the industrial monitoring sites are established in close proximity to their facilities and are usually required as part of a provincial permit to operate. In some jurisdictions industry also collaborates with government (and NGOs) to operate regional monitoring networks. Most of the air quality measurement data presented in this chapter are from the National Air Pollution Surveillance (NAPS) network. However, another important national network is the Canadian Air and Precipitation Monitoring Network (CAPMoN) and some data from this program are also used in this chapter.

The National Air Pollution Surveillance (NAPS) Program is a joint federal, provincial, territorial and municipal initiative. The purpose of this Program is to coordinate the collection of air quality data from existing provincial, territorial and municipal air quality monitoring networks and provide a unified Canada-wide air quality data base. Sites designated as NAPS sites are selected by Environment Canada and the Province or Territory in which the site is located, primarily to support national air quality programs like the CWS and the national ambient air quality objectives NAAQOs. Several provinces, territories and regional governments operate additional monitoring stations to meet the requirements of their own ambient monitoring programs. Ambient air pollution data from most of these sites are also submitted to the NAPS Canada-wide database at the discretion of each monitoring agency.

Table 3.1 provides 2009 information on each of the main pollutants monitored, such as number of operating sites and measurement frequency. The associated provincial/territorial/regional monitoring networks reporting data to the NAPS

Table 3.1 Overview of the main NAPS and CAPMoN ambient air pollutant measurements across Canada for 2009. A complete listing of observations by NAPS site code can be found at http://www.etc-cte.ec.gc.ca/NapsData/Web-Information

Pollutant	Time resolution (h)	Measurement method	No. of Stns	Start year	Comments
O_3	1	UV light absorption	212		Reporting to CWAQD[f]
O_3	1	UV light absorption	16		CAPMoN
NO_2	1	chemiluminescence	147		
NO	1	chemiluminescence	147		Reported nationally since 1986
SO_2	1	pulse-fluorescence ultraviolet (UV) adsorption	124		Reporting to CWAQD[f]
SO_2	24	integrated open-face filterpack	13		CAPMoN multiple PM constituents
$PM_{2.5}$	1	Tapered element oscillating microbalance (TEOM)[a]	134		Urban
$PM_{2.5}$	1	TEOM with filter dynamic measurement system (FDMS)	11		Urban
$PM_{2.5}$	1	Beta attenuation (BAM)	51		Urban
$PM_{2.5}$[c]	24	dichotomous sampler or FRM sampler[b]	41		Every 3rd or 6th day 0-0 LST
PM_{coarse}[c]	24	dichotomous sampler	30		Every 3rd or 6th day 0-0 LST
$PM_{2.5}$	24	Speciation sampler[d]	13	2002	Every 3rd day 0-0 LST
VOC[e]	24	Summa™ polished canisters	37		Urban Every 6th day 0-0 LST
VOC[e]	4	Summa™ polished canisters	14		Rural Every 6th day 12–16 LST

The Province of Québec, while not a signatory to the CWS, has undertaken analogous efforts as those covered by the CWS and has also developed working inter-jurisdictional arrangements for many provisions of the CWS.

[a] Beginning in 2002 many TEOM instruments in the NAPS $PM_{2.5}$ network were fitted with a sample equilibration system (SES). The SES incorporates a low-particle-loss Nafion dryer allowing for conditioning of the PM sample stream to a lower humidity and temperature level. Unless indicated otherwise, all $PM_{2.5}$ mass data in this chapter are from TEOM-SES instruments operated at 30 °C or TEOMs operating at 40 °C

[b] FRM = EPA Federal Reference Method

[c] The pre and post weighed Teflon filters from the dichotomous and FRM samplers are analyzed for elements using energy dispersive X-ray fluorescence (EDXRF) and for anions and cations using ion chromatography (IC)

[d] This particle speciation program was designed for more accurate and complete measurement of all the important components of $PM_{2.5}$. Along with the major ions (e.g., sulphate, nitrate, ammonium) and mass and trace elements from the co-located dichotomous sampler, this includes organic carbon (OC), elemental carbon (EC), gaseous ammonia and nitric acid. By 2006 this program included 14 (4 rural and 10 urban) sites, which are a subset of the $PM_{2.5}$ dichotomous sampler locations. A complementary short-term program was initiated at 5 CAPMoN sites following the same methods

[e] Content of canister analyzed for more than 100 different C_2 to C_{12} volatile organic carbon (VOC) species; Beginning in 2003 a number of polar species including α-pinene, β-pinene, δ-limonene and camphene began to be quantified

[f] Stations reporting to the Canada-wide Air Quality Database (CWAQD) include NAPS designated sites, CAPMoN stations and some additional provincial stations not designated as "NAPS"

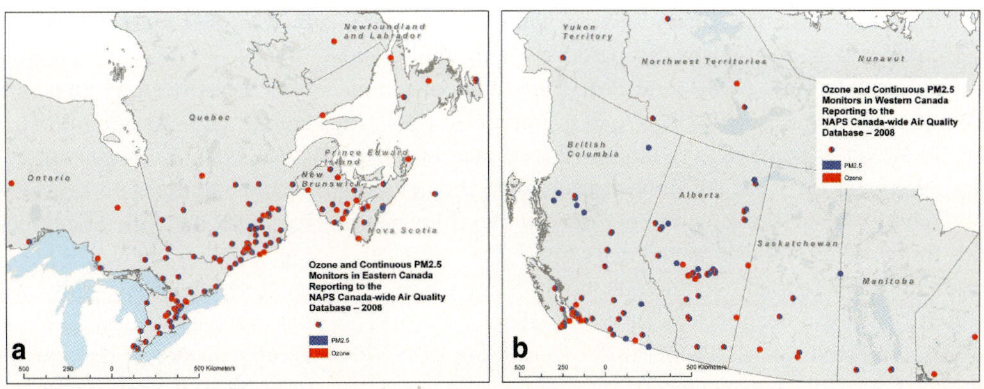

Fig. 3.1 Location of NAPS ozone and $PM_{2.5}$ monitoring sites in **a** eastern Canada in 2008, **b** western Canada in 2008

database in 2009 consisted of approximately 325 monitoring stations in over 215 communities. Although NAPS is primarily an urban monitoring network, 82 of the monitoring stations are located in rural areas. On a pollutant-by-pollutant basis there are approximately 831 continuous monitors for sulphur dioxide (SO_2), carbon monoxide (CO), nitrogen dioxide (NO_2), ozone (O_3), and particulate matter (PM), and over 113 air samplers measuring components of PM and volatile organic compounds (VOCs). Maps showing NAPS $PM_{2.5}$ and ozone monitoring sites are provided in Fig. 3.1a, b.

CAPMoN

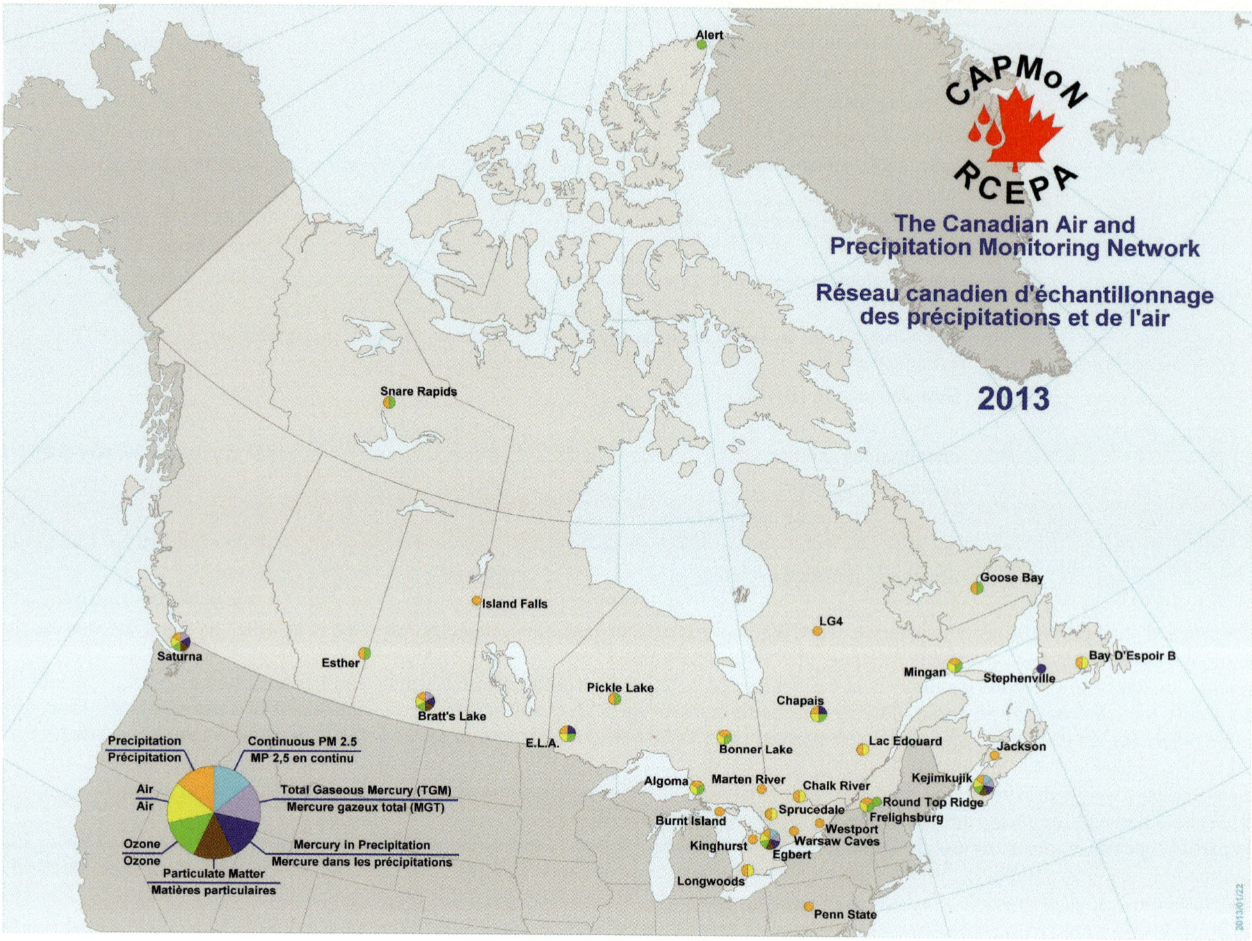

The Canadian Air and Precipitation Monitoring Network (CAPMoN) was initially motivated by the Acid Rain issue, but has grown to include a range of other measurements. Four types of measurements are made, namely: the chemical composition of precipitation (major ions and cations), the chemical composition of atmospheric particles (both acidic and basic), particle mass ($PM_{2.5}$ and PM_{10}), and the concentration of selected gases including O_3, SO_2, nitric acid (HNO_3), $NO/NO_2/NO_y$[1], peroxyacetyl nitrate (PAN), mercury (Hg) and ammonia (NH_3). CAPMoN measurement methods include both size-selective and non-size-selective filter methods for particle composition and mass, specialized denuders, impregnated filters and continuous monitors for gases, and wet-only deposition collectors for precipitation chemistry.

The objectives of CAPMoN differ from those of NAPS in that CAPMoN measurements provide data for research into: (1) regional-scale spatial and temporal variations of air pollutants and deposition, (2) long range transport of air pollutants (including transboundary transport), (3) atmospheric processes, and chemical transport model evaluation. To meet these objectives and to best complement NAPS, the CAPMoN sites are located in rural and remote areas.

[1] $NO_y = NO_z + NO_x$ and $NO_z = HNO_3 + HONO + 2N_2O_5 + HO_2NO_2 + PAN + NO_3$ + organic nitrates (not NH_3)

3.2.1 Measurement of $PM_{2.5}$

NAPS has been measuring the fine (<2.5 μm—$PM_{2.5}$) and coarse (2.5–10 μm—$PM_{10-2.5}$) fractions of particle mass since 1984. As of 2009 there were 27 dichotomous samplers and an additional 13 FRM $PM_{2.5}$ samplers (Table 3.1) operating in the network. Continuous or real-time particle monitoring began in the NAPS network in 1995, and the number of instruments grew rapidly with over 185 instruments reporting to the network in 2009. The use of these instruments has greatly increased the spatial and temporal coverage of the network. It is well-known that measurements obtained using the Tapered Element Oscillating Microbalance (TEOM) are biased low under certain ambient conditions due to the loss of semi-volatile chemical constituents (Allen et al. 1997; Brook and Dann 1998; Dann et al. 2006). Unless otherwise stated all $PM_{2.5}$ observations from TEOMs that are reported in this chapter have been corrected for this bias.

Table 3.2 Air toxics (Hazardous Air Pollutants) exceeding or approaching health-based guidelines in Canada

Air toxic/HAP	Volatility Class and Sampling Details
PAHs[a]	Semivolatile organic compounds (SVOCs) (gas phase and particles combined)
PCDD/Fs[b]	*Samples collected for 24 h every 6, 12 or 24 days in southern Canada or for 7 days every week in the Arctic*[c] *using high-volume filter-sorbent samplers* *Samples analysed by gas chromatography/mass spectrometry or high performance liquid chromatography with fluorescence detection*
Acrolein	Volatile organic compounds (VOCs) (gas phase only)
Acrylonitrile	*see VOCs in Table* 3.1
Benzene	
1,3-Butadiene	
Carbon tetrachloride	
Chloroform	
1,4-Dichlorobenzene	
1,2-Dichloroethane	
Dichloromethane	
Ethylene oxide	
Formaldehyde	
Naphthalene[d]	
Perchloroethylene	
Styrene	
Trichloroethylene	
Vinyl chloride	
Xylenes	
Arsenic	Metals and trace elements (particles only)
Cadmium	*Fine and coarse particulate matter samples collected for 24 h every 3 or 6 days using dichotomous samplers*
Lead	*Samples analysed by energy dispersive x-ray fluorescence and/or inductively coupled plasma/mass spectrometry*
Manganese	
Nickel	

[a] polycyclic aromatic hydrocarbons represented by benzo[a]pyrene
[b] polychlorinated dibenzo-*p*-dioxins and polychlorinated dibenzofurans
[c] PAHs only; one sample per month analysed and remainder of samples archived
[d] naphthalene is commonly listed as a PAH but is sampled with methods used to collect VOCs due to its high volatility

Adjustment of TEOM measurements In Canada, the manual, 24-hour, filter-based, gravimetric method (Method No.: 8.06/1.0/M) has been adopted as the NAPS Reference Method (RM) for $PM_{2.5}$ measurements. Data quality objectives (DQOs) have also been established for comparison of continuous $PM_{2.5}$ instruments with the RM monitors. At a number of locations in Canada, TEOMs, Beta-Attenuation Monitors (BAMs) and TEOM-FDMS (filter dynamic measurement system) units have been co-located with RM filter-based samplers. Consequently, numerous comparisons of continuous and manual $PM_{2.5}$ measurement methods have been conducted. These show that TEOM mass measurements are lower than mass measured by the NAPS reference method due to the volatilization of semi-volatile compounds from the TEOMs. This is caused by heating of the sampled air to 30 or 40 °C in the TEOM to remove unwanted water. This loss is typically larger in the winter than in the summer because of the larger differences between the TEOM filter and ambient temperatures in winter (Environment Canada 2004). Losses also appear to occur after very high concentration events are observed, such as during biomass burning. In this case, TEOMs often report negative concentrations for a short time (i.e., a few hours) after the event while semi-volatile material that rapidly accumulated on the filter during the event evaporates. TEOMs operated at 40 °C can over-report summer $PM_{2.5}$ concentrations when the ambient dew point is higher than approximately 20 °C, because of elevated relative humidity conditions by the filter, which can enhance water uptake on the particles.

The relationship between the TEOM and the RM in the cold season is quite consistent across Canada (Environment Canada 2004). Thus, correction or adjustment equations have been developed and perform reasonably well at reducing the bias in reported average $PM_{2.5}$ concentrations. The effects of such adjustments on the observed levels are discussed in Sect. 3.5.

3.2.2 Measurements of Hazardous Air Pollutants (Air Toxics)

Table 3.2 lists air toxics that exceeded or approached health-based guidelines at Canadian ambient air concentrations in recent years (*adapted from* Galarneau and Dann 2011). These

HAPs are measured routinely in Canadian air though NAPS and some are also measured in the Great Lakes region by the Integrated Atmospheric Deposition Network (IADN) and in the Canadian Arctic through the National Contaminants Program (NCP). It is important to note that federal ambient air quality standards are not available for the vast majority of HAPs, though some provinces have devised standards applicable in their jurisdictions. Standards from the province of Ontario have been used in this chapter.

3.3 The Unperturbed Airmass Over Canada: Background O_3 and $PM_{2.5}$

The existence of a background level of O_3 in the atmosphere is well established. Background O_3 is defined as the ambient level resulting from anthropogenic and natural emissions outside North America and natural sources within North America (CCME 2006). Background O_3 concentrations are variable, with variability due to geographic location, elevation, time of year, meteorology and long-range transport influences. Geographically, background O_3 levels are higher in the northern hemisphere, with notably large enhancements observed during the spring season. Elevation is particularly important, especially at sites with exposure to the free troposphere.

Unlike most air pollutants, background O_3 can exert a significant influence on ambient O_3 levels. There are four sources which contribute to background O_3: (1) downward mixing from the stratosphere to the troposphere, (2) transport of O_3 from outside North America, (3) in-situ photochemical production from natural precursors (biogenic methane, NO_X, VOC) and (4) intercontinental transport of O_3 precursors. Much of the background O_3 affecting surface levels resides in the free troposphere. Downward transfer from the free troposphere to the surface occurs relatively frequently and is facilitated by vertical motion caused by convective and frontal activity. A combination of processes appears to contribute to this dynamic repository, including in situ O_3 production, downward fluxes from the stratosphere and intercontinental transport. There is currently some controversy around the magnitude of the influence of stratospheric O_3 on surface background levels (Lefohn et al. 2001; Fiore et al. 2003).

The spring O_3 maximum is a northern hemisphere phenomenon and is a defining feature of the annual cycle at remote and rural sites (Monks 2000). There appear to be several mechanisms responsible for the spring maximum, including UV-enhanced photochemistry in the free troposphere (Dibb et al. 2003), stratospheric-tropospheric exchange (Diem 2004; Tarasick et al. 2005) and enhanced hemispheric transport (Jaffe et al. 2010). This observation of springtime O_3 maxima is contrary to the notion that O_3 peaks during the summer, when local photochemistry is usually greatest. As shown in Fig. 3.2, Alberta sites experience March to May peaks in

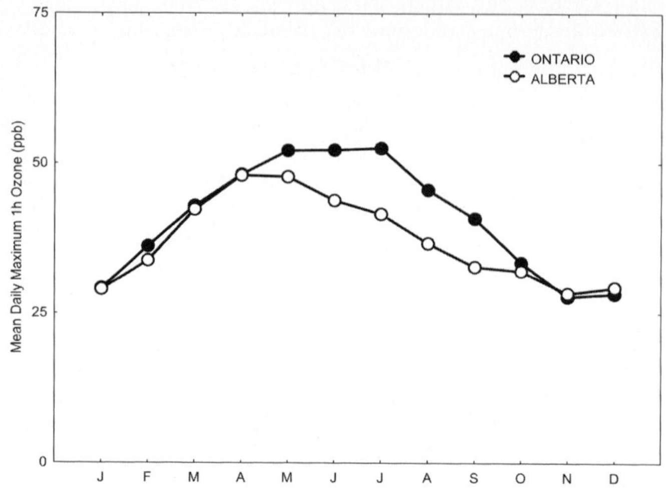

Fig. 3.2 Mean daily maximum ozone (ppb) by month (2006–2008) for southern Ontario and Alberta sites

mean daily maximum O_3 whereas the Ontario sites experienced the highest mean daily maximum O_3 in the months of June, July or August. Assuming that the spring peak is a background feature affecting all areas, the excess ozone in Ontario for the May-September period reflects the photochemical creation of ozone in the region related to emissions from human activities. Figure 3.2 also shows that during the months of January-April and October-December mean ozone concentrations are essentially the same in the two regions and hence are related to continental background conditions.

"Background" particulate matter (PM) is defined here as PM arising from local natural (non-anthropogenic) emissions of PM and PM precursors, as well as PM and precursors transported into an airshed from afar. This long range transport PM may be both natural and anthropogenic in origin (McKendry 2006), but comes from outside North America. Examples of natural emissions of PM and its precursors include wind-blown dust, sea spray, volcanic eruptions, lightning, forest fires, wild animals and plants.

Few values of background $PM_{2.5}$ are available in the scientific literature. Background annual median $PM_{2.5}$ concentrations in Canada are both location- and time-dependent and have been estimated to range from 1.2 $\mu g\ m^{-3}$ at a continental site in northern Alberta to 4.2 $\mu g\ m^{-3}$ at an agricultural site near Montréal, QC (unpublished analysis of NAPS data). Of considerable relevance to Canada are estimates from McKendry (2006) who reported a mean background concentration in air masses arriving in British Columbia from north Pacific trajectories to be 1.5–2 $\mu g\ m^{-3}$. Lower background $PM_{2.5}$ values, generally less than 1.0 $\mu g\ m^{-3}$ for northern portions of the U.S., were used to estimate the overall public health burden due to $PM_{2.5}$ in the U.S. (Fann et al. 2011). These were derived from air quality model simulations that excluded anthropogenic emissions as opposed to from ambient observations.

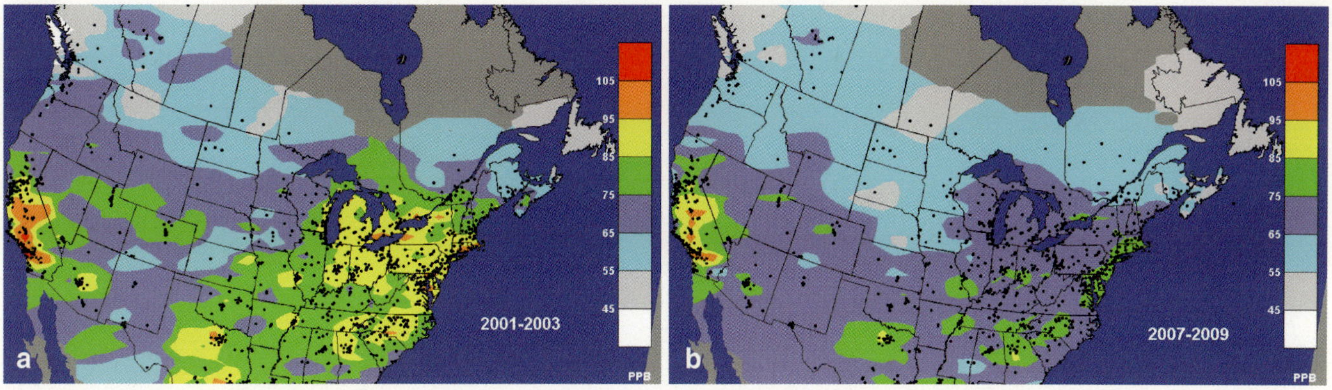

Fig. 3.3 **a**, **b** Comparison of the spatial distribution of the 3-year-average 4th highest daily maximum 8-hr ozone concentration in ppb for 2001–2003 vs. 2007–2009. The figure includes data from the United States Environmental Protection Agency (U.S. EPA) Air Quality System (AQS) database. Annual statistics for the measurement sites for these two time periods were spatially interpolated using inverse distance weighting

3.4 Observations of Ozone Concentrations Across Canada

Ozone data have been analyzed extensively throughout the past to illustrate seasonal, diurnal and day-of-the-week patterns and episodic O_3 concentrations in different regions of Canada (Fuentes and Dann 1993; Pryor et al. 1995; Environment Canada 1997). A complete analysis of ozone in terms of the Canada-wide Standard can be found in the Government of Canada Five-year Progress Report on Canada-wide Standards for Particulate Matter and Ozone (Government of Canada 2007).

For the period 2003–2005, at least 40 % of the Canadian population (approximately 13 million) lived in communities with levels above the CWS. Most of these were located in Ontario and Québec. Outside these two provinces, only one community in British Columbia and one non-urban area in Atlantic Canada had levels above the CWS. With the exception of Saskatchewan, Manitoba and the territories, all other regions had at least one location with levels within 10 % of the CWS. Provinces and territories reported data for a total of 70 communities (CMA, CA and RSA) representing more than 70 % of the Canadian population.

The spatial pattern of peak O_3 concentrations, as defined by the CWS metric, is shown for 2001–2003 and for 2007–2009 in Fig. 3.3a, b. From the earlier part of the last decade to the latter part there was a dramatic drop in O_3, including throughout the United States (U.S.). However, through 2009 southern Ontario was still experiencing levels above the CWS. The maps in Fig. 3.3a, b show that these high levels are part of a large area that encompasses the eastern U.S. These high levels are due, in part, to Ontario emissions and in part to transboundary transport of O_3 and ozone precursors from the U. S. Evidence of the impact of local Canadian sources is best exemplified by noting the small area of higher levels downwind of Toronto (i.e., to the east) in both time periods (Fig. 3.3a, b).

Compared to southern Ontario, western and eastern Canada have had lower levels of O_3, with limited evidence of CWS exceedances. Slightly elevated levels can be seen around Edmonton, Alberta and there are parts of the Lower Fraser Valley (LFV) of British Columbia and southern Atlantic Canada with locally elevated levels. Due to their location downwind of the populated areas of southwestern British Columbia and northwestern Washington, parts of the eastern LFV (e.g., Hope) may continue to be near CWS exceedance levels. However, the scale of the maps do not allow for such local maxima to be visible.

3.4.1 Temporal Variations in Ozone Concentrations

Ozone is well known to be present in higher concentrations during the warmer months and during the afternoon. In terms of seasonal behaviour, however, much of Canada is not as strongly influenced by local to regional scale sources and strong summertime photochemical O_3 production events as expected. Out of 125 sites examined for the 2001–2005 period only 46 actually experienced the highest mean daily maximum O_3 in the months of June, July or August. At the remaining sites, which are classified as being more remote rural and background sites, the maxima in O_3 concentrations occur in the springtime. All of the sites where O_3 peaks in June-August, except Hope, British Columbia, were located in southern Ontario and southern Québec. In these areas, regional-scale photochemical O_3 production and long-range transport of pollutants are relatively more important and thus, summer-time maximum O_3 concentrations are significantly higher than those in other areas of the country.

The relative frequency of days per month during 2006–2010 with eight hour average O_3 greater than 65 ppb is shown for selected geographic regions in Fig. 3.4. Only 8 sites outside of Ontario and Québec met the criteria of inclu-

Fig. 3.4 Average days per month with ozone concentrations greater than 65 ppb for the years 2006–2010. (Only sites averaging at least one day per year >65 ppb are included in the plot)

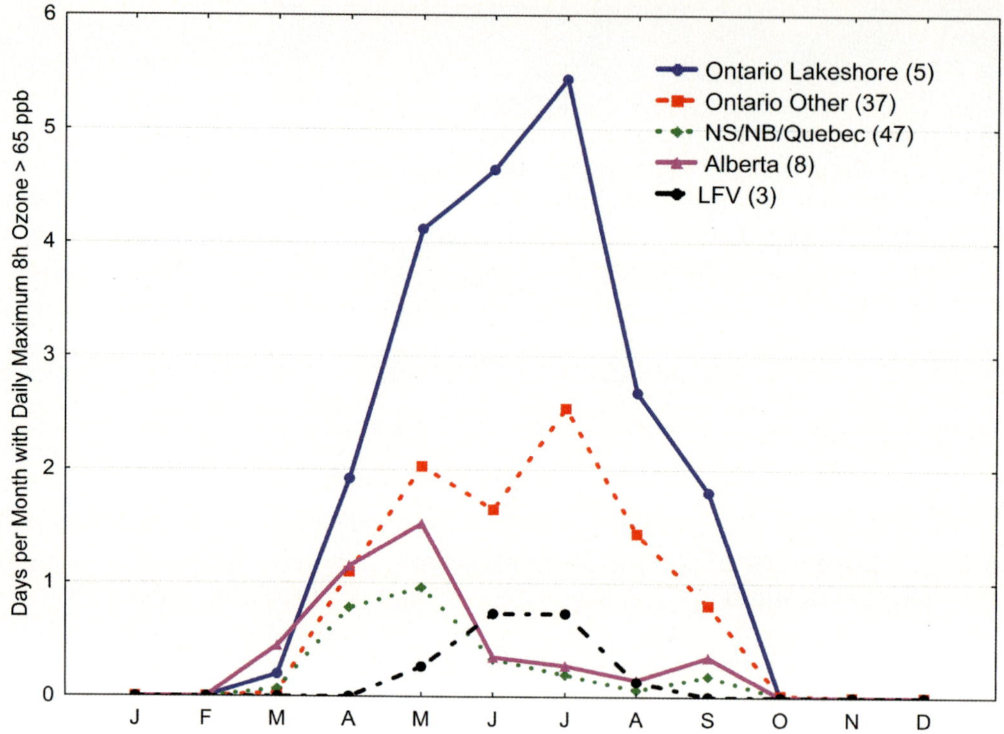

sion in the figure of having at least one day >65 ppb. Sites in Ontario near the Great Lakes shorelines experienced the highest frequency of days and thus these are shown separately on the figure. It can be seen that for all areas O_3 mixing ratios greater than 65 ppb can be experienced in any of the months from March through October, but the greatest frequencies occur for June through September. The most noticeable spring peak was in the eastern part of the country (NS/NB/Québec).

Through the course of the day, greater sun intensity in the afternoon typically leads to higher O_3 levels at this time. In addition, the O_3 precursor emissions have strong temporal variations on several time scales, which may influence the diurnal pattern in O_3. For example, the main source of NO in urban areas—traffic—shows a strong diurnal cycle with a maximum generally in the daylight hours. Superimposed on this is a further modulation, especially in urban areas, related to the maximum traffic flow in the morning and evening rush hours. In addition, the diurnal variation of emissions is different on weekends to that occurring on weekdays. The number of heavy duty diesel vehicles is also substantially reduced on Saturdays and even further on Sundays. These changes in precursor emission patterns, especially during the daylight hours have led to substantial differences between weekdays and weekends in urban diurnal O_3 profiles (Environment Canada 1997; Fuentes and Dann 1993; Pryor et al. 1995). For the 2001–2005 time period 65 out of 72 urban sites in Canada experienced a "weekend effect" while 32 out of 44 rural sites showed lower O_3 or no difference on weekends—this included all rural sites in southern Ontario.

3.4.2 Trends in Ozone

During the past several decades attempts have been made to reduce O_3 by addressing precursor emissions in the U.S. and Canada from a variety of mobile and stationary sources. NARSTO (2011) recently examined North America changes in O_3 in response to these controls to assess the extent to which management actions have resulted in the intended benefits. In some places in Canada the results have been favourable, but not in all situations, due to the complexity of the atmospheric chemistry and growth in some regions. In addition, trends are difficult to detect due to the large influence of meteorology. Trends in O_3 vary markedly depending on site location (urban, rural, background) and on which statistic is being examined (Jenkin 2008; Ainslie and Steyn 2007).

Year to year changes in daily maximum 8h O_3 (April to September only) at different percentile levels (10th, 25th, 50th, 75th, 95th and maximum) are shown for a group of urban and rural sites for the period 1990–2010 in Figs. 3.5 and 3.6, respectively. The urban sites shown are distributed across Canada but the rural sites are primarily located in Ontario and Québec with 3 sites in Atlantic Canada and 2 sites in British Columbia. Trends in the higher end of the O_3 distribution tends to reflect the impact from precursor emissions at closer distances to the measurement sites, while the average or lower portions of the distribution generally represent the background O_3 (Lelieveld et al. 2004).

Figure 3.5 shows trend results for the urban sites with a statistically significant (at the 95th confidence level) positive trend for the 10th and 25th percentiles indicating a rise in urban background levels. These results illustrate the effect

Fig. 3.5 Trend in daily maximum 8 h ozone 10th, 25th, 50th, 75th, 90th, 95th percentiles and maximum for Canadian urban sites for 1990–2010. Regression lines marked with ** are statistically significant at the 95th percentile confidence level

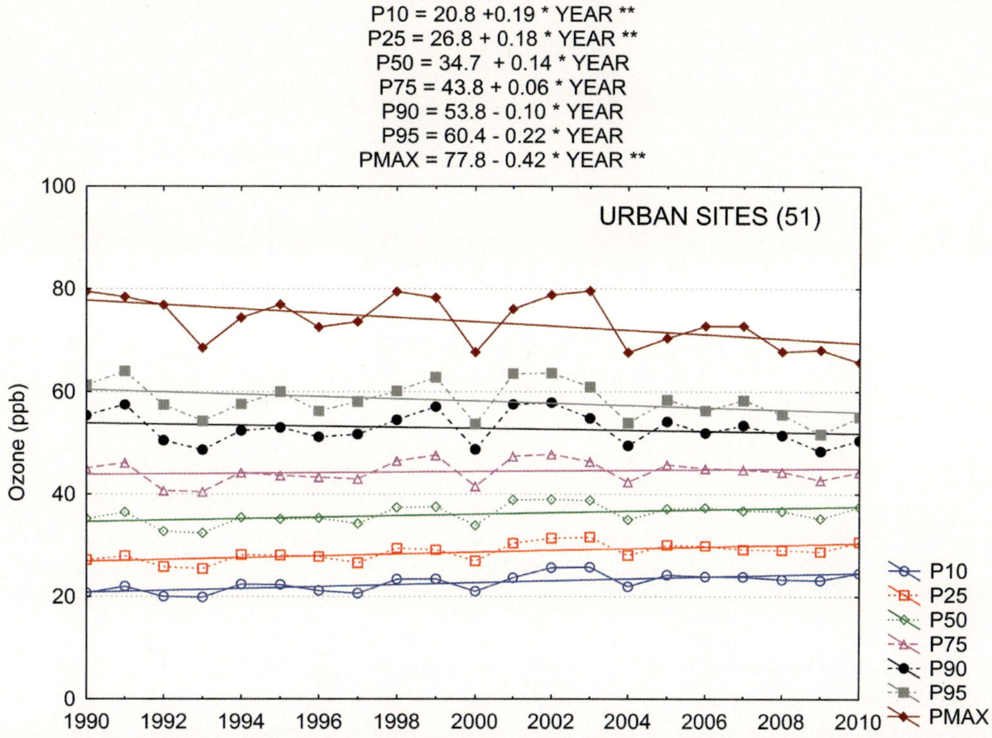

Fig. 3.6 Trend in daily maximum 8 h ozone 10th, 25th, 50th, 75th, 90th, 95th percentiles and maximum for Canadian rural sites for 1990–2010. Regression lines marked with ** are statistically significant at the 95th percentile confidence level

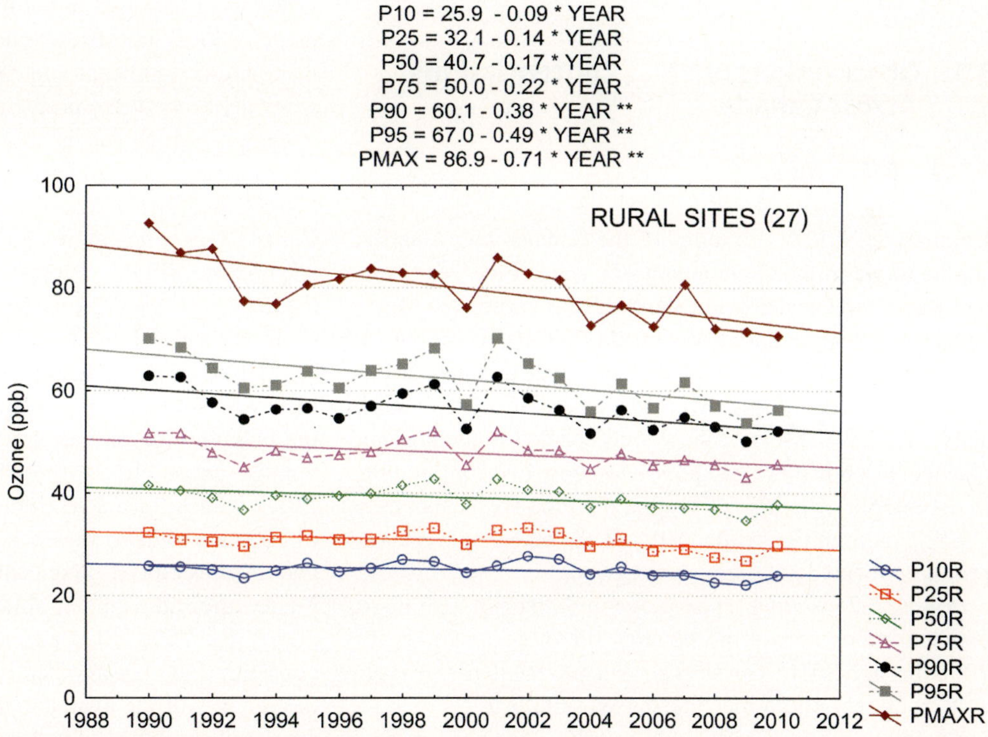

of local NO reductions in reducing O_3 scavenging leading to a net upward trend in lower percentile and average ozone concentrations. In contrast, a significant negative slope was detected for the maximum urban O_3 values, while for the 90th and 95th percentiles there were non-significant decreases. This behaviour suggests that O_3 arising from emission sources that are more local to Canadian cities has decreased during the past 20 years.

For the rural sites, Fig. 3.6 shows that there have been decreases (i.e., negative slopes) throughout the O_3 distribution, but only the maximum, 95th and 90th percentiles of 8 h O_3 had statistically significant decreases. Thus, region-

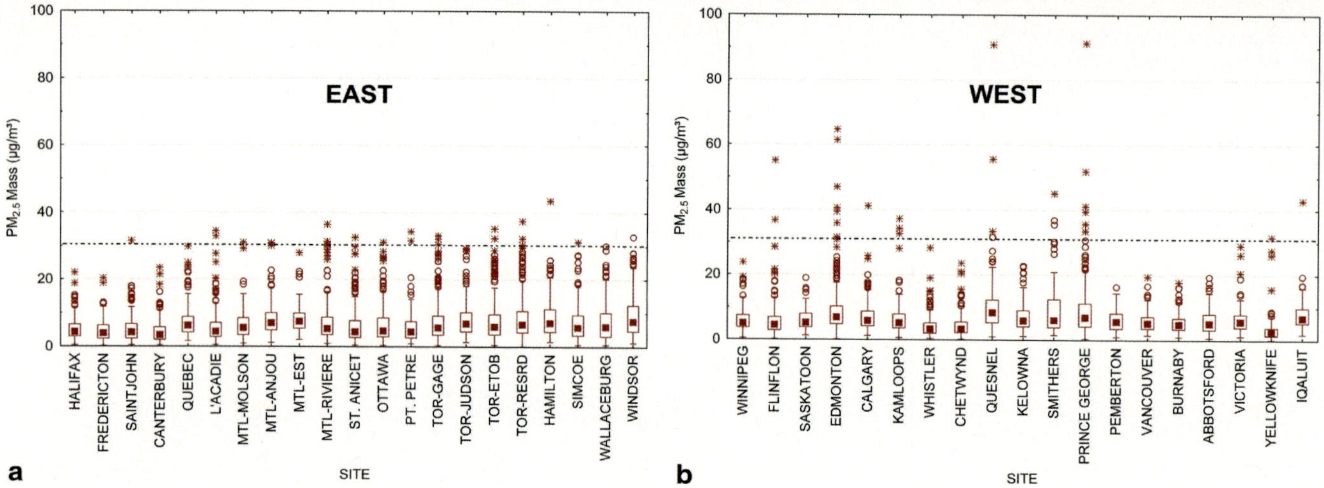

Fig. 3.7 Comparison of 24 h $PM_{2.5}$ mass concentrations from filter-based samplers: 2008–2010. The statistics provided are the median, 2nd, 25th, 75th and 98th percentiles and outliers and extremes. Only sites with a minimum of 100 samples collected over this period are included and the sites are arranged from eastern Canada to the west

al scale O_3 has responded to the past emission reductions implemented in Canada and the U.S. Importantly, there is no evidence from this group of rural sites (predominantly in Ontario and Québec) that there has been an increase in background O_3.

3.5 Observations of $PM_{2.5}$ Concentrations Across Canada

3.5.1 $PM_{2.5}$ Mass

An analysis of $PM_{2.5}$ in terms of the Canada-wide standard can be found in the Government of Canada Five-year Progress Report on Canada-wide Standards for Particulate Matter and Ozone (Government of Canada 2007). For the period 2003–2005, at least 30 % of the Canadian population (approximately 10 million) lived in communities with levels above the CWS. Most of these were located in Ontario and Québec. Outside these two provinces, only two communities in the interior of British Columbia reported levels above the CWS. Communities within 10 % of the standard were also primarily located in Ontario and Québec.

Figure 3.7 provides a box-plot comparison of 24 h $PM_{2.5}$ concentrations across the country for more recent years: 2008 through 2010. The sites shown represent all the available, routinely operating filter-based samplers. As seen for 2008–2010, there were more sites with 24 h observations greater than 30 µg m $^{-3}$ in eastern Canada compared to the west. However, there continue to be locations in the interior of BC (e.g., Prince George) with relatively high means and peak 24 h concentrations. Edmonton is also seen to have experienced a number of days above the 30 µg m^{-3} threshold.

Similar to O_3, the conditions over southern Ontario and southern Québec are part of a high concentration Canada-U.S. airshed. Throughout this large area, annual mean $PM_{2.5}$ exceeded 8 µg m^{-3} in 2004–2006. In central British Columbia, where there are also high concentrations, it is likely that these levels were confined to the more populated valleys where a range of local industries, residential wood smoke and traffic make more significant contributions. The winter months are generally when these problems occur due to stronger valley inversions, which trap the local emissions, and due to a greater need for supplemental heating by wood-burning appliances.

As Table 3.1 indicates there are many more locations in Canada reporting hourly $PM_{2.5}$ from continuous samplers such as the TEOM. These data are valuable for enhancing the density of the network such that it is more comparable to O_3 monitoring and thus, provides greater population coverage. However, as indicated above, there are potential problems with some of the available instruments. To demonstrate the potential importance of the TEOM instrument's loss of volatile mass, Fig. 3.8 provides the annual 98th percentile $PM_{2.5}$ values for a selection of TEOM trend (at least 8 out of 10 years of data) sites based upon adjusted (transformed) and unadjusted $PM_{2.5}$ concentrations. Values are 2–5 µg m^{-3} higher after adjustment, although both measures show that during the 2000–2009 period there was a net decrease in concentration. As another example, adjusting $PM_{2.5}$ values among all the available TEOM sites across the country among the measurements taken during 2004–2006 increased the number of sites surpassing the CWS by 27. Clearly, in order to obtain a proper understanding of the current state of $PM_{2.5}$ levels over as much of Canada as possible it critical that the continuous monitors, which are necessary to provide daily sampling frequency, produce measurements that are consistent with the reference method.

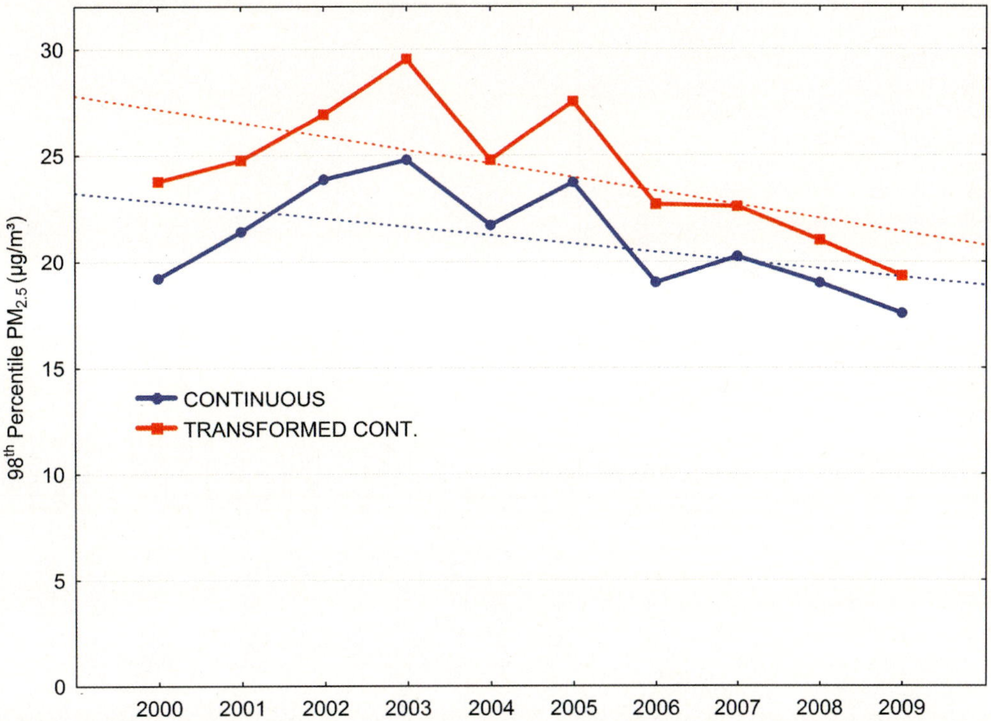

Fig. 3.8 Yearly variation of 98th percentile $PM_{2.5}$ for national TEOM trend sites 2000–2009. Adjusted and unadjusted concentrations are shown to provide an indication of the amount of mass that is estimated to be lost due to the sample heating inside the TEOM instrument

Current Status of Continuous $PM_{2.5}$ Measurement Across Canada Due to reasons described above the TEOM continuous $PM_{2.5}$ instruments deployed across the Canadian monitoring networks since the mid-1990's do not meet the data quality objectives (DQOs) that have been established for comparison with NAPS reference method samplers (Allen 2010). However, NAPS managers have determined that instruments approved as U.S. EPA Class III federal equivalency methods (FEM) do meet the NAPS DQOs. Thus, monitoring agencies across Canada have either transitioned to or are in the process of transitioning to $PM_{2.5}$ FEM instruments in their monitoring networks (Allen 2010).

There are primarily three makes of Class III FEM instruments currently being deployed across Canada: the Met One BAM1020; the Thermo TEOM 1400ab-FDMS (to be discontinued); and the Thermo 5030 Sharp. The U.S. EPA conducted an assessment of $PM_{2.5}$ FEMs compared to collocated federal reference method (FRM) samplers. For this analysis, 61 Met One BAM stations and 17 Thermo 8500 FDMS stations were examined. Based on these comparisons it was found that both FEMs predicted annual means similar to the FRM or higher by 2–5 µg m^{-3}. The bias for both instruments appeared to be highest for the spring/summer seasons (U.S. EPA 2011). The Canadian Council of Ministers of the Environment (CCME) has recommended that all jurisdictions strive to deploy new Class III FEM instruments of their choice by December 31, 2012.

Ultimately with greater geographic coverage of $PM_{2.5}$ monitoring using consistent daily measurement techniques (i.e., RM or FEM) it may be possible to generate $PM_{2.5}$ maps such as shown in Fig. 3.3a, b for O_3. However, the current network does not have the density of monitors for this purpose and differences in measurement techniques among the available sites add to the problem. In addition, $PM_{2.5}$ differs from O_3 in that there can be significant primary emissions of $PM_{2.5}$. This can lead to greater uncertainty in the interpolation necessary to produce maps, particularly in areas of significant local sources and limited monitoring.

Satellite-based estimates of surface $PM_{2.5}$ concentrations are proving to be useful in filing the surface monitoring gap (van Donkelaar et al. 2010), but limitations remain (Hoff and Christopher 2009). This includes the limited temporal coverage, lost observations due to clouds and the columnar nature of satellite observations. Nonetheless, satellite-derived data represent an important emerging technology for air pollutant monitoring. The national level exposure assessment used by Crouse et al. (2012) to successfully quantify the effects on mortality rates of long term $PM_{2.5}$ exposure across Canada relied on observations derived from satellite. Lee et al. (2011) showed what level of information on surface NO_2 patterns over southern Ontario could be derived by combining satellite and routine monitoring data and recently McLinden et al. (2012) used satellite data to study NO_2 and SO_2 patterns over the oil sands region of Alberta.

3.5.2 Temporal Variations in $PM_{2.5}$ Concentrations

Unlike O_3, which has a strong seasonality, $PM_{2.5}$ varies less through the year. Winter and summer levels of $PM_{2.5}$ mass

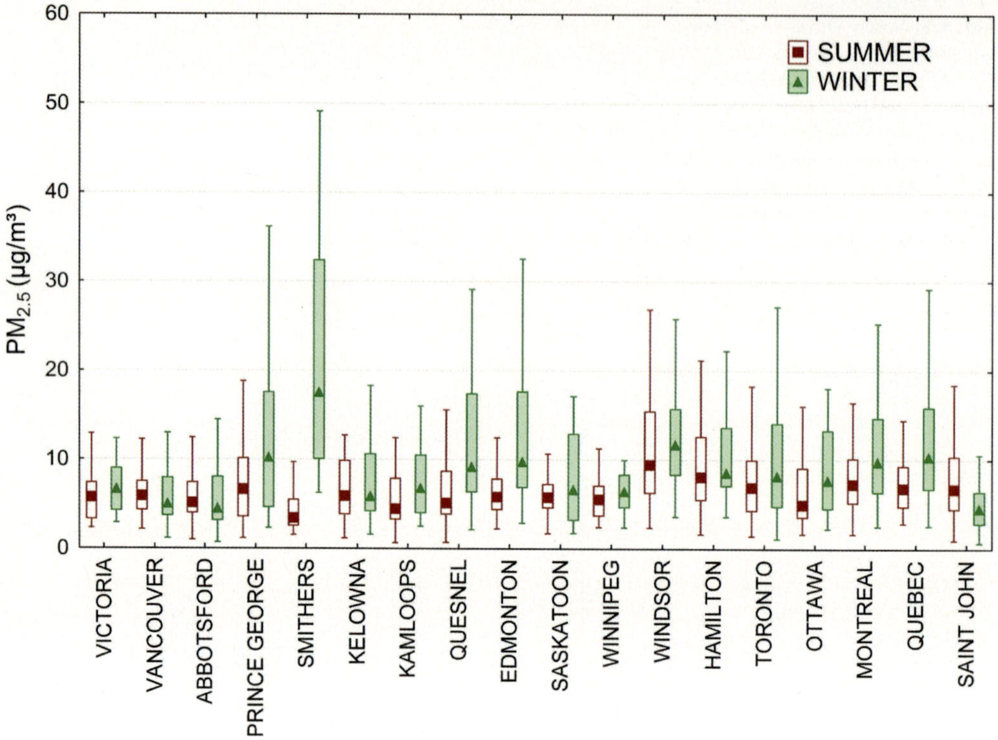

Fig. 3.9 Winter (December, January, February) vs. summer (June, July, August) concentrations of $PM_{2.5}$ mass (2007–2009)—filter-based samplers for selected urban areas (median, 25th and 75th percentile, non-outlier maximum and minimum). The time periods correspond to those in Fig. 3.12

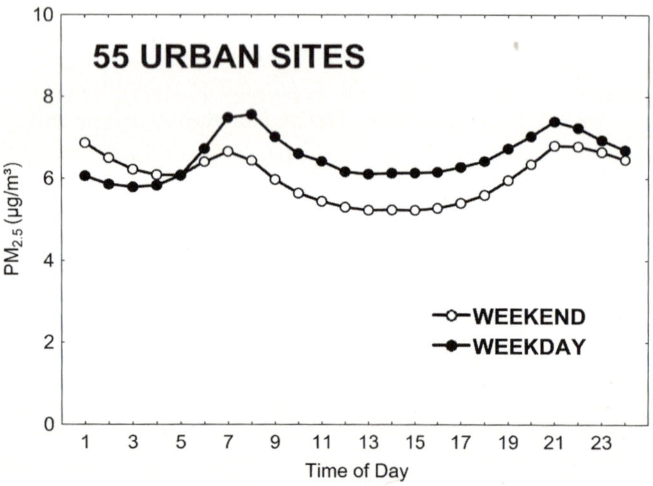

Fig. 3.10 Diurnal variation in $PM_{2.5}$ mass—weekday vs. weekend—TEOM samplers (2007–2009)

are compared for 2008–2010 in Fig. 3.9. Overall there were more sites with higher winter means and peaks.

Using composite hourly data from the TEOM network the diurnal variability of $PM_{2.5}$ mass for weekdays versus weekends is illustrated in Fig. 3.10. This figure clearly shows the impact of transportation sources on $PM_{2.5}$ concentrations with an approximate 25 % decrease in the morning peak and a 17 % decrease in the daily mean on weekends as compared to weekdays.

3.5.3 Recent Trends in $PM_{2.5}$

The lack of long-term $PM_{2.5}$ mass data at rural/remote monitoring sites in Canada precludes any analyses of long-term trends outside of cities. Figure 3.11 shows how the annual mean $PM_{2.5}$ mass composited across multiple urban dichotomous sampler sites changed during the last decade. In eastern Canada the peak year was 2002 with a level of 13 µg m^{-3} and since then $PM_{2.5}$ decreased to below 9 µg m^{-3}, on average. During this same ten year period the $PM_{2.5}$ also decreased among the western Canadian sites, but only by 2 µg m^{-3}. As a result, in 2008 the mean was 6 µg m^{-3} and thus, the difference between the east and west urban areas has diminished.

Also shown in Fig. 3.11 is the annual mean sulphate concentration (SO_4^{2-}). This $PM_{2.5}$ constituent is one of the main contributors to mass (see next section) and during the past 20 years there have been considerable North American reductions in the emissions of SO_2 in order to lower acid deposition rates and vehicular emissions of $PM_{2.5}$. In eastern Canada sulphate was variable from 1999 to 2005 at around 11 µg m^{-3}. However, decreases in the last three years have led to a composite average concentration of 2.2 µg m^{-3} in 2008. The rate of reduction of $PM_{2.5}$ during this time period was very similar to the rate observed for sulphate suggesting that the SO_2 controls had a beneficial impact on $PM_{2.5}$ in the east. Sulphate was a smaller fraction of the $PM_{2.5}$ in the west and its concentration has remained steady at about 1 µg m^{-3} from 2001 to 2008.

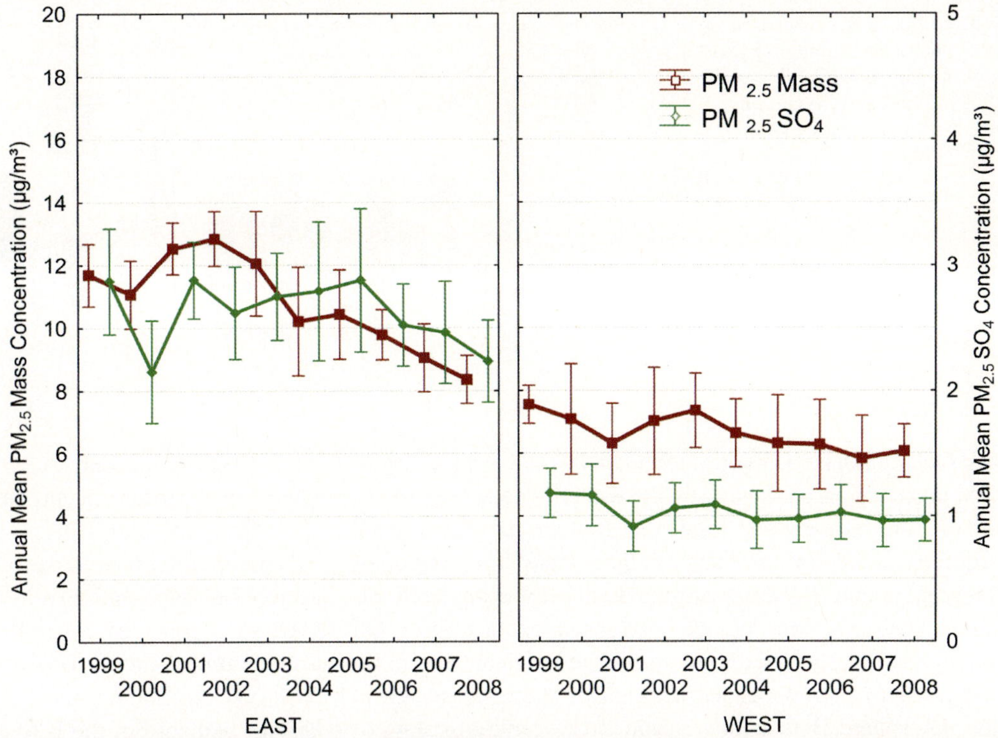

Fig. 3.11 Yearly variation in mean $PM_{2.5}$ mass and sulphate (SO_4) from filter based network urban trend sites east and west of the Ontario/Manitoba border (1999–2008). Composite means and the 90th percent confidence interval around the mean are plotted

3.5.4 Chemical Composition of $PM_{2.5}$

$PM_{2.5}$ is a complex mixture of inorganic and organic chemical compounds. The main contributors to the mass are the ions of sulphate, nitrate and ammonium and organic carbon (OC), which consists of thousands of different species and not all have been identified. Another important component of the $PM_{2.5}$ is elemental carbon (EC), which is often referred to as black carbon (BC). Here we use EC because of the approach to determine the concentration reported, which is based upon the thermal-optical approach (Chow et al. 1993). A small, but consistent contribution to $PM_{2.5}$ is also made by a large number of trace elements or metals such as iron, aluminum, silicon, calcium, zinc, manganese, titanium, nickel and arsenic.

Sulphate and major elements of $PM_{2.5}$ mass from the dichotomous sampler network have been determined since 1986. These data have been previously described (Brook et al. 1997, 1999; Brook and Dann 1999). Monitoring of complete fine particle speciation began at Canadian network sites in 2000 (Lee et al. 2003). The number of sites started to grow beyond Toronto and Burnaby South (greater Vancouver area) in 2003. In this speciation network, which currently consists of 11 sites (Table 3.1), the major components of $PM_{2.5}$ are reported. These measurements were recently discussed by Dabek-Zlotorzynska et al. (2011), including the seasonal variation in each of the major chemical species making up the $PM_{2.5}$.

Complete fine particle speciation allows the mass to be reconstructed from the chemical species. Dabek-Zlotorzynska et al. (2011) described the reconstruction approach used for the Canadian $PM_{2.5}$ network. This involves grouping the chemical composition measurements according to ammonium sulphate ($(NH_4)_2SO_4$ or ASO_4), ammonium nitrate (NH_4NO_3 or ANO_3)), organic matter (OM), which is taken as $1.6 \times OC$ at urban sites and $1.8 \times OC$ at rural sites, elemental carbon (EC), crustal material and other oxidized metals (SOIL) and sodium chloride (NaCl). Particle-bound water (PBW) can also be estimated from the measurements. Figure 3.12a, b shows the reconstructed mass for the warm (April to September) and cold (October to March) seasons of 2007–2009.

On average the reconstructed mass was in agreement with the measured mass. Combined $(NH_4)_2SO_4$ and NH_4NO_3 account for 32–43 % of total $PM_{2.5}$ mass at the eastern sites during the summer and 36–59 % in the winter. Thus, secondary compounds (i.e., those that form in the atmosphere from emissions of gaseous precursors) are a major contributor. For the western sites $(NH_4)_2SO_4$ and NH_4NO_3 account for 18–30 % of mass in summer and 22–45 % in winter, also indicating that gaseous precursor emissions are quite important to $PM_{2.5}$. Figure 3.12a, b also shows that the relative amounts of $(NH_4)_2SO_4$ and NH_4NO_3 change from summer to winter. Nitrate species increase dramatically at most locations with the highest concentrations in Edmonton and Windsor.

At the eastern sites OM is the next most important contributor to mass after the two inorganic secondary species. In contrast, in Edmonton and at the British Columbia sites, OM contributes the most. OM contributions to $PM_{2.5}$ mass for urban and rural sites range from 31 to 52 % in summer and 23 to 58 % in winter. Interior BC locations, such as Quesnel

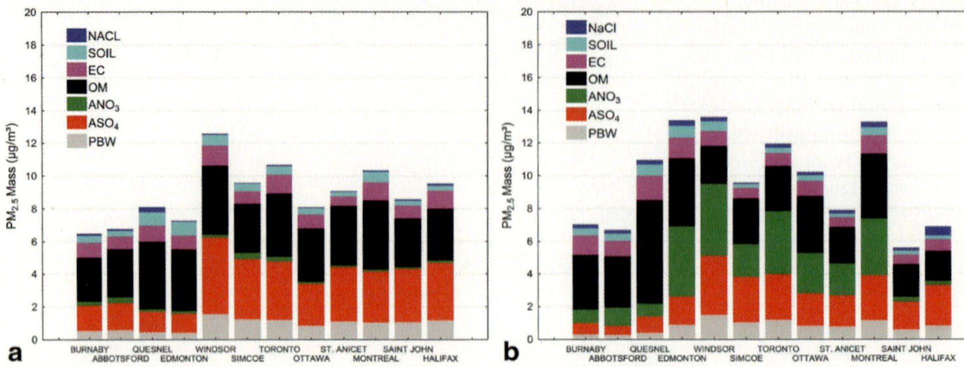

Fig. 3.12 **a, b** Reconstructed $PM_{2.5}$ mass by major component and site for (a) June, July, August (2007–2009) and (b) December, January, February, March (2007–2009)

and Golden (Dabek-Zlotorzynska et al. 2011), have the largest OM percentages, particularly in the winter. Wood combustion, either from industry and residential heating, as well as from wildfires, is the key contributor in these areas. The fraction of OM that is due to secondary formation, originating from gas phase VOC emissions, has not been accurately quantified. On average, it is suspected to be a significant fraction and also increases away from populated areas and/or as the aerosol age in the atmosphere. Both anthropogenic and biogenic emissions of VOCs are important contributors to the secondary OM (e.g., de Gouw et al. 2008; Slowik et al. 2010, 2011; Liggio et al. 2010).

The average $PM_{2.5}$ chemical composition in Fig. 3.12a, b changes during high mass concentration episodes due to greater contributions from certain sources or atmospheric formation processes. For the ten highest days in summer, $PM_{2.5}$ mass at eastern sites is primarily composed of $(NH_4)2SO_4$ and OM (70–80 %), while at the western sites other than Golden, NH_4NO_3 is also an important contributor. For the highest concentration days in winter NH_4NO_3 and OM are the primary contributors to mass at almost all the sites, but $(NH_4)2SO_4$ is also an important contributor at the eastern sites and in particular in Halifax.

Among major cities, trace metals (four selected toxic elements) have the highest concentrations in Halifax. However, Fig. 3.13 shows that among the locations monitored these metals were by far the highest in the small, base metal smelting community of Flin Flon, Manitoba, where arsenic was very high during the monitoring period (2007–2009). In terms of cities, these four metals are also relatively high in Montreal, followed by Windsor, Saint John and Burnaby. Both the east and west coast and Montreal have high nickel and vanadium levels. These are due to use of oil for power and heating (Halifax), heating (Montreal) and due to ship traffic (all four locations).

3.6 Elevated $PM_{2.5}$ and O_3 (SMOG) Events

Smog episodes are an important feature of Canadian air quality because the response by the government is to issue an air quality advisory. This leads to greater public awareness, affects their perception of the problem and focuses attention on the actions needed to avoid such events. High O_3 and $PM_{2.5}$ events often occur during multi-day, regional scale episodes, particularly in eastern Canada. The meteorological conditions contributing to pollutant build up may become established over Ontario and then propagate eastward. However, such episodes are relatively uncommon over Atlantic Canada unless the air flow is such that pollutants that have built up over Ontario, the U.S. Midwest and particularly the eastern seaboard of the U.S. are transported to the region to combine with local emissions.

Defining an episode is subjective in that to generate frequency statistics a threshold concentration needs to be defined. The appropriate threshold is a matter of perception and ideally should be set relative to the prevailing conditions in each region of the country. For example, the threshold level for high $PM_{2.5}$ in Ontario is well above the level that noticeably impacts visibility in the Lower Fraser Valley. In this chapter the CWS metrics for $PM_{2.5}$ and O_3 are used to identify episodes, but again, this may misrepresent how local populations perceive the occurrence of episodes. Thus, there is a need for local authorities and stakeholders to determine their own threshold to best meet their needs of public awareness and health and welfare protection.

Historically the greatest frequency of regional-scale episodes $PM_{2.5}$ have occurred in Ontario followed by Québec and for both these regions high $PM_{2.5}$ values often persisted for several days. Regional scale episodes of $PM_{2.5}$ were infrequent in the Prairies with the most notable regional scale $PM_{2.5}$ event associated with the August 2003 forest fires in the southern British Columbia interior. The relative frequency of days per month with $PM_{2.5}$ concentrations greater than 30 $\mu g\ m^{-3}$ is shown in Fig. 3.14 for the five year period 2005–2009. $PM_{2.5}$ concentrations greater than 30 $\mu g\ m^{-3}$ occurred in all months of the year. However, in British Columbia these events were largely a cold season phenomenon and were most often due to local emissions being trapped under inversions in the interior valleys. A similar seasonal behaviour occurred in Alberta with strong inversions limiting vertical mixing over large emission areas (e. g., cities) being responsible. The summertime occurrences of $PM_{2.5}$ episodes in western Canada were usually associated with forest fires

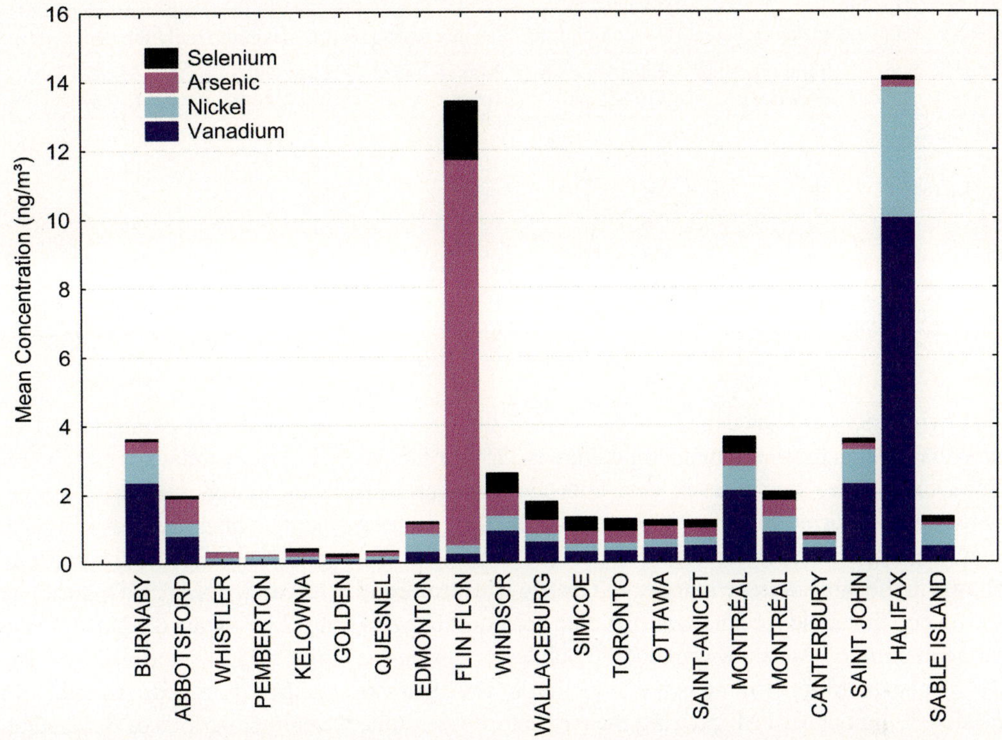

Fig. 3.13 Mean concentrations (ng m^{-3}) of vanadium, nickel, arsenic and selenium at NAPS sites (2007–2009). Measurements are of water soluble $PM_{2.5}$ by ICP-MS

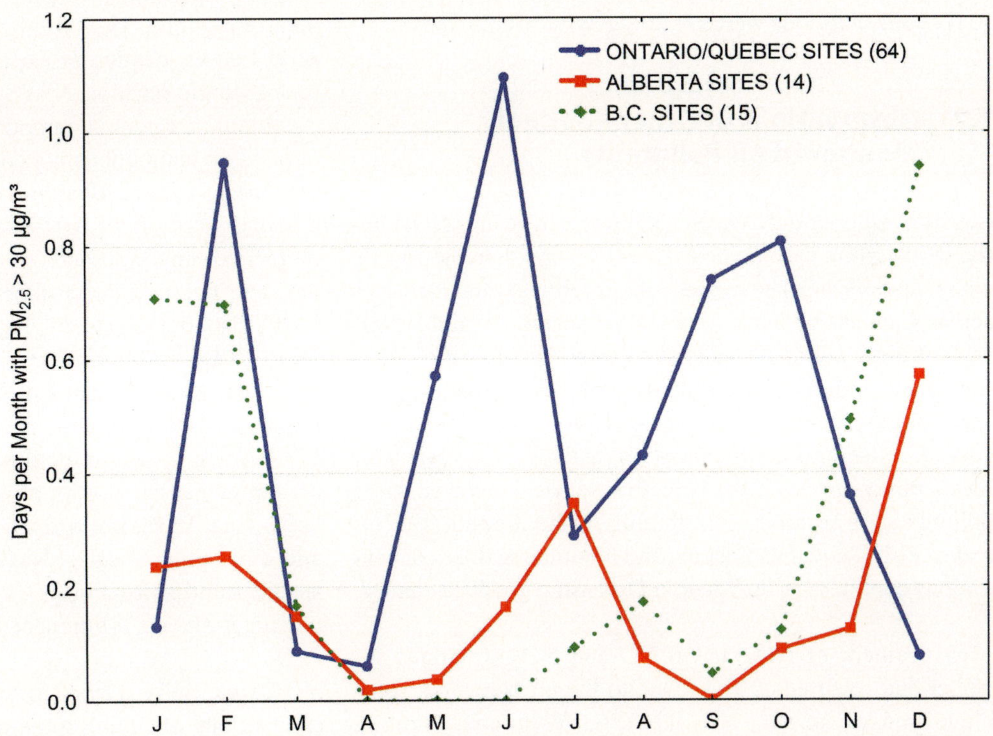

Fig. 3.14 Average days per month with $PM_{2.5}$ concentrations greater than 30 µg m^{-3} for the years 2005–2009. Results have been adjusted for cold-season TEOM mass loss. Only sites averaging at least one day per year >30 µg m^{-3} were used to prepare this graph, which resulted in no days in Atlantic Canada, Manitoba and Saskatchewan. The numbers of sites included are shown in brackets in the legend

(e.g., August 2003). As expected, Ontario/Québec sites experienced the greatest overall frequency of days >30 µg m^{-3} with the peak month being June followed by February and October. Thus, a $PM_{2.5}$ episode can occur in any season.

The greatest frequency of regional-scale episodes of O_3 also has occurred in Ontario followed by Québec. For both these regions high ozone values often persisted for several days and were highly likely to be associated with $PM_{2.5}$ values greater than the CWS metric of 30 µg m^{-3}. This situation has changed over the past ten years, however, and to illustrate this Table 3.3 provides a comparison of a regional scale episode that occurred in Ontario in 2003 contrasted with one that occurred in 2010.

Table 3.3 Comparison of a Regional Scale Ozone and $PM_{2.5}$ Episode in Ontario for two heat-wave periods in 2003 and 2010. (A regional scale episode is defined as days where 33 % of monitoring sites in a region record maximum daily 8 h ozone greater than 65 ppb)

Episode date	Total reporting sites (O_3)	Sites with 8 h $O_3 > 65$ ppb	Max. 8 h O_3 (ppb)	Sites with $PM_{2.5} > 30$ µg/m³	Max 24 h $PM_{2.5}$ (µg/m³)	Max. temp Toronto (°C)	Max. temp Ottawa (°C)
24-Jun-2003	56	36	112	8	33	32.6	32.5
25-Jun-2003	56	36	123	28	46	34.2	33.0
26-Jun-2003	56	35	108	28	55	33.3	32.3
05-Jul-2010	50	29	82	1	32	33.8	33.7
06-Jul-2010	50	32	99	1	31	33.1	33.6
07-Jul-2010	50	26	89	3	33	32.0	33.8
08-Jul-2010	50	26	97	2	38	33.3	34.5

Environment Canada defines a heat wave as three consecutive days when the maximum temperature is 32 °C or higher. For Ottawa there was a seven year time gap between heat waves that occurred in June 2003 and July 2010. It is these two heat wave periods that are compared in Table 3.3. As shown in the table, a large number of Ontario sites exceeded the 65 ppb threshold for 8 h O_3 during both episodes. However the maximum O_3 values were 20–30 ppb lower during the 2010 episode. More striking is the reduction in sites that exceeded 30 µg m^{-3} of $PM_{2.5}$ during the two episodes (despite the same number of reporting sites) with over 25 sites exceeding this value in 2003 and only 1–3 in 2010. Maximum $PM_{2.5}$ concentrations were 15–20 µg m^{-3} lower in 2010.

3.7 Observations of Other Routinely Monitored Air Pollutants

Levels of sulphur dioxide (SO_2), nitrogen dioxide (NO_2), nitric oxide (NO), carbon monoxide (CO) and volatile organic compounds (VOCs) have been monitored at several locations across Canada for many years (>20 years). At high levels each of these pollutants can have human health and/or environmental impacts. Consequently ambient air quality standards or objectives have been in place for many years to encourage reductions in concentrations and in recent years their levels rarely approach the objective values. However, these pollutants are reactive, contributing to the formation of O_3 and/or $PM_{2.5}$, and thus tracking and minimizing their concentrations continues to be important to ensure good air quality.

Measurement of NO_2 Often NO and NO_2 are added together and reported as NO_x. Due to the measurement technique used to determine NO_2 at NAPS sites the concentration reported can experience a small interference from other oxidized nitrogen compounds, namely nitric acid, peroxyacetyl nitrate, organic nitrates and particle nitrate. When these species make it through the inlet of the measurement system and are converted to NO before being detected they cause an overestimation of NO_2 (and NO_X). The magnitude of the overestimation is dependent on the extent of photochemical processing of these other nitrogen species during the course of atmospheric transport to the measurement site. In general, the overestimate of NO_2 increases as the distance from emission sources increases.

The magnitude of this positive interference has been estimated at a few rural locations in Canada as part of CAPMoN research (NARSTO 2011) and have also been discussed by Lee et al. (2011) for southwestern Ontario. At a rural site ~75 km northwest of Toronto (Egbert) the interference ranges from approximately 4 % in the winter to 39 % during summer periods of elevated photochemical activity. It should be noted that most Canadian NO_X measurement sites are located in urban environments which are impacted mostly by local emissions. Thus the positive interference seen at the rural Egbert location is expected to be closer to the worst case scenario for most NAPS sites.

Volatile organic compounds, or VOCs, are generally defined as compounds containing at least one carbon atom (excluding carbon dioxide and carbon monoxide) and with a vapour pressure of 0.01 kPa or greater at 25 °C. Although there are many thousands of organic compounds in the natural and polluted troposphere that meet the definition of a VOC, most measurement programs have concentrated on the 50 to 150 most abundant C2 to C12 hydrocarbons consisting of the general formula CxHy and on C2 to C6 carbonyls (compounds that contain the structural element R2C = O). In this chapter, total non-methane hydrocarbons (total NMHC) are defined as the sum of all identified C2 to C12 hydrocarbons. Total VOCs are defined as total NMHC plus carbonyls and other polar species. NMHC also includes a number of species emitted from biogenic sources including isoprene, α-pinene, β-pinene, δ-limonene and camphene.

VOC Reactivity Ozone formation includes a complex array of reactions involving the atmospheric oxidation of VOCs. In this process, individual VOCs differ in their efficiency towards O_3 formation. Therefore, a scale in which each compound is ranked according to its potential to form O_3 can help guide certain emission regulations. The use of the rate coefficient for the reaction of OH + VOC (*k*OH) as a measure of the reactivity of a VOC is one common ranking approach. While this approach does not take into account the complex chemical reactions that follow the initial reaction between

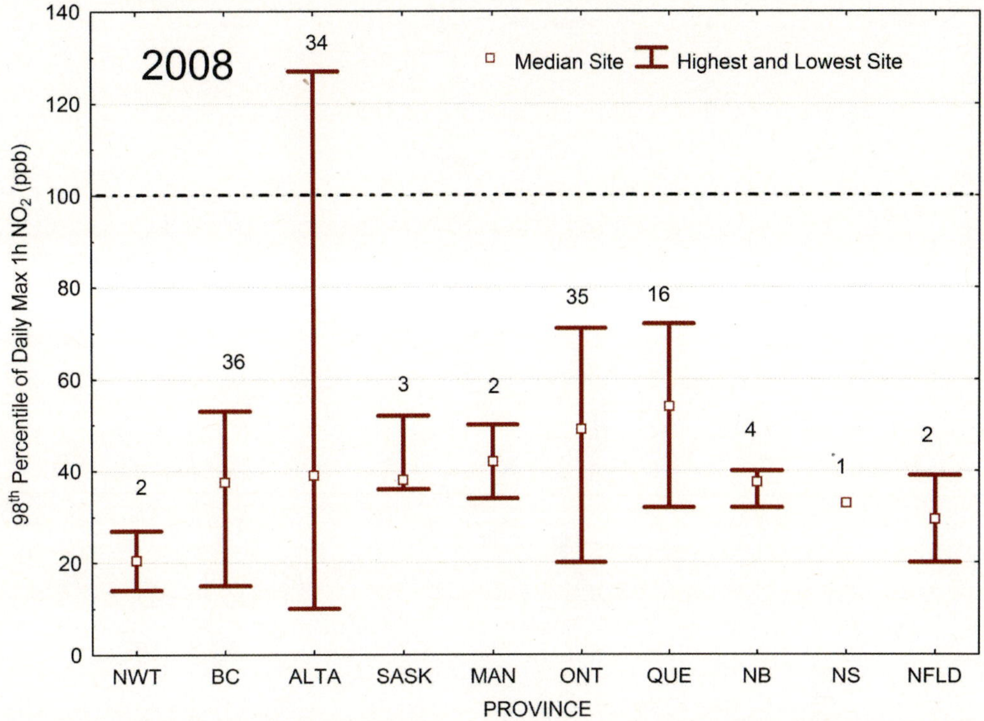

Fig. 3.15 Provincial comparison of 98th percentile of daily maximum 1 h NO_2 concentrations (ppb) for 2008. The plotting point represents the median of all sites and the whiskers represent highest and lowest site. The number of reporting sites in each province is also provided

the VOC and OH, it is useful to consider the *k*OH reactivity scale under conditions where the production of O_3 is largely limited by the supply of NO_X rather than VOCs. These sort of conditions prevail at locations removed from major source areas of VOCs, e.g., in rural and remote areas. To help with this limitation Chameides et al. (1992) introduced a procedure for scaling VOC-based *k*OH that involves scaling a VOC by multiplying its concentration (in ppbC) by the ratio of its rate coefficient with OH and the rate coefficient of the reaction between propylene and OH. This adjusted VOC concentration is called its Propy-Equivalent concentration and also helps rank VOCs according to their impact on O_3.

3.7.1 Urban NO and NO_2 Concentrations

Figure 3.15 summarizes the NO_2 concentrations by province for 2008. This figure shows the median and range among urban sites of the 98th percentile values of the daily hourly maximum. The median was highest among the Ontario and Québec sites at about 55 ppb. Western Canadian cities from Manitoba westward and New Brunswick had similar median 98th percentiles at near 40 ppb. However, the location with the highest 98th percentile was in Alberta as was the site with the lowest value. Nova Scotia, Newfoundland and the Northwest Territories all had lower values. Annual mean NO concentrations ranging from 22 to 32 ppb are found at centre city or roadway sites in the larger metropolitan areas. At centre city or roadway sites in the larger metropolitan areas NO can reach high levels surpassing 150 ppb and explaining 50–70 % of measured NO_X. At suburban sites and commercial sites in smaller urban centres, annual mean NO concentrations are typically in the range of 10 to 20 ppb while NO_2-to-NO ratios are in the range of 1 to 4.

Monthly mean NO and NO_2 concentrations for selected sites are presented in Fig. 3.16 averaged over the years 2003–2005. Both NO and NO_2 concentrations are lower in the summer months and NO_2/NO ratios are higher. The majority of urban sites show a strong seasonal cycle in NO, with maximum concentrations experienced in the winter months. The winter maximum is a result of three factors: increased emissions in the winter (primarily from fuel combustion); reduced atmospheric dispersion and a shallower mixed layer in winter; and less photochemical activity, resulting in slower destruction of NO_X. Figure 3.16 shows that these factors contribute to the largest seasonal variability and highest wintertime NO concentrations in the country in Calgary. The NO_2 seasonal cycle had a smaller amplitude compared to NO. The minimum was generally recorded in midsummer, due to faster conversion of NO_2 to other more oxidized nitrogen species (e.g., nitric acid) and lower NO levels.

3.7.2 Levels of Volatile Organic Compounds (VOCs)

VOC samples are routinely collected at 37 urban and 13 rural sites across Canada with five sites located near refineries. In Fig. 3.17, these sites are categorized according to these three settings and are compared using mean, 10th and 90th percen-

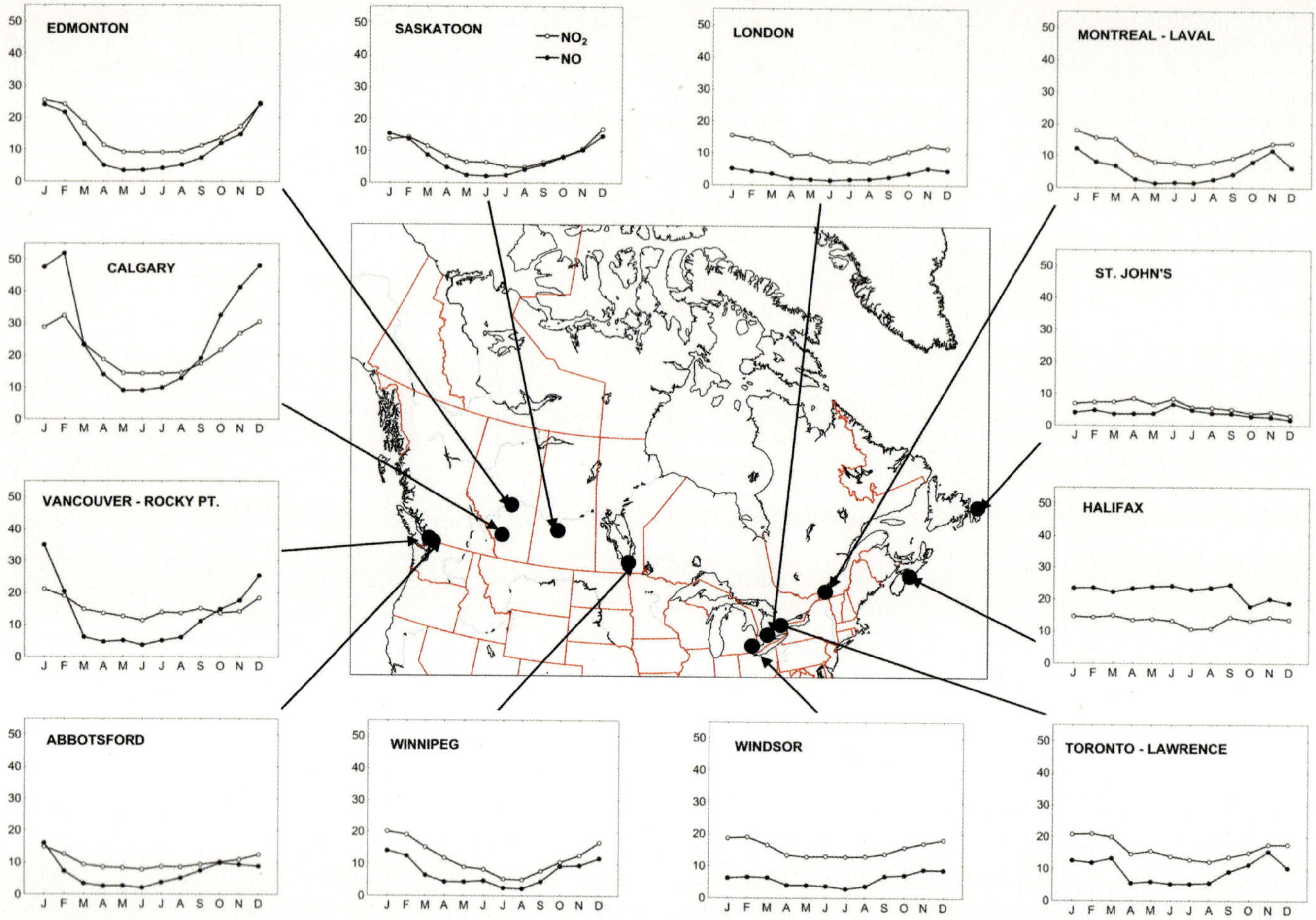

Fig. 3.16 Mean NO (*open circles*) and NO_2 (*solid circles*) by month averaged over the years 2007–2009. (Mean NO and NO_2 concentration (ppb) shown on y-axis)

Fig. 3.17 Mean, 10th and 90th percentile non-methane hydrocarbon concentrations (ppbC) for May to September in 2007–2009

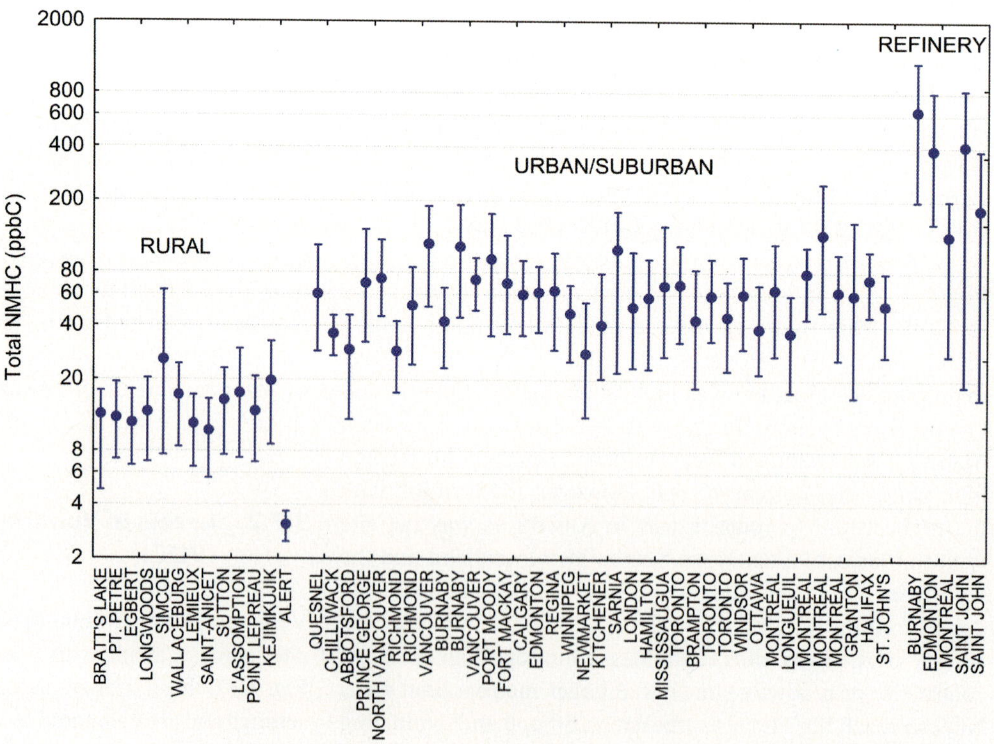

Table 3.4 Ten most abundant NMHC Species as a percentage of total carbon and as a percent of $NMHC_{prop}$—Urban and Rural Sites (Summer Only) 2007–2009

Rank	Compound	Mean (ppbC)	% of total	Compound	Mean (ppbC)	% of total
Urban—sorted by concentration				*Urban—sorted by* $NMHC_{prop}$		
1	Toluene	5.6	9.5	*m* and *p*-Xylene	2.7	15.3
2	Isopentane	4.9	8.4	Toluene	1.9	10.9
3	Propane	4.7	8.0	Ethylene	1.5	8.6
4	Ethane	3.7	6.3	Isoprene	1.0	5.7
5	Butane	3.5	5.9	Propylene	0.8	4.5
6	*m* and *p*-Xylene	2.9	5.0	Isopentane	0.7	4.0
7	Pentane	2.2	3.8	*o*-Xylene	0.6	3.4
8	Isobutane	2.2	3.7	1,2,4-Trimethylbenzene	0.5	3.1
9	Ethylene	1.9	3.3	3-Ethyltoluene	0.4	2.3
10	2-Methylpentane	1.3	2.2	Butane	0.4	2.3
All species		*59.0*			*17.6*	
Rural—Sorted by Concentration				*Rural—sorted by* $NMHC_{prop}$		
1	Ethane	2.3	16.3	Isoprene	1.3	30.0
2	Isoprene	1.4	10.0	Ethylene	0.3	6.9
3	Propane	1.2	8.9	Toluene	0.3	6.4
4	Isopentane	0.8	5.9	α-Pinene	0.3	6.2
5	Toluene	0.8	5.8	m and p-Xylene	0.2	5.2
6	α-Pinene	0.7	5.1	Propylene	0.2	4.3
7	Butane	0.7	4.7	Isopentane	0.1	2.8
8	Pentane	0.5	3.2	β-Pinene	0.1	2.4
9	Ethylene	0.4	2.7	1-Butene/Isobutene	0.1	2.1
10	β-Pinene	0.4	2.6	δ-Limonene	0.1	1.7
All species		*14.0*			*4.3*	

tile NMHC concentrations. Not surprisingly concentrations were much lower at the regionally representative and remote sites compared to the rural-urban impact sites. Higher levels are apparent at locations situated in the downtown core (e.g., Vancouver-Robson) and (Montréal-Maisoneuve) as well as in cities near petrochemical industry (Sarnia).

Mean concentrations varied by almost three orders of magnitude from the lowest site (Alert—3.1 ppbC) to the highest site (Burnaby-Eton/Madison—633 ppbC). The Burnaby site is located approximately 1 km from a Chevron refinery. The differences among sites are not the same for all individual compounds. Some of the NMHC species with relatively long atmospheric life times such as ethane are found in similar concentrations at urban and rural sites. Biogenic species are usually more abundant at forested, rural sites (e.g., Kejimkujik). For the rural sites the proximity of a large source regions plays a role. For example, NMHC concentrations are relatively large at such sites as Simcoe, Wallaceburg and l'Assomption which are relatively close to large urban areas.

Table 3.4 provides a list of the 10 most abundant NMHC species as a percent of total carbon and as a percent of reactivity weighted total carbon $NMHC_{prop}$ for the urban and rural site categories. The most abundant species on a percent of carbon basis in the urban area are toluene, isopentane, propane, butane and m and *p*-xylene. In contrast, the species *m* and *p*-xylene, toluene, ethylene, propylene and isopentane are the most important contributors to $NMHC_{prop}$ (i.e., important for O_3 formation). The ranking of major species measured in NMHC mixtures has remained stable during the past 20 years of measurement (Environment Canada 1997), particularly at the urban sites, which reflects the large contribution from transportation sources. Species profiles for the refinery impact sites (not shown) are similar to urban sites except for a greater abundance of C3-C5 alkanes (propane, butanes and pentanes). There also systematic differences between the profiles among refinery sites due to differences in industrial activities. For example, the Saint John, NB, site records high levels of MTBE and propylene while the Burnaby site shows high levels of 2,2-dimethylbutane. At the rural sites the importance of biogenic NMHC species, such as isoprene and α-pinene, is relatively large. In terms of $NMHC_{prop}$ isoprene is the dominant species.

3.7.3 Sulphur Dioxide Concentrations

Sulphur dioxide is a major precursor of $PM_{2.5}$ and has been routinely monitored across Canada for decades. Ambient SO_2 concentrations across Canada exhibit a seasonal cycle with higher urban SO_2 concentrations in the winter than in the summer. As shown below in Fig. 3.22, concentrations have continued to decrease in recent years, but sites with

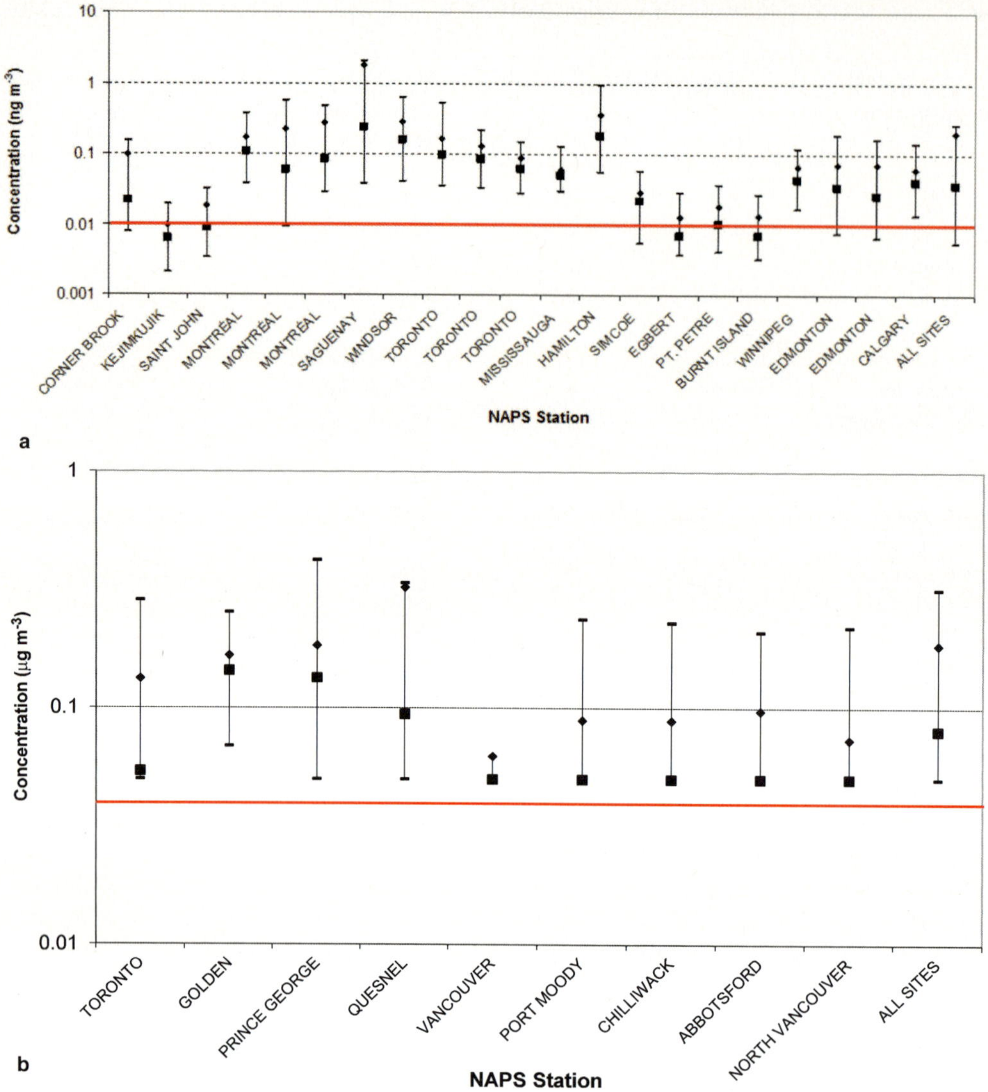

Fig. 3.18 Concentrations of polycyclic aromatic hydrocarbons (PAHs), as represented by benzo[a]pyrene (**a**), and ethylene oxide (**b**) at NAPS sites over 2003–2009. ■ Median concentration. — 10th and 90th percentile concentrations. ♦ Mean concentration. — Annual average ambient air quality criterion (AAQC) from Ontario

nearby point sources (usually non-ferrous smelters) continue to experience the highest SO_2 concentrations. However, the observed mean annual concentrations across the country are still below 10 ppb at all but a few sites, such as the industrially-impacted Temiscaming, Québec, and Trail, BC, locations. The lowest concentrations are found at western and northern sites with no nearby industrial emissions. Some rural sites in Ontario, such as Egbert and Longwoods, continue to experience higher SO_2 levels than a number of the western urban sites. This is because of higher emissions of SO_2 within the region, including the U.S. Midwest.

3.7.4 Levels of Hazardous Air Pollutants

Typical concentrations of the HAPS listed in Table 3.2 vary over a wide range depending upon compound (e.g., from femtograms (10^{-15} g) to micrograms (10^{-6} g) per cubic meter). Similarly, the concentrations at which no adverse effects are expected vary by compound.

Two of the HAPS listed in Table 3.2 exceeded their respective annual Ontario ambient air quality criteria (AAQC) by substantial margins in recent years: polycyclic aromatic hydrocarbons (PAHs, as represented by benzo[a]pyrene) and ethylene oxide. Their concentration distributions among NAPS sites are shown in Figs. 3.18a, b and compared to their AAQCs. Figure 3.18a shows that PAH concentrations varied widely by location. Sparsely populated areas, such as Kejimkujik National Park and Burnt Island (Lake Huron), had median concentrations below Ontario's annual AAQC. Conversely, most urban areas had median concentrations that exceeded the AAQC due to the impact of local mobile and industrial sources. All NAPS sites that measured ethylene oxide had median concentrations that were above the

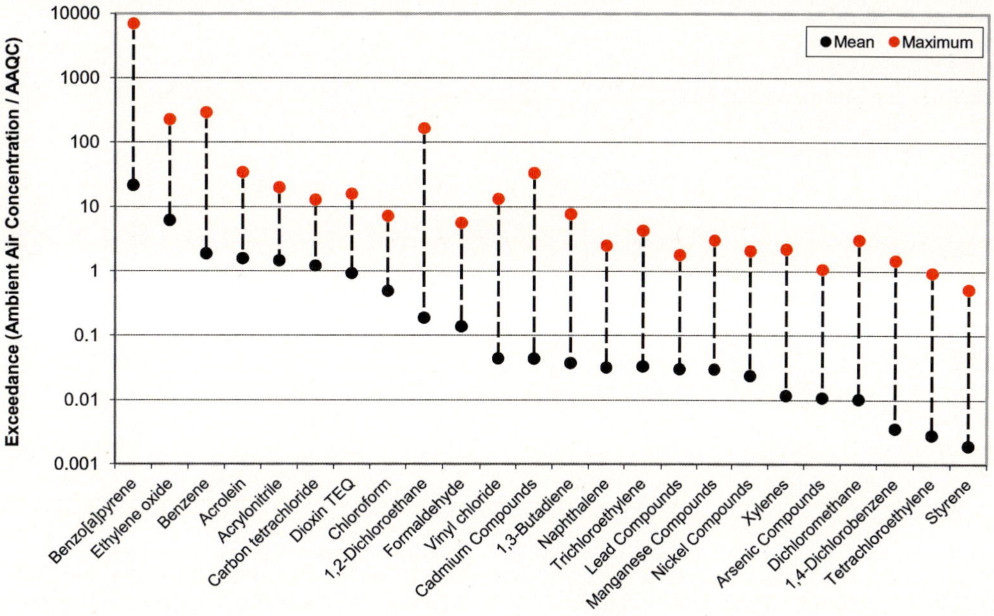

Fig. 3.19 Air toxics in Canada with maximum 24-Hour exceedance greater than 0.1 over 2003–2008. Note that not all pollutants are measured at all stations and/or in all years. Benzo[a]pyrene is a surrogate compound representing the PAHs. 24-hour AAQCs used to calculate exceedances

Ontario AAQC (Fig. 3.18b). However, this HAP was only measured in British Columbia and at a single site in Ontario. It is possible that improved spatial coverage in ethylene oxide measurements would reveal areas with concentrations below the AAQC and others further above it (i.e., hotspots). More measurements are needed to determine the spatial variability of ethylene oxide in Canada.

The range in concentrations among the HAPS in Table 3.2 is large, and a useful approach to summarizing their levels and potential impact is to express their concentrations relative to their respective Ontario AAQCs. Compounds that exceeded or approached their relevant AAQCs in Canada in recent years are shown in Fig. 3.19. A plotted value > 1 indicates that the concentration (mean or maximum Canadian value) was above the Ontario AAQC. The mean Canadian concentrations of six air toxics (PAHs as represented by benzo[a]pyrene, ethylene oxide, benzene, acrolein, acrylonitrile, and carbon tetrachloride) exceeded their respective annual Ontario AAQCs. A further sixteen air toxics exceeded their AAQCs at their maximum recorded concentrations. In other words, there was at least one observation among the limited number of measurement sites in Canada that was above the AAQC for those compounds. Continued investigations of HAPs across Canada will be useful to understanding the risks posed to Canadians by air toxics. However, careful attention must be paid to setting science priorities given the large list of HAPs that are potentially relevant in the Canadian context (Galarneau and Dann 2011).

3.7.5 Trends in NO_2, NO, VOCs and SO_2

Recent trends in annual mean NO and NO_2 are shown in Fig. 3.20 and trends in May-September VOCs at urban, rural and refinery impacted sites are provided in Fig. 3.21. Between 1999 and 2008 annual mean NO decreased by 48 %, NO_2 by 26 % and summer mean VOCs by 46 % at these Canadian urban sites. The trends were consistent, on a site by site basis, with essentially all urban sites in Canada, recording similar decreases in ambient levels. Interestingly, given the reductions in urban NO_2 during the period, the levels in eastern and western Canadian cities are now comparable. In contrast, mean NO is higher in the cities in western Canada. As a result, the ratio of composite annual mean NO_2 to NO increased from 1.1 to 1.5 during the time period. Rural sites experienced a decline in anthropogenic VOCs similar to the urban sites with a 49 % decrease in mean summer concentration between 1999 and 2008. The reductions in ambient air NO_X and NMHC also matches the reduction in on-road transportation sector emissions (Government of Canada 2007). In contrast, Fig. 3.21 shows that mean and 90th percent confidence limits for the VOCs at the six sites located near refineries have increased during the past decade. This upward trend was largely driven by higher concentrations measured at the Edmonton and Vancouver monitoring sites.

In Fig. 3.22 annual mean SO_2 trends are shown separately for 12 urban and 9 industrial influence sites for the 1999–2008 time period. Composite annual mean SO_2 concentrations decreased by approximately 52 % at both groups of sites. Mean levels in locations impacted by industrial emissions now range from 3 to 9 ppb, while at most other urban locations they average less than 2 ppb.

3.8 Summary

There are over 300 air quality monitoring locations across Canada from coast to coast to coast. Ozone and fine particulate matter ($PM_{2.5}$) are the most frequently measured pollut-

Fig. 3.20 Yearly variation in annual mean NO and NO_2 from Canadian urban trend sites (1999–2008). Composite means and the 90th percent confidence interval around the mean are plotted

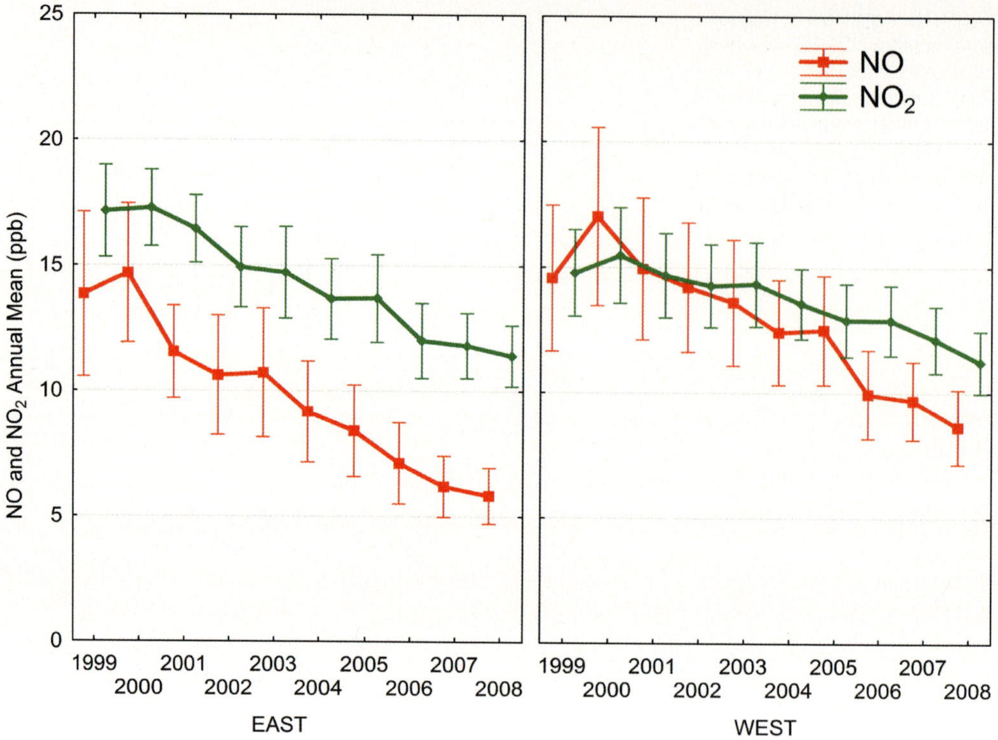

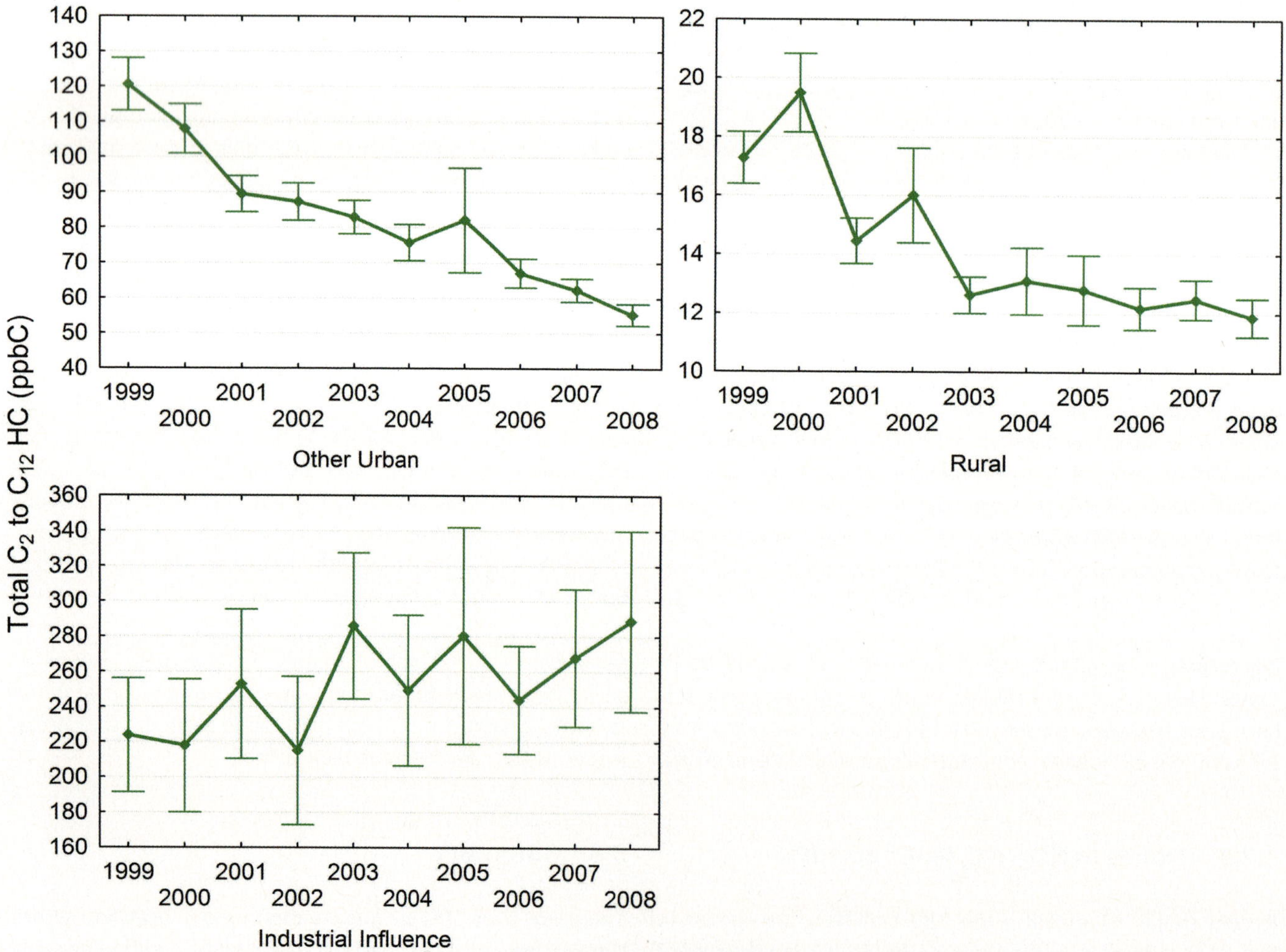

Fig. 3.21 Yearly variation in non-biogenic C_2 to C_{12} hydrocarbons at industrial influence, other urban and rural sites (May to September only). Composite means and the 90th percent confidence interval around the mean are plotted. Independent scales are used for each plot

Fig. 3.22 Yearly variation in annual mean SO_2 from Canadian industrial influence and other trend sites (1999–2008). Composite means and the 90th percent confidence interval around the mean are plotted

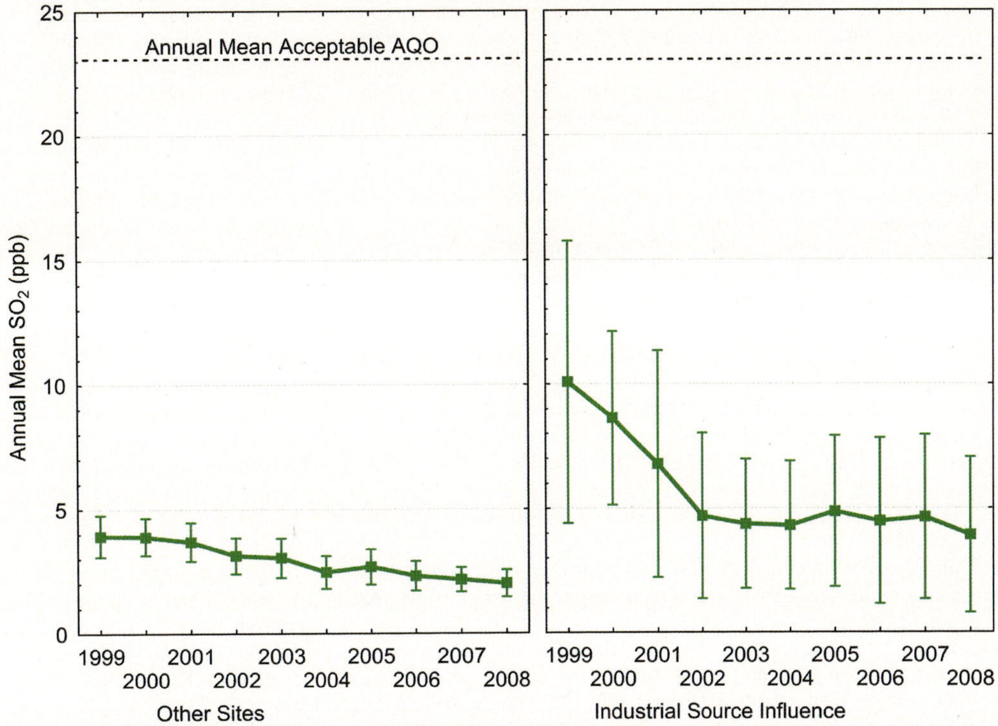

ants with the majority of locations situated where population density is high or where there is considerable industrial activity. Nitrogen oxides and sulphur dioxide are also measured at a considerable number of locations in Canada while volatile organic compounds (VOC) and other air toxics are measured at a relatively small number of places. In this chapter, national scale patterns for these air pollutant concentrations have been presented along with information on their temporal variations, such as diurnal, seasonal and year to year changes. The composition of fine particulate matter ($PM_{2.5}$) and volatile organic compounds (VOC) has also been determined and described in this chapter.

The concentration measurements indicate that there are areas where the current Canadian standards or provincial guidelines are exceeded for ozone, $PM_{2.5}$ and toxics. Ozone in southern Ontario and southern Québec has been the most problematic although there has been some improvement in recent years. Similarly, on a national scale $PM_{2.5}$ has decreased indicating that, due to the range of emission reduction strategies implemented in the past decade, air quality is improving. There are exceptions, such as VOCs in areas of industrial influence and, although not a focus in this chapter, local areas of industrial growth such as NO_2 over the oil sands region (McLinden et al. 2012). Thus, there remains a need to monitor air quality, especially in regions of projected industrial or population growth in order to maintain current levels and to identify options for improvement and to inform adaptive management approaches. Furthermore, current health research indicates that pollutants such as $PM_{2.5}$ have effects on the population at low levels. Thus, air quality remains an important public health issue to track; nationally, regionally and in particular at the local scale where areas of concentrated emissions due to industrial activity or population behaviour (e.g., traffic, wood burning) can lead to higher population exposures.

In the context of the past 30–40 years, Canadian air quality has been an environmental management success story. Levels throughout the populated regions of the country are considerably lower than in the 1970's and 1980's. The 1990's also brought reductions and this chapter shows that even up through the 2000's improvements in air quality have been realized.

Acknowledgements This chapter would not have been possible without the hard work and leadership from Ewa Dabek-Zlotorzynska, Daniel Wang, Robert Vet, Mike Shaw and many staff associated with the Environment Canada analytical laboratories, the Provincial Environment Ministries and field site operators.

References

Ainslie B, Steyn DG (2007) Spatiotemporal trends in episodic ozone pollution in the lower fraser valley, B.C. in relation to meso-scale atmospheric circulation patterns and emissions. J Appl Meteorol 46(10):1631–1644

Allen G, Sioutas C, Koutrakis P, Reiss R, Lurmann FW, Roberts PT (1997) Evaluation of the TEOM method for measurement of ambient particulate mass in urban areas. J Air Waste Manag Assoc 47:682–689

Allen G (2010) Evaluation of transformation methods for adjustment of historical TEOM® data in the NAPS network—Report for the Analysis and Air Quality Section. Environment Canada

Brook J, Dann T (1998) Contribution of Nitrate and Carbonaceous species to PM_{10} and $PM_{2.5}$ observed in Canadian cities. J Air Waste Manag Assoc 49:193–199

Brook J, Dann T (1999) Contribution of Nitrate and Carbonaceous species to PM_{10} and $PM_{2.5}$ observed in Canadian cities. J Air Waste Manag Assoc 49:193–199

Brook JR, Dann T, Burnett RT (1997) The relationship among TSP, PM_{10}, $PM_{2.5}$ and inorganic constituents of atmospheric particulate matter at multiple Canadian locations. J Air Waste Manag Assoc 47:2–19

Brook JR, Dann TF, Bonvalot Y (1999) Observations and interpretations from the Canadian fine particle monitoring program. J Air Waste Manag Assoc 49:PM-35–44

Brook RD, Rajagopalan S, Pope CA III, Brook JR, Bhatnagar A, Diez-Roux AV, Holguin F, Hong Y, Luepker RV, Mittleman MA, Peters A, Siscovick D, Smith SC Jr, Whitsel L, Kaufman JD (2010) Particulate matter air pollution and cardiovascular disease: an update to the scientific statement from the American Heart Association. Circulation 121:2331–2378

Burnett RT, Brook JR, Yung WT, Dales RE, Krewski D (1997) Association between ozone and hospitalization for respiratory diseases in 16 Canadian cities. Envir Res 72:24–31

Canada (1995) Toxic Substances Management Policy. Government of Canada, Ottawa, ON. ISBN 0-662-61860-2, Cat. no. En40-499/1–1995

CCME (Canadian Council of Ministers of the Environment) (2006) Canada wide standards for particulate matter and ozone: five year report: 2000–2005

Chameides WL, Fehsenfeld F, Rodgers MO, Cardelino C, Martinez J, Parrish D, Lonneman W, Lawson DR, Rasmussen RA, Zimmerman P, Greenberg J, Middleton P, Wang T (1992) O3 precursor relationships in the ambient atmosphere. J Geophys Res 97:6037–6055

Chow JC, Watson JG, Pritchett LC, Pierson WR, Frazier CA, Purcell RG (1993) The DRI thermal/optical reflectance carbon analysis system: description, evaluation and application in U.S. air quality studies. Atmos Environ 27A:1185–1201

Crouse DL, Peters PA, Donkelaar A van, Goldberg MS, Villeneuve PJ, Brion O, Khan S, Atari DO, Jerrett M, Pope CA III, Brauer M, Brook JR, Martin RV, Stieb D, Burnett RT (2012) Risk of cardiovascular mortality in relation to long-term exposure to low concentrations of fine particulate matter: a Canadian National-level Cohort Study. Environ Health Perspect 120:708–714

Dabek-Zlotorzynska E, Dann TF, Martinelango PK, Celo V, Brook JR, Mathieu D, Ding L, Austin C (2011) Canadian National Air Pollution Surveillance (NAPS) PM2.5 speciation program: methodology and PM2.5 chemical composition for the years 2003–2008. Atmos Environ 45:673–686. doi:10.1016/j.atmosenv.2010.10.024

Dann T, White L, Biron A (2006) Performance of continuous PM2.5 monitors at a monitoring site in Ottawa, Canada. EPA National Air Monitoring Conference. November 2006

de Gouw JA, Brock CA, Atlas EL, Bates TS, Fehsenfeld FC, Goldan PD, Holloway JS, Kuster WC, Lerner BM, Matthew BM, Middlebrook AM, Onasch TB, Peltier RE, Quinn PK, Senff CJ, Stohl A, Sullivan AP, Trainer M, Warneke C, Weber RJ, Williams EJ (2008) Sources of particulate matter in the northeastern United States in summer: 1. direct emissions and secondary formation of organic matter in urban plumes. J Geophys Res 113:D08301. doi:10.1029/2007JD009243

Dibb JE, Talbot RW, Scheuer E, Seid G, DeBell L, Lefer B, Ridley B (2003) Stratospheric influence on the northern North American free troposphere during TOPSE: 7Be as a stratospheric tracer. J Geophys Res 108:11.1–11.8. doi:10.1029/2001JD001347

Diem J (2004) Explanations for the spring peak in ground-level ozone in the southwestern United States. Phys Geogr 25:105–129

Environment Canada (1997) Ground Level Ozone and Its Precursors in Canada. (1980–1993). Canadian 1996 NOX/VOC Science Assessment: Report of the Data Analysis Working Group. (Edited by T. Dann and P. Summers)

Environment Canada (2004) Performance of Continuous PM2.5 Monitors at Canadian Monitoring Locations. NAPS Managers Technical Working Group on PM Measurement Technology, November 2004

Fann N, Lamson AD, Anenberg SC, Wesson K, Risley D, Hubbell BJ (2011) Estimating the national public health burden associated with exposure to ambient PM2.5 and ozone. Risk Anal. doi:19.1111/j.1539-6924.2011.0160

Fiore A, Jacob DJ, Liu H, Yantosca RM, Fairlie TD, Li Q (2003) Variability in surface ozone background over the United States: Implications for air quality policy. J Geophys Res 108:19.1–19.16. doi:10.1029/2003JD003855

Fuentes JD, Dann TF (1993) Ground-Level Ozone in Canada during 1980 to 1990. Report: ARD-93-010. Atmospheric Environment Service, Downsview, Ontario

Galarneau E, Dann T (2011) Air toxics in Canada (ATiC): preliminary scoping report. Environment Canada Departmental Report, Toronto, ON, p 23

Government of Canada (2007) Five-year progress report: Canada-wide standards for particulate matter and ozone. January 2007

Hoff RM, Christopher SA (2009) Critical review—remote sensing of particulate pollution from space: have we reached the promised land? A Critical Review. J Air Waste Manag Assoc 59:645–675. doi:10.3155/1047-3289.59.6.645

Jaffe DA, Parrish D, Goldstein A, Price H, Harris J (2010) Increasing background ozone during spring on the west coast of North America. J Geophys Res 30. doi:10.1029/2003GL017024

Jerrett M, Burnett RT, Pope CA III, Ito K, Thurston G, Krewski D, Shi Y, Calle E, Thun M (2009) Long-term ozone exposure and Mortality. N Engl J Med 360:1085–1095

Jenkin ME (2008) Trends in ozone concentration distributions in the UK since 1990: local, regional and global influences. Atmos Environ 42:5434–5445

Lee PK, Brook JR, Dabek-Zlotorzynska E, Mabury S (2003) Identification of the major sources contributing to PM2.5 observed in Toronto. Envir Sci Technol 37:4831–4840

Lee CJ, Brook JR, Evans GJ, Martin RV, Mihele C (2011) Novel application of satellite and in-situ measurements to map surface-level NO2 in the Great Lakes region. Atmos Chem Phys 11:11761–11775

Lefohn AS, Oltmans SJ, Dann T, Singh HB (2001) Present day variability of background ozone in the lower troposphere. J Geophys Res 9:9945–9958

Lelieveld J, Aardenne J van, Fischer H, Reus M de, Williams J, Winkler P (2004) Increasing ozone over the Atlantic Ocean. Science 304:1483

Liggio J, Li SM, Vlasenko A, Sjostedt S, Chang R, Shantz N, Abbatt J, Slowik JG, Bottenheim JW, Brickell PC, Stroud C, Leaitch WR (2010) Primary and secondary organic aerosols in urban air masses intercepted at a rural site. J Geophys Res 115:D21305. doi:10.1029/2010JD014426

McKendry IG (2006) Background concentrations of PM2.5 and Ozone in British Columbia, Canada. University of British Columbia report prepared for the British Columbia Ministry of the Environment. Vancouver, B.C

McLinden CA, Fioletov V, Boersma KF, Krotkov N, Sioris CE, Veefkind JP, Yang K (2012) Air quality over the Canadian oil sands: a first assessment using satellite observations. Geophysical Research Letters (in press)

Monks PS (2000) A review of the observations and origins of the spring ozone maximum. Atmos Environ 34:3545–3561

NARSTO (2011) Technical challenges of multipollutant air quality management. Hidy et al. editors. Springer. doi:10.1007/978-94-007-0304-9

Pope CA III, Dockery DW (2006) Health effects of fine particulate air pollution: lines that connect. J Air Waste Manag Assoc 56(6):709–742

Pope CA III, Burnett RT, Thurston GD, Thun MJ, Calle EE, Krewski D, Godleski JJ (2004) Cardiovascular mortality and long-term exposure to particulate air pollution: epidemiological evidence of general pathophysiological pathways of disease. Circulation 109:71–77

Pryor SC, McKendry IG, Steyn DG (1995) Synoptic-Scale meteorological variability and surface ozone mixing ratios in Vancouver, British Columbia. J Appl Meteorol 34:1824–1833

Slowik JG, Brook J, Chang RY-W, Evans GJ, Hayden K, Jeong C-H, Li S-M, Liggio J, Liu PSK, McGuire M, Mihele C, Sjostedt S, Vlasenko A, Abbatt JPD (2011) Photochemical processing of organic aerosol at nearby continental sites: contrast between urban plumes and regional aerosol. Atmos Chem Phys 11:2991–3006

Stieb DM, Burnett RT, Beverige RC, Brook JR (1996) Association between ozone and asthma emergency department visits in Saint John, New Brunswick, Canada. Envir Health Persp 104:1354–1360

Tarasick DW, Fioletov VE, Wardle DI, Kerr JB, Davies J (2005) Changes in the vertical distribution of ozone over Canada from ozone sondes: 1980–2001. J Geophys Res 110:D02304

US EPA (2010) The US Clean Air Act Amendments of 1990. List of hazardous air pollutants. US EPA Technology Transfer Network. http://www.epa.gov/ttnatw01/orig189.html

US EPA (2011) Assessment of $PM_{2.5}$ FEMs Compared to Collocated FRMs. PM NAAQS Docket, EPA—HQ—OAR—2007—0492 (April, 2011) www.epa.gov/ttn/naaqs/standards/pm/data/HanleyandReff040711.pdf

US EPA (2013) National air toxics assessments. US EPA Technology Transfer Network. http://www.epa.gov/ttn/atw/natamain/

van Donkelaar A, Martin RV, Brauer M, Kahn R, Levy R, Verduzco C et al (2010) Global estimates of ambient fine particulate matter concentrations from satellite-based aerosol optical depth: development and application. Environ Health Perspect 118(6):847–855

Long-Range Transport of Air Pollutants and Regional and Global Air Quality Modelling

4

Michael D. Moran, Ashu Dastoor and Gilles Morneau

Abstract

The long-range transport of air pollutants from distant sources affects air quality in many parts of Canada. During long-range transport, meteorological conditions will vary and there will be sufficient time for chemical transformations to occur, allowing secondary pollutants such as ozone to form, and for the removal of pollutants by dry or wet processes. Regional and global air quality (AQ) models are the only tools available that can take into account the details of both pollutant emissions and this complex set of atmospheric processes and then predict the pollutant concentration and deposition fields that will result. This capability allows AQ models to provide important information and guidance for AQ management.

This chapter begins by reviewing the nature of long-range transport and its importance in Canada. Next, the capabilities of AQ models and some possible applications of these models to AQ management are discussed, the formulation and architecture of these models is described, a summary of the various regional and global AQ models that have been applied over the past three decades to support AQ management in Canada is provided, and some of the metrics that have been employed to quantify AQ impacts are reviewed. Lastly, three recent Canadian case studies are presented to provide concrete examples of how AQ models can be applied to AQ management.

Keywords

Long-range transport · Air-quality models · AURAMS · GEM-MACH · GRAHM · Biodiesel · Mercury · Trans-Pacific transport · Chemical weather · Air-quality forecasting

4.1 Introduction

Air pollutants that are emitted directly to the atmosphere are called primary pollutants. Air pollutants that do not have primary sources but instead are created by chemical reactions in the atmosphere are called secondary pollutants. Those chemical reactions may take place in clear air (gas-phase chemistry), in clouds within cloud droplets (aqueous-phase chemistry), or on particle surfaces (heterogeneous chemistry). Pollutants such as nitrogen dioxide (NO_2) and sulphur dioxide (SO_2) are primary pollutants. Ozone (O_3), on the other hand, is a secondary pollutant. Particulate matter (PM) is more complicated because it can be directly emitted (e.g., combustion processes, fugitive dust, sea salt) but it can also be formed through chemical reactions (e.g., particle sulphate and nitrate, secondary organic matter). This means that PM can be considered both a primary and a secondary pollutant.

Several later chapters in this book discuss *local* emissions of pollutants from different types of sources, including industry, transportation, and agriculture. However, pollutants emit-

M. D. Moran (✉)
Environment Canada, Toronto, Canada
e-mail: mike.moran@ec.gc.ca

A. Dastoor · G. Morneau
Environment Canada, Montreal, Canada
e-mail: ashu.dastoor@ec.gc.ca

G. Morneau
e-mail: gilles.morneau@ec.gc.ca

E. Taylor, A. McMillan (eds.), *Air Quality Management,* DOI 10.1007/978-94-007-7557-2_4,

ted by *distant* sources can travel long distances and then affect local air quality (AQ) as well, since this is determined by the contributions of all sources, both near and far. The contribution of pollutants from distant sources is often referred to as long-range transport, which has been defined as transport of air pollutants over distances of 100 km or more (OECD 2012). For such distances travel times will range from hours to days and meteorological conditions are likely to vary in both time and space. Moreover, during long-range transport, pollutants or their precursors may also undergo chemical transformations and removal processes. This means that atmospheric chemistry and both primary and secondary pollutants need to be considered when long-range transport is an issue.

This chapter begins by looking at the nature of long-range transport, some of its physical and chemical characteristics, and its importance in Canada. Next, regional and global AQ models are examined: these are the AQ management tools best able to represent and quantify the contributions of both long-range transport and local emissions of air pollutants to local air quality. The capabilities and some possible applications of these models to AQ management are then reviewed, followed by a description of the formulation and architecture of these models, including the differences between air quality and dispersion models, Eulerian and Lagrangian models, regional and global models, and off-line and on-line models. A summary of the different regional and global AQ models that have been applied over the past three decades to support AQ management in Canada is then given and some of the metrics that have been employed to quantify AQ impacts are reviewed. The chapter concludes with three recent Canadian case studies to provide concrete examples of how AQ models can be applied to AQ management. The first case study considers the application of a regional AQ model to estimate the impact on human health of substituting biodiesel fuel in place of fossil diesel fuel in heavy-duty diesel vehicles. The second case study looks at the application of a global AQ model to study one episode of long-range transport of Asian emissions of mercury to western North America. And the third case study describes Environment Canada's use since 2001 of operational regional AQ models to make routine short-term national AQ forecasts in support of the national Air Quality Health Index (AQHI) program and as a complement to the national weather forecasts that it has made for decades.

4.2 Importance of Long-Range Transport of Air Pollutants

The term "long-range transport" refers to the transport by the wind of air pollutants or their precursors from the areas where they were emitted to other locations at downwind distances of 100 km or more. In cities, ambient pollutant levels are typically due to a combination of local sources of these pollutants and long-range transport. In rural or remote regions, on the other hand, where local emissions are often small or non-existent, long-range transport is usually the dominant source of the pollutants present in these regions.

Secondary pollutants have a close connection to long-range transport because both long-range transport and chemical reactions take time to occur, and hence secondary pollutants can be formed during long-range transport. This implies that unless the pollutant of interest is both a primary pollutant and chemically inert, then long-range transport and atmospheric chemistry need to be considered together. The time required for the formation of secondary pollutants also means that the spatial distribution of secondary pollutants is smoother than the spatial distribution of primary pollutants due to the atmospheric dispersion that happens at the same time as atmospheric chemistry.

One important factor for long-range transport is the atmospheric lifetime or residence time of a pollutant. The atmospheric lifetime of a pollutant is the average time that a pollutant molecule will remain in the atmosphere before it is removed. This value is determined by the rate at which it is removed from the atmosphere independent of its abundance in the atmosphere (e.g., Seinfeld and Pandis 2006). The three primary removal mechanisms in the atmosphere are removal by chemical reactions, removal by direct transfer to the Earth's surface (dry deposition), and removal by precipitation (wet deposition). A pollutant that is neither very reactive nor very soluble in water will thus have a longer atmospheric lifetime than a pollutant that is either reactive or soluble because all three removal mechanisms will be less effective. For example, due to their individual chemical and physical properties, NO_2 has a lifetime of one to two days near the Earth's surface but about 2 weeks in the free troposphere above the planetary boundary layer and below the stratosphere, SO_2 has a lifetime of 2 days, PM has a lifetime on the order of one week, O_3 has a lifetime of 25 days (in the troposphere, longer in the stratosphere), carbon monoxide (CO) has a lifetime of one to 3 months, and elemental mercury (Hg) has a lifetime of 1 year (e.g., Summers and Fricke 1989; Seinfeld and Pandis 2006).

Clearly, a pollutant with a longer atmospheric lifetime can be transported over greater distances than a pollutant with a shorter lifetime. It requires about 1 week for a pollutant to be well-mixed vertically in the troposphere, that is, from the Earth's surface up to the bottom of the stratosphere, but it requires about 1 year for a pollutant to be well-mixed horizontally around the globe (e.g., Seinfeld and Pandis 2006). This is the reason that elemental mercury is considered to be a global pollutant whereas other pollutants that have shorter atmospheric lifetimes are only considered to be local or regional pollutants.

The long-range transport of air pollutants is a very common phenomenon. For example, Chap. 3 considers the role of long-range transport of ozone and $PM_{2.5}$ (i.e., fine particulate matter smaller than 2.5 μm in aerodynamic diameter) when

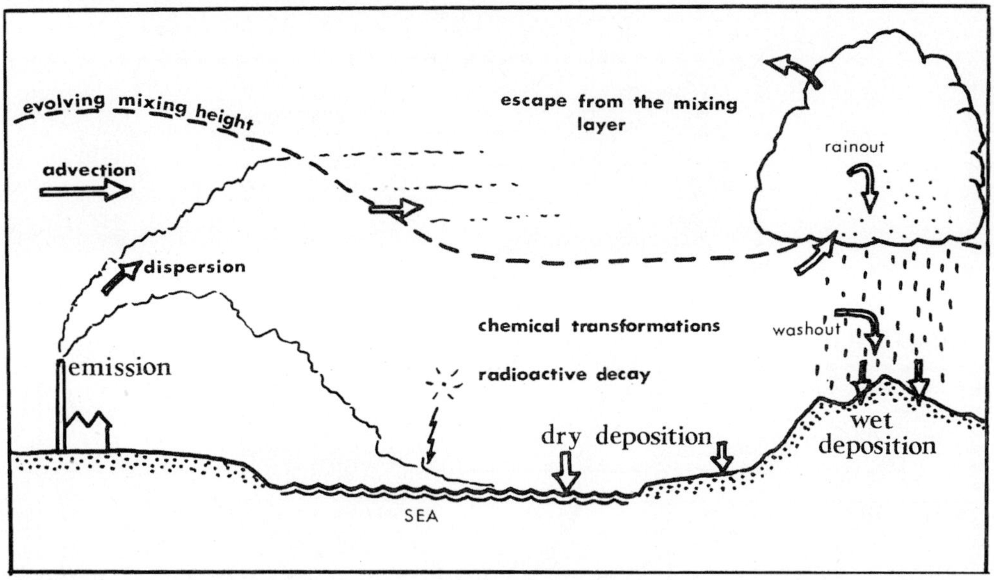

Fig. 4.1 Some of the processes affecting pollutant concentrations in the atmosphere. (Reproduced with permission from Ellis Horwood Ltd, Pasquill and Smith 1983)

analyzing and interpreting Canadian measurements of these species. Many studies of acid rain and photochemical smog in Canada have shown that long-range transport plays an important role in controlling pollutant levels in eastern Canada, including Ontario (e.g., Zeng and Hopke 1989; Keeler et al. 1990; Brankov et al. 2003; Gálvez 2007). One study suggested that long-range transport from the U.S. is responsible for between 56 % and 83 % of the non-sea-salt $SO_4^{=}$and NO_3^{-} wet deposition received in eastern Canada annually (Vet and Ro 2008). Another study estimated that long-range transport contributes 50–60 % of summertime ozone in southern Ontario (Yap et al. 1988). A third study estimated that long-range transport is responsible for 65–70 % of summertime $PM_{2.5}$ in the greater Toronto area (Brook et al. 2002), while a fourth study estimated that U.S. emissions are responsible for approximately 55 % of health and environmental damages in Ontario due to ground-level ozone and fine PM (Ontario Ministry of the Environment 2005). Conversely, Brankov et al. (2003) examined the potential for an emissions trading program between the province of Ontario and neighbouring New York state given the two-way exchange of pollutants between the two jurisdictions. Vet and Ro (2008) reviewed several studies that estimated the extent to which the long-range transport of Canadian SO_2 and NO_x emissions contributed to acid deposition in the eastern U.S. And Bouchet et al. (2013) presented AQ model results that estimated the contribution of emissions from Ontario to pollutant levels in neighbouring provinces and states along with similar results for emissions from British Columbia, the Prairie provinces, Quebec, and the Maritime provinces.

Pollutants from both human and natural sources can also be transported even longer distances, for example, from Asia across the Pacific Ocean to British Columbia (e.g., Leaitch et al. 2009; Walker et al. 2010; Sect. 4.6.2). Other examples of intercontinental transport include the transport of pollutants from mid-latitudes to the Arctic (e.g., Arctic haze), the transport of radionuclides from accidental releases such as the Chernobyl and Fukushima reactor disasters, and the transport of pollutants such as PM, CO, and SO_2 from natural sources such as major volcanic eruptions and large wildfires (e.g., O'Neill et al. 2006; Shindell et al. 2008; Durnford et al. 2010). Recently an intercomparison of global AQ models carried out by the United Nations Economic Commission for Europe focused on the transport of O_3 and PM and their precursors between continents, including expected health impacts from this long-range transport (Sanderson et al. 2008; Shindell et al. 2008; Anenberg et al. 2009; Fiore et al. 2009).

4.3 Role of Regional and Global AQ Models in Canadian AQ Management

Air quality models were introduced in Chap. 1 as important tools for AQ management. Some but not all of these models can be used to simulate long-range transport. The reason for this difference is that when longer distances (e.g., regional, continental, and global scales) and time periods (e.g., days to months) are modelled, more atmospheric processes, including chemistry and removal processes, have time to play a role. In order to represent the effects of all of these processes on pollutant concentrations, AQ models that are capable of treating long-range transport must be more complex and able to represent all of these processes (unless the pollutants of interest are inert primary pollutants). Atmospheric dispersion models, on the other hand, are simpler models that consider emissions, transport, and diffusion but usually not other processes such as chemistry.

Figure 4.1 shows a cartoon of most of the major processes that play a role in determining pollutant concentrations in the atmosphere. Starting from the left-hand side of the figure, we

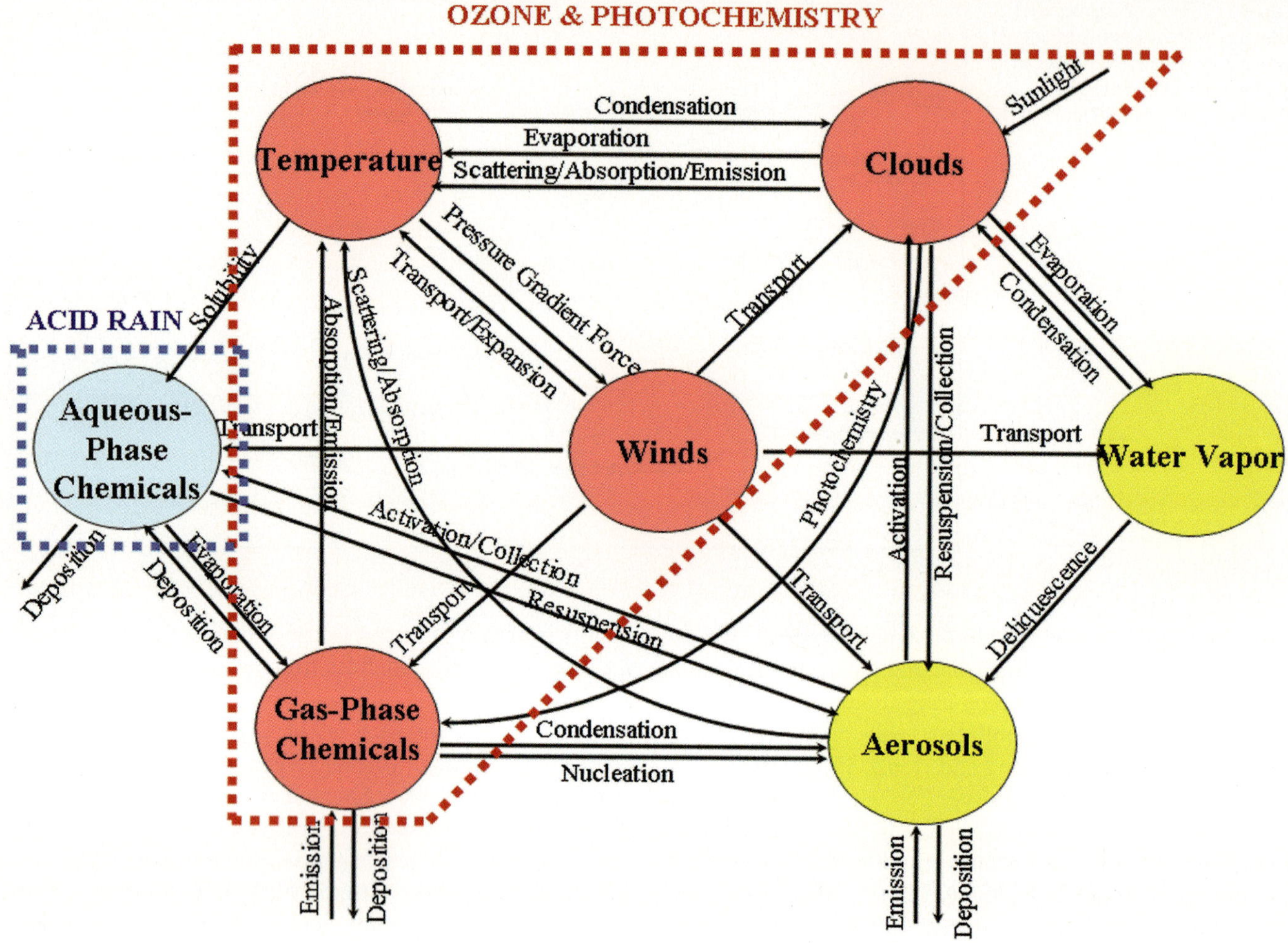

Fig. 4.2 Schematic diagram showing the physical and chemical processes linking a number of important atmospheric components. The four components coloured in red and their connecting processes control atmospheric ozone concentrations; the five components coloured in red or blue and their connecting processes control acid deposition; and the seven components coloured in red, blue, or yellow and their connecting processes control atmospheric particulate concentrations. (Adapted from Peters et al. 1995)

see pollutants being emitted into the atmosphere, being transported by the wind (advection), spreading out by turbulent mixing (horizontal and vertical diffusion), entering clouds, being transformed by chemical reactions either in clear air (gas-phase chemistry) or inside cloud droplets (aqueous-phase chemistry) or on particle surfaces (heterogeneous chemistry), and being removed from the atmosphere by contact with the Earth's surface (dry deposition) or by precipitation (wet deposition). Note that the combination of advection and diffusion, two of the processes shown in this figure, is often referred to as "dispersion" (e.g., Moran 2000).

Figure 4.2 depicts this same system from the perspective of atmospheric components and relationships: it shows seven major atmospheric-chemistry components such as winds, clouds, and different chemical phases and the "web" created by the numerous processes that link them together. An AQ model that is used to predict atmospheric ozone (O_3) concentrations must represent all four of the atmospheric components in this figure that are coloured in red and all of the processes connecting them. If acid deposition is also to be simulated, then aqueous-phase chemistry (marked in blue) and additional processes such as wet deposition must also be represented by the model. And if atmospheric particulate matter (PM) is to be modelled, then all of the atmospheric components and processes shown in this figure should be modelled because they all affect PM concentrations in some way.

Models that do represent all or most of the components and processes shown in Figs. 4.1 and 4.2 are usually referred to as AQ models or chemical transport models (CTMs), but they may also be referred to as first-principles models, because they represent fundamental physical and chemical processes, and as emissions-based or source-oriented models, because they need values of the emissions of various pollutants or their precursors as inputs in order to predict the atmospheric concentrations of these species. An atmospheric dispersion model, on the other hand, is also an emissions-based model, but it neglects atmospheric chemistry and can only predict the concentrations of chemically inert species.

The great strength of AQ models is their ability to quantify the relationship between emissions of pollutants to the atmosphere and the resulting concentrations of these pollutants in the atmosphere. An AQ model needs to be provided with a detailed description of both pollutant emissions and meteorological conditions: with this information the model can then calculate the atmospheric pollutant concentrations that would occur as a result. This means that AQ models can be used to predict current AQ conditions using information about current pollutant emissions levels, but they can also be used to simulate "what if" scenarios to examine what the impact of different future (or past) emissions would be on atmospheric concentrations. By contrast, AQ measurements (see Chap. 3) can be used to quantify the same relationship for *current* emissions, but they cannot provide information on what would happen to pollutant levels if emissions of these pollutants or their precursors were to change. And since AQ models are prognostic models, that is, they can predict changes in time, then they can also quantify the impact of *meteorological* changes on pollutant concentrations either for current weather conditions or for a changing climate (e.g., Kelly et al. 2012). This capability is important because short-term variations in air pollutant concentrations, that is, variations over time scales of minutes to days, are controlled more by changes in weather than by changes in emissions.

Given the above capabilities, AQ models can contribute to AQ management in a number of ways. These include the following AQ policy-, forecast-, and research-related applications:

- provide a quantitative link between emissions to the atmosphere and ambient air concentrations and deposition;
- simulate impact of "what if" future emissions scenarios and historical emissions scenarios;
- estimate emissions changes required to attain AQ objectives or standards;
- permit a synthesis of our best understanding of all processes relevant to air quality;
- quantify importance of various processes to air quality;
- test scientific hypotheses about air quality;
- "fill in" gaps in AQ monitoring networks in both space and time;
- help to optimize design of AQ monitoring networks;
- calculate source attributions for different emissions sectors or geographic regions;
- help to design and to support AQ field campaigns;
- evaluate quality of emissions inventories;
- support objective analyses of measurements and model predictions (i.e., "data fusion"); and
- deliver real-time AQ forecasts for the next few days.

Further guidance about carrying out some of these applications can be found in Ellis (1988), Trujillo-Ventura and Ellis (1991), Seigneur and Moran (2004), Brook et al. (2008), and Zhang et al. (2012b, c), among others.

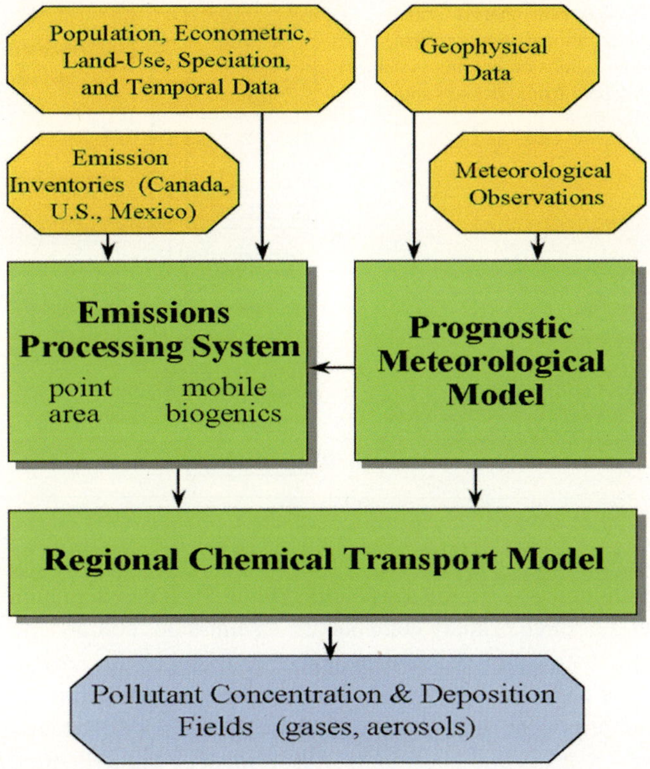

Fig. 4.3 Conceptual diagram of the components of an AQ modelling system. (Seigneur and Moran 2004)

Note that it is important that any guidance for using predictions from AQ models should acknowledge and address the unavoidable fact that these predictions will have errors and uncertainties. The nature and magnitude of AQ model errors and uncertainties can be quantified through performance evaluations against AQ measurement data sets, comparisons with other AQ models, and sensitivity tests (see next section). Information on model errors and uncertainties must then be considered when model results are used to support AQ management.

The next section describes how an AQ model or, more accurately, an AQ modelling *system* actually works and provides some information on sources of model errors and uncertainties.

4.4 Overview of Regional and Global AQ Modelling Systems

As discussed in the previous section, AQ models need both detailed meteorological and emissions inputs in order to function. The preparation of such time-varying input fields are very demanding tasks in and of themselves. In fact, when the generation of these required input fields is considered together with the AQ model, it is more accurate to talk about

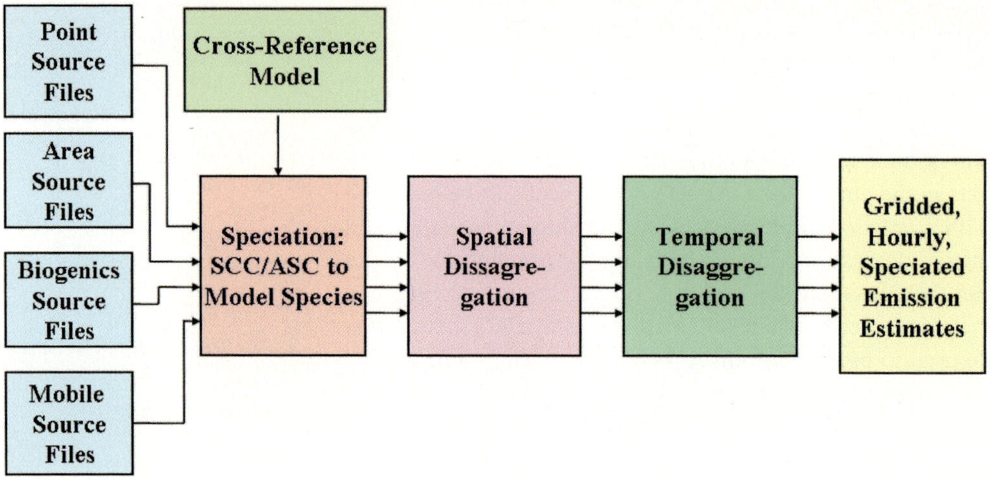

Fig. 4.4 Flowchart of primary computational steps performed by an emissions processing system. (Adapted from Dickson and Oliver 1991)

an AQ modelling system. Figure 4.3 shows a conceptual diagram of a generic AQ modelling system. Such a system consists of three primary components: an emissions component; a meteorological component; and an AQ component. Each of these components is subject to errors and uncertainties that can propagate through the rest of the system, so the issue of model errors and uncertainty must always be borne in mind when analyzing and considering model predictions. In the rest of this section each of these three components of an AQ modelling system is discussed in turn.

4.4.1 Emissions Inventories and Emissions Processing Systems

The emissions inputs required by an AQ model are typically provided by an emissions processing system, which in turn obtains information about emissions from one or more emissions inventories. Emissions inventories have already been discussed in some detail in Chap. 13. Emissions inventories of so-called criteria air contaminants (CACs) usually contain values of *annual* emissions for a particular base year reported by source sector (e.g., industrial, residential, and commercial stationary sources; on-road and off-road mobile sources) and by *jurisdiction* (e.g., province, state, county, census division) for a small number of *criteria* air pollutants (SO_2, NO_x, VOC, CO, NH_3, $PM_{2.5}$, and PM_{10}, where the acronym VOC stands for volatile organic compounds, NH_3 is the chemical formula for ammonia, and PM_{10} is fine particulate matter smaller than 10 μm in aerodynamic diameter). For example, summaries of several Canadian national CAC emissions inventories can be accessed on the internet at http://www.ec.gc.ca/inrp-npri/. (Note that other emissions inventories exist for toxic chemicals and for greenhouse gases.)

In many cases the emissions values contained in emissions inventories may have considerable uncertainty associated with them since very few emissions are measured directly (the major exception in North America being large U.S. power plants, which are required by law and by eligibility requirements for emissions trading programs to have instruments to measure instantaneous emissions of SO_2 and NO_x from their smokestacks (e.g., http://www.epa.gov/airmarkets/emissions/index.html). NARSTO (2005) contains an extensive review and analysis of emissions inventory uncertainties and the uncertainty and sensitivity analysis methods that have been applied to examine and to quantify uncertainties associated with inventory compilation and inventory processing.

The reason that emissions inventories cannot be used directly to provide emissions inputs for AQ models is that there is a major mismatch between their contents versus the level of detail about emissions needed by an AQ model. In contrast to a typical inventory, with annual emissions reported by jurisdiction for a small number of pollutants, AQ models often need emissions fields for every hour of every day of the year, and they need to know emission for each model grid cell for the larger set of model chemical species (e.g., NO, NO_2, individual VOC species).

The role of the emissions processing system "box" shown in Fig. 4.3 is thus to transform the emissions values contained in emissions inventories into the form needed by AQ models. Figure 4.4 shows the three primary processing steps that must be performed at a minimum by any emissions processing system.

- The first processing step shown in Fig. 4.4 is the chemical speciation step. In this step those criteria pollutants from the inventory that are composed of a group or aggregate of two or more individual chemical species or components are separated, recombined if necessary, and then mapped onto corresponding model species. For example, NO_x emissions are separated into emissions of its two component species, NO and NO_2. Bulk VOC emissions, on the other hand, may represent the combined total of

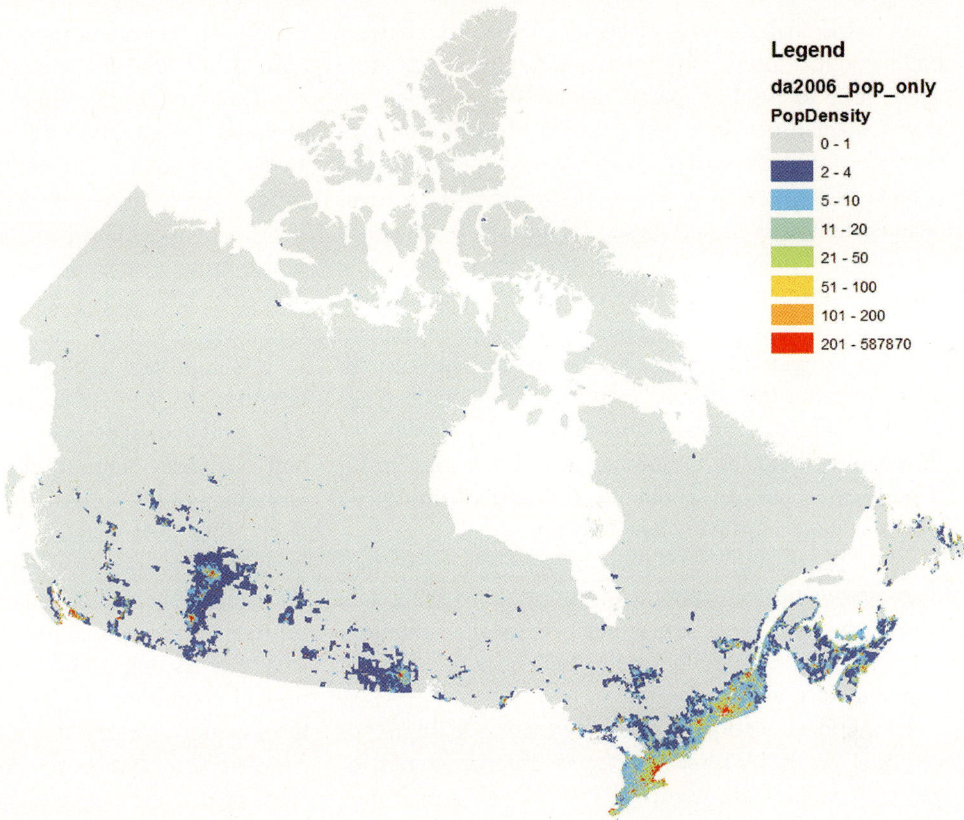

Fig. 4.5 Map of population density for Canada based on 2006 census statistics (units of persons km^{-2})

thousands of individual volatile organic compounds. In this step bulk VOC emissions are first speciated into these individual compounds, then recombined into a set of model VOC species that are often composed of 10–100 or more individual VOC species (e.g., Makar et al. 2003).

Note that this step is not required for the other three gas-phase criteria air pollutants (SO_2, CO, NH_3) since they are already individual chemical species. Bulk $PM_{2.5}$ and PM_{10} emissions, on the other hand, may or may not be speciated into their primary chemical components, depending on the AQ model chosen.

The actual chemical speciation is performed using a library of NO_x, VOC, and PM speciation profiles, where each profile is associated with one or more emissions source sectors or source types. For example, separate VOC speciation profiles are available for coal-fired power plants, heavy-duty diesel vehicles, and solvent evaporation (USEPA 2012; Reff et al. 2009). However, it must be noted that the library of speciation profiles contains many fewer profiles than there are different source types, so profiles are not always a perfect match for the source types being speciated. As well, there are uncertainties associated with the speciation profiles themselves: some are based on only a few measurements, so that representativeness is an issue, while others contain a non-negligible and unspeciated "unknown" component due to limitations of the instrument or analytical method used to prepare the speciation profile.

Note that some AQ models employ a more detailed representation of the PM size distribution, which may require the bulk $PM_{2.5}$ and PM_{10} emissions to be distributed amongst smaller size sections or modes using a library of size-distribution profiles in an analogous manner to the chemical speciation step.

- The second processing step shown in Fig. 4.4 is the spatial disaggregation or spatial allocation step. In this step emissions from jurisdictions are mapped to and allocated amongst model grid cells. Depending upon the size of a jurisdiction or a model grid cell, emissions from a single jurisdiction may be mapped to multiple model grid cells (e.g., Quebec emissions vs. 15 km horizontal grid spacing) or emissions from multiple jurisdictions may be mapped to a single model grid cell (e.g., eastern U.S. counties vs. $2° \times 2°$ horizontal grid spacing).

Spatial allocation is performed using a library of spatial surrogate fields, where each surrogate field is associated with one or more emissions source sectors or source types. These "gridding surrogates" are called surrogates because they are only proxies for actual emissions distributions, which are not known. Spatial surrogate fields include such quantities as population density, road networks, shipping lanes, and land use (e.g., cropland, water, deciduous forest). Population density is the most commonly used spatial surrogate field. Figure 4.5 shows a map of population density for Canada. This information can be used to build a separate spatial surrogate field for *each* Canadian province or territory with which to allocate emissions from each province to an AQ model grid.

Note that this step is not required for point sources such as individual smokestacks, which can be assigned uniquely to single grid cells based on their geographic location. Note also that like the chemical speciation step, there are uncertainties associated with the spatial allocation step. In particular, there are far fewer spatial surrogates available (about 100) than there are source types (about 5,000), so for many source types the spatial surrogate used is not an optimal proxy, which may result in non-representative spatial distributions of emissions being calculated for many source types.

- The third processing step shown in Fig. 4.4 is the temporal disaggregation or temporal allocation step. In this step, depending on the inventory, annual, seasonal, or monthly emissions are allocated in time to each hour of each day of the model simulation period.

  Temporal allocation is performed using a library of temporal profiles. Typically, three different types of temporal profiles are used: (a) diurnal, or hour-of-day, profiles, which allocate daily emissions to each hour of the day; (b) weekly, or day-of-week, profiles, which allocate weekly emissions to each day of the week, and (c) monthly, or month-of-year, profiles, which allocate annual emissions to each month of the year. For some source types, the associated temporal profiles may be uniform: for example, for base-load power plants, the diurnal and weekly profiles may be "flat", that is, the same for each hour of the day and each day of the week. For other source types, the temporal profiles may be strongly non-uniform: for example, the diurnal profile associated with light-duty gasoline vehicles typically has two peaks in 24 h that coincide with morning and evening rush hours.

  There are also uncertainties associated with the temporal allocation step. Again, the available library of temporal profiles is considerably smaller than the number of source types contained in the emissions inventory, which can lead to representativeness issues when a single profile is used for multiple source types. There may also be representativeness issues *within* a single source type if there is a lot of variation within the source type (e.g., using a single diurnal profile to allocate emissions from commercial meat cooking when different restaurants are open for different periods of a day). Special situations can also arise when standard temporal profiles do not apply, such as facility shutdowns for re-tooling or vacations, "upset" conditions such as stack flaring or leaks that are sporadic and unscheduled, and public holidays.

Figure 4.6 shows a sample NO total emissions field from on-road motor vehicles for a particular hour (11 a.m. EDT: Eastern Daylight Saving Time) of a particular day of the week (Friday) and month of the year (August) that was generated by the Sparse Matrix Operator Kernel Emissions (SMOKE) processing system (see CEP 2012) for a 2.5 km model grid located over northeastern North America. The locations of major urban centres such as Chicago, Detroit, Toronto, and Montreal are readily identifiable as are major and minor highways.

It is important to note that AQ models require knowledge of emissions from *all* source types, including both anthropogenic and natural sources, in order to be both comprehensive and complete. Natural emissions sources include biogenic emissions (VOC emissions from vegetation and NO emissions from soil), sea-salt emissions, wildfire and prescribed burning emissions, wind-blown dust emissions, and lightning emissions. Wildfire emissions are sometimes included in detailed emissions inventories whereas the other natural source types, which are all strongly dependent on meteorological conditions, are usually either modelled directly by the AQ model or neglected.

More information on emissions inventories and emissions processing can be found in Dickson and Oliver (1991), Houyoux et al. (2000), Makar et al. (2003), NARSTO (2005), Pouliot et al. (2012), and Zhang et al. (2012a). Note that NARSTO (2005) provides a good review of the uncertainties associated with emissions inventories, but little work has been done to date on uncertainties associated with emissions processing.

4.4.2 Prognostic Meteorological Models

As shown in Fig. 4.3 the meteorological inputs required by an AQ model are usually provided by a meteorological (or numerical weather prediction) model. Meteorological models are mathematical models that integrate a set of coupled, time-dependent partial differential equations corresponding to conservation equations for momentum, mass, energy, and moisture (e.g., Haltiner and Williams 1980; Seaman 2000). In mathematics this type of problem, in this case predicting the weather, is known as an initial-value problem, and the momentum, mass, energy, and moisture fields that must be specified at the start of the integration are known as initial conditions. The set of conservation equations itself is often referred to as the governing equations. A meteorological model is thus prognostic because it can predict future weather conditions from a detailed description of current weather conditions (i.e., initial conditions). Meteorological quantities predicted by meteorological models include three-dimensional fields of wind speed and wind direction, temperature, humidity, pressure, turbulence intensity, shortwave and

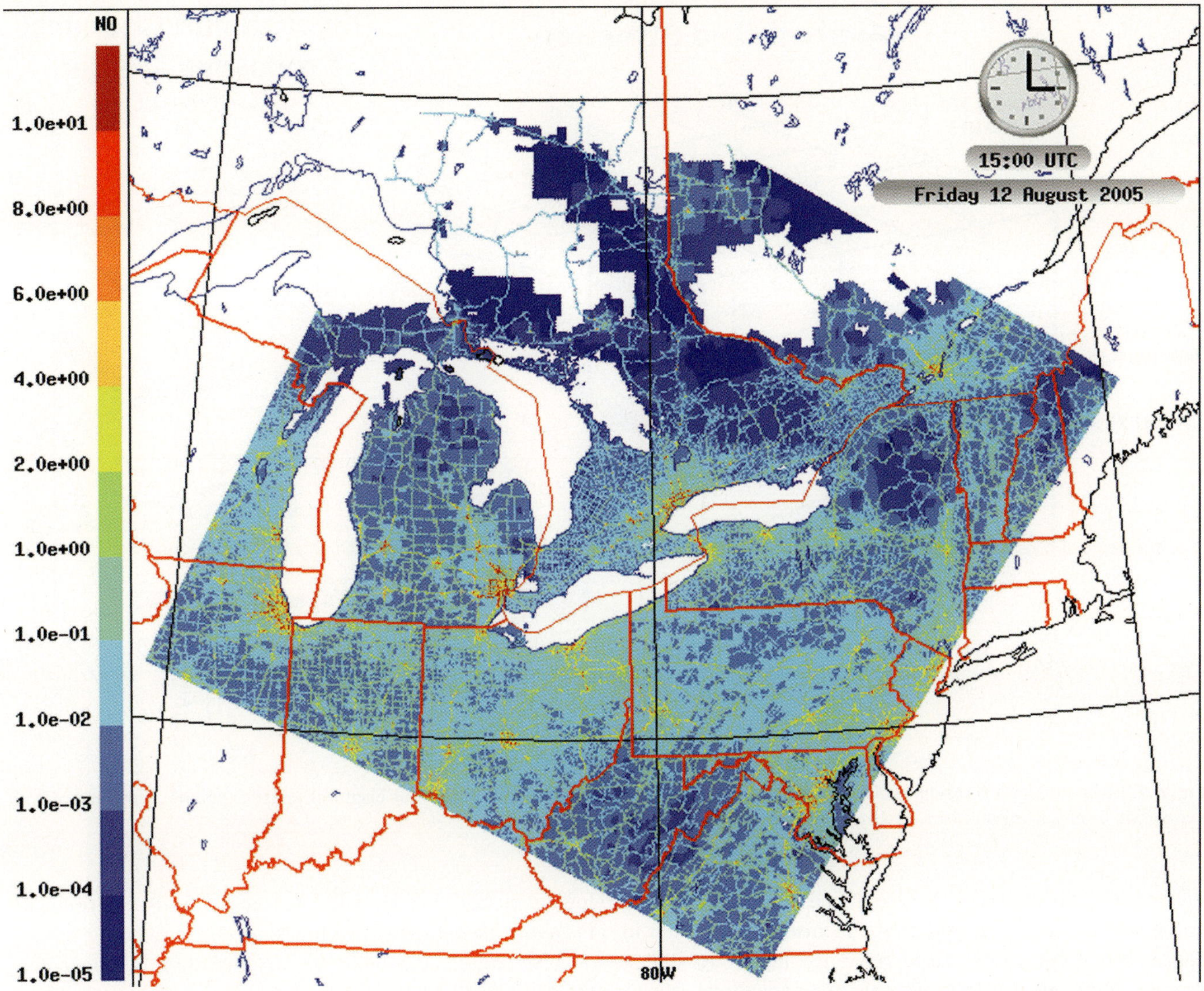

Fig. 4.6 Plot of NO emissions fluxes (g s^{-1} $cell^{-1}$) from on-road motor vehicles at 11 a.m. EDT on a Friday in August for northeastern North America on a polar-stereographic grid with 2.5 km grid spacing. Note that the contour spacing is logarithmic

longwave radiation, clouds, and precipitation, and also their variation in time. All of these meteorological quantities must also be known in order to predict air quality (see Fig. 4.2).

The mathematical system of governing equations that must be solved to forecast the weather is so complicated that it cannot be solved analytically; instead, it can only be solved numerically and with the aid of computers. Even so, of necessity the governing equations describe a simplified version of the real world since some physical processes in the atmosphere, such as radiation, turbulent mixing, cloud dynamics and microphysics, and land-surface processes, are too complicated to be described in detail and occur at very fine time and space scales. Instead, these processes are "parameterized" using simplified conceptual models and descriptions that can be solved more easily but that are still sufficiently realistic to represent the atmospheric impacts of these processes with reasonable accuracy at the time and space scales being considered (e.g., Haltiner and Williams 1980; Lee and Hong 2005). Note that an analogous approach is often taken in classical physics, as in the cases of the ideal gas laws vs. the kinetic theory of gases and continuum mechanics vs. the atomic theory of materials.

One common numerical method to solve the governing equations for weather prediction, the finite difference method, requires the portion of the atmosphere being modelled to be discretized using a geometric grid or lattice composed of three-dimensional grid boxes. Such a grid is illustrated in Fig. 4.7. The grid may span the entire atmosphere, in which case the meteorological model is called a global model, or only a portion of the atmosphere (e.g., over North America), in which case the meteorological model is called a regional model. The fourth dimension, time, is also discretized: the time integration of the governing equations is performed with finite steps called time steps and the values of different

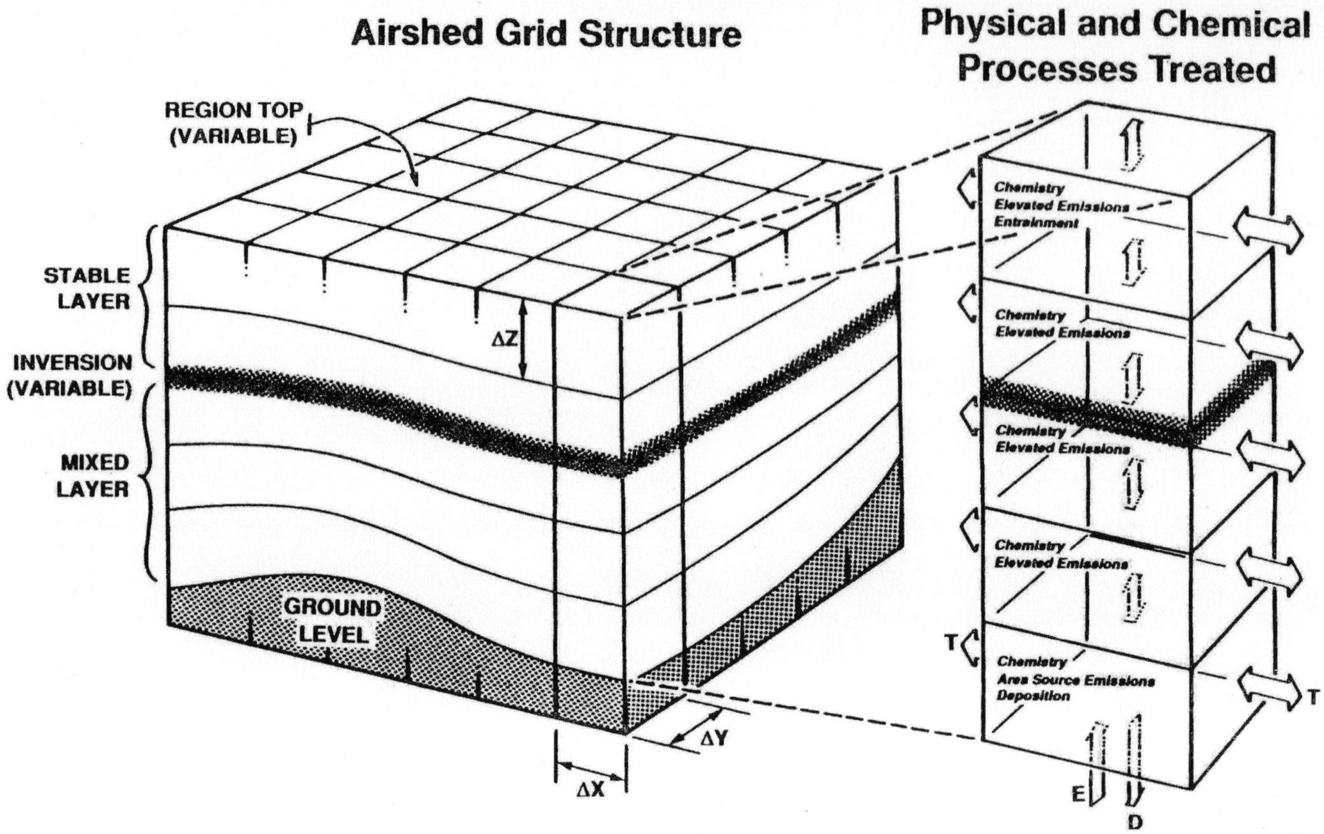

Fig. 4.7 Illustration of a three-dimensional grid used with an Eulerian meteorological model or chemical transport model. (Reproduced with permission from Lewis Publishers, Tesche and McNally 1991)

meteorological quantities are determined only at grid points and discrete times. The grid points themselves are fixed in space, consistent with a fixed reference frame. Note that numerical models that solve the governing equations using a fixed reference frame and fixed grid are called Eulerian models (after an 18th century Swiss mathematician, Leonhard Euler), whereas numerical models that use a moving reference frame are called Lagrangian models (after an 18th century French-Italian mathematician, Joseph-Louis Lagrange).

The discretization of the spatial domain shown in Fig. 4.7 suggests a fundamental limitation of using the finite difference method to solve the governing equations for meteorological models (and AQ models), which is that it imposes a finite "resolution" in both space and time. This means that the models do not provide any information at locations in between grid points or at times in between time steps. This limitation can be addressed by using smaller grid-box sizes and smaller time steps, but then more calculations must be performed for the same size model domain and forecast period, which will increase the amount of computer time required for the model integration. In an extreme case, trying to increase model resolution by too large a factor for a short-term meteorological forecast may mean that the period of interest has already come and gone by the time the meteorological model has completed its forecast for that period with the available computer resources.

Meteorological models are an important component of the AQ modelling system shown in Fig. 4.3 because meteorology is the dominant modulator of air quality through its influence on emissions, transport and diffusion, chemistry, and removal processes (see Fig. 4.2). Uncertainties in meteorological predictions thus translate directly into uncertainties in AQ predictions. Uncertainties and errors in meteorological model predictions can be caused by errors in the initial conditions, the numerical techniques, and the physical parameterizations used by the model, and by discretization errors due to the finite sizes of the model grid spacing and time step. More information about the role of meteorological models in AQ modelling systems and meteorological model performance may be found in Pielke and Uliasz (1998), Seaman (2000), Smyth et al. (2006a), and Vautard et al. (2012). More information about AQ model sensitivity to meteorological inputs may be found in Smyth et al. (2006a), Appel et al. (2010), and Vautard et al. (2012).

4.4.3 Chemical Transport Models

Figure 4.3 shows the outputs from emissions processing systems and meteorological models, namely gridded emissions fields and meteorological fields, being supplied as inputs to AQ models. But how are AQ models themselves constructed? Like meteorological models, all AQ models are mathematical models. The governing equations for an AQ model are a set of coupled, time-dependent partial differential equations corresponding to conservation-of-mass equations for each chemical species of interest. According to Seigneur and Moran (2004), these models "consist of mathematical representations of the relevant physical and chemical atmospheric processes that are solved with numerical algorithms to obtain pollutant concentrations as a function of space and time for a given set of pollutant emissions and meteorological conditions". Furthermore, those mathematical representations are themselves almost always simplifications of the real world, in part because some chemical processes, such as gas-phase chemistry, aqueous-phase chemistry, heterogeneous chemistry, and secondary organic aerosol formation, are too complicated to predict directly and instead must be "parameterizated" (i.e., treated using simplified descriptions of the chemical processes that are analogous to the physical parameterizations that must be used by meteorological models to represent some physical processes).

When the governing equations of a time-dependent Eulerian AQ model are solved (i.e., integrated in time), chemical and physical transformations are treated *in situ* within each grid cell and transport and diffusion processes move chemical species between grid cells. This is analogous to the treatment in meteorological models of water vapour, itself a chemical species. However, the need for AQ models to represent time-varying emissions of pollutants from both human activities and natural sources from the Earth's surface (i.e., bottom boundary) to the atmosphere constitutes a fundamental difference between AQ models and meteorological models. Whereas the mathematical model solved to predict the weather is classified as an initial-value problem, the inclusion of emissions, which in effect represent a chemical forcing term, means that the mathematical model solved to predict air quality is classified as an initial-value/boundary-value problem.

The treatment of grid domain boundaries is also the key difference between regional and global AQ (and meteorological) models. While both types of models have a bottom boundary (Earth's surface) and a top boundary (a selected level in the atmosphere, sometimes in the stratosphere, sometimes higher), regional models also have lateral boundaries (i.e., the sides of model grid domains) whereas global models do not. Since there must be flows at some model lateral boundaries from the atmosphere external to the regional model domain into the model domain, it is necessary to specify the concentrations of pollutants entering the model grid at these inflow boundaries in some way. Different approaches are used by different regional AQ models (e.g., Brost 1988; Samaali et al. 2009; Schere et al. 2012), but whatever the approach used, pollutant concentrations in model interiors are often very sensitive to the imposed chemical boundary values. As a consequence the treatment of these chemical lateral boundary conditions is a key aspect of all regional AQ models.

Note that Fig. 4.3 shows an AQ modelling system with a meteorological model and an AQ model as separate components. In this configuration, the meteorological model supplies meteorological fields to the AQ model and hence the meteorological model must be run before the AQ model is run so that the meteorological fields needed by the AQ model will be available in advance. For this configuration the AQ model is called an "off-line" model since it is run independently of the meteorological model. In fact, the AQ model might be run multiple times for the same set of meteorological fields. This can be cost-effective computationally if a number of emissions scenarios are to be considered but the meteorology is assumed not to change so that the meteorological model need only be run once. An off-line AQ model might also be run several times using meteorological fields from different meteorological models in order to investigate sensitivity to meteorological inputs (e.g., Smyth et al. 2006a; Appel et al. 2010; Vautard et al. 2012).

The off-line configuration, however, also has some drawbacks, such as high input/output costs, mass inconsistencies caused by grid and time-step differences, and an inability to account for pollutant feedbacks to meteorology (e.g., Grell et al. 2004; Zhang 2008; Gong et al. 2013). A different strategy is to integrate the meteorological model and AQ model together (i.e., combine the meteorological and AQ components in Fig. 4.3); this configuration is sometimes called an on-line AQ model and sometimes a chemical weather prediction model. A number of such on-line AQ models have been developed recently, including the Canadian GEM-AQ, GEM-MACH, and GRAHM models that are described in the next section.

More information about AQ models may be found in Peters et al. (1995), Jacobson (1999), Russell and Dennis (2000), Seigneur and Moran (2004), and Seinfeld and Pandis (2006). AQ model errors can be quantified and assessed through comparison to AQ measurements in what are called model performance evaluations. More information about and examples of such model performance evaluations can be found in Fox (1981), Clark et al. (1989), Sirois et al. (1999), McKeen et al. (2005, 2007, 2009), Smyth et al. (2006b), Appel et al. (2007, 2008), Moran et al. (2008, 2011), Stroud et al. (2008), and Solazzo et al. (2012a, b).

Errors and uncertainties in AQ model predictions can arise from a number of sources. One obvious source is errors in model inputs, particularly in emissions and meteorological fields. Other sources are equivalent to other error sources in meteorological models: that is, errors in the numerical techniques and in the physical and chemical parameterizations used by the model and discretization errors due to the finite sizes of the model grid spacing and time step. Parameterization errors might be due to the algorithm used (e.g., treatment of VOC species in a gas-phase chemistry mechanism) but can also be due to the choice of parameter values required by the parameterization (e.g., the rate constants used in the gas-phase chemistry mechanism). More information about the sources of AQ model error and AQ model sensitivity and uncertainty can be found in Russell and Dennis (2000), Dunker et al. (2002), Hakami et al. (2003), Seigneur and Moran (2004), Zhang et al. (2005), and Koo et al. (2009), among others.

4.5 Regional and Global AQ Models and Impact Metrics That Have Been Used in Canada

4.5.1 Regional AQ Models

Regional AQ models have been used in Canada by the federal government and by some provincial governments since the 1980s for AQ management. The first of these models were developed to study and to manage regional acid deposition, starting with Environment Canada's AES (Atmospheric Environment Service: renamed Meteorological Service of Canada in 1999) Lagrangian Sulphur Model (ALSM) (Olson et al. 1982, 1989; Olson and Oikawa 1989; Clark et al. 1989). Other regional acid deposition models included the AES Lagrangian Nitrogen Model (ALNM) (Olson et al. 1990, 1992: used by Environment Canada (EC)), the Acid Deposition and Oxidants Model (ADOM) (Venkatram et al. 1988; Fung et al. 1991; Karamchandani and Venkatram 1992: used by the province of Ontario and EC), and the REgional Lagrangian Acid Deposition model (RELAD) (Cheng et al. 1995; Cheng and Angle 1996; McDonald et al. 1996: used by the province of Alberta). All of these AQ models were off-line models, and various means were used to provide their emissions and meteorological input files. Reviews of these models and their applications can be found in Environment Canada (1998) and Moran (2005). Note that all of these early models have now been retired as newer and more comprehensive regional AQ models have been developed and come into service.

Beginning in the 1990s regional AQ models were also developed to study and manage photochemical oxidants, in particular, ozone. ADOM predicted ozone as well as acid deposition, and another off-line Eulerian model, the Canadian Hemispheric and Regional Ozone and NO_x System (CHRONOS), was developed by EC specifically to predict ozone (Pudykiewicz et al. 1997; Sirois et al. 1999). Emissions files for CHRONOS were prepared using the Canadian Emissions Processing System or CEPS (Moran et al. 1997; Scholtz et al. 1999) while meteorological files for CHRONOS were produced by EC's Global Environmental Multiscale (GEM) meteorological model (Côté et al. 1998a, b). Another off-line Eulerian regional AQ model, the Community Multiscale Air Quality (CMAQ) model, which was developed by the U.S. Environmental Protection Agency (e.g., Byun and Schere 2006; Appel et al. 2007, 2008), was applied by the National Research Council of Canada and by EC to predict ozone in Canada (Smyth et al. 2006b; Fox and Kellerhals 2007). The CMAQ model uses emissions files prepared by the SMOKE emissions processing system (CEP 2012) and can use meteorological files prepared by GEM or by other meteorological models such as the MM5 (Fifth-Generation Penn State/NCAR Mesoscale Model) and WRF (Weather Research and Forecasting) models (e.g., Smyth et al. 2006a; Appel et al. 2010).

Another pollutant of concern to health officials and policymakers is particulate matter (PM). As noted in the discussion of Fig. 4.2, PM is a more complex pollutant than ozone, and so prediction of PM requires the use of more complex AQ models. The U.S. EPA's CMAQ model was developed to predict both ozone and PM and has been applied in Canada to simulate PM (Jiang et al. 2006; Smyth et al. 2006b). Using the CHRONOS model as a starting point, Environment Canada developed a new off-line Eulerian regional AQ modelling system called AURAMS (A Unified Regional Air quality Modelling System) to predict both ozone and PM (e.g., Moran et al. 1998; Gong et al. 2006; Smyth et al. 2009; Makar et al. 2009; Park et al. 2010; Levy et al. 2010). The AURAMS chemical transport model uses emissions files prepared by the SMOKE emissions processing system (CEP 2012) and meteorological files prepared by the GEM meteorological model (Côté et al. 1998a, b). AURAMS has also been used to model acid deposition (Moran et al. 2008), and a recent AQ management application of AURAMS is discussed in Sect. 4.6.1 Recently, Gong et al. (2013) provided a summary of the status, performance, and recent developments for CHRONOS, AURAMS, and CMAQ, and Bouchet et al. (2013) provided an overview of the use of these three models for AQ management in Canada between 2000 and 2007.

4.5.2 Global AQ Models

For some AQ problems of interest, global AQ models are more suitable tools than regional models because they cover the entire globe and do not require chemical lateral boundary conditions to be specified. Two examples of such problems include modelling long-lived pollutants such as mercury or persistent organic pollutants (POPs), which can circle the globe many times, and quantifying the contribution of intercontinental transport, such as the trans-Pacific transport of pollutants from Asia to North America or the transport of pollutants from mid-latitude regions to polar regions (e.g., Arctic haze).

The Global-Regional Atmospheric Heavy Metal (GRAHM) model is a global on-line Eulerian model developed by EC to model atmospheric mercury (e.g., Dastoor and Larocque 2004; Dastoor et al. 2008; Durnford et al. 2010). To build GRAHM, parameterizations of mercury chemistry and removal were embedded in the GEM meteorological model, which can employ either a regional or a global grid, and mercury emissions fields are input. GRAHM is described in more detail and a recent AQ management application of GRAHM is discussed in Sect. 4.6.2.

A global on-line PM chemistry version of the GEM model called GEM-AQ was developed through a collaboration between York University (Ontario) and EC. GEM-AQ has been used to study the hemispheric and global transport of both ozone and PM (e.g., O'Neill et al. 2006; Kaminski et al. 2008; Reidmiller et al. 2009; Gong et al. 2012). GEM-AQ was also extended by adding modules for the treatment of gas-particle partitioning and atmosphere-soil and atmosphere-water exchange of semi-volatile persistent organic pollutants (POPs) to create a special air toxics version called GEM-POPs (Gong et al. 2007; Huang et al. 2007).

More recently, EC merged a newer version of the GEM meteorological model with the AURAMS regional AQ model to create a new on-line multiscale AQ model called GEM-MACH (Global Environmental Multiscale—Modelling Air quality and CHemistry) (Talbot et al. 2008; Anselmo et al. 2010; Moran et al. 2010, 2011). Because GEM can easily be configured to run for either a regional or global domain, GEM-MACH can also be run as either a regional or global AQ model. As described in Sect. 4.6.3, a regional version of GEM-MACH is now being used by EC to produce operational 48 h AQ forecasts for Canada.

4.5.3 Impact Metrics

Regional and global AQ models can predict concentration, dry deposition, and wet deposition fields for many chemical species. Additional calculations can then be performed after the AQ model simulation has finished (i.e., a post-processing step) to characterize predicted AQ effects or impacts in different ways in order to match policy requirements or questions. These impacts typically fall into one of two categories: (1) human health effects; and (2) environmental effects.

To assess the human health effects of AQ (see Chap. 7), one common approach is to compare predicted AQ concentrations against AQ standards for pollutants such as ozone and $PM_{2.5}$ (for some examples, see Bouchet et al. 2013). A second approach is to combine predicted pollutant concentrations into a single index such as the national Air Quality Health Index (AQHI) discussed in Chap. 18. And a third approach is to estimate total health impacts by estimating population exposure to selected pollutants, where predicted pollutant fields are combined with a population density field (Fig. 4.5) using one of a number of different exposure models (e.g., Seigneur and Moran 2004).

While health effects can be either acute (i.e., short-term) or chronic (long-term), environmental impacts from air pollutants are generally chronic, that is, long-term and cumulative. To quantify the environmental impacts of air pollutants, one commonly-used approach in the case of ozone is to calculate the cumulative exposure of agricultural crops or other vegetation such as forests to ozone over a season or other period. One such vegetation exposure index, called AOT40 (Accumulated dose of ozone Over a Threshold of 40 ppb), has been used extensively in Europe. AOT40 has been used by EC to assess crop-yield impacts, where hourly ozone concentration increments above 40 ppbv are accumulated between 5:30 and 21:00 local time from May 1 to July 31. Another cumulative exposure index, referred to as SUM60, has also been used in Canada. It is defined as the sum of hourly ozone concentration increments above 60 ppbv between 08:00 and 19:59 local standard time for a 3-month or 3-day period, and it has been used for characterizing both season-long and short-term effects (see CCME 2003).

Another common approach in the case of acid deposition is to compare a model-predicted deposition field with the corresponding critical-load field for acid deposition or eutrophication. The acid-deposition critical load (ADCL) field, for example, is a quantitative measure of the acid buffering capacity of an ecosystem. It provides an objective metric that can be used to determine both the spatial extent of a region subject to damaging levels of acid deposition and the degree of acidification. An ADCL field will typically vary geographically, since some locations will have a lower acid buffering capacity than others and hence will be more sensitive to acid deposition. If the difference between the annual atmospheric total (= wet + dry) acid deposition to an ecosystem predicted by the regional AQ model and the ecosystem's ADCL value is positive, then that difference is termed an exceedance because acid deposition to that ecosystem is predicted to be larger than the ecosystem's available acid buffering capacity and hence the ecosystem is becoming more acidic with time.

As discussed by Jeffries and Ouimet (2005), Canada-wide ADCL fields have been developed for two types of ecosystems: freshwater aquatic ecosystems and upland forest ecosystems. Figure 4.8 shows a plot of the national ADCL field for upland forest ecosystems. The most sensitive forest ecosystems can be seen in the coastal mountains of British Columbia, in northern Alberta, Saskatchewan, Manitoba,

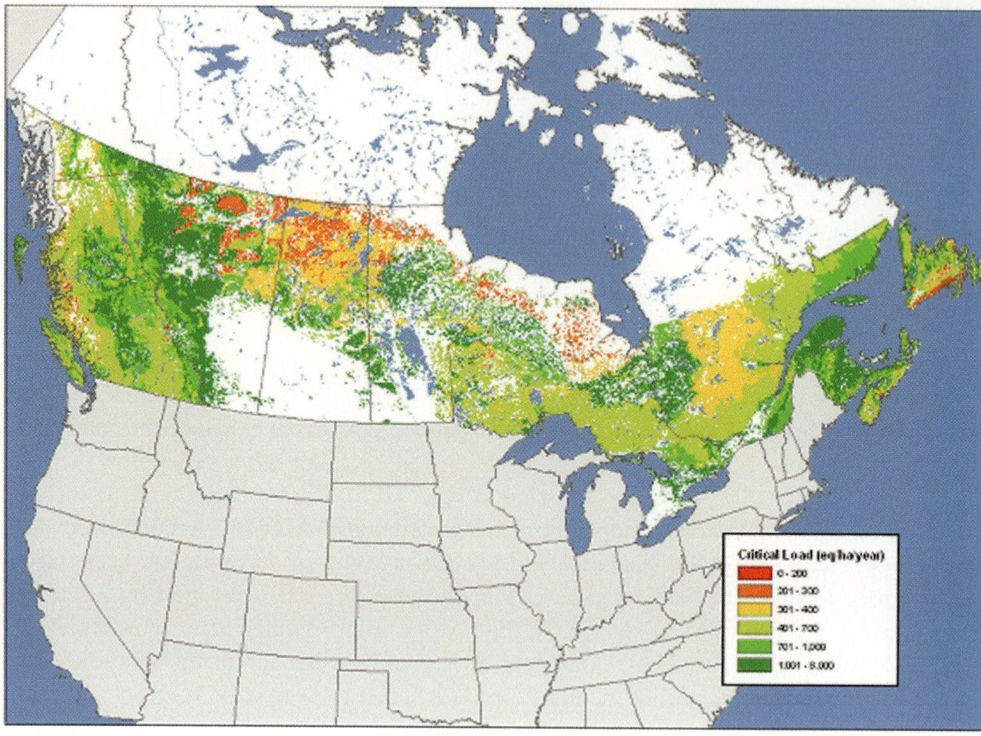

Fig. 4.8 National distribution of acid deposition critical loads for upland forest soils in the southern half of Canada. (CCME 2008)

and Ontario, and in Nova Scotia and Newfoundland. Some sensitive aquatic ecosystems, on the other hand, are located in other parts of Canada (Jeffries and Ouimet 2005; CCME 2008). Examples of the use of ADCL fields to quantify the environmental impacts of sulphur and nitrogen deposition fields in Canada can be found in Moran et al. (2008) and Bouchet et al. (2013).

4.6 Examples of Recent Canadian Applications of Regional and Global AQ Models

The AQ models reviewed in Sect. 4.5 have been used to support and guide AQ management in Canada for decades. Some examples of such applications include Olson et al. (1982, 1983, 1990, 1992), Cheng et al. (1995), McDonald et al. (1996), Environment Canada (1998), Brankov et al. (2003), Dastoor and Larocque (2004), Moran (2005), Smyth et al. (2006b), Gálvez (2007), Huang et al. (2007), Moran et al. (2008), Makar et al. (2009), Durnford and Dastoor (2010), Levy et al. (2010), Park et al. (2010), Cho et al. (2012a, b), and Kelly et al. (2012). Bouchet et al. (2013) recently provided an extensive review of applications of the AURAMS, CHRONOS, and CMAQ regional AQ models performed in Canada between 2000 and 2007 that were concerned with ozone and $PM_{2.5}$.

The following three case studies provide additional examples of more recent AQ model applications in Canada in support of AQ management. The first case study describes an application of the regional off-line AURAMS model to assess the potential impact on ambient AQ and human health if biodiesel fuel blends were to become widely used in Canada. The second case study describes an application of the global on-line GRAHM model to simulate a spring 2004 episode of trans-Pacific transport of mercury emitted in East Asia to western North America. And the third case study describes the application of a regional configuration of the on-line GEM-MACH model to operational national AQ forecasting in Canada.

4.6.1 Estimated Impacts of Biodiesel Fuel Use on Ambient Pollutant Concentrations

AQ models can provide guidance to policymakers on the AQ impacts that may result from proposed technological changes. Recently, Health Canada wished to assess the potential impact on ambient AQ and human health if biodiesel fuel blends were to become widely used in Canada. AQ models are useful to investigate this type of question since they are capable of simulating AQ for different emissions scenarios. In this kind of assessment study, it is customary to use the model first to estimate the air quality for the base case without the proposed change and then again for the scenario with the changed emissions. The calculated difference in AQ between the scenario and the base case will then provide a measure of the impact of the use of biodiesel on air quality from which the health impacts and benefits can be estimated. This section summarizes the AQ modelling study that was

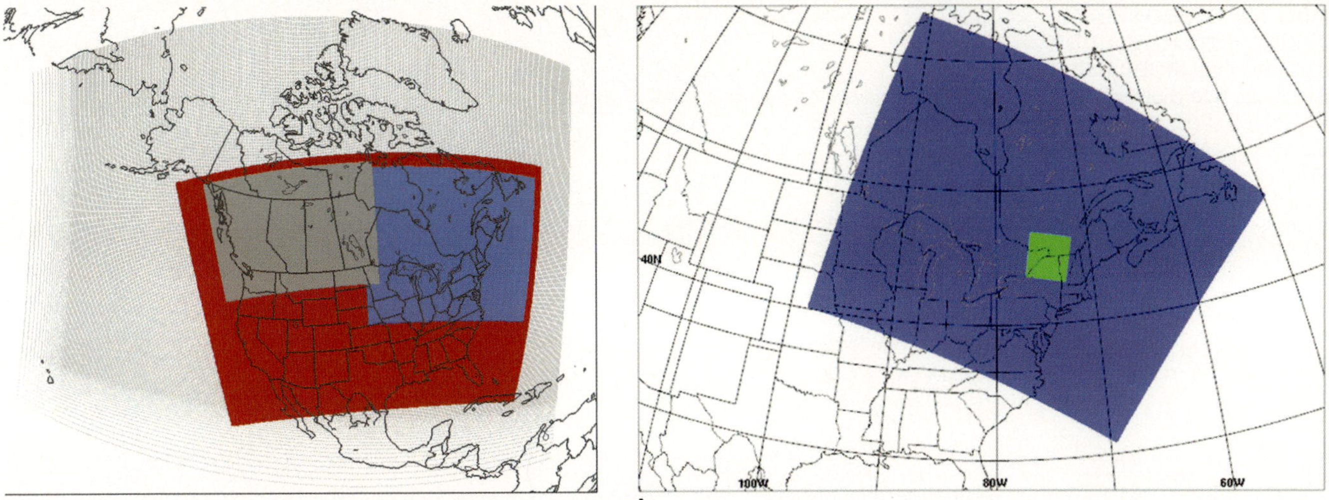

a **b**

Fig. 4.9 Location of the model grid domains: (*left*) Meteorological outer domain in light grey, AURAMS continental 45 km domain in red and nested Canadian western and eastern regional 22.5 km domains in darker grey and blue, respectively; (*right*) High-resolution (3 km grid spacing) domain covering the Montreal region is shown in green nested in the eastern regional 22.5 km domain in blue

Table 4.1 List of AURAMS biodiesel scenarios. (Source: Health Canada 2012)

Years	Fuel	Continental domain	Eastern 22.5 km domain	Western 22.5 km domain	High resolution
2006 and 2020	B0	Annual	Annual	Annual	June 12–23
	B5	Annual	Annual	Annual	June 12–23
	B20 summer, B0 winter	B20: May 1st to Sept 30th, otherwise B0	B20: May 1st to Sept 30th, otherwise B0	B20: May 1st to Sept 30th, otherwise B0	B20: June 12–23

carried out for Health Canada and its results. Further information about the study, and in particular how health benefits were estimated, can be found in the complete assessment published by Health Canada (2012).

Biodiesel is produced from vegetable oil and animal fats by reaction with an alcohol (generally methanol). Biodiesel is blended with ultra-low sulphur diesel (ULSD) and the blends are denoted by the letter B followed by the percentage of biodiesel by volume. For example, B5 denotes a blend containing 5% biodiesel and 95% ULSD. The scope of the modelling study was to assess the impact of switching to B5 and B20 blends on emissions from the Canadian heavy-duty diesel vehicle (HDDV) fleet and on ambient air quality in 2006 and 2020. The reasons to limit the study scope to the HDDV fleet are that this vehicle class accounts for 94% of diesel fuel use in Canada and emissions measurement data for biodiesel blends mostly apply to this class.

AQ modelling was performed with the AURAMS regional AQ model using a grid-cascade configuration, where the model was first run for a continental North American domain with horizontal grid spacing of 45 km, then for two smaller overlapping regional domains with 22.5 km grid spacing that were nested within the coarser continental domain, and then for one even smaller high-resolution domain covering the Montreal region at 3 km. The AURAMS integrations over the continental domain were performed first to produce initial and boundary conditions for the two regional domains. The reason to choose this approach is that it allows a higher-resolution treatment of a region of interest for lower computational cost while still accounting for the contribution of long-range transport. The locations of the four AQ model domains are shown in Fig. 4.9 and the list of emissions scenarios that were considered using this nested model configuration is shown in Table 4.1.

For all scenarios in 2006 and 2020, meteorological fields for 2006 were obtained from Environment Canada's GEM weather forecast model and were used for all of the 2006 and 2020 AURAMS CTM simulations. GEM was run for the entire year 2006 over North America with a grid spacing of approximately 15 km. The GEM model was also run in cascade mode on a smaller domain with a grid spacing of about 2.5 km to generate higher-resolution meteorological fields in order to drive the high-resolution AQ domain. The meteorological fields were then interpolated onto each AQ model grid. The locations of the meteorological model domains are also illustrated in Fig. 4.9.

Anthropogenic emissions of seven criteria air contaminants (CO, NO_x, SO_x, VOC, NH_3, $PM_{2.5}$ and PM_{10}) for 2006 were obtained from the 2006 Canadian national CAC emissions inventory and the 2005 U.S. national CAC emissions inventory.

Table 4.2 Percent change in Canadian fleet-average HDDV emissions from MOBILE6.2C for B5 and B20 compared to the reference fuel (ULSD) in 2006 and 2020. (Source: Health Canada 2012)

Pollutants	B5		B20	
	2006	2020	2006	2020
1,3-Butadiene/Acetaldehyde/Acrolein/Formaldehyde	−4	−1	−18	−3
Benzo[a]pyrene	−3	−2	−14	−9
Benzene	−4	−1	−18	−4
Carbon monoxide	−3	−2	−11	−7
Elemental carbon/Organic carbon	−3	−2	−13	−9
NO_x	+1	+1	+4	+3
$PM_{2.5}$ (exhaust)	−3	−2	−13	−9
SO_2, NH_3	0	0	0	0
Toluene/Ethylbenzene/Xylene	−4	−1	−18	−3
Total VOC, Total hydrocarbon	−4	−1	−18	−3

The Canadian 2020 emissions projection was obtained from Environment Canada's Economic Analysis division, whereas for the United States the latest available 2020 emissions projection at the time of the study was the one prepared in 2004 by the U.S. Environmental Protection Agency for the Clean Air Interstate Rule impact study. These two inventories were processed with the SMOKE emissions processing system (CEP 2012) to do spatial allocation onto the AURAMS grids, temporal allocation, and chemical speciation for the ADOM-II gas-phase mechanism (Stroud et al. 2008).

Canadian on-road mobile emissions, which were the key emissions sector of interest in this study, were computed using the MOBILE6.2C emissions-factor model, which is the "Canadianized" version of the U.S. EPA MOBILE6.2 model. MOBILE6.2C was modified for this study in order to account for renewable fuels in the HDDV class based on available measurement data. Although the modified version can handle five separate biodiesel feedstocks, it was assumed for this study that canola would be the only feedstock used by the Canadian on-road fleet. The ambient meteorological conditions used by MOBILE6.2C to generate the 2006 and 2020 on-road mobile emissions were monthly averages for 2006 in order to be consistent with the 2006 base-year meteorology that was used for the AQ modelling. In Canada, on-road mobile emissions were estimated at the provincial level except where motor-vehicle inspection/maintenance (I/M) programs were in place (British Columbia and Ontario), in which case they were estimated at the sub-provincial level. For the high-resolution AURAMS domain around Montreal, a traffic-demand model was used to generate vehicle flow and average speed on every road segment; these road-segment-based data were then used as input to MOBILE6.2C to generate the emissions at high resolution (Noriega et al. 2007).

Table 4.2 illustrates the changes in CAC emissions and some specific VOCs and PAHs that are expected to result from the use of biodiesel blends relative to the reference fuel (ULSD) for the whole country. In summary, as biodiesel concentration increases in fuel, emissions of CO, PM_{10}, $PM_{2.5}$ and VOCs from the on-road transportation sector are expected to decrease whereas NO_x emissions are expected to increase. It should be noted that the impact of biodiesel on HDDV emissions is expected to be smaller in 2020 than in 2006 due to the assumption that newer vehicles have been introduced into the vehicle fleet and older vehicles retired. Moreover, when these emissions changes are scaled by the corresponding *total* Canadian anthropogenic emissions, the relative changes in emissions are very small. For example in 2006, the B20 scenario produced a decrease of 0.1 % in total Canadian anthropogenic VOC emissions, a decrease of 0.3 % in total Canadian anthropogenic $PM_{2.5}$ emissions, and an increase of 0.5 % in total Canadian anthropogenic NO_x emissions.

The statistical metrics that were used for human-health impact assessment were the 1-hour daily maximum and the 24 h daily averages for CO, NO_2, PM_{10}, and $PM_{2.5}$ concentrations for all of 2006 as well as the daily maximum 8 h running average and 1 h daily maximum for O_3 concentration during the summer of 2006. Results for the summer average of daily maximum 8 h O_3 concentration and the annual average $PM_{2.5}$ concentration were as follows.

Figures 4.10 and 4.11 show the predicted impact of B20 biodiesel on daily maximum 8 h ozone concentration in 2006. One noticeable feature in Fig. 4.10 is that ozone is predicted to decrease in major urban centres due to the increase in NO_x emissions. These cities were under a high-NO_x emissions regime in 2006, and thus any additional contribution of NO_x led to further scavenging of ozone. On the other hand, the increase in NO_x emissions led to an increase in ozone outside major cities that is not cancelled by the reductions in VOC emissions. The same pattern is seen in Fig. 4.11 on the high-resolution Montreal domain, where ozone scavenging by NO is noticeable in downtown Montreal and along major highways.

Figures 4.12 and 4.13 illustrate the impact of B20 biodiesel on annual average $PM_{2.5}$ concentration in 2006. Since $PM_{2.5}$ primary emissions decreased and SO_2 emissions remained unchanged (see Table 4.2), annual $PM_{2.5}$ ambient concentrations were predicted to decrease in major cities.

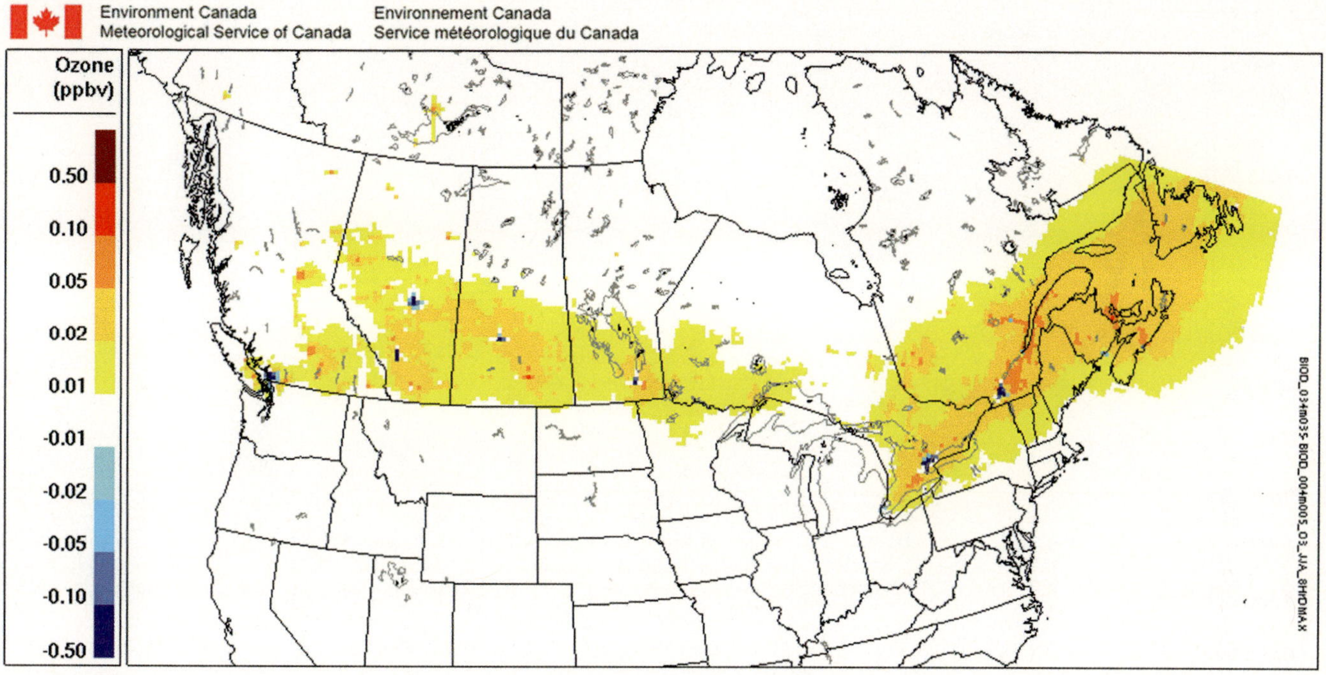

Fig. 4.10 Change in summer average (June-August) daily maximum 8 h ozone concentration (ppbv) on a portion of the AURAMS 45 km continental grid when using a B20 biodiesel fuel blend compared to ULSD in Canada in 2006. (Health Canada 2012)

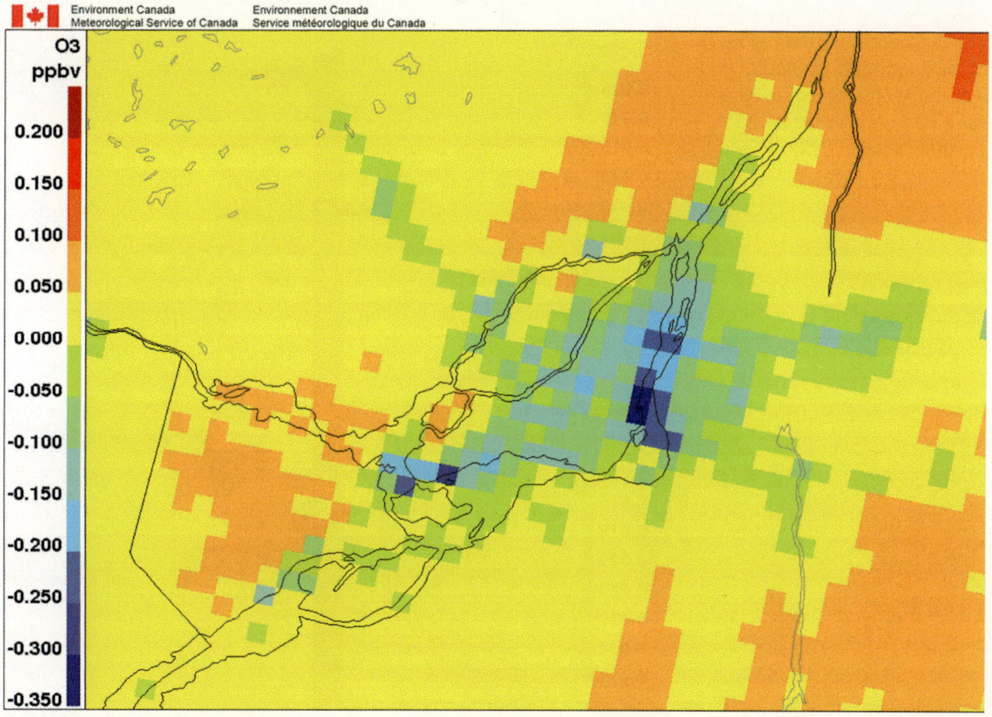

Fig. 4.11 Change in the two-week average (12–23 June 2006) of the daily maximum 8 h ozone concentration (ppbv) on the 3 km high-resolution grid over the Montreal region when using B20 compared to ULSD. (Health Canada 2012)

The NO_x emissions increase produced some increase in $PM_{2.5}$ concentration in some rural areas around industrialized regions (i.e., Lower Fraser Valley, Calgary-Edmonton corridor and Windsor-Quebec City corridor) that is mostly a wintertime contribution to the annual average.

In summary, the B20 and B5 scenarios, when applied to 2006 emission levels, produced small but significant changes in predicted air quality that were focused in major urban centres and surrounding areas (Lower Fraser Valley, central Alberta, southern Ontario, and southern Quebec). B5 results

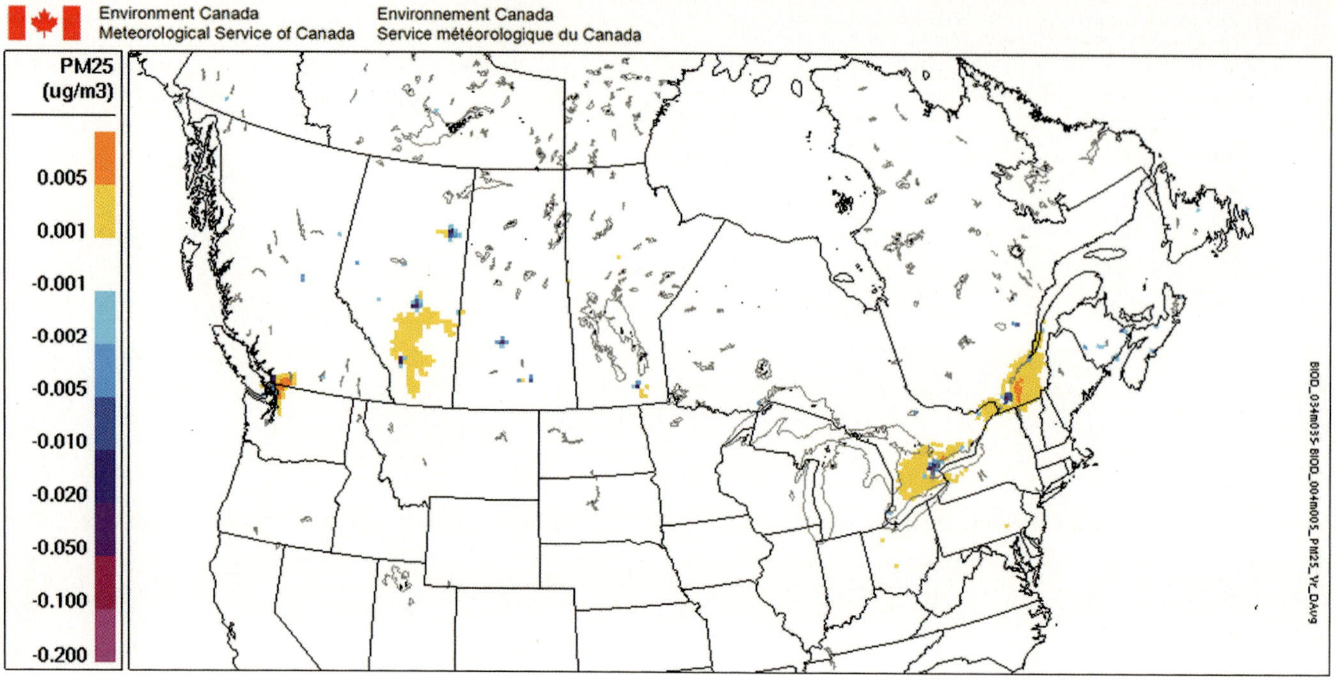

Fig. 4.12 Change in annual average $PM_{2.5}$ concentration (μgm^{-3}) on a portion of the AURAMS 45 km continental grid when using B20 compared to ULSD in Canada in 2006. (Health Canada 2012)

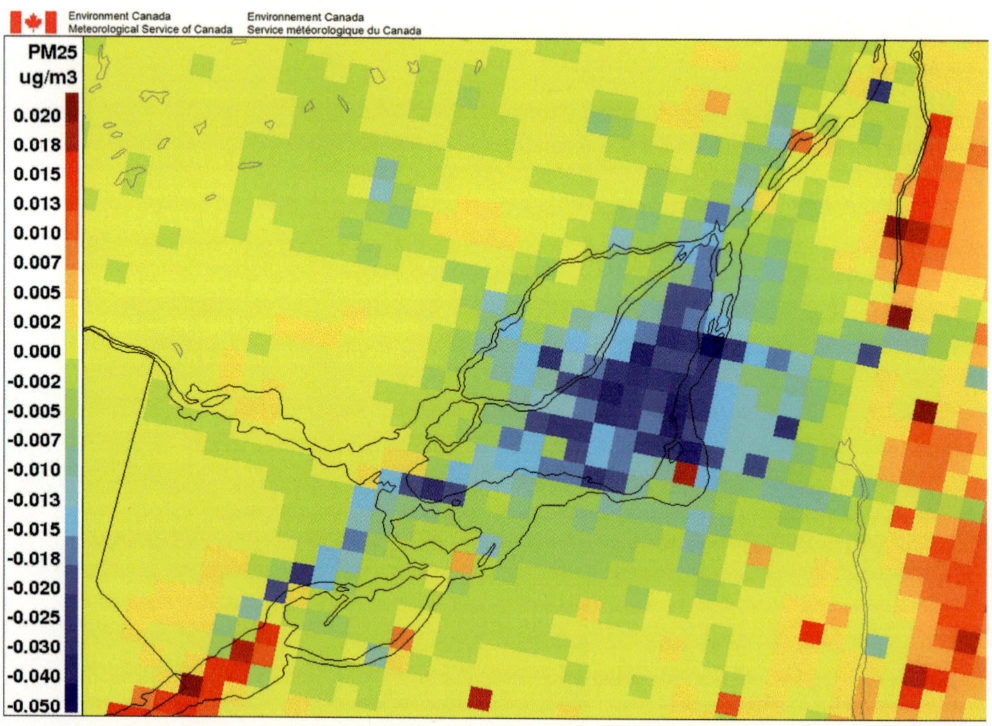

Fig. 4.13 Change in the two-week average (12–23 June 2006) of $PM_{2.5}$ concentration ($\mu g\ m^{-3}$) on the 3 km high-resolution grid over the Montreal region when using B20 compared to ULSD. (Health Canada 2012)

(not shown) had the same pattern as B20 but with much smaller amplitude, consistent with the smaller B5 emissions changes (Table 4.2).

The results for the 2020 projections show that changes in predicted air quality had the same general pattern as the 2006 results but the differences were much smaller. In particular, the model results showed that ozone will continue to decrease in the same major urban centres as in 2006 due to NO_x emissions increases, but with a smaller footprint. Differences between the scenario and the base case for 2020 are often close to the limit of model accuracy. Generally speaking, though, impacts of B20 in 2020 were comparable to impacts of B5 in 2006, and impacts of B5 in 2020 (not shown) affected a very limited area, even in urban centres.

4.6.2 Tracking Long-Range Transport of Mercury to Canada

Mercury (Hg) is a natural element that is ubiquitous in the environment and found in virtually all geological media. Mercury is released to the atmosphere in nature by two primary source processes: the weathering of rocks and volcanic/geothermal activities. It is naturally present in coal and other fossil fuels, minerals, and metal ores, including zinc, copper, and gold, and thus can be emitted to the atmosphere from such human activities as fossil-fuel combustion and ore processing. Mercury is also released to the environment from biomass burning. In the pre-industrial period (i.e., before the mid–18th century), the natural release of mercury to the atmosphere was generally in balance with the natural processes that removed it from the atmosphere (Sunderland and Mason 2007). Since the Industrial Revolution, however, anthropogenic activities have mobilized vast quantities of mercury from the Earth's crust and redistributed it to the more biologically active and mobile media of the environment (surface soils, atmosphere, lakes, rivers, oceans). Analysis of mercury concentrations in lake sediments shows that current global mercury deposition rates are approximately 2–5 times higher than those during pre-industrial times (Lamborg et al. 2002; Biester et al. 2007; Lindberg et al. 2007).

Currently, human activities such as the combustion of fossil fuels, the extraction of metals from ores, as well as the use and disposal of consumer products containing mercury (e.g., batteries and light bulbs) are responsible for the release of roughly 1,900 tonnes of Hg annually to the atmosphere (Pacyna et al. 2010). As much as 46 % of global anthropogenic emissions of Hg result from the combustion of fossil fuels, mainly from coal-fired power plants and industrial and residential heating. Metal production, cement manufacturing, and artisanal gold production are also large contributors to global Hg emissions. Asia is by far the region with the largest emissions of Hg to the atmosphere, with about 65 % of the total anthropogenic emissions, followed by North America and Europe. Deposited Hg is reduced via photochemical and biochemical processes and may then be re-emitted back to the atmosphere from soil and vegetation as well as from the ocean surface. Re-emission is a significant source of Hg to the atmosphere in comparison to primary emissions. Available estimates of natural emissions and re-emissions are within the range 2,000–5,000 tonnes per year and are thus of the same magnitude as or larger than anthropogenic emissions (Pirrone et al. 2010).

Most of the Hg released to the atmosphere is in its elemental gaseous form, Hg^0, with smaller amounts emitted in both gaseous and particulate oxidized forms, Hg^{II}. Hg^0, which is insoluble, has a long lifetime in the atmosphere (~ 1 year) and is transported globally. However, Hg^0 is slowly oxidized in the atmosphere to highly soluble Hg^{II}, which is rapidly removed from the atmosphere by wet and dry deposition. Atmospheric transport followed by deposition and re-emission is the main pathway for the global dispersal of Hg. Once deposited to the Earth's surface, mercury can be transformed into a highly toxic organic form, methyl-mercury (MeHg), which can bio-accumulate and bio-magnify through the food chain. MeHg exposure via consumption of fish leads to adverse neurological and developmental effects in humans and wildlife (Champoux et al. 2006; Clarkson and Magos 2006). The impacts of mercury are most significant in the Arctic region, which acts as a sink for long-range transport of Hg emissions (AMAP 2011).

Since the 1970s, Canada has reduced its domestic emissions of mercury to the atmospheric by approximately 90 % and currently contributes less than 0.4 % of global anthropogenic emissions (Environment Canada 2010). Despite continuing efforts to regulate anthropogenic mercury emissions in some regions of the world such as North America and Europe, however, mercury contamination remains a concern in Canada due to the long-range transport of mercury. Currently, a large number of fish consumption advisories in Canada are due to mercury (Environment Canada 2010). Monitoring data indicate that mercury contamination in Canadian Arctic biota has generally been increasing during the past 30 years, and high levels of mercury are found in the tissue of aboriginal people who rely on traditional "country food" such as marine mammals and fish (AMAP 2011). Mercury is identified as a toxic substance under the 1999 Canadian Environmental Protection Act and it is included in the List of Toxic Substances in Schedule I of the Act.

Long-Range Transport of Mercury

The majority of anthropogenic mercury emissions come from mid latitudes of the Northern Hemisphere. Mercury is primarily transported within the Northern Hemisphere; hemispheric exchange is not as significant, as evidenced by the measured Hg concentration gradient between the Northern and Southern Hemispheres (Lamborg et al. 2002). Current mercury monitoring networks are sparse and are unevenly distributed; as a consequence, the measurement data have a limited capability to detect signs of intercontinental transport. Atmospheric AQ models thus provide a powerful tool for investigating the long-range transport of mercury. Global and hemispheric mercury models have been used to examine the role of long-range transport of atmospheric mercury in regional mercury contamination (Christensen et al. 2004; Travnikov 2005; Dastoor et al. 2008; Holmes

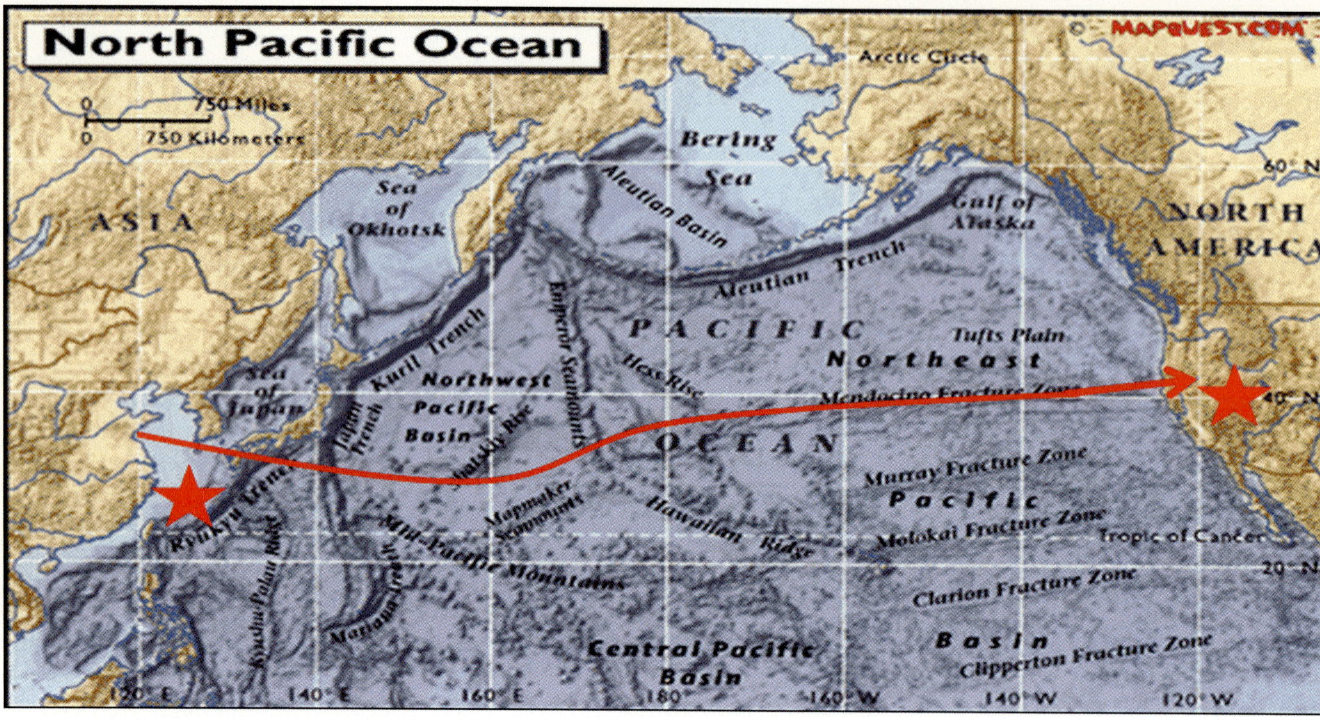

Fig. 4.14 Transport pathway from East Asia to North America for this episode. The left star marks the location of the Hedo Station monitoring site and the right star marks the location of the Mt. Bachelor Observatory monitoring site

et al. 2010; HTAP, 2010). Environment Canada's GRAHM model, which was introduced in Sect. 4.5.2, has been applied to study the impact of long-range transport on Hg concentration and deposition in Canada (Durnford et al. 2010).

Durnford et al. (2010) analyzed long-range transport events from the four major global mercury source regions, East Asia, Europe, North America, and Russia, into Canada and the Arctic. East Asia was found to be the largest source of Hg in Canada, contributing to the most long-range transport events at all Arctic sites combined (43 %), followed by Russia (27 %), North America (16 %), and Europe (14 %). Long-range transport from major source regions was found to occur principally in the mid-troposphere. Measurements coupled with modelling analyses were used to detect industrial plumes of Hg transported out of East Asia during the ACE-Asia campaign (Friedli et al. 2004; Pan et al. 2006, 2008, 2010). The outflow of Hg to the Pacific Ocean from East Asian anthropogenic emissions was estimated to be 681–714 Mg yr^{1} by Pan et al. (2010). Lifting of mercury plumes from the planetary boundary layer to the free troposphere above followed by rapid horizontal transport in the free troposphere is the typical long-range-transport pathway for Asian Hg outflows to the Pacific Ocean. The transported plumes then often reach the Earth's surface through atmospheric subsidence over another continent. Through this pathway, Asian Hg plumes can be transported relatively undiluted to the west coast of North America in about one week. These plumes are mainly emitted from large point sources that inject mercury to higher elevations due to high smokestack exit temperatures and subsequent plume rise (Friedli et al. 2004). Springtime has been found to be the most active period for trans-Pacific transport (e.g., Reidmiller et al. 2009; Durnford et al. 2010). At Reifel Island, B.C., a site on the west coast of Canada, Durnford et al. (2010) found that Asian mercury explains about 60 % of the variability in spring and summer, but less than 25 % of the variability in autumn and winter.

Jaffe et al. (2005) conducted a field campaign in spring 2004 to investigate the outflow of Asian mercury across the Pacific Ocean to North America. They simultaneously measured air concentrations of Hg^0 in the western Pacific at Hedo Station, Okinawa, Japan, which is often located downwind of major Hg emissions from China, and at the Mt. Bachelor Observatory in central Oregon, U.S. (see Fig. 4.14). The Mt. Bachelor Observatory is located at an elevation of ~2.7 km above sea level (ASL) and usually measures free-tropospheric air. During this period, the observed mean Hg^0 concentration at Hedo Station was found to be 2.04 ng m^{-3}, which is higher than the Northern Hemisphere background value of 1.5–1.8 ng m^{-3}, due to the impact of Asian outflows (Fig. 4.15a). Several episodes of enhanced Hg^0 concentrations were observed at Hedo Station with peak concentrations reaching 2.5 ng m^{-3}. The largest episode was observed around April 19 and lasted about 3 days with peak concentrations of 3.5 ng m^{-3}. They then analyzed the Hg^0 measurements at Mt. Bachelor to identify any episodes of

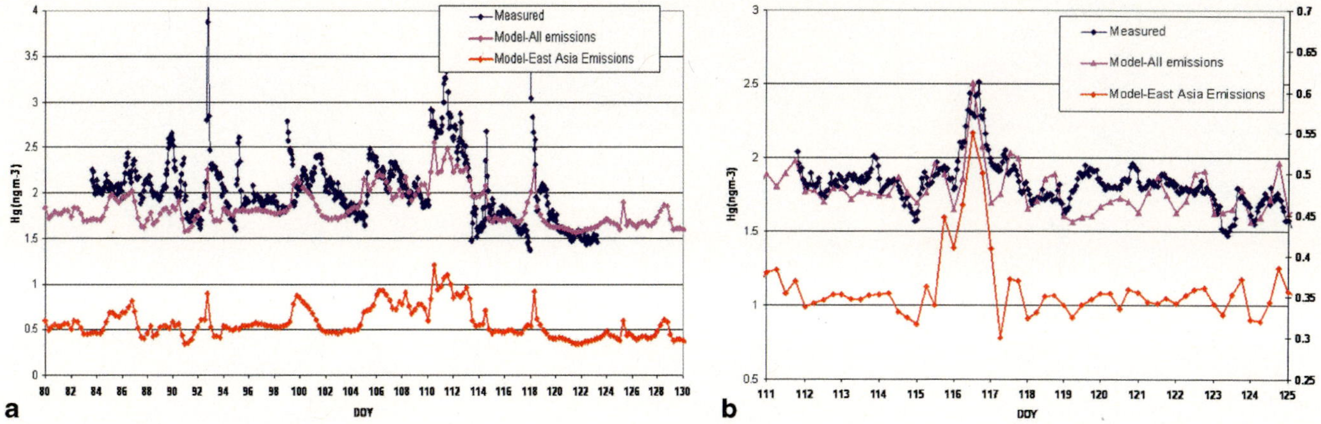

Fig. 4.15 Elemental mercury concentrations (ng m^{-3}) at (**a**) Hedo Station, Okinawa, Japan as observed (Jaffe et al. 2005; blue) and simulated by GRAHM using all global emissions (*purple*) vs. only East Asian emissions (*red*) and (**b**) Mt. Bachelor Observatory, Oregon, U.S. as observed (Jaffe et al. 2005; blue) and simulated by GRAHM using all global emissions (*purple*) vs. only East Asian emissions (*red*). Note that concentrations for the East Asian simulation use the scale on the right

long-range transport of Hg^0 that were also observed at the Hedo Station site. One large episode was observed at Mt. Bachelor around April 25, where the peak total Hg^0 concentrations reached ~2.5 ng m^{-3}, which is ~0.7 ng m^{-3} above the background value. The authors used observed ratios of Hg^0 to CO concentration as a chemical signature to confirm that the source of the episode was outflow from East Asia. The Hg^0/CO ratio at Mt. Bachelor on April 25 was found to be very similar to the Hg^0/CO ratio measured at Hedo Station on April 19. Using measured Hg^0/CO ratio and CO emissions estimates, they inferred mercury emissions from Asia to be double the anthropogenic Asian Hg emission estimates of Pacyna et al. (2010), suggesting that this global inventory underestimates Asian emissions of mercury.

GRAHM was applied to investigate the mechanism and the origin of the observed episode of mercury at Mt. Bachelor on April 25, 2004. At each model time step, mercury emissions were injected into the atmosphere, meteorological processes were simulated, and three atmospheric mercury species (gaseous Hg^0, gaseous Hg^{II} and particulate Hg^{II}) were transported, transformed chemically and deposited. The primary tropospheric gaseous oxidants of Hg^0 represented in GRAHM are O_3 and OH, whereas in the polar regions and the marine boundary layer, atomic bromine is the major oxidant and this reaction was also included in GRAHM. The representation of dry deposition of the three Hg species was based on the standard resistance approach (Zhang et al. 2003). Gaseous and particulate Hg^{II} species were also scavenged in GRAHM by hydrometeors both in and below clouds, but elemental Hg is insoluble and was not scavenged. The anthropogenic emissions of Hg included in GRAHM were based on the 2005 global anthropogenic mercury emissions inventory developed by Pacyna et al. (2010). Global emissions from natural sources and re-emissions of previously deposited mercury (from land and oceans) were based on the global Hg budgets of Mason and Sheu (2002). Seasonally-varying natural emissions were spatially distributed according to the natural enrichment of mercury in the Earth's crust at different locations, and re-emissions were spatially distributed according to historic Hg deposition patterns. In the presence of sunlight, a significant portion of deposited mercury in snowpacks is known to be rapidly reduced and revolatilized back to the atmosphere. A multi-layer snowpack and melt-water parameterization was used in GRAHM to treat Hg chemistry in snow and fluxes of Hg between air and snow (Durnford et al. 2012). The model was first integrated for several years to establish a steady state between the emissions, atmospheric mercury concentrations, and deposition. The performance of the GRAHM model has been evaluated previously using measured surface air concentrations of Hg^0 and wet deposition fluxes (HTAP 2010; Durnford et al. 2012).

A base model simulation was performed using global emissions from all sources from all regions. Four control simulations were also performed in which anthropogenic emissions from only one source region were included in each model run. The four major source regions considered were East Asia, Europe, Russia, and North America. Time series of Hg^0 concentrations at Mt. Bachelor Observatory from each of the control simulations were then compared to the time series from the base simulation to detect the region of origin of the April 25 episode at Mt. Bachelor. The spatial resolution of the model runs was inceased incrementally to simulate the intensity of the episode better. A model horizontal grid spacing of 0.25° × 0.25° latitude-longitude was found to be successful in reproducing the transport, timing, and magnitude of the observed episode at Mt. Bachelor with good accuracy.

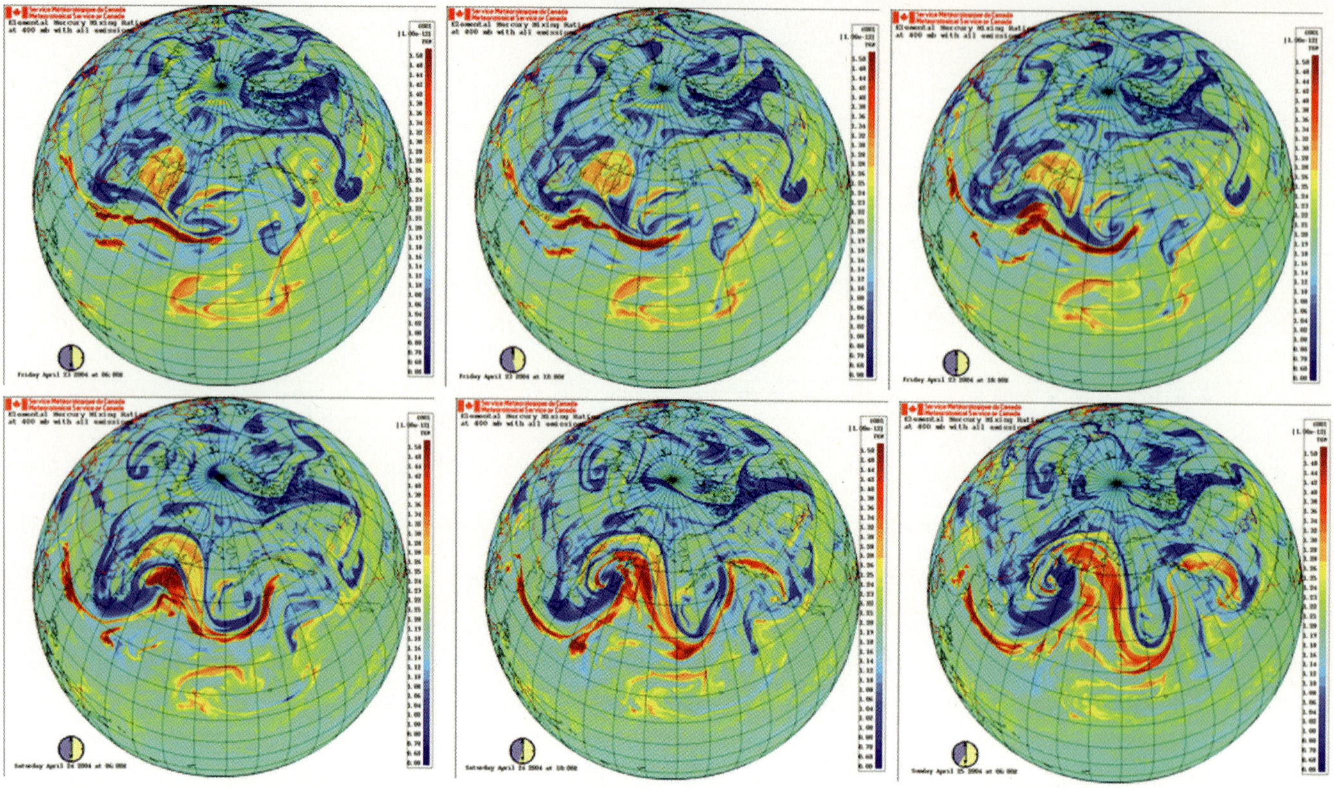

Fig. 4.16 Model-simulated transport of Hg^0 shown as instantaneous mass mixing ratios (ng kg^{-1}) at 500 hPa from April 23, 6 UTC to April 25, 18 UTC at 12-hour intervals (panels are ordered in time from left to right, top row first)

Figure 4.16 shows snapshots of Hg^0 mass mixing ratio distributions at ~500 hPa 12 hours apart from April 23, 0600 Universal Time Coordinated (UTC) to April 25, 1800 UTC. Most of the mercury transport was found to occur between 750–400 hPa and took approximately seven days to cross the Pacific Ocean and then descend over western North America through a deep anticyclonic system. The origin of the episode was confirmed by the control simulation that used only East Asian anthropogenic emissions. The mercury-rich air mass from East Asia reached northern British Columbia in Canada first and then traveled southward on the lee side of the Rockies, impacting western Canada and the western U.S. This is a significant pathway for the episodic transport of Hg from Asia into northern and western Canada as previously described by Durnford et al. (2010).

Time series of surface air concentrations of Hg^0 from the base simulation and the control simulation with only East Asian anthropogenic emissions along with the observed time series are presented in Fig. 4.15 for the Hedo Station and Mt. Bachelor Observatory sites. Excellent correlation between measured and simulated Hg^0 concentrations from the "all emissions" base run as well as the "East Asian emissions only" run suggests that Hg^0 concentrations at Hedo Station are strongly influenced by East Asian emissions. The model simulation shows that the Hg^0 episode on April 25, 2004 observed at Mt. Bachelor had earlier impacted the Hedo Station site around April 19, 2004. The enhancement of Hg^0 at the Hedo Station site is underpredicted by the model in both the base and control runs (~1 ng m^{-3} observed vs. ~0.5 ng m^{-3} in the model). These results are thus consistent with the conclusion in Jaffe et al. (2005) that East Asian Hg emissions are underestimated in the global emissions inventory developed by Pacyna et al. (2010).

Elevated Hg^0 concentrations at Mt. Bachelor are clearly seen around April 25 in Fig. 4.15b. The Hg^0 concentration is enhanced by ~0.25 ng m^{-3} above background in the "East Asian emissions only" control simulation compared to the observed enhancement of ~0.7 ng m^{-3}, consistent with the underprediction at Hedo Station. However, the model simulation using all emissions simulates the peak in line with the observations. The base simulation includes re-emissions of Hg from Asia in addition to anthropogenic emissions, leading to the conclusion that the contribution from the re-emissions of Hg from Asia compensates for the underprediction of the episode in the "all emissions" model simulation. The uncertainty in re-emission estimates is large; it is thus possible that this source is overestimated in the model. In conclusion, this case study provides both observational and modelling evidence for the trans-Pacific transport of mercury to Canada and points to East Asia as the source region of the emissions in this particular episode.

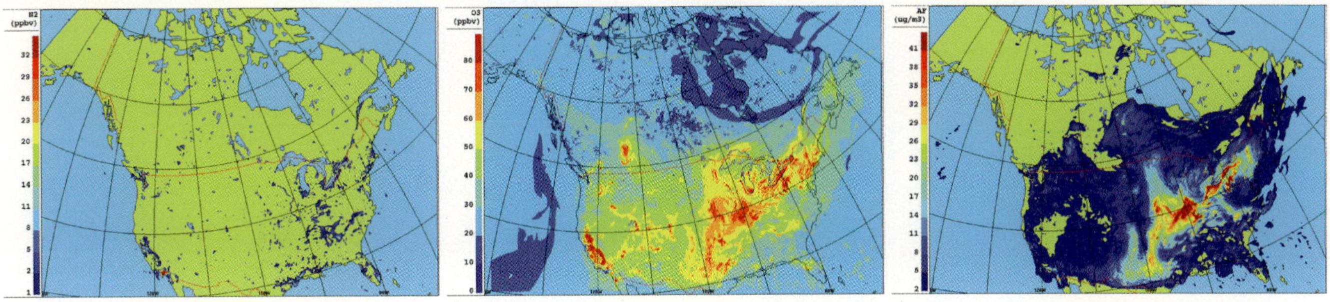

a b c

Fig. 4.17 Plots of instantaneous (**a**) NO_2, (**b**) O_3, and (**c**) $PM_{2.5}$ surface concentration fields predicted by the 10 km version of GEM-MACH at 4 p.m. EDT, 12 July 2012. Units are ppbv, ppbv, and μg m^{-3}, respectively

4.6.3 Chemical Weather Forecasting for Canada

By the late 1990s, computers had improved to the point where it started to become possible to apply regional AQ models in real time for the prediction of next-day air quality. This development opened up the possibility of routine AQ forecasts and public-health advisories and warnings analogous to the weather forecasts, advisories, and warnings that have been issued for decades (e.g., Kukkonen et al. 2012; Zhang et al. 2012b, c). The requirements for such routine AQ forecasts include the availability of a reliable numerical meteorological forecast, emissions fields that are up-to-date and day-specific, and computational resources that allow a multi-day AQ forecast to be performed in one or two hours, but if these requirements can be met, then real-time AQ forecasting provides a valuable tool for short-term AQ management and public guidance.

Environment Canada has been issuing national two-day forecasts of ozone since 2001 and of $PM_{2.5}$ and PM_{10} since 2003. The first operational AQ forecast model at EC was CHRONOS (Sect. 4.5.1; Pudykiewicz et al. 2003; McHenry et al. 2004; McKeen et al. 2005, 2007, 2009). This off-line regional AQ model was run operationally from 2001 to 2009 on a North American grid with 21 km horizontal grid spacing and 24 terrain-following vertical levels extending from the Earth's surface to 6 km above ground level (AGL). A 3600-s advective time step was used, and 48-hour forecasts were made once per day starting at 00 UTC.

Recently, as discussed in Sect. 4.5.2, an on-line multiscale chemistry model called GEM-MACH has been developed at EC (Talbot et al. 2008; Anselmo et al. 2010; Moran et al. 2010, 2011). In Nov. 2009 a limited-area version of GEM-MACH, GEM-MACH15, replaced CHRONOS as EC's operational regional AQ forecast model. In this initial implementation, GEM-MACH15 used a continental-scale North American rotated latitude-longitude grid with 15 km horizontal grid spacing and 58 vertical sigma-pressure hybrid levels spanning the atmosphere from the Earth's surface to 0.1 hPa. A 450-s meteorological time step and 900-s chemistry time step were selected. Another significant change from CHRONOS was that some additional process representations were included and several other process parameterizations were upgraded using process representations from the AURAMS CTM (Anselmo et al. 2010). The number of 48-hour forecasts was also increased to two per day, one beginning at 00 UTC and one beginning at 12 UTC. Results from a two-year performance evaluation of this version of GEM-MACH have been reported by Moran et al. (2011).

More recently, in Oct. 2012, a new operational version of GEM-MACH called GEM-MACH10 was implemented at EC with 10 km horizontal grid spacing and 80 vertical hybrid levels to 0.1 hPa (Moran et al. 2013). The meteorological time step was also decreased from 450 to 300 s for consistency with the reduced grid spacing. GEM-MACH10 uses hourly emissions input files that were generated by processing the 2006 Canadian national CAC emissions inventory, a 2012 projected U.S. CAC emissions inventory, and the 1999 Mexican CAC emissions inventory with the SMOKE emissions processing system (CEP 2012). The emissions files input by GEM-MACH10 contain 30 different chemical species; GEM-MACH10 in turn outputs predictions of the concentrations of 59 chemical species, both gas-phase species and PM chemical and size-fraction components. The three most important output species are NO_2, O_3, and $PM_{2.5}$ because these species are the three component species considered in the Canadian national Air Quality Health Index (AQHI: see Chap. 18).

Figure 4.17 shows an example of GEM-MACH hourly forecast fields on the new 10 km model grid for these three species. The afternoon NO_2 surface concentration field (Fig. 4.17a) is very patchy because only values greater than 1 ppbv are coloured. Most of the areas of high NO_2 concentration are associated with urban centres in Canada and the U.S. The highest NO_2 values at the time of this plot (4 p.m. EDT) are predicted to occur in Los Angeles, California. The O_3 spatial pattern (Fig. 4.17b) also displays large spatial variations but is somewhat smoother than the NO_2 pattern. The largest areas of elevated O_3 predicted for this

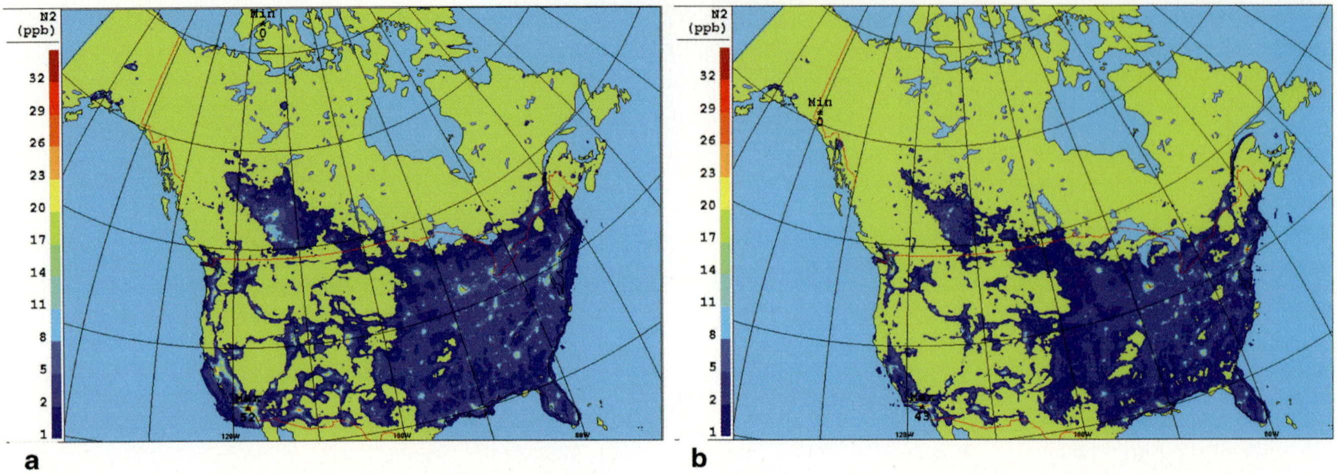

Fig. 4.18 Mean NO_2 surface concentration field (ppbv) predicted by the 15 km version of GEM-MACH for a two-month (**a**) winter 2012 period (1 Jan.–2 Mar.; 62 days) and (**b**) summer 2012 period (22 June–2 Sept.: 73 days)

time can be seen over the U.S. midwest, southern Great Lakes region, and northeastern U.S., with isolated areas of elevated O_3 over southern Alberta and southern California. This pattern also suggests that air with higher O_3 concentrations is being transported from the main North American landmass eastward to the Canadian Maritimes and the Atlantic Ocean. The $PM_{2.5}$ spatial pattern (Fig. 4.17c) is more like the O_3 pattern than the NO_2 pattern, consistent with $PM_{2.5}$'s dual nature as both a primary and secondary pollutant. In fact the broad-scale $PM_{2.5}$ and O_3 patterns display some noticeable similarities over the U.S. midwest and the southern Great Lakes.

It is clear from Fig. 4.17 that the GEM-MACH model domain covers most of North America and not just Canada. The reason for this is that it is important to include CAC emissions from the continental U.S. in the domain because of the importance on many days of long-range transport from the U.S. to Canada combined with the fact that U.S. CAC emissions are five to 10 times larger in total than Canadian CAC emissions (e.g., Moran and Makar 2001; Canada-U.S. Air Quality Committee 2004).

Figure 4.18 shows the predicted mean hourly NO_2 surface concentration field over North America for a winter 2012 period and a summer 2012 period. Note that only locations with NO_2 concentrations greater than 1 ppbv have been coloured. It is clear from this figure that NO_2 levels vary widely geographically and tend to be higher in the winter than the summer, consistent with the reduced vertical mixing near the Earth's surface expected in the cold season. Average fields like these two-month mean fields are smoother than instantaneous fields (i.e., "snapshots") due to the averaging of day-to-day meteorological fluctuations that result in NO_2 emissions being transported in different directions by different wind patterns (contrast Fig. 4.18b with the instantaneous NO_2 concentration field shown in Fig. 4.17a). Mean spatial concentration patterns tend to emphasize locations with high NO_2 emissions, with local maxima in the NO_2 concentration field (i.e., "hot spots") centred over these high emissions areas. As a consequence, large urban centres such as Vancouver, Calgary, Edmonton, Chicago, Detroit, Toronto, Montreal, and New York City are readily identifiable in this Fig. 4.6 (and compare this figure with the NO emissions plot shown in Fig. 4.6).

While Fig. 4.17 showed plots of surface pollutant concentration fields in space predicted by GEM-MACH at one instant in time, Fig. 4.19 shows two examples of predicted changes of ozone concentration in time over a 48-h period in the summer at a single location in space, in this case, a 10 km by 10 km surface grid cell located over midtown Montreal. For both of the two-day periods shown, the model predicted daytime O_3 peaks very well but significantly underpredicted nighttime minimum values. From the public-health perspective, though, maximum values are of greater interest than minimum values.

The last steps in the forecast chain are the dissemination and communication of forecasts. Once a GEM-MACH AQ forecast run has finished, the results are disseminated to AQ forecasters in Environment Canada and in some provinces and municipalities. As well, the AQ forecasts are made available to the public in several ways. One route is through the local broadcast media as AQHI forecasts. Another route is by the Internet. Daily national GEM-MACH forecasts of O_3 and $PM_{2.5}$ fields can be accessed by the public at the EC Weatheroffice website (see http://weather.gc.ca/aqfm/index_e.html), and local AQHI forecasts based on GEM-MACH forecasts can be accessed at the same website from the forecasts for individual cities (see http://weather.gc.ca/canada_e.html) or from the EC AQHI webpage http://www.ec.gc.ca/cas-aqhi/ (see also Chaps. 6 and 18).

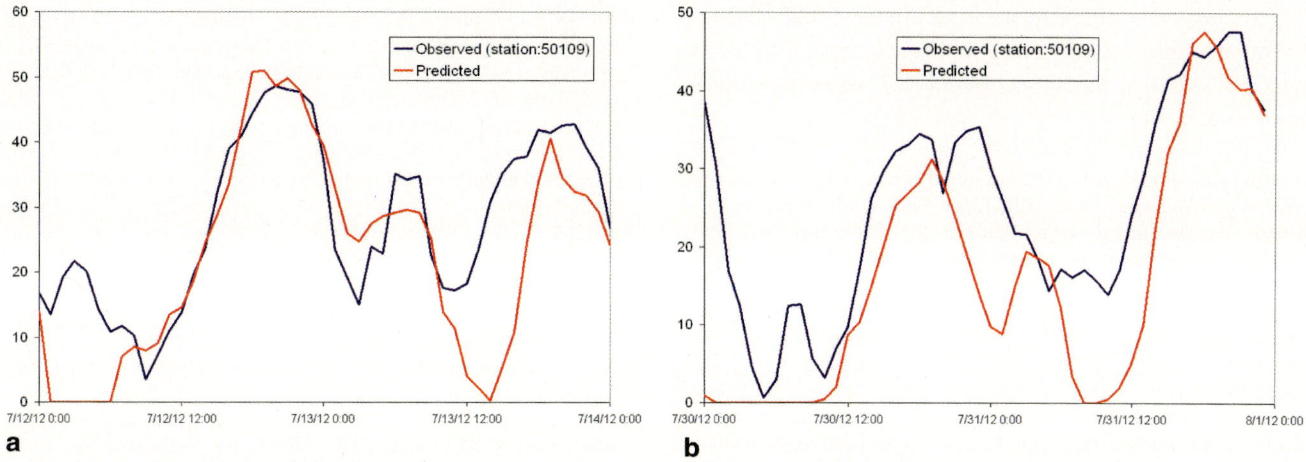

Fig. 4.19 Time series of hourly O_3 surface concentration (ppbv) for a grid box located over Montreal from the 00 UTC 48-hour forecast made by 10 km version of GEM-MACH (plotted in red) vs. the measurements from a surface monitor (no. 50109) located in the same grid box (plotted in blue) for two periods: (**a**) 12–13 July 2012; and (**b**) 30–31 July 2012. Time is in UTC units (where 12 UTC=08 EDT)

4.7 Summary

Air pollutants that are emitted directly to the atmosphere are called primary pollutants. Air pollutants that do not have emissions sources but instead are created by chemical reactions are called secondary pollutants. The long-range transport of air pollutants, where the term "long range" refers to distances of 100 km or more, has an important influence on air quality in many parts of Canada and a number of examples have been provided in this chapter. Long-range transport also involves a complex set of atmospheric processes. During long-range transport, meteorological conditions will vary and there will be sufficient time both for chemical transformations to occur, which may allow secondary pollutants such as ozone to form, and for the removal of pollutants by dry or wet processes. The chemical reactions may occur in clear air, inside water droplets in clouds or fog, and on aerosol particle surfaces.

Regional and global AQ models are the only tools available that can take into account both pollutant emissions and the complex set of atmospheric processes associated with long-range transport and then predict the pollutant concentration and deposition fields that will result. This capability allows AQ models to provide important information and guidance in support of AQ management, although model errors and uncertainties must also be considered. A long list of potential applications of AQ models to AQ management has been provided in this chapter.

The formulation and architecture of regional and global AQ models has also been described, including the differences between AQ and dispersion models, Eulerian and Lagrangian models, regional and global models, and off-line and on-line models. It was also noted that it is more accurate to talk about an AQ modelling system than an AQ model, where such a system consists of three primary components: an emissions component; a meteorological component; and an AQ component. Details about each of these components and their individual uncertainties were provided.

A summary of the different regional and global AQ models and their companion meteorological models and emissions processing systems that have been applied over the past three decades to support AQ management in Canada was also provided. The first of these were the following regional Lagrangian and Eulerian acid deposition models: ALSM, ALNM, RELAD, and ADOM. Then, as more attention began to be paid to photochemical smog, especially ozone and PM, new regional Eulerian AQ models were developed or adopted, including CHRONOS, AURAMS, CMAQ, and GEM-MACH. A few global AQ models have also been developed and employed in Canada to study such global pollutants as mercury, ozone, and persistent organic pollutants, including GRAHM, GEM/POPs, GEM-AQ, and GEM-MACH.

In order to characterize the health and environmental impacts of air pollutants, the output fields from AQ models are often processed to calculate additional statistical or chemical metrics. Some examples of these metrics, including exceedances of ambient AQ standards, of AQ indexes, and of acid-deposition critical loads and the calculation of population or crop exposures, have been provided.

Finally, three recent Canadian case studies were summarized to provide concrete examples of how AQ models can be applied to AQ management. The first case study described an application of the regional off-line AURAMS model to assess the potential impact on ambient AQ and human health if biodiesel fuel blends were to become widely used in Canada. The second case study described an application of the global on-line GRAHM model to simulate a spring 2004 episode of trans-Pacific transport of mercury emitted in East

Asia to western North America. And the third case study described an application of a regional configuration of the online GEM-MACH model to operational national-scale AQ forecasting in Canada.

Acknowledgements Ashu Dastoor would like to thank Daniel Jaffe, University of Washington-Bothell, for providing the 2004 Hg campaign measurement data and for helpful discussions. The authors would also like to acknowledge the many contributions to the three case studies by a number of colleagues, in particular Mathieu Rouleau of Health Canada and Didier Davignon, Jack Chen, Mehrez Samaali, Brett Taylor, and Marc Besner of EC for their contributions to the biodiesel study, Sylvain Ménard, Radenko Pavlovic, Hugo Landry, Sylvie Gravel, Alexander Kallaur, Samuel Gilbert, Jack Chen, Paul Makar, Wanmin Gong, Craig Stroud, Sunling Gong, Paul-André Beaulieu, and David Anselmo of EC for their help with the development and implementation of GEM-MACH, and Radenko Pavlovic for his help with the GEM-MACH figures. The careful review of a draft version of this chapter by Wanmin Gong of EC is also much appreciated.

References

AMAP (2011) AMAP Assessment 2011: Mercury in the Arctic. Arctic Monitoring and Assessment Programme (AMAP), Oslo, Norway, xiv + pp. 193

Anenberg SC, West JJ, Fiore AM, Jaffe DA, Prather MJ, Bergmann D, Cuvelier K, Dentener FJ, Duncan BN, Gauss M, Hess P, Jonson JE, Lupu A, Mackenzie IA, Marmer E, Park RJ, Sanderson MG, Schultz M, Shindell DT, Szopa S, Vivanco MG, Wild O, Zeng G (2009) Intercontinental impacts of ozone pollution on human mortality. Environ Sci Technol 43:6482–6487

Anselmo D, Moran MD, Ménard S, Bouchet V, Makar P, Gong W, Kallaur A, Beaulieu P-A, Landry H, Stroud C, Huang P, Gong S, Talbot D (2010) A new Canadian air quality forecast model: GEM-MACH15. Proc. 12th AMS Conf. on Atmos. Chem., Jan. 17–21, Atlanta, GA, American Meteorological Society, Boston, MA, 6 pp. (see http://ams.confex.com/ams/pdfpapers/165388.pdf)

Appel KW, Gilliland AB, Sarwar G, Gilliam RC (2007) Evaluation of the Community Multiscale Air Quality (CMAQ) model version 4.5: Sensitivities impacting model performance. Part I—Ozone. Atmos Environ 41:9603–9615

Appel KW, Bhave PV, Gilliland AB, Sarwar G, Roselle SJ (2008) Evaluation of the community multiscale air quality (CMAQ) model version 4.5: Sensitivities impacting model performance. Part II—particulate matter. Atmos Environ 42:6057–6066

Appel KW, Roselle SJ, Gilliam RC, Pleim JE (2010) Sensitivity of the Community Multiscale Air Quality (CMAQ) model v4.7 results for the eastern United States to MM5 and WRF meteorological drivers. Geosci Model Dev 3:169–188

Biester H, Bindler R, Martínez-Cortizas A, Engstrom DR (2007) Modeling the past atmospheric deposition of mercury using natural archives. Environ Sci Technol 41:4851–4860

Bouchet V, Buset K, Bloxam R, Carou S, Cousineau S, di Cenzo C, Davignon D, Farrell C, Jiang W, Kellerhals M, Makar PA, Morneau G, Pelletier S, Smyth S, Vingarzan R (2013) Model scenarios and applications. In: Canadian Smog Science Assessment. Vol. 1. Atmospheric Science and Environmental Effects. Environment Canada and Health Canada (Available upon request)

Brankov E, Henry RF, Civerolo KL, Hao W, Rao ST, Misra PK, Bloxam R, Reid N (2003) Assessing the effects of transboundary ozone pollution between Ontario, Canada and New York, USA. Environ Pollu 123:403–411

Brook JR, Lillyman CD, Shepherd MF, Mamedov A (2002) Regional transport and urban contributions to fine particle concentrations in southeastern Canada. J Air Waste Manage Assoc 52:855–866

Brook JR, Moran MD, Pennell W, Craig L (2008) Elements of air quality management: atmospheric science tools for developing effective policy. In: Gurjar BR, Molina LT, Ojha CSP (eds) Air pollution: health and environmental impacts. CRC Press, Boca Raton, Florida, pp 363–418

Brost RA (1988) The sensitivity of input parameters of atmospheric concentrations simulated by a regional chemical model. J Geophys Res 93:2371–2387

Byun D, Schere KL (2006) Review of the governing equations, computational algorithms, and other components of the Models-3 Community Multiscale Air Quality (CMAQ) modeling system. Appl Mech Rev 59:51–76

Canada-U.S. Air Quality Committee (2004) Canada-United States transboundary PM science assessment. EC Catalogue No. En56–203/2004E, ISBN 0-662-38678-7, Environment Canada, Toronto, pp. 129 (http://publications.gc.ca/collections/Collection/En56-203-2004E.pdf)

CCME (2003) Effects of ozone on vegetation: Update in support of the Canada-wide Standards for particulate matter and ozone. Report prepared for Canadian Council of Ministers of the Environment, Winnipeg, Manitoba, March, pp. 77 (http://www.ccme.ca)

CCME (2008) A national picture of acid deposition critical loads for forest soils in Canada. Publication PN 1412, Canadian Council of Ministers of the Environment, Winnipeg, Manitoba (www.ccme.ca/assets/pdf/national_picture_acid_deposition_pn1412.pdf)

CEP (2012) Sparse Matrix Operator Kernel Emission (SMOKE) modeling system. Carolina Environmental Program, University of North Carolina, Chapel Hill, North Carolina (see http://www.smoke-model.org/index.cfm)

Champoux L, Masse DC, Evers D, Lane OP, Plante M, Timmermans STA (2006) Assessment of mercury exposure and potential effects on common loons (Gavia immer) in Quebec. Hydrobiologia 567:263–274

Cheng L, Angle RP (1996) Model-calculated interannual variability of concentration, deposition and transboundary transport of anthropogenic sulphur and nitrogen in Alberta. Atmos Environ 30:4021–4030

Cheng L, Angle RP, Peake E, Sandhu HS (1995) Effective acidity modelling to establish acidic deposition objectives and manage emissions. Atmos Environ 29:383–392

Christensen JH, Brandt J, Frohn LM, Skov H (2004) Modelling of mercury in the Arctic with the Danish Eulerian Hemispheric Model. Atmos Chem Phys 4:2251–2257

Clark TL, Voldner EC, Dennis RL, Seilkop SK, Alvo M, Olson MP (1989) The evaluation of long-term sulfur deposition models. Atmos Environ 23:2267–2288

Clarkson TW, Magos L (2006) The toxicology of mercury and its chemical compounds. Critical Rev Toxicol 36:609–662

Côté J, Gravel S, Méthot A, Patoine A, Roch M, Staniforth A (1998a) The operational CMC/MRB Global Environmental Multiscale (GEM) model. Part 1: Design considerations and formulation. Mon Wea Rev 126:1373–1395

Côté J, Desmarais J-G, Gravel S, Méthot A, Patoine A, Roch M, Staniforth A (1998b) The operational CMC-MRB Global Environment Multiscale (GEM) model. Part II: Results. Mon Wea Rev 126:1397–1418

Dastoor AP, Larocque Y (2004) Global circulation of atmospheric mercury: a modelling study. Atmos Environ 38:147–161

Dastoor AP, Davignon D, Theys N, Roozendael MV, Steffen A, Ariya PA (2008) Modeling dynamic exchange of gaseous elemental mercury at polar sunrise. Environ Sci Technol 42:5183–5188

Dickson RJ, Oliver WR (1991) Emissions models for regional air quality studies. Environ Sci Technol 25:1533–1535

Dunker AM, Yarwood G, Ortmann JP, Wilson GM (2002) Comparison of source apportionment and source sensitivity of ozone in a three-dimensional air quality model. Environ Sci Technol 36:2953–2964

Durnford D, Dastoor A, Figueras-Nieto D, Ryjkov A (2010) Long range transport of mercury to the Arctic and across Canada. Atmos Chem Phys 10:6063–6086

Durnford D, Dastoor A, Ryzhkov A, Poissant L, Pilote M, Figueras-Nieto D (2012) How relevant is the deposition of mercury onto snowpacks? Part 2: a modeling study. Atmos Chem Phys 12:9251–9274. doi:10.5194/acp-12-9251-2012

Ellis JH (1988) Acid rain control strategies. Environ Sci Technol 22:1248–1255

Environment Canada (1998) 1997 Canadian acid rain assessment. Vol 2: atmospheric science assessment report. ISBN 0-662-2598-6, February, Downsview, pp. 296

Environment Canada (2010) Risk management strategy for mercury. Available from http://www.ec.gc.ca/mercure-mercury (Accessed Nov. 6 2012)

Fiore AM, Dentener FJ, Wild O, Cuvelier C, Schultz MG, Hess P, Textor C, Schulz M, Doherty RM, Horowitz LW, MacKenzie IA, Sanderson MG, Shindell DT, Stevenson DS, Szopa S, Van Dingenen R, Zeng G, Atherton C, Bergmann D, Bey I, Carmichael G, Collins WJ, Duncan BN, Faluvegi G, Folberth G, Gauss M, Gong S, Hauglustaine D, Holloway T, Isaksen ISA, Jacob DJ, Jonson JE, Kaminski JW, Keating TJ, Lupu A, Manner E, Montanaro V, Park RJ, Pitari G, Pringle KJ, Pyle JA, Schroeder S, Vivanco MG, Wind P, Wojcik G, Wu S, Zuber A (2009) Multimodel estimates of intercontinental source-receptor relationships for ozone pollution. J Geophys Res Atmos 114 D04301

Fox DG (1981) Judging air quality model performance. Bull Amer Meteorol Soc 65:599–609

Fox D, Kellerhals M (2007) Modelling of ozone levels in Alberta: base case, sectoral contributions and a future scenario. Environment Canada Internal Report, pp. 83 (Available from Air Quality Science Unit, Prairie and Northern Region, Environment Canada, Edmonton, Alberta)

Friedli HR, Radke LF, Prescott R, Li P, Woo J-H, Carmichael GR (2004) Mercury in the atmosphere around Japan, Korea, and China as observed during the 2001 ACE-Asia field campaign: Measurements, distributions, sources, and implications. J Geophys Res 109:D19S25

Fung CS, Misra PK, Bloxam R, Wong S (1991) A numerical experiment on the relative importance of H_2O_2 and O_3 in aqueous conversion of SO_2 to SO_4. Atmos Environ 25A:411–423

Gálvez O (2007) Synoptic-scale transport of ozone into southern Ontario. Atmos Environ 41:8579–8595

Gong S, Huang P, Zhao TL, Sahsuvar L, Barrie LA, Kaminski JW, Li Y-F, Niu T (2007) GEM/POPs: A global 3-D dynamic model for semi-volatile persistent organic pollutants—Part 1: model description and evaluations of air concentrations. Atmos Chem Phys 7:4001–4013

Gong S, Lavoué D, Zhao TL, Huang P, Kaminski JW (2012) GEM-AQ/EC, an on-line global multi-scale chemical weather modelling system: Model development and evaluation of global aerosol climatology. Atmos Chem Phys 12:8237–8256

Gong W, Farrell C, Makar PA, Ménard R, Moran MD, Morneau G, Stroud C (2013) Chemical transport models: model description and evaluation. In: canadian smog science assessment. Vol. 1. atmospheric science and environmental effects. Environment Canada and Health Canada (Available upon request)

Grell GA, Peckham SE, Schmitz R, McKeen SA (2004) Online versus offline air quality modeling on cloud-resolving scales. Geophys Res Lett 31:1–4

Hakami A, Odman MT, Russell AG (2003) High-order, direct sensitivity analysis of multidimensional air quality models. Environ Sci Technol 37:2442–2452

Haltiner GJ, Williams RT (1980) Numerical prediction and dynamic meteorology, 2nd edition. John Wiley & Sons, New York, pp 477

Health Canada (2012) Human health risk assessment for biodiesel production, distribution and use in Canada. Health Canada publications, Ottawa, Ontario, 227 pp + Apps

Holmes CD, Jacob DJ, Corbitt ES, Mao J, Yang X, Talbot R, Slemr F (2010) Global atmospheric model for mercury including oxidation by bromine atoms. Atmos Chem Phys 10:12037–12057

Houyoux MR, Vukovich M, Coats Jr CJ, Wheeler NJM (2000) Emission inventory development and processing for the Seasonal Model for Regional Air Quality (SMRAQ) project. J Geophys Res 105:9079–9090

HTAP (2010) Chapter 4: Global and Regional Modeling, Part B: Mercury, Air Pollution Studies No. 18, Hemispheric Transport of Air Pollution. Available from http://www.htap.org/

Huang P, Gong S, Zhao TL, Neary L, Barrie LA (2007) GEM/POPs: a global 3-D dynamic model for semi-volatile persistent organic pollutants—Part 2: Global transports and budgets of PCBs. Atmos Chem Phys 7:4015–4025

Jacobson MZ (1999) Fundamentals of atmospheric modeling. Cambridge University Press, Cambridge, pp. 656

Jaffe D, Prestbo E, Swartzendruber P, Weiss-Penzias P, Kato S, Takami A, Hatakeyama S, Yoshizumi K (2005) Export of atmospheric mercury from Asia. Atmos Environ 38:3029–3038

Jeffries DS, Ouimet R (2005) Critical loads: Are they being exceeded? Chapter 8, 2004 Canadian Acid Deposition Science Assessment, ISBN 0-662-68662-4, Environment Canada, Downsview, 341–369

Jiang W, Smyth S, Giroux E, Roth H, Yin D (2006) Differences between CMAQ fine mode particle and $PM_{2.5}$ concentrations and their impact on model performance evaluation in the lower Fraser valley. Atmos Environ 40:4973–4985

Kaminski JW, Neary L, Struzewska J, McConnell JC, Lupu A, Jarosz J, Toyota K, Gong S, Côté J, Liu X, Chance K, Richter A (2008) GEM-AQ, an on-line global multiscale chemical weather modelling system: Model description and evaluation of gas phase chemistry processes. Atmos Chem Phys 8:3255–3281, 2008.

Karamchandani PK, Venkatram A (1992) The role of non-precipitating clouds in producing ambient sulfate during summer: results from simulations with the Acid Deposition and Oxidant Model (ADOM). Atmos Environ 26A:1041–1052

Keeler GJ, Spengler JD, Koutrakis P, Allen GA, Raizenne M, Stern B (1990) Transported acid aerosols measured in southern Ontario. Atmos Environ 24A:2935–2950

Kelly J, Makar PA, Plummer DA (2012) Projections of mid-century summer air-quality for North America: Effects of changes in climate and precursor emissions. Atmos Chem Phys 12:5367–5390

Koo B, Wilson GM, Morris RE, Dunker AM, Yarwood G (2009) Comparison of source apportionment and sensitivity analysis in a particulate matter air quality model. Environ Sci Technol 43:6669–6675

Kukkonen J, Olsson T, Schultz DM, Baklanov A, Klein T, Miranda AI, Monteiro A, Hirtl M, Tarvainen V, Boy M, Peuch V-H, Poupkou A, Kioutsioukis I, Finardi S, Sofiev M, Sokhi R, Lehtinen KEJ, Karatzas K, San JR, Astitha M, Kallos G, Schaap M, Reimer E, Jakobs H, Eben K (2012) A review of operational, regional-scale, chemical weather forecasting models in Europe. Atmos Chem Phys 12:1–87

Lamborg CH, Fitzgerald WF, O'Donnell J, Torgersen T (2002) A non-steady-state compartment model of global-scale mercury biochemistry with inter-hemispheric atmospheric gradients. Geochim Cosmochim Ac 66:1105–1118

Leaitch WR, MacDonald AM, Anlauf KG, Liu PSK, Toom-Sauntry D, Li S-M, Liggio J, Hayden KL, Wasey M, Russell LM, Takahama S, Liu S, Van Donkelaar A, Duck T, Martin RV, Zhang Q, Sun Y, McKendry I, Shantz NC, Cubison M (2009) Evidence for Asian dust effects from aerosol plume measurements during INTEX-B 2006 near Whistler, BC. Atmos Chem Phys 9:3523–3546

Lee T-Y, Hong S-Y (2005) Physical parameterization in next-generation NWP models. Bull Amer Meteor Soc 86:1615–1618
Levy I, Makar PA, Sills D, Zhang J, Hayden KL, Mihele C, Narayan J, Moran MD, Sjostedt S, Brook J (2010) Unraveling the complex local-scale flows influencing ozone patterns in the southern Great Lakes of North America. Atmos Chem Phys 10:10895–10915
Lindberg S, Bullock R, Ebinghaus R, Engstrom D, Feng X, Fitzgerald W, Pirrone N, Prestbo E, Seigneur C (2007) A synthesis of progress and uncertainties in attributing the sources of mercury in deposition. Ambio 36:19–32
Makar PA, Moran MD, Scholtz MT, Taylor A (2003) Speciation of volatile organic compound emissions for regional air quality modelling of particulate matter and ozone. J Geophys Res 108:4041. doi:10.1029/2001JD000797
Makar PA, Moran MD, Zheng Q, Cousineau S, Sassi M, Duhamel A, Besner M, Davignon D, Crevier L-P, Bouchet VS (2009) Modelling the impacts of ammonia emissions reductions on North American air quality. Atmos Chem Phys 9:7183–7212
Mason RP, Sheu G-R (2002) The role of the ocean in the global mercury cycle. Global Biogeo Cycles 16:40–41, 10.1029/2001GB001440
McDonald KM, Cheng L, Olson MP, Angle R (1996) A comparison of box and plume model calculations for sulphur deposition and flux in Alberta, Canada. Atmos Environ 30:2969–2980
McHenry JN, Ryan WF, Seaman NL, Coats Jr CJ, Pudykiewicz J, Arunachalam S, Vukovich JM (2004) A real-time Eulerian photochemical model forecast system. Bull Amer Meteor Soc 85:525–548
McKeen S, Wilczak J, Grell G, Djalalova I, Peckham S, Hsie E-Y, Gong W, Bouchet V, Ménard S, Moffet R, McHenry J, McQueen J, Tang Y, Carmichael GR, Pagowski M, Chan A, Dye T (2005) Assessment of an ensemble of seven real-time ozone forecasts over Eastern North America during the summer of 2004. J Geophys Res 110, D21307, doi:10.1029/2005JD005858, pp. 16
McKeen S, Chung SH, Wilczak J, Grell G, Djalalova I, Peckham S, Gong W, Bouchet V, Moffet R, Tang Y, Carmichael GR, Mathur R, Yu S (2007) Evaluation of several real-time $PM_{2.5}$ forecast models using data collected during the ICARTT/NEAQS 2004 field study. J Geophys Res 112 D10S20, doi:10.1029/2006JD007608:20 pp
McKeen S, Grell G, Peckham S, Wilczak J, Djalalova I, Hsie E-Y, Frost G, Peischl J, Schwarz J, Spackman R, Holloway J, De Gouw J, Warneke C, Gong W, Bouchet V, Gaudreault S, Racine J, McHenry J, McQueen J, Lee P, Tang Y, Carmichael GR, Mathur R (2009) An evaluation of real-time air quality forecasts and their urban emissions over eastern Texas during the summer of 2006 Second Texas Air Quality Study field study. J Geophys Res D 114, D00F11, pp. 26
Moran MD (2000) Basic aspects of mesoscale atmospheric dispersion. In: Boybeyi Z (ed) Mesoscale atmospheric dispersion. WIT Press, Southampton, 27–119
Moran MD (2005) Current and proposed emission control programs: How will acid deposition be affected? Chapter 4, 2004 Canadian Acid Deposition Science Assessment, ISBN 0–662-68662-4, Environment Canada, Downsview, 99–162
Moran MD, Makar PA (2001) The nature of the source emissions. In: Precursor contributions to ambient fine particulate matter in Canada. Meteorological Service of Canada report, Catalogue No. En56-167/2001E, Environment Canada, Downsview, Ontario, 33–84
Moran MD, Scholtz MT, Slama CF, Dorkalam A, Taylor A, Ting NS, Davies D, Makar PA, Venkatesh S (1997) An overview of CEPS1.0: Version 1.0 of the Canadian Emissions Processing System for regional-scale air quality models. Proc. 7th AWMA Emission Inventory Symp., Oct. 28–30, Research Triangle Park, North Carolina, Air & Waste Management Association, Pittsburgh
Moran MD, Dastoor A, Gong S-L, Gong W, Makar PA (1998) Conceptual design for the AES regional particulate-matter model/unified air quality model. Unpublished report, October, Environment Canada, Downsview, Ontario, 100 pp
Moran MD, Zheng Q, Pavlovic R, Cousineau S, Bouchet VS, Sassi M, Makar PA, Gong W, Stroud C (2008) Predicted acid deposition critical-load exceedances across Canada from a one-year simulation with a regional particulate-matter model. 15th Conf on Applications of Air Pollution Meteorology, Jan. 21–24, New Orleans, American Meteorological Society, Boston, pp. 20 (Available from weblink http://ams.confex.com/ams/pdfpapers/132916.pdf)
Moran MD, Ménard S, Talbot D, Huang P, Makar P, Gong W, Landry H, Gong S, Gravel S, Crevier L-P, Kallaur A (2010) Particulate-matter forecasting with GEM-MACH15, a new Canadian operational air quality forecast model. In: Air Pollution Modeling and its Application XX, Steyn DG, Rao ST (eds) Springer, Dordrecht, 289–292
Moran MD, Ménard S, Pavlovic R, Landry H, Beaulieu P-A, Gilbert S, Chen J, Makar PA, Gong W, Stroud C, Kallaur A, Robichaud A, Gong S (2011) Two years of operational AQ forecasting with GEM-MACH15: A look back and a look ahead. 10th CMAS Conference, 24–26 Oct., Chapel Hill, North Carolina, 7 pp. (see http://www.cmascenter.org/conference/2011/abstracts/moran_two_years_2011.pdf)
Moran M, Menard S, Gravel S, Pavlovic R, Anselmo D (2013) RAQDPS version 1.5.0: A new version of the CMC operational Regional Air Quality Deterministic Prediction System. CMC technical note, Canadian Meteorological Centre, Dorval, Quebec, Oct. 23 pp. (http://collaboration.cmc.ec.gc.ca/cmc/cmoi/product_guide/docs/lib/op_systems/doc_opchanges/technote_rdaqps_20121003_e.pdf)
NARSTO (2005) Improving emission inventories for effective air quality management across North America: A NARSTO assessment. NARSTO-05-001, Oak Ridge, TN (Available from http://www.narsto.org/emission_inventory_1)
Noriega Y, Florian M, Morneau G (2007) Computing mobile emissions for the Montreal Area. In: Air Pollution XV, Wessex Institute of Technology, UK
OECD (2012) Organisation for Economic Co-operation and Development, http://stats.oecd.org/glossary/detail.asp?ID=1558 (Accessed Nov. 6 2012)
Olson MP, Oikawa KK (1989) Interannual variability of transboundary sulphur flux. Atmos Environ 23:333–340
Olson MP, Voldner EC, Oikawa KK (1982) A computed sulphur budget for the eastern Canadian provinces. Water, Air. Soil Pollu 18:139–155
Olson MP, Voldner EC, Oikawa KK (1983) Transfer matrices from the AES-LRT model. Atmos-Ocean 31:344–361
Olson MP, Bottenheim JW, Oikawa KK (1990) Nitrogen budget for eastern Canada. Atmos Environ 24A:897–901
Olson MP, Bottenheim JW, Oikawa KK (1992) Nitrogen source-receptor matrices and model results for eastern Canada. Atmos Environ 26A:2323–2340
O'Neill NT, Campanelli M, Lupu A, Thulasiraman S, Reid JS, Aubé M, Neary L, Kaminski JW, McConnell JC (2006) Evaluation of the GEM-AQ air quality model during the Québec smoke event of 2002: Analysis of extensive and intensive optical disparities. Atmos Environ 40:3737–3749
Ontario Ministry of the Environment (2005) Transboundary air pollution in Ontario. Report prepared by D Yap, N Reid, G De Brou, R Bloxam, June, pp. 132. Available from http://www.ene.gov.on.ca/environment/en/resources/STD01_076512.html
Pacyna EG, Pacyna JM, Sundseth K, Munthe J, Kindbom K, Wilson S, Steenhuisen F, Maxson P (2010) Global emission of mercury to the atmosphere from anthropogenic sources in 2005 and projections to 2020. Atmos Environ 44:2487–2499
Pan L, Woo J-H, Carmichael GR, Tang Y, Friedli HR, Radke LF (2006) Regional distribution and emissions of mercury in east Asia: A modeling analysis of Asian Pacific Regional Aerosol Characterization Experiment (ACE-Asia) observations. J Geophys Res 111:D07109
Pan L, Carmichael GR, Adhikary B, Tang Y, Streets D, Woo J-H, Friedli HR and Radke LF (2008) A regional analysis of the fate and transport of mercury in East Asia and an assessment of major uncertainties. Atmos Environ 42:1144–1159

Pan L, Lin C-J, Carmichael GR, Streets DG, Tang Y, Woo J-H, Shetty SK, Chu H-W, Ho TC, Friedli HR, Feng X (2010) Study of atmospheric mercury budget in East Asia using STEM-Hg modeling system. Sci Total Environ 408:3277–3291

Park SH, Gong S, Gong W, Makar PA, Moran MD, Stroud CA, Zhang J (2010) Relative impact of wind-blown dust vs. anthropogenic fugitive dust in $PM_{2.5}$ on air quality in North America. J Geophys Res 115 D16210, doi:10.1029/2009JD013144:13 pp

Pasquill F, Smith FB (1983) Atmospheric diffusion, 3rd edition. Ellis Horwood Ltd, Chichester, 437 pp

Peters LK, Berkowitz CM, Carmichael GR, Easter RC, Fairweather G, Ghan SJ, Hales JM, Leung LR, Pennell WR, Potra FA, Saylor RD, Tsang TT (1995) The current state and future direction of Eulerian models in simulating the tropospheric chemistry and transport of trace species: a review. Atmos Environ 29:189–222

Pielke RA, Uliasz M (1998) Use of meteorological models as input to regional and mesoscale air quality models – limitations and strengths. Atmos Environ 32:1455–1466

Pirrone N, Cinnirella S, Streets DG, Feng X, Mukherjee AB, Leaner J, Telmer K, Mason R, Friedli HR, Finkelman RB, Stracher G (2010) Global mercury emissions to the atmosphere from anthropogenic and natural sources. Atmos Chem Phys 10:5951–5964

Pouliot G, Pierce T, Denier van der Gon H, Schaap M, Moran M, Nopmongcol U (2012) Comparing emission inventories and model-ready emission datasets between Europe and North America for the AQMEII project. Atmos Environ 53:4–14

Pudykiewicz JA, Kallaur A, Smolarkiewicz PK (1997) Semi-Lagrangian modelling of tropospheric ozone. Tellus 49B:231–248

Pudykiewicz JA, Kallaur A, Moffet R, Bouchet VS, Jean M, Makar PA, Moran MD, Gong W, Venkatesh S (2003) Operational air quality forecasting in Canada: numerical model guidance for ground-level ozone and particulate matter. 5th Amer Meteor Soc Conf on Atmospheric Chemistry, Feb 9–13, Long Beach, California, pp. 6 (see https://ams.confex.com/ams/pdfpapers/54490.pdf)

Reff A, Bhave PV, Simon H, Pace TG, Pouliot GA, Mobley JD, Houyoux M (2009) Emissions inventory of $PM_{2.5}$ trace elements across the United States. Environ Sci Technol 43:5790–5796

Reidmiller DR, Fiore AM, Jaffe DA, Bergmann D, Cuvelier C, Dentener FJ, Duncan BN, Folberth G, Gauss M, Gong S, Hess P, Jonson JE, Keating T, Lupu A, Marmer E, Park R, Schultz MG, Shindell DT, Szopa S, Vivanco MG, Wild O, Zuber A (2009) The influence of foreign vs. North American emissions on surface ozone in the US. Atmos Chem Phys 9:5027–5042

Russell A, Dennis R (2000) NARSTO critical review of photochemical models and modeling. Atmos Environ 34:2283–2324

Samaali M, Moran MD, Bouchet VS, Pavlovic R, Cousineau S, Sassi M (2009) On the influence of chemical initial and boundary conditions on annual regional air quality model simulations for North America. Atmos Environ 43:4873–4885

Sanderson MG, Dentener FJ, Fiore AM, Cuvelier C, Keating TJ, Zuber A, Atherton CS, Bergmann DJ, Diehl T, Doherty RM, Duncan BN, Hess P, Horowitz LW, Jacob DJ, Jonson J-E, Kaminski JW, Lupu A, MacKenzie IA, Mancini E, Marmer E, Park R, Pitari G, Prather MJ, Pringle KJ, Schroeder S, Schultz MG, Shindell DT, Szopa S, Wild O, Wind P (2008) A multi-model study of the hemispheric transport and deposition of oxidised nitrogen. Geophys Res Lett 35. doi:10.1029/2008GL035389L17815

Schere K, Flemming J, Vautard R, Chemel C, Colette A, Hogrefe C, Bessagnet B, Meleux F, Mathur R, Roselle S, Hu R-M, Sokhi RS, Rao ST, Galmarini S (2012) Trace gas/aerosol boundary concentrations and their impacts on continental-scale AQMEII modeling domains. Atmos Environ 53:38–50

Scholtz MT, Taylor A, Ivanoff A, Moran MD, Davies D, Makar PA, Venkatesh S, Cheung P, Barton J (1999) Application of the Canadian Emissions Processing System, Version 1.0 (CEPS1.0): four case studies. Proc. 9th AWMA Emission Inventory Symp., AWMA Book VIP-93, Oct. 26–28, Raleigh, North Carolina, Air & Waste Management Association, Pittsburgh, 456–468

Seaman NL (2000) Meteorological modeling for air-quality assessments. Atmos Environ 34:2231–2259

Seigneur C, Moran MD (2004) Using models to estimate particle concentration. In: McMurry P, Shepherd M, Vickery J (eds) Particulate matter science for policy makers: A NARSTO assessment. Cambridge University Press, Cambridge, 283–323

Seinfeld JH, Pandis SN (2006) Atmospheric chemistry and physics—from air pollution to climate change, 2nd edition. John Wiley & Sons, New York, pp 1326

Shindell DT, Chin M, Dentener F, Doherty RM, Faluvegi G, Fiore AM, Hess P, Koch DM, MacKenzie IA, Sanderson MG, Schultz MG, Schulz M, Stevenson DS, Teich H, Textor C, Wild O, Bergmann DJ, Bey I, Bian H, Cuvelier C, Duncan BN, Folberth G, Horowitz LW, Jonson J, Kaminski JW, Marmer E, Park R, Pringle KJ, Schroeder S, Szopa S, Takemura T, Zeng G, Keating TJ, Zuber A (2008) A multi-model assessment of pollution transport to the Arctic. Atmos Chem Phys 8:5353–5372

Sirois A, Pudykiewicz JA, Kallaur A (1999) A comparison between simulated and observed ozone mixing ratios in eastern North America. J Geophys Res 104:21397–21423

Smyth S, Yin D, Roth H, Jiang W, Moran MD, Crevier L-P (2006a) The impact of GEM and MM5 meteorology on CMAQ air quality modeling results in eastern Canada and the northeastern United States. J Appl Meteorol 45:1525–1541

Smyth SC, Jiang W, Yin D, Roth H, Giroux É (2006b) Evaluation of CMAQ O_3 and $PM_{2.5}$ performance using Pacific 2001 measurement data. Atmos Environ 40:2735–2749

Smyth SC, Jiang W, Roth H, Moran MD, Makar PA, Yang F, Bouchet VS, Landry H (2009) A comparative performance evaluation of the AURAMS and CMAQ air quality modelling systems. Atmos Environ 43:1059–1070

Solazzo E, Bianconi R, Vautard R, Appel KW, Moran MD, Hogrefe C, Bessagnet B, Brandt J, Christensen JH, Chemel C, Coll I, Denier van der Gon H, Ferreira J, Forkel R, Francis XV, Grell G, Grossi P, Hansen AB, Jeričević A, Kraljević L, Miranda AI, Nopmongcol U, Pirovano G, Prank M, Riccio A, Sartelet KN, Schaap M, Silver JD, Sokhi RS, Vira J, Werhahn J, Wolke R, Yarwood G, Zhang J, Rao ST, Galmarini S (2012a) Model evaluation and ensemble modelling of surface-level ozone in Europe and North America in the context of AQMEII. Atmos Environ 53:60–74

Solazzo E, Bianconi R, Pirovano G, Matthias V, Vautard R, Moran MD, Appel KW, Bessagnet B, Brandt J, Christensen JH, Chemel C, Coll I, Ferreira J, Forkel R, Francis XV, Grell G, Grossi P, Hansen AB, Hogrefe C, Miranda AI, Nopmongcol U, Prank M, Sartelet KN, Schaap M, Silver JD, Sokhi RS, Vira J, Werhahn J, Wolke R, Yarwood G, Zhang J, Rao ST, Galmarini S (2012b) Operational model evaluation for particulate matter in Europe and North America in the context of AQMEII. Atmos Environ 53:75–92

Stroud CA, Morneau G, Makar PA, Moran MD, Gong W, Pabla B, Zhang J, Bouchet VS, Fox D, Venkatesh S (2008) OH-Reactivity of volatile organic compounds at urban and rural sites across Canada: evaluation of air quality model predictions using speciated VOC measurements. Atmos Environ 42:7746–7756

Summers PW, Fricke W (1989) Atmospheric decay distances and times for sulphur and nitrogen oxides estimated from air and precipitation monitoring in eastern Canada. Tellus B 41:286–295

Sunderland EM, Mason RP (2007) Human impacts on open ocean mercury concentrations. Global Biogeochem Cycles 21, GB4022

Talbot D, Moran MD, Bouchet V, Crevier L-P, Ménard S, Kallaur A, Team GEM-MACH (2008) Development of a new Canadian operational air quality forecast model. In: Air Pollution Modelling and its Application XIX, Borrego C, Miranda AI (eds) Springer, Dordrecht, 470–478

Tesche TW, McNally D (1991) A three-dimensional photochemical-aerosol model for episodic and long-term simulation: formulation and initial application in the Los Angeles Basin. In: Atmospheric Chemistry: Models and Predictions for Air Quality and Chemistry, Sloane CS, Tesche TW (eds) Lewis Publishers, Chelsea, Michigan, 63–82

Travnikov O (2005) Contribution of the intercontinental atmospheric transport to mercury pollution in the Northern Hemisphere. Atmos Environ 39:7541–7548

Trujillo-Ventura A, Ellis JH (1991) Multiobjective air pollution monitoring network design. Atmos Environ 25A:469–479

USEPA (2012) SPECIATE version 4.3 webpage, U.S. Environmental Protection Agency, Research Triangle Park, North Carolina (see http://www.epa.gov/ttn/chief/software/speciate/index.html)

Vautard R, Moran MD, Solazzo E, Gilliam RC, Matthias V, Bianconi R, Chemel C, Ferreira J, Geyer B, Hansen AB, Jericevic A, Prank M, Segers A, Silver JD, Werhahn J, Wolke R, Rao ST, Galmarini S (2012) Evaluation of the meteorological forcing used for the Air Quality Model Evaluation International Initiative (AQMEII) air quality simulations. Atmos Environ 53:15–37

Venkatram A, Karamchandani PK, Misra PK (1988) Testing a comprehensive acid deposition model. Atmos Environ 22:737–747

Vet R, Ro C-U (2008) Contribution of Canada–United States transboundary transport to wet deposition of sulphur and nitrogen oxides—A mass balance approach. Atmos Environ 42:2518–2529

Walker TW, Martin RV, Van Donkelaar A, Leaitch WR, MacDonald AM, Anlauf KG, Cohen RC, Bertram TH, Huey LG, Avery MA, Weinheimer AJ, Flocke FM, Tarasick DW, Thompson AM, Streets DG, Liu X (2010) Trans-Pacific transport of reactive nitrogen and ozone to Canada during spring. Atmos Chem Phys 10:8353–8372

Yap D, Ning DT, Dong W (1988) An assessment of source contributions to the ozone concentrations in southern Ontario, 1979–1985. Atmos Environ 22:1161–1168

Zeng Y, Hopke PK (1989) A study of the sources of acid precipitation in Ontario, Canada. Atmos Environ 23:1499–1509

Zhang Y (2008) Online-coupled meteorology and chemistry models: History, current status, and outlook. Atmos Chem Phys 8:2895–2932

Zhang L, Brook JR, Vet R (2003) A revised parameterization for gaseous dry deposition in air-quality models. Atmos Chem Phys 3:2607–2082

Zhang Y, Vijayaraghavan K, Seigneur C (2005) Evaluation of three probing techniques in a three-dimensional air quality model. J Geophys Res 110. D02305, doi:10.1029/2004JD005248

Zhang J, Zheng Q, Moran MD, Gordon M, Liggio J, Makar P, Stroud C, Taylor B (2012a) Improvements to SMOKE processing of Canadian on-road mobile emissions. 20th Emissions Inventory Conference, 13–16 Aug., Tampa, Florida, pp. 15 (http://www.epa.gov/ttn/chief/conference/ei20/session1/jzhang.pdf)

Zhang Y, Bocquet M, Mallet V, Seigneur C, Baklanov A (2012b) Real-time air quality forecasting 1. History, techniques, and current status. Atmos Environ 60:632–655

Zhang Y, Bocquet M, Mallet V, Seigneur C, Baklanov A (2012c) Real-time air quality forecasting 2. State of the science, current research needs, and future prospects. Atmos Environ 60:656–676

Modelling the Dispersion of Pollutants: Two Case Studies

5

Bob Humphries and Tyler Abel

Abstract

This Chapter will describe air quality modelling in and around an urban area and a mixed urban/rural area, using two case studies. One case study is a power plant that was proposed near Nanaimo, BC and the other case study is a major road development in the Metro Vancouver area. In the interest of public health and environmental protection, government agencies monitor current ambient air quality and regularly assess the impact that human activities can have on these interests. Any major project that will potentially impact the ambient air quality must be assessed before the project can go forward. Air quality models are one of the tools used for such assessments. The chapter begins with a general discussion of dispersion models, the types of models, where one can access approved models, and the input data requirements of dispersion models. This chapter includes excerpts from "Guidelines for Air Quality Dispersion Modelling in British Columbia" (BC MoE, http://www.elp.gov.bc.ca/epd/bcairquality/reports/air_disp_model_08.html, 2008) and "A Primer on the Guidelines for Air Quality Dispersion Modelling in British Columbia" (BC MoE, http://www.elp.gov.bc.ca/epd/bcairquality/reports/aq_disp_model_06_primer.html, 2006).

Keywords

Dispersion model · Screening dispersion model · Refined dispersion model · Advanced model · Point source · Area source · Volume source · CALPUFF · AERMOD · CALINE 3 · SCREEN3 · Dispersion model domain · Dispersion model receptors · Modelling acid deposition · Criteria air contaminants · Gaussian plume

5.1 What is a Dispersion Model?

Air quality dispersion modelling uses mathematical equations and numerical methods to describe the physical and chemical processes that affect air pollutants (gases or particles) as they react and move through the air after being emitted from sources such as industrial plants, vehicular traffic, sewage lagoons, etc. The equations are used to estimate or predict the downwind concentrations of the emitted pollutants, which provide useful information for the air permitting and environmental assessment process. In the past these equations were used in manual calculations but have now been converted into algorithms used in computer programs generally referred to as dispersion models.

There are many types of dispersions models available and many ways to apply them. The United States Environmental Protection Agency (US EPA) lists over two-dozen preferred or recommended models as well as many alternative models that may be used for assessing air emission sources. The US EPA Support Center for Regulatory Atmospheric Modeling (SCRAM) categorizes air quality models into three categories: dispersion models, photochemical models, and receptor models.

Dispersion models, which are the focus of this chapter, are typically used in the air permitting or environmental assessment process to estimate the concentration of pollutants at specified ground-level locations surrounding an

B. Humphries (✉) · T. Abel
Levelton Consultants Ltd., Richmond, Canada
e-mail: bhumphries@levelton.com

T. Abel
e-mail: tabel@levelton.com

E. Taylor, A. McMillan (eds.), *Air Quality Management*, DOI 10.1007/978-94-007-7557-2_5,

emissions source; referred to as receptors. Photochemical models are typically used in regulatory or policy assessments to simulate the impacts from all sources by estimating pollutant concentrations and deposition of both inert and chemically reactive pollutants over large spatial scales. Receptor models are observational techniques which use the chemical and physical characteristics of gases and particles measured at source and receptor to both identify the presence of and to quantify source contributions to receptor concentrations.

Each jurisdiction has its own set of preferred or recommended models, but most will follow the lead of the US EPA. All US EPA preferred/recommended and alternative models are available for download from their SCRAM web site[1]. A few of the models listed on the US EPA website are:

- AERMOD Modeling System—is a steady-state plume model that has some Gaussian plume characteristics, but also contains new or improved algorithms that incorporates air dispersion based on planetary boundary layer turbulence structure and scaling concepts, including treatment of both surface and elevated sources, and both simple and complex terrain.
- CALPUFF Modeling System—is a non-steady-state puff dispersion model that simulates the effects of time- and space-varying meteorological conditions on pollution transport, transformation, and removal. CALPUFF can be applied for long-range transport and for complex terrain.
- CALINE3—is a steady-state Gaussian dispersion model designed to determine air pollution concentrations at receptor locations downwind of highways located in relatively uncomplicated terrain.
- SCREEN3—is a single source Gaussian plume model which provides maximum ground-level concentrations for point, area, flare, and volume sources, as well as concentrations in the cavity zone, and concentrations due to inversion break-up and shoreline fumigation.
- Community Multi-scale Air Quality (CMAQ)—the EPA's CMAQ modeling system is supported by the Community Modeling and Analysis System (CMAS) Center. The CMAQ model includes state-of-the-science capabilities for conducting urban to regional scale simulations of multiple air quality issues, including tropospheric ozone, fine particles, toxics, acid deposition, and visibility degradation.

There is no one single model that is able to handle all geophysical, atmospheric and source situations as well as a range of applications. The available models can be roughly categorized as Screening, Refined and Advanced (BC MoE 2006).

- A *Screening* model provides a quick way to calculate and "flag" the "worst case" concentration that might possibly occur. It requires very little input as it has a built-in set of meteorological conditions and is relatively simple to use. This type of model is useful to identify those sources, which may require more focused effort (either for control actions, or more detailed modelling to confirm whether or not they are a concern).
- A *Refined* model includes more rigorous treatment of meteorology, dispersion, and atmospheric processes than a Screening model and therefore requires more input and expertise to run. It needs hourly measures of meteorology over a period (a year is common) that represent the conditions experienced by the emission and thus makes predictions that are site specific and more detailed. The output consists of concentration predictions for a range of time averages (typically 1 h to annual) at specified locations. These provide a rich dataset of information from which to understand the air quality impacts that reflect the meteorological conditions experienced at the emission source.
- An *Advanced* model includes comprehensive treatments of the physics and chemistry of emissions in the atmosphere, and thus requires considerable expertise and computer resources to set-up, run and interpret the results.

Of the models that are listed above, SCREEN3, as its name implies is a screening model; AERMOD, CALPUFF and CALINE3 are refined models; and CMAQ is an advanced model.

Detailed descriptions of dispersion models are provided in several of the references provided at the end of this chapter (Turner 1994; Zannetti 2003; Beychok 1994). However, it is useful to understand in simple terms what is meant by Gaussian, steady-state plume, and non-steady-state plume models.

The term Gaussian plume model assumes that a pollutant plume is carried downwind from its emission source by a mean wind and that concentrations in the plume can be approximated by assuming that the highest concentrations occur on the horizontal and vertical midlines of the plume, with the distribution about these midlines characterized by Gaussian or bell-shaped concentration profiles. Figure 5.1 show a Gaussian distribution, while Fig. 5.2 shows a plume where the concentration of pollutants varies according to a Gaussian distribution.

Gaussian-plume models assume steady-state conditions where the Gaussian-plume dispersion formulae do not depend on time, although they do represent an ensemble time average, typically an hour. The meteorological conditions are assumed to remain constant during the dispersion from source to receptor, which is effectively instantaneous. Emissions and meteorological conditions can vary from hour to hour but the model calculations in each hour are independent of those in other hours.

The assumptions of steady state used in models results in several limitations. First, steady state models assume that for each hour the plume extends instantaneously out to infinity. Concentrations may then be found at points too distant for

[1] US EPA Support Center for Regulatory Atmospheric Modeling (SCRAM) http://www.epa.gov/ttn/scram/. Accesssed May 2013.

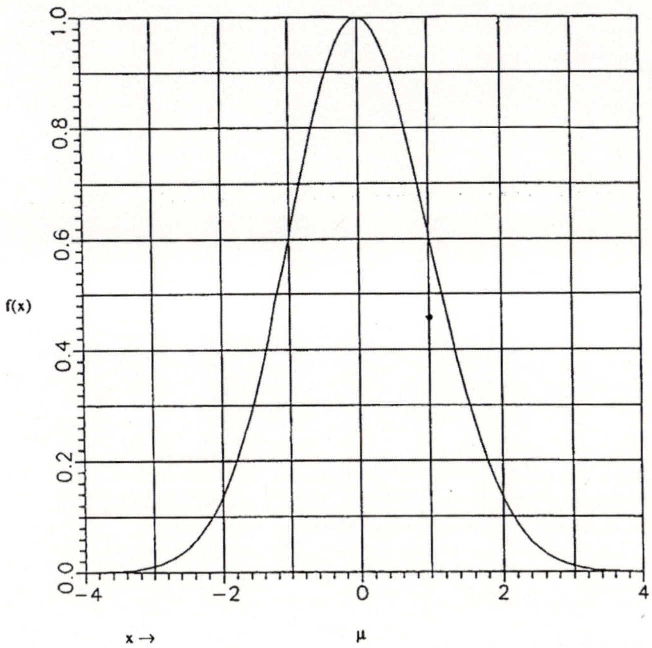

Fig. 5.1 A Gaussian distribution or normal distribution (Turner 1994)

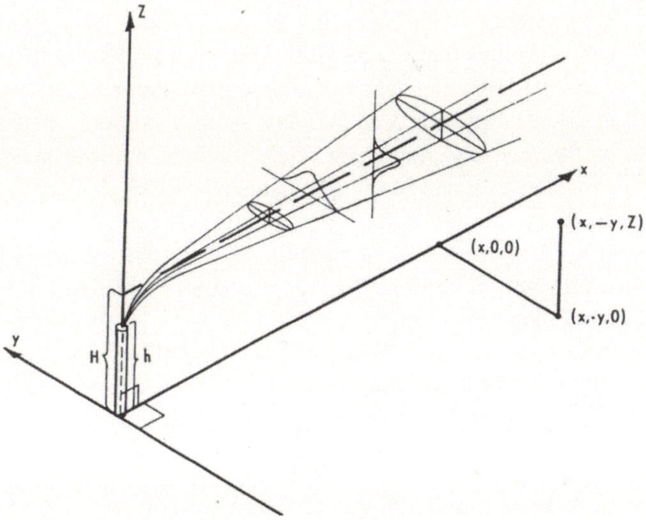

Fig. 5.2 Coordinate system showing Gaussian distributions of pollutant concentration in the horizontal and vertical

emitted pollutants to have reached them in an hour. Secondly, these models do not handle calm or light wind conditions. Consequently these models are mainly used to estimate pollutant concentrations within 50 km of the source and for wind conditions above 1 m/s. Finally, steady-state models assume straight-line plume trajectories with spatially uniform meteorological conditions. These assumptions combined with the independent treatment of each time step's, predicted concentration make steady-state models less than ideal for application in areas with complex terrain or multi-hour atmospheric processes (i.e. inversions and fumigation).

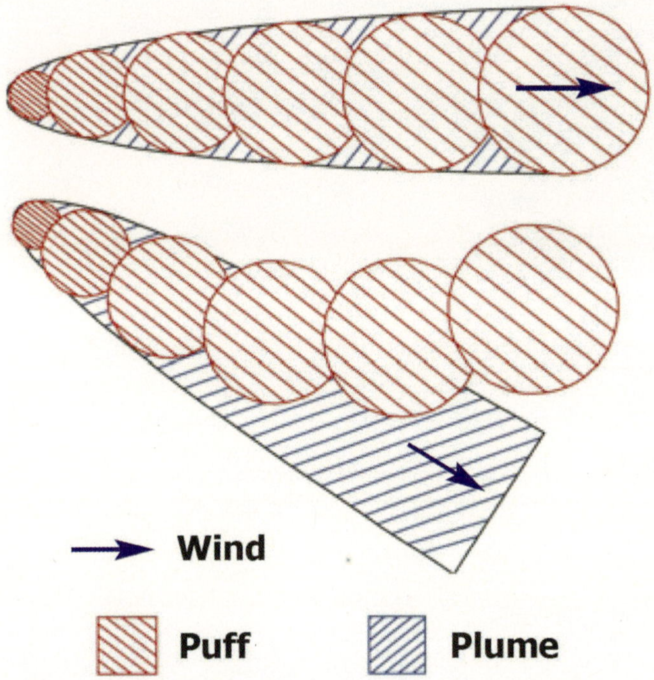

Fig. 5.3 A schematic depicting the tracking differences of a puff and a plume model (from Lakes Environment CALPUFF View brochure)

A non-steady-state puff dispersion model assumes that a continuously emitted plume or instantaneous cloud of pollutants can be simulated by the release of a series of puffs or discrete packets of pollutants that will be carried in a time- and space-varying wind field. The puffs are assumed to have Gaussian or bell-shaped concentration profiles in their vertical and horizontal planes'. The model keeps track of the movement and dispersion of each "puff" from hour-to-hour on a grid point system, allowing for pollutant transport and dispersion over multiple hours to be treated more realistically. In Fig. 5.3 a steady wind from the West shows that both the puffs (in red) and the steady state plume (in blue) lie along the same line. However when the wind shifts to come from the northwest the individual puffs shift with the wind but only the most recent puffs line up along the new wind direction.

5.2 Dispersion Model Inputs

Air dispersion models require emission information in order to predict downwind concentrations of the contaminant. It follows that uncertainties and errors in the emissions estimates will be reflected in the modelling results (i.e. the model will not magically correct errors or reduce uncertainty) so considerable effort is required to determine the following source information: source type (point, area, volume, line) and dimensions, physical height above ground where the emission occurs (stack, roof vent, road); exit temperature and speed of the effluent; and the mass emission rate.

Fig. 5.4 Examples of point sources showing a rendition of the proposed VIGP power plant plus the existing pulp mill at the right edge of the picture

Fig. 5.6 Fugitive dust from stockpile of sand. This is an example of a volume source because the source of the wind blown dust comes from the sides and the top of the stockpile

Fig. 5.5 A field of lava in Hawaii. The gaseous emissions would be treated as an area source

Source parameters, such as exit temperature, exit velocity and emission rate tend to vary over time. These variations may be due to normal process changes or abnormal conditions such as start-up, shutdown and upset conditions. Although the latter situation may occur very infrequently, the resulting emission rates could lead to serious air quality problems. These variations, whether under normal or abnormal operations, should be considered as part of the air quality assessment and modelled accordingly.

Emission Sources— can be simplified into four types based on geometry: point, area, volume, and line sources.

- A point source is a stationary, specific point of origin where contaminants are emitted into the atmosphere (such as a stack, exhaust vent, etc.). The source parameters normally required for point sources include the UTM or grid coordinates, release height (i.e., stack height), exit velocity, stack diameter, exit temperature, and mass emission rates of the contaminants of concern. A flare is considered to be a point source, but special treatment is required. See Fig. 5.4 for examples of point sources.
- An area source is an emission into the atmosphere that is distributed over a stationary spatial area. Examples of area sources include settling ponds, feedlots, and even urban regions that include multiple point sources (which combined together act as an area source). Parameters normally required for area sources include the coordinates of the area perimeter, the release height, and the mass emission flux rate of the pollutants of concern (i.e., mass emission rate per unit of area, g/s m^2). Figure 5.5 shows emissions from a lava field in Hawaii. This emission source would be treated as an area source for dispersion modelling.
- A volume source is an emission to the atmosphere that has an initial width and depth at a stationary release point. An example of a volume source is dust emission from an aggregate storage pile. Parameters normally required for volume sources include the coordinates and initial dimensions of the release and the mass emission rates. Figure 5.6 shows an example of fugitive dust emissions from a stockpile of sand.
- A line source is an emission to the atmosphere that is distributed over a line. Examples of line sources include conveyor belts, roadways, and rail lines. Parameters normally required for line sources include the dimensions of the line, release height, and the mass emission rates. Figure 5.7 is an example of a roadway line source.

Meteorological Data— should provide a dispersion model with a description of the atmospheric conditions that control the transport and mixing of the contaminant. Refined dispersion models require a time series of hourly sequential meteorological data. The ideal meteorological input data

Fig. 5.7 Although not visible, these automobiles and trucks are emitting pollutants. Such roadway emissions are treated as a line source

would come from a specially designed and sited meteorological collection program and would represent the conditions that determine the behaviour of the modelled emissions. However, such data is not always available, and sometimes not practical to collect. When this is the case, one may use:

- measurements at a different location that represent the key features of the conditions at the site,
- data produced by a prognostic, mesoscale meteorological model, or
- screening meteorological data that represents all possible realistic combinations of wind speed, stability class, and mixing heights. As such, the data may not reflect the actual meteorological conditions at the location of interest, and can only be used to model one-hour averages (unless averaging time adjustments are made for longer time-averages.

The Model Domain— is the area within which model predictions are made. The domain will generally be greater for large buoyant sources (e.g., tall stacks with buoyant emissions) where a domain of 50 by 50 km centred on the stack may be required for a flat terrain area. For shorter stacks, a smaller domain may be appropriate (e.g., 10 km by 10 km). The domain may be altered to include sensitive receptor areas such as a community or recreation area. The shape of the domain does not have to be square if there are features such as a valley, which could constrain the plume.

Receptors— are the locations within the model domain where the concentration/deposition predictions are calculated. Some models allow the user to select polar or Cartesian receptor grids. A polar grid is most applicable when evaluating a single source (i.e., a single stack). The polar grid is typically selected to correspond to 36 equally spaced radials with a 10-degree separation. A Cartesian grid is better for evaluating multiple source facilities, area sources or for evaluating multiple facilities.
Since the number of receptors impacts model run-times, modellers typically select receptors that are more densely located in areas where the maximum concentrations are expected. For example receptors may have 20 m along the plant boundary and the spacing could increase with distance from the source to 1,000 m beyond 5 km.

In some cases, it may be desirable to identify sensitive receptors that should be included as discrete receptors. Sensitive receptors may include individual residences, residential areas, schools, hospitals, campgrounds, parks, recreational areas, commercial day care and seniors' centres, sensitive ecosystems, etc.

Many models allow the user to select the receptor height, called flagpole receptors. Where ground level concentrations are of interest the receptor height is 0 m, but if the interest is in the concentration at the height of an average person, including children, the receptor height could be set to 1.5 m. Elevated receptors would be used for determining concentrations at treetop heights or at the heights of apartment building balconies.

Geophysical Data— is required when the core or preprocessing models require the geophysical setting of the source(s) to be defined. Depending on the model, this can be a simple switch to indicate whether the surrounding area is urban or rural, or can include detailed topographical and land use information with corresponding values of surface roughness, Albedo[2] and Bowen[3] ratios.

Existing Air Quality— dispersion models can predict the incremental change in concentrations due to a new source or a change to an existing source. There is always an existing level (or "background") due to sources not included in the model predictions. Some of these sources may be so far away or so diverse and complex that it is not possible to include them all in the model. Hence the existing background ambient concentration of pollutants of interest is needed to compare with or add to the modelled concentrations. If the background is high, new emissions may cause ambient concentrations to exceed regulatory objectives. In these areas, new emission sources may be required to implement more rigorous emission control technologies or techniques.

In cases where, background concentrations of pollutants of interest are not well known, there may be a requirement

[2] Albedo is the fraction of solar energy (shortwave radiation) reflected from the Earth back into space.

[3] Bowen ratio for any moist surface is the ratio of heat energy used for sensible heating (conduction and convection) to the heat energy used for latent heating (evaporation of water or sublimation of snow).

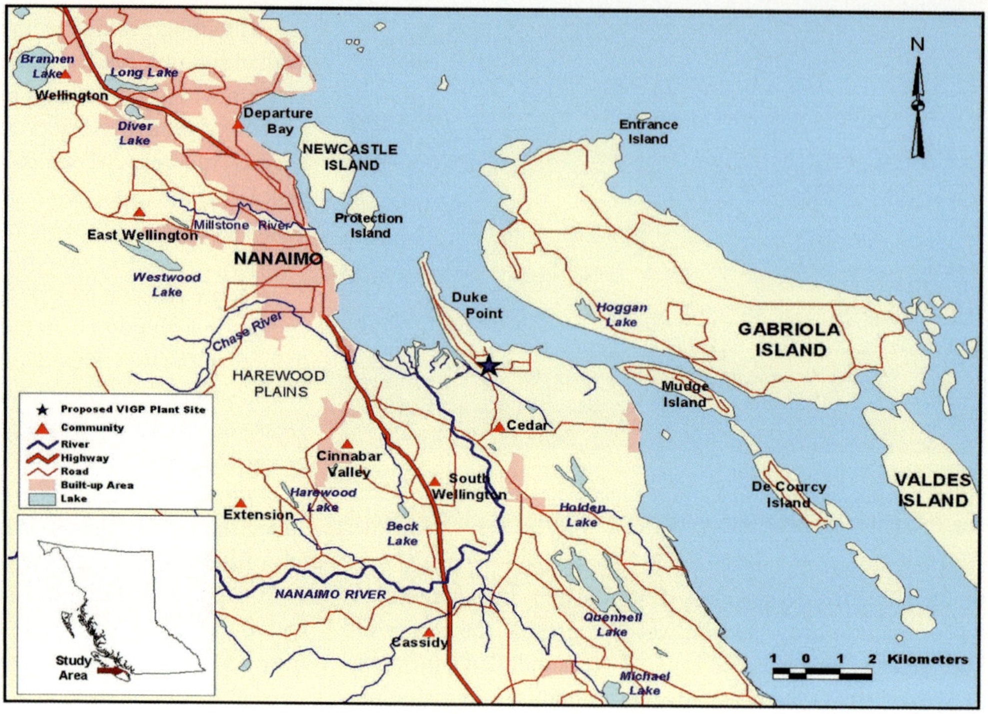

Fig. 5.8 General location of the proposed VIGP plant site

to install ambient monitoring equipment to obtain measurements of ambient concentrations.

In order to understand the application of dispersion models, two cases studies are provided. Each case has its own unique modelling requirements.

5.3 Case Study 1—Use of CALPUFF to Assess Air Quality Impacts from a Proposed Power Plant

On November 27, 2003 the government of British Columbia issued an Environmental Assessment Certificate for the Vancouver Island Generation Project (VIGP) to be built at Duke Point near Nanaimo, British Columbia. Although the project received the required approvals, certificates and permits, the facility has not been developed. However, the air quality assessment of the project serves as a good example of how air dispersion models are used in the environmental assessment process. The assessment included predictions of future ambient concentrations of emitted pollutants, determination of potential fog events resulting from emissions of water vapour, assessment of acidic deposition and determination of visibility impacts from the plant's emitted plume.

5.3.1 The Project

Vancouver Island Energy Corporation (VIEC), a wholly owned subsidiary of BC Hydro, had proposed to develop a $ 370 million power generation project based on advanced, combined-cycle gas turbine technology capable of producing nominally 265 MW (increased to 295 MW with duct-firing) of electricity for delivery to the BC Hydro power grid. The plant was to use state of the art technology and would have been fuelled with natural gas to generate electricity at high efficiency and with low emissions. The site proposed for the power plant was 10.1 ha in area and located at Duke Point near Nanaimo, BC (Fig. 5.8). Figure 5.4 shows a rendition of the VIGP facility.

The major air emission related components of the power plant (Fig. 5.9) were to consist of:

- a GE 7FA natural gas fired combustion turbine generator equipped with dry low NO_x (and low CO) emission combustors;
- a heat recovery steam generator (HRSG) with duct burners that would provide additional heat input and steam generation during peak electricity demand periods;
- a selective catalytic reduction (SCR) unit with an aqueous ammonia storage and handling system to reduce NO_x emissions to the atmosphere;
- a single condensing steam turbine generator (STG); and
- a surface condenser, cooling water circuit, and wet cooling tower to condense the exhaust steam from the STG and vent this heat to the atmosphere.

5.3.2 Assessment Tasks

The first step in an air quality assessment is to meet with the appropriate regulatory agencies and agree on a terms of reference (TOR). In this case meetings were held with the BC

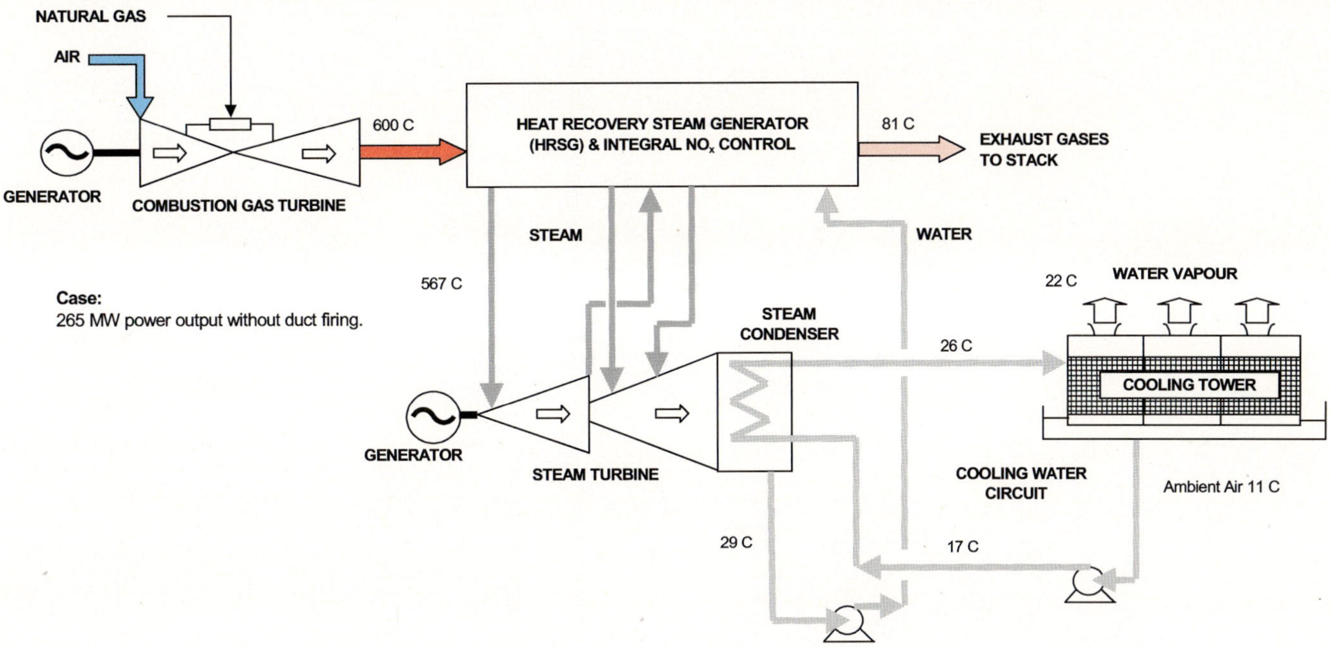

Fig. 5.9 Simplified process schematic diagram and typical stream Temperatures for the Vancouver Island generation project

Environmental Assessment Office (BC EAO) and the BC Ministry of Environment (BC MoE). All related documentation can be found on the BC EAO web site (http://www.eao.gov.bc.ca). The following key points, which impact the dispersion modelling, were contained within the agreed upon TOR:

- Carry out baseline studies to determine the existing background ambient air quality concentrations for $PM_{2.5}$, PM_{10}, O_3, and NO_2. In addition to the existing ambient air quality monitoring stations operated by the BC MoE, VIGP installed two additional monitoring stations in the Nanaimo area to provide additional observations of PM_{10}. O_3 and NO_2. Locations and pollutants monitored at these stations are presented in Tables 5.1 and 5.2 and shown in Fig. 5.10.
- Examine available meteorological and climatological data to ensure the meteorological data used in the dispersion model represented the climatology of the region.
- Identify major point sources of emissions in the airshed.
- Estimate emissions of concern from existing point sources.
- Model air quality impacts for two assessment areas. Assess local impacts for the Nanaimo region using a square area of about 50 km per side centred on the VIGP. Assess long-range air quality impacts in the southern Georgia Basin using an area of 80 km (N-S) by 104 km (E-W) centred over the southern tip of Gabriola Island.
- Compare predicted concentrations with federal and provincial ambient air quality objectives and determine the acute and chronic risks to human health.
- Determine the impact of acidic deposition caused by emissions of SO_2 and NO_x (oxides of nitrogen).
- Determine the characteristics of the brown plume that would be formed by secondary particulate matter ($(NH_4)_2SO_4$ and NH_4NO_3) from precursor emissions of NO_x and SO_x, and primary emissions of $PM_{2.5}$.
- Because of the complexity of the terrain in the region, the requirement to model out to the mainland, Vancouver and the International border plus the influence of the ocean and the requirement to model acid deposition, the CALPUFF model was selected for this project (Figs. 5.11 and 5.12). Note the CALPUFF modelling domain was chosen to be slightly smaller than the CALMET modelling domain, to ensure the elimination of any numerical "edge" effects.
- The Fog model was selected for modeling water vapour emissions. The algorithms in the Fog model have been incorporated in recent versions of CALPUFF.
- Model the impact of the water vapour plume from the cooling tower on the local highway to the ferry terminal.

5.3.3 Modelling Methodology for VIGP

Meteorology and Climatology Meteorology is one of the most important and complex data sets required by the CALPUFF model. The meteorological information required by CALPUFF is provided by CALMET, which is a diagnostic computer model that produces detailed three-dimensional fields of meteorological parameters based on surface, marine

Table 5.1 BC Government ambient air monitoring station locations and pollutants monitored within the long range transport domain

Station Information					Pollutant Monitoring Period					
Station Name	Station Number	Location	Latitude	Longitude	PM_{10}	$PM_{2.5}$	O_3	NO_2	CO	SO_2
Harmac Cedar Woobank	E225377	1624 Woobank Road	49 06 78	123 50 96	7/97–07/02	–	–	–	–	–
Nanaimo Labieux Road	E229797	2080A Labieux Road	49 12 03	123 59 38	1/98–9/98	12/97–07/02	12/97–07/02	–	–	–
Duncan Deykin Avenue	E234670	6364 Deykin Ave	48 48 10	123 38 42	11/98–07/02	–	–	–	–	–
Duncan Koksilah	E242620	5300 Island Highway	48 45 29	123 41 20	–	–	9/00–12/01	–	–	–
Duncan Mobile	E223361	5700 Menzies Road	48 54 46	123 46 53	7/96–10/96	–	7/96–10/96	7/96–10/96	–	–
Duncan Mobile Transfer Station	E242166	3900 Drinkwater Road	48 48 13	123 41 03	7/00–5/01	7/00–5/01	7/00–5/01	7/00–5/01	7/00–5/01	–
Victoria Topaz	E231866	923 Topaz	48 25 45	123 21 30	–	5/98–07/02	5/98–07/02	5/98–07/02	5/98–07/02	5/98–07/02
Vancouver Kitsilano	0310175	2550 West 10th Ave	49 15 40	123 09 49	12/93–07/02	–	8/80–07/02	8/80–07/02	5/82–07/02	8/80–07/02
Vancouver International Airport#2	E232246	3153 Templeton St	49 11 11	123 09 09	2/98–07/02	3/99–07/02	2/98–07/02	2/98–07/02	2/98–07/02	–
Richmond South	E207417	Williams and Aragon Road	49 08 30	123 06 29	10/93–07/02	–	7/86–07/02	7/86–07/02	7/86–07/02	7/86–07/02

Table 5.2 VIGP station locations and pollutants monitored in the Nanaimo Area

Station Information				Pollutant Monitoring Period					
Station Name	Location	Latitude	Longitude	PM_{10}	$PM_{2.5}$	O_3	NO_2	CO	SO_2
Harmac-Cedar (HC)	1624 Woobank Road	49 06 78	123 50 96	–	–	2/02–07/02	2/02–07/02	–	–
Department of National Defence	Nanaimo Lakes Road and Nanaimo Parkway	49 09 03	123 57 55	2/02–07/02	–	2/02–07/02	2/02–07/02	–	–

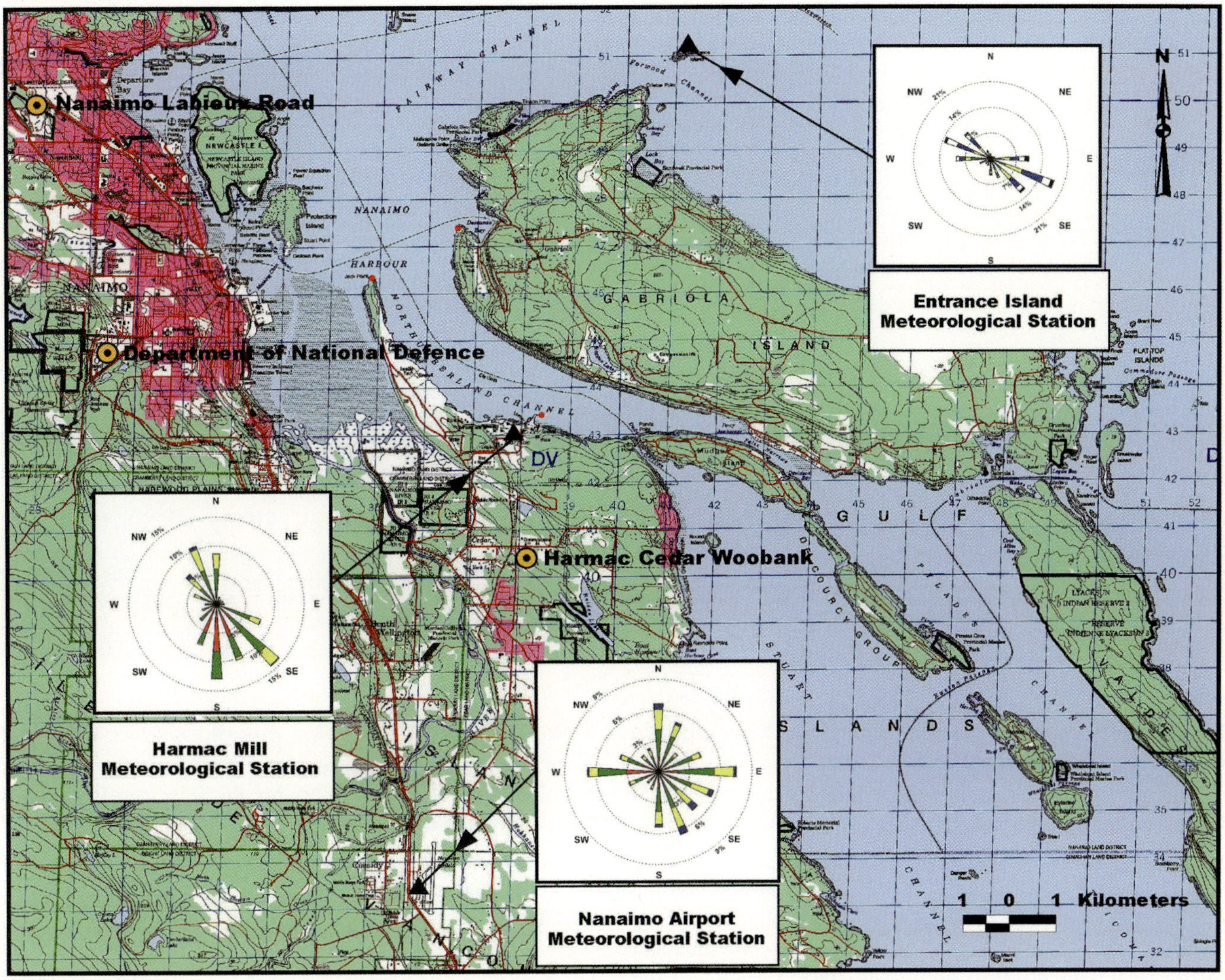

Fig. 5.10 Location of air quality and meteorological monitoring stations in the study area (VIGP Stations note by *orange circles*)

(optional) and upper air weather measurements, digital land use data, terrain data, and prognostic meteorological data (optional). Figure 5.13 shows the location of the surface and marine based meteorological stations that were used in this study. In this application, CALMET was executed for two entire model years, using meteorological input data from five surface stations, two upper air stations, and one marine buoy station. Digital terrain and land use data covering the model domain were included in the CALMET input data set. In addition, three-dimensional prognostic meteorological data produced by two different mesoscale meteorological models were used to improve the performance of the CALMET model.

The British Columbia and Alberta Governments commissioned a project where the fifth generation NCAR/Penn State Mesoscale Model (MM5) was executed at a 20-km horizontal grid resolution for Western Canada for the entire year of 1995. However, at a 20-km grid resolution, it is difficult for

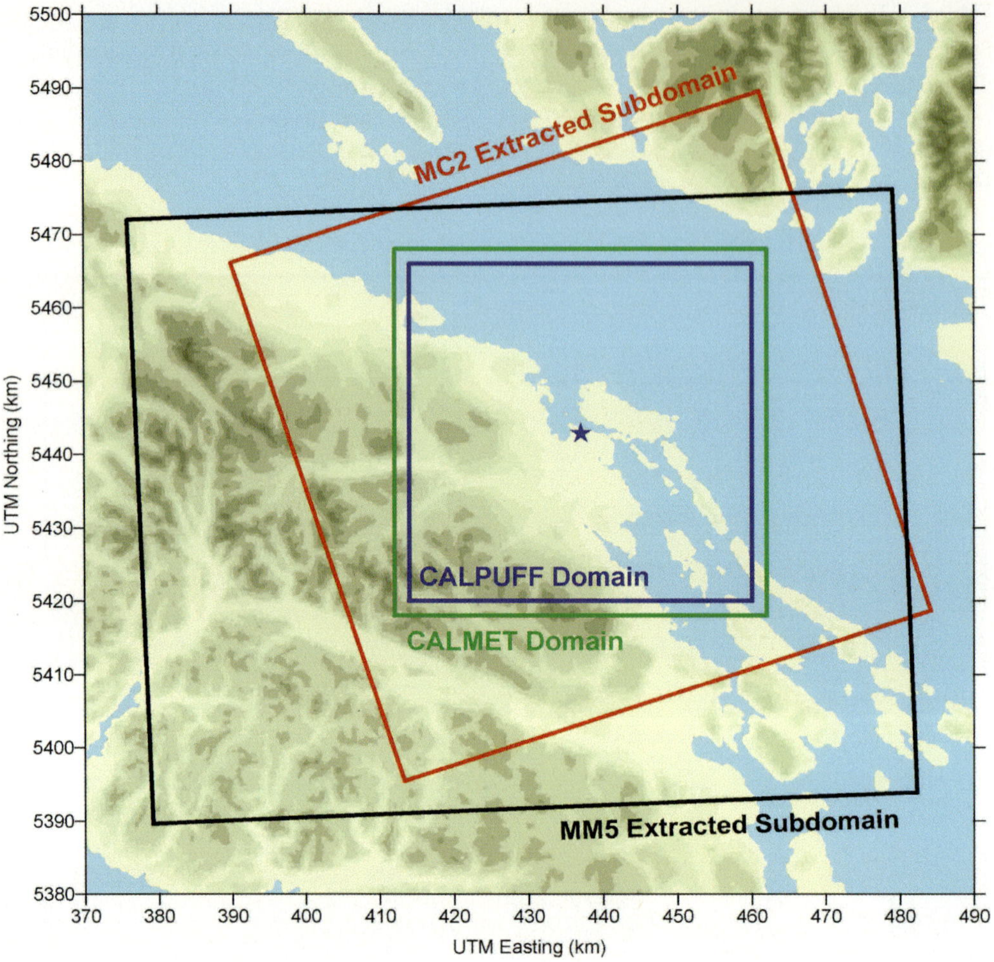

Fig. 5.11 Regional topographical map displaying the model domain boundaries. (The star shows the location of the VIGP)

the prognostic model to capture the complex topography in the Georgia basin. It was therefore decided, for the purposes of this application, to augment the prognostic meteorological input data set by generating another 12 months of prognostic data at a higher grid resolution. The University of British Columbia (UBC) was commissioned to run the MC2 model at a 3.3-km resolution for a one-year period starting in July 2000. Hence, the two model years that were used in the local impacts assessment: 1995 and 2000–2001. Because of the MC2 prognostic data set's better resolution, long range modelling only used the 2000–2001 data set.

A comparison of the meteorology in the region with the climate normals from Nanaimo Airport showed that the 1995 and 2000–2001 data from the four surface stations are climatologically representative of the region. For example, Figs. 5.14 and 5.15 show the temperature normals and the temperature observations for 1995 and the 2000–2001 period for the Nanaimo Airport. The Meteorological Service of Canada (MSC) provided the data. The temperature trends and values for 1995 and 2000–2001 are similar to the 30-year normals.

Note in Fig. 5.10 the wind rose plots, which show the frequency of the time that the wind is blowing from a given direction. The wind rose plots for the Entrance Island and Harmac Mill stations show the influence of the Strait of Georgia, which forces winds to align up and down the strait. The Nanaimo Airport station is further inland and is clearly not affected as much by the strait.

5.3.4 Air Quality Baseline for the Power Plant

Ambient Air Quality The existing air quality was determined by analyzing the data from the BC MoE monitoring stations as well as the two stations that VIGP installed. Tables 5.1 and 5.2 summarize the pollutants monitored and the time period that was used in the analysis. Analysis of the data indicated that at the time of the assessment (2002) air quality in Nanaimo was relatively good, with the frequency of exceeding various objectives and standards ranging from low to none. Nanaimo air quality is typical of the east coast of Vancouver Island, and cleaner than larger urban centres

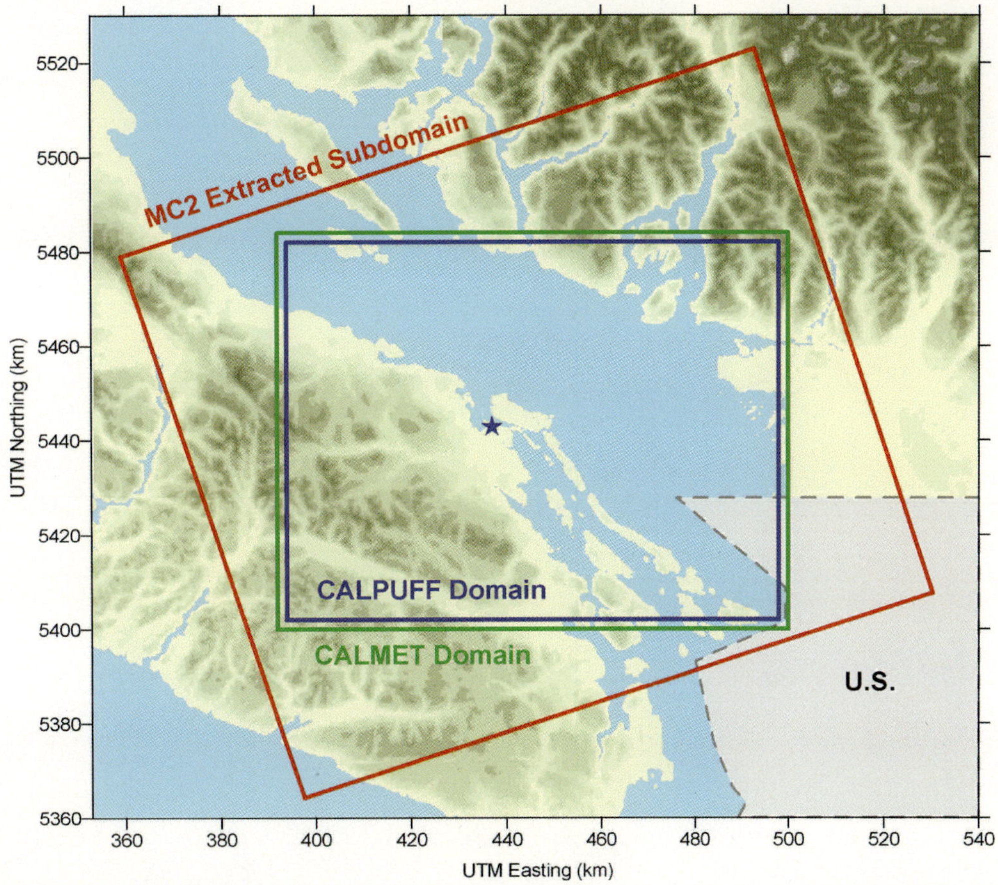

Fig. 5.12 Topographical map displaying the long-range model domain boundaries. (The star shows the location of the VIGP)

with respect to particulate matter. It did however experience higher ozone levels than Victoria and Vancouver, which may be an indirect result of having low NO_x concentrations, as NO_x can scavenge ozone from the air. The baseline levels were based on the 98th percentiles for 1-h and 24-h values and the average for the annual values:

- PM_{10}—24-h value was 23 μg/m^3 and annual average was 11 μg/m^3,
- $PM_{2.5}$—24-h value was 12 μg/m^3 and annual average was 4.4 μg/m^3,
- NO_2—1-hourly value was 35 μg/m^3, 24-h value was 19 μg/m^3, and period average was 11 μg/m^3.

The baseline, referred to as the background ambient air quality concentrations for each pollutant was added to the modelling results as part of the assessment. The ambient air quality objectives of interest for the two case studies are shown in Table 5.3.

In addition to analysis of the existing air quality data and the emission inventory, emissions from the existing Harmac mill were modelled with CALPUFF. The mill was included because it was a significant nearby emission source. By modelling both facilities the cumulative air quality impacts of the existing mill and the proposed power plant could be evaluated.

5.3.5 Dispersion Modelling of the Power Plant

CALMET Files for CALPUFF The first step was to gather together and format for CALMET the upper air and surface meteorological data, topographic data, land use information, and the three-dimensional fields of meteorological data generated by prognostic models (MM5 and MC2). The CALMET model was then run and the three dimensional wind fields generated by the model were reviewed to ensure the results were realistic. In this application, various combinations of weighting parameters and surface station maximum radii of influence were tested for CALMET to produce final wind fields that are representative of the meteorological wind fields in the area. Within the CALMET modelling domain, a square horizontal grid at a one-kilometre resolution was used. (Appendix E of the assessment provides more details on the process of refining the CALMET model results.)

Emission Data for CALUFF The main emission source for the power plant was the exhaust stack for the gas turbine (see Fig. 5.9). As noted in the discussion of the baseline study, the existing emissions from the Harmac Mill were modelled with CALPUFF. The stack parameters and emission rates for

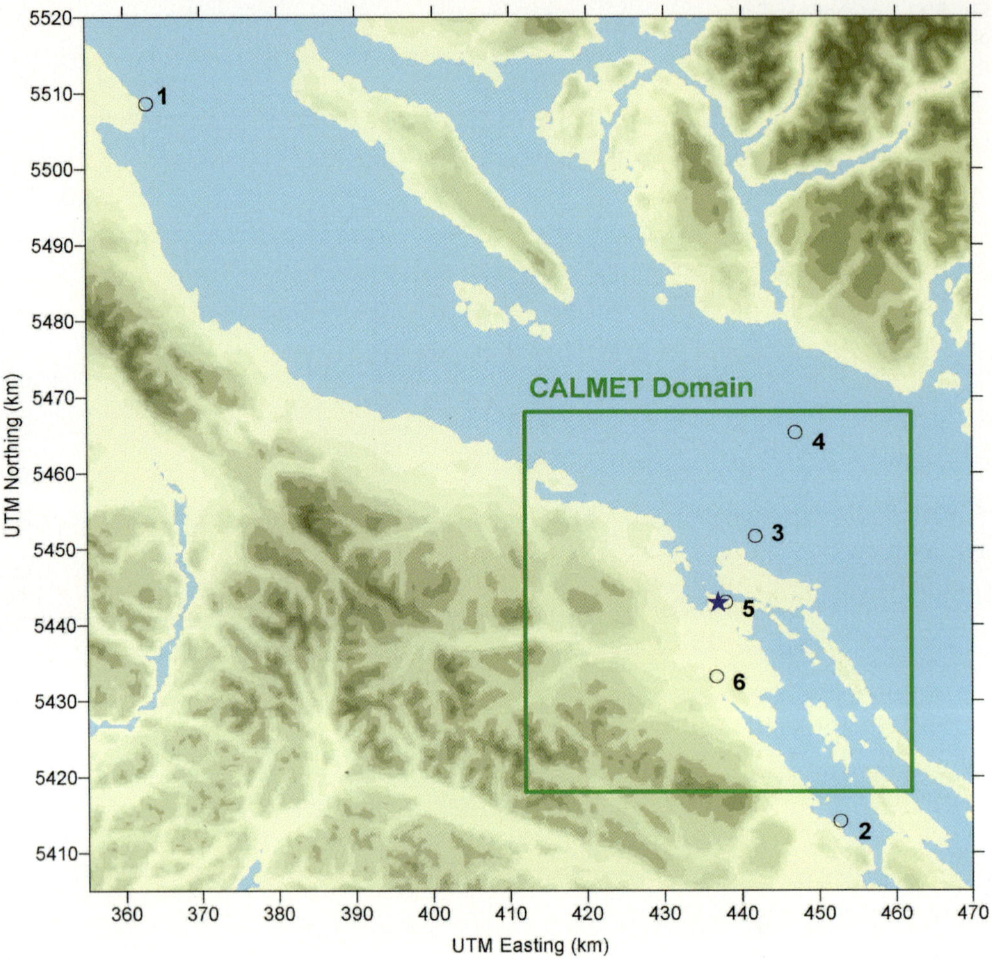

Fig. 5.13 Locations of surface and marine meteorological stations (*1* Comox Airport; *2* Crofton Mill; *3* Entrance Island; *4* Halibut Bank Buoy; *5* Harmac Pacific Mill; *6* Nanaimo Airport)

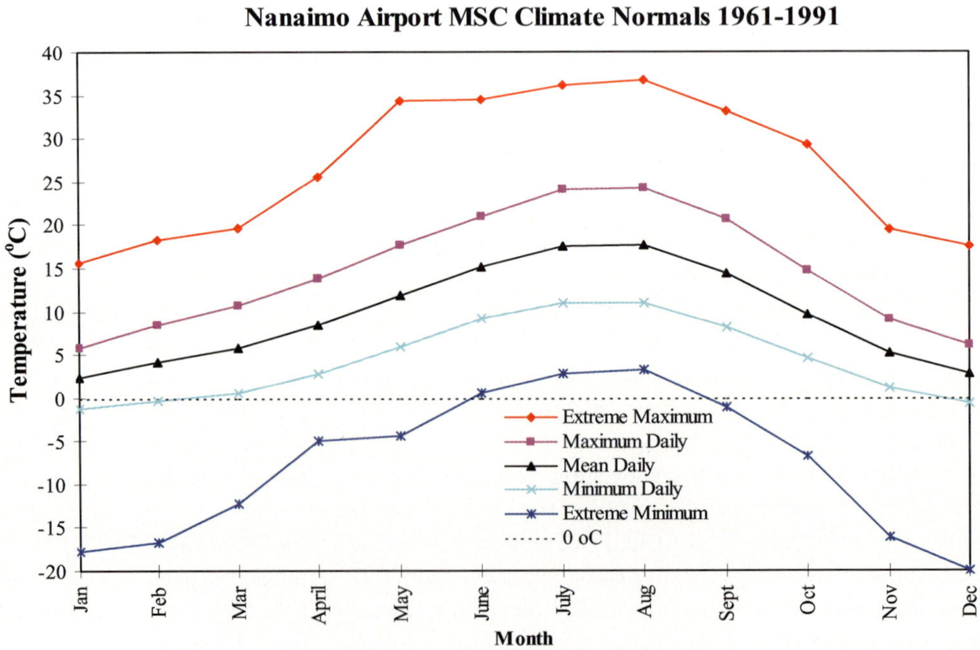

Fig. 5.14 Temperature normals for Nanaimo Airport (1961–1991)

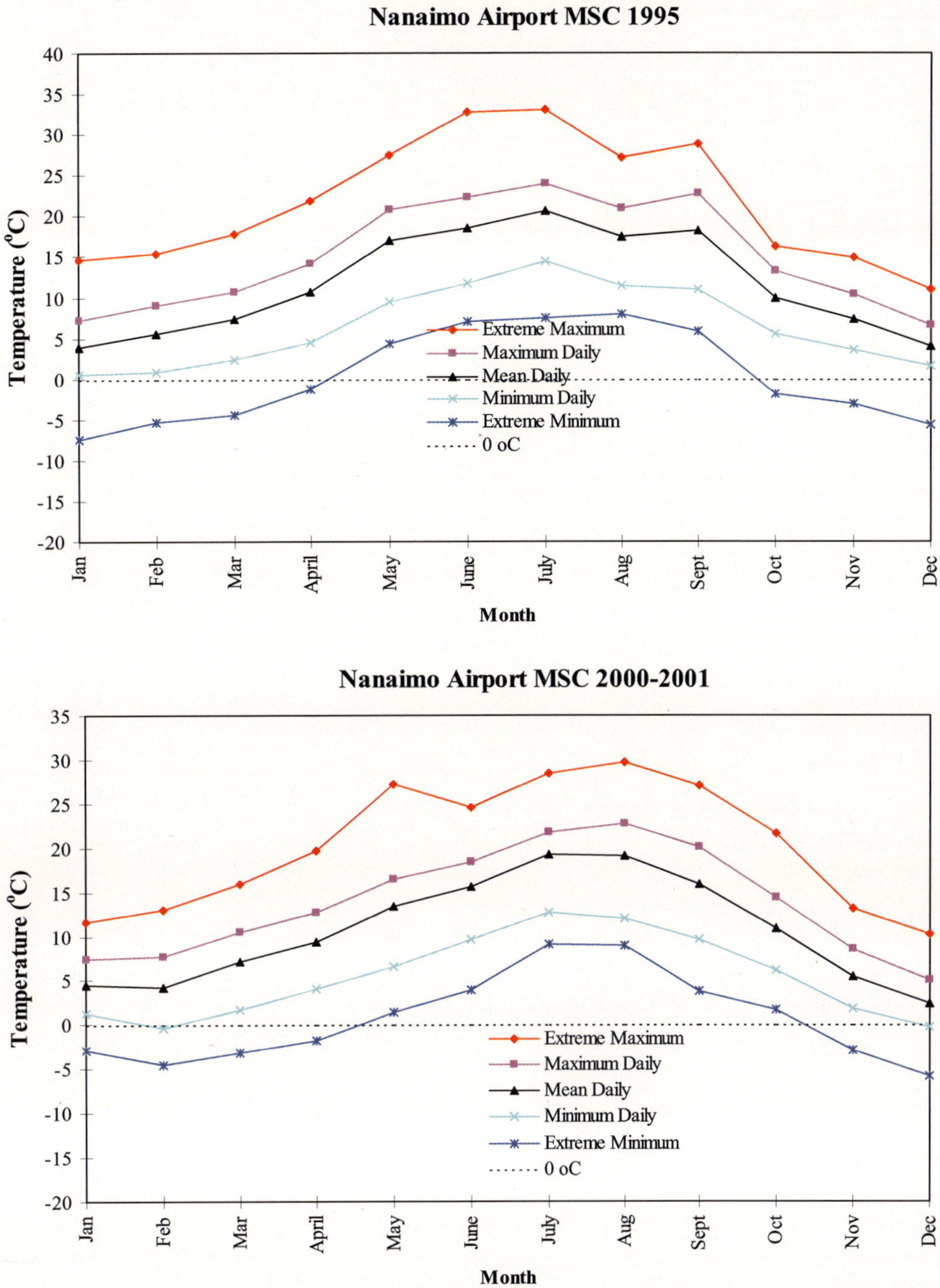

Fig. 5.15 1995 and 2000–2001 temperature observations for Nanaimo Airport

the criteria air contaminants (CACs) for the power plant and the mill are shown in Table 5.4.

Receptor Grid CALPUFF and other dispersion models calculate concentrations at points along a grid, which are defined by the dispersion modeller. Figure 5.16 shows the nested receptor grid used for the local modelling. The hourly concentrations are calculated at each point on the grid for every hour of available data.

Model Results The CALPUFF model was run once for the Harmac Mill sources and once for the Power Plant. Post processing software allowed for the two model runs to be added together in order to assess cumulative air quality impacts. The output files from CALPUFF were processed by contouring software, which showed the spatial distribution of the ambient concentrations of the emitted pollutants. Only a few examples of the output are provided here for illustration. Table 5.5 shows the maximum predicted concentrations for the CACs for the Mill plus VIGP emissions and for each source alone.

Table 5.3 Summary of ambient air quality objectives for B.C. and Canada

Parameter	British Columbia Objective			Federal Objective			Metro Vancouver
	Level A[a] (μg/m^3)	Level B[a] (μg/m^3)	Level C[a] (μg/m^3)	Maximum[b] Desirable (μg/m^3)	Maximum[b] Acceptable (μg/m^3)	Maximum[b] Tolerable (μg/m^3)	Objective (μg/m^3)
Nitrogen Dioxide (NO_2)							
1-h Maximum	–	–	–	–	400	1,000	200
24-h Maximum	–	–	–	–	200	300	–
Annual Mean	–	–	–	60	100	–	40
Carbon Monoxide (CO)							
1-h Maximum	14,300	28,000	35,000	15,000	35,000	–	30,000
8-h Maximum	5,500	11,000	14,300	6,000	15,000	20,000	10,000
Sulphur Dioxide (SO_2)							
1-h Maximum	450	900	900–1,300	450	900	–	450
24-h Maximum	160	260	360	150	300	800	125
Annual Mean	25	50	80	30	60	–	30
PM_{10}							
24-h Maximum	–	50	–	–	–	–	50
Annual Mean	–	–	–	–	–	–	20
$PM_{2.5}$							
24-h Maximum	–	–	–	–	30[c]	–	25
Annual Mean	–	–	–	–	–	–	12
Ozone (O_3)							
1-h Maximum	–	–	–	100	160	300	–
24-h Maximum	–	–	–	30	50	–	126[d]
Annual Mean	–	–	–	–	30	–	–

[a] Concentrations given at 20 °C, 101.3 kPa, dry basis
[b] Concentrations given at 25 °C, 101.3 kPa, dry basis
[c] Canada Wide Standard
[d] 8-h average

Table 5.4 Modelling stack parameters for emissions from VIGP at 100 % load and Harmac Mill sources

Emission Sources	Source Modelling Characteristics									
	Inside Stack Diameter (m)	Stack Height (m)	Exit Temp. (K)	Exit Velocity (m/s)	CO (g/s)	NO_x (g/s)	PM_{10}[a] (g/s)	$PM_{2.5}$[a] (g/s)	SO_x (g/s)	VOC (g/s)
Modelled VIGP Emission Scenarios:										
1. With Duct Burning	5.46	45.72	348.67	20.86	6.96	3.28	0.289	0.289	0.35	0.53
2. Without Duct Burning	5.46	45.72	354.56	21.15	4.64	2.97	0.264	0.264	0.32	0.41
3. Without duct burning, and without SCR.	5.46	45.72	354.56	21.15	4.64	7.63	0.264	0.264	0.32	0.41
Existing Harmac Mill:										
Power Boilers 1,2 & 9	4.22	76.2	440	11.01	26.92	2.99	2.25	1.94	2.34	0.64
Recovery Boilers 4 & 5	3.70	76.2	478	12.20	35.19	1.59	7.06	6.08	54.13	2.56
Recovery Boiler 6	3.61	76.2	489	15.89	40.26	2.84	2.74	2.37	7.53	2.93
East Lime Kiln	1.51	30.48	483	9.04	0.34	0.36	0.10	0.07	0.68	0.02
West Lime Kiln	1.51	30.48	472	13.80	0.58	0.69	0.08	0.06	2.02	0.04
Dissolving Tank 4	1.07	25.73	339	5.12	–	–	0.14	0.13	–	–
Dissolving Tank 5	1.21	30.68	351	7.23	–	–	0.42	0.37	–	–
Dissolving Tank 6	1.79	52.73	344	7.98	–	–	2.09	1.86	–	–

[a] Filterable particulate

Fig. 5.16 CALPUFF nested receptor grid. (The image has been cropped and does not display the entire CALPUFF domain.)

The contour plots for the maximum predicted 1-h NO_2 concentration for VIGP and for VIGP plus the Harmac Mill are shown in Figs. 5.17 and 5.18 (note the red star in the two figures marks the location of the Mill and the blue star the location of VIGP) the location. Clearly the greatest concentrations occur close to the sources. One interesting plot (Fig. 5.19) shows the net increase in the 24-h $PM_{2.5}$ concentrations. This plot shows that the power plant was predicted to increase concentrations by less than 1 μg/m^3 and typically less than 0.1 μg/m^3.

Models like CALPUFF provide the ability to look at difference plots, 98th percentile plots, 1-h, 8-h, 24-h, and annual or period averages. The post processing programs of the CALPUFF model suite as well as the tools provided by various plots programs allow one to combine sources, calculate various time averages, and calculate various percentile values.

Long Range Modelling for VIGP Because the city of Vancouver and the United States/Canada border were within 50 km of the power plant CALMET and CALPUFF were run for long-range modelling Domain. The CALMET long-range domain covered an 84 (N-S) by 108 (E-W) kilometre area centred on the southern part of Gabriola Island. The emissions of interest were NO_2, $PM_{2.5}$, and SO_2. The modelling results indicated that VIGP would have had no measurable impact on the air quality of the southern portion of the Strait of Georgia and, in particular, the Lower Fraser Valley airshed.

Acidic Deposition Effects of VIGP Acidic deposition occurs through wet and dry deposition processes. Wet acid deposition results from the dissolution of sulphur dioxide (SO_2) and oxides of nitrogen (NO_x) in water present in the atmosphere (cloud/fog droplets, rain, snow). The result is the formation of dilute sulphuric

Table 5.5 Predicted concentrations from the Harmac Mill and VIGP together for the combined 1995 and 2000–2001 modelling period

Pollutant	Averaging Period	Predicted Concentration (μg/m³)				Strictest Ambient Objective (μg/m³)
		Harmac Mill Plus VIGP		VIGP Alone	Harmac Alone	
		Maximum	98th Percentile	Maximum	Maximum	
NO_2	1-hour	123.2	27.6	32.1	123.2	400
	24-hour	31.1	16.0	12.7	31.1	200
	Annual	3.1		0.8	3.1	60
PM10	24-hour	58.6	37.3	1.1	58.6	50
	Annual	8.6		0.07	8.6	
$PM_{2.5}$	24-hour	52.2	33.0	1.1	52.2	30 (98th percentile)
	Annual	7.5		0.07	7.5	
SO_2	1-hour	515.5	158.8	3.4	515.5	450
	24-hour	136.9	116.7	1.4	136.9	150
	Annual	24.8		0.09	24.8	25
CO	1-hour	753.0	178.2	68.2	753.0	14300
	8-hour	201.8	156.9	31	201.8	5500
	Annual	28.5			28.4	
VOC	1-hour	44.6	11.4	5.2	44.6	
	24-hour	10.7	8.9	2.1	10.7	
	Annual	1.8		0.1	1.8	

VIGP is assumed to be operating at 295 MW power output with duct firing

acid (H_2SO_4) and nitric acid (HNO_3). This acidic atmospheric moisture then deposits on soil, vegetation, water, and structures, at which point it can affect the receiving environment. Dry acid deposition refers to the direct deposition of acidic gases and particles onto receiving surfaces.

Since the proposed facility would emit NO_x and, to a much lesser extent, SO_x, an assessment was conducted of the potential for VIGP emissions to increase local acidic deposition rates and affect water quality and aquatic habitats.

The modelling predicted maximum acid deposition rates in the Nanaimo area resulting from worst-case VIGP emissions. The model takes into account the various possible fates of emitted SO_2 and NO_x, local meteorology and climatology, as well as the dispersion of NO_x and SO_x emissions in the atmosphere. Modelling results showed that the increase in acidic deposition from VIGP emissions would be very small, and would have been dispersed over a large area, as shown in the plot of results for 2000/2001 (Fig. 5.20).

Miscellaneous Modelling Results Brown Plume—The potential for VIGP emissions to affect visibility in the Nanaimo region was assessed using the CALPUFF model outputs and the visibility option of the CALPOST postprocessor utility. Given CALPUFF model outputs of hourly concentration of sulphates and/or nitrates and/or other particulate matter resulting from emission sources, the CALPOST utility computes and summarizes the corresponding light extinction coefficient (b_{ext}), relative to the background light extinction (Scire et al. 2000). The light extinction coefficient includes both scattering and absorption components and is a measure of the attenuation of light over a unit distance, expressed in inverse megameters (1/Mm) where a megameter is one million meters, or 1,000 km. Another measure derived from b_{ext} is also computed to indicate perceived visibility on a linear scale. The deciview (dv) is defined as

$$dv = 10 \ln\left[\frac{b_{ext}}{10}\right]$$

Here, a reference extinction coefficient of 10 (1/Mm) in the denominator of the logarithm corresponds to a pristine environment (with only Rayleigh scattering from clean dry air exclusive) and a deciview of zero.

CALPUFF was run to produce hourly concentration estimates of secondary particulate matter ($(NH_4)_2SO_4$ and NH_4NO_3) from precursor emissions of NO_x and SO_x, and primary emissions of $PM_{2.5}$ (see Appendix E of assessment). Results of these runs were summed and then processed with the CALPOST visibility algorithms using the EPA criteria for assessment of impacts on CLASS 1 visibility sensitive areas as background.

The CALPOST postprocessor assesses visibility changes on both a 24-h and a run-length basis. The 24-h assessment gives the maximum change in both extinction coefficient and deciview predicted for any of the modelled receptors for each day in the model run. This is presented as the number of days that show a maximum visibility change greater than some threshold criteria (i.e. a Δ deciview greater than 1). The model results showed that there were no receptors that

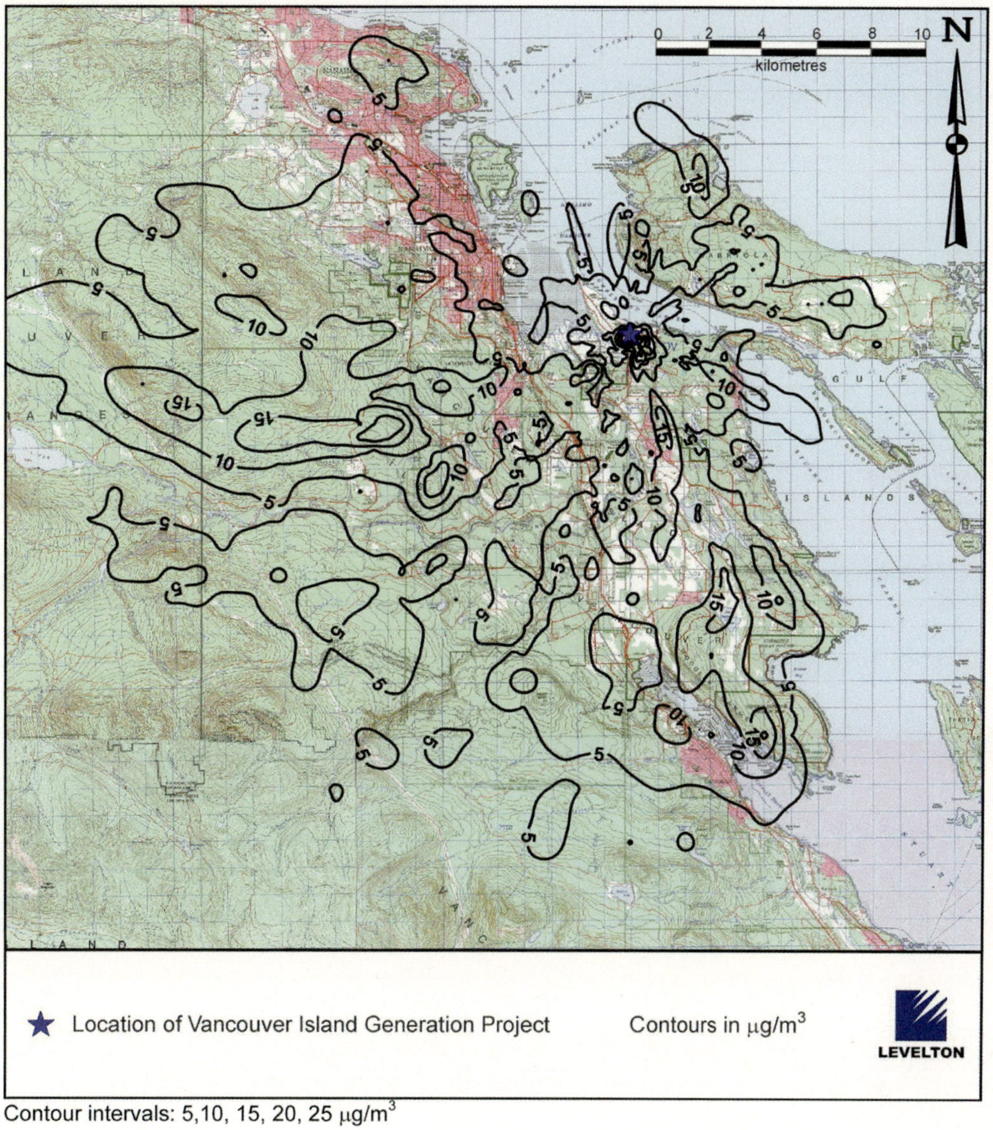

Fig. 5.17 Maximum predicted 1-h NO_2 concentration for VIGP for the combined 1995 and 2000–2001 modelling period

showed an annual extinction change greater than 1 % and there were no receptors that showed a change in Deciview of more than 0.10.

Water Vapour Plume—Plume Visibility and the potential for ground-level fog occurrence were assessed using the FOG model (Scire 1997). FOG is a Gaussian plume based model designed explicitly for dispersion of water vapour. It contains equations for calculating ambient water content and saturation mixing ratio from hourly measurements of temperature and relative humidity in addition to a plume based atmospheric dispersion module. The model uses a terrain following 'plume path coefficient' to simulate complex terrain effects on plume height. The reader is referred to Scire (1997) for details on the mathematical implementation of these features. The algorithms of the FOG model have been incorporated into CALPUFF.

The FOG model was used to examine the water vapour plume from the cooling towers. VIGP's cooling tower was predicted to produce ground level fog on a maximum of 1.3 % of hours modelled for the winter emissions scenario. This amounts to 114 h out of the year modelled. For the summer emissions load 0.44 % of modelled hours produced ground level fog, equal to 39 h. For both scenarios, the extent of areas with fog potential was limited to a small proportion of the model domain. Furthermore, there were no occurrences of ground-based icing or ice plumes in any of the model runs. These results provided an indication that the VIGP facility would have minimal impact on driving conditions along the local highway to the ferry terminal.

Start-up and Partial Load Assessment—NO_x and CO emissions from a gas turbine can increase by a factor of 10–100 under partial load. The impact of increased NO_x and CO emissions under partial load was assessed using SCREEN3.

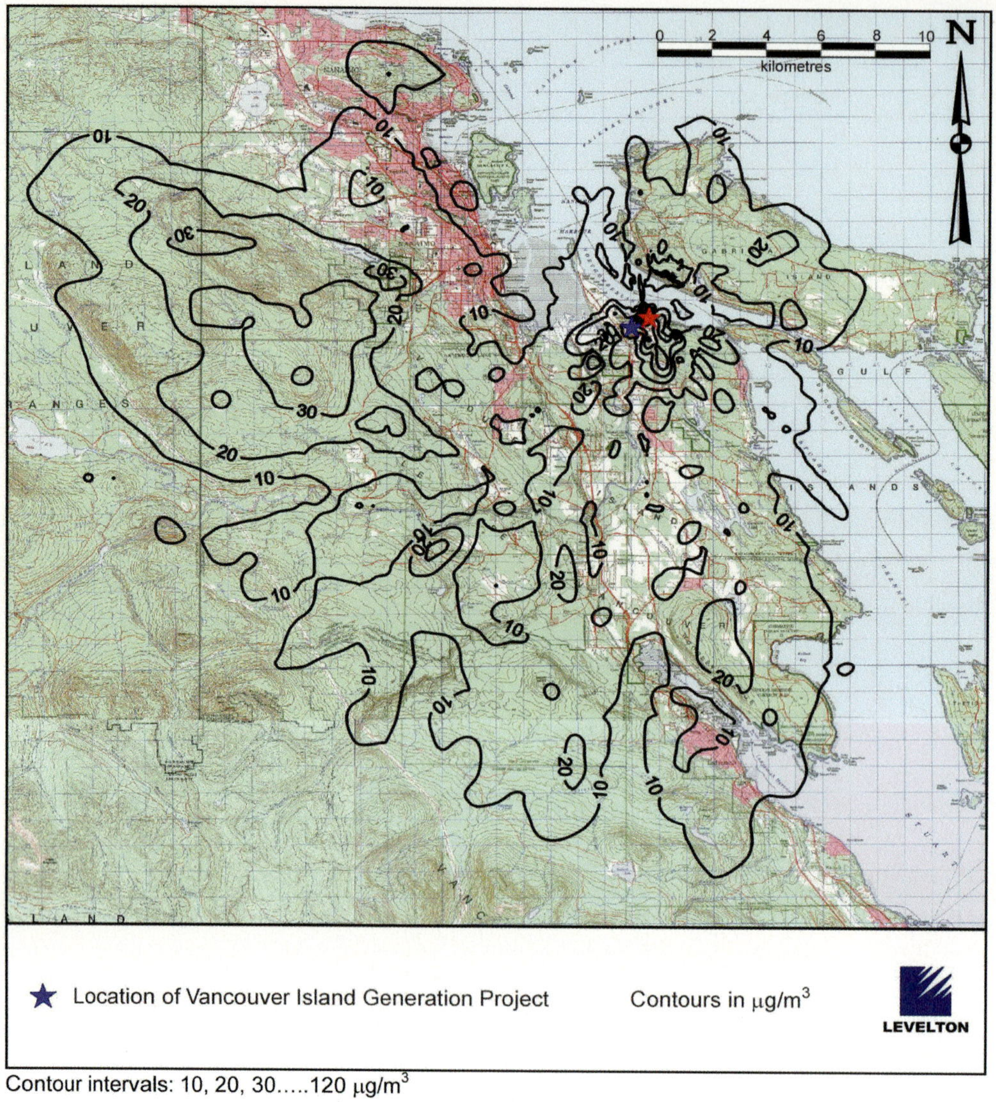

Fig. 5.18 Maximum predicted 1-h NO_2 concentration for VIGP and the Harmac Mill together for the combined 1995 and 2000–2001 modelling period

Although a simple model, SCREEN3 was a good choice for this analysis because one could quickly compare the SCREEN3 results for base load and for warm and cold start-up scenarios.

The SCREEN 3 predicted maximum 1-h NO_x and CO concentrations for VIGP under start-up and partial load conditions (Table 5.6) were similar in magnitude to the values predicted from normal operation. The screening of these additional scenarios provided an indication that impacts under start-up and partial load conditions were acceptable given that the higher emission rates under these scenarios would occur over a period of 2 h or less.

Non-criteria Pollutants from VIGP—The maximum ambient concentrations of non-criteria pollutants (e.g. Acetaldehyde and Acrolien) emitted from the VIGP combustion processes were predicted using the same modelling methods described previously for the CACs. These predicted values were used in the public health impact assessment, demonstrating that dispersion models are useful tools for a variety of environmental impacts studies.

5.3.6 Conclusion Regarding the Dispersion Modelling of a Power Plant

The preceding case study provides an overview of an air quality assessment of a proposed power plant. Dispersion modelling was used as a tool to:

- Predict ambient concentrations of pollutants emitted from the proposed power plant. Predicted concentrations were then compared to government objectives and used in a health impact assessment to assess the impacts of the plant.
- Predict acid deposition
- Assess impacts to visibility.

Fig. 5.19 Net increase in the maximum predicted 24-h $PM_{2.5}$ concentration for VIGP plus the Harmac Mill Compared to that for the Harmac Mill Alone

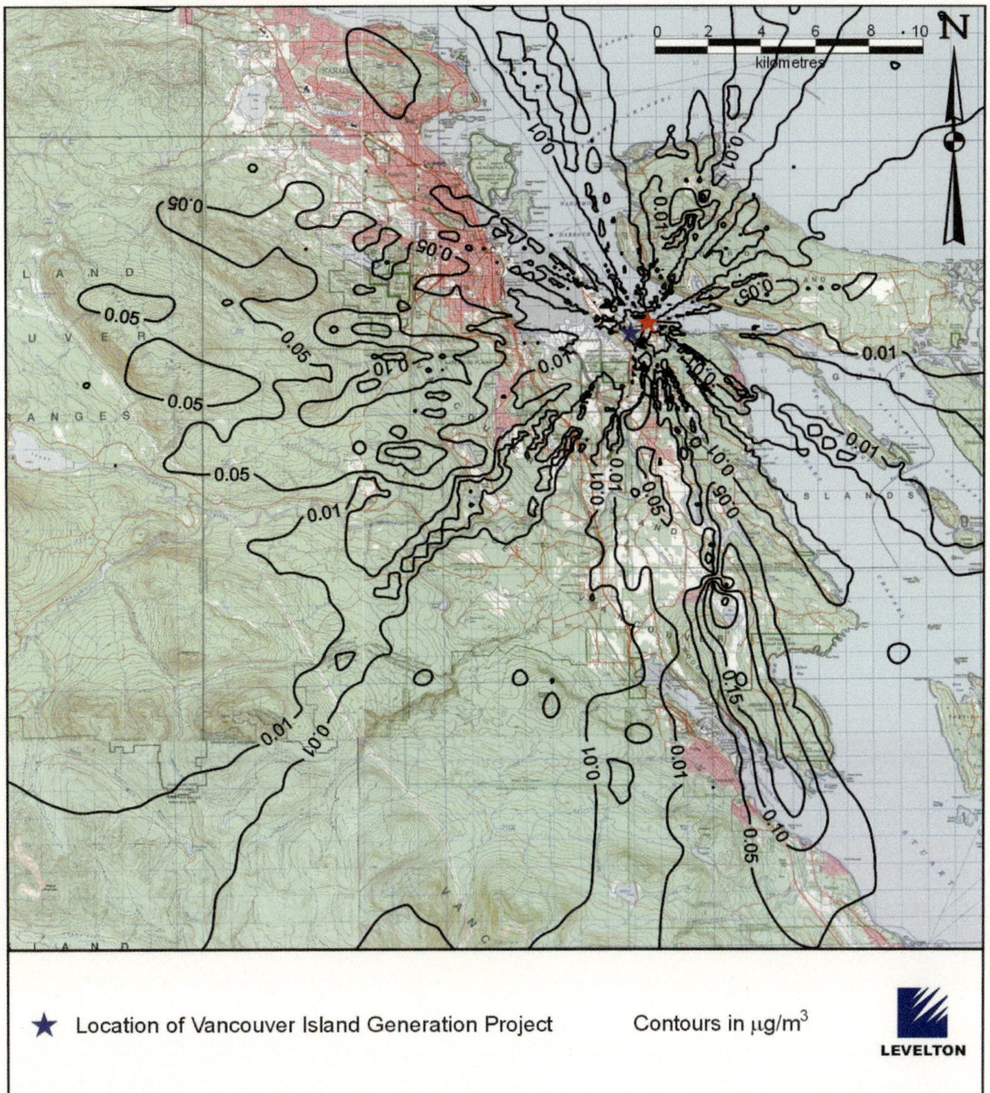

Contour intervals: 0.01, 0.05, 0.1, 0.15, 0.2, 0.25 μg/m³

- Provide guidance on the impact of the water vapour plume on ground level visibility and icing conditions.
- Screen various emissions scenarios within the complex air quality assessment for quick comparison of potential impacts.

5.4 Case Study 2—Use of CALINE to Assess the Air Quality Impacts from the South Fraser Perimeter Road (SFPR)

5.4.1 The Project

The South Fraser Perimeter Road (SFPR) project consisted of building a new highway, and upgrading existing roads to highway standard over an approximately 40 km route on the south side of the Fraser River between Delta, BC and Surrey, BC (Fig. 5.21). Part of the overall environmental assessment for this project included an air quality assessment to identify the impacts that vehicle emission could have on the local environment with an emphasis on human health impacts. The assessment for this project can be downloaded from the BC EAO web site (http://www.eao.gov.bc.ca). The SFPR is part of the Gateway program for which more information is available at: www.gatewayprogram.bc.ca.

5.4.2 Assessment Tasks

Prior to undertaking the studies to define baseline air quality and to assess potential impacts of the proposed SFPR, a preliminary issues identification process was conducted. This identified the key issues anticipated to be of concern for the project, and used the list of issues identified in other

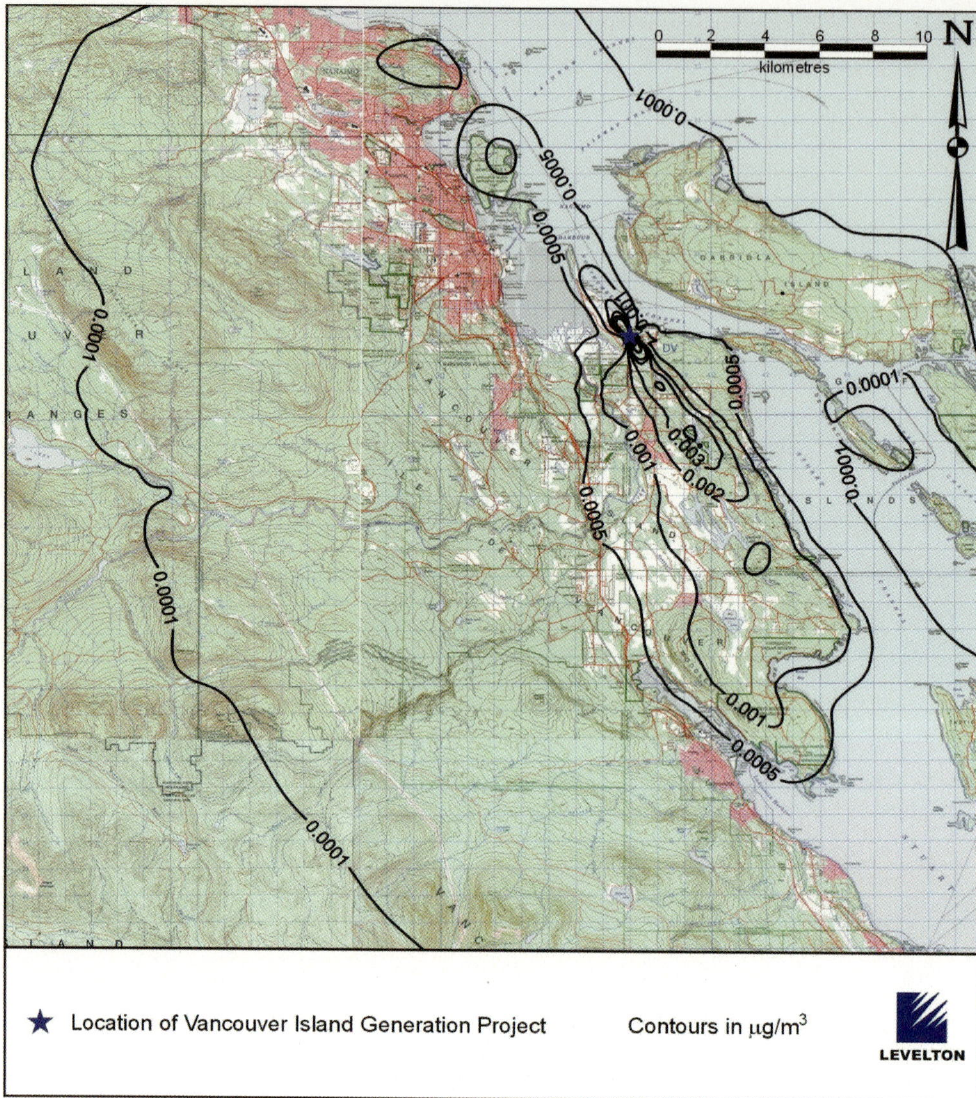

Contour intervals: 0.0001, 0.005, 0.001, 0.002, 0.003, 0,004, 0.005 and 0.01 kmol H^+eq/ha/yr.

Fig. 5.20 Maximum predicted acidic deposition rates from VIGP emissions for 2000/2001

Table 5.6 Summary of SCREEN3 model Results for start-up scenarios

Scenario	Maximum 1-hr NO_2 Concentration ($\mu g/m^3$)	Maximum 1-hr CO Concentration ($\mu g/m^3$)	Distance from Stack to Maximum (m)
Base Load, without Duct Burning	32.6	20	330
Hot Start	78	389	300
Warm Start	74	735	300
Cold Start	113	866	300

transportation projects within the province and stakeholder (Metro Vancouver and Fraser Health) issues from specific SFPR discussions.

The following is a list of the key air quality issues for the proposed SFPR:

- Baseline air quality—summary of baseline air quality information collected at appropriate locations and justification for locations representing the baseline;
- Potential overall impact of the project on air quality in the project area, including potential cumulative effects;
- Identification of those locations within the project area that are most sensitive to vehicle tailpipe and road dust emissions;
- Amount and type of air contaminants associated with the project;
- Potential impact of diesel exhaust;
- Environmental standards used to evaluate air quality modelling results;
- Influence of local weather patterns on potential air quality impacts;
- Local health issues arising from potential air quality impacts;
- Greenhouse gas emissions—impact of the SFPR on total GHG emissions;
- Residual air quality impacts—if residual impacts are identified determine cumulative effects;

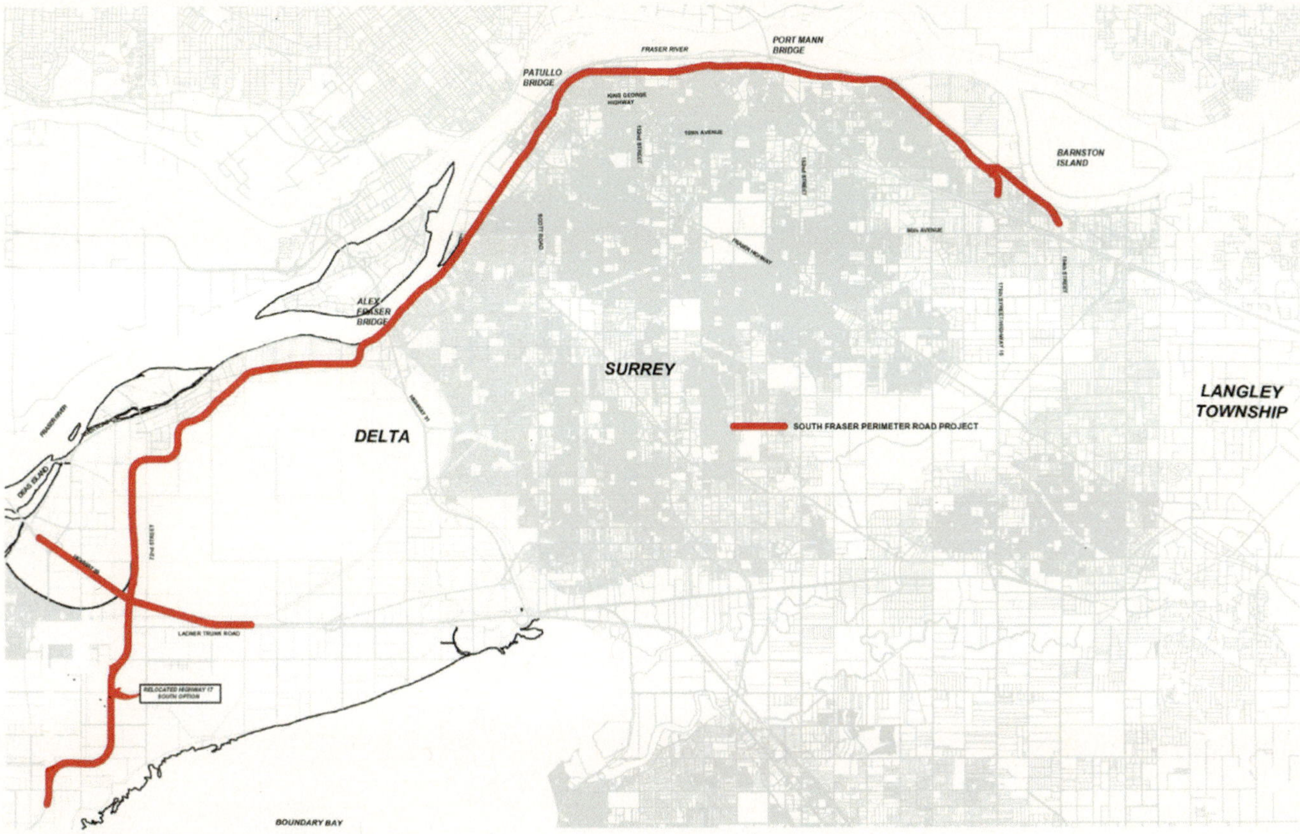

Fig. 5.21 South Fraser Perimeter road site map

- Mitigation—identify mitigation measures for emissions impacting air quality; and
- Road dust—determine level of impact from road dust emissions.

All of these points will not be addressed in this chapter only those related to dispersion modelling.

5.4.3 Modelling Methodology for SFPR

Model Selection The modelling approach used in the assessment was developed in consultation with Metro Vancouver and involved the use of a proven dispersion model for road vehicle impacts, together with a reliable meteorological model for determining the wind fields near SFPR.

Predictions of ambient concentrations resulting from vehicle exhaust on highways in previous assessments in B.C. had commonly used the CALINE 3 model. CALINE 3 is a steady-state Gaussian dispersion model designed to determine air pollution concentrations at receptor locations downwind of highways located in relatively uncomplicated terrain. Other models such as CALPUFF, or the Industrial Source Complex model (ISC3), a Gaussian plume model, can handle a variety of emission source types, but do not have explicit algorithms to address road sources.

CALINE3 divides individual highway links into a series of elements from which incremental concentrations are computed and then summed to form a total concentration estimate for a particular receptor location. The receptor distance is measured along a perpendicular from the receptor to the highway centerline, Fig. 5.22. The model does not treat plume rise.

Study Area Boundaries Almost all air quality impacts from vehicles operating on roadways occur within the first 1 km from the road, and the maximum impact tends to occur within 100 m. Therefore, the modelling conducted for this study focused on areas within 1 km from the edge of the proposed roadway, plus sections of major crossroads.

Ambient Air Quality Metro Vancouver operates an extensive network of ambient air quality monitoring stations (Fig. 5.23). Data from six monitoring stations (T13, T15, T17, T18 T20 and T30) were used for conducting the ambient air quality analyses for the region surrounding SFPR. The background air quality was set as the 98th percentile

Fig. 5.22 CALINE3 models the dispersion of pollutants downwind of roadway links by partitioning the link into elements, represented by a series of finite line sources (from US EPA Unabridged User's Guide)

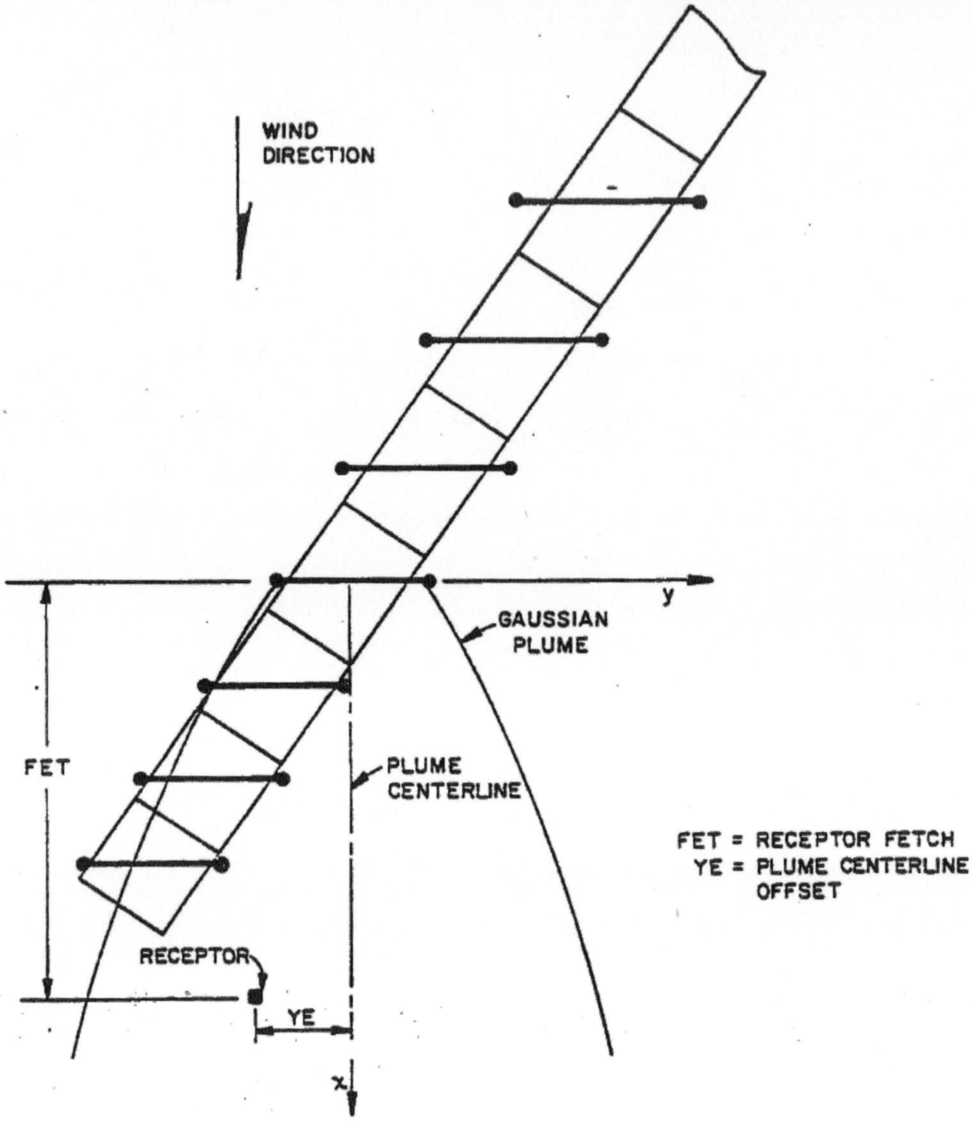

of the data for the period of interest averaged over the years of available data and the values are shown in Table 5.7. The background values were all below the Metro Vancouver objectives listed in Table 5.3.

Meteorology and Climatology The six Metro Vancouver ambient air quality monitoring stations surrounding the SFPR also measure various meteorological parameters such as wind speed, wind direction, and temperature. Data from these stations as well as meteorological data from the Vancouver Airport were examined to determine the climatology of the SFPR area. In addition, the data from the six Metro Vancouver stations were used together with meteorological models to determine the representative winds in the SFPR project area. This local meteorological data, although providing good geographical coverage, was not sufficient on its own to determine wind speed and direction along the entire road alignment.

CALMET was used to provide a better estimate of the meteorology in the vicinity of SFPR than was available from any of the stand-alone monitoring stations. Model simulations were initialised using prognostic outputs from the MC2 mesoscale model that was used in the VIGP assessment, and observational meteorological input data from sixteen surface stations and two upper air stations. As for the VIGP case, CALMET was run for just over one full year, from June 13, 2000 to July 7, 2001.

Representative hourly meteorology for each road section was extracted and formatted for CALINE from the resultant three-dimensional fields. The section specific extracted meteorology was then subsequently used to drive the CALINE simulations on which the air-quality assessment was based.

The year of data used in the model were compared with long-term climate data from the Vancouver airport and the

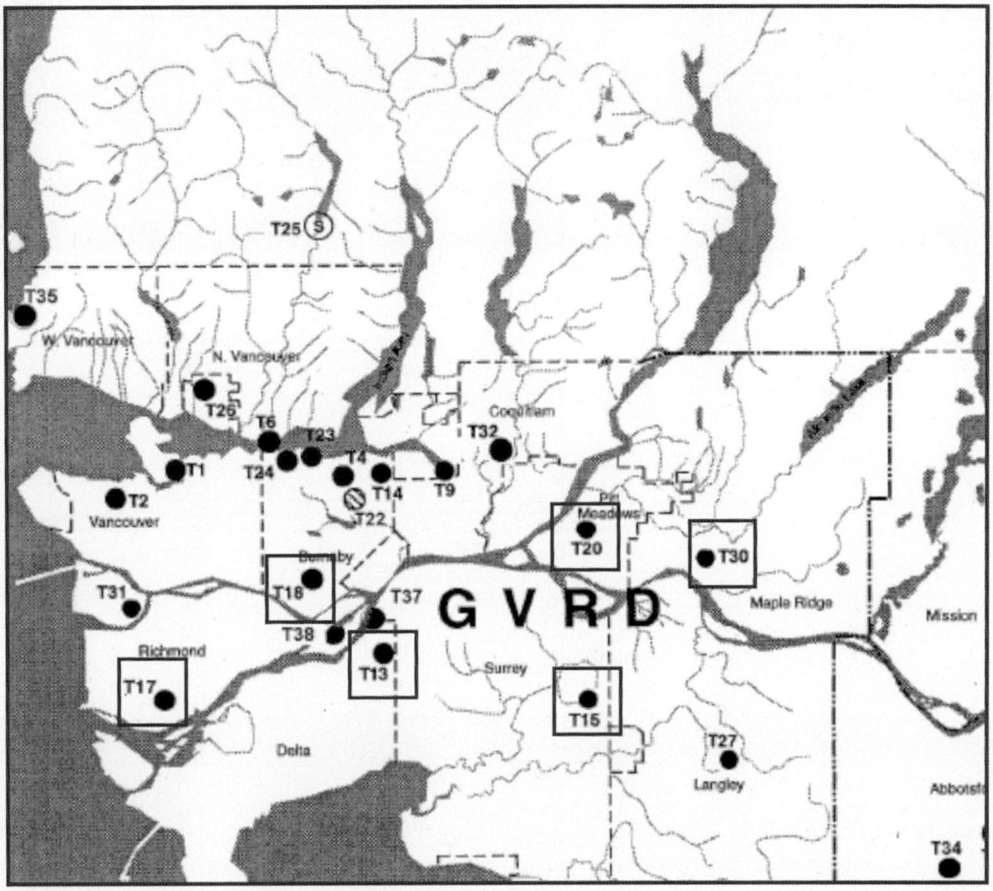

Fig. 5.23 The GVRD ambient air quality monitoring network, squares identify the stations used in this study (figure extracted from GVRD 2003)

Table 5.7 Background values of CO, NO_2, PM_{10}, $PM_{2.5}$ and SO_2 for the SFPR study area

	CO			NO_2			PM_{10}		$PM_{2.5}$		SO_2		
	1-hr	8-hr	yr	1-hr	24-hr	yr	24-hr	yr	24-hr	yr	1-hr	24-hr	yr
Baseline Value (μg/m^3)	1706	1515	551	70	56	30	27	13	15	5	12	9	3

Metro Vancouver stations to ensure the year of data was representative of the climatology of the region.

Vehicular Emissions Vehicle emissions were determined using the MOBILE6.2C model. It estimates vehicle emission factors for:

- hydrocarbon (HC);
- carbon monoxide (CO);
- oxides of nitrogen (NO_x);
- exhaust particulate matter, which consists of several components including,
- tire wear particulate matter; and
- brake wear particulate matter.
- sulphur dioxide (SO_2), ammonia (NH_3);
- six hazardous air contaminant (HAP); and
- carbon dioxide (CO_2).

Vehicles modelled included gasoline and diesel fuelled highway motor vehicles, and motorcycles. The model accounts for the emission impacts of factors such as changes in vehicle emission standards, changes in vehicle populations and activity, and variation in local conditions such as temperature, humidity, and fuel quality. MOBILE 6.2C was the latest U.S. EPA model that was modified for application in Canada to reflect the base emission rate and deterioration rate of Canadian vehicles. Vehicle emission factors used for the modelling take into account improvements in vehicle emission systems that are expected to occur between 2003 and 2021.

Traffic volumes and emission rates are based on the results from the Paramics model. The Paramics model is a micro-simulation model of traffic operations, which models the behaviour of each vehicle on the modelled network. The

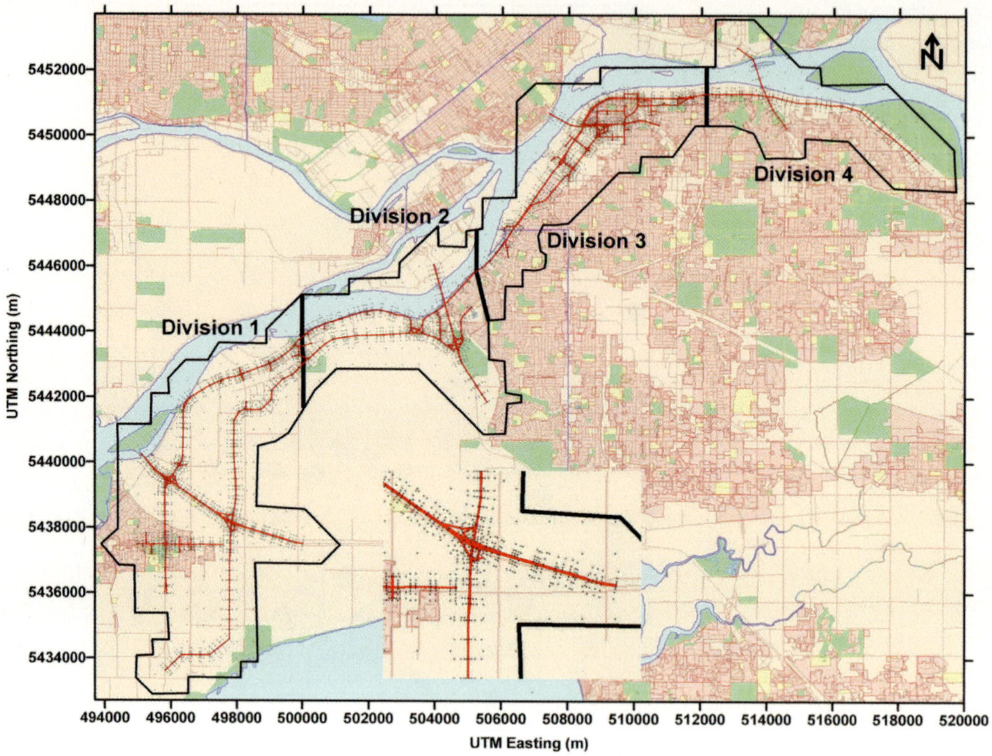

Fig. 5.24 Example of receptors used in the CALINE model. The area inside the black polygon defines the study area for the assessment of local air quality for the SFPR. The southern part of Division 1 has been enlarged to show some of the detail of the receptor grid

road network for SFPR was coded in detail using Paramics, which included all traffic control measures (e.g. traffic signals a, stop signs). The results of the Paramics model are summarized by hour. Since the Paramics model tracks the operating characteristics of each vehicle in the modelled network, it is possible to calculate vehicle emissions if emission rates are available. For the purposes of the SFPR air quality analysis, rates developed from Mobile 6.2C were applied to estimate vehicle emissions. The rates are a function of vehicle speed and the second-by-second speed was used to calculate emissions by vehicles over the simulation period. These emissions are aggregated and reported by hour by vehicle class for each segment in the Paramics network.

Traffic Forecasts were available from the Gateway Strategic Transport (GST) model. Forecasts from the GST model provided an indication of conditions in the future based on a set of assumptions including the Metro Vancouver Growth Management Strategy v 4.0 (modified) Land Use forecast.

Two traffic rates were provided for use in the CALINE3 model. One was the AM peak hour volume (AM PH), which provided rush hour traffic volumes in units of vehicles per hour. The AM PH value was used in the CALINE3 model for all available hours of meteorological data to determine the maximum 1-h ambient concentrations. However, the AM PH is mainly valid for peak traffic hours and would be overly conservative if used to determine the 24-h average and annual average concentrations. Therefore, to determine results from other averaging periods other than one hour a scaling factor was applied during the post processing. The scaling factor for all scenario's and division's was 0.56. This factor represents an "average hour" during the week. In the tables and results that follow, the 1-h concentrations are based on AM PH estimates and the 24-h and annual average values are based on the average hour during the week.

Receptor Grid This CALINE computer model predicts hourly ambient concentrations at each receptor. Receptors are a grid of points related to locations on a map, usually uniformly spaced, where the model calculates ambient concentrations of contaminants emitted from the vehicles on the road. The CALINE 3 model was run for a receptor grid extending a kilometre from SFPR and the cross streets. Compare the CALINE receptor grid shown in Fig. 5.24 with the CALPUFF receptor grid shown in Fig. 5.16. CALMET was used to provide an estimate of the wind fields in the immediate vicinity of the SFPR. Meteorological data was extracted from CALMET and the data that was determined to be most representative for each division was used the for the CALINE modelling.

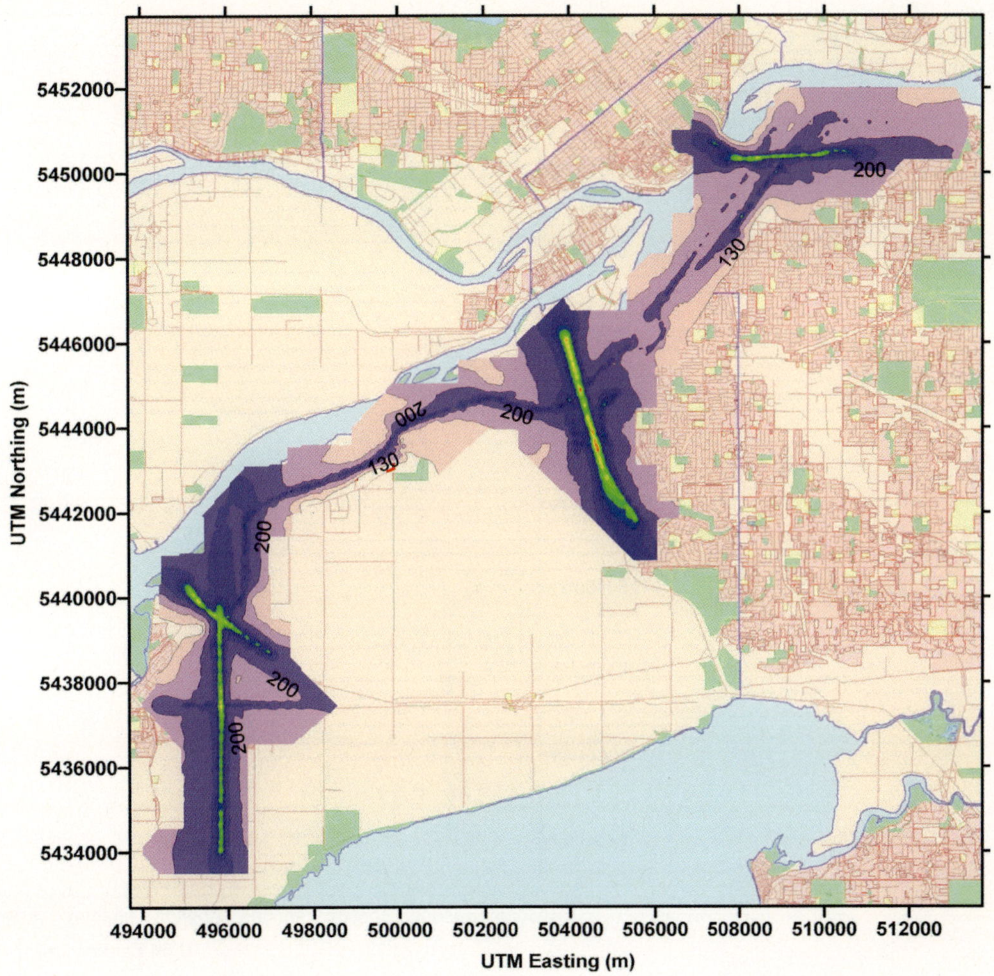

Fig. 5.25 Maximum 1-h NO_X concentration for existing roads in 2003 along the proposed SFPR route (*red*) plus several cross streets (*blue*). The contour lines are 100, 130, 200, and then increase by 100–1,200 μg/m³

5.4.4 Modelling Results for SFPR

The CALINE model used to analyse the impact of the SFPR on local air quality was run for the following scenarios:

- Existing roads in 2003: air quality impact on the area in the vicinity of the proposed SFPR due to present vehicular traffic load;
- Existing roads in 2021: air quality impact on the area in the vicinity of the proposed SFPR due to increase in vehicular traffic from 2003 to 2021 assuming no modifications are made to the existing roads in the SFPR corridor;
- SFPR Gateway in 2021: air quality impact in the vicinity of the proposed SFPR with a new corridor to the east of Ladner starting at the intersection of Deltaport Way and existing Highway 17.

The contour plots of the 1-h NO_x concentrations are presented for illustration of the model results for the CALINE model for a road network like the SFPR. In the original assessment, the maximum NO_2 concentrations were determined by assuming they were equal to 100% of the NO_x concentration (a conservative estimate), and then converted by using ozone limiting to convert the NO_x to NO_2 (a common technique), and finally by using the ambient measurement of NO/NO_2 ratios to convert NO_x to NO_2. The background value was used with the model results for final comparison to air quality objectives. For example, in the contour plots for 1-h maximum NO_x (assuming 100% conversion), the 130 μg/m³ contour represents the concentration that exceeds the GVRD 1-hr guideline when the background concentration is included.

Figure 5.25 shows the model calculated maximum 1-h NO_x concentrations along the SFPR route for 2003. For the most part, the highest concentrations above the ambient air quality objectives occurred on the major highways that intersect the proposed SFPR. For the existing roads in 2003, 1-h NO_x concentrations exceeded the GVRD (200 μg/m³) objective along the edge of the proposed SFPR roadway and along all the major cross streets.

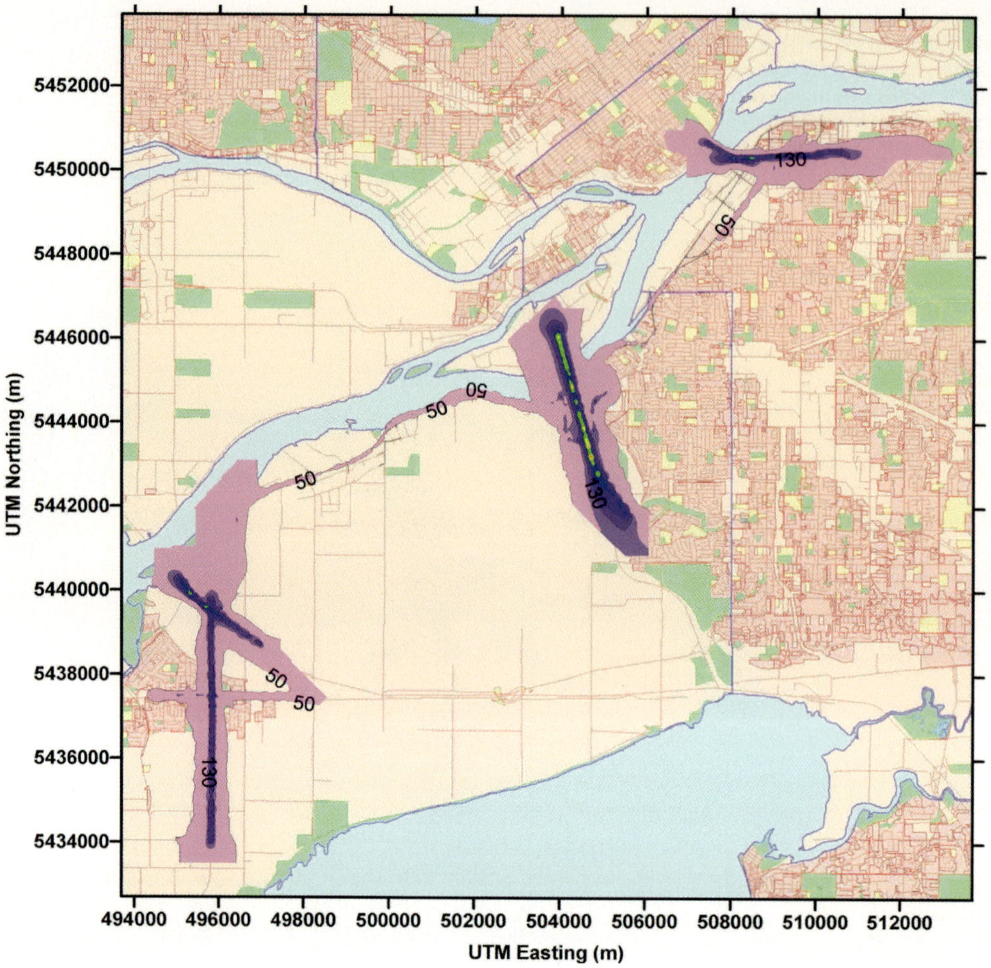

Fig. 5.26 Maximum 1-h NO_X for the existing roads for 2021 (without SFPR) along the proposed SFPR route plus several cross streets. The contours are 50, 100, 130, 150, 200, and then increment by 50 μg/m³

Figure 5.26 shows the model calculated 1-h maximum NO_x concentrations for 2021 without the SFPR. The concentrations decreased and fewer objectives were exceeded in 2021 for the existing road network. Improvements in the efficiency of vehicles and the reduced sulphur content of fuels are the likely reason for fewer air quality objectives being exceeded.

Emissions from traffic on the SFPR in 2021 were predicted to exceed the GVRD maximum 1-h objective for NO_x at only a few locations along the SFPR within 50 m of the roadway (Fig. 5.27) plus other areas are along the major cross streets. Figure 5.27 also shows the Port Mann Bridge in the upper right corner, which was not shown in the previous two figures.

While comparing the model predicted concentration of various pollutants to ambient air quality objectives is useful, it does not provide a clear picture of human health impacts. However, it is possible to use the model output to assess the acute (short-term) and chronic (long-term) non-cancer health effects as well as the cancer health impacts.

Figure 5.28 shows the steps taken in the health impact assessment. Figure 5.29 shows the spatial distribution of the predicted acute respiratory hazard quotients in the existing roads in 2003 and 2021 and the 2021 SFPR scenario. The health effects threshold is 1.0 for all affected organ systems.

5.5 Summary

The two case studies show that dispersion models are useful tools for use in air quality impact assessments. There are many types of models available for use; some are very specialized and only useful for specific applications. In case study 1 the CALPUFF model suite helped to assess the impacts of stack emissions on local and long range ambient air quality and human health as well as to assess the potential impact of acid disposition on the local area. On the other hand, the Fog model helped assess the impact of water vapour emissions on visibility and icing on

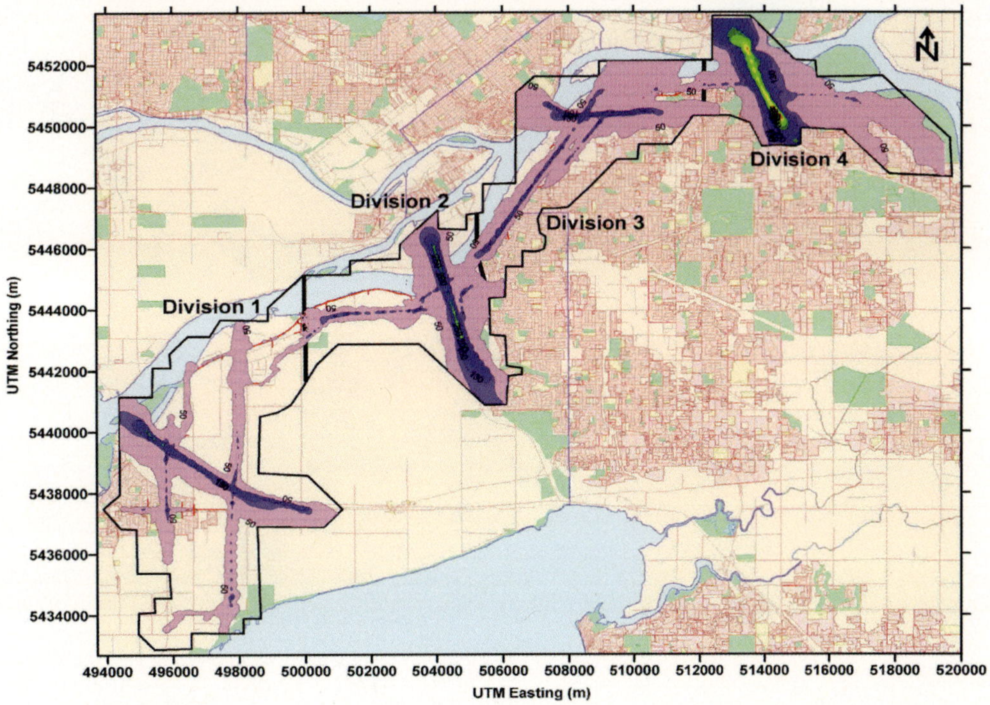

Fig. 5.27 Maximum 1-h NO_X concentration for the SFPR in 2021, along the SFPR plus several cross streets for 2021. The contours are 50, 100, 130, 150, 200, and then increment by 50 μg/m³

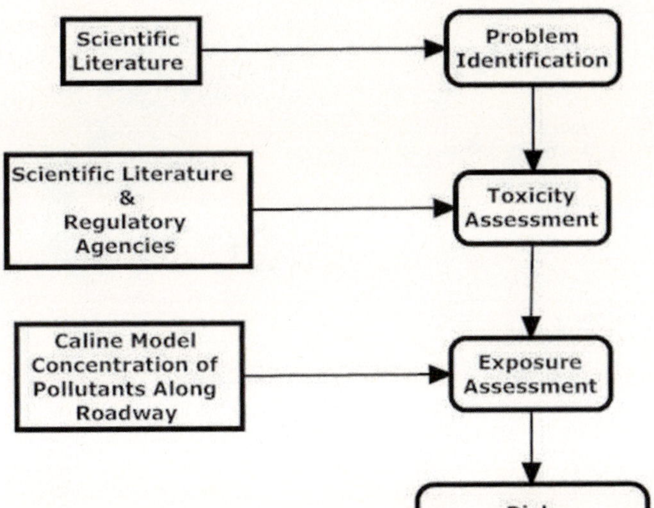

Fig. 5.28 Flow diagram showing the steps in the SFPR health impact assessment

roads near the power plant. The SCREEN3 model helped assess the dispersion of pollutants emitted during the start-up and shutdown of a gas turbine. Case study 2 relied on the CALINE model, which is a line model specifically developed to assess the impacts of roadway emissions on the local environment.

In both case the main objective was to estimate the effect that air pollutant emissions would have on human health. All the models used in these case studies required information about the meteorology and topography of the study areas plus the emission sources. Although the models may differ there are usually commonalities in the required input data and the final application of the model results.

Fig. 5.29 Spatial distribution of the predicted acute respiratory hazard quotients in the existing roads in 2003 and 2021 and the 2021 SFPR (Gateway) scenario

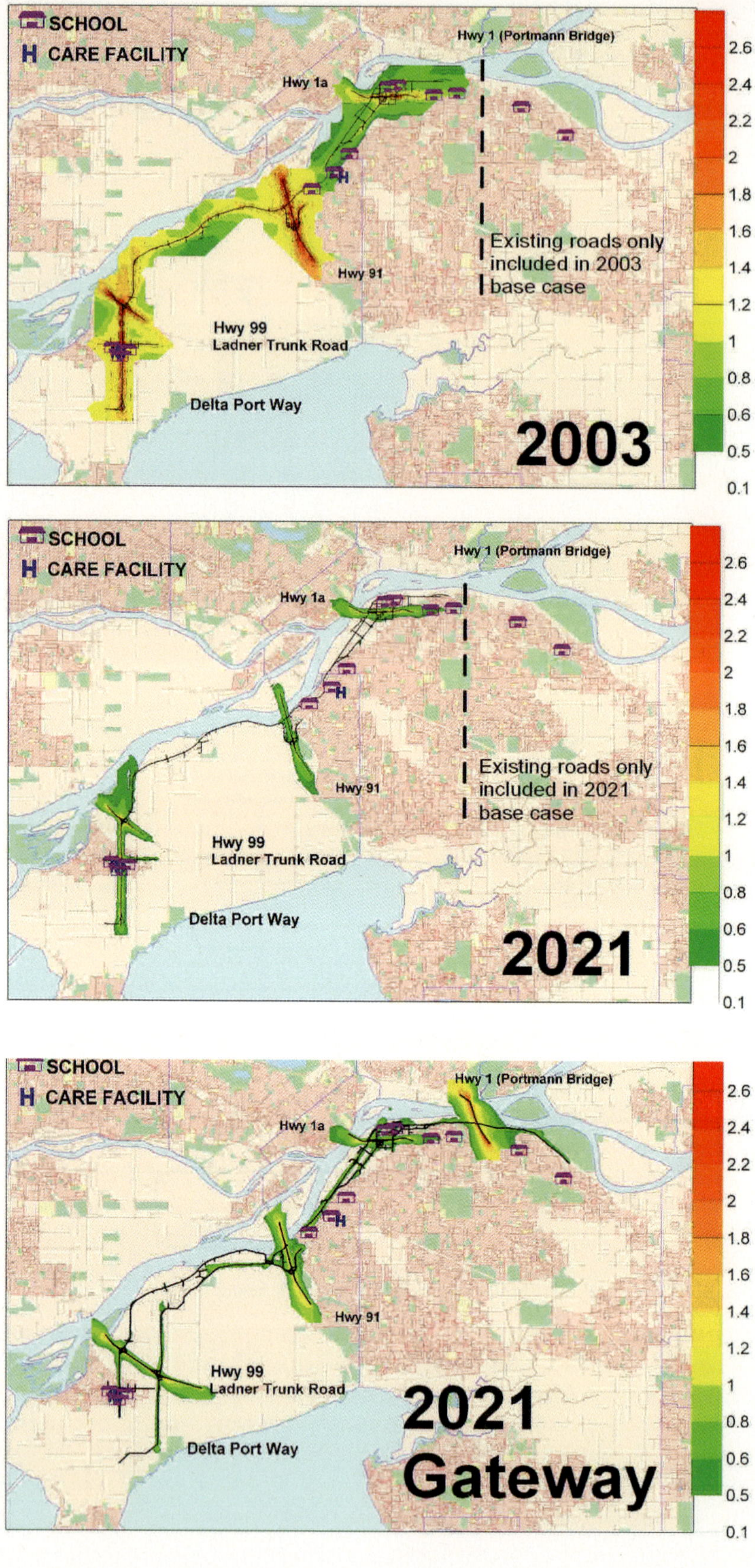

References

BC MoE (2006) A primer on the guidelines for air quality dispersion modelling in British Columbia. http://www.elp.gov.bc.ca/epd/bcairquality/reports/aq_disp_model_06_primer.html. Accessed May 2013

BC MoE (2008) Guidelines for air quality dispersion modelling in British Columbia. http://www.elp.gov.bc.ca/epd/bcairquality/reports/air_disp_model_08.html. Accessed May 2013

Beychok M (1994) Fundamentals of stack gas dispersion, 3rd edn, Irvine, CA. CALINE3 1979, Unabridged User's Guide, US EPA. http://www.epa.gov/ttn/scram/dispersion_prefrec.htm#caline3. Accessed May 2013

GVRD (2003) Lower Fraser Valley ambient air quality report, August 2004

Scire JS (1997) Technical description and user's guide for the FOG model, Earth Tech, Inc, Concord, MA

Scire JS, Strimaitis DG, Yamartino RJ (2000) "A users guide for the CALPUFF dispersion model" Version 5, Earth Tech, Inc. January 2000

South Fraser Perimeter Road (SFPR) (2006) BC Environmental Assessment Office. http://a100.gov.bc.ca/appsdata/epic/html/deploy/epic_project_home_196.html. Accessed May 2013

Turner DB (1994) Workbook of atmospheric dispersion estimates—an introduction to dispersion modeling, 2nd edn. Lewis Publishers, Boca Raton, FL

Vancouver Island Generation Project (VIGP) (2003) BC Environmental Assessment Office. http://a100.gov.bc.ca/appsdata/epic/html/deploy/epic_project_home_195.html. Accessed May 2013

Zannetti P (ed) (2003) Air quality modeling—theories, methodologies, computational techniques, and available databases and software. Volume I—Fundamentals, EnviroComp Institute and Air & Waste Management Association, Publishers

Air Quality Forecasting

6

Brian Bukoski and Eric Taylor

Abstract

Concentrations of air pollutants can change rapidly over hours and days, sometimes reaching levels that can affect human health, particularly within the vulnerable population. Short term air quality forecasts that are reliable and sufficiently accurate can help the public protect their health on a daily basis by providing advice to help lower their health risks. This advice could include limiting short-term exposure to air pollution or reducing physical activity levels.

Short term changes in air quality are a function of both pollutant emissions and the changing state of the atmosphere. Predicting pollutant concentrations on an hourly and daily basis requires not only knowledge of the type and rate of emissions but also an understanding of how pollutants may move, change and disperse in the atmosphere over time. This chapter gives an air quality meteorologist's perspective of the forecasting process, focussing on the atmospheric mechanisms that affect air quality. It summarizes the steps needed to produce a forecast, including the analysis of pollutant concentrations and atmospheric conditions as well as the prediction of their evolution over time.

Keywords

Bluesky · Ambient · Weather prediction · Pollutant emissions · Atmospheric stability · Dispersion modelling · Photochemistry · Air quality forecasting

6.1 Introduction

Air pollutants are solid, liquid or gaseous substances suspended in the atmosphere that can harm human health, as described in the chapter by David Stieb in this book, and can also impact the environment and impair visibility. Pollutant concentrations are a function of their emission rate, movement, dispersion and transformation in the atmosphere and their eventual removal.

Providing short term forecasts of air quality to the public can reduce air pollution-related health risk by providing information with which they can make lifestyle changes to reduce exposure. This is particularly true for vulnerable populations such as infants (and their caregivers) and those with pre-existing medical conditions.

Ozone, particulate matter, lead, carbon monoxide, nitrogen dioxide, sulfur dioxide, toxics and volatile organic compounds are some of the major pollutants affecting health and the environment in Canada. Emission sources are many and varied and include motor vehicles, industry, forest fires and domestic heating. At source, these pollutants may be highly concentrated and harmful to human health, but soon after

B. Bukoski (✉)
Environment Canada, Winnipeg, Canada
e-mail: brian.bukoski@ec.gc.ca

E. Taylor
British Columbia Ministry of Environment, Victoria, BC, Canada
e-mail: eric.taylor@gov.bc.ca

E. Taylor, A. McMillan (eds.), *Air Quality Management,* DOI 10.1007/978-94-007-7557-2_6,

being emitted, these *primary*[1] pollutants become diluted and dispersed in the atmosphere by turbulent mixing. This generally reduces their concentrations and, as a result, their risk to human health. This dilution is particularly rapid where winds are strong and atmospheric turbulence is high. Occasionally, turbulence can actually lead to increases in ground level concentrations when pollutants at higher elevations from such sources as tall industrial stacks or distant forest fires are turbulently mixed down to the surface.

Ambient pollutant concentrations[2] can increase over time if their emission or production rate is higher than the atmosphere's ability to disperse and dilute them. This usually occurs during so-called *stable* atmospheric conditions when the wind and atmospheric turbulence in the surface layer is weak or non-existent. Pollutants emitted into a stable layer are then trapped near the ground, allowing for gradually increasing pollutant concentrations. Identifying these stable conditions is an important part of air quality forecasting and is discussed in more detail later in this chapter.

Some pollutants also undergo photochemical transformation resulting in new, so-called *secondary,* pollutants, such as ozone, being formed. This can sometimes occur even when local emissions are relatively low. For example, as primary pollutants from Metro Vancouver in the western part of the Lower Fraser Valley of British Columbia are blown eastward, photochemical reactions can gradually convert these pollutants into ozone in the far eastern end of the valley, where local emissions are minimal.

6.2 Air Quality Forecasting Essentials

The forecasting of air pollutant concentrations ("air quality") for a particular community or region therefore requires knowledge of the following three processes.

- *Emissions* refers to the types of pollutants that are released by each pollutant source into the atmosphere, their release rates, and the location, elevation and areal extent of the important fixed and mobile sources. Emissions rates vary temporally and spatially, sometimes being dependent on temperature and other environmental characteristics. For example, emissions from domestic wood stoves can increase as outside temperatures fall.
- *Transport and Dispersion.* This refers to the horizontal movement (or lack thereof) of pollutants by winds at all levels and their spreading in three dimensions over time through atmospheric turbulence and diffusion. These processes vary, sometimes rapidly, as atmospheric structures, like cold fronts, move through an area.
- *Transformation* is the chemical and physical changes that pollutants undergo as they move through the atmosphere to the receptor (e.g. a community). These transformations include deposition of solid and liquid particles on surfaces either directly or through precipitation and the physical and chemical reactions that they undergo. These reactions depend on the presence of other airborne pollutants as well as on temperature, solar radiation and other meteorological variables. Sometimes transformation can change pollutant concentrations in a relatively predictable way, such as when ozone is formed by photochemical processes during the day and destroyed overnight. These kind of patterns can persist for several days.

6.3 Outline of the Air Quality Forecasting Process

6.3.1 Forecasting Steps

Forecasting outdoor air quality is still in its early stages. Air quality forecasts made in 2010 by computer models were at the same point that weather forecasting computer models were in the early 1990s (Mike Howe, personal communication 2010). This is largely because significantly more computing power is required for both air quality and weather forecast modelling than for weather forecasting alone. The good news is that computing power continues to increase rapidly, allowing models to better simulate the complex interactions between the atmosphere and air pollutants.

Since the concentrations of pollutants depend largely on the ability of the atmosphere to disperse and dilute them, air quality forecasting relies fundamentally on an accurate assessment of the current and future state of atmospheric conditions. There are three main steps in the air quality and weather forecasting process:

- *Analysis* of the recent air quality and meteorological conditions.
- *Diagnosis* of the current atmospheric processes to identify their potential to either transport and disperse pollutants or allow their concentrations to rise.
- *Prognosis* of how atmospheric processes, air pollutant transformations and transport and dispersal processes will evolve over time.

Air quality forecasts are typically made by an experienced air quality meteorologist at a government or private weather office. The meteorologist can incorporate knowledge of atmospheric changes, chemical and physical pollutant transformation processes and recent pollutant concentration trends in developing the forecast. Air quality computer models are an essential tool in this process, particularly for longer term

[1] *Primary* pollutants are emitted directly into the atmosphere where they can be converted to *secondary* pollutants through chemical and physical processes

[2] "Ambient" air is outside air to which the public is exposed. Generally it applies to air that is far from emission sources

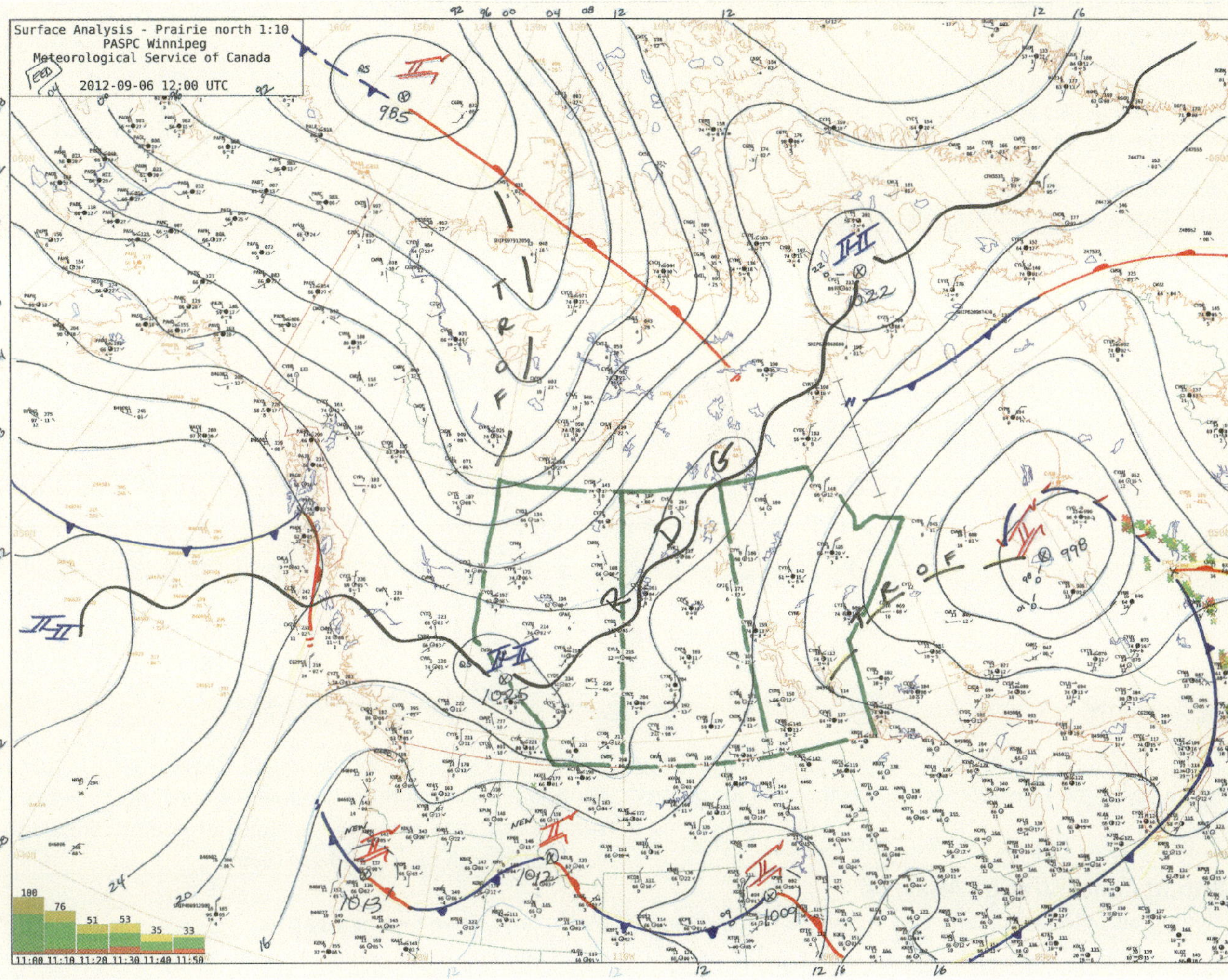

Fig. 6.1 Surface analysis of the weather situation over central Canada. The *solid black* lines are isobars (lines of equal atmospheric pressure). Tightly packed isobars suggest strong pressure gradients and therefore strong surface winds that can disperse pollutants. More stagnant conditions exist under high pressure areas ("*H*" on map) with very weak pressure gradients. Stagnant conditions promote the accumulation of air pollutants. (Image courtesy of Environment Canada's Meteorological Services)

(2–4 day) forecasts. The following sections deal with each step in the forecast process.

6.3.2 Data and Analysis

Meteorological Data

Measurements of hourly meteorological parameters are collected in real time from a variety of federal, provincial and municipal monitoring stations over a wide geographical area (see chapter on air quality monitoring by Jeff Brooks et al). These synoptic data[3] are quickly plotted on weather maps that are then analyzed by a meteorologist to help in the understanding of the spatial pattern of temperature, atmospheric pressure, atmospheric stability, winds and other parameters important to air quality forecasting. Since meteorological data is measured both at the earth's surface and aloft[4], these maps can also represent conditions at various levels in the atmosphere.

An example of a surface synoptic weather map is in Fig. 6.1. It shows areas of low and high atmospheric pressure, often associated with good and poor dispersion of pollutants, respectively. Dispersion is also aided by brisk winds. These are caused by a strong gradient of atmospheric pressure, usually identified by closely spaced isobars[5], represented by the black lines on the map in Fig. 6.1.

Data from individual surface meteorological monitoring stations can also be graphed as a time series to show how

[3] Synoptic meteorological data are measurements that have been made at the same time over a large area.

[4] Upper air meteorological data is measured directly by balloon-borne radiosonde instruments, as well as remotely by satellite

[5] Isobars are lines connecting points of equal atmospheric sea level pressure

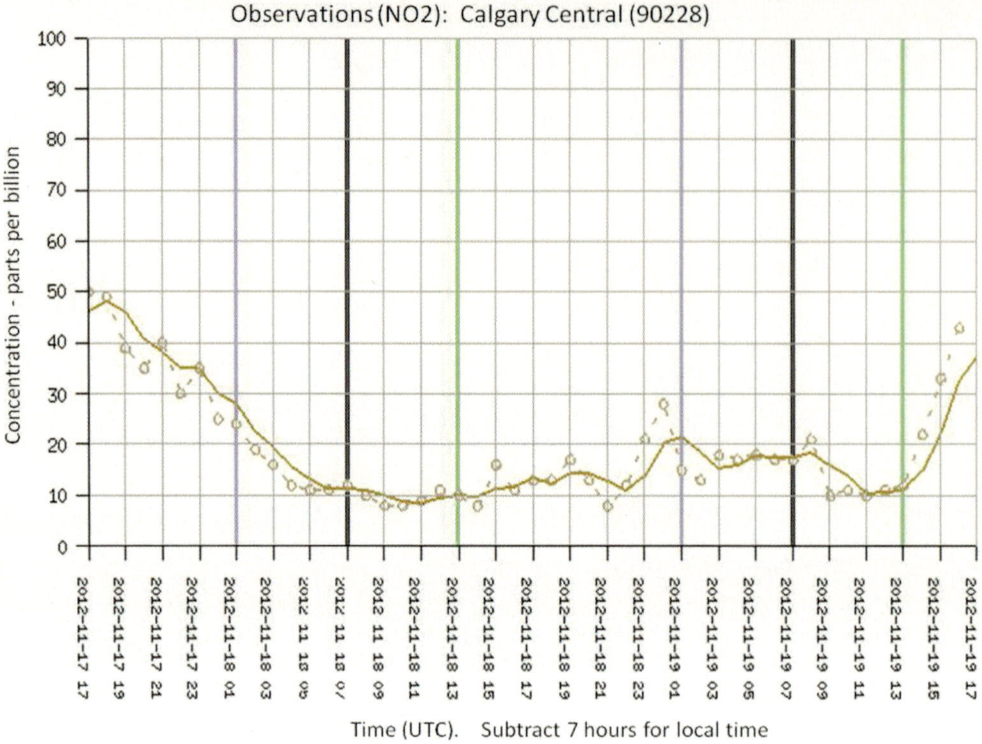

Fig. 6.2 A forty-eight hour time series of observed nitrogen dioxide (NO_2) concentrations at Calgary, Alberta. *Circles* joined by *dotted line* are hourly concentrations. The *solid line* is a three hour moving average. (Graph courtesy of Environment Canada's Meteorological Service)

weather conditions are changing from hour to hour and day to day, helping the forecaster relate weather and air quality changes.

Pollutant Concentration Analysis

Analyzing air pollutant data in near real-time requires rapid access to accurate measurements of major pollutants. Air quality monitoring equipment can measure concentrations of particulates ($PM_{2.5}$[6] and PM_{10}), ozone, nitrogen oxides, sulfur oxides, hydrogen sulfide and carbon monoxide, which can then be quickly transmitted to air quality forecasters. However, in contrast to meteorological monitoring, which often occurs at airports in rural settings, air quality in Canada is generally measured more intensively in urban areas due to the relatively high number of pollutant sources in close proximity to populated areas that can increase human exposure to pollutants. Because of this uneven monitoring pattern, it is often not useful to analyze maps of air pollutant concentrations that encompass large rural areas in the same manner as one would analyze maps of meteorological data, since the urban areas generally have much higher pollutant concentrations. Important exceptions to this are when pollutants are distributed widely over rural areas, often as a result of transport from distant sources such as heavily industrialized regions or extensive and persistent forest fires. Another exception is where pollutant sources of concern are located in rural areas, such as wells, pipelines or compressor stations associated with the oil and gas sector.

Pollutant Time Series Graphs

Pollutant time series graphs are used operationally to analyze recent changes in surface concentrations of each pollutant. For example, Figs. 6.2 and 6.3 show time series of hourly nitrogen dioxide concentrations in Calgary and ozone in Prince George. Pollutant concentrations during degraded air quality episodes over several days often repeat a similar diurnal pattern under long-lasting, stable atmospheric conditions. These analyses can help the air quality meteorologist in the prediction of future pollutant concentrations.

Analysis of Remotely-sensed Air Quality Information

Satellite imagery can help in the assessment and predicted movement of some air pollutants that are visible from space, including forest fire smoke, dust storms, pollution plumes, and ash from volcanic eruptions (Fig. 6.4). Coupling this information with surface and upper level wind information can be used to predict short term movement and dispersion of pollutants. Longer term predictions benefit from programs like AERONET (AErosol RObotic NETwork) that use ground based remote sensors and air pollutant dispersion models to predict the long range transport of pollutants (NASA 2012).

[6] Microscopic aerosols less than 2.5 μm in diameter, often composed of multiple chemicals and usually formed through combustion or other chemical reactions.

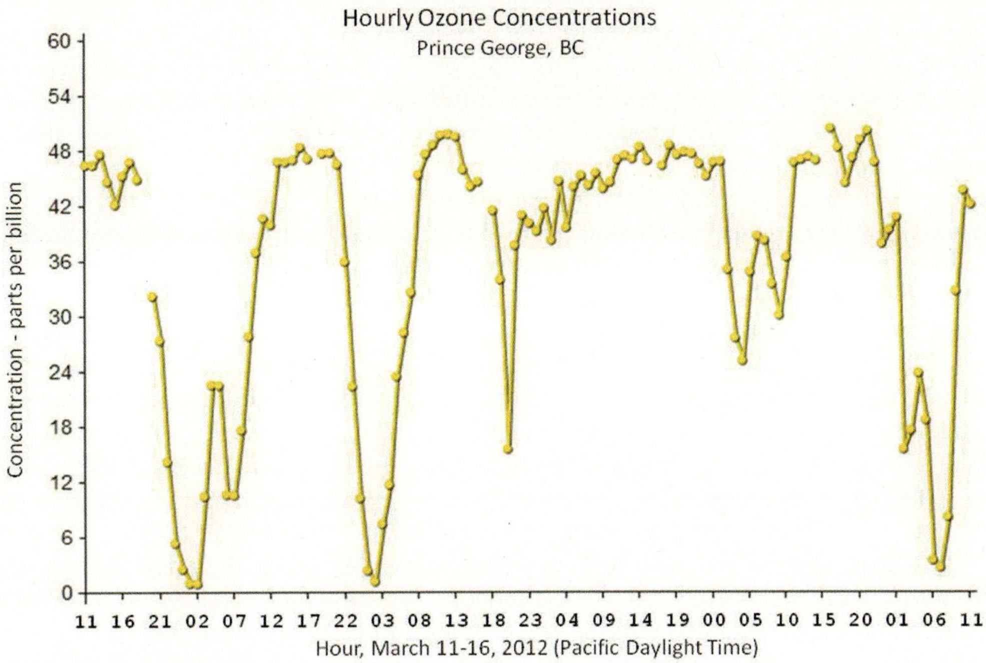

Fig. 6.3 Five day hourly ozone time series at Prince George, British Columbia from a publicly accessible website. (Used with permission of BC Ministry of Environment)

Fig. 6.4 High resolution satellite imagery of forest fire smoke from a polar orbiting satellite. Post processing has identified forest fire "hotspots" in Northern Quebec southeast of James Bay. Smoke (partially transparent *light grey area*) is being blown southward into Ontario and the northern United States. (Courtesy of National Aeronautics and Space Administration (NASA))

6.3.3 Diagnosis of Air Quality and Meteorological Situation

Stable Atmospheric Layers and Poor Dispersion

An important part of forecasting air quality is the ability to diagnose the location of *stable* layers of the atmosphere where pollutants can be trapped near the earth's surface. Stable layers develop when winds are light and where a relatively cooler, and therefore heavier, layer of the atmosphere has formed at the surface. These stable conditions can occur at any time but often form overnight as surface temperatures fall. They also are common when a ridge of high pressure dominates an area, causing temperatures to rise aloft, thus leaving a relatively cool layer at the surface.

A stable layer has limited ability to mix and dilute pollutants upwards or horizontally. Pollutants continually emitted into, or produced in, such a stable layer tend to remain trapped there, leading to increasing concentrations over hours and days. In valley locations, concentrations climb even more rapidly since the stable surface layer is confined horizontally by surrounding hills. Low lying areas are prone to these kinds of stagnant episodes and are particularly common in the deep valleys of western Canada (British Columbia, Yukon, and western areas of the Northwest Territories and Alberta). Stable layers that develop in low lying areas in urban centres can result in high rates of human exposure to air pollutants.

Air quality improves when the stable layer is weakened by brisk winds, often caused by the approach of a frontal system or a similar atmospheric disturbance, or by sufficient solar heating which creates turbulent mixing that disperses and therefore dilutes the pollutants.

Stable surface layers and other phenomena that lead to degraded air quality can be identified from vertical atmospheric profiles of temperature, humidity and wind speed that are obtained from radiosonde soundings. These radiosonde balloons are routinely sent aloft every 12 h from hundreds of locations around the world. Vertical profile information can be augmented with information from satellite and aircraft-based monitors.

Other Types of Diagnosis

Diagnosis involves determining a cause and effect relationship between the changing atmospheric phenomena, changing pollutant emissions and changing air pollutant concentrations. For example, sharp increases in emissions from local sources can cause air quality to be degraded even when there are no significant stable layers at the surface. This can occur, for example, with new large smoke emissions from wildfires or prescribed forest burning or if there has been an "upset" in a local industrial source that has resulted in a rapid increase in pollutant releases. Alternatively, changing wind direction at the surface and aloft can also bring increased pollutant concentrations to a community from both nearby and distant sources.

6.3.4 Prognosis

Prognosis Tools

Tools used for prediction of the future concentration of pollutants range from simple extrapolation of atmospheric features to the use of complex air quality models that take into account emissions, transport, dispersion and chemical transformation of air pollutants.

Simple short term qualitative predictions (1–12 h) can be made by an air quality forecaster projecting the movement of key atmospheric features and estimating the resulting air quality changes in a community. The forecaster generally makes these predictions based on an understanding of how pollutant concentrations have changed in the recent past under similar atmospheric conditions. This kind of qualitative forecast is produced frequently in winter for $PM_{2.5}$ in small valley communities in British Columbia, where stable atmospheric conditions in fall and winter are commonly accompanied by degraded air quality from smoke emissions from domestic wood heating and industry. Similarly, day to day forecasts of ozone are made in major cities in summer when persistent stable layers allow the gradual buildup of pollutants under clear skies. A passing cold front or other disturbance signals the end of these kinds of ozone episodes.

At the small time and space scales described above, computer models are not generally used operationally to predict air quality. However, air quality models are very important for predicting pollutant concentrations in the medium to longer term such as in the next 24–72 h, and over larger geographical areas. These models are based on several types of computer programs:

- weather forecast computer programs, often covering an entire continent or the globe
- databases and programs that estimate emission rates over larger areas (see the chapter on emissions inventory by Warren McCormick in this book),
- programs that simulate air pollutant dispersion as well as chemical and physical transformation of pollutants.

Air quality models have been developed on the global, regional and local scale, and are used to predict concentrations based on estimated emission rates of pollutants and/or their precursors. These air quality models are routinely run in North America and Europe to provide daily air quality predictions for use by environmental and health agencies and the general public.

Environment Canada has developed a comprehensive air quality computer model containing a full description of atmospheric chemistry and meteorological processes. This is an in-line model, meaning that the meteorology is integrated in-step with the chemistry. The model uses an emissions inventory to assess the source of tropospheric ozone precursors. The current version of the model (called GEM-MACH) is run twice a day over North America and is used to provide guidance for the production of air quality forecasts for Canada. The model produces air quality forecast information that shows predicted concentrations of various pollutants (such as ground level ozone, nitrogen dioxide, sulfur dioxide, and fine particulate matter) at various time steps through the forecast period. The forecast information includes predictions of the three components of Canada's Air Quality Health Index (Environment Canada 2012), described elsewhere in this book. Some of these processes are further explained in the section later in this chapter on predicting concentrations of the pollutants used in the Air Quality Health Index.

Environment Canada continues to carry out modelling research on the transport, dispersion, transformation and deposition of pollutants by the atmosphere, including improving the spatial resolution of the model (Environment Canada 2012). More information on air quality modelling can be found in the chapter in this book by Mike Moran (Fig. 6.5).

The Canadian BlueSky Wildfire Smoke Forecasting System

Canada has 10 % of the world's forests, and significant areas of Canadian forests are burned by wildfires annually. For example, 3.2 million hectares were lost due to fire in 2010. Forest fire smoke from these fires can result in elevated concentrations of fine particulate matter ($PM_{2.5}$) and unhealthy gases for hundreds or even thousands of kilometres

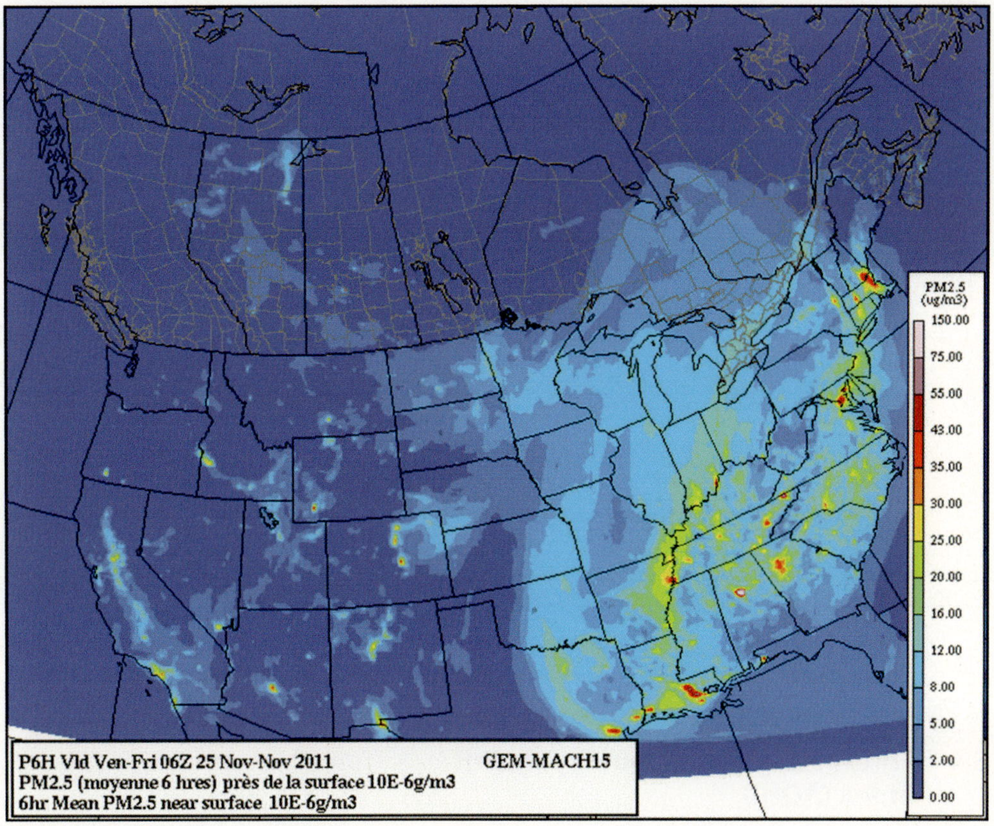

Fig. 6.5 A typical output of the GEM-MACH air quality model that combines a meteorological model with an air pollutant dispersion model. This forecast map shows an estimate of $PM_{2.5}$ surface concentrations 6 h into the future. (Courtesy of Environment Canada's Meteorological Services)

downwind. This smoke can contain harmful chemical compounds. (World Health Organization 2005; Naeher et al. 2007; BC Environment 2012).

Because of this health threat, a specialized air quality model that predicts $PM_{2.5}$ concentrations from wildfire smoke is operating in western and central Canada during the late spring and summer. It is based on the BlueSky Smoke Modelling framework developed by the U.S. Forest Service AirFire Team (Larkin et al. 2008). It consists of data and models of fuel consumption, pollutant emissions, fire locations and intensity, weather forecasts, and smoke dispersion, all linked into a single system. Hourly forecasts for up to two days are generated for ground-level concentrations of $PM_{2.5}$ from wildfires, prescribed burns, and agricultural fires. The output is available from the web (http://www.bcairquality.ca/bluesky/) and is visualized in the form of colours and animations over a region. Figure 6.6 shows a prediction of the Bluesky Smoke Forecasting System.

The Canadian BlueSky system, currently still experimental, covers most of Canada west of Lake Superior as well as a 400 km-wide northern strip of the continental United States. The system requires the participation of a wide variety of organizations that provide timely data and advice. This includes the current and future state of the atmosphere, fire locations, fire intensity and growth and smoke emissions from each fire. These organizations include several provincial and federal government environmental and health agencies, the University of British Columbia and the U.S. Forest Service (BC Environment 2012b; Sakiyama 2012).

"Playground" is a prototype for an alternate version of Bluesky that is based on a similar system operating in the U.S. Playground is currently designed to estimate smoke dispersal from "prescribed" forestry burns and is now being tested at the University of British Columbia. Prescribed burns are purposely ignited by government agencies under controlled situations for hazard reduction and ecosystem restoration. Smoke amounts from these burns are generally much lower than that from large wildfires, though nearby communities can still be negatively affected by smoke. In order to minimize the smoke impacts and to provide notification to people in the potential affected areas, Playground is a web-based tool used to inform burn managers on the smoke consequences of a burn before it actually is ignited. Based on fuel characteristics, burn location and start time, Playground will forecast the resulting spatial distribution and magnitude of smoke concentrations up to two days into the future. Based on this, burn managers can assess whether a burn should occur at the proposed time, or delayed until there are more favourable atmospheric and burning conditions where the smoke impacts would be minimal (Steve Sakiyama, personal communication).

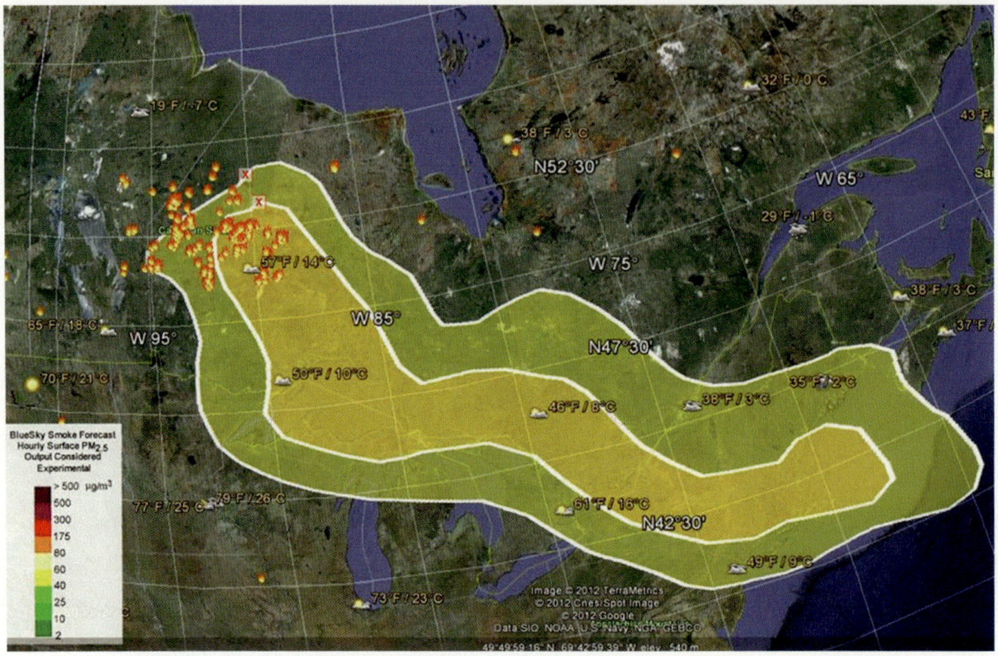

Fig. 6.6 A prediction by the BlueSky forest fire smoke dispersion model showing the expected extent of the smoke plume and the estimated surface $PM_{2.5}$ concentrations from scores of forest fires located southwest of Hudson Bay in northern Ontario. (Reproduced with permission from the BC Ministry of Environment)

6.4 Forecasting the Air Quality Health Index (AQHI)

6.4.1 Predicting the Three Components of the AQHI

One example of the process to predict air quality is the routine forecasting of the three individual pollutants needed for Canada's Air Quality Health Index (AQHI). The AQHI is a function of the concentration of three pollutants: ozone (O_3), nitrogen dioxide (NO_2) and particulate matter ($PM_{2.5}$ or PM_{10}) (Environment Canada 2011). Forecasting the AQHI requires predictions of each of these pollutants (see chapter on the AQHI in this book for more information).

Forecasting NO_2 and ozone necessitates an understanding of their relationships. In urban areas, the primary source of nitrogen dioxide (NO_2), a brown gas commonly associated with urban smog, is the rapid oxidation of nitrogen oxide (NO). NO is produced primarily from fuel combustion, mostly from vehicle traffic and industrial activities.

NO_2, besides being a key part of the AQHI, is also an important precursor to the production of ozone, which itself is a major component of the AQHI. When NO_2 is exposed to sunlight, ozone formation occurs as a result of its photolysis in the presence of other gases such as volatile organic compounds. Photolysis of NO_2 therefore usually results in rising ozone concentrations during the day. Once formed, ozone reacts with NO to regenerate NO_2 (Seinfeld and Pandis 1998). This latter reaction generally leads to relatively higher NO_2 concentrations and lower ozone concentrations at night when the photolytic destruction of NO_2 has ended.

Figure 6.7 shows an example of the hourly concentrations of, and the relationships between, NO, NO_2 and ozone.

Forecasting daytime ozone requires information on current concentrations of ozone, NO and NO_2, their concentrations patterns on the previous day and the expected NO emissions. More important though are the atmospheric changes that are likely to occur, since concentrations of all pollutants are closely related to low level stability, surface wind speeds, turbulence, precipitation (or lack of it) and the amount of solar radiation. In many areas the identification of long range transport in the atmosphere of ozone and its precursors from distant sources is also important for local ozone prediction.

Primary sources of $PM_{2.5}$ include industrial activities, including forestry, wildfires, domestic heating by wood stoves, long range transport of particulates from outside a region and emissions from local traffic. Secondary sources of $PM_{2.5}$ include photochemical reactions, aggregation of smaller particles, and condensation of gas phase compounds onto existing aerosol particles. As with ozone and NO_2, an accurate forecast of atmospheric parameters is often important, particularly in winter.

PM10 can also be used in calculating the AQHI in areas prone to dust. PM10 includes all particles smaller than 10 μm in size.

6.4.2 Producing the Predicted Air Quality Health Index

The Air Quality Health Index (AQHI) is a public health protection tool. It is designed to help people make decisions

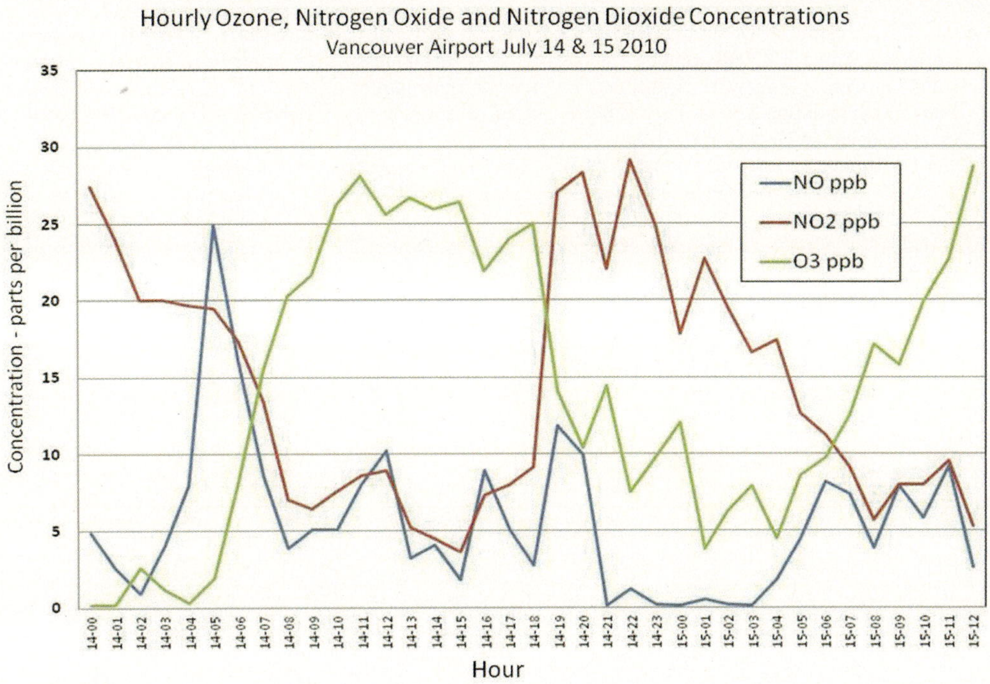

Fig. 6.7 Typical hourly concentrations of nitrogen oxide, nitrogen dioxide and ozone at an urban centre in summer. Vehicle emissions at morning and evening rush hours can briefly increase NO concentrations (*blue line*). Atmospheric chemical reactions can lead to increasing ozone concentrations (*green line*) during the day and decreasing concentrations overnight. NO_2 concentrations (*brown line*) increase in the evening as ozone is destroyed

about lifestyle to protect their health. Two key pieces of advice are included in the AQHI message during degraded air quality episodes:

- Limit short-term exposure to air pollution
- Reduce activity levels

This advice is particularly important for those vulnerable to higher concentrations of air pollutants. More information on this advice is contained in the chapter in this book on the AQHI.

AQHI forecasts are based on the predicted values of its constituent pollutants, ozone, NO_2 and $PM_{2.5}$. In practice, air quality models provide a first estimate of the concentrations of three pollutants. The final forecast of the AQHI is based on a modification of these concentrations where necessary by an air quality meteorologist.

6.5 Conclusion

Accurate, timely and useful short term air quality forecasts are a tool that can lower health risk from air pollution. This is particularly true for the very young, the elderly and those with respiratory or cardio-vascular disease that are vulnerable to degraded air quality. Since the atmosphere plays the dominant role in changing ambient concentrations, air quality forecasters need a sound understanding of how the atmosphere moves, disperses and concentrates particles and gases and how these are transformed over time, both chemically and physically. Real time measurements of both pollutant concentrations and weather variables are needed, along with estimates of emissions. Of particularly importance is the ability of a forecaster to identify geographical areas that are, or will, experience stable conditions in the low levels of the atmosphere, since this is where emissions can increase pollutant concentrations over time, raising health risks. This situation can occur in winter when stable surface layers develop during cold, stagnant weather events if domestic wood heaters are emitting large amounts of $PM_{2.5}$ into the community airshed. Stable conditions in summer can also concentrate pollutants such as NO_2 and other pollutants associated with traffic. During sunny conditions, this can result in solar radiation triggering the creation of ground level ozone through photochemical processes. Air quality forecasters rely increasingly on complex air quality computer models from the Canadian Meteorological Centre to develop their forecast products. These models simulate the dispersion, transport and transformation of pollutants to provide estimates of future ambient pollutant concentrations. They are particularly useful when pollutants are transported over long distances.

References

Environment BC (2012) Air quality and your health. http://www.bcairquality.ca/health/. Accessed 24 May 2012

Environment BC (2012b) Smoke forecasts for Western Canada. http://www.bcairquality.ca/bluesky/. Accessed 24 May 2012

Environment Canada (2012) Air quality model forecasts. http://www.weatheroffice.gc.ca/aqfm/index_e.html and http://www.ec.gc.ca/air-sc-r/default.asp?lang=En&n=1C816B9D-1. Accessed 8 May 2012

Environment Canada (2011, June 30) Air quality health index. http://www.airhealth.ca. Accessed 24 May 2012

Larkin NK, Sullivan D, O'Neill S, Raffuse S, Solomon R, Krull C, Rorig K, Strand T (2008) Smoke forecasting using the bluesky smoke modeling framework. 15th joint conference on the applications of air pollution meteorology with the air and waste management association. New Orleans, LA

Naeher LP, Brauer M, Lipsett M, Leikoff JT, Simpson CD, Koenig JQ, Smith KR (2007) Woodsmoke health effects: a review. Inhal Toxicol 19:27–106

NASA (2012) AERONET. http://aeronet.gsfc.nasa.gov/. Accessed 20 Oct 2012

Sakiyama S (2012) The bluesky Western Canada wildfire smoke forecasting system. http://www.bcairquality.ca/bluesky/BlueSky-Smoke-Forecasts-for-Western-Canada.pdf. Accessed 24 May 2012

Seinfeld JH, Pandis SN (1998) Atmospheric chemistry and physics. Wiley, USA

World Health Organization (2005) Air quality guidelines, global update. Chapter 10. Particulate matter. World Health Organization report, ISBN 92 890 2192 6. http://www.who.int/phe/health_topics/outdoorair_aqg/en/. Accessed 25 May 2012

Part II

Air Quality Impacts

Air Quality Impacts on Health

7

David M. Stieb and Ling Liu

Abstract

In this chapter, we provide a broad overview of the evidence linking air pollution with adverse health effects, focusing on "criteria" pollutants (CO, NO_2, O_3, SO_2 and particulate matter). We begin with a brief historical background including a description of air pollution events which gave rise to modern air pollution control legislation. We then describe types of research evidence, their advantages and disadvantages and their role in understanding effects on human health. Following this is a review of factors which determine vulnerability to effects of air pollution and which influence exposure and dose. Next we summarize the most influential studies from the worldwide scientific literature (restricted to effects observed in humans), as well as highlighting a number of Canadian studies. Finally, we briefly discuss policy implications and identify gaps in the evidence which must be addressed by future research in order to ensure that exposure and health risks of air pollution are understood accurately and can be reduced effectively.

Keywords

Air pollution exposure · Air pollution epidemiology · Respiratory · Cardiovascular · Traffic · Woodsmoke

7.1 Introduction

A series of severe air pollution episodes in the 1930s–1960s in the Meuse Valley, Belgium, Donora, Pennsylvania and London, England, drew attention to the link between emissions from industry, home heating and transportation sources, weather conditions, and effects on health. Pollutant concentrations at that time were much higher, and associations with adverse health impacts were more readily apparent compared to current conditions in Canada. As seen in Fig. 7.1, thousands of deaths occurred in London, England immediately following a severe episode of increased SO_2 and particulate matter in December 1952[1] (Bell and Davis 2001). As a result of these severe episodes, clean air legislation was implemented in many jurisdictions which resulted in dramatic improvements in air quality in many wealthy industrialized countries. Still, a series of studies conducted in the 1980s and 1990s revealed that despite these improvements, associations between air pollution and adverse health impacts could still be detected even at these lower levels of exposure. Studies in Canada have been particularly informative with respect to evaluating effects at low levels of exposure. Increasing attention has also been paid to comparatively high levels of exposure and impacts in middle and low-income countries (Prüss-Üstün and Corvalán 2006).

D. M. Stieb (✉)
Environmental Health Science and Research Bureau,
Health Canada, 3rd floor, 4595 Canada Way, Burnaby,
BC, V5G 1J9, Canada
e-mail: dave.stieb@hc-sc.gc.ca

L. Liu
Environmental Health Science and Research Bureau, Health Canada,
50 Colombine Driveway, Tunney's Pasture (Building # 8), A.L. 0801A,
K1A 0K9 Ottawa, ON, Canada
e-mail: ling.liu@hc-sc.gc.ca

[1] Interestingly, Dr. David Bates, a pioneer of research on air pollution and health in Canada, was practising medicine in London at the time of the London fog episode. He recounts his experiences and puts the event in historical perspective in an informative commentary (Bates 2002).

E. Taylor, A. McMillan (eds.), *Air Quality Management,* DOI 10.1007/978-94-007-7557-2_7,

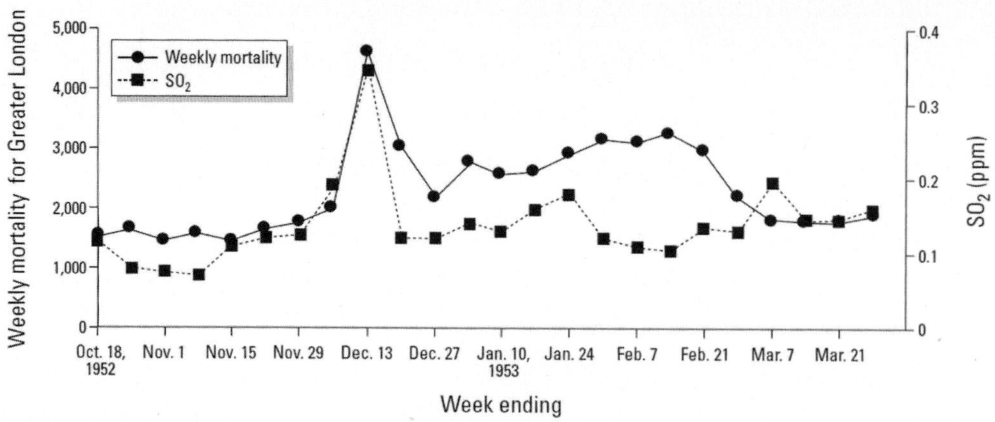

Fig. 7.1 Approximate weekly mortality and SO_2 concentrations for Greater London, 1952–1953. (Reproduced with permission from Environmental Health Perspectives, Bell and Davis 2001)

7.2 Types of Evidence of Adverse Health Effects

Evidence linking air pollution with adverse health effects comes from essentially three complementary forms of research: toxicology, human clinical studies and epidemiology (Boyes et al. 2007).

Toxicological studies employ whole animals (*in vivo* studies) or animal cells (*in vitro* studies), high doses over short or long durations in small homogeneous samples, and close control of experimental conditions. The primary limitations of toxicological studies are related to extrapolation between species, and from high to low dose. Incorporating evidence from toxicological studies into assessments of the risk to humans typically involves applying safety factors to account for differences between species.

Human clinical studies are similar to toxicological studies in that they involve close control of experimental conditions and generally small numbers of subjects who are often pre-screened (eg. healthy non-smokers). Durations are short (usually a few hours) and doses are lower than in toxicological studies but often higher than in epidemiological studies. Health outcomes observed are mild and reversible. The main limitations of human clinical studies again relate to extrapolation: from small homogeneous samples in controlled conditions to the general population in real-world conditions and from acute to chronic exposures. In the specific context of air pollution research, this type of study is conducted in a laboratory with controlled atmospheric conditions. Such experiments document the health effects that result from breathing known concentrations of pollutants. The highly controlled environment allows researchers to identify responses to individual pollutants and characterize an exposure-response relationship. The double-blind protocol, which uses a cross-over design with filtered clean air as control, is optimal. In such experiments, the investigator, the technician and the subjects are unaware of the exposure conditions, that is, whether the exposure on a single day involves a pollutant or control substance. Such a strategy eliminates potential observer bias and avoids relaying clues to participants regarding anticipated responses.

Epidemiological studies are generally observational and employ varying sample sizes of diverse individuals. The degree of confidence in epidemiological studies is greatest when information on exposure and other known or suspected risk factors for the health outcome is collected at the outset, prior to the occurrence of the health outcome of interest rather than after the fact (prospectively rather than retrospectively).

Exposures and effects can be both acute and chronic, but exposures tend to be characterized poorly compared to toxicological or human clinical studies. In epidemiological studies of air pollution, exposure is often based on routinely available monitoring data from government monitoring sites. Because people live at different distances from these sites and move around their communities for work, school and other activities in the course of their daily routines, relying on fixed monitoring sites does not reflect their exposure with complete accuracy. The effect of this exposure measurement error is that the true effect of air pollution may actually be larger than that detected in epidemiological studies. The primary concern with epidemiological studies is causal attribution in that it is often difficult to account for all factors which could potentially obscure the relationship between exposure and response.

Prospective cohort studies provide the strongest evidence that air pollution causes adverse health effects. In these studies, participants are recruited at the outset of the study, a variety of information is collected on their air pollution exposure as well as health status and risk factors such as smoking, alcohol consumption, diet and occupation, and they are followed sometimes for many years, periodically determining their health status over time.

Another commonly applied study design is the time-series study. In this type of study, the frequency of health outcomes such as deaths, hospital admissions or emergency department visits is observed from day to day together with daily air pollution concentrations. The relationship between health outcomes and air pollution is examined, adjusting for factors like weather which also vary from day to day, and "filtering" out other factors like day of the week and season which can affect both air pollution concentrations and health outcomes. Advantages of this type of study include the low cost because it does not require collection of new data, and the fact that individual risk factors like smoking do not confound the results because they do not vary from day-to-day within the population. Case-crossover studies employ a related design in which exposure conditions on the day of an event like a hospital admission (the "case" period) are compared with conditions on another day before or after the event (the "control" period).

A sort of hybrid of cohort and time-series studies is the panel study in which groups of people are examined as they engage in daily activities in their usual environment where pollutant levels may be closely monitored from day to day. In some panel studies, participants wear small air samplers to collect data on air pollution at the personal level. In these studies, participants may be either healthy or have existing disease(s) or health conditions; they can be of all ages, and their medical histories, socioeconomic status and lifestyles are known. Their daily activity patterns, episodes of illness, physiological and biochemical measures may be closely monitored.

With increasing computerization of health outcome and other data, record linkage has emerged as a potentially powerful new tool in conducting epidemiological studies. By linking records for a health outcome like death, with records for hospital admissions, physician office visits, or prescription medication, or with other public records which provide information on personal characteristics or place of residence, it has been possible for investigators to more fully understand an individual's health status prior to an event like death or hospital admission, to improve exposure classification, or to more fully account for other risk factors like socioeconomic status. Record linkage permits this without the cost of collecting entirely new data.

Controlled toxicological and clinical studies assess the effects of exposure of precisely controlled concentrations of pollutants, usually alone, but sometimes in combination. Real world exposures, on the other hand, involve complex mixtures which cannot be readily reproduced in a laboratory. In most cases, epidemiological studies rely on data on a small number of routinely monitored pollutants collected at fixed location regulatory monitoring sites. These data effectively constitute a convenient indicator of the true pollutant mix, and observed associations based on these data should be interpreted with caution—they do not necessarily reflect a causal association with the measured pollutant. Particulate matter is itself a complex mixture which is poorly understood in terms of which constituents are associated with the greatest toxicity. It is not readily resuspended in a laboratory setting to simulate real world exposures, and there is currently only one facility in Canada located in Toronto capable of concentrating particulate matter from outdoor air for the purposes of controlled clinical studies. Another facility in Vancouver permits evaluation of the effects of controlled exposure to diluted vehicle exhaust.

Although each type of evidence has limitations, taken together they constitute a complementary weight of evidence. Human clinical studies and toxicological studies often uncover physiological mechanisms which explain effects observed in epidemiological studies. In practice, results from epidemiological studies, particularly those employing data from central site air quality monitors used to measure compliance with standards, are most directly applicable to the development of standards, quantification of public health impacts of air pollution (Toronto Public Health 2007), and assessment of the benefits of proposed regulatory programs (Government of Canada 2007). Unique applications of results of epidemiologic studies in Canada include development of the Air Quality Health Index public communication tool (see Chap. 18) and evaluation of changes in health risk over time (Shin et al. 2008).

7.3 Vulnerability, Exposure and Dose

Not surprisingly, some individuals are more likely than others to experience adverse health effects from air pollution exposure. Children, the elderly and those with respiratory or cardiovascular disease and diabetes appear to be more susceptible to adverse effects. These groups together constitute approximately 40 % of the population (source: 2006 census and 2005 Canadian Community Health Survey). Children have immature immune and respiratory defences while the elderly are more likely to be physiologically compromised and have a greater prevalence of chronic disease. Individuals with diabetes are at greater risk of cardiovascular disease. Lower socioeconomic status may also result in increased vulnerability (Sacks et al. 2011). Certain genetic variants (polymorphisms) which influence blood clotting and response to oxidative stress may also put some individuals at greater risk as a result of gene-environment interactions (Peters et al. 2009; Hayes and McLellan 1999; Romieu et al. 2004; Schwartz et al. 2005).

The probability of an adverse health effect due to air pollution exposure also depends on the amount of pollutant inhaled, or dose, which is the product of exposure (concentration of pollutant per volume of air) and inhalation rate. Exposure to outdoor air pollution is governed partly by time activity patterns in that exposure is increased when outdoor activity corresponds to periods of higher ambient concentrations. Since individuals spend the vast majority of their time indoors (Fig. 7.2; Leech et al. 1996), factors such as opening windows which increase penetration of pollutants indoors, increase outdoor pollutant exposure. Conversely,

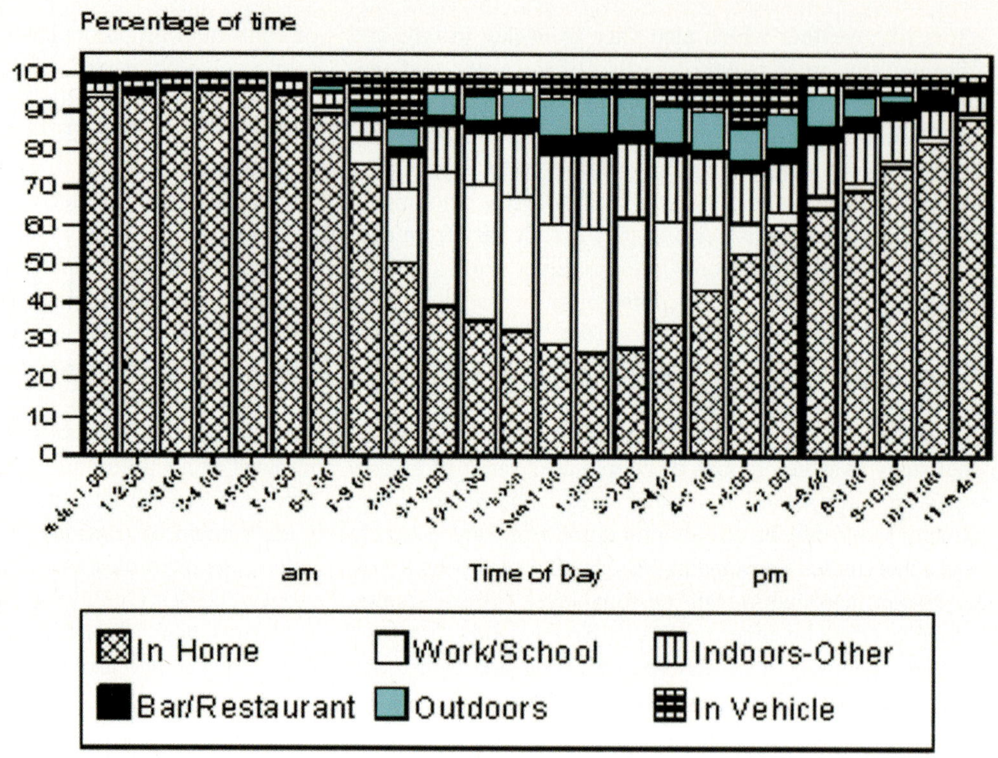

Fig. 7.2 Percent of time in selected environments by time of day. (Reprinted with permission from Chronic Diseases in Canada, Leech et al. 1996)

air conditioning reduces exchange with outdoor air, reducing exposure. Dose is increased with increased inhalation rate, as in those engaged in strenuous physical activity for work or recreation. Inhalation rate varies several-fold over the range from sedentary to intense physical activity (Fig. 7.3; USEPA 2011a). Children are more likely to be physically active outdoors, increasing their exposure, and also inhale at a greater rate per kg of body weight, increasing their relative dose.

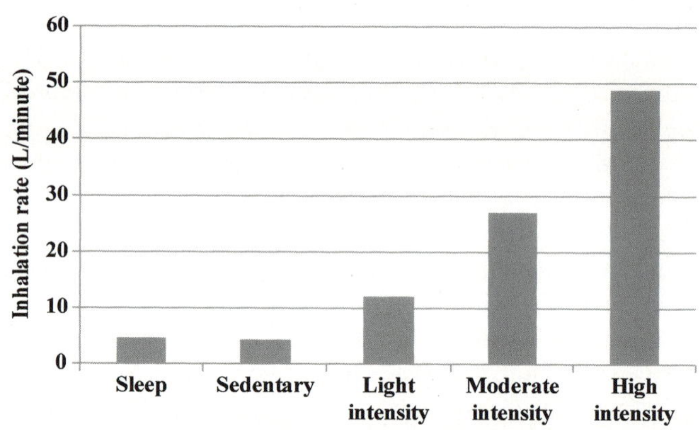

Fig. 7.3 Inhalation rate by level of activity. (USEPA 2011a)

7.4 Summary of Key Studies

7.4.1 Short-term Health Effects

Hundreds of studies have been completed examining short term associations between air pollution and health. This is largely attributable to the ready availability of administrative data (eg. mortality, hospital admission) originally collected for other purposes but suitable for examining associations with air pollution. Several meta-analyses pooling results across individual studies have also been completed (Bell et al. 2005; Ito et al. 2005; Levy et al. 2005; Stieb et al. 2002). Perhaps the most influential time-series study is the National Morbidity, Mortality and Air Pollution Study (NMMAPS) comprising an analysis of air pollution and mortality in 95 US cities (Samet et al. 2000a, b) in which particulate matter exhibited the most consistent associations with mortality. The Air Pollution and Health in Europe and North America (APHENA) study was an international collaboration involving analysis of air pollution and mortality on two continents. Significant associations of PM_{10} with mortality were observed, although interestingly, associations were larger in Canada compared to Europe and the US (Fig. 7.4; Samoli et al. 2008). This may indicate that air pollution is most potent at lower concentrations and becomes less potent as concentrations increase. This is consistent with the relative potency of long term exposure to particulate matter in ambient air, environmental tobacco smoke and from personal smoking (Pope et al. 2009, 2011).

A number of Canadian time-series studies have also been conducted (see also Table 7.1). David Bates and colleagues conducted some of the earliest Canadian studies of this kind

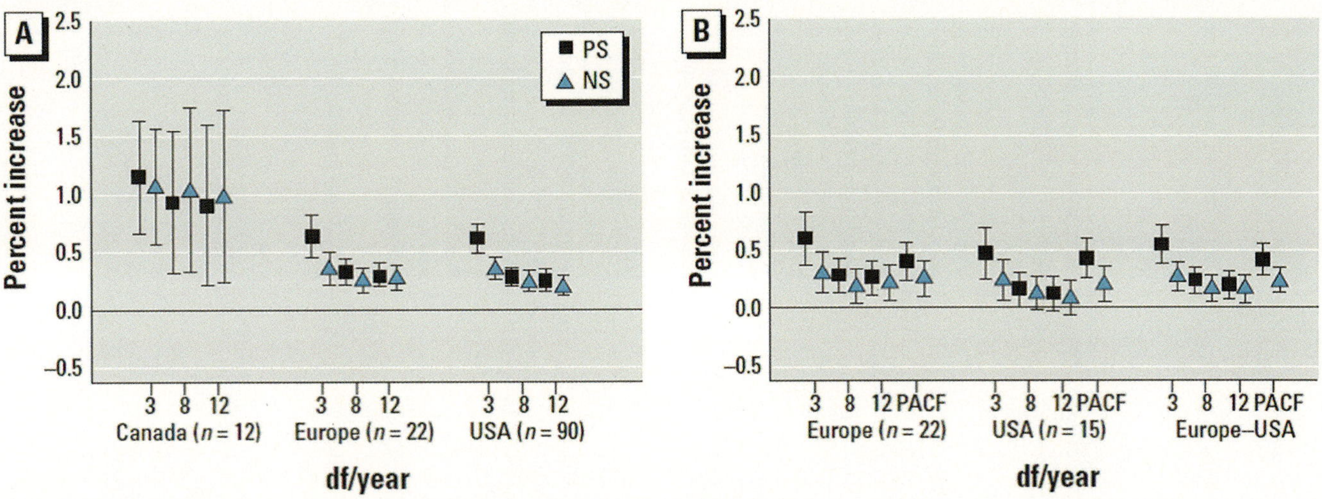

Fig. 7.4 Percent increase in mortality per 10 μg/m^3 increase in PM_{10} (error bars indicate the degree of uncertainty) comparing Canada, the US and Europe according to alternative statistical models (NS and PS; 3, 8, 12; and PACF); n refers to number of cities. (Reproduced with permission from Environmental Health Perspectives, Samoli et al. 2008)

using hospital admission and emergency department visit data. In an analysis of acute care hospitals in Southern Ontario, they found that summertime respiratory hospital admissions were associated with ozone and sulphate concentrations and termed this the "acid summer haze effect" (Bates and Sizto 1987). In contrast, in a study of asthma emergency visits in Vancouver, SO_2 was associated with asthma visits in winter (Bates et al. 1990). Rick Burnett and colleagues subsequently expanded these datasets to include a larger number of cities and applied newer statistical methods. Like the original study by Bates they found associations between ozone and sulphate and respiratory hospital admissions in Southern Ontario (Burnett et al. 1994). Associations of $PM_{2.5}$ with both respiratory and cardiovascular admissions were also reported in a related analysis (Burnett et al. 1995). Since then a number of Canadian investigators have published analyses of the short-term associations between air pollution and adverse health effects using administrative data (eg. Burnett et al. 1997a, 1997b, 1999, 2004; Smargiassi et al. 2009; Stieb et al. 2009; Vedal et al. 2003; Villeneuve et al. 2006). In particular, Goldberg et al. have exploited a unique capacity to link mortality and morbidity data (records for hospital admissions, physician office visits, and prescription medication use) in Quebec. In this way, they have been able to identify health conditions such as acute lower respiratory disease and heart failure which appear to predispose people to dying in relation to air pollution exposure (Goldberg et al. 2001, 2003).

Studies of acute effects of air pollution are not limited to mortality and hospital admissions or emergency department visits evaluated using administrative data. A number of panel studies have examined the association between day to day changes in air pollution concentration and respiratory symptoms and lung function. Ward and Ayres, for example, reviewed 22 panel studies examining the association of PM_{10} and $PM_{2.5}$ with lung function measures and respiratory symptoms in children. Average effects across the studies showed that both PM_{10} and $PM_{2.5}$ were associated with reduced lung function (Fig. 7.5) and increased reporting of respiratory symptoms. Considerable variability was noted among studies and there was some evidence that results were biased in favour of studies showing positive effects (publication bias). In a recent panel study in Windsor, Ontario, a group of asthmatic children (grades 4–8) was followed over one month, and their pulmonary function and lung inflammation tested repeatedly. Among these asthmatic children, particulate matter, NO_2 and SO_2 were significantly associated with reduced pulmonary function (Dales et al. 2009a; Liu et al. 2008) and markers of increased oxidative stress (Liu et al. 2008).

Although the respiratory system is the first point of contact with air pollution, studies have also demonstrated a variety of cardiovascular effects, including acute and chronic effects on blood vessels (eg. blood pressure), clotting, metabolism (eg. diabetes), and heart rate and rhythm. These were recently summarized in a review published by the American Heart Association (Fig. 7.6). Panel studies have demonstrated that air pollution is associated with increased heart rate and decreased heart rate variability (HRV—a measure of nervous system control of heart rate) in adults, especially in the elderly with pre-existing cardiovascular conditions (Whitsel et al. 2009; Chahine et al. 2007; Park et al. 2005). This adverse effect was reduced in subjects with higher intakes (above median levels) of vitamins B6, B12, or methionine (Baccarelli et al. 2008). In a study done in Vancouver, researchers observed only weak associations between particle concentration and heart rhythm abnormalities and with

Table 7.1 Summary of selected Canadian studies

Reference	Design	Outcome	Population	Location	Time period	Pollutants	Summary of findings
Time series studies of mortality							
Samoli et al. (2008)	Time-series	Mortality	General population	Calgary, Edmonton, Halifax, Hamilton, Montreal, Ottawa, Quebec, Saint John, Toronto, Windsor, Winnipeg, Vancouver (+112 US and European cities)	1987–1996	PM_{10}, O_3	Larger risk in Canadian than US/European cities
Burnett et al. (2004)	Time-series	Mortality	General population	Calgary, Edmonton, Halifax, Hamilton, Montreal, Ottawa, Quebec, Saint John, Toronto, Windsor, Winnipeg, Vancouver	1981–1999	CO, NO_2. O_3, PM_{10}, $PM_{2.5}$, SO_2	Mortality was most strongly associated with NO_2
Vedal et al. (2003)	Time-series	Mortality	General population	Vancouver	1994–1996	CO, NO_2. O_3, PM_{10}, SO_2	Mortality was most strongly associated with O_3 in summer, and NO_2 in winter
Goldberg et al. (2001)	Time-series	Mortality	General population	Montreal	1984–1993	CO, COH, NO_2. O_3, SO_2, SO_4	Particle measures were associated with mortality only in individuals with acute lower respiratory disease or chronic coronary artery disease or heart failure
Goldberg et al. (2003)	Time-series	Mortality	General population	Montreal	1984–1993	CO, COH, NO_2. O_3, SO_2	COH, NO_2 and SO_2 were associated with mortality in individuals with pre-existing heart failure
Time series studies of cardiac or respiratory morbidity							
Bates and Sizto (1987)	Time-series	Respiratory hospital admissions	General population	Southern Ontario	1974–1983	COH, NO_2, O_3, SO_2 SO_4	Summer admissions associated with O_3, SO_4
Bates et al. (1990)	Time-series	Asthma emergency department visits	General population	Vancouver	1984–1986	COH, NO_2, O_3, SO_2 SO_4	Winter visits associated with SO_2
Burnett et al. (1994)	Time-series	Respiratory hospital admissions	General population	Southern Ontario	1983–1988	O_3, SO_4	Summer admissions associated with O_3, SO_4
Burnett et al. (1995)	Time-series	Cardiac, respiratory hospital admissions	General population	Southern Ontario	1983–1988	O_3, SO_4	Admissions associated with SO_4
Burnett et al. (1997a)	Time-series	Heart failure hospital admissions	>65 years	Calgary, Edmonton, Hamilton, London, Montreal, Ottawa, Toronto, Windsor, Winnipeg, Vancouver	1981–1991	CO, COH, NO_2, O_3, SO_2	Admissions associated with CO
Burnett et al. (1997b)	Time-series	Respiratory hospital admissions	General population	Calgary, Edmonton, Halifax, Hamilton, London, Montreal, Ottawa, Quebec, Regina, Saint John, Saskatoon, Toronto, Windsor, Winnipeg, Vancouver, Victoria	1981–1991	CO, COH, NO_2, O_3, SO_2	Admissions associated with O_3

Table 7.1 (continued)

Reference	Design	Outcome	Population	Location	Time period	Pollutants	Summary of findings
Burnett et al (1999)	Time-series	Cardiac, respiratory, cerebrovascular, peripheral vascular hospital admissions	General population	Toronto	1980–1994	CO, NO_2. O_3, PM_{10}, $PM_{2.5}$, $PM_{10-2.5}$, SO_2	Cardiac and respiratory admissions associated with mix of CO, NO_2, O_3, $PM_{2.5}$, $PM_{10-2.5}$ and SO_2
Smargiassi et al. (2009)	Case-cross-over	Asthma emergency department visits, hospital admissions	Children 2–4 years	Montreal	1996–2004	SO_2	SO_2 from petroleum refinery stack emissions was associated with asthma episodes in young children
Stieb et al. (2009)	Time-series	Cardiac and respiratory emergency department visits	General population	Edmonton, Montreal, Ottawa, Saint John, Toronto, Vancouver	1992–2003	CO, NO_2. O_3, PM_{10}, $PM_{2.5}$, SO_2	CO and NO_2 were most strongly associated with cardiac visits and O_3 was most strongly associated with respiratory visits
Villeneuve et al. (2006)	Case-cross-over	Stroke emergency department visits	General population	Edmonton	1992–2002	CO, NO_2. O_3, PM_{10}, $PM_{2.5}$, SO_2	CO and NO_2 were most strongly associated with stroke in the warm season
Panel studies							
Dales et al. (2009a)	Panel	Lung function	Children 9–14 years with asthma	Windsor	2005	NO_2, O_3, $PM_{2.5}$, SO_2	$PM_{2.5}$ was most strongly associated with reduced lung function
Liu et al. (2008)	Panel	Lung function, inflammatory and oxidative stress markers	Children 9–14 years with asthma	Windsor	2005	NO_2, O_3, $PM_{2.5}$, SO_2	NO_2, $PM_{2.5}$, and SO_2 were associated with lung function and oxidative stress in children with asthma, particularly those not taking inhaled steroids
Goldberg et al. (2008)	Panel	Oxygen saturation and heart rate	Heart failure patients	Montreal	2002–2003	O_3, NO_2, SO_2, $PM_{2.5}$	SO_2 was associated with reduced oxygen saturation
Goldberg et al. (2009)	Panel	Shortness of breath and rating of general health	Heart failure patients	Montreal	2002–2003	O_3, NO_2, SO_2, $PM_{2.5}$	O_3 was associated with rating of general health
Brauer et al. (2001)	Panel	Lung function, heart monitoring, blood pressure	Chronic obstructive pulmonary disease patients	Vancouver		CO, NO_2. O_3, PM_{10}, $PM_{2.5}$, SO_2	Weak associations were observed between PM and heart rhythm disturbances and blood pressure
Liu et al. (2007b)	Panel	Heart rate, blood pressure, vascular function, oxidative stress markers	Diabetic patients	Windsor	2005	PM_{10}	PM_{10} was associated with oxidative stress and vascular function
Liu et al. (2009)	Panel	Heart rate, blood pressure, vascular function and markers of inflammation and oxidative stress	Seniors	Windsor	2007	$PM_{2.5}$, black carbon	$PM_{2.5}$ and black carbon were associated with blood pressure, heart rate, brachial artery diameter and mediators of vascular function and oxidative stress

Table 7.1 (continued)

Reference	Design	Outcome	Population	Location	Time period	Pollutants	Summary of findings
Controlled human exposure studies							
Brook et al. (2009)	Controlled exposure	Blood pressure, heart rate variability, vascular reactivity	Healthy adults	Toronto		O_3, $PM_{2.5}$	$PM_{2.5}$ was associated with increased blood pressure and reduced heart rate variability and vascular function
Sivagangabalan et al. (2011)	Controlled exposure	Heart electrical activity	Healthy adults	Toronto		O_3, $PM_{2.5}$	O_3 together with $PM_{2.5}$ affected electrical activity in the heart
Brook et al. (2002)	Controlled exposure	Vascular function	Healthy adults	Toronto		O_3, $PM_{2.5}$	O_3 together with $PM_{2.5}$ affected vascular function
Urch et al. (2005)	Controlled exposure	Blood pressure	Healthy adults	Toronto		O_3, $PM_{2.5}$	O_3 together with $PM_{2.5}$ increased blood pressure
Urch et al. (2004)	Controlled exposure	Vascular function	Healthy adults	Toronto		O_3, $PM_{2.5}$	Organic and elemental carbon components of $PM_{2.5}$ were associated with vascular function
Studies of non-cardiorespiratory outcomes							
Brauer et al. (2008)	Cohort	Low birth weight, preterm birth	Newborns	Vancouver	1999–2002	Black carbon, CO, NO, NO_2. O_3, PM_{10}, $PM_{2.5}$, SO_2	Proximity to highways and pollutants other than O_3 were associated with birth weight outcomes and preterm birth
Dugandzic et al. (2006)	Cohort	Low birth weight	Newborns	Nova Scotia	1988–2000	O_3, PM_{10}, SO_2	PM_{10} and SO_2 were associated with low birth weight
Liu et al. (2003)	Cohort	Intrauterine growth restriction, low birth weight, preterm birth	Newborns	Vancouver	1985–1998	CO, NO_2. O_3, SO_2	Low birth weight was associated with SO_2, intrauterine growth restriction was associated with CO, NO_2 and SO_2 and preterm birth was associated with CO and SO_2.
Liu et al. (2007)	Cohort	Intrauterine growth restriction	Newborns	Calgary, Edmonton, Montreal	1985–2000	CO, NO_2. O_3, $PM_{2.5}$, SO_2	CO, NO_2 and $PM_{2.5}$ were associated with intrauterine growth restriction
Szyszkowicz et al. (2009a)	Time-series	Emergency department visits for headache	General population	Edmonton	1992–2002	CO, NO_2. O_3, $PM_{2.5}$, SO_2	NO_2, $PM_{2.5}$ and SO_2 were associated with emergency department visits for headache
Szyszkowicz et al. (2009b)	Time-series	Emergency department visits for headache	General population	Edmonton, Halifax, Ottawa, Toronto, Vancouver	1992–2003	CO, NO_2. O_3, PM_{10}, $PM_{2.5}$, SO_2	NO_2, $PM_{2.5}$ and SO_2 were associated with emergency department visits for headache

Table 7.1 (continued)

Reference	Design	Outcome	Population	Location	Time period	Pollutants	Summary of findings
Szyszkowicz et al. (2009c)	Time-series	Emergency department visits for headache	General population	Vancouver	1999–2003	CO, NO_2. O_3, PM_{10}, $PM_{2.5}$, SO_2	SO_2 was associated with emergency department visits for migraine
Kaplan et al. (2009)	Case-cross-over	Hospital admissions for appendicitis	General population	Calgary	1999–2006	CO, NO_2. O_3, PM_{10}, $PM_{2.5}$, SO_2	CO, NO_2. O_3, PM_{10}, and SO_2 were associated with appendicitis hospital admissions during summer months
Studies of long term exposure and mortality							
Crouse et al. (2012)	Cohort	Mortality	≥25 years	Canada	1987–2001	$PM_{2.5}$	Long term $PM_{2.5}$ exposure was associated with mortality, particularly from ischemic heart disease
Jerrett et al. (2009b)	Cohort	Mortality	Adult respiratory clinic patients	Toronto	1992–2002	NO_2, O_3, $PM_{2.5}$	NO_2 was associated with mortality
Gan et al. (2010)	Cohort	Mortality	45–85 years	Vancouver	1994–2002	Proximity to major roads	Proximity to major roads was associated with increased risk of ischemic heart disease mortality, and those who moved further from roads reduced their risk
Villeneuve et al. (2012)	Cohort	Mortality	>35 years	Ontario	1982–2004	NO_2, proximity to green space	NO_2 and traffic were associated with increased mortality and proximity to green space with reduced mortality
Studies of traffic							
Clark et al. (2010)	Case-control	Asthma	<4 years	British Columbia	1999–2003	CO, NO, NO_2, PM_{10}, SO_2, black carbon, proximity to point sources	CO, NO, NO_2, PM_{10}, SO_2, black carbon, proximity to point sources were associated with asthma diagnosis; traffic-related pollutants exhibited strongest associations
Dell et al. (2009)	Case-control	Asthma	Grade 1 and 2 school children	Toronto	1998–2006	NO_2, $PM_{2.5}$	Early life exposure to NO_2 was associated with the prevalence of asthma
Jerrett et al. (2009b)	See above						
Gan et al. (2010)	See above						
Crouse et al. (2010)	Case-control	Breast cancer	Post-menopausal women	Montreal	1985–1997	NO_2	NO_2 was associated with incidence of breast cancer
Smargiassi et al. (2006)	Case-control	Respiratory hospital admissions	≥60 years	Montreal	2001–2002	Proximity to road traffic	Proximity to road traffic increased the risk of respiratory hospital admission

Table 7.1 (continued)

Reference	Design	Outcome	Population	Location	Time period	Pollutants	Summary of findings
Dales et al. (2008)	Cross-sectional	Lung function	9–11 years	Windsor	2005	$PM_{2.5}$, proximity to roads	Proximity to roads and $PM_{2.5}$ were associated with airway inflammation and lung function
Dales et al. (2009b)	Cross-sectional	Respiratory symptoms in children	Grade 1–8 school children	Windsor	2004	Proximity to roads	Proximity to roads was associated with prevalence of wheeze and asthma
Weichenthal et al. (2011)	Panel	Lung and cardiovascular function	Healthy adults	Ottawa	2010	$PM_{0.1}$, $PM_{2.5}$, black carbon, volatile orgaic compounds	Measures of traffic pollution affected nervous system control of the heart
Studies of woodsmoke							
Henderson et al. (2011)	Cohort	Respiratory, cardiovascular physician visits, hospital admissions	General population	British Columbia	2003	PM_{10}	PM_{10} from forest fire smoke was associated with respiratory physician visits and hospital admissions
Barn et al. (2008)	Panel	Infiltration of forest fire and woodsmoke	General population	British Columbia		$PM_{2.5}$	Use of HEPA air cleaner reduced infiltration indoors of $PM_{2.5}$ from forest fires and residential woodsmoke
Wheeler et al. (2011)	Panel	Infiltration of woodsmoke	General population	Nova Scotia	2009–2010	$PM_{2.5}$	Use of HEPA air cleaner reduced infiltration indoors of $PM_{2.5}$ from residential woodsmoke
Allen et al. (2011)	Panel	Vascular function, markers of inflammation and oxidative stress	Healthy adults	British Columbia	2008–2009	$PM_{2.5}$	Use of HEPA air cleaner reduced particle concentrations, improved vascular function and reduced inflammatory markers

Fig. 7.5 Change in lung function measure (peak expiratory flow) in L/min per μg/m³ increase in PM_{10} and $PM_{2.5}$ (error bars indicate the degree of uncertainty and "symptomatic" indicates individuals with pre-existing respiratory disease). (Reproduced from, Ward and Ayres 2004, with permission from BMJ Publishing Group Ltd.)

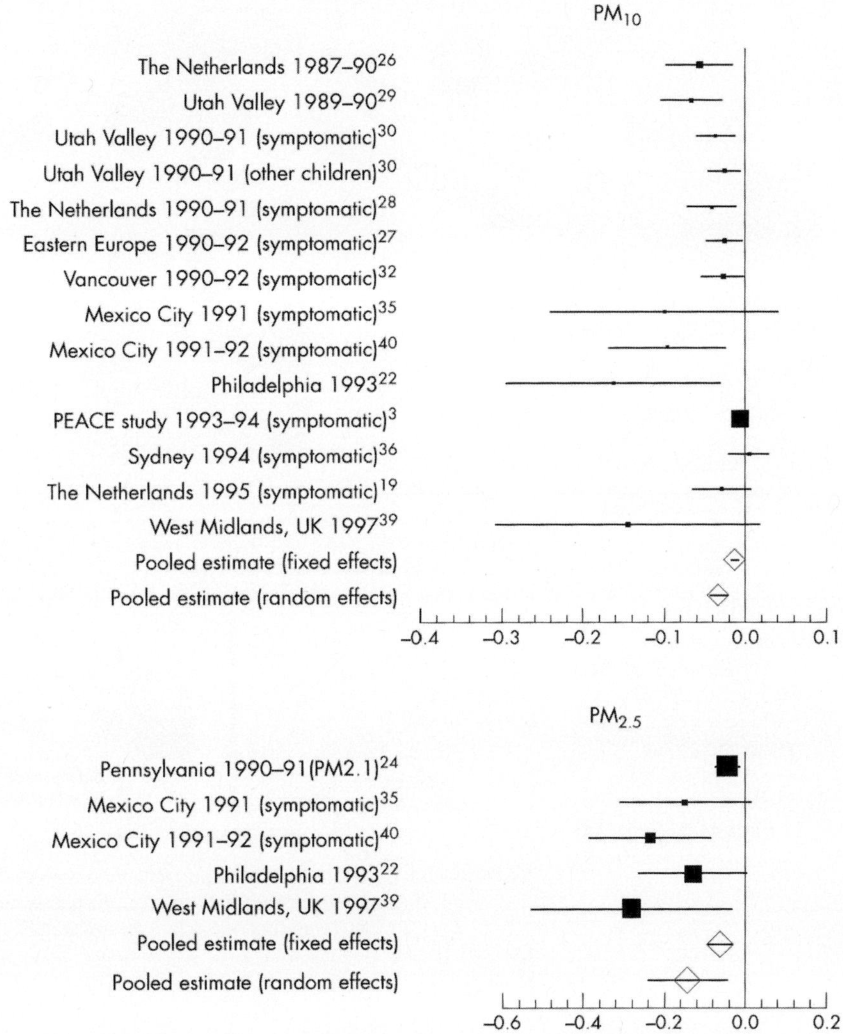

decreased blood pressure among 16 chronic obstructive pulmonary disease patients (Brauer et al. 2001). They did not observe consistent associations between particles and HRV. Goldberg et al. (2008, 2009) found some evidence of associations between air pollution and physiological parameters and self-rated health in heart failure patients.

Panel studies have also documented associations between air pollution and measures of increased blood clotting such as plasma viscosity and fibrinogen concentrations in blood (Seaton et al. 1999; Peters et al. 1997). Elderly people and people who have diabetes mellitus appear to be more vulnerable to air pollution-related changes in cardiovascular physiology. In a study conducted in Boston, particulate matter was significantly associated with decreased ability of blood vessels to dilate among patients with diabetes but not those without diabetes (O'Neill et al. 2005). In two studies conducted in Windsor, Ontario, seniors (>65 years old) and people with diabetes were followed over a month, and their blood pressure, heart rate and brachial artery diameter were measured several times. Participants were found to have elevated blood pressure and heart rate, and reduced brachial artery diameter, on high pollution days (Liu et al. 2007b, 2009).

These observations have been corroborated by several experimental studies where volunteers were exposed to known amount of air pollutant(s) in a facility with a controlled atmosphere. In Toronto, healthy volunteers were exposed to concentrated ambient fine particles drawn from a downtown street for two hours at a concentration similar to those observed on a smoggy day (Brook et al. 2009; Sivagangabalan et al. 2011). The researchers found that exposure to particles and ozone (Sivagangabalan et al. 2011) resulted in a significant reduction in HRV (Brook et al. 2009). A small reduction in HRV was also observed in Los Angeles, where volunteers inhaled ambient coarse particles for 2 h (Gong et al. 2004). Controlled exposure studies have also demonstrated significant increases in blood pressure and reduction in brachial artery diameter or vascular function (Brook et al. 2002, 2009; Urch et al. 2004, 2005). In a study using controlled exposure to diluted diesel exhausts for 1 h at concentrations occurring in urban road traffic, results showed significant increases in

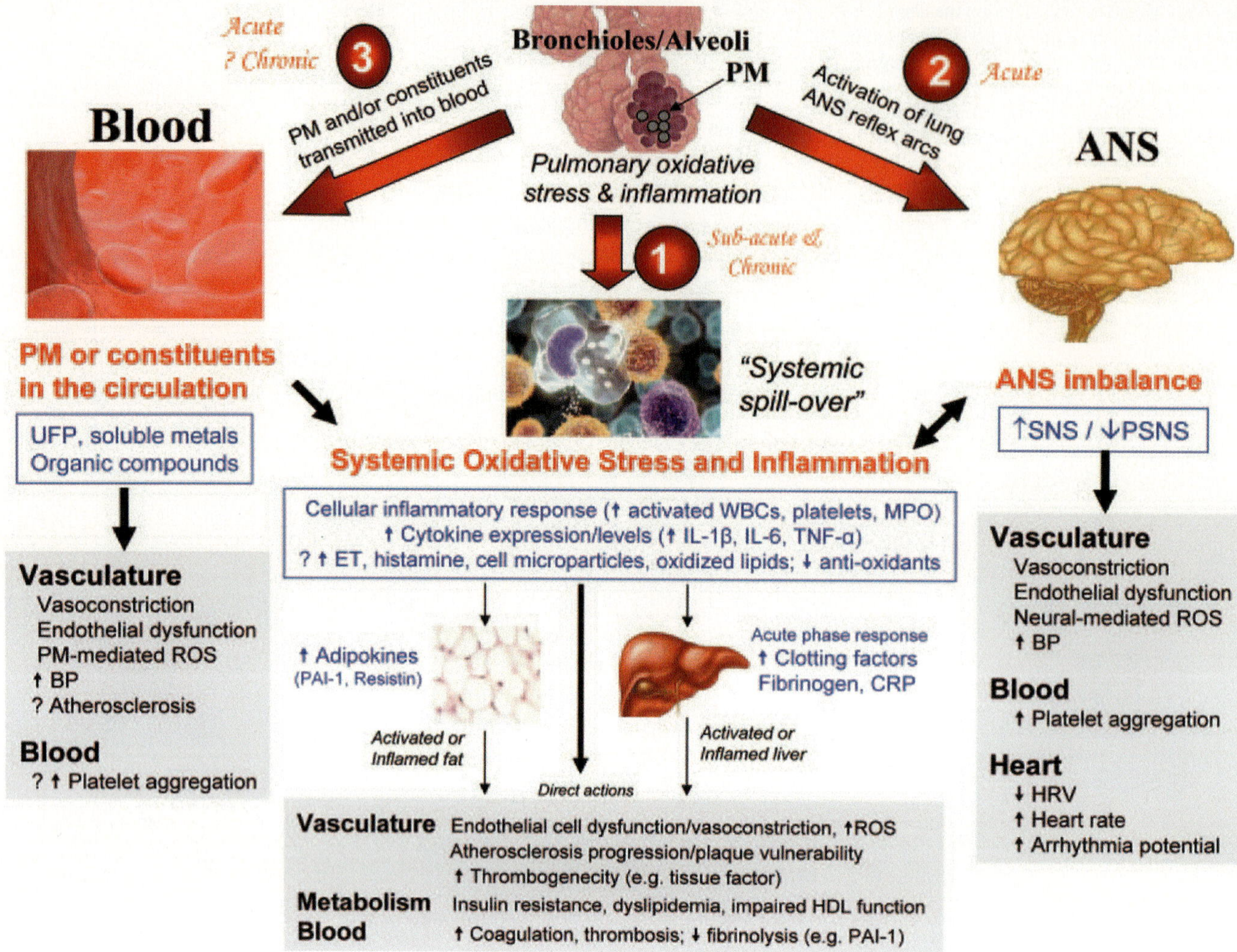

Fig. 7.6 Summary of physiological mechanisms through which air pollution is believed to contribute to cardiovascular health effects. *PM* particulate matter, *ANS* autonomic nervous system, *SNS* sympathetic nervous system, *PSNS* parasympathetic nervous system, *ROS* reactive oxygen species, *BP* blood pressure, *HRV* heart rate variability, *UFP* ultrafine particles, *WBC* white blood cell, *MPO* myeloperoxidase, *IL* interleukin, *TNF* tumour necrosis factor, *ET* endothelin, *CRP* C-reactive protein, *HDL* high density lipoprotein), *PAI-1* Plasminogen Activator Inhibitor Type 1. (Reprinted with Permission, Brook et al. 2010)

heart rate and exercise-induced abnormalities on electrocardiograms, and reduction in the acute release of factors mediating blood clotting in patients with stable heart disease (Mills et al. 2007).

7.4.2 Physiological Mechanisms

Studies have also shown changes in biological markers for inflammation and oxidative stress, in response to elevated pollution levels. Some of these biomarkers play an important role in regulating cardiovascular function inside the body. Oxidative stress is defined by the imbalance in levels of reactive oxygen species and antioxidants. Oxidative stress has emerged as a potential mechanism implicated in the development, progression and cell dysfunction of several diseases, including diabetes complications (Kawano et al. 1999; King and Brownlee 1996). Oxidative stress can result in initiation of the process of plaque formation on blood vessel walls (Gonzalez and Selwyn 2003), a critical step for developing atherosclerosis.

Panel studies have shown significant associations between outdoor or indoor fine particulate matter and markers of oxidative stress in diabetic patients (Liu et al. 2007b) and elderly people (Liu et al. 2009). Other studies have also shown an association between inflammatory markers and exposure to particulate matter that was present inside automobiles (Riediker et al. 2004) and in ambient air (Peters et al. 2001; Pope et al. 2004b; Rückerl et al. 2006). Some of these inflammatory markers have been linked to oxidative stress (Abramson et al. 2005) and reduced vascular function (Vita et al. 2004), and are thought to be a risk factor for atherosclerosis (Verma et al. 2002) and diabetes mellitus (Pradhan et al. 2001).

Calderón-Garcidueñas et al. reported elevated concentrations of mediators of blood vessel constriction in children who were chronically exposed to air pollution in Mexico City compared to children living in a less polluted city (Calderón-Garcidueñas et al. 2007). Haak et al. observed elevated levels of these substances as well as increased heart rate and blood pressure shortly after cigarette smoking (Haak et al. 1994). Liu et al. observed a positive association between these mediators and indoor fine particulate pollutants in elderly people living in nursing homes in Windsor, Ontario (Liu et al. 2009).

Delfino et al. studied associations of multiple air pollutants with systemic inflammation in a group of elderly subjects who had a history of coronary artery disease and lived in retirement communities in Los Angeles. They found significant positive associations for a host of inflammatory markers with outdoor and/or indoor concentrations of particulate matter, elemental carbon, organic carbon, carbon monoxide, and nitrogen dioxide from the current-day and multiday averages (Delfino et al. 2008).

There are also findings of immunological changes in association with particulate pollution (Leonardi et al. 2000). Leonardi et al. conducted a survey within the framework of the Central European Study of Air Quality and Respiratory Health (CESAR) to measure a panel of immune biomarkers in children of 17 countries (Leonardi et al. 2000). These biomarkers, including various lymphocytes and immunoglobulins, detect subtle modifications of the immune system. The results suggest that long-term exposure to airborne fine particles may lead to inflammation of the airways and activation of the cellular immune system.

Results from studies using controlled exposure to diluted diesel exhausts showed inflammatory changes in the airways of healthy and asthmatic subjects (Nordenhäll et al. 2000; Salvi et al. 2000; Stenfors et al. 2004), together with evidence of oxidative stress (Pourazar et al. 2005). Controlled inhalation of diesel exhaust impaired the response to agents used to dilate blood vessels (vasodilators) and increased levels of factors mediating blood coagulation (Mills et al. 2005). The concentrations of air pollutants used in these experimental studies often occur in occupational settings, or have been experienced by urban population in road traffic.

7.4.3 Non-cardiorespiratory Health Effects

In addition to the now very large body of evidence of effects of air pollution on the respiratory and cardiovascular systems, evidence is accumulating of effects on other body systems. Approximately 60 studies have been conducted around the world examining the association between air pollution, birth weight and preterm birth. These include studies in four Canadian cities (Brauer et al. 2008; Liu et al. 2003, 2007) and the province of Nova Scotia (Dugandzic et al. 2006). Most studies have utilized administrative birth record data together with data from fixed monitoring sites. In recent meta-analyses pooling results among several studies, Sapkota et al. (2012) reported increased risks of low birth weight and preterm birth in relation to $PM_{2.5}$, and Stieb et al. (2012) reported reduced birth weight and increased risk of low birth weight and preterm birth associated with CO, NO_2, PM_{10} and $PM_{2.5}$. Parker et al. (2011) recently reported preliminary results from a coordinated international analysis from 14 centres in 9 countries and reported significant associations between PM_{10} and birth weight as well as low birth weight in several centers although there was substantial variability between centers.

Studies have also detected associations between air pollution and outcomes such as headache (Dales et al. 2009c; Larrieu et al. 2009; Mukamal et al. 2009; Szyszkowicz et al. 2009a, b, c), seizures (Cakmak et al. 2010) and gastrointestinal disease (Kaplan et al. 2009, 2010). Whether these outcomes reflect some underlying general physiological mechanism such as inflammation is unclear at this point. However, replication of these associations in multicity studies in a variety of geographic areas is needed before they can be considered to reflect causal associations.

7.4.4 Long-term Health Effects

Because examining chronic effects entails following subjects over extended periods, which is expensive and time consuming, these studies are less numerous than those of acute effects. Two of the most influential long-term studies are the Harvard Six Cities study and the American Cancer Society cohort study. The Six Cities Study established a cohort of approximately 8,000 adults in cities selected to contrast exposure at varying levels of particulate air pollution. A significant linear trend in risk of mortality was observed across the cities in relation to several measures of particulate exposure (Fig. 7.7; Dockery et al. 1993), which was confirmed in extended follow-up (Laden et al. 2006). The American Cancer Society cohort study examined whether results from the six cities study could be replicated in the much larger (>500,000 subjects) Cancer Prevention Study cohort. Significant associations of particulate matter with mortality were observed (Fig. 7.8; Pope et al. 1995) and continued to be observed in extended follow-up (Pope et al. 2002, 2004). An association of long-term ozone exposure with respiratory mortality has also been observed in this cohort (Jerrett et al. 2009a). A Canadian team of investigators, led by Dr. Daniel Krewski, made a key contribution to the interpretation of these studies by conducting an extensive audit, replication of findings and re-analysis, which supported the validity of the findings and their importance in regulatory decision making (Health Effects Institute 2000).

Fig. 7.7 Estimated ratio of mortality from the Harvard Six Cities study, comparing each city to Portage, Wisconsin (*P*). Other cities are Topeka, Kansas (*T*), Watertown, Massachusetts (*W*), St. Louis, Missouri (*L*), Harriman, Tennesse (*H*), and Steubenville, Ohio (*S*). Adjusted for other risk factors and plotted by air pollution level. (Reprinted with permission of the Massachusetts Medical Society. Copyright © 2012 Massachusetts Medical Society, Dockery et al.1993)

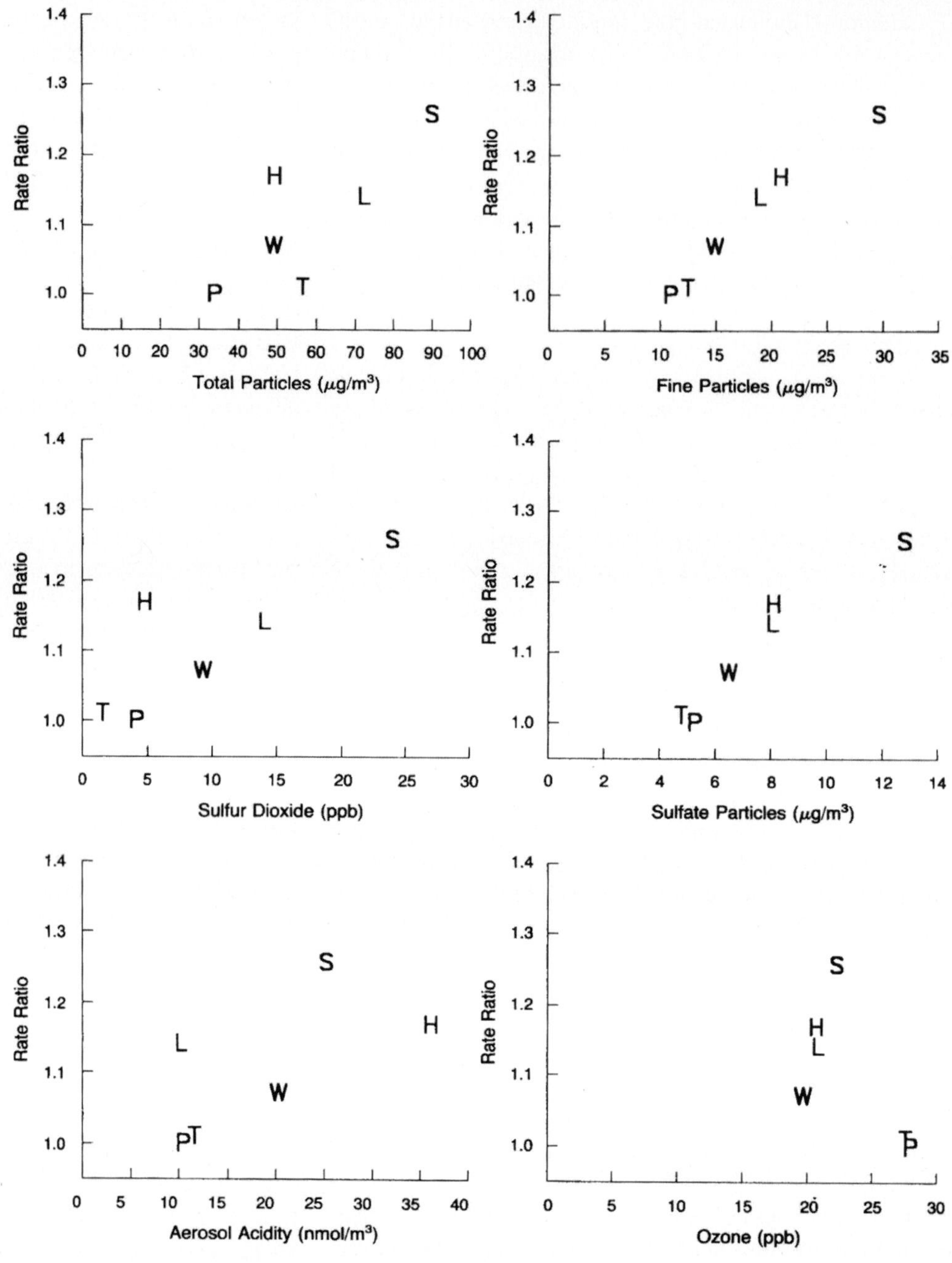

A number of other cohort studies of particulate air pollution and mortality have appeared subsequently from various locations in the US and Europe (Abbey et al. 1999; Beelen et al. 2008; Chen et al. 2005; Eftim et al. 2008; Filleul et al. 2005; Gehring et al. 2006; Lipfert et al. 2006; Miller et al. 2007; Naess et al. 2007; Nafstad et al. 2004; Puett et al. 2008; 2009; Zeger et al. 2008). Two large Canadian cohort studies using linked data and exposures based on remote sensing satellite technology have also demonstrated associations of long term exposure to $PM_{2.5}$ (Crouse et al. 2012) and traffic-related pollution (Villeneuve et al. 2012) with mortality. Estimated $PM_{2.5}$ exposures in Canada from the Crouse study are shown in Fig. 7.9.

Relatively few studies have been conducted of chronic exposure to air pollution and respiratory morbidity. Two examples are the Adventist Health Study on Smog (AHSMOG) and the Southern California Children's Health Study (CHS). The AHSMOG study examined the association between air pollution exposure and development of chronic lung disease in a cohort of over 6,000 non-smoking adults in California. Associations were observed between $PM_{2.5}$ and PM_{10} and development of chronic bronchitis symptoms (Abbey et al. 1995a, b), between ozone, PM_{10}, SO_2 and lung cancer incidence (Beeson et al. 1998), and between $PM_{2.5}$ and mortality (Chen et al. 2005). The CHS recruited over 3,000 children beginning at age 10 and fol-

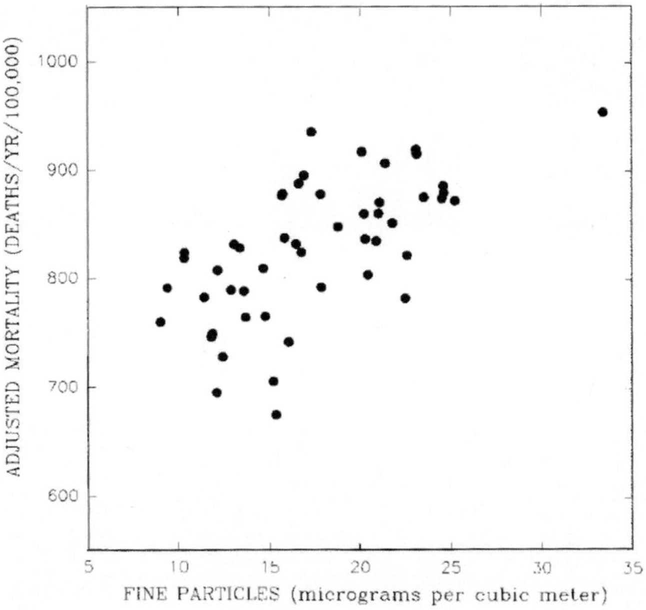

Fig. 7.8 Death rates for 1980 plotted against fine particle concentrations for 1979 to 1983 (adjusted for age, sex and race) from the American Cancer Society cohort study. (Reprinted with permission of the American Thoracic Society. Copyright © 2012 American Thoracic Society, Pope et al. 1995)

lowed them over several years prospectively evaluating changes in lung function and the development of new cases of asthma, among other outcomes. Findings have included a greater risk of developing asthma among children participating in team sports in high ozone communities (substantially higher than concentrations in Canada) (McConnell et al. 2002), reduced lung growth in relation to nitrogen dioxide, acid vapour and elemental carbon (Gauderman et al. 2004) as well as proximity to traffic (Gauderman et al. 2007), greater impacts of air pollution on lung function in families with high stress (Islam et al. 2011), and increased susceptibility to the effects of ozone among certain genetic variants (Islam et al. 2009).

Long-term exposure to current levels of air pollutants in ambient air may also contribute to the development of atherosclerosis, leading to changes that play a major role in cardiovascular morbidity and mortality later in life. Researchers from the University of Southern California examined the association between residential ambient fine particulate matter and thickness of a portion of the carotid artery wall, a quantitative measure of generalized atherosclerosis that correlates well with all of the major cardiovascular risk factors, with coronary artery atherosclerosis, and with clinical cardiovascular events (Mack et al. 2000). Long-term exposure to fine particulate matter and proximity to roadways was significantly associated with an increase in wall thickness (Kunzli et al. 2005, 2010). Mixed results have been reported in the Multi-Ethnic Study of Atherosclerosis (MESA)—a multicity study of nearly 7,000 subjects examining the progression of cardiovascular disease over time. While $PM_{2.5}$ exposure was associated with vascular abnormalities on digital photographs of the retina (Adar et al. 2010), it was not associated with measures of arterial stiffness (O'Neill et al. 2011).

7.4.5 Impacts of Specific Sources

One particular area of recent interest is the identification of health effects of specific pollutant sources such as traffic. The Health Effects Institute (HEI) recently completed a review of the literature in this area (HEI 2010). They concluded that the health of large numbers of people living within 300–500 m of major roadways is potentially affected by traffic pollution in many major cities. With respect to health impacts, they determined that there was sufficient evidence for a causal association between traffic pollutant exposure and exacerbation of asthma, together with suggestive but not conclusive evidence of associations with cardiovascular morbidity and mortality and onset of asthma. A recent meta-analysis provided further support for the existence of a causal association between traffic related air pollution and incidence of asthma in children (Anderson et al. 2013). It has recently been estimated that approximately one third of Canadians live within 100 m of a major road or 500 m of a highway (Brauer et al. 2012), and that approximately one third of elementary schools in 10 Canadian cities are located within 200 m of a major road (Amram et al. 2011).

Several European studies have reported associations between NO_2, a traffic marker, and mortality (Filleul et al. 2005; Gehring et al. 2006; Hoek et al. 2002; Naess et al. 2007; Nafstad et al. 2004). Brauer et al. (2002) and Gehring et al. (2010) found that following a cohort of approximately 4,000 children in the Netherlands since birth, traffic-related air pollution was associated with the incidence of asthma indicating that it may cause development of asthma in children. Similar associations have also been observed in Canada. An analysis in British Columbia found that there was an association between traffic pollution exposure and incidence of asthma, using linked birth records and health claims utilization data (Clark et al. 2010). In 2004–2006, a series of epidemiologic studies were conducted in Windsor, Ontario, to investigate air pollution from various sources and the effects on children's respiratory system. In Windsor, the major source of air pollution is traffic and in particular the Ambassador Bridge connecting Windsor, Canada with Detroit, US is the busiest border crossing point in Canada. A cross-sectional questionnaire survey was conducted on all school children of grades 1–8 on their family health history, living conditions and respiratory symptoms. A second survey was conducted on grades 4–6 children to test pulmonary function and exhaled nitric oxide, a biomarker of airway inflamma-

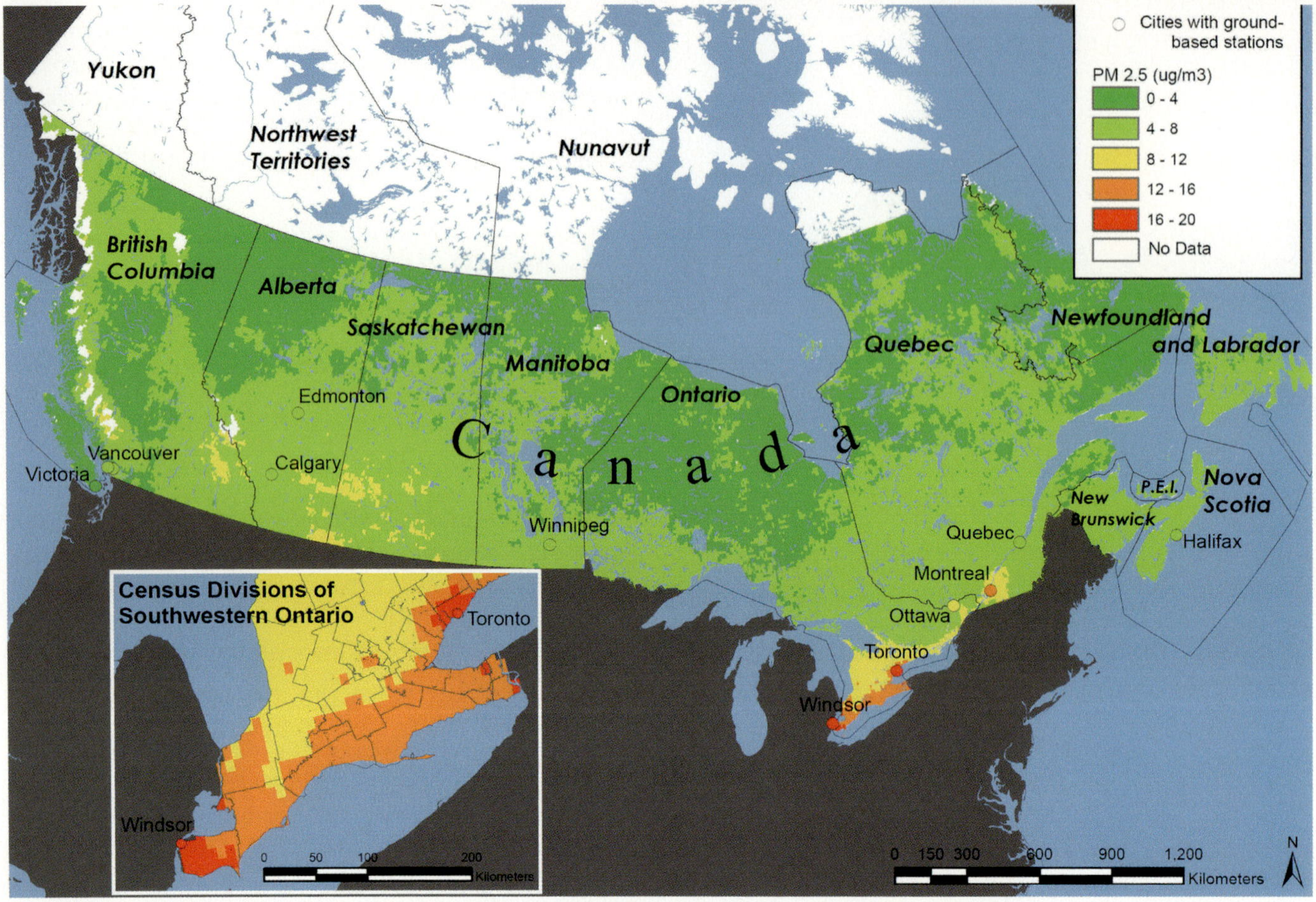

Fig. 7.9 Mean satellite-derived estimates of $PM_{2.5}$ across Canada, 2001–2006 and mean ground-based concentrations in 11 cities. (Reproduced with permission from Environmental Health Perspectives, Crouse et al. 2012)

tion. These studies reported that after adjusting for age, sex and socioeconomic status, living near roadways was consistently significantly associated with increased asthma symptoms (Dales et al. 2009b) and exhaled nitric oxide (Dales et al. 2008). A study in Toronto schoolchildren also found that early life exposures to traffic pollutants was associated with onset of asthma in childhood (Dell et al. 2009). With respect to other health outcomes, modelled exposure to NO_2 as a marker for traffic-related pollution (Fig. 7.10) or proximity to roads was associated with all-cause and circulatory mortality in a cohort of respiratory clinic patients in Toronto (Jerrett et al., 2009b), and with respiratory hospital admission in the elderly (Smargiassi et al. 2006), and breast cancer (Crouse et al. 2010) in Montreal.

Woodsmoke from wildfires and from heating has also recently emerged as an area of particular interest, both in high and low income countries. In the latter, indoor biomass burning has in fact been identified as one of the leading causes of mortality (Prüss-Üstün and Corvalán 2006). Woodsmoke contains numerous pollutants including carbon monoxide, nitrogen dioxide, fine particulate matter and a number of known carcinogens. Toxicological and epidemiological studies have documented adverse health impacts following exposure (Naeher et al. 2007). In some parts of Canada, woodsmoke from residential heating has been estimated to account for a significant majority of particulate matter emissions (Fig. 7.11). Forest fire smoke is also an important source. Fires in the interior of British Columbia in 2003 resulted in very high concentrations of PM_{10} and associations with respiratory morbidity were detected (Henderson et al. 2011). (Fig. 7.12)

7.4.6 Studies of Interventions to Control Air Pollution

Several studies have examined the impact on health of changes in pollutant concentrations resulting from policy interventions or plant closures. These are powerful tools in establishing causal relationships between exposure and response because they are considered analogous to true experiments, the strongest form of evidence, where investigators change an exposure and observe the response.

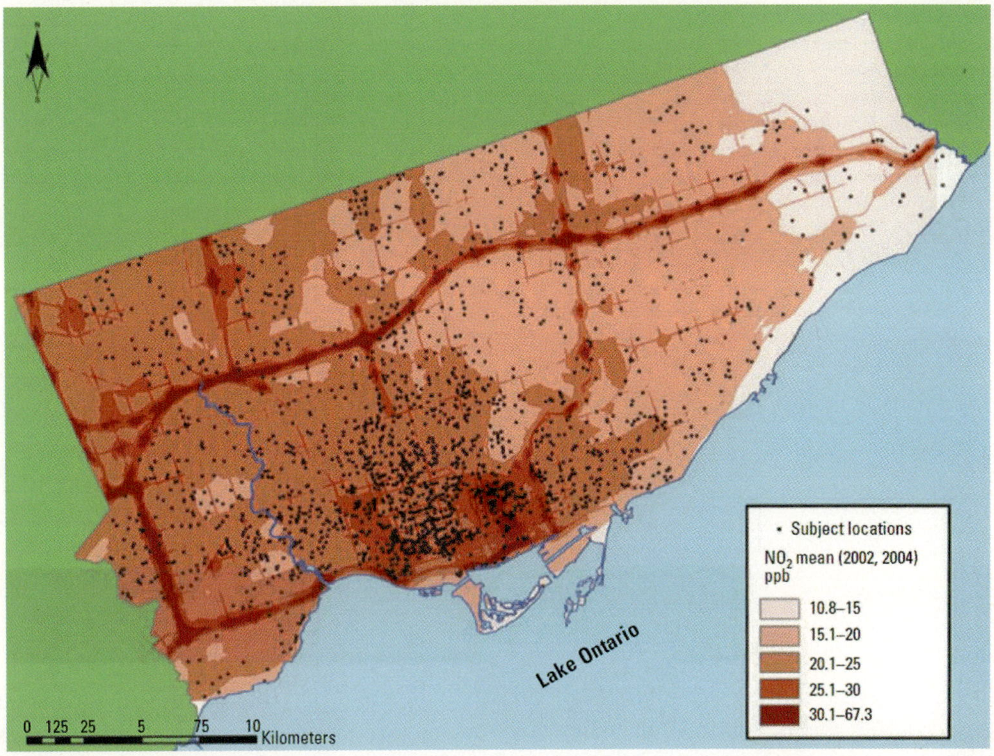

Fig. 7.10 Modelled concentrations of NO_2 in Toronto based on monitoring campaigns in 2002 and 2004. Approximate subject coordinates are shifted to protect privacy. (Reproduced with permission from Environmental Health Perspectives, Jerrett et al. 2009b)

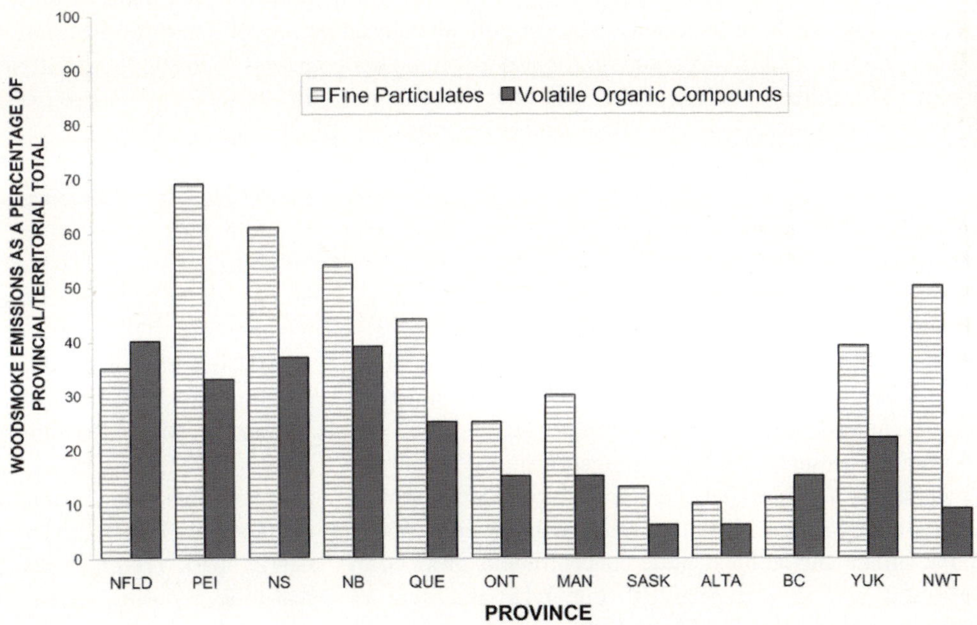

Fig. 7.11 Percent contribution of woodsmoke to fine particulate matter and volatile organic compounds by province. (Reproduced from, Naeher et al. 2007, with permission of TAYLOR & FRANCIS INC.)

Policy Interventions and Plant Closures

Perhaps one of the best known natural experimental studies on air pollution is the Utah Valley steel mill study (Pope 1989). In 1989, Pope observed that during a period of closure of a local steel mill in Utah Valley in 1986–1987, hospital admissions due to incidence of severe respiratory diseases decreased significantly compared with the periods before and after the steel mill was closed (Pope 1989). Statistical analysis indicated that the reduction of these hospital admissions could be attributed to lower levels of PM_{10} during the closure of the steel mill (Fig. 7.13). In Utah Valley, the main source of particulate pollution was the steel mill, which when in operation, contributed approximately 50–80 % of the PM_{10} pollution. A reduction in preterm birth was also observed during

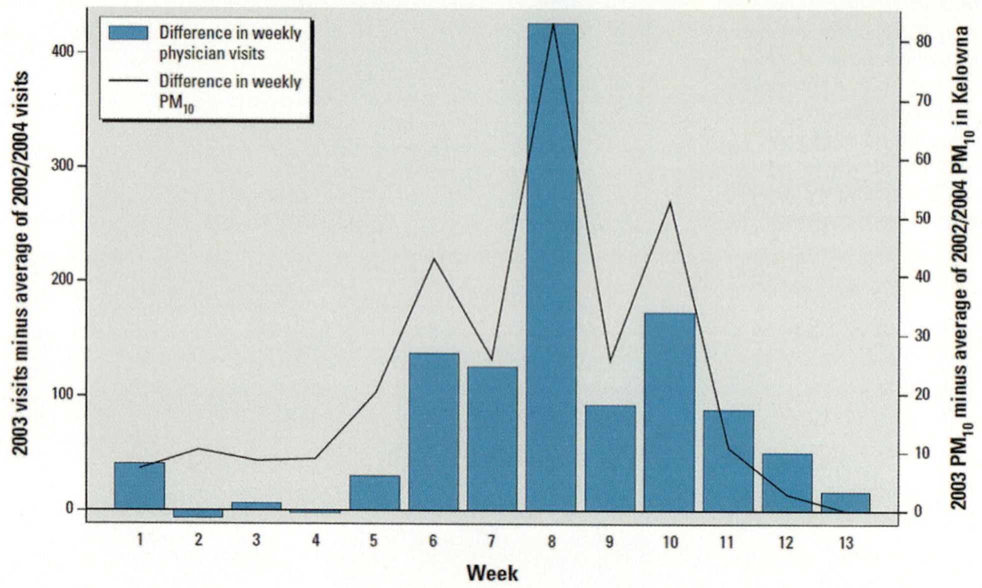

Fig. 7.12 Differences in weekly physician visits and PM_{10} concentrations between 2003 forest fires and average of 2002, 2004 in Kelowna, British Columbia. (Reproduced with permission from Environmental Health Perspectives, Henderson et al. 2011)

the mill closure (Parker et al. 2008) (Fig. 7.14). Particles collected during this period in Utah Valley were later used to study their toxicity on human subjects. Significantly higher concentrations of iron, copper, zinc, lead, nickel and vanadium were found in particles collected when the steel mill was open than in those collected when the mill was closed (Ghio and Devlin 2001). Exposure to the water-soluble portion of PM_{10} collected before closure and after reopening of the steel mill provoked a greater inflammatory response relative to the PM extract acquired during the plant shutdown (Ghio and Devlin 2001). PM_{10} collected when the steel mill was open weakened the immune defence capacity in human macrophages in the lungs (Soukup et al. 2000). These experimental studies provide a biological explanation for the epidemiological findings of an association between particles and hospital admissions in Utah Valley (Pope 1989). They also suggest that metals in particulate matter play an important role in adversely affecting respiratory health.

Another persuasive study examined changes in ambient concentrations of black smoke and sulfur dioxide following a ban on coal sales in Dublin in 1990 (Clancy et al. 2002). Black smoke and sulfur dioxide concentrations dropped by 70% and over 30% respectively (Fig. 7.15) and were accompanied by 10% and 16% reductions in cardiovascular and respiratory deaths (Fig. 7.16), after controlling for trends in deaths elsewhere in Ireland. In a study in Libby, Montana, investigators examined changes in ambient concentrations of particulate matter following a woodstove change out program in which old woodstoves were replaced with new ones with improved emission controls. They observed a 28% reduction in woodsmoke source $PM_{2.5}$ following the replacement of 1200 woodstoves (Ward et al. 2010). Significant reductions in ambient air pollution concentrations were also observed in relation to air pollution controls implemented during the Beijing Olympics, and these were associated with improvements in levels of biomarkers of inflammation, oxidative stress and thrombosis (Huang et al. 2012; Rich et al. 2012). Finally, a study of the impact of reduced traffic congestion at toll plazas after the introduction of electronic passes found 11% and 12% reductions in preterm birth and low birth weight respectively in the surrounding area (Currie and Walker 2009).

Individual-level Actions

In addition to societal or policy level interventions to control air pollution, there is also evidence that actions by individuals can lead to reductions in exposure and the risk of adverse health effects (Giles et al. 2011).

McCreanor et al. conducted a randomized crossover study on volunteers who had mild to moderate asthma in London, UK, to investigate the adverse health effects of diesel traffic emissions (McCreanor et al. 2007). Each participant walked for two hours along a London street with heavy diesel traffic (Oxford Street) and, on a separate occasion, through a nearby park (Hyde Park). They determined participants' real-time exposure as well as physiological and immunologic measurements. Participants had significantly higher exposures to particulate pollution and nitrogen dioxide on Oxford Street than in Hyde Park. Walking for two hours on Oxford Street resulted in significant reduction in pulmonary function, accompanied by increases in markers of inflammation. Walking in Hyde Park resulted in much smaller changes in pulmonary function. These changes were associated most consistently with exposures to ultrafine particles and elemental carbon. This study demonstrates that reducing short-term exposure by changing the location of physical ac-

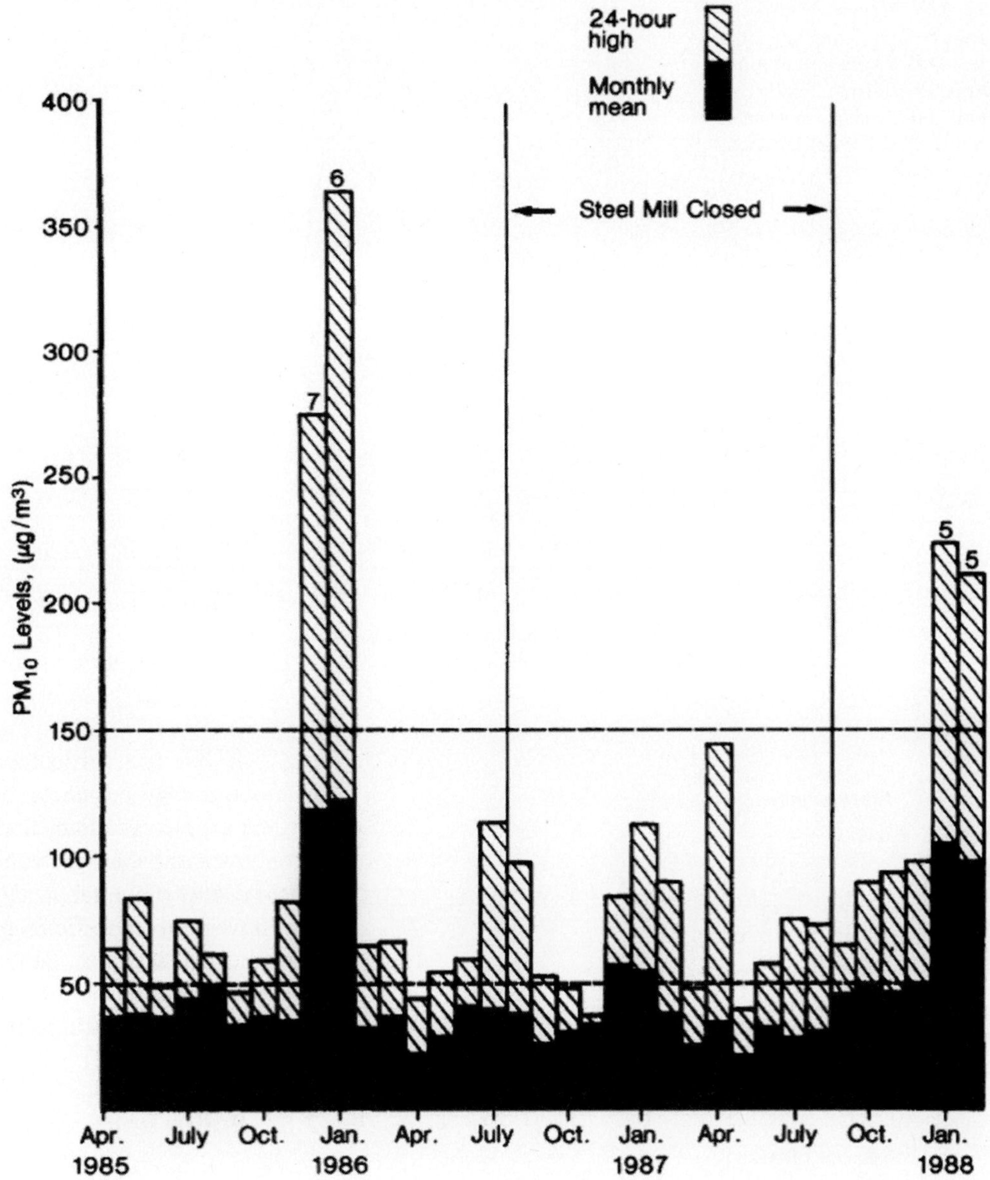

Fig. 7.13 PM_{10} concentrations before, during and after Utah Valley steel mill closure. (Reprinted with permission of the American Public Health Association. Copyright © 2012 American Public Health Association. Pope 1989)

tivity can reduce health risk, and this is supported in a recent study of cardiovascular effects of traffic pollution among cyclists in Ottawa (Weichenthal et al. 2011). In relation to long-term exposure to traffic-related air pollution, Gan et al. (2010) reported that individuals in Vancouver who moved further from road traffic had a reduced risk of heart disease mortality.

Studies have also documented reduced exposure to both traffic pollution and woodsmoke particles with the use of portable indoor air cleaner devices employing HEPA (high-efficiency particulate air) filters (Bräuner et al. 2008; Barn et al. 2008; Wheeler et al. 2011), as well as improved vascular and reduced inflammatory markers (Allen et al. 2011)[2].

[2]Note that air cleaning devices which generate ozone are not recommended http://www.epa.gov/iaq/pubs/ozonegen.html.

To evaluate whether acute effects of air pollution on lung function could be attenuated by antioxidant vitamin supplementation, Romieu et al. conducted several randomized trials using a double-blind crossover design on street workers (Romieu et al. 1998) or children with asthma (Romieu et al. 2002) in Mexico City. Participants took a daily supplement of vitamin E and C or placebo for 6–7 months, and pulmonary function was tested. For street workers and children with moderate and severe asthma, ozone exposure resulted in reduced lung function measures in the placebo group but not in the supplement group (Romieu et al. 1998, 2002). The researchers also ran a randomized trial, treating older subjects (>60 years) with fish oil that contains omega-3 polyunsaturated fatty acids or soy oil as control for 6 months (Romieu et al. 2005). During the study, they measured participants' heart rate variability (HRV) repeatedly. In the group receiv-

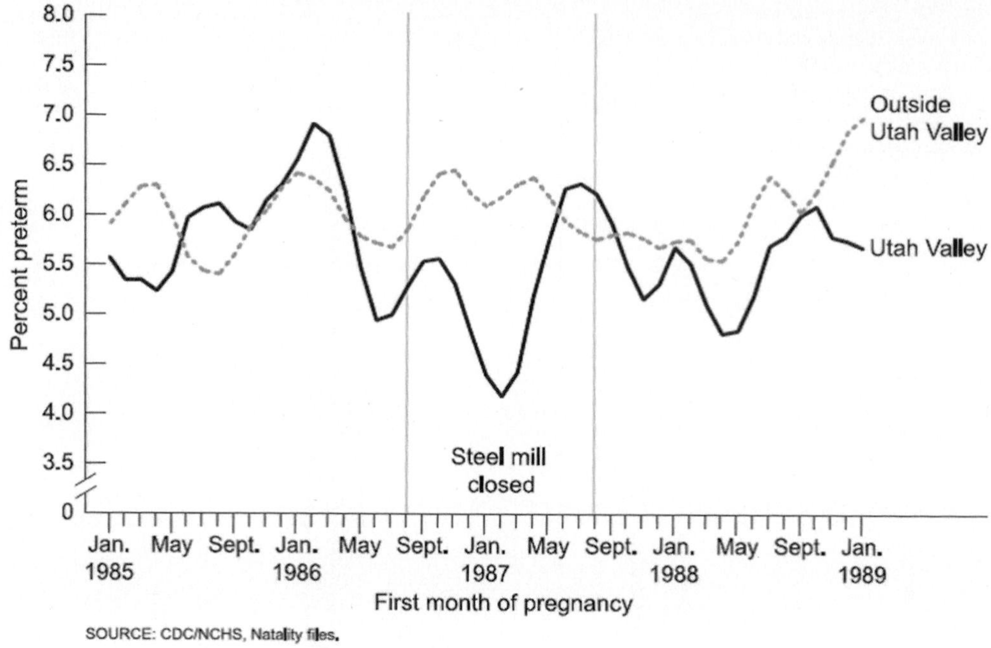

Fig. 7.14 Prevalence of preterm births before, during and after Utah Valley steel mill closure. (Reproduced from, Parker et al. 2008, with permission of LIPPINCOTT WILLIAMS & WILKINS)

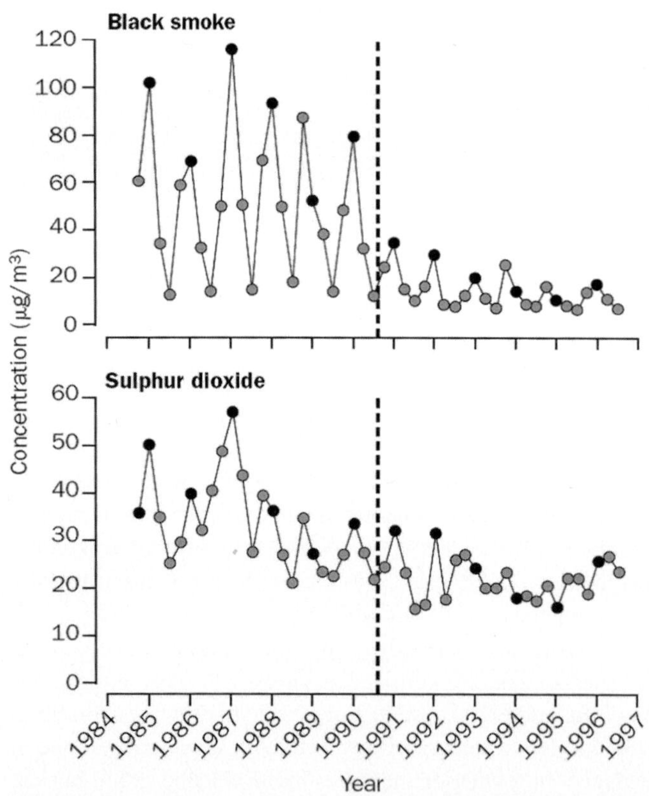

Fig. 7.15 Concentrations of black smoke and sulphur dioxide in winter (*black dots*) and summer (*grey dots*) before and after Dublin coal ban (*vertical line*). (Reprinted from The Lancet, 360, Clancy et al. 2002, Copyright (2002), with permission from Elsevier.)

ing fish oil, the particulate pollution-associated adverse changes in HRV became significantly less pronounced during the supplementation. Supplementation with soy oil did not show significant protective effect against air pollution. These results suggest that supplementation with antioxidants may have protective effects against air pollution-induced adverse health outcomes, although at this time there is insufficient evidence to routinely recommend supplementation as a means of reducing air pollution impacts.

7.5 Policy Implications

7.5.1 Multi-pollutant Mixtures

As indicated earlier, one of the challenges facing research on health effects of air pollution is evaluating the effects of complex mixtures. The weight of toxicological, clinical and epidemiological evidence points most strongly to particulate matter and ozone in terms of the existence of causal associations with adverse health effects at current levels of exposure. At the same time, pollutants such as NO_2 appear to be an effective marker for specific sources like traffic. Recognition of the reality of pollutant mixtures also has regulatory implications. Reducing ambient concentrations of an individual pollutant may not produce the anticipated health benefits if the pollutant is simply serving as a marker for the true toxic component or source, but is not causally associated with adverse health impacts. This recognition

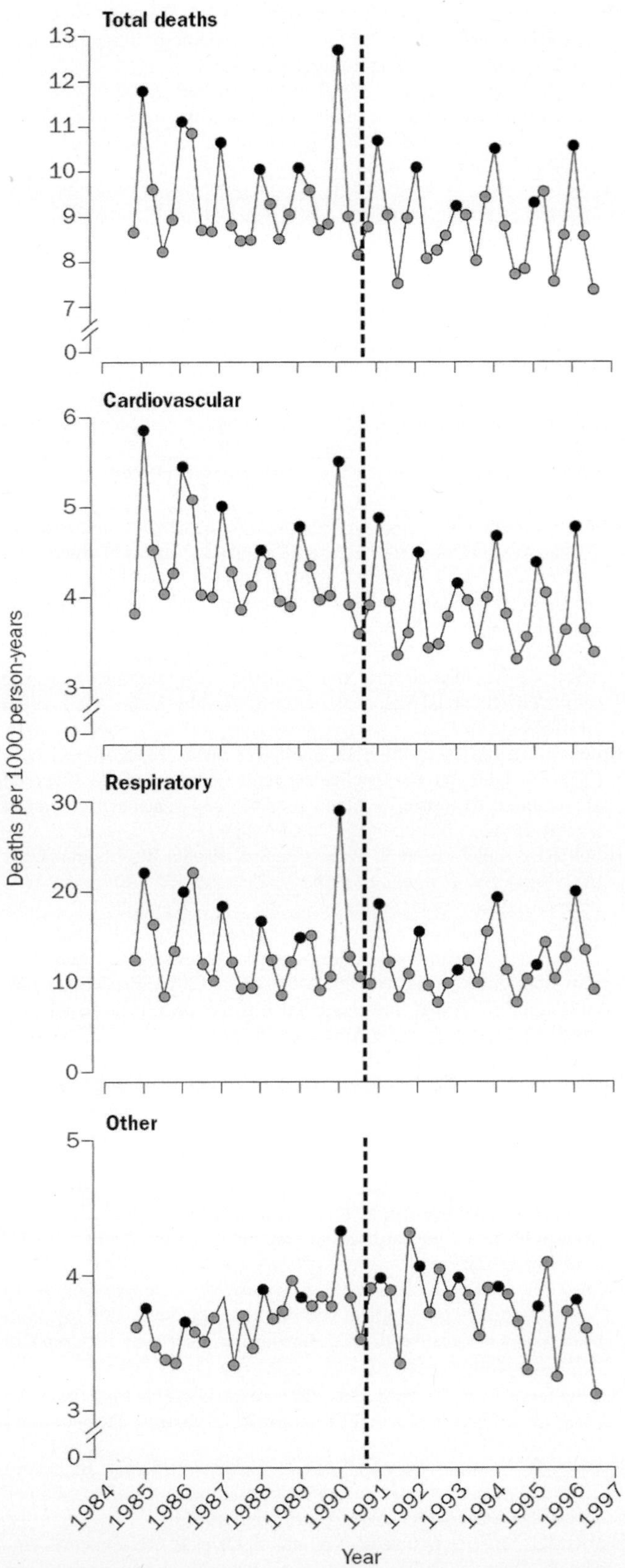

Fig. 7.16 Age-adjusted (standardized) death rates in winter (*black dots*) and summer (*grey dots*) before and after Dublin coal ban (*vertical line*). (Reprinted from The Lancet, 360, Clancy et al. 2002, Copyright (2002) with permission from Elsevier.)

has recently led to calls for a multi-pollutant approach to air quality management (Dominici et al. 2010; Mauderly et al. 2010), although the process for achieving this remains unclear.

7.5.2 Priorities for Future Research

Considerable progress has been made over the years in improving our understanding of the adverse health effects of air pollution, in order to guide policies to reduce emissions, ambient concentrations and health risks. Nonetheless, important gaps in our understanding remain. Identified research priorities include: understanding the effects of mixtures (including improved statistical analysis methods), in particular mixtures originating from key sources like traffic; examination of particulate matter as a mixture to identify key components; understanding effects of long-term exposure, of ultrafine particles, of new fuels and engine technologies, and of interactions of air pollution with climate change; and improved ability to measure the benefits of policy interventions to improve air quality, including both large scale policies such as air quality standards and natural experiments such as air quality changes during major events like Olympic games (Health Effects Institute 2010; National Research Council 2004; USEPA 2011b).

7.6 Conclusions

Although air quality has improved considerably since the severe air pollution episodes of the mid 20th century, adverse health effects can still be detected with a variety of research methods. The probability of experiencing effects depends on personal factors which mediate susceptibility as well as behaviors which determine exposure and dose. Current evidence indicates that air pollution causes both acute and chronic health effects on the respiratory and cardiovascular systems. There is also suggestive evidence air pollution contributes to adverse health effects on other body systems. In an effort to identify the most effective means of reducing health risks, increasing attention is being paid to addressing air quality management from a multi-pollutant perspective, and isolating sources of air pollution which have the greatest potency in causing adverse health effects. Traffic-related air pollution and woodsmoke are sources of particular interest. Studies of air pollution interventions have shown that reducing air pollution results in reduced health risks, thus providing powerful support for the value of continuing to improve air quality.

Acknowledgments The authors thank Melanie Fortune for assistance in summarizing key studies and Dr. Michael Brauer for helpful comments.

References

Abbey DE, Ostro BE, Petersen F, Burchette RJ (1995a) Chronic respiratory symptoms associated with estimated long-term ambient concentrations of fine particulates less than 2.5 µm in aerodynamic diameter ($PM_{2.5}$) and other air pollutants. J Expo Anal Environ Epidemiol 5:137–159

Abbey DE, Hwang BL, Burchette RJ, Vancuren T, Mills PK (1995b) Estimated long-term ambient concentrations of PM_{10} and development of respiratory symptoms in a nonsmoking population. Arch Environ Health 50:139–152

Abbey DE, Nishino N, McDonnell WF, Burchette RJ, Knutsen SF, Beeson WL, Yang JX (1999) Long-term inhalable particles and other air pollutants related to mortality in nonsmokers. Am J Respir Crit Care Med 159:373–382

Abramson JL, Hooper WC, Jones DP, Ashfaq S, Rhodes SD, Weintraub WS et al (2005) Association between novel oxidative stress markers and C-reactive protein among adults without clinical coronary heart disease. Atherosclerosis 178:115–121

Adar SD, Klein R, Klein BE, Szpiro AA, Cotch MF, Wong TY, O'Neill MS, Shrager S, Barr RG, Siscovick DS, Daviglus ML, Sampson PD, Kaufman JD (2010) Air pollution and the microvasculature: a cross-sectional assessment of in vivo retinal images in the population-based multi-ethnic study of atherosclerosis (MESA). PLoS Med 7:e1000372

Allen RW, Carlsten C, Karlen B, Leckie S, Eeden S van, Vedal S, Wong I, Brauer M (2011) An air filter intervention study of endothelial function among healthy adults in a woodsmoke-impacted community. Am J Respir Crit Care Med 183:1222–1230

Amram O, Abernethy R, Brauer M, Davies H, Allen RW (2011) Proximity of public elementary schools to major roads in Canadian urban areas. Int J Health Geogr 10:68

Anderson HR, Favarato G, Atkinson RW. (2013) Long-term exposure to air pollution and the incidence of asthma: meta-analysis of cohort studies. Air Qual Atmos Health 6:47-56

Baccarelli A, Cassano PA, Litonjua A, Park SK, Suh H, Sparrow D et al (2008) Cardiac autonomic dysfunction: effects from particulate air pollution and protection by dietary methyl nutrients and metabolic polymorphisms. Circulation 117:1802–1809

Barn P, Larson T, Noullett M, Kennedy S, Copes R, Brauer M (2008) Infiltration of forest fire and residential wood smoke: an evaluation of air cleaner effectiveness. J Expo Sci Environ Epidemiol 18:503–511

Bates DV (2002) A half century later: recollections of the London fog. Environ Health Perspect 110:A735

Bates DV, Sizto R (1987) Air pollution and hospital admissions in Southern Ontario: the acid summer haze effect. Environ Res 43:317–331

Bates DV, Baker-Anderson M, Sizto R (1990) Asthma attack periodicity: a study of hospital emergency visits in Vancouver. Environ Res 51:51–70

Beelen R, Hoek G, Brandt AP van den, Goldbohm RA, Fischer P, Schouten LJ, Jerrett M, Hughes E, Armstrong B, Brunekreef B (2008) Long-term effects of traffic-related air pollution on mortality in a Dutch cohort (NLCS-AIR Study). Environ Health Perspect 116:196–202

Beeson WL, Abbey DE, Knutsen SF (1998) Long-term concentrations of ambient air pollutants and incident lung cancer in California adults: results from the AHSMOG study. Adventist health study on smog. Environ Health Perspect 106:813–822

Bell ML, Davis DL (2001) Reassessment of the lethal London fog of 1952: novel indicators of acute and chronic consequences of acute exposure to air pollution. Environ Health Perspect 109(Suppl 3):389–394

Bell ML, Dominici F, Samet JM (2005) A meta-analysis of time-series studies of ozone and mortality with comparison to the national morbidity, mortality, and air pollution study. Epidemiology 16:436–445

Boyes WK, Moser VC, Geller AM, Benignus VA, Bushnell PJ, Kamel F (2007) Integrating epidemiology and toxicology in neurotoxicity risk assessment. Hum Exp Toxicol 4:283–293

Brauer M, Ebelt ST, Fisher TV, Brumm J, Petkau AJ, Vedal S (2001) Exposure of chronic obstructive pulmonary disease patients to particles: respiratory and cardiovascular health effects. J Expo Anal Environ Epidemiol 11:490–500

Brauer M, Hoek G, Van Vliet P, Meliefste K, Fischer PH, Wijga A, Koopman LP, Neijens HJ, Gerritsen J, Kerkhof M, Heinrich J, Bellander T, Brunekreef B (2002) Air pollution from traffic and the development of respiratory infections and asthmatic and allergic symptoms in children. Am J Respir Crit Care Med 166:1092–1098

Brauer M, Lencar C, Tamburic L, Koehoorn M, Demers P, Karr C (2008) A cohort study of traffic-related air pollution impacts on birth outcomes. Environ Health Perspect 116:680–686

Brauer M, Reynolds C, Hystad P (2012) Traffic-related air pollution and health: a Canadian perspective on scientific evidence and potential exposure mitigation strategies. Report for Water, Air and Climate Change Bureau, Health Canada. University of British Columbia, Vancouver, BC

Bräuner EV, Forchhammer L, Møller P, Barregard L, Gunnarsen L, Afshari A, Wåhlin P, Glasius M, Dragsted LO, Basu S, Raaschou-Nielsen O, Loft S (2008) Indoor particles affect vascular function in the aged: an air filtration-based intervention study. Am J Respir Crit Care Med 177:419–425

Brook RD, Brook JR, Urch B, Vincent R, Rajagopalan S, Silverman F (2002) Inhalation of fine particulate air pollution and ozone causes acute arterial vasoconstriction in healthy adults. Circulation 105:1534–1536

Brook RD, Urch B, Dvonch JT, Bard RL, Speck M, Keeler G et al (2009) Insights into the mechanisms and mediators of the effects of air pollution exposure on blood pressure and vascular function in healthy humans. Hypertension 54:659–667

Brook RD, Rajagopalan S, Pope CA 3rd, Brook JR, Bhatnagar A, Diez-Roux AV, Holguin F, Hong Y, Luepker RV, Mittleman MA, Peters A, Siscovick D, Smith SC Jr, Whitsel L, Kaufman JD, American Heart Association Council on Epidemiology and Prevention, Council on the Kidney in Cardiovascular Disease, and Council on Nutrition, Physical Activity and Metabolism (2010) Particulate matter air pollution and cardiovascular disease: an update to the scientific statement from the American Heart Association. Circulation 121:2331–2378

Burnett RT, Dales RE, Raizenne ME, Krewski D, Summers PW, Roberts GR, Raad-Young M, Dann T, Brook J (1994) Effects of low ambient levels of ozone and sulfates on the frequency of respiratory admissions to Ontario hospitals. Environ Res 65:172–194

Burnett RT, Dales R, Krewski D, Vincent R, Dann T, Brook JR (1995) Associations between ambient particulate sulfate and admissions to Ontario hospitals for cardiac and respiratory diseases. Am J Epidemiol 142:15–22

Burnett RT, Dales RE, Brook JR, Raizenne ME, Krewski D (1997a) Association between ambient carbon monoxide levels and hospitalizations for congestive heart failure in the elderly in 10 Canadian cities. Epidemiology 8:162–167

Burnett RT, Brook JR, Yung WT, Dales RE, Krewski D (1997b) Association between ozone and hospitalization for respiratory diseases in 16 Canadian cities. Environ Res 72:24–31

Burnett RT, Smith-Doiron M, Stieb D, Cakmak S, Brook JR (1999) Effects of particulate and gaseous air pollution on cardiorespiratory hospitalizations. Arch Environ Health 54:130–139

Burnett RT, Stieb D, Brook JR, Cakmak S, Dales R, Raizenne M, Vincent R, Dann T (2004) Associations between short-term changes in nitrogen dioxide and mortality in Canadian cities. Arch Environ Health 59:228–236

Cakmak S, Dales RE, Vidal CB (2010) Air pollution and hospitalization for epilepsy in Chile. Environ Int 36:501–505

Calderón-Garciduеñas L, Vincent R, Mora-Tiscareno A, Franco-Lira M, Henriquez-Roldan C, Barragan-Mejia G et al (2007) Elevated plasma endothelin-1 and pulmonary arterial pressure in children exposed to air pollution. Environ Health Perspect 115:1248–1253

Chahine T, Baccarelli A, Litonjua A, Wright RO, Suh H, Gold DR et al (2007) Particulate air pollution, oxidative stress genes, and heart rate variability in an elderly cohort. Environ Health Perspect 115:1617–1622

Chen LH, Knutsen SF, Shavlik D, Beeson LW, Petersen F, Ghamsary M, Abbey D (2005) The association between fatal coronary heart disease and ambient particulate air pollution: are females at greater risk? Environ Health Perspect 113:1723–1729

Clancy L, Goodman P, Sinclair H, Dockery DW (2002) Effect of air-pollution control on death rates in Dublin, Ireland: an intervention study. Lancet 360:1210–1214

Clark NA, Demers PA, Karr CJ, Koehoorn M, Lencar C, Tamburic L, Brauer M (2010) Effect of early life exposure to air pollution on development of childhood asthma. Environ Health Perspect 118:284–290

Crouse DL, Goldberg MS, Ross NA, Chen H, Labrèche F (2010) Postmenopausal breast cancer is associated with exposure to traffic-related air pollution in Montreal, Canada: a case-control study. Environ Health Perspect 118:1578–1583

Crouse DL, Peters PA, Donkelaar A van, Goldberg MS, Villeneuve PJ, Brion O, Khan S, Atari DO, Jerrett M, Pope CA, Brauer M, Brook JR, Martin RV, Stieb D, Burnett RT (2012) Risk of non-accidental and cardiovascular mortality in relation to long-term exposure to low concentrations of fine particulate matter: a Canadian National-level cohort study. Environ Health Perspect 120:708–714

Currie J, Walker R (2009) Traffic congestion and infant health: Evidence from E-ZPass. National Bureau of Economic Research Working Paper No. 15413

Dales RE, Wheeler A, Mahmud M, Frescura AM, Smith-Doiron M, Nethery E et al (2008) The influence of living near roadways on spirometry and exhaled nitric oxide in elementary schoolchildren. Environ Health Perspect 116:1423–1427

Dales R, Chen L, Frescura AM, Liu L, Villeneuve PJ (2009a) Acute effects of outdoor air pollution on forced expiratory volume in 1 s: a panel study of schoolchildren with asthma. Eur Respir J 34:316–323

Dales R, Wheeler AJ, Mahmud M, Frescura AM, Liu L (2009b) The influence of neighborhood roadways on respiratory symptoms among elementary schoolchildren. J Occup Environ Med 51:654–660

Dales RE, Cakmak S, Vidal CB (2009c) Air pollution and hospitalization for headache in Chile. Am J Epidemiol 170:1057–1066

Delfino RJ, Staimer N, Tjoa T, Polidori A, Arhami M, Gillen DL et al (2008) Circulating biomarkers of inflammation, antioxidant activity, and platelet activation are associated with primary combustion aerosols in subjects with coronary artery disease. Environ Health Perspect 116:898–906

Dell SD, Foty R, To T, Beckerman BB, Jerret MJ, Stieb DM (2009) Birth and cumulative lifetime exposures to traffic-related air pollutants (TRAP) are associated with asthma in Toronto school children. Abstract #4212. European Respiratory Conference. Vienna, Austria. 747S

Dockery DW, Pope CA 3rd, Xu X, Spengler JD, Ware JH, Fay ME, Ferris BG Jr, Speizer FE (1993) An association between air pollution and mortality in six U.S. cities. N Engl J Med 329:1753–1759

Dominici F, Peng RD, Barr CD, Bell ML (2010) Protecting human health from air pollution: shifting from a single-pollutant to a multi-pollutant approach. Epidemiology 21:187–194

Dugandzic R, Dodds L, Stieb D, Smith-Doiron M (2006) The association between low level exposures to ambient air pollution and term low birth weight: a retrospective cohort study. Environ Health 5:3

Eftim SE, Samet JM, Janes H, McDermott A, Dominici F (2008) Fine particulate matter and mortality: a comparison of the six cities and American Cancer Society cohorts with a medicare cohort. Epidemiology 19:209–216

Filleul L, Rondeau V, Vandentorren S, Le Moual N, Cantagrel A, Annesi-Maesano I, Charpin D, Declercq C, Neukirch F, Paris C, Vervloet D, Brochard P, Tessier J-F, Kauffmann F, Baldi I (2005) Twenty five year mortality and air pollution: results from the French PAARC survey. Occup Environ Med 62:453–460

Gan WQ, Tamburic L, Davies HW, Demers PA, Koehoorn M, Brauer M (2010) Changes in residential proximity to road traffic and the risk of death from coronary heart disease. Epidemiology 21:642–649

Gauderman WJ, Avol E, Gilliland F, Vora H, Thomas D, Berhane K, McConnell R, Kuenzli N, Lurmann F, Rappaport E, Margolis H, Bates D, Peters J (2004) The effect of air pollution on lung development from 10 to 18 years of age. N Engl J Med 351:1057–1067

Gauderman WJ, Vora H, McConnell R, Berhane K, Gilliland F, Thomas D, Lurmann F, Avol E, Kunzli N, Jerrett M, Peters J (2007) Effect of exposure to traffic on lung development from 10 to 18 years of age: a cohort study. Lancet 369:571–577

Gehring U, Heinrich J, Kra¨mer U, Grote V, Hochadel M, Sugiri D, Kraft M, Rauchfuss K, Eberwein HG, Wichmann H-E (2006) Long-term exposure to ambient air pollution and cardiopulmonary mortality in women. Epidemiol 17:545–551

Gehring U, Wijga AH, Brauer M, Fischer P, Jongste JC de, Kerkhof M, Oldenwening M, Smit HA, Brunekreef B (2010) Traffic-related air pollution and the development of asthma and allergies during the first 8 years of life. Am J Respir Crit Care Med 181:596–603

Ghio AJ, Devlin RB (2001) Inflammatory lung injury after bronchial instillation of air pollution particles. Am J Respir Crit Care Med 164:704–708

Giles LV, Barn P, Künzli N, Romieu I, Mittleman MA, Eeden S van, Allen R, Carlsten C, Stieb D, Noonan C, Smargiassi A, Kaufman JD, Hajat S, Kosatsky T, Brauer M (2011) From good intentions to proven interventions: effectiveness of actions to reduce the health impacts of air pollution. Environ Health Perspect 119:29–36

Goldberg MS, Burnett RT, Bailar JC 3rd, Tamblyn R, Ernst P, Flegel K, Brook J, Bonvalot Y, Singh R, Valois MF, Vincent R (2001) Identification of persons with cardiorespiratory conditions who are at risk of dying from the acute effects of ambient air particles. Environ Health Perspect 109:S487–S494

Goldberg MS, Burnett RT, Valois MF, Flegel K, Bailar JC 3rd, Brook J, Vincent R, Radon K (2003) Associations between ambient air pollution and daily mortality among persons with congestive heart failure. Environ Res 91:8–20

Goldberg MS, Giannetti N, Burnett RT, Mayo NE, Valois MF, Brophy JM (2008) A panel study in congestive heart failure to estimate the short-term effects from personal factors and environmental conditions on oxygen saturation and pulse rate. Occup Environ Med 65:659–666

Goldberg MS, Giannetti N, Burnett RT, Mayo NE, Valois MF, Brophy JM (2009) Shortness of breath at night and health status in congestive heart failure: effects of environmental conditions and health-related and dietary factors. Environ Res 109:166–174

Gong H Jr, Linn WS, Terrell SL, Clark KW, Geller MD, Anderson KR et al (2004) Altered heart-rate variability in asthmatic and healthy volunteers exposed to concentrated ambient coarse particles. Inhal Toxicol 16:335–343

Gonzalez MA, Selwyn AP (2003) Endothelial function, inflammation, and prognosis in cardiovascular disease. Am J Med 115:99S–106S

Government of Canada (2007) Regulatory framework for air emissions. Government of Canada, Ottawa

Haak T, Jungmann E, Raab C, Usadel KH (1994) Elevated endothelin-1 levels after cigarette smoking. Metabolism 43:267–269

Hayes JD, McLellan LI (1999) Glutathione and glutathione-dependent enzymes represent a co-ordinately regulated defence against oxidative stress. Free Radic Res 31:273–300

Health Effects Institute (HEI) (2000) Reanalysis of the Harvard six cities study and the American cancer society study of particulate air pollution and mortality: a special report of the institute's particle epidemiology reanalysis project. Health Effects Institute, Cambridge MA

Health Effects Institute (HEI) Panel on the Health Effects of Traffic-Related Air Pollution (2010) Traffic-related air pollution: a critical review of the literature on emissions, exposure, and health effects. HEI Special Report 17. Health Effects Institute, Boston, MA

Henderson SB, Brauer M, Macnab YC, Kennedy SM (2011) Three measures of forest fire smoke exposure and their associations with respiratory and cardiovascular health outcomes in a population-based cohort. Environ Health Perspect 119:1266–1271

Hoek G, Brunekreef B, Goldbohm S, Fischer P, Brandt PA van den (2002) Association between mortality and indicators of traffic-related air pollution in the Netherlands: a cohort study. Lancet 360:1203–129

Huang W, Wang G, Lu SE, Kipen H, Wang Y, Hu M, Lin W, Rich D, Ohman-Strickland P, Diehl SR, Zhu P, Tong J, Gong J, Zhu T, Zhang J (2012) Inflammatory and oxidative stress responses of healthy young adults to changes in air quality during the Beijing Olympics. Am J Respir Crit Care Med. (Epub ahead of print)

Islam T, Berhane K, McConnell R, Gauderman WJ, Avol E, Peters JM, Gilliland FD (2009) Glutathione-S-transferase (GST) P1, GSTM1, exercise, ozone and asthma incidence in school children. Thorax 64:197–202

Islam T, Urman R, Gauderman WJ, Milam J, Lurmann F, Shankardass K, Avol E, Gilliland F, McConnell R (2011) Parental stress increases the detrimental effect of traffic exposure on children's lung function. Am J Respir Crit Care Med 14. (Epub ahead of print)

Ito K, De Leon SF, Lippmann M (2005) Associations between ozone and daily mortality: analysis and meta-analysis. Epidemiology 16:446–457

Jerrett M, Burnett RT, Pope CA 3rd, Ito K, Thurston G, Krewski D, Shi Y, Calle E, Thun M (2009a) Long-term ozone exposure and mortality. N Engl J Med 360:1085–1095

Jerrett M, Finkelstein MM, Brook JR, Arain MA, Kanaroglou P, Stieb DM, Gilbert NL, Verma D, Finkelstein N, Chapman KR, Sears MR (2009b) A cohort study of traffic-related air pollution and mortality in Toronto, Ontario, Canada. Environ Health Perspect 117:772–777

Kaplan GG, Dixon E, Panaccione R, Fong A, Chen L, Szyszkowicz M, Wheeler A, MacLean A, Buie WD, Leung T, Heitman SJ, Villeneuve PJ (2009) Effect of ambient air pollution on the incidence of appendicitis. CMAJ 181:591–597

Kaplan GG, Hubbard J, Korzenik J, Sands BE, Panaccione R, Ghosh S, Wheeler AJ, Villeneuve PJ (2010) The inflammatory bowel diseases and ambient air pollution: a novel association. Am J Gastroenterol 105:2412–2419

Kawano H, Motoyama T, Hirashima O, Hirai N, Miyao Y, Sakamoto T et al (1999) Hyperglycemia rapidly suppresses flow-mediated endothelium-dependent vasodilation of brachial artery. J Am Coll Cardio 31:146–154

King GL, Brownlee M (1996) The cellular and molecular mechanisms of diabetic complications. Endocrinol Metab Clin North Am 25:255–270

Künzli N, Jerrett M, Mack WJ, Beckerman B, LaBree L, Gilliland F, Thomas D, Peters J, Hodis HN (2005) Ambient air pollution and atherosclerosis in Los Angeles. Environ Health Perspect 113:201–206

Künzli N, Jerrett M, Garcia-Esteban R, Basagaña X, Beckermann B, Gilliland F, Medina M, Peters J, Hodis HN, Mack WJ (2010) Ambient air pollution and the progression of atherosclerosis in adults. PLoS One 5:e9096

Laden F, Schwartz J, Speizer FE, Dockery DW (2006) Reduction in fine particulate air pollution and mortality: extended follow-up of the Harvard Six Cities study. Am J Respir Crit Care Med 173:667–672

Larrieu S, Lefranc A, Gault G, Chatignoux E, Couvy F, Jouves B, Filleul L (2009) Are the short-term effects of air pollution restricted to cardiorespiratory diseases? Am J Epidemiol 189:1201–1208

Leech JA, Wilby K, McMullen E, Laporte K (1996) The Canadian human activity pattern survey: report of methods and population surveyed. Chronic Dis Can 17:118–123

Leonardi GS, Houthuijs D, Steerenberg PA, Fletcher T, Armstrong B, Antova T (2000) Immune biomarkers in relation to exposure to particulate matter: a cross-sectional survey in 17 cities of central Europe. Inhal Toxicol 12:1–14

Levy JI, Chemerynski SM, Sarnat JA (2005) Ozone exposure and mortality: an empiric Bayes metaregression analysis. Epidemiology 16:458–468

Lipfert FW, Baty JD, Miller JP, Wyzga RE (2006) $PM_{2.5}$ Constituents and related air quality variables as predictors of survival in a cohort of U.S. Military Veterans. Inhal Toxicol 18:645–657

Liu S, Krewski D, Shi Y, Chen Y, Burnett RT (2003) Association between gaseous ambient air pollutants and adverse pregnancy outcomes in Vancouver, Canada. Environ Health Perspect 111:1773–8

Liu S, Krewski D, Shi Y, Chen Y, Burnett RT (2007a) Association between maternal exposure to ambient air pollutants during pregnancy and fetal growth restriction. J Expo Sci Environ Epidemiol 17:426–432

Liu L, Ruddy T, Dalipaj M, Szyszkowicz M, You H, Poon R et al (2007b) Influence of personal exposure to particulate air pollution on cardiovascular physiology and biomarkers of inflammation and oxidative stress in subjects with diabetes. J Occup Environ Med 49:258–265

Liu L, Poon R, Chen L, Frescura AM, Montuschi P, Ciabattoni G et al (2008) Acute effects of air pollution on pulmonary function, airway inflammation and oxidative stress in asthmatic children. Environ Health Perspect 117:668–674

Liu L, Ruddy T, Dalipaj M, Poon R, Szyszkowicz M, You H et al (2009) Effects of indoor, outdoor, and personal exposure to particulate air pollution on cardiovascular physiology and systemic mediators in seniors. J Occup Environ Med 51:1088–98

Mack WJ, LaBree L, Liu C-R, Liu C-H, Selzer RH, Hodis HN (2000) Correlations between measures of atherosclerosis change using carotid ultrasonography and coronary angiography. Atherosclerosis 150:371–379

Mauderly JL, Burnett RT, Castillejos M, Ozkaynak H, Samet JM, Stieb DM, Vedal S, Wyzga RE (2010) Is the air pollution health research community prepared to support a multipollutant air quality management framework? Inhal Toxicol 22(S1):1–19

McConnell R, Berhane K, Gilliland F, London SJ, Islam T, Gauderman WJ, Avol E, Margolis HG, Peters JM (2002) Asthma in exercising children exposed to ozone: a cohort study. Lancet 359:386–391

McCreanor J, Cullinan P, Nieuwenhuijsen MJ, Stewart-Evans J, Malliarou E, Jarup L et al (2007) Respiratory effects of exposure to diesel traffic in persons with asthma. New Engl J Med 357:2348–2358

Miller KA, Siscovick DS, Sheppard L, Shepherd K, Sullivan JH, Anderson GL, Kaufman JD (2007) Long-term exposure to air pollution and incidence of cardiovascular events in women. N Engl J Med 356:447–458

Mills NL, Törnqvist H, Robinson SD, Gonzalez M, Darnley K, MacNee W et al (2005) Diesel exhaust inhalation causes vascular dysfunction and impaired endogenous fibrinolysis. Circulation 112:3930–3936

Mills NL, Törnqvist H, Gonzalez ML, Vink E, Robinson SD, Söderberg S et al (2007) Ischemic and thrombotic effects of dilute diesel-exhaust inhalation in men with coronary heart disease. New Engl J Med 357:1075–1082

Mukamal KJ, Wellenius GA, Suh HH, Mittleman MA (2009) Weather and air pollution as triggers of severe headaches. Neurology 72:922–927

Naeher LP, Brauer M, Lipsett M, Zelikoff JT, Simpson CD, Koenig JQ, Smith KR (2007) Woodsmoke health effects: a review. Inhal Toxicol 19:67–106

Næss Ø, Nafstad P, Aamodt G, Claussen B, Rosland P (2007) Relation between concentration of air pollution and cause-specific mortality: four-year exposures to nitrogen dioxide and particulate matter pollutants in 470 neighborhoods in Oslo, Norway. Am J Epidemiol 165:435–443

Nafstad P, Håheim LL, Wisløff T, Gram F, Oftedal B, Holme I, Hjermann I, Leren P (2004) Urban air pollution and mortality in a cohort of Norwegian men. Environ Health Perspect 112:610–615

National Research Council, Committee on Research Priorities for Airborne Particulate Matter (2004) Research priorities for airborne particulate matter: IV. Continuing Research Progress. National Academies Press, Washington, DC

Nordenhäll C, Pourazar J, Blomberg A, Levin J-O, Sandström T, Ädelroth E (2000) Airway inflammation following exposure to diesel exhaust: a study of time kinetics using induced sputum. Eur Respir J 15:1046–1051

O'Neill MS, Veves A, Zanobetti A, Sarnat JA, Gold DR, Economides PA et al (2005) Diabetes enhances vulnerability to particulate air pollution-associated impairment in vascular reactivity and endothelial function. Circulation 111:2913–2920

O'Neill MS, Diez-Roux AV, Auchincloss AH, Shen M, Lima JA, Polak JF, Barr RG, Kaufman J, Jacobs DR Jr (2011) Long-term exposure to airborne particles and arterial stiffness: the multi-ethnic study of atherosclerosis (MESA). Environ Health Perspect 119:844–851

Park SK, O'Neill MS, Vokonas P, Sparrow D, Schwartz J (2005) Effects of air pollution on heart rate variability: the VA normative aging study. Environ Health Perspect 113:304–309

Parker JD, Mendola P, Woodruff TJ (2008) Preterm birth after the Utah Valley Steel Mill closure: a natural experiment. Epidemiology 19:820–823

Parker JD, Rich DQ, Glinianaia SV, Leem JH, Wartenberg D, Bell ML, Bonzini M, Brauer M, Darrow L, Gehring U, Gouveia N, Grillo P, Ha E, Hooven EH van den, Jalaludin B, Jesdale BM, Lepeule J, Morello-Frosch R, Morgan GG, Slama R, Pierik FH, Pesatori AC, Sathyanarayana S, Seo J, Strickland M, Tamburic L, Woodruff TJ (2011) The International Collaboration on air pollution and pregnancy outcomes: initial results. Environ Health Perspect 119:1023–1028

Peters A, Doring A, Wichmann H-E, Koenig W (1997) Increased plasma viscosity during an air pollution episode: a link to mortality? Lancet 349:1582–1587

Peters A, Fröhlich M, Döring A, Immervoll T, Wichmann H-E, Hutchinson WL et al (2001) Particulate air pollution is associated with an acute phase response in men. European Heart J 22:1198–1204

Peters A, Greven S, Heid IM, Baldari F, Breitner S, Bellander T et al (2009) Fibrinogen genes modify the fibrinogen response to ambient particulate matter. Am J Respir Crit Care Med 179:484–491

Pope CA 3rd (1989) Respiratory disease associated with community air pollution and a steel mill, Utah Valley. Am J Public Health 79:623–628

Pope CA 3rd, Thun MJ, Namboodiri MM, Dockery DW, Evans JS, Speizer FE, Heath CW Jr (1995) Particulate air pollution as a predictor of mortality in a prospective study of U.S. adults. Am J Respir Crit Care Med 151:669–674

Pope CA 3rd, Burnett RT, Thun MJ, Calle EE, Krewski D, Ito K, Thurston GD (2002) Lung cancer, cardiopulmonary mortality, and long-term exposure to fine particulate air pollution. JAMA 287:1132–1141

Pope CA 3rd, Burnett RT, Thurston GD, Thun MJ, Calle EE, Krewski D, Godleski JJ (2004a) Cardiovascular mortality and long-term exposure to particulate air pollution: epidemiological evidence of general pathophysiological pathways of disease. Circulation 109:71–77

Pope CA 3rd, Hansen ML, Long RW, Nielsen KR, Eatough NL, Wilson WE et al (2004b) Ambient particulate air pollution, heart rate variability, and blood markers of inflammation in a panel of elderly subjects. Environ Health Perspect 112:339–345

Pope CA 3rd, Burnett RT, Krewski D, Jerrett M, Shi Y, Calle EE, Thun MJ (2009) Cardiovascular mortality and exposure to airborne fine particulate matter and cigarette smoke: shape of the exposure-response relationship. Circulation 120:941–948

Pope CA 3rd, Burnett RT, Turner MC, Cohen A, Krewski D, Jerrett M, Gapstur SM, Thun MJ (2011) Lung cancer and cardiovascular disease mortality associated with ambient air pollution and cigarette smoke: shape of the exposure-response relationships. Environ Health Perspect 119:1616–1621

Pourazar J, Mudway IS, Samet JM, Helleday R, Blomberg A, Wilson SJ et al (2005) Diesel exhaust activates redox-sensitive transcription factors and kinases in human airways. Am J Physiol Lung Cell Mol Physiol 289: L724–L730

Pradhan AD, Manson JE, Rifai N, Buring JE, Ridker PM (2001) C-reactive protein, interleukin 6, and risk of developing type 2 diabetes mellitus. JAMA 286:327–334

Prüss-Üstün A, Corvalán C (2006) Preventing disease through healthy environments. Towards an estimate of the environmental burden of disease. WHO, Geneva

Puett RC, Schwartz J, Hart JE, Yanosky JD, Speizer FE, Suh H, Paciorek CJ, Neas LM, Laden F (2008) Chronic particulate exposure, mortality, and coronary heart disease in the nurses' health study. Am J Epidemiol 168:1161–1168

Puett RC, Hart JE, Yanosky JD, Paciorek C, Schwartz J, Suh H, Speizer FE, Laden F (2009) Chronic fine and coarse particulate exposure, mortality and coronary heart disease in the nurses' health study. Environ Health Perspect 117:1697–1701

Rich DQ, Kipen HM, Huang W, Wang G, Wang Y, Zhu P, Ohman-Strickland P, Hu M, Philipp C, Diehl SR, Lu SE, Tong J, Gong J, Thomas D, Zhu T, Zhang JJ (2012) Association between changes in air pollution levels during the Beijing Olympics and biomarkers of inflammation and thrombosis in healthy young adults. JAMA 16:2068–78

Riediker M, Cascio WE, Griggs TR, Herbst MC, Bromberg PA, Neas L et al (2004) Particulate matter exposure in cars is associated with cardiovascular effects in healthy young men. Am J Respir Crit Care Med 169:934–940

Romieu I, Meneses F, Ramirez M, Ruiz S, Padilla RP, Sienra JJ et al (1998) Antioxidant supplementation and respiratory functions among workers exposed to high levels of ozone. Am J Respir Crit Care Med 158:226–232

Romieu I, Sienra-Monge JJ, Ramirez M, Tellez-Rojo MM, Moreno-Macias H, Reyes-Ruiz NI et al (2002) Antioxidant supplementation and lung functions among children with asthma exposed to high levels of air pollutants. Am J Respir Crit Care Med 166:703–709

Romieu I, Sienra-Monge JJ, Ramirez-Aguilar M, Moreno-Macias H, Reyes-Ruiz NI, Estela dR-N et al (2004) Genetic polymorphism of GSTM1 and antioxidant supplementation influence lung function in relation to ozone exposure in asthmatic children in Mexico City. Thorax 59:8–10

Romieu I, Tellez-Rojo MM, Lazo M, Manzano-Patino A, Cortez-Lugo M, Julien P et al (2005) Omega-3 fatty acid prevents heart rate variability reductions associated with particulate matter. Am J Respir Crit Care Med 172:1534–1540

Rückerl R, Ibald-Mulli A, Koenig W, Schneider A, Woelke G, Cyrys J et al (2006) Air pollution and markers of inflammation and coagulation in patients with coronary heart disease. Am J Respir Crit Care Med 173:432–441

Sacks JD, Stanek LW, Luben TJ, Johns DO, Buckley BJ et al (2011) Particulate matter–induced health effects: who Is susceptible? Environ Health Perspect 119:446–454

Salvi SS, Nordenhall C, Blomberg A, Rudell B, Pourazar J, Kelly FJ et al (2000) Acute exposure to diesel exhaust increases IL-8 and GRO-α production in healthy human airways. Am J Respir Crit Care Med 161:550–557

Samet JM, Dominici F, Curriero FC, Coursac I, Zeger SL (2000a) Fine particulate air pollution and mortality in 20 U.S. cities, 1987–1994. N Engl J Med 343:1742–1749

Samet JM, Zeger S, Dominici F et al (2000b) The national morbidity, mortality, and air pollution study (NMMAPS). 2. Morbidity and mortality from air pollution in the United States. Cambridge, Mass.: Health Effects Institute (See http://healtheffects.org/pubs/samet2.pdf.)

Samoli E, Peng R, Ramsay T, Pipikou M, Touloumi G, Dominici F, Burnett R, Cohen A, Krewski D, Samet J, Katsouyanni K (2008) Acute effects of ambient particulate matter on mortality in Europe and North America: results from the APHENA study. Environ Health Perspect 116:1480–1486

Sapkota A, Chelikowsky AP, Nachman KE, Cohen AJ, Ritz B (2012) Exposure to particulate matter and adverse birth outcomes: a comprehensive review and meta-analysis. Air Qual Atmos Health 5:369–381

Schwartz J, Park SK, O'Neill MS, Vokonas PS, Sparrow D, Weiss S et al (2005) Glutathione-S-transferase M1, obesity, statins, and autonomic effects of particles. Gene-by-drug-by-environment interaction. Am J Respir Crit Care Med 172:1529–1533

Seaton A, Soutar A, Crawford V, Elton R, McNerlan S, Cherrie J et al (1999) Particulate air pollution and the blood. Thorax 54:1027–1032

Shin HH, Stieb DM, Jessiman B, Goldberg MS, Brion O, Brook J, Ramsay T, Burnett RT (2008) A temporal, multicity model to estimate the effects of short-term exposure to ambient air pollution on health. Environ Health Perspect 116:1147–1153

Sivagangabalan G, Spears D, Masse S, Urch B, Brook RD, Silverman F et al (2011) The effect of air pollution on spatial dispersion of myocardial repolarization in healthy human volunteers. J Am Coll Cardio 57:198–206

Smargiassi A, Berrada K, Fortier I, Kosatsky T (2006) Traffic intensity, dwelling value, and hospital admissions for respiratory disease among the elderly in Montreal (Canada): a case-control analysis. J Epidemiol Community Health 60:507–512

Smargiassi A, Kosatsky T, Hicks J, Plante C, Armstrong B, Villeneuve PJ, Goudreau S (2009) Risk of asthmatic episodes in children exposed to sulfur dioxide stack emissions from a refinery point source in Montreal, Canada. Environ Health Perspect 117:653–659

Soukup JM, Ghio AJ, Becker S (2000) Soluble components of Utah Valley particulate pollution alter alveolar macrophage function *in vivo* and *in vitro*. Inhal Toxicol 12:401–414

Stenfors N, Nordenhäll C, Salvi SS, Mudway I, Söderberg M, Blomberg A et al (2004) Different airway inflammatory responses in asthmatic and healthy humans exposed to diesel. Eur Respir J 23:82–86

Stieb DM, Judek S, Burnett RT (2002) Meta-analysis of time-series studies of air pollution and mortality: effects of gases and particles and the influence of cause of death, age, and season. J Air Waste Manag Assoc 52:470–484

Stieb DM, Szyszkowicz M, Rowe BH, Leech JA (2009) Air pollution and emergency department visits for cardiac and respiratory conditions: a multi-city time-series analysis. Environ Health 8:25

Stieb DM, Chen L, Eshoul M, Judek S (2012) Ambient air pollution, birth weight and preterm birth: a systematic review and meta-analysis. Environ Res 117:100–111

Szyszkowicz M, Stieb DM, Rowe BH (2009a) Air pollution and daily ED visits for migraine and headache in Edmonton, Canada. Am J Emerg Med 27:391–396

Szyszkowicz M, Kaplan GG, Grafstein E, Rowe BH (2009b) Emergency department visits for migraine and headache: a multi-city study. Int J Occup Med Environ Health 22:235–242

Szyszkowicz M, Rowe BH, Kaplan GG (2009c) Ambient sulphur dioxide exposure and emergency department visits for migraine in Vancouver, Canada. Int J Occup Med Environ Health 22:7–12

Toronto Public Health (2007) Air pollution burden of illness from traffic in Toronto—problems and solutions. Toronto Public Health, Toronto

Urch B, Brook JR, Wasserstein D, Brook RD, Rajagopalan S, Corey P et al (2004) Relative contributions of $PM_{2.5}$ chemical constituents to acute arterial vasoconstriction in humans. Inhal Toxicol 16:345–352

Urch B, Silverman F, Corey P, Brook JR, Lukic KZ, Rajagopalan S et al (2005) Acute blood pressure responses in healthy adults during controlled air pollution exposures. Environ Health Perspect 113:1052–1055

USEPA (2011a) Exposure factors handbook: Edition Washington, DC: 2011. EPA/600/R-09/052F

USEPA (2011b) Clean air research. Research Priorities. http://www.epa.gov/airscience/research.htm. Accessed 19 Oct 2011

Vedal S, Brauer M, White R, Petkau J (2003) Air pollution and daily mortality in a city with low levels of pollution. Environ Health Perspect 111:45–52

Verma S, Li S-H, Badiwala MV, Weisel RD, Fedak PWM, Li R-K et al (2002) Endothelin antagonism and interleukin-6 inhibition attenuate the proatherogenic effects of C-reactive protein. Circulation 105:1890–1896

Villeneuve PJ, Chen L, Stieb D, Rowe BH (2006) Associations between outdoor air pollution and emergency department visits for stroke in Edmonton, Canada. Eur J Epidemiol 21:689–700

Villeneuve PJ, Jerrett M, G Su J, Burnett RT, Chen H, Wheeler AJ, Goldberg MS (2012) A cohort study relating urban green space with mortality in Ontario, Canada. Environ Res. (Epub ahead of print)

Vita JA, Keaney JF Jr, Larson MG, Keyes MJ, Massaro JM, Lipinska I et al (2004) Brachial artery vasodilator function and systemic inflammation in the Framingham Offspring Study. Circulation 110:3604–3609

Ward DJ, Ayres JG (2004) Particulate air pollution and panel studies in children: a systematic review. Occup Environ Med 61:e13

Ward TJ, Palmer CP, Noonan CW (2010) Fine particulate matter source apportionment following a large woodstove changeout program in Libby, Montana. J Air Waste Manag Assoc 60:688–693

Weichenthal S, Kulka R, Dubeau A, Martin C, Wang D, Dales R (2011) Traffic-related air pollution and acute changes in heart rate variability and respiratory function in urban cyclists. Environ Health Perspect 119:1373–1378

Wheeler AJ, Gibson MD, Ward T, Allen RW, Guernsey JR, Seaboyer M, Kutcha J, Gould R, Stieb D (2011) Reductions in residential wood smoke concentrations and infiltration efficiency using electrostatic air cleaner interventions. 12th International Conference on Indoor Air Quality and Climate. June 5–10, 2011, Austin, TX, USA

Whitsel EA, Quibrera PM, Christ SL, Liao D, Prineas RJ, Anderson GL et al (2009) Heart rate variability, ambient particulate matter air pollution, and glucose homeostasis: the environmental epidemiology of arrhythmogenesis in the women's health initiative. Am J Epidemiol 169:693–703

Zeger SL, Dominici F, McDermott A, Samet JM (2008) Mortality in the medicare population and chronic exposure to fine particulate air pollution in urban centers (2000–2005). Environ Health Perspect 116:1614–1619

Visual Air Quality Management

8

Eric Taylor

Abstract

Airborne particulate matter and certain gases can degrade visual air quality, reducing the public's ability to enjoy views of distant mountains and other scenic landmarks. Improving visual air quality can have economic, health and social benefits.

Visual air quality can be assessed qualitatively using human perception of a scenic view as well as quantitatively by measuring the degree by which air pollutants interfere with light travelling through the atmosphere. Significant research has been undertaken in Canada since 1980 to further our understanding of visual air quality degradation, particularly in British Columbia. This work is laying the foundation for a numerical index to communicate visual air quality in British Columbia and Canada and for a strategy to protect and improve visual air quality in the future.

Keywords

Visibility · Visual air quality · Vista · Smog · Aerosol · Extinction coefficient · Photons · Glacier · Real estate · Tourism · Film · First nations · Health · Agriculture · Measurement · Perception · Deciviews · Visual range · BCVCC · Air quality index · Visibility protection framework

8.1 Introduction to Visual Air Quality

Canada is known for its natural environment and unspoiled wilderness. Its beautiful vistas are highly valued and are an important aspect of the quality of life. They are also important to tourism, real estate, business and other sectors since stunning, natural scenery makes Canada a more attractive place to work, live and visit. Unfortunately, air pollution is a threat to this scenic beauty, particularly in mountainous and coastal areas where people expect to enjoy distant landmarks.

Visual air quality is a measure of how the air "looks". Distant objects and landmarks can be vivid in colour and clarity when the atmosphere is clean, but they can appear hazy and washed out when air quality deteriorates. Poor visual air quality occurs when sunlight interacts with gases, mainly nitrogen dioxide, and fine (microscopic) airborne particles called "fine particulate matter" or $PM_{2.5}$[1]. The latter can be made up of many chemical species, each of which interacts with light differently. These particles and gases can make objects and vistas appear indistinct and even disappear. The higher the concentration of these pollutants, the more visual air quality is obscured. Humidity can also affect visual air quality since some chemical species are hygroscopic and attract water, causing an increase in particle size which results in changes in how they interfere with light.

The main goal of air quality management in Canada has justifiably been to reduce the risk of air pollution to human health, so managing visual air quality is in its infancy. However, visual air quality can be compared to the "canary in the coal mine"—a warning of the threat to health of degrading air quality if pollutants are not suitably managed.

This chapter, like others in this book, is focussed on the Canadian experience. However, there is an emphasis on work that has been done recently in the province of British

E. Taylor (✉)
British Columbia Ministry of Environment, Victoria, BC, Canada
e-mail: eric.taylor@gov.bc.ca

[1] $PM_{2.5}$ is airborne particulate matter with a diameter less than 2.5 µm (millionths of a metre), much smaller than the width of a human hair.

E. Taylor, A. McMillan (eds.), *Air Quality Management,* DOI 10.1007/978-94-007-7557-2_8,

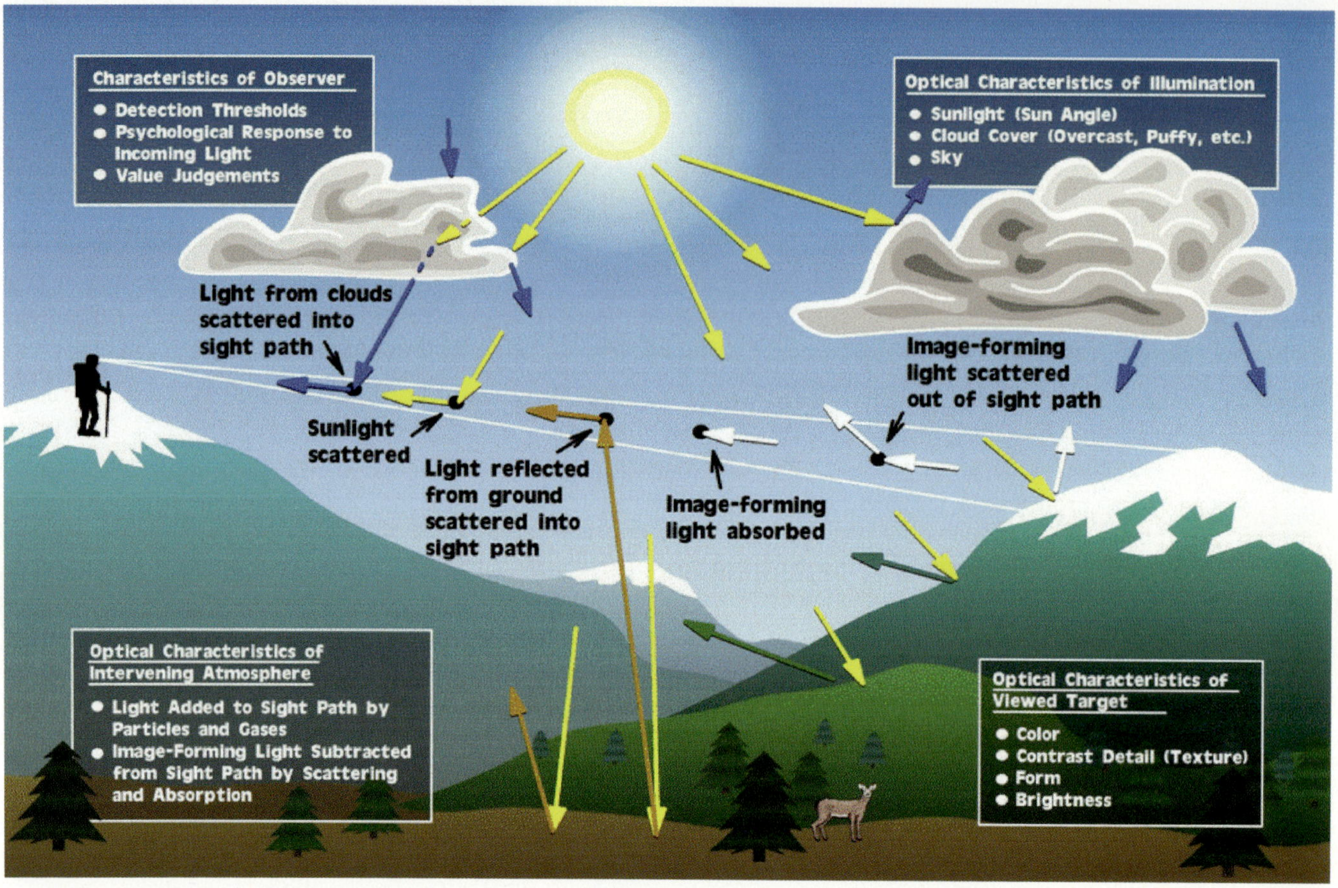

Fig. 8.1 An illustration showing the various mechanisms through which particles and certain gases can degrade visual air quality. (Malm 1999)

Columbia. Residents there enjoy and take pride in the abundance of magnificent views of the natural mountainous environment. BC is also where significant activity is now leading to the development of a visual air quality program that aims to protect and improve the views of these natural wonders.

Note that in this chapter the terms "visibility" and "visual air quality" refer to the same phenomenon and are used interchangeably.

8.2 The Impact of Air Pollutants on Visual Air Quality

This section summarizes our understanding of how air pollution impacts visual air quality and lessens the aesthetic value of scenic vistas and draws largely from Malm (1999).

In urban areas, visual air quality can manifest as a yellow or brown layer of smog that appears following several days of warm, clear weather. In rural and wilderness areas it most often appears as a uniform white haze that washes out or obscures distant landscape features. In both instances the cause is primarily fine particles that interfere with visible light, resulting in the loss of scene information to the eye. Visual air quality is primarily degraded when microscopic particles sizes are between 0.1 and 1 μm[2], about the same length as the wavelengths of visible light. This size of particle unfortunately also tends to be associated with man-made pollutant emissions.

Some light is ***scattered*** when light photons[3] collide with airborne particles or certain gases, particularly nitrogen dioxide, and become dispersed in all directions. This scattered light, as well as the additional ***air light*** that is scattered in the direction of the observer, reduces the amount of image-forming information received by the observer. Air light occurs when sunlight, light from clouds, and ground-reflected light all impinge on and scatter from particulates and certain gases located in the sight path. Some of this scattered light remains in the sight path, and at times it can become so bright that the image essentially disappears. Some light photons are also ***absorbed*** by gases and dark or coloured airborne particles, further limiting the image information transmitted to the observer.

Scattering, absorption and other important factors involved in observing a scenic vista are outlined in Fig. 8.1.

[2] A micron is one millionth of a metre.

[3] Photons are "particles" of light. Light has the characteristics of both a wave and a particle.

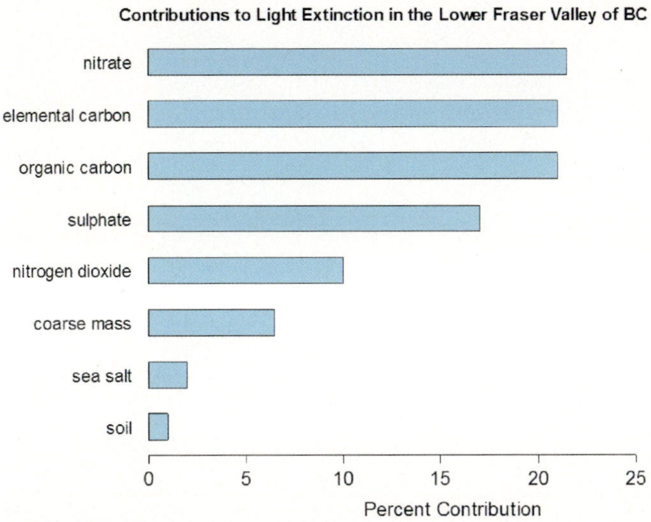

Fig. 8.2 Annual calculated average contributions to total extinction by particle and gas species at the Burnaby South and Abbotsford air quality monitoring sites. Based on data from 2003 to 2008. (So et al. 2011)

Fig. 8.3 Top: View of Toronto, Canada on a day with excellent air quality and low ex-tinction coefficient. Bottom: Same view artificially degraded assuming poor air quality and high extinction coefficient. Top photo courtesy Raymond Hoff. Bottom image digi-tally enhanced by Yahya Golestani and Bill Malm

A final important factor in seeing and appreciating a scenic vista are the characteristics of the human observer (Malm et al. 2011).

Several coefficients have been defined to better quantify the processes related to visual air quality. These coefficients are simply the *fraction* of the light that is scattered or absorbed by particles or gases. Very clean air scatters and absorbs almost no light, so these coefficients are close to zero. Highly polluted air scatters and absorbs a large portion of incoming light and so the coefficients can be much higher, approaching one. The fraction of light that is scattered is called the *scattering coefficient*; the fraction absorbed, the *absorption coefficient*. The *extinction coefficient* is the sum of the scattering and absorbtion coefficients of both partculates and gas. Units are per unit distance, which could be any metric (centimetres, kilometres, megametres etc.).

As a result of these scattering and absorption processes, image-forming information from an object is reduced as it passes through the atmosphere to the human observer. This process is summarized in the following equations that use the coefficients described above:

$$\text{Scattering coefficient} = \text{sum of scattering coefficients of particles and gases}$$

$$\text{Absorption coefficient} = \text{sum of absorption coefficients of particles and gases}$$

$$\text{Extinction coefficient} = \text{sum of scattering and absorption coefficients}$$

Typical values for extinction coefficients range from about 0.02 (clean air) to 0.30 (polluted air) per kilometre. This means that only 2 % of photons are being scattered or absorbed, per kilometre, for clean air while 30 % of photons are being scattered or absorbed per kilometre for polluted air. This does not take into account the effect of additional air light being scattered from all directions into the sight view of the observer and further deteriorating the image information.

The concentrations of air pollutants is related to their capacity to absorb or scatter light. Knowledge of these concentrations can help in the management of visual air quality. These relationships vary between different locations since they depend on relative humidity and chemical and physical characteristics of the fine particulate. Particles are composed of varying amounts of different constituents, each of which have different abilities to extinguish light. For example, particles containing sulphates, nitrates and elemental and organic carbon are particularly adept at interfering with light, being responsible for about 75 % of light extinction in the Lower Fraser Valley of BC (Fig. 8.2).

Figure 8.3 compares two scenes of the skyline of Toronto with differing extinction coefficients. The top scene has excellent visual air quality with a very low extinction coefficient (0.025 km^{-1}). This is a relatively rare event (top 0.2 % of days with low humidity). The bottom scene has degraded air quality and a correspondingly high extinction coefficient (0.316 km^{-1}) and would be in the bottom 8 % of dry days for visual air quality. The visual air quality in the bottom scene has been artificially degraded using computer simulation.

Fig. 8.4 Air pollutants occasionally restrict visibility of distant landmarks in the Lower Fraser Valley (LFV) of British Columbia. See text for details. Image courtesy of Environment Canada

8.3 The Importance of Visual Air Quality[4]

Most are aware that haze, smoke and smog produced by human activities can reduce the ability of an observer to enjoy scenic views. We also sense that this could have, at a minimum, subtle impacts on our well-being. The public also associates degraded visibility with *unhealthy* air quality. Unfortunately, negative impressions of degraded visual air quality can be left with residents and visitors alike. Tourists, employers, potential employees planning to relocate and retirees who are considering moving to a clean, safe area may look elsewhere.

Figure 8.4 is an example of air pollutants degrading visibility in Metro Vancouver and the Lower Fraser Valley (LFV) in British Columbia, Canada. This image was taken from a height of about 1000 m on a mountain just north of Vancouver. A white haze in the lower levels of the atmosphere lays over the LFV, blocking out views of eastern Vancouver suburbs located about 40 km from the camera. The peak of Mount Baker, a glacier-capped 3300 m volcano in Washington state, is about100 km from the camera (see map in Fig. 8.10). The volcanic peak was clearly visible above the haze layer but was likely obscured when viewed from Vancouver, which is close to sea level. Residents of the Lower Fraser Valley and Metro Vancouver rate Mount Baker and its adjacent mountains as valuable panoramas. Indeed, clear, unlimited visibility days have in the past been referred to as "Baker Days" by local residents.

Although visual air quality has no separate market value (i.e. it can't be bought or sold), the following studies have attempted to estimate its value.

8.3.1 Residents' Willingness-to-pay for 'Good Visual Air Quality'

By its very nature, visual air quality is perception-based, so the most common method of estimating its value is by determining the viewer's "willingness to pay" for clearer vistas. One estimate of this can be made using a survey whereby individuals are asked to value improvements in both local

[4] This section draws largély from the 2011 draft report Economic Impacts of Visibility Degradation in the Lower Fraser Valley, BC, A Discussion Paper by the BC Visibility Coordinating Committee (Bates-Frymel et al. 2011).

Fig. 8.5 Moraine Lake, Banff National Park, Alberta and other scenic vistas make Canada an important tourist destination. Air pollution can degrade such scenes, negatively impacting tourism. Photo courtesy Tour Smart Ltd, UK

visual air quality and health conditions separately. In an example of such a survey, residents of Metro Vancouver and the Lower Fraser Valley were presented with photographs of summer views with differing degrees of degraded visual air quality and potential costs to each household of improving local visual air quality. From this survey, it was estimated that a reasonable value/cost for visual air quality improvement (10 % improvement in visual range[5]) would be about $ 40 per household in the Lower Fraser Valley (LFV), or $ 37 million annually (2011 dollars) (Bates-Frymel et al. 2011). This value does not consider the spin-off health benefits. This result suggests that residents highly value their visual air quality and do not want it degraded.

Similar surveys in the US, Korea and New Zealand have produced comparable results.

8.3.2 Real Estate/Housing Values

Good visual air quality can result in a higher perceived quality of life in a city, which in turn impacts the ability to attract residents, workers, and businesses to a region. Conversely, decisions to avoid moving to or investing in areas with poor visual air quality could negatively impact the real estate and housing industries. Therefore, an economic benefit to good visual air quality could be the "avoided losses" for the real estate and housing industries.

For example, a study in Los Angeles estimated that the visibility impact on housing prices is a small but significant part of the total house price (Beron et al. 2001). In another study in Metro Vancouver, this translated into a "real estate visibility valuation" benefit of $ 208 million annually in 2009 dollars. Also, spin-off benefits related to maintaining or improving good visual air quality on the real estate and the housing industries, such as professional services, taxes, increased employment and increased capital expenditures, have been estimated to be an additional $ 24 million annually in Metro Vancouver (Bates-Frymel et al. 2011).

8.3.3 Tourism Industry Impacts

Tourism contributes as much to Canada's wealth as agriculture, fisheries and forestry combined (Industry Canada 2011). In 2008, tourism activity, including visits to wilderness areas like the Rocky Mountains (Fig. 8.5), generated over $ 74 billion in revenues. This represented 2 % of Canada's gross domestic product (GDP) and directly employed over 660,000 Canadians. Tourism is crucial to the bottom-line of key industry sectors and is the primary source of revenues for passenger airlines and accommodation properties. Tourism also supports restaurants and the entertainment industry, including performing arts, cultural and sporting events, and a variety of small- and medium-sized businesses.

[5] Visual Range is the distance at which a given standard object can be seen and identified with the unaided eye.

Degraded visual air quality has been shown to impact tourism. Visitor surveys demonstrate that tourists, when confronted with poor visual air quality at their destination, may tend to shorten their stay, go elsewhere or not return in the future. In a 1999 survey, it was concluded that one poor visual air quality event could result in revenue losses of up to $ 9 million in the Lower Mainland/Fraser Valley area of BC (McNeill and Roberge 2000). In 2009 these losses would have increased to $ 18 million per poor visual air quality event.

Further, Tourism Vancouver has stated that "any deterioration of visual air quality is a deterioration of our brand, our reputation and the visitor experience" and "there could be a tipping point at which poor visual air quality will erode our competitiveness".

8.3.4 Impacts on Film Industry

Canada has some of the largest film and television service production centres in North America. In B.C., for example, a total of 239 productions were filmed in 2009 contributing $1.3 billion to the province's economy and providing an estimated 20,000 direct and 15,000 indirect jobs (Bates-Frymel et al. 2011). Film production companies find a broad appeal for filming in Canada due to its variety of location resources (e.g. urban, wilderness, mountains, coastlines etc) and diversity of the county's climate and geography. Degraded visual air quality could reduce this appeal.

8.3.5 First Nations Cultural Impacts

One study found that First Nations[6] peoples of the Lower Fraser Valley have distinct concerns over declining visual air quality (Carlson 2009). Through interviews, it was determined that aboriginal peoples have a spiritual need to see the mountains, see *from* the mountains and also be *seen by* the "mountain mother" who watches over the people and the returning sockeye salmon. Impaired views of the landscape threaten this spiritual need and disrupt the flow of knowledge transfer across generations, tribes and cultures through the following:

- First Nations have a history of sharing cultural knowledge and history about mountains and celestial features during times of travelling. Landscape features served as geographic markers and help to develop a shared sense of the environment. But as haze and urban smog impair views of these features, it has become increasingly difficult to transmit this knowledge.
- The Coast Salish society Stó:lõ people in the Pacific Northwest not only need to see the mountains, but must also be able to see *from* them. Smog and haze prevents aboriginal leaders from seeing the extent of their territory and therefore undermines their ability to fulfill their ancient obligations. It restricts a tribe's ability to assert control over a region.
- First Nations are also concerned about the ability of the mountains to see *them*. These concerns stem from the belief that local mountains were once human beings who were transformed into stone, but retain the living spirit of their original human.

The study recommended creating opportunities for First Nations to have a direct role in helping to improve visual air quality and that First Nations economic ventures related to cultural tourism be assessed in light of degraded visual air quality.

8.3.6 Health Co-Benefits

Canadian standards for fine particulate matter are designed to minimize health impacts, not to improve or protect visual air quality. As a result, even when pollutant concentrations are low enough to meet air quality standards, visual air quality can still be impaired by fine particulates, particularly those with diameters less than 1 micron. However, it has also been shown that human health continues to be impacted even by low concentrations of particulates (Pope 2006). Therefore, continued reductions in fine particulate to protect and improve visual air quality will also yield human health co-benefits[7].

8.3.7 Potential Impacts on Agriculture[8]

Air pollutants reduce the amount of solar radiation reaching crops and thus interfere with photosynthesis. Therefore, while not quantified, haze and smog could also be having a negative effect on crop growth in Canada. One study estimated that improvements to visual air quality in China could have the potential to increase crop yields there by 5–30 % (Chameides et al. 1999). One source of visibility degradation in rural areas is ammonia from agricultural activities, including manure spreading on fields. Ammonia has been shown to be an important contributor to the white haze that degrades visibility in agricultural areas like the lower Fraser Valley of British Columbia.

[6] First Nations is a term that collectively refers to various Aboriginal peoples in Canada who are neither Inuit or Métis (Statistics Canada 2006).

[7] More details on health impacts of air quality are available in the chapter in this book by Dave Stieb.

[8] A detailed review of the issue of agriculture and air quality can be found in the chapter by Shabtai Bittman et al. in this book.

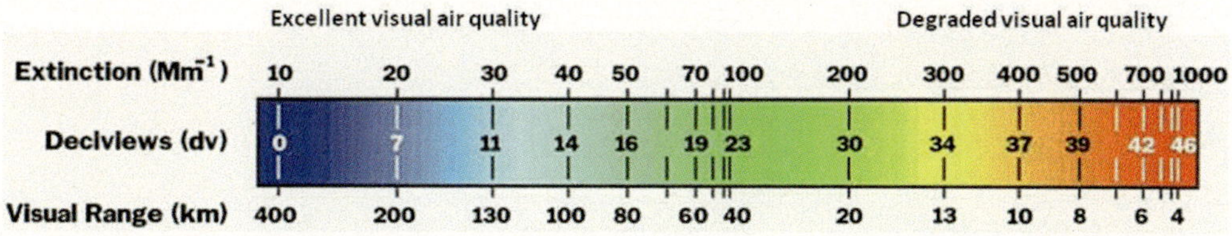

Fig. 8.6 Comparisons of extinction (Mm-1), deciviews (dv), and visual range (km). For example, extinction of 10 Mm-1 corresponds to about 400 km visual range and 0.0 dv, while extinction of 1,000 Mm-1 is about 4 km visual range and 46 dv. Figure courtesy of the US Environmental Protection Agency (US EPA 2011)

8.4 Measuring Visual Air Quality

Visual air quality can be either *quantitatively* measured using physical parameters such as extinction coefficients, or *qualitatively* by assessing how visual air quality is perceived by a human observer. This section discusses these two approaches.

8.4.1 Quantitative Measurement of Visual Air Quality Degradation

Visibility (or visual air quality) has been defined as "the greatest distance at which an observer can just see a black object viewed against the horizon sky" (Malm 1999). While visual air quality can most simply be described as a distance, there are infinite nuances to appreciate in the visual environment. Clear air compliments the awe-inspiring wilderness vistas for which much of Canada is famous. From a scenic point of view, visual air quality is not how far a person can see, but rather the ability of an observer to clearly see and appreciate the many and varied elements in each vista. Also, urban dwellers are conditioned to associate urban haze with air pollution, to the extent that they will not believe that air quality is improved until they see a difference. The *clarity* of the air is a key aspect of its quality.

There is a need, however, to measure visual air quality *quantitatively* to better understand how visual air quality varies with location, pollutant type and concentration and to identify visual air quality trends over time. This can help in the establishment of a Visual Air Quality Rating and a visual air quality goal (Sakiyama and Kellerhals 2011). Quantitative measurements can also provide information on background levels of visual air quality and can be used to compare visual air quality between different locations.

The following are some common visual air quality concepts and metrics.

Atmospheric extinction coefficient: Described earlier, this is the fraction of light that is scattered or absorbed per unit distance. It is the sum of the scattering and absorption coefficients.

Just Noticeable Change: This concept refers to identifying where or when a threshold of visual air quality is crossed in space or time. This threshold can be noticed when a layered haze forms a contrasting edge to a more transparent haze in the same scene. It could also be noticed if visual air quality is changing over time and a landmark just disappears (or just appears) as air quality concentrations increase (or decrease). (Malm et al. 2011).

Contrast: This is the ratio of the brightness or radiance of an object to that of its background. "Inherent contrast" is the contrast directly adjacent to the scenic element of interest, while the contrast at some distance is called the "apparent contrast". The "equivalent contrast" of a digital image is a mathematical assessment that compares more favourably to the contrast detected by the human visual system. (Malm et al. 2011)

Visual Range: This is the distance at which a given standard object can be seen and identified with the unaided eye. Visual range in the context of visual air quality protection is estimated by the U.S. National Weather Service (NWS) by viewing a series of landscape features at a variety of distances and recording the most distant feature that can be seen (Malm 1999).

Deciview: A related objective measurement of visual air quality is the *deciview*, a metric based on light extinction, and therefore, air pollution concentrations. One deciview is defined as one incremental change in *perceived* visual air quality. It is associated with approximately a 10% change in light extinction. The deciview scale is near zero for a pristine atmosphere and increases as visual air quality degrades.

Expressed in terms of extinction coefficient (b_{ext}) or in terms of visual range (Vr), the deciview scale is:

$$\text{deciview} = 10\ln(b_{ext})/0.01km^{-1} = 10\ln(391km/vr)$$

Where:

Ln is the natural logarithmic function
b_{ext} is the light extinction coefficient
Vr is Visual Range expressed in kilometres
Figure 8.6 illustrates the relationship between three objective measures of visual air quality: the extinction coefficient; the deciview and visual range (in kilometres).

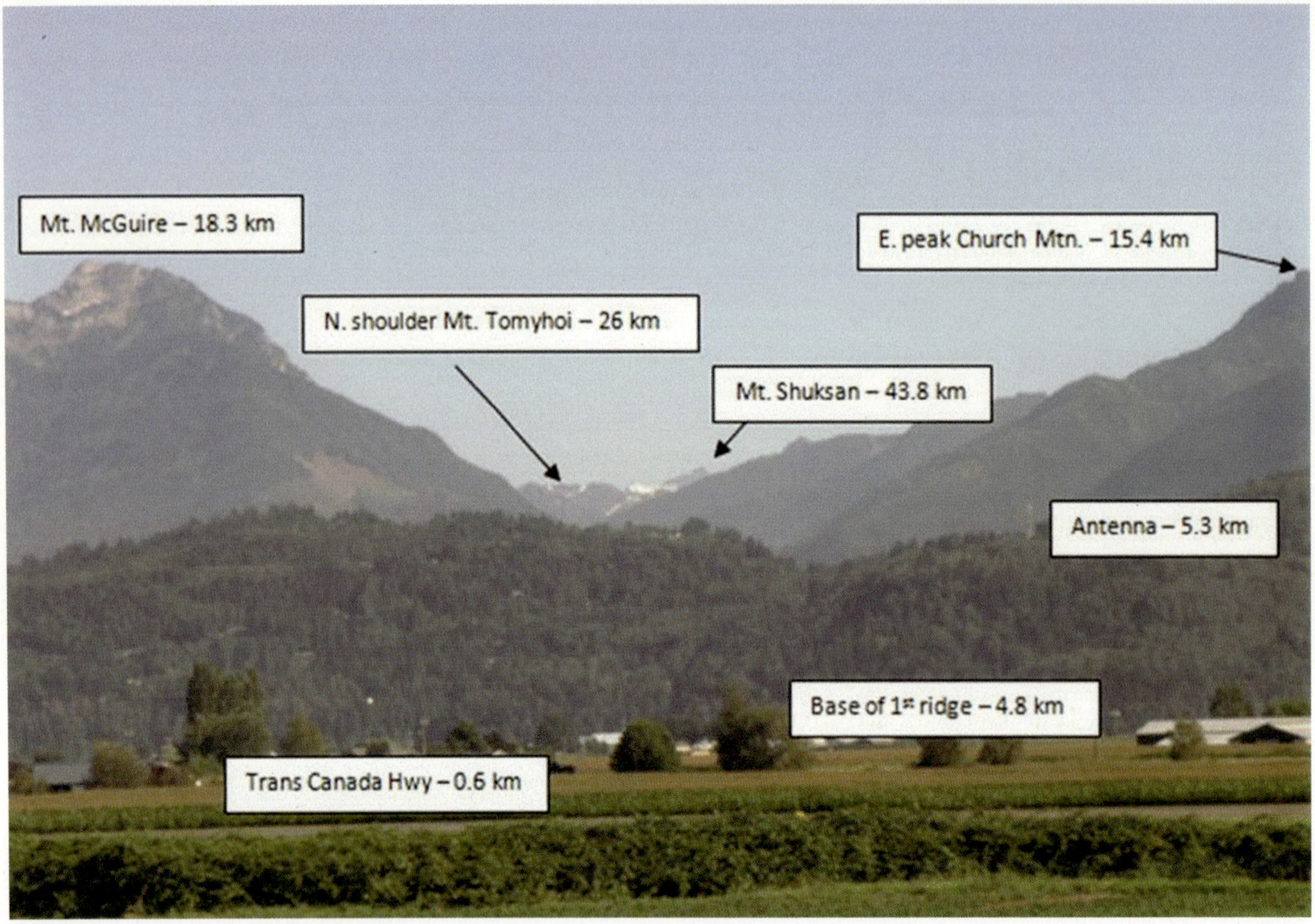

Fig. 8.7 An example of one or many images used in assessing an observer's perception of visual air quality. The labels on the image are usually removed when showing images to observers during perception studies. Image courtesy of BC Ministry of Environment

8.4.2 Measuring the Human Perception of Visual Air Quality

"Extinction", "deciviews" and "visual range", although scientifically useful, are meaningless, esoteric terms to the public unless they can be related to human ratings of visual air quality (Latimer 1979). In other words, although visual air quality improvement can be proven through quantifiable measures, the public may not perceive it to be any different. Thus the goal of a visual air quality management program must consider the public's perception of the visual conditions (Malm 1999).

Perception studies assess how human observers value a series of scenic images. These images are generally captured by automated cameras co-located with monitoring equipment that continually measures light scattering, absorption, extinction and sometimes pollutant concentrations. Absorption and scattering is measured by a transmissometer, while scattering alone is measured by a nephelometer.

In perception studies, groups of observers are asked to review photographs of the same scenes with differing levels of visual degradation or optical clarity (Fig. 8.7). Groups are provided with identical verbal instructions. Observers base their assessment on their own perceptions of these images, including whether they feel that the scenes are vivid, colourful and sharp and whether they can detect distant objects and landmarks. Survey participants are usually asked to indicate, for each photograph, whether they consider the scene to be unacceptably degraded—i.e. "how much haze or visibility degradation is too much?" Participants can also be asked to rate a scene on a given scale, sometimes called a Visual Air Quality (VAQ) scale.

An example of a perception study was during the 1993 REVEAL study in the lower Fraser Valley of British Columbia (discussed later in this chapter). Automated cameras were used to obtain sequences of images of distant vistas at different times of the day and under differing conditions of humidity, cloud and light extinctions. Similar studies were completed in subsequent years, including in 2011, in order to build up a database of perception and extinction information. Central to the success of these visual air quality perception studies was the selection of suitable images

(Sakiyama and Kellerhals 2011). For example, in the 1993 REVEAL study, the views portrayed were consistent from slide to slide such that a change of view angle was not misinterpreted as a change in visual air quality. Strict criteria were used in selecting slides: the relative humidity was less than 75 %, the image was captured between 1000 and 1600 local time and cloud cover was less than 5/10 or more than 7/10 of the sky.

Similar perception studies have been carried out in Washington, DC, Denver and Phoenix (Malm et al. 2011). Unfortunately, these studies showed that when quantitative measures of visual air quality such as extinction, visual range, or deciviews were compared to perceived preference ratings, there was not a single indicator that represented acceptable levels of visual air quality for the varied settings investigated. (Malm et al. 2011). However, a comparison of *perceived* visual air quality to *quantitative measurements* of visual air quality does still provide a valuable basis for the development of visual air quality indices and standards.

8.5 Visual Air Quality Research in Canada

8.5.1 Motivation for Research

Those jurisdictions around the world that manage visual air quality do so in much the same way that they manage air pollutants to protect human health. The chief difference is that the management of visual air quality focuses on reducing the concentrations of the specific pollutants, namely fine particulates and selected gases, that extinguish light.

Visual air quality research in Canada was initiated in response to long-standing visual air quality and health concerns and the Canada/US Air Quality Agreement. This was signed in 1990 and has an important visual air quality component (see text box)[9]:

Text box 8.1. The visibility protection clause contained in the US-Canada Air Quality Agreement

Prevention of Air Quality Deterioration and Visibility Protection

Recognizing the importance of preventing significant air quality deterioration and protecting visibility, particularly for international parks, national, state, and provincial parks, and designated wilderness areas:

A. For the United States:

Requirement that the United States maintain means for preventing significant air quality deterioration and protecting visibility, to the extent required by Part C of Title I of the Clean Air Act, with respect to sources that could cause significant transboundary air pollution.

B. For Canada:

Requirement that Canada, by January 1, 1995, develop and implement means affording levels of prevention of significant air quality deterioration and protection of visibility comparable to those in paragraph A above, with respect to sources that could cause significant transboundary air pollution.

Canada and the US agreed to prevent air quality deterioration and protect visual air quality from sources that could cause significant transboundary effects. Part of Canada's actions resulting from this agreement was a commitment to monitor visual air quality near the US border. The intent of these measurements was to provide optical, physical and chemical information in Canada to assess whether U.S. visual air quality research results could apply to Canada.

The visual air quality components of the Agreement grew out of the U.S. Congress enacting the Clean Air Act (CAA) in 1977. Class I areas, described as parks or wilderness areas greater than 5,000 acres, were labelled "protected environments" where visual air quality is a critical resource for human appreciation of the natural area. This resulted in the U.S. Environmental Protection Agency announcing a major effort to improve air quality in national parks and wilderness areas. The Regional Haze Rule followed, calling for state and federal agencies to work together to improve visual air quality in 156 national parks and wilderness areas such as the Grand Canyon, Yosemite, the Great Smokies and Shenandoah. The rule requires the states, in coordination with the Environmental Protection Agency, the National Park Service, U.S. Fish and Wildlife Service, the U.S. Forest Service, and other interested parties, to develop and implement air quality protection plans to reduce the pollution that causes visual air quality impairment. Canada's responsibility under the Agreement was to be "equivalent" to the United States efforts to protect visual air quality (Jordan 2006).

Two border areas were of particular U.S. concern when it came to visual air quality (Hoff et al. 1997):

1. In 1932, Waterton Lakes National Park (Alberta, Canada) was combined with the Glacier National Park (Montana, USA) to form the world's first International Peace Park (Fig. 8.8). Initial visual air quality measurements were made by the Canadian government in Waterton National Park and nearby Esther, Alberta in the 1990s as part of Canada's obligations under the US/Canada Air Quality Agreement. These measurements suggested that visual air quality was more degraded in the US park than in the Canadian park.
2. To assess the impact of pollutants on visual air quality in the Acadia National Park in Maine, visibility measurements were made in St. Andrews, New Brunswick, about 60 km east of Acadia National Park.

[9] A detailed analysis of the Canada/US Air Quality Agreement can be found in Chap. 16 by Jean Melious in this book.

Fig. 8.8 Situated on the border between the US and Canada and offering outstanding scenery, the Waterton-Glacier International Peace Park is exceptionally rich in prairie, forest, alpine and glacial features. (Photo courtesy Waterton Park Information Services)

Another motivation for research into visual air quality in Canada was the creation of the *Visibility Task Force* (Stubbs 1993 personal communication). This was a collaboration of federal, provincial and regional government environmental agencies and Tourism BC. This technical and policy advisory group was established to address visual air quality degradation and protection issues. An international workshop titled "Protecting Visibility in Western Canada and the Pacific Northwest" was held at Harrison Hot Springs in March 1993. A Task Force Report outlined key issues and interests of different agencies and recommended the creation of the Visibility Management Steering Group to champion visual air quality science, policy and management.

8.5.2 Canadian Visual Air Quality Research Prior to 2000

REVEAL and Pacific93

An opportunity for research on visual air quality in western Canada appeared during the summer of 1993 with *Pacific 93*, an intensive multi-agency atmospheric field campaign examining air quality and meteorology in the Lower Fraser Valley of BC. Taking advantage of the infrastructure and air quality experts available for *Pacific 93*, the *Regional Visibility Experimental Assessment in the Lower Fraser Valley* (REVEAL) was initiated. This study was a collaborative effort between the BC Ministry of Environment, Metro Vancouver, Environment Canada and the Fraser Valley Regional District to characterize summertime visual air quality and ambient aerosol loadings in south-western B.C. (Pryor and Steyn 1994). It was the first study of this kind in Canada.

A significant number of air quality and visibility monitors as well as cameras were deployed throughout Metro Vancouver and the Lower Fraser Valley. Locations of these monitoring sites are in Fig. 8.9.

Highlights of the conclusions of the 1993 REVEAL research were (Pryor 2011):

- Visual air quality, aerosol concentrations, and aerosol composition are highly variable in the time and space in the Lower Fraser Valley.
- Organic matter dominates fine particulate mass, accounting for 35–46% of the mean fine aerosol mass in the Lower Fraser Valley (LFV),
- Nitrate and sulphate dominate particle light scattering by particles (accounting for 55–67% of mean scattering by particles) in the LFV.
- The severity of visual air quality impairment in the LFV, despite comparatively low fine mass concentrations, was attributed to the high mass scattering efficiencies of the ambient aerosols (due in part to the relative abundance of ammonia and ammonium).
- Direct vehicle emissions, secondary aerosols resulting from atmospheric photochemical reactions and soil/road dust were the primary sources of fine aerosols in the LFV.
- Visual air quality conditions are frequently degraded below publically defined 'acceptable levels'.

REVEAL II and Related Studies

Visual air quality perception studies began in 1993/94 in REVEAL II, a follow-up to the initial REVEAL study. Several groups of observers were recruited for these studies during which they reviewed photographic images where the scene characteristics remained the same while the pollutant concentrations changed.

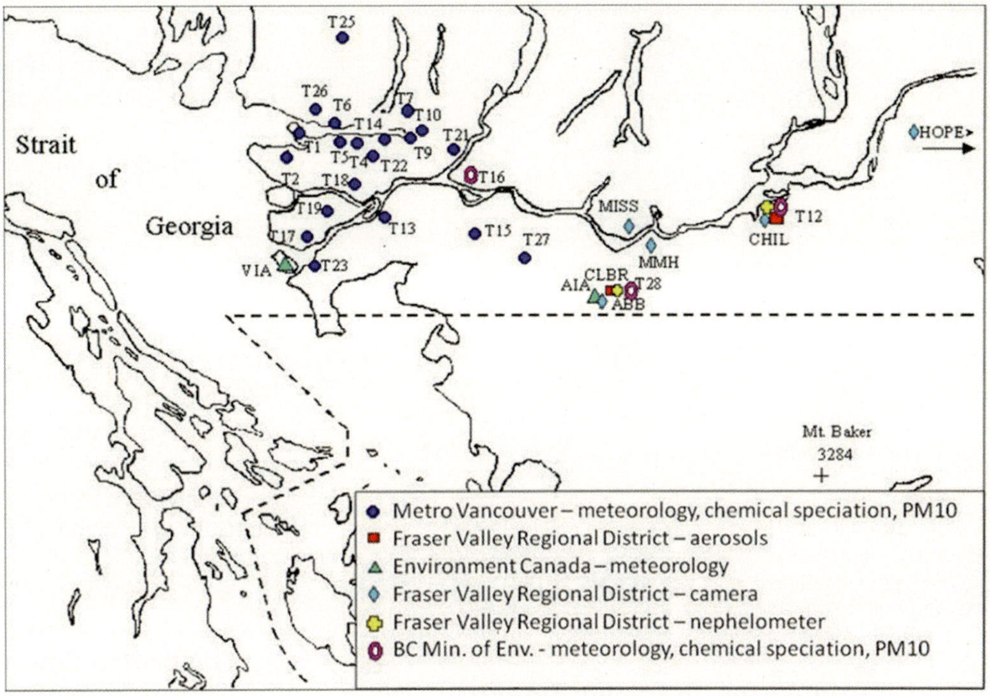

Fig. 8.9 Air quality, meteorology, light extinction and automated visibility camera monitoring sites operated during REVEAL and REVEAL II studies in Metro Vancouver and the Lower Fraser Valley in the mid-1990s

Following REVEAL II, an international Air Issues Workshop triggered by the Canada/US Air Quality Agreement and co-sponsored by Parks Canada and the U.S. National Parks Service provided recommendations on visual air quality science and policy to Canadians working on visual air quality issues. This resulted in the U.S. Parks Service assisting Parks Canada with visual air quality monitoring in the Kootenay National Park Visibility Project with the use of photographic and light scattering monitoring to distinguish local and long distance causes of visual air quality degradation.

8.5.3 Visual Air Quality Research 2000–2005

Visual air quality initiatives re-emerged in the early 2000's as concerns about air quality in urban areas re-emerged. A comprehensive visual air quality monitoring program was undertaken in Metro Vancouver and in the Lower Fraser Valley, including $PM_{2.5}$ speciation[10] monitoring at two locations, specifically nitrate and sulphate speciation, light scattering and extinction (at four locations) and photographic information (at six locations). This resulted in a great deal of visual air quality-related data being collected as a basis for future visual air quality work.

A Serendipitous Experiment

Ammonia has been recognized as a leading cause of degraded visual air quality, particularly the "white haze", in the Lower Fraser Valley (LFV). A unique opportunity to assess ammonia's influence occurred in 2004 when an avian influenza epidemic struck the poultry sector in the LFV, resulting in a cull of nineteen million birds. The agricultural sector contributes approximately 76 % of total atmospheric ammonia in the LFV. Results of field studies indicate that in spite of an approximate 67 % decline in atmospheric ammonia concentrations in response to the poultry cull, the effect on average $PM_{2.5}$ levels was relatively small (less than a 6 % decline). There was also no improvement in average visual air quality during this period, based on extinction measurements (So and Vingarzan 2010).

These results suggest that due to the overabundance of ammonia in the LFV, control of ammonia emissions alone is likely to have limited value at producing significant valley-wide reductions in average PM or improvements in visual air quality. Recommendations from this research included a multi-pollutant reduction approach which would simultaneously address the principal precursors of fine particulate matter (nitrogen oxides, sulphates and volatile organic compounds), in addition to ammonia, as a strategy for reducing fine particulate matter and improving visual air quality in the LFV (Vingarzan 2011 personal communication).

[10] Speciation identifies the type and quantity of chemical species such as nitrates or sulphates in particulates.

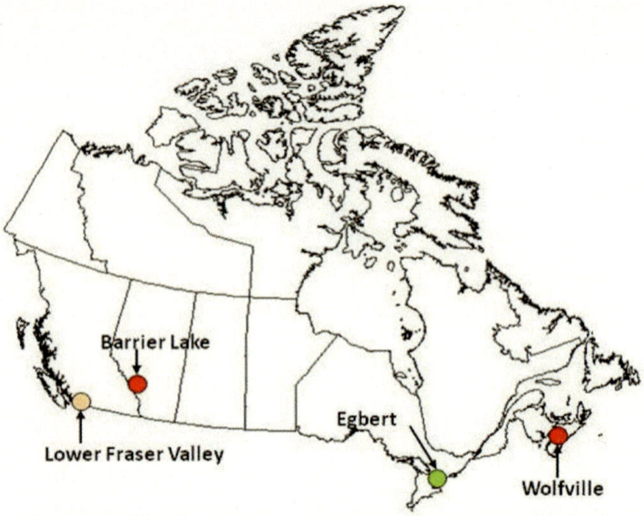

Fig. 8.10 Environment Canada and its partners continue to operate visual air quality monitors in four areas of Canada. (Source: Environment Canada)

8.5.4 Recent Visual Air Quality Research

Environment Canada and its partners continue to operate visual air quality monitoring equipment in three other locations (Fig. 8.10) besides the Lower Fraser Valley of BC:

1. Barrier Lake, Alberta, in the scenic Rocky Mountains;
2. Egbert, Ontario, an important air quality monitoring site north of Toronto
3. Wolfville, Nova Scotia

These sites measure light extinction, light scattering and pollutant concentrations. Cameras at some of the sites capture hourly images of distant landmarks.

Environment Canada's 2008 Smog Assessment also contains an analysis of recent visual air quality information based on data from Metro Vancouver, Victoria, Saturna Island, Kelowna, Golden and Prince George. Photochemical models that were traditionally used to model ozone and $PM_{2.5}$ have recently been used with this information to estimate visual air quality under different future emission scenarios.

Recent research has also shown that, in general in the Lower Fraser Valley:

- Visual air quality is observed to degrade when $PM_{2.5}$ concentrations reach 7–8 μg/m^3(micrograms per cubic metre), becomes significantly degraded above 10–14 μg/m^3 and severely degraded above 20 μg/m^3
- Visual air quality is most degraded in the fall. On a diurnal basis, it is most degraded in the hours just before dawn, on average.
- Visual air quality improved sharply after 1970, based on an analysis of visibility at local airports,. This coincided with the introduction of air quality regulations in British Columbia.
- Vehicle emissions, gasoline and solvent evaporation and ammonia from agricultural emissions have the highest impact on visual air quality.
- During the most severely degraded days, organic and elemental carbon and nitrate, all contained in particulate matter, were responsible for the highest amount of light extinction.
- Ammonia emissions cause a white haze in the summer. A 10 % reduction in the visual range through this white haze would require a massive 60 % reduction in ammonia emissions.
- The implementation of Metro Vancouver's 2015 Air Quality Management Plan will not result in detectable improvements in visual air quality.

(So and Vingarzan 2010) (So et al. 2011) (Vingarzan 2010) (Vingarzan 2011 personal communication) (Vingarzan and So 2011)

8.6 The British Columbia Visibility Coordinating Committee

Agencies in British Columbia are currently the most active in visual air quality-related activities in Canada. These activities may form the basis of strategies to protect and improve visual air quality in BC and elsewhere Canada. As described above, recent work has consisted of visual perception studies, visual air quality monitoring, developing visual air quality goals, formulating a Visual Air Quality Rating and using dispersion modelling to link emissions, ambient concentrations and visual air quality. This work will form the basis of visual air quality management programs or policies. Indeed, improved visual air quality is already a goal of some air quality management plans in British Columbia. The current management approach continues to focus on managing air quality to meet health-based standards, with an expectation that visual air quality will be protected and improved as a consequence.

As a result of this and other work carried out since 1993 investigating visual air quality in the Lower Fraser Valley, as well as the announcement of the 1999 *Regional Haze Rule* by the US EPA, the *British Columbia Visibility Coordinating Committee* (BCVCC) was formed in 2006. The BCVCC was created to address visual air quality management issues in BC. It is a collaborative involving government agencies at all levels, including Environment Canada, Health Canada, BC Ministry of Environment, Metro Vancouver, the Fraser Valley Regional District and the City of Kelowna. It coordinates and provides direction on visual air quality management in the province. The Vision of the BCVCC is:

> The ability to see our natural and urban environment is an important part of the British Columbia experience, thus the BCVCC is committed to achieving clean air and pristine visibility for the health and enjoyment of present and future generations.

Fig. 8.11 The BC Visibility Coordinating Committee web site. Real-time images allow the public to assess visual air quality at several locations. URL is http://clearairbc.ca

The BCVCC website (Fig. 8.11) promotes the protection and improvement of visual air quality, providing real time images and information on visual air quality at several locations in Metro Vancouver and the Lower Fraser Valley.

8.7 A Proposed "Visual Air Quality Rating"

One of the goals of the BCVCC is to develop a Visual Air Quality Rating, a metric that can be used as a real-time public reporting tool. It attempts to reflect the human perception of visual air quality (Sakiyama and Kellerhals 2011). This necessitates developing a relationship between visual air quality *perception* and measurements of *light extinction.* This relationship can also help in the establishment of a visual air quality goal and in the development of a suitable metric to determine progress toward that goal. This would be an important step in developing a comprehensive visual air quality management program (Ely et al. 1991). Several jurisdictions around the world have developed these kinds of visual air quality ratings. This section will focus on the Visual Air Quality Rating being developed for Metro Vancouver and Lower Fraser Valley.

The basis of the Visual Air Quality Rating is the relationship between the participant responses in perception studies and an objective measure of optical clarity such as deciviews. Participant's answers to two distinct questions were used in helping to determine this relationship in the lower Fraser Valley:

1. How would you rate the visual air quality (VAQ) in this scene on a scale of 1 to 7?
2. Is the visual air quality in this scene unacceptably degraded?

The graph in Fig. 8.12 is based on the answers to question 1 and summarizes the relationship between the average visual air quality and deciviews, a metric for light extinction. This relationship supports the hypothesis that there is a linear relationship between human perception of visual air quality and light extinction (in deciviews) by air pollutants. This would lead to the assumption, for example, that a doubling of light extinction would lead to a halving of people's rating of the visual air quality of a scene, on average.

Figure 8.13 is based on the average responses to question 2 above and summarizes the relationship between the light extinction and the fraction of observers who assessed the visual air quality to be "acceptable". The data has been fitted to a non-linear logit function. Similar relationships have been found in other locations (BBC Research & Consulting 2003).

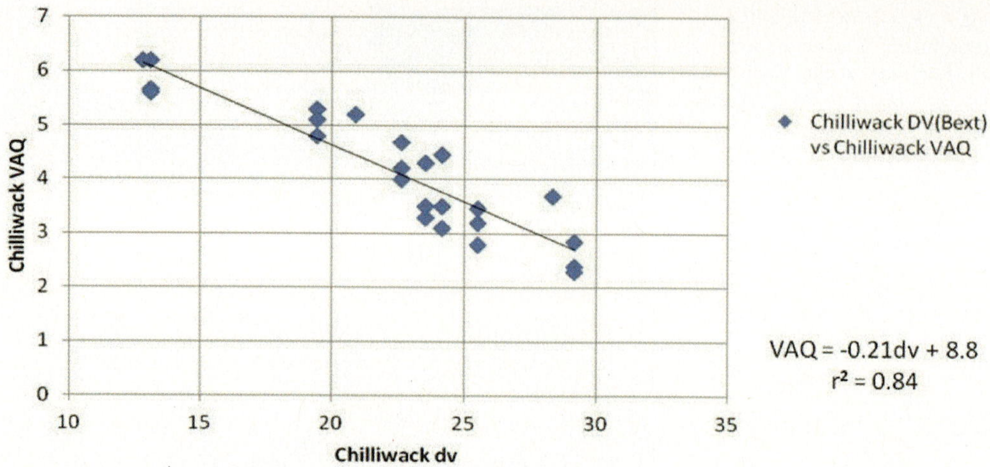

Fig. 8.12 Plot of the average perception of the Visual Air Quality (VAQ) on a scale of 1 to 7 versus the light extinction measured in deciviews (dv) near Chilliwack, a community in the lower Fraser Valley of British Columbia. A linear relationship appears to exist (Sakiyama and Kellerhals 2011)

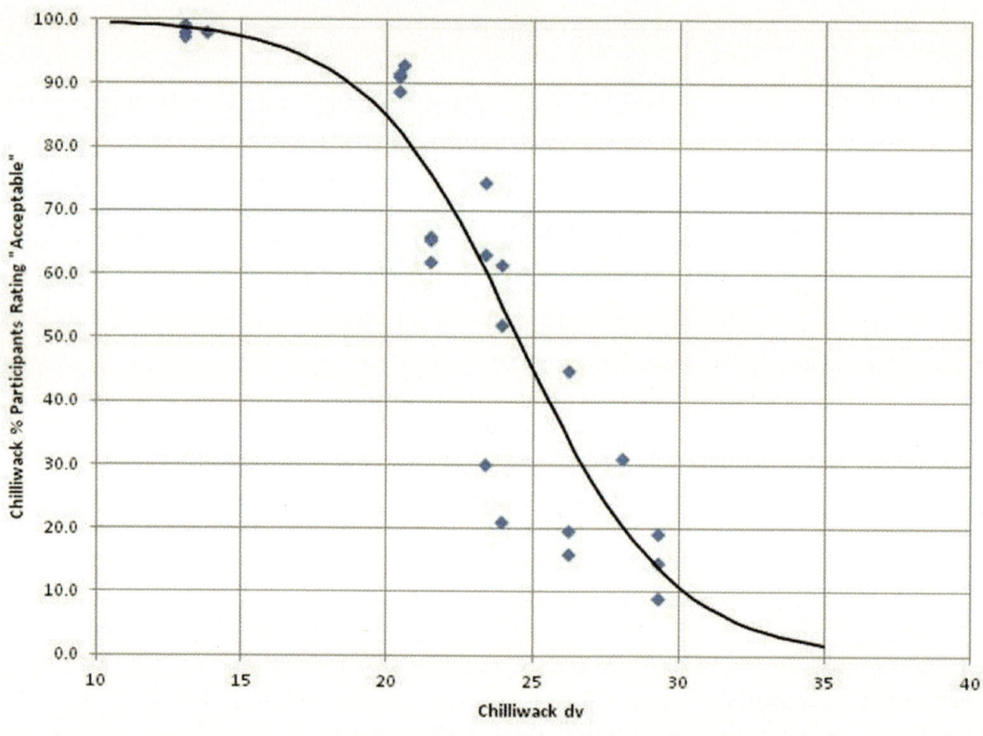

Fig. 8.13 Plot of the percent of participants in perception surveys that felt that the visual air quality of a scene at Chilliwack was "acceptable" versus the light extinction measured in deciviews. The relationship exists, but appears to be non-linear (Sakiyama and Kellerhals 2011)

Using the graphs in Figs. 8.12 and 8.13 and other information, a preliminary Visual Air Quality Rating has been proposed (Table 8.1) that relates the visual air quality, based on perception studies, to the light extinction, measured in deciviews.

The final version of the Visual Air Quality Rating will be used as an indicator of real time visual air quality in Metro Vancouver and the lower Fraser Valley and as a basis for a visual air quality goal.

8.8 A Visual Air Quality Protection Framework

This section describes the work being planned to identify the sources of pollutants that degrade visual air quality and outlines a strategy to reduce these emissions to protect and improve visual air quality (BCVCC 2011).

Air pollutants that impair visual air quality come from a variety of sources[11] including fires, agriculture, automobiles, ships, industry, manufacturing activities and natural sources.

[11] Chapters by Steve Sakiyama and Warren McCormick provide detailed information on air pollutant emissions in Canada.

Table 8.1 Preliminary Visual Air Quality Rating and Category (Sakiyama and Kellerhals 2011) and the average annual occurrence of each category at two sites in the lower Fraser Valley (So et al. 2011)

Visual air quality category	Visual air quality rating	Deciview	Percent scene acceptability	Average occurrence of visibility at two sites in the Lower Fraser Valley 2003–2008
Excellent	5	<17	>95	62%
Good	4	17–22	70–95	30%
Fair	3	22–25	40–70	7%
Poor	2	25–29	10–40	1%
Very Poor	1	>29	<10	0%

Some haze-causing particles are directly emitted to the air from such sources, while other pollutants react together to form fine particulate matter in the air downwind from the source. Reductions in particulate levels can be achieved by reducing emissions of primary particulates as well as pollutants such as nitrogen oxides, sulphur oxides and ammonia which can react together in the presence of light to form secondary particulates in the atmosphere. (Nitrogen dioxide also directly impairs visual air quality by producing a brown haze). The pie chart in Fig. 8.14 identifies the relative contribution of different emission sources that cause smog and degraded visual air quality in Metro Vancouver and the Lower Fraser Valley.

Canada has an abundance of mountainous wilderness, providing spectacular scenery that is highly valued by residents and visitors alike. Mountainous viewscapes near large urban areas with multiple and large emission sources are therefore particularly at risk of visual air quality degradation and in need of a management plan. Unfortunately it is estimated that noticeable improvements in particulate levels, and hence visual air quality, will not be achieved with the implementation of existing air quality management programs (Vingarzan and So 2011), at least in some urban areas. More will need to be done.

In 2010 the BCVCC created a *BC Visibility Protection Framework* that began the process of developing recommendations for air quality management plans that will achieve the BCVCC Vision. This Framework is a series of adaptive management steps where the application of scientific information and explicit feedback mechanisms are applied to refine and improve certain management decisions over time.

The BC Visibility Protection Framework is founded on the following principles.

- Based on science
- Involves an engaged public and stakeholders
- Actions based on priority and reasonable time frames
- Includes the concept of continuous improvement
- Adaptable for local situations
- Collaborative in development and implementation through the BCVCC
- Linked to air pollution-related health risk and economic impact policies
- Harmonized and leveraged with existing air quality management plans

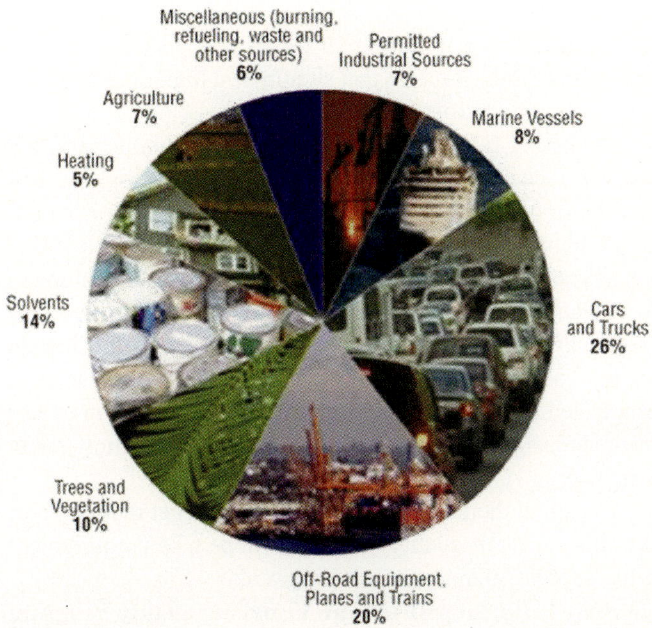

Fig. 8.14 The contributions of various emission sources to air pollution in Metro Vancouver and the lower Fraser Valley (Metro Vancouver 2011). Many of these pollutants also impact visual air quality

- Incorporates a precautionary approach: Prevention of future problems

The four steps of the BC Visibility Protection Framework are:

Step 1: Initiate Process, Analyse Data and Track Progress Visual air quality monitoring and analysis is a crucial first step in the Framework. It involves the measurement and analysis of visual air quality relevant parameters. These include scene perception values, optical extinction, scattering and absorption coefficients and aerosol and gas concentrations. These are needed to support the visual air quality metrics that will be the used as the basis of a standard and index.

Step 2: Set/Reassess Visual air quality Goals (Form and Metric) The second step is to develop and later improve a Visual Air Quality Rating and goals, based on a scientific assessment of all appropriate data. The aim is to mitigate the visual air quality impacts of various air pollutants to a level which is acceptable by a date determined with input from stakeholders.

Step 3: Develop/Adjust Science and Social Science Programs Broad programs are needed to reduce emissions so as to improve and protect visual air quality. Step 3 identifies the emissions that contribute to degraded visual air quality for a given area over a given time period and assesses how much they need to be reduced to meet visual air quality goals and standards. An inventory will be needed of the sources and relative contribution of pollutants to visual air quality impairment. This inventory will include historical trends of these specific emissions and forecasts of visual air quality-affected pollutants. This step will also will require identifying those sources that should be a high priority target for emission controls.

Predicting future visual air quality is a daunting task, but estimates can be made using photochemical atmospheric dispersion models. This modelling helps to establish relationships between pollutant emissions and their possible effect on visual air quality, both in the past and in the future. It will also provide predictions of the relative contribution of pollutants to visual air quality impairment and the impact of new air quality related regulations and policies, such as improved fuels and emission control technology. It may also provide insight into how human health and the economy may improve as visual air quality improves.

Engaging the public and stakeholders to get their support for this visual air quality initiative will be a key requirement for a successful management program.

Step 4: Implement/Update Visual air quality Program Step 4 is the implementation of the visual air quality program. Emissions controls will be a high priority. This will include technology (e.g. new equipment, cleaner processes and fuels), economic instruments, and planning (bike lanes, densification) to control/reduce the emissions of visual air quality relevant emissions. These improvements provide the means to achieve visual air quality goals by reducing visual air quality-related emissions, generally by building upon existing air quality management plans that already include emission control programs.

An important part of Step 4 will be government-promulgated policies to protect and improve visual air quality. The purpose of these policies will be to provide authority and public accountability to take action on visual air quality management.

8.9 Conclusion

The ability to see and enjoy the spectacular natural environment is an important part of life in Canada. Air pollution can rob residents and visitors alike of the opportunity to benefit from these magnificent vistas. Canadians need to work together to protect and improve visual air quality throughout the country, but particularly in those areas that are blessed with outstanding natural beauty—mountains, lakes, coastlines and wilderness areas. Improving visual air quality will benefit Canadians economically, socially and from a health perspective.

Protecting and improving visual air quality has been an important issue in Canada, since the 1980s. Studies of people's perception of visual air quality have been combined with specialized monitoring of air pollutants and their ability to interfere with light. These studies have laid the foundation for the creation of a Visual Air Quality Rating for the Lower Fraser Valley of British Columbia as well as visual air quality goals and standards that can be used as yardsticks to measure improvements in visual air quality over the next decades.

The importance of visual air quality management continues. The creation of the B.C. Visibility Coordinating Committee, a collaborative of concerned government agencies, has provided a focus for visual air quality science, communication and management initiatives. The science of visual air quality impairment is well known, but more research is needed to understand how pollutant emissions should best be managed to protect and improve visual air quality.

A new BC Visibility Protection Framework being developed by the BCVCC will provide a broad road map for British Columbia, and possibly for the rest of Canada.

References

Bates-Frymel L et al (2011) Economics impacts of visibility degradation in the lower Fraser Valley BC—A discussion paper by the BC visibility coordinating committee. (in preparation)

BBC Research & Consulting (2003) Phoenix area visibility survey. Arizona department of environmental quality, p 9

BCVCC (2011) A visibility protection framework for British Columbia. (in preparation)

Beron K, Murdoch J, Thayer M (2001) The benefits of visibility improvement: new evidence from the Los Angeles metropolitan area. J Real Estate Financ 22:(2/3)319–337

Carlson KT (2009) Mountains that see, and that need to be seen: aboriginal perspectives on degraded visibility associated with air pollution in the BC lower mainland and Fraser Valley-a traditional knowledge study. Prepared for Environment Canada

Chameides WL, Yu H, Liu SC, Bergin M, Zhou X, Mearns L, Wang G, Kiang CS, Saylor RD, Luo C, Huang Y, Steiner A, Giorgi F (1999) Case study of the effects of atmospheric aerosols and regional haze on agriculture: an opportunity to enhance crop yields in China through emission controls? Proceedings of the national academy of sciences. 96:13626–13633

Ely DW, Leary JT, Stewart TR, Ross DM (1991) The establishment of the Denver visibility standard. presented at the 84th annual meeting and exhibition of the A&WMA, Vancouver, BC. June 1991

Hoff Ray et al (1997) Recent visibility measurements in Canada. Presentation to the air & waste management association 90th annual meeting

Industry Canada (2011) Federal tourism strategy. http://www.ic.gc.ca/eic/site/dsib-tour.nsf/eng/h_qq00000.html. Accessed September 2011

Jordan NS (2006) Analysis of air pollutant transport to class I areas using multiple satellite products and in-situ ground based measure-

ments. Proceedings of the annual meeting of the American Meteorological Society
Latimer DA (1979) Analysis of scenic degradation caused by air pollution. Proceedings of the workshop in visibility values. Ft. Collins, Colorado. USDA general technical report WO-18
Malm WC (1999) Introduction to visibility, air resources division, national park service, cooperative institute for research in the atmosphere, NPS visibility program, Colorado State University, Fort Collins CO
Malm WC, Molenar JV, Pitchford ML, Deck LB (2011) Which visibility indicators best represent a population's preference for a level of air quality? Presented at the air & waste management 104th annual conference, Orlando, June 21–24, 2011
McNeill R, Roberge A (2000) The impact of visual air quality on tourism revenues in Greater Vancouver and the lower Fraser Valley. Environment Canada
Metro Vancouver (2011) Clear air—Your guide to visibility in Metro Vancouver and the Lower Fraser Valley. http://clearairbc.ca/visibility/Pages/EmissionSources.aspx. Accessed October 2011
Pope III A, Dockery DW (2006) Health effects of fine particulate air pollution: lines that connect. J Air Waste Manage Assoc 56:709–742
Pryor S (2011) Reveal II home page http://www.indiana.edu/~reveal2/index.html. Accessed January 2012
Pryor S, Steyn D (1994) Visibility and ambient aerosols in Southwestern British Columbia during REVEAL. Government of BC
Sakiyama S, Kellerhals M (2011) The development of a visibility index for the Lower Fraser Valley of British Columbia. BC ministry of environment, Victoria, Canada. (in preparation)
So R, Vingarzan R (2010) Analysis of visibility in the Lower Fraser Valley of British Columbia. Environment Canada, Meteorological services of Canada
So R, Vingarzan R, Teakles A (2011) Analysis of visibility in the Lower Fraser Valley of British Columbia. Draft, December 2011. Environment Canada
Statistics Canada (2006) Aboriginal identity. Government of Canada. Retrieved 2009-10-08
US EPA (2011) Visibility—nature and sources of the problem. http://www.epa.gov/ttn/naaqs/ozone/ozonetech/airtrends/vis.html. Accessed October 2011
Vingarzan R (2010) Lower Fraser Valley visibility pilot science summary and recommendations. Environment Canada
Vingarzan R, So R (2011) Statistical modelling of visibility in the Lower Fraser Valley of British Columbia (in preparation)

Part III

Management of Emissions

Industrial Emissions Management

9

Randolph P. Angle

Abstract

Both direct and indirect controls are used in managing industrial emissions in Canada. Regulatory restraints are supplemented with economic incentives, partnerships and recognition programs. The federal government manages emissions concurrently with 10 provincial and 3 territorial governments. In two of the ten provinces, responsibility for air management in an urban metropolitan area has been delegated to a municipal government. Most Canadian jurisdictions use a source permitting system in which a number of air management tools are applied. The source permitting system processes information about industrial facility design, existing ambient air quality, source standards, ambient standards, and dispersion model results together with regulatory agency policies to set emission limits, establish operating conditions, and determine the self-monitoring required for performance feedback. Additional feedback is provided through the regulator's compliance program of education, prevention, verification and enforcement. Airshed management plans, environmental assessments, research results and emissions inventories may contribute to any of the system's components.

Keywords

Industrial emissions · Emission management · Source permitting · Air quality regulations · Emission standards

9.1 Introduction

Preventing or reducing emissions to the atmosphere from industrial operations requires both engineering and sociological knowledge. Engineering determines what can be achieved technically. This information finds its way into emission standards and other strategies for managing emissions. Whether technical solutions will be applied is a decision made by the owners and operators of a facility. The social sciences provide various ways of influencing these decisions. Generally society can exercise control over its members through either direct (also known as external or formal) or indirect (also known as internal or informal) means (Encyclopedia of Sociology 2001).

Direct social control is expressed using law enforcement mechanisms and other formal sanctions such as fines and imprisonment as determined through legislation by elected representatives. Specific legal requirements may be imposed on emission sources along with corresponding penalties if there is a failure to comply. Alternatively or in combination, financial incentives may be provided to motivate appropriate environmental behavior –preventing or reducing emissions, or for performing better than the legal requirement. The combination "carrot and stick" approach can be seen in Ontario's emission trading system, and in British Columbia's emission fees.

Indirect social controls aim to internalize values and norms, that is, have the recipient accept new attitudes and display appropriate behaviors in conformity with societal expectations. In the air management field, this means persuading the owners and operators of industrial facilities

R. P. Angle (✉)
R. Angle Consulting, Edmonton, Alberta, Canada
e-mail: rangle2009@gmail.com

E. Taylor, A. McMillan (eds.), *Air Quality Management,* DOI 10.1007/978-94-007-7557-2_9,

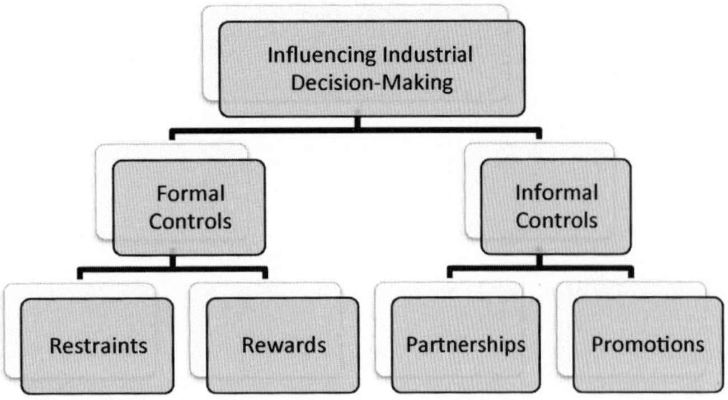

Fig. 9.1 Direct and indirect means of influencing industrial decision-making

that emission controls are an ordinary part of doing business. This requires long-term, sustained effort in providing information, education, undertaking joint projects, and supporting research, technology development and technology deployment. Many jurisdictions complement their regulatory activities with various types of informational outreach and partnerships with industry.

Figure 9.1 shows the four main ways that societies and their governments influence human environmental behavior through direct controls (restraints and rewards) and indirect controls (partnerships and promotions). Some specific examples for air quality management have been described in the chapter on processes and tools. An extensive compilation of policy instruments can be found in Marbek and Amec (2007). A different taxonomy of approaches is given by de Nevers et al. (1976) and Elsom (1992).

9.2 Direct Controls

Canada is a federal state where legislative power has been divided between a federal government and provincial governments. The Canadian constitution allocates specific powers to the provinces and remaining powers to the federal government. Because the constitution predates the emergence of environmental concerns, authority for environmental legislation is not explicitly allocated, and therefore derives from the other powers. This had understandably led to some friction between federal and provincial governments when both have chosen to regulate in the same environmental area. Regulated parties have taken this perceived problem to the Supreme Court, who has ruled that jurisdiction is concurrent; regulated parties must follow both laws, and in any areas of direct conflict, the federal law is paramount (Deimann 1998).

9.2.1 Legislation

Parliament passed the federal Clean Air Act on June 23, 1971. The Act had three primary objectives: "promote a uniform approach (to air pollution management) across Canada, make provisions for the mechanisms and institutions needed to ensure that all measures to control air pollution can be taken, and delineate a leadership role for the federal government" (Environment Canada 1973). In 1988 the Clean Air Act and other environmental legislation was consolidated into the Canadian Environmental Protection Act (CEPA) as part of a framework for the management and control of toxic substances throughout their life cycle. CEPA also provided additional management tools, many of which would be categorized as indirect controls. CEPA was also administered jointly by Environment Canada and Health Canada, in contrast to the Clean Air Act which was administered solely by Environment Canada. CEPA also included provisions for a mandatory review after five years, and in 1999 a revised CEPA was passed with the intent to make pollution prevention the cornerstone of national efforts to reduce toxic substances in the environment. After a substance has been declared "toxic" for the purposes of CEPA, the federal government has a number of instruments (Fig. 9.2) at its disposal to impose preventative and control actions.

Table 9.1 shows some examples for each of the instrument types under the Canadian environmental Protection Act.

Each province and territory also has air pollution legislation and regulations. In the Montreal and Vancouver metropolitan areas, the provinces have delegated administrative responsibility to municipal governments, whose direct controls take the form of bylaws. Table 9.2 gives the main air pollution legislation and examples of statutory regulations (bylaws for municipalities) in the 15 Canadian jurisdictions.

Provincial and territorial legislation generally provides authority to make statutory regulations and create various types of quasi-regulations related to the tools outlined in the chapter on processes and tools. They also authorize the operation of a *source permit system* which is central to air management in most jurisdictions.

9.2.2 Emission Standards

Emission standards, as defined in the chapter on processes and tools, are an important component of all Canadian air management systems. They appear in statutory regulations and in various quasi-regulations such as guidelines, mandatory codes of practice and source permits. While it is possible to derive emission standards for industrial sources from considerations of ambient air quality, general reduction targets, or aesthetic requirements (Stern 1976), in Canada such standards are generally determined from industrial process and equipment considerations. Because it is much more expensive to retrofit pollution control devices into existing facilities than to install them in new facilities, there are often different requirements for new and existing facilities. As technology improves over time, it becomes possible to

Fig. 9.2 Management instruments under the Canadian Environmental Protection Act

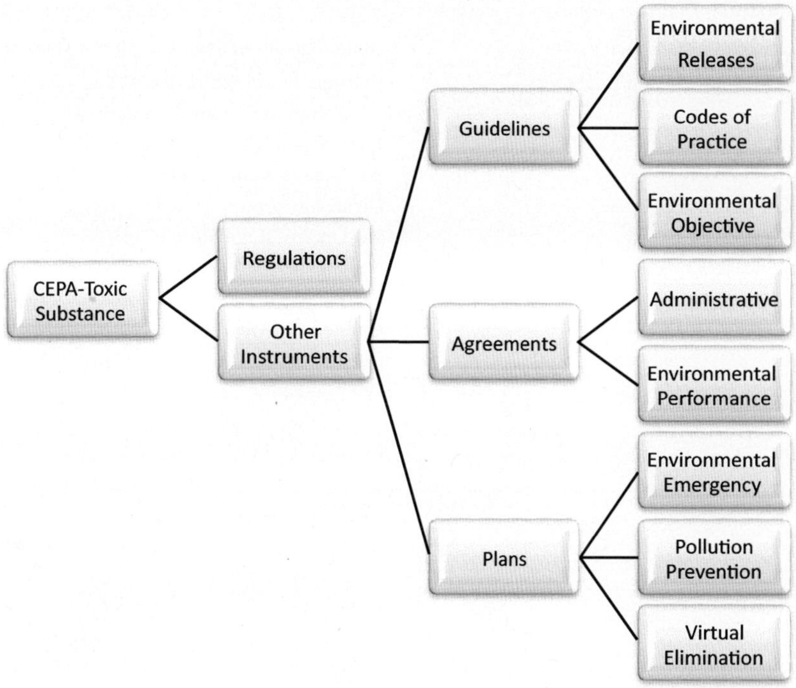

Table 9.1 Examples of CEPA instruments for industrial emissions

Type of CEPA instrument	Examples
Regulations	Off-Road compression ignition engine emission
	Off-Road small spark-ignition engine emission
	Volatile organic compound (VOC) concentration limits for architectural coatings
Guidelines	New source emission guidelines for thermal electricity generation
	Guidelines for the reduction of ethylene oxide releases from sterilization applications
	Environmental code of practice for integrated steel mills
	Code of practice for the reduction of dichloromethane emissions from the use of paint strippers in commercial furniture refinishing and other stripping applications
Agreements	Environmental performance agreement concerning atmospheric emissions of polycyclic aromatic hydrocarbons-Rio Tinto Alcan
	Canada-Wide standards
	MOU between the government of Canada and the Canadian chemical producer's association
Plans	Pollution prevention plan for acrylonitrile
	Environmental emergency regulations
	Virtual elimination list

set more stringent emission standards and limits. Each new source can be progressively cleaner than its predecessor. The older facilities are often "grandfathered", that is, subject only to the standards in place at the time of their construction, to provide certainty to the operators. Significant expansions or modifications usually trigger upgrades to meet current standards. Sometimes, emissions standards incorporate a schedule for upgrading the older installations at regular intervals.

As an example, Vale Inco spent nearly $ 1 billion since 1986 to reduce SO_2 emissions at its Sudbury operations to meet Ontario's regulatory limits. Through improvements in milling and smelting technologies, as shown in Fig. 9.3 emissions of SO_2 have decreased by more than 75%. Vale has embarked on a new $ 2 billion initiative to further reduce SO_2 emission to approximately 45,000 t per year by 2015. Vale Inco Sudbury has also developed an advanced emissions reduction program using sophisticated meteorological instrumentation, dispersion modelling and a community monitoring network to reduce smelter production when conditions indicate that SO_2 emissions may impact surrounding communities.

Because costs increase at an ever increasing rate as complete containment is approached, there is a challenge in determining the degree of control that should be required. Consequently the concept of "best practical technology" has been introduced, where "practical" implies both technological and economic feasibility. A variety of terms have been introduced to characterize this balancing of costs and controls:

Table 9.2 Main air pollution legislation and example regulations in Canadian air quality jurisdictions

Canadian jurisdiction	Main legislation and examples of statutory regulations
Newfoundland and Labrador	Environmental Protection Act
	Air Pollution Control Regulations
Nova Scotia	Environment Act
	Air Quality Regulations
New Brunswick	Clean Air Act
	Air Quality Regulation 97–133
	Administrative Penalties Regulation
	Appeal Regulation
	Ozone Depleting Substances and Other Halocarbons Regulation
	Public Participation Regulation
Prince Edward Island	Environmental Protection Act
	Air Quality Regulations
Quebec	Environment Quality Act (Loi sur la qualité de l'environnement)
	Quality of the atmosphere Regulation (Règlement sur la qualité de l'atmosphère)
	Hot mix asphalt plants Regulation
	Mandatory reporting of certain emissions of contaminants into the atmosphere Regulation
	Pulp and paper mills Regulation
Ontario	Ontario Environmental Protection Act
	Ontario Environmental Bill of Rights
	Airborne Contaminant Discharge Monitoring and Reporting
	Air Pollution—Local Air Quality
	Boilers
	Cessation of Coal Use—Atikokan, Lambton, Nanticoke and Thunder Bay Generating Stations
	Emissions Trading
	Environmental Penalties
Manitoba	Environment Act
	Classes of Development Regulation
	Licensing Procedures Regulation
Saskatchewan	Clean Air Act
	Clean Air Regulations
	Potash Refining Air Emissions Regulations
Alberta	Environmental Protection and Enhancement Act
	Substance Release Regulation
	Activities Designation Regulation
	Administrative Penalty Regulation
	Approvals and Registrations Procedure Regulation
	Emissions Trading Regulation
	Mercury Emissions from Coal-fired Power Plants Regulation
	Release Reporting Regulation
British Columbia	Waste Management Act
	Environmental Management Act
	Asphalt Plant Regulation
	Oil and Gas Waste Regulation
	Permit Fees Regulation
	Waste Discharge Regulation
	Agricultural Waste Control Regulation
	Gasoline Vapour Control Regulation
	Open Burning Smoke Control Regulation
	Wood Residue Burner and Incinerator Regulation
	Code of Practice for Concrete and Concrete Products Industry
Yukon	Environment Act
	Air Emissions Regulations
Northwest Territories	Environmental Protection Act
	Environmental Rights Act
	Asphalt Paving Industry Emission Regulations
	Spill Contingency Planning And Reporting Regulations
	Used Oil And Waste Fuel Management Regulations

Table 9.2 (continued)

Canadian jurisdiction	Main legislation and examples of statutory regulations
Nunavut	Environmental Protection Act
	Environmental Rights Act
	Asphalt Paving Industry Emission Regulations
	Spill Contingency Planning And Reporting Regulations
Communauté métropolitaine de Montréal (Montreal Metropolitan Community)	Règlement 2001–2010
Metro Vancouver	Air Quality Management Bylaw 1082
	Air Quality Management Fees Bylaw 1083
	Concrete and Concrete Products Industries Emission Bylaw 1084
	Gasoline Distribution Emission Bylaw 1085, 1129
	Automotive Refinishing Emission Bylaw 1086
	Boilers and Process Heaters Emission Bylaw 1087
	Agricultural Boilers Emission Bylaw 1098
	Non-Road Diesel Engine Emission Bylaw 1137

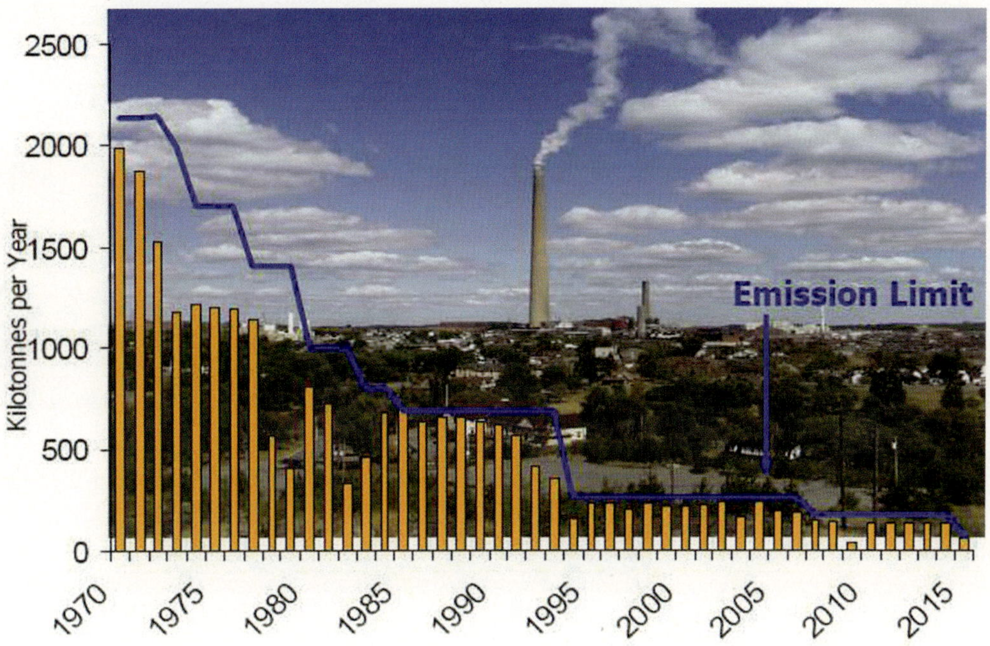

Fig. 9.3 Vale Inco Sudbury emissions from 1970–2015. (Courtesy Vale Inco)

BATNEEC (best available techniques not entailing excessive costs), BATEA (best available technology economically achievable), BPT (best practicable technology), BADT (best available demonstrated technology), or BACTEA (best available control technology economically achievable).

The procedure for developing an emission standard can be represented as six steps (after Ontario Ministry of Environment 2005a):

- Identify all potential control technologies (clearinghouses, journal articles, vendor information, permits elsewhere, technical reviews)
- Evaluate technical feasibility (use at comparable facilities elsewhere, pilot scale tests, experimental installations)
- Rank control technologies by effectiveness
- Determine control costs and emission reductions
- Review requirements in other jurisdictions
- Select the technology that best balances effectiveness and cost

Usually the affected industry will be consulted and sometimes environmental interest or community stakeholder groups also participate in the standard setting process (for example, CASA 2003). Many regulators use websites to make regulatory proposals available to stakeholders for comment (for example, British Columbia posts intention papers; Ontario operates an Environmental Registry under its Environmental Bill of Rights; and the federal government maintains a CEPA Environmental Registry).

Emission standards are generally set for an entire industrial sector or for common equipment across sectors. Because of the effort required to develop sector emission standards,

Table 9.3 Examples of CCME emissions standards

Emissions standard	Year	Form of standard
National emission guidelines for stationary combustion turbines	1991	g/GJ for oxides of nitrogen and sulphur dioxide depending on fuel type, turbine use and size
National emission guideline for commercial/industrial boilers and heaters	1998	g/GJ for oxides of nitrogen depending on capacity and fuel type
National emission guidelines for cement kilns	1998	kg NO_2/tonne of clinker by fuel and kiln type
New source performance standards and guidelines for the reduction of volatile organic compound emissions from Canadian automotive OEM coating facilities	1995	g/m^2 of Volatile Organic Compounds for different types of vehicle
Recommended CCME standards and guidelines for the reduction of VOC emissions from Canadian industrial maintenance coatings	2002	g/L of VOC in different types of coatings
Canada-Wide standards for mercury emissions from coal-fired electric power generation plants	2006	% capture in coal burned for different coal types

Fig. 9.4 Enhanced electrostatic precipitator at poplar river power station near Coronach, Saskatchewan. (Photo provided by and property of SaskPower)

Canadian jurisdictions often work together under the Canadian Council of Ministers of the Environment (CCME) to develop emission standards for industry sectors that have installations in many jurisdictions, or for types of equipment that can be found in any industry. Table 9.3 summarizes some of these standards. CCME standards are not binding on any of the participating jurisdictions. They come into force when a jurisdiction adopts them through regulation, policy, or permit.

Industry response to such standards can take many forms. For example, SaskPower's Emissions Control Research Facility explored various mercury-capture technologies before finding the best one to meet the requirements of the Canada-Wide Standard for mercury emissions. Key to the work was the development of in-house ability for analyzing mercury in coal and ash streams. At the Poplar River Power Station (Fig. 9.4) powdered activated carbon is injected into the flue gas to absorb mercury prior to particulate removal by the electrostatic precipitators. This was the first Canadian facility of its kind to do so.

While the CCME has not updated many of its emissions standards, a recent initiative around a comprehensive national air management system (Steering Committee 2010) will lead to the development of new emissions standards in the form of *Base-level Industrial Emission Requirements* (*BLIERs*). These requirements are based on what leading jurisdictions inside or outside Canada are requiring of industry in areas with air quality meeting ambient standards, adjusted for Canadian circumstances. More stringent industrial requirements could be imposed where needed to reduce air pollution to acceptable levels. BLIERs initially will focus on nitrogen oxides, sulphur dioxide, volatile organic compounds, and particulate matter. BLIERS are anticipated for 13 industry sectors: aluminum and alumina, base metal smelting, cement, chemicals, electricity, iron ore pellets, iron and steel, oil sands, petroleum refineries, pipelines, potash, pulp and paper, and upstream oil and gas, and two types of common equipment: boilers and heaters, combustion turbines. BLIERs are to be set as quantitative performance requirements, with qualitative requirements being adopted only in sectors where quantitative standards are not feasible. BLIERs would apply to new facilities immediately and existing facilities on some defined schedule. BLIERS are to be set under a federally led, time-limited, federal/provincial/territorial consensus process, with stakeholder involvement, and will be reviewed regularly to ensure they reflect technological improvements.

In a country as large and diverse as Canada, the existence and importance of each sector will vary widely. If a national standard does not exist for a particular sector and a jurisdiction has many installations, it may choose to invest in developing its own sector-wide standard, or it may simply adopt the standard of a similar jurisdiction elsewhere. Because of a large oil and gas industry that extracts sulphur containing material, Alberta has requirements of 70–99.8 % removal of the contained sulphur depending upon the size of the facility (ERCB 2001). Removing the sulphur from a fuel prevents it from entering the atmosphere as sulphur dioxide when the fuel is combusted. With forest products being a major industry, British Columbia has emission guidelines for medium-density fibreboard (British Columbia Ministry of Environment 2008b). Newfoundland and Labrador has operating requirements for used oil combustion (Dutton and Lawrence 2006). Many jurisdictions place emission standards for a number of industries into a single regulation.

If a jurisdiction has only a few installations of an industry type, it may not invest in developing sector-wide standards. Instead it will tailor a specific *emission limit* for an individual permit following much the same procedure as outlined for developing a standard, or by using the standard from a similar jurisdiction. For simple activities of lower environmental risk, some jurisdictions use a *Code of Practice* to articulate operating requirements (for example, asphalt paving plants in Alberta, concrete and concrete products in British Columbia, abrasive blasting in Newfoundland and Labrador) rather than issue a permit. Such Codes of Practice are legally enforceable when established under the authority of legislation or statutory regulations. They are not to be confused with the codes of practice produced by some industry associations as guidance to good conduct of members.

9.2.3 Penalties

Failure of industry to comply with regulations usually results in financial penalties. For example, in Alberta for a major offence a corporation may be fined up to $ 1,000,000. Lesser offences may be subject to administrative penalties of up to $ 5000 per day. In Quebec the most serious contraventions can receive a fine of up to $ 500,000 on first offence, $ 1,200,000 on second offence and $ 1,500,000 on subsequent offences.

Ontario determines penalties from a formula:

$$\text{Penalty} = A + (B - C)$$

where "A" is the monetary benefit received by the regulated person as a result of the contravention, "B" is the gravity of the contravention, "C" is a reduction to the gravity component for a variety of extenuating circumstances including prevention or mitigation efforts, presence of an environmental management system, and agreed actions with the Ministry.

The penalties under the Canadian Environmental Protection Act include: (a) a fine of up to $ 1 million a day for each day of an offence, (b) imprisonment of corporate officials for up to for three years, or (c) both a fine and imprisonment. Sentencing criteria for the court includes the cost to remedy the damage done to the environment, and violators may also have to pay for clean-up costs or forfeit any profits earned as a result of an offence.

9.2.4 Rewards

Rewards for good industrial performers are not as visible as the restraints. The economic instruments described in the chapter on processes and tools provide financial incentives to control or reduce emissions (or disincentives for not controlling emissions).

Permit fees are set by British Columbia in direct proportion to the authorized emissions. The rates are higher for substances that pose more relative risk. This provides an incentive for an industrial operator to request a lower emissions limit especially for the more costly emissions. Until 2008 Ontario provided a *tax deduction* for pollution control equipment. Similarly, the federal government in the past has allowed *accelerated depreciation* for capital expenditures on pollution control equipment. In 2011 incentives existed for systems that produced heat and/or electric power efficiently from fossil fuels or from alternative renewable energy sources. At different times in the past governments have also provided grants or loan guarantees to assist industry in upgrading equipment and technology for improved environmental performance

In 2005 Ontario expanded its *emission trading* program for oxides of nitrogen and sulphur dioxide, allowing capped electricity and industrial emitters to purchase excess allowances from other capped emitters or to purchase emission reduction credits from uncapped emitters who voluntarily reduce their emissions. Emission trading systems impart a market value to emissions and encourage the lowest cost reductions of emissions. Alberta has a limited trading program for its electricity producers. While national emission trading has been largely successful in the United States (e.g. Burtraw and Szambelan 2009), Canada does not have a similar system. Although long discussed (for example, Nichols 1993) the small number of potential participants together with regional differences and other difficulties have not made this an attractive option. Recently, the potential for cross-border emissions trading between the United States and Canada has been explored (Border Air Quality Strategy 2005).

9.3 Indirect Controls

Indirect controls often take the form of partnerships to undertake some beneficial activity or promotional campaigns (social marketing) to educate, convince and motivate action.

9.3.1 Recognition Programs

From 1995 to 2000 Environment Canada operated the first nationwide voluntary pollution reduction initiative, the Accelerated Reduction/Elimination of Toxics (ARET) program. Some 318 separate industrial facilities, representing 171 corporations from eight major industrial sectors, participated in the program which targeted 117 toxic substances. ARET was guided by a Stakeholders Committee comprising representatives from industry, health and professional associations, and federal and provincial governments. A 50% reduction target for 87 substances was met by ARET program participants in 1997, three years ahead of schedule, and a reduction of 72% was achieved by the end of 2000. A 61% reduction of 30 toxic, persistent, and bioaccumulative substances was achieved, somewhat less than the 90% target.

The National Pollutant Release Inventory (NPRI) is Canada's legislated, publicly accessible inventory of pollutant releases and transfers. It comprises information reported by facilities to Environment Canada under the Canadian Environmental Protection Act, together with air pollutant emission estimates for facilities not required to report and non-industrial sources such as motor vehicles, residential heating, forest fires and agriculture. Public access to the NPRI motivates industry to prevent and reduce pollutant releases so that they avoid the negative publicity, reputation risk and potential market repercussions of appearing on anyone's list of the "worst polluters".

Ontario has used recognition programs for many years. Originally called Cooperative Agreements, the Ontario Environmental Leaders (OEL) Program was a non-regulatory vehicle providing incentives to environmental leaders who commit to meeting "beyond compliance" environmental targets. The program was intended to motivate continuous improvement. Entry requirements include having an environmental management system (EMS), a good environmental compliance record and committing to beyond compliance environmental targets. Facilities also develop and implement a communications and outreach plan and annually report on their progress. In return, participants were eligible for a range of incentives from the Ministry, including enhanced permits, guaranteed turn-around time for amendments, technical assistance, single window access to the Ministry with a dedicated customer service representative, faster decisions on applications for new technologies, greater regulatory certainty, public acknowledgement in appropriate Ministry communications, and recognition on the Ministry's web site. The Ontario Environmental Leaders program is being phased out and replaced with the Minister's Award for Environmental Excellence. The new program will recognize and showcase the dedicated environmental efforts of a broader range of participants including business, non-government organizations, institutions, communities and individuals.

Similarly, Alberta operates a good performance recognition program called EnviroVista. It encourages industry to go beyond just compliance with the terms of its permit by providing a number of incentives: facility name, company name and company logo are featured on the government website; company names appear in printed material published by the department and in presentations by the senior department staff; participants are issued a certificate and have the right to use the EnviroVista logo on their printed and electronic materials, signs and on other products; participants will have a say in the design of any future refinements to the Program; participants who commits to a Stewardship Agreement receive a simplified permit that provides facilities with operational, regulatory and administrative flexibility.

From 1997 to 2009 the Canadian Council of Ministers of Environment (CCME) sponsored the Pollution Prevention Awards Program to give national recognition to companies and organizations showing cutting-edge accomplishment and leadership in preventing pollution at the source, rather than cleaning it up or treating it later. Over 60 companies have been recognized for their pollution prevention achievements over the life of the program.

9.3.2 Performance Agreements

Environment Canada has used Environmental Performance Agreements to achieve specified environmental results. They may be negotiated with a single company, multiple companies, regional industry associations, a sector association or a number of sector associations. Other government agencies (federal, provincial, territorial or municipal) and third parties (non-government organizations) may also be parties to such agreements. The Environmental Performance Agreement stipulates clear and measurable performance standards and includes effective accountability mechanisms. Environment Canada assumes certain obligations such as facilitating

Fig. 9.5 Nova Chemicals Sarnia-Lambton operations. (Courtesy Nova Chemicals)

information exchange and performance monitoring, public recognition of good performance or relief for participating parties from other management and control tools. Similar undertakings may also be provided by other federal, provincial or territorial agencies that are parties to an Environmental Performance Agreement. In 2011 Environment Canada had the following agreements: Canadian Paint and Coatings Association, E.I. DuPont Canada Company, Perfluorinated Carboxylic Acids (PFCAs) and their Precursors, Railway Association of Canada, Rio Tinto Alcan, and Vinyl Industry. Previous agreements existed with Alcoa Ltd., Algoma Steel Inc., Automotive Parts Manufacturers' Association, Canadian Chemical Producers' Association, Dofasco Inc., Dow Chemical Canada Inc., Refractory Ceramic Fibre Industry, and Specialty Graphic Imaging Association.

One of the earliest agreements was the 2001 Memorandum of Understanding with the Canadian Chemical Producers Association (CCPA) to reduce the release of chemical substances through voluntary, non-regulatory action under the CCPA Responsible Care® program. Figure 9.5 shows three chemical production sites operated by one of the member companies. The target was a 25% reduction in VOC emissions by 2002, based on a 1997 base year, focusing action on the 25 companies that had the largest VOC emissions. Signatories to the Memorandum were: Environment Canada, Health Canada, Industry Canada, CCPA, Ontario Ministry of the Environment and Alberta Environment (the two provinces where the majority of Canada's chemical production takes place). Non-governmental organizations, specifically Pollution Probe and STOP, as well as members of the CCPA's National Advisory Panel, were active partners in the negotiation of this MOU.

At the end of 2002 VOC emissions had been reduced by 25.1%. The MOU also provided for action on other substances of concern to parties to the agreement. At the completion of the agreement in 2005, emissions of CEPA-Toxic substances had declined by 75% since 1992. The members of Responsible Care® reduced their overall emissions by 87%, reduced releases of known and probable carcinogens by 94%, eliminated 98% of emissions of 14 high-priority substances targeted by Canada's Chemicals Management Plan, and decreased ozone-depleting discharges by 66% since 2000.

9.3.3 Partnerships

Partnerships with industry have also been used to develop emission reduction strategies. In 2001 the Canadian Petroleum Products Institute approached provincial and federal environment and energy departments with a proposal to regulate air emissions from Canadian petroleum refineries in a way that would stimulate innovation but preserve or enhance the competitiveness of the Canadian petroleum refining industry. In 2002, CCME took up the offer and started a development process involving all levels of government, industry, and non-governmental environmental and health organizations.

The National Framework for Petroleum Refinery Emission Reductions (CCME 2005) was premised on two basic concepts: 1) establish multi-pollutant refinery-wide emissions caps, rather than individual sources at the refinery; and 2) make the caps "performance-based" rather than "prescriptive", that is, dictating specific technologies for the required emission reductions. The National Framework for Petroleum Refinery Emissions Reductions provided the principles and methods for doing this, in a system with four key elements:

- A methodology for prioritizing and setting emission caps,
- A strategy to monitor and report on refinery emissions and reductions,
- A 10-year plan to keep the framework tools updated, measure performance and report on progress, and
- Expectations for jurisdictional management of refineries.

Since 1994 major policies around industrial emissions in Alberta are developed through a multi-stakeholder partnership of industry, government and environmental/health public interest groups known as The Clean Air Strategic Alliance (CASA). CASA operates a comprehensive air management system that screens and scopes issues, sets priorities and delegates tasks, designs and approves actions plans, and implements and evaluates the plans. Consensus decision-making is used throughout. Since formation, among its many accomplishments (CASA 2010), CASA has developed emissions management frameworks for solution gas flaring, sulphur dioxide sources, coal-fired electricity generators, and confined feeding operations.

Flaring is a combustion process used in the oil and gas industry to dispose of small volumes of natural gas that are technically difficult and uneconomic to conserve. Figure 9.6 shows a typical flare.In 1998 the Clean Air Strategic Alliance, consisting of industry, government, and environmental non-government organizations, established a solution gas management framework. In 2009 the volume of solution gas flared was 80.4 % below the 1996 flaring baseline.

Another area of partnership between government and industry is research. Background studies undertaken for initiatives of the Canadian Council of Ministers of the Environment are often co-funded by industry associations. In the past, the Canadian Association of Petroleum Producers has supported a large Acid Deposition Research Program (Legge and Krupa 1990) and a major study of 33,000 cattle in 205 herds in three provinces over two years to examine the association between exposure to emissions from the petroleum industry and animal health and productivity (WISSA 2006). Other partners in the latter study were the provincial governments of Alberta, British Columbia, Manitoba and Saskatchewan, the Alberta Cattle Commission, and Environment Canada.

Public Pressure

As environmental awareness has grown, the public has become much more involved in air quality issues, both pressuring governments for regulation and pressuring industries directly for improvements. At one time receiving a legal permit to operate, as described in the next section, was viewed as meeting corporate social obligations because the applicable regulations were assumed to represent societal expectations (Gunningham et al. 2002). Industries would not generally go beyond compliance unless there was a clear financial benefit in doing so. The past success of community groups and non-government organizations in forcing ever more stringent regulations has led many industrial corporations to realize that meeting legal obligations is not sufficient to meet societal expectations. To be successful a corporation must also possess a *social license* from the local community and other stakeholders. The social license may demand measures that would not normally be undertaken using standard financial justifications.

Fig. 9.6 A typical flare used in the oil and gas industry. (Courtesy Clean Air Strategic Alliance)

For example, in response to complaints from local residents and from nearby North Dakota, the Saskatchewan government, owner of SaskPower, invested $ 70 million to reduce particulate emissions at Boundary Dam Power Station. Electrostatic precipitators (ESPs) were installed over a five year period with completion in 2003. Better than 99.8 % of the particulates are being removed. Visible particulate plumes have been eliminated completely from the Boundary Dam Power Station (Fig. 9.7).

In the words of one Canadian CEO:

> We have to balance our primary interests as commercial enterprises with the wider interests of the societies of which we are a part. And, we must realize that our health as a company is very much tied to the overall well-being of the world around us. (Eric Newell, Chairman & CEO, Syncrude Ltd. (1999) 8th International Conference on Thinking. July 6, 1999)

The social license is monitored and enforced by a variety of community groups using social mechanisms such as adverse publicity and consumer boycotts. If a corporation does not respond appropriately, then stakeholders can pressure regulatory agencies to change the legal license. As noted below,

Fig. 9.7 Boundary dam power station near Estevan, Saskatchewan. (Photo provided by and property of SaskPower)

source permitting procedures usually provide some means of public input and legal requirements for public reporting provide information that community groups can use.

The concept of social license is closely related to the concept of *corporate social responsibility* (CSR), which is also called corporate responsibility, corporate accountability, corporate ethics, corporate citizenship, corporate stewardship, responsible entrepreneurship, "triple bottom line," "responsible competitiveness" and "corporate sustainability" (Hohnen and Potts 2007). CSR is understood to be the way firms integrate social, environmental and economic concerns into their values, culture, decision making, strategy and operations in a transparent and accountable manner. There are many different perspectives on CSR and the relationship between a corporation and its stakeholders (Branco and Rodriques 2007).

9.4 Source Permitting

Most Canadian provincial, territorial and municipal jurisdictions use a permit system (enabled under legislation) to manage industrial emissions. Such systems allow for the flexibility in applying combinations of various controls and generally aim to: (1) minimize source emissions, and (2) disperse the residual emissions so that ambient air concentrations are below the ambient air quality standards. Jurisdictions also commonly apply the principles of pollution prevention, continuous improvement and polluter pay. A permit may be called an *approval, authorization, or license* depending on the jurisdiction.

Permitting can be viewed as a system where a number of inputs are transformed into an output in the form of a Permit as Fig. 9.8 depicts. The inputs consist of specific information about emission standards, ambient air quality standards, existing ambient air quality, the applicant's emissions, and the results of dispersion modelling. Processing consists of regulatory agency deliberations, and the output is the Permit with its emission limits and operating conditions. Feedback is provided by industry self-monitoring of its emissions and impacts on air quality, and by the regulator's compliance assessment. If necessary, adjustments to permit limits and operating conditions may be made, either soon after a problem is detected or at the time of permit renewal. Permits are renewed typically at 5-year intervals. Alberta permits can be up to 10-years and British Columbia permits are indefinite. Depending on the situation, permits may be issued for much shorter times. The setting for the permitting system is the social milieu in which it operates. Many types of general societal information could conceivably affect the system components, but the most relevant would likely be research results, environmental assessments, and emission inventories.

9.4.1 Specific Information Inputs

An applicant for a Permit must supply detailed information about the industrial operation; typically there are application forms and application guides (for example, Ontario Ministry of Environment 2005b; Alberta Environment 1999; British

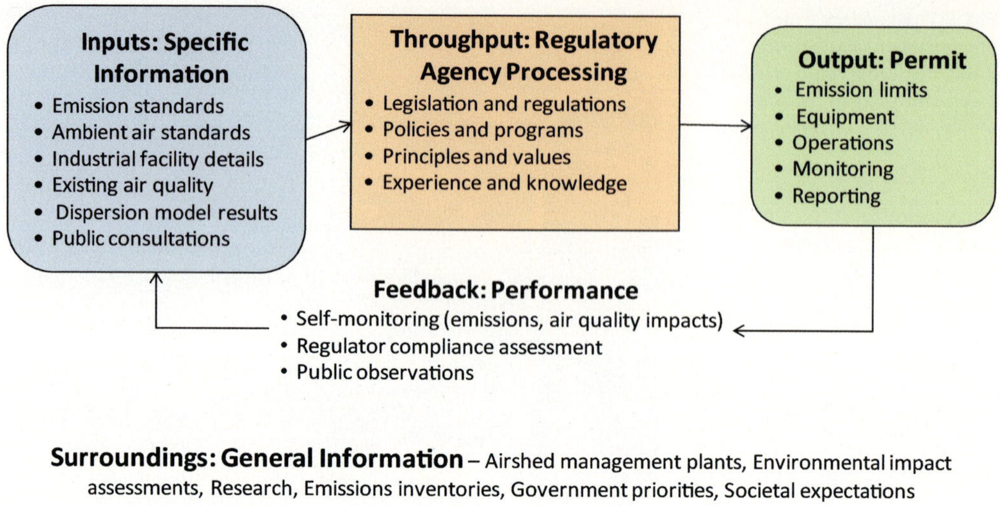

Fig. 9.8 Source permitting as a dynamic system

Columbia Ministry of Environment 2010). The information would include: industrial processes that will be in operation, pollution control equipment that will be installed, sources of emissions, substances that will be emitted, the rates of emission, characteristics of the emissions (height, temperature, velocity, diameter of stack), nearby buildings, predicted ambient air concentrations at the point of maximum and at sensitive receptors in the vicinity, existing ambient air quality in the area, demonstration that resulting air quality is below the ambient air quality standards, and demonstration that any relevant source emission standards have been met or bettered.

Plume dispersion models are tools that link emissions to ambient concentrations. These computer models are used to determine the required source characteristics (such as stack height or emission rates) so that any residual substances emitted will not exceed the ambient air quality standards. Modelling is also used in the siting of ambient air monitoring stations in the vicinity of industrial facilities, and can take into account the cumulative impact of all other sources emitting similar substances in the area. Because of the complexities inherent in dispersion modelling several jurisdictions have provided guidelines for conducting the work: British Columbia (British Columbia Ministry of Environment 2008a), Ontario Ministry of Environment (2009), Quebec (Leduc 2005) Newfoundland and Labrador (Lawrence 2006), and Alberta (Idriss and Spurrell 2009).

Ambient air quality monitoring is conducted across Canada through a cooperative arrangement between the federal and provincial/territorial governments (Environment Canada 2004). These data are available to determine existing air quality in urban areas and some rural locations. If no such data are available in the vicinity of the proposed industrial facility, the applicant may be asked to conduct some initial monitoring prior to startup, or to provide model estimates of the existing air quality. Ambient air quality data at the site may have been collected if a facility has undergone an earlier environmental assessment.

Jurisdictions may require a proponent to conduct *public consultation* as part of the permit application process. This usually includes notification in local newspapers. Proponents may have to demonstrate that they have addressed stakeholder concerns when they submit their application documents. There may also be a requirement to consult with First Nations whose rights may be impacted by a new or modified emission source.

9.4.2 Regulator Processing

The regulatory agency takes all of the input information and integrates it, along with considerations of legislation, regulations, policies, programs, principles, values, public concerns, previous experience with the type of operation or the operator, and knowledge about the technology and current industry best practices in leading jurisdictions. If an emission standard exists, then the capacity of the process or equipment can be used to determine the allowable emissions. For example, suppose an applicant wishes to install a 25 MW gas turbine. The CCME emission standard for oxides of nitrogen is 140 g/GJ. Converting, 25 MW equals 90 GJ/h and multiplied by 140 g/GJ yields the allowable emission of 12.6 kg/h. If the turbine is operated in an efficient cogeneration cycle, making 77 GJ/h. of thermal energy available, then the emission standard provides an additional allowance of $40\ g/GJ \times 77\ GJ/h = 3.1\ kg/h$. Such calculations can become quite complicated.

The dispersion modelling work of the applicant is reviewed, and perhaps checked by running the same or a similar model with the input data provided by the applicant. The influence of terrain, local meteorology, and other nearby sources will be evaluated. The regulator will review existing air quality to ensure that the additional emissions will not exceed the ambient standards. If the incremental addition is too large, then better control may be required. If there is an emission trading regime, regional plan or other government policies that could influence the emissions limit, the relevant documentation is reviewed and the implications determined.

9.4.3 The Permit

The result of the regulatory agency's deliberations is a document herein called a *permit*. Such a permit could contain any number of provisions, but as a minimum, would specify the emission limits for each source or for the facility as a whole, and list the required pollution control equipment and technologies. It might contain specifications about the emission points (stack or vent designs) and operational procedures required to minimize emissions. Many jurisdictions impose emission monitoring and reporting requirements and some include ambient air quality monitoring as well.

Point source emissions can be measured in two ways: manual stack surveys and continuous emission monitoring. Manual stack surveys are short duration tests, usually consisting of three one-hour periods, in which effluent gas samples are collected from the stack. These surveys are conducted by specially trained stack sampling personnel in accordance with prescribed methods (for example, Lawrence 2001; Alberta Environment 1995). If emissions are large, then continuous emission monitoring instruments are permanently installed on the stack and operated in accordance with prescribed procedures (for example, Ontario Ministry of Environment 2001; Alberta Environment 1998). The nature of the pollutant, available technology, and cost will determine the final requirement.

An industrial installation can also have emissions, particularly of volatile organic compounds (VOCs) from leaking valves, flanges, sampling connections, pumps, pipes and compressors. These are commonly called *fugitive emissions*. Industries, especially organic chemical plants, may be required to implement programs which will detect such leaks so that prompt corrective action can be taken. Fugitive emissions are also common during oil and gas extraction and processing.

Some industries may be required to conduct ambient air quality impact monitoring for some or many of the substances released. The permit would specify the number of monitoring stations, frequency and duration of monitoring or sampling, measuring or sampling techniques, and analytical methods. In *perimeter monitoring* discrete samples of substances are taken at various locations along the property boundary of the industrial facility for specified periods, or remote sensors are installed along the boundary lines to provide line-averaged concentrations around the entire facility. In *compliance ambient monitoring* continuous monitors are located at the point of predicted maximum ground level concentration, and/or points with high long-term average concentrations, and/or points with sensitive receptors and/or other considerations. A number of jurisdictions provide guidance on the conduct of such monitoring: Saskatchewan Environment (2007), Newfoundland and Labrador (Redmond 2010), Ontario Ministry of Environment (2008), and Alberta Environment (2006).

Industry may be required to submit monitoring reports monthly, quarterly or annually to the regulatory agency and sometimes to the public through a website. These reports summarize ambient and source monitoring data and the quality assurance and quality control measures that have been performed. The reports can outline problems which may have arisen, and corrective actions taken. Immediate reporting may be required for accidental or emergency releases to the atmosphere.

Permits also reflect negotiated agreements between the regulator and the industrial operator for improvements in performance, upgrades in technology and reductions in emissions. For example, in 2009 Syncrude Canada (Fig. 9.9) completed the installation of sulphur scrubbing technology to reduce stack emissions by about 60 %. A lime-based spray dryer absorber was retrofitted to the original two cokers at a cost of approximately $ 1.6 billion. Similarly, in 1997 Suncor Energy had completed improvements to the upgrader sulphur plant at its oil sands plant (Fig. 9.10). The 190 million $ installation of a flue gas desulphurization unit reduced sulphur dioxide emissions by 75 %.

9.4.4 Performance Feedback

Both the regulatory agency and the industrial operator will receive feedback in three ways: from the industry compulsory self-monitoring, from the regulator's industrial compliance program, and from observations by the general public. Many regulators make their permits available online so that interested stakeholders can view the contents and become aware of the requirements. Regulators generally operate some means for citizens to register complaints or report any observed non-compliance.

The regulator's compliance program may have three components (for example, Alberta Environment 2005): (1) *education* to promote compliance by raising awareness of environmental requirements, how to comply with those requirements, and the consequences of non-compliance; (2) *prevention* to identify and address potential problems before they cause environmental damage; and (3) *enforcement* to ensure that there are consequences for non-compliance, and to remedy problems and mitigate any damage.

Fig. 9.9 Syncrude Canada oil sands plant near Ft. McMurray, Alberta. (Courtesy Syncrude Canada)

Fig. 9.10 Suncor Energy Oil Sands Plant near Ft. McMurray, Alberta. (Courtesy Suncor Energy)

Educational activities are varied and might include:

- Guidance information, including fact sheets and brochures, website pages, media advertisements, and articles
- Displays, newsletters, direct mailing materials, publications
- Participation in meetings, seminars, conferences, workshops, tradeshows
- News releases, enforcement summaries, annual compliance assessment and enforcement reports

Prevention activities typically fall into three areas:

1. *Inspections* are site and field evaluations of a regulated activity to verify that specified requirements are being met. Inspectors meet with facility personnel, check on adherence to permit conditions, review monitoring data

and procedures, point out any areas of non-compliance and take samples, if necessary.

2. *Audits* are site and field examinations to verify that designated methods and procedures for data collection meet all quality assurance/quality control criteria (for example, British Columbia Ministry of Environment 2009). They may be conducted on a facility's manual source (stack) emission surveys, on continuous emission monitoring equipment and procedures, and on ambient air monitoring equipment and methods of sample collection and calibration.
3. *Performance reviews* are assessments of compulsory monitoring reports, notices, records and other required submissions.

Enforcement actions are generally proportionate to the nature of the contravention and its potential impact on human health and the environment, as well as the operator's likelihood of compliance in the future. For example, escalating enforcement actions could include:

a) *Written warning* for minor contraventions and first-time offenders
b) Administrative penalty (monetary fine) for minor offenses
c) *Environmental order* commanding that steps be taken to fix environmental problems
d) *Prosecution* for a major offence, with a trial before a judge to obtain punitive fines and/or imprisonment. The courts may opt for *creative sentencing* that compels the offender to do something constructive for the community such as donate funds to an educational institution for related research, perform a community service, or educate others through training programs.
e) *Suspension or cancellation of permit* for a very serious offence (thereby removing the regulated party's legal ability to operate a business)

9.4.5 General Information

The permitting system exists within a broader set of circumstances that may interact with any of the components. General information that may be relevant to an air quality decision might be found in airshed management plans, environmental impact assessments, research results, emissions inventories, or other public and government sources.

Airshed management plans (as described in the chapter on airshed planning) provide goals and actions for improving air quality in regions affected by the cumulative impact of a multitude of activities and emission sources, both regulated and unregulated.

An *environmental assessment* for the proposed industrial installation may be required under federal and/or provincial/territorial legislation. Such assessments identify possible adverse environmental effects (such as wildlife or water quality), propose measures to mitigate these effects, and provide an opportunity for those who may be affected by the installation, including First Nations, to provide input and advice.

Many government agencies provide financial support for *research* that may provide insights relevant to the application either in terms of technology that may be applied or in terms of environmental risks that may need consideration. The British Columbia Clean Air Research Fund aims to support transformative, internationally significant research of strategic importance for the management of air quality in British Columbia. Ontario's New Environmental Technologies Evaluation Program (NETE) assists in the development, marketing and application of new technologies for air pollution prevention and control. Ontario's Best in Science (BIS) Program invests in science that supports the development of good public policy and improved environmental outcomes by investigating emerging issues and technologies. Collaborative research is funded either through a competitive contracts process for specific projects identified by the Ministry, or through a *grants* process for projects proposed by research institutions. Alberta supports basic and applied research in clean coal technology under the auspices of a central government-wide agency called Alberta Innovates. Environment Canada invests directly in atmospheric science and technology (Environment Canada 2007). Federal granting agencies support numerous university research projects that may provide useful results.

Emission inventories are useful to the regulator in a variety of policy development and policy evaluation activities. For permitting their principal values lies in providing: total emissions and trends in the jurisdiction, total emissions in a specific industry sector, total emissions in a specific geographic area, and typical emissions from similar facilities elsewhere. The emission data for nearby and possibly more distant industrial facilities are used in dispersion models for predicting pollutant concentrations in the ambient air.

9.5 Conclusion

Canadian air quality jurisdictions have applied a number of direct controls to industrial emitters in the form of emission standards, guidelines, codes of practice, mandatory plans, penalties and economic incentives. They have also used a number of indirect controls in the form of recognition programs, performance agreements, and partnerships. Public pressure and social license to operate have also been influential in motivating industry to reduce emissions. To manage industrial emissions the federal government relies primarily on statutory regulations and quasi-regulations while provincial, territorial and municipal governments rely primarily on a source permitting system.

References

Alberta Environment (1995) Alberta stack sampling code. Alberta Environmental Protection, Edmonton

Alberta Environment (1998) Continuous emission monitoring system (CEMS) Code. Alberta Environmental Protection, Edmonton

Alberta Environment (1999) A guide to content of industrial approval applications. Alberta Environment, Edmonton

Alberta Environment (2005) Alberta environment compliance assurance. Alberta Environment, Edmonton

Alberta Environment (2006) 2006 amendments to the air monitoring directive, 1989 (AMD 2006). Alberta Environment, Edmonton

Alberta Environment (2009) Air management in Alberta. Alberta Environment, Edmonton

Border Air Quality Strategy (2005) Canada-United States emissions cap and trading feasibility study. Canada-United States Air Quality Agreement, Ottawa and Washington

Branco MC, Rodrigues LL (2007) Positioning stakeholder theory within the debate on corporate social responsibility. Electron J Bus Ethics Organ Stud 12(1):5–15

British Columbia Ministry of Environment (2008a) Guidelines for air quality dispersion modelling in British Columbia. British Columbia Ministry of Environment, Victoria

British Columbia Ministry of Environment (2008b) Provincial medium density fibreboard (MDF) emission guidelines. British Columbia Ministry of Environment, Victoria

British Columbia Ministry of Environment (2009) Standard auditing procedure for continuous emission monitoring and ambient air monitoring instruments. British Columbia Ministry of Environment, Victoria

British Columbia Ministry of Environment (2010) Guidance on applications for authorizations under the environmental management act. British Columbia Ministry of Environment, Victoria

Burtraw D, Szambelan SJ (2009) U.S. emissions trading markets for SO_2 and NO_x. Resources for the Future, Washington

CASA (2003) An emissions management framework for the Alberta Electricity Sector report to stakeholders. Clean Air Strategic Alliance, Edmonton

CASA (2010) Partnering for success. Clean Air Strategic Alliance, Edmonton

CCME (2005) National framework for petroleum refinery emissions reductions. Canadian Council of Ministers of the Environment, Winnipeg

de Nevers NH, Neligan RE, Slater HH (1976) Air quality management, pollution control strategies, modeling and evaluation. In: Stern AC (ed) Air pollution, vol 5, 3rd edn. Academic Press, New York

Deimann S (1998) R. v. Hydro-Qu6bec: Federal environmental regulation as criminal law. McGill Law J 43:923–952

Dutton JJ, Lawrence B (2006) Implementation guide for reviewing and approving used oil combustion equipment installation and operation proposals. Newfoundland and Labrador Department of Environment & Conservation, St. John's

Elsom DM (1992) Atmospheric pollution: a global problem, 2nd edn. Blackwell, Oxford

Encyclopedia of Sociology. 2001. The Gale Group, Farmington Hills, Michigan

Environment Canada (2007) Backgrounder benefits and costs of the regulatory framework for air emissions. Environment Canada, Ottawa

Environment Canada (2007) Environment Canada's science plan. Environment Canada's technology role: a supplement to environment Canada's science plan. Environment Canada, Ottawa

Environment Canada (2004) National air pollution surveillance network quality assurance and quality control guidelines, Ottawa

Environment Canada (1973) The clean air act annual report 1972–1973. Environment Canada, Ottawa

ERCB (2001) Interim directive ID 2001–3 sulphur recovery guidelines for the province of Alberta. Energy Resources Conservation Board, Calgary

Gunningham N, Kagan RA, Thornton D (2002) Social license and environmental protection: why businesses go beyond compliance. London School of Economics, London

Hohnen P, Potts J (2007) Corporate social responsibility: an implementation guide for business. International Institute for Sustainable Development, Winnipeg

Idriss A, Spurrell F (2009) Air quality model guideline. Alberta Environment, Edmonton

Lawrence B (2001) Procedural guide for source emission testing. Newfoundland and Labrador Department of Environment & Conservation, St. John's

Lawrence B (2006) Guideline for plume dispersion modelling. Newfoundland and Labrador Department of Environment & Conservation, St. John's

Leduc R (2005) Guide de la Modélisation de la Dispersion Atmosphérique. Ministère du Développement durable, de l'Environnement et des Parcs, Quebec

Legge AH, Krupa SV (eds) (1990) Acidic deposition: sulphur and nitrogen oxides—the Alberta Government/Industry Acid Deposition Research Program (ADRP). Lewis Publishers, Chelsea

Marbek and Amec (2007) Air quality management policy tools leading practice research. Alberta Environment, Edmonton

Nichols AL (1993) An emission trading program for sulphur dioxide sources in Canada. Canadian Council of Ministers of the Environment, Winnipeg

Ontario Ministry of Environment (2001) Guideline for the installation and operation of continuous emission monitoring systems (CEMS) and their use for reporting under the provisions of regulation O.Reg. 127/01. Ontario Ministry of Environment, Toronto

Ontario Ministry of Environment (2005a) Guideline for identification of best available control technology—economically achievable (BACTEA) for Ontario regulation 194/05 "Industry Emissions—nitrogen oxides and sulphur dioxide." Ontario Ministry of Environment, Toronto

Ontario Ministry of Environment (2005b) Guide to applying for approval (air and noise). Ontario Ministry of Environment, Toronto

Ontario Ministry of Environment (2008) Operations manual for air quality monitoring in Ontario. Ontario Ministry of Environment, Toronto

Ontario Ministry of Environment (2009) Air dispersion modelling guideline for Ontario. Ontario Ministry of Environment, Toronto

Redmond R (2010) Guidelines for ambient air monitoring. Newfoundland and Labrador Department of Environment & Conservation, St. John's

Saskatchewan Environment (2007) Air monitoring directive for Saskatchewan, DRAFT. Saskatchewan Environment, Regina

Steering Committee (2010) Comprehensive air management system: a proposed framework to improve air quality management. Canadian Council of Ministers of the Environment, Winnipeg

Stern AC (1976) Emission standards for stationary sources. In: Stern AC (ed) Air pollution, vol 5, 3rd edn. Academic Press, New York

WISSA (2006) Western Canada study on animal health effects associated with exposure to emissions from oil and natural gas field facilities. Western Interprovincial Scientific Studies Association, Calgary

Transportation Emissions: Sources and Regulation

10

Deniz Karman, Greg Rideout, Wendy Bailey, Andrew Green and Peter Eggleton

Abstract

The transportation sector accounts for a significant fraction of Canada's criteria air contaminant (CAC) and greenhouse gas (GHG) emissions. These emissions are reviewed in this chapter separately for on-road vehicles, off-road vehicles, aviation, marine and rail sub-sectors. Each subsection aims to quantify the emissions, identify the tools and methodologies for estimating emission inventories, identify the regulations and regulatory authorities for managing the emissions, and address interactions between emission regulations and fuel regulations where appropriate. The complex emission regulations governing the transportation sector are presented in detail. The alignment of emission regulations with those in the United States and other international organizations is discussed for different sub-sectors. The relationship between CAC and GHG emissions is examined in terms of the past trends with time and projections into the future.

Keywords

Aviation emissions · Heavy-duty vehicle emissions · Light-duty vehicle emissions · Locomotive emissions · Marine vessel emissions · On-road vehicle emissions · Off-road vehicles and equipment emissions · Sulphur in fuels

10.1 Introduction

The transportation sector accounts for a significant fraction of Canada's criteria air contaminant (CAC) and greenhouse gas (GHG) emissions. These emissions arise mostly from the use of fossil fuel combustion as the dominant source of primary energy in transportation. Canada is not unique in this respect; World energy statistics (IEA 2010) highlight the large and increasing share of transportation in world oil consumption while its share of other fossil fuels (coal and gas) are much lower as shown in Fig. 10.1. The electrification of transportation, particularly of road transport, is potentially an important opportunity for reducing emissions by substituting renewable sources of primary energy such as hydro, solar, wind, and nuclear energy. However, transportation currently accounts for a very small fraction of electricity use. Conversely, electricity represents a very small fraction of the end use energy in the transportation sector.

Statistics and projections by Natural Resources Canada (NRCan 2006) of Canadian GHG emissions in Fig. 10.2 reflect the dominance of fossil fuels in transportation and the importance of the sector relative to others. Transportation also accounts for significant shares of criteria air contaminants (Fig. 10.3).

D. Karman (✉)
Carleton University, Ottawa, Canada
e-mail: deniz_karman@carleton.ca

G. Rideout
Environment Canada, Ottawa, Canada
e-mail: greg.rideout@ec.gc.ca

W. Bailey
Transport Canada, Ottawa, Canada
e-mail: wendy.bailey@tc.gc.ca

A. Green
Lynnwood, Washington, U.S.A.
e-mail: green@alum.mit.edu

P. Eggleton
TELLIGENCE Group Enr., St-Lambert, Quebec, Canada
e-mail: peter.eggleton@videotron.ca

E. Taylor, A. McMillan (eds.), *Air Quality Management,* DOI 10.1007/978-94-007-7557-2_10,

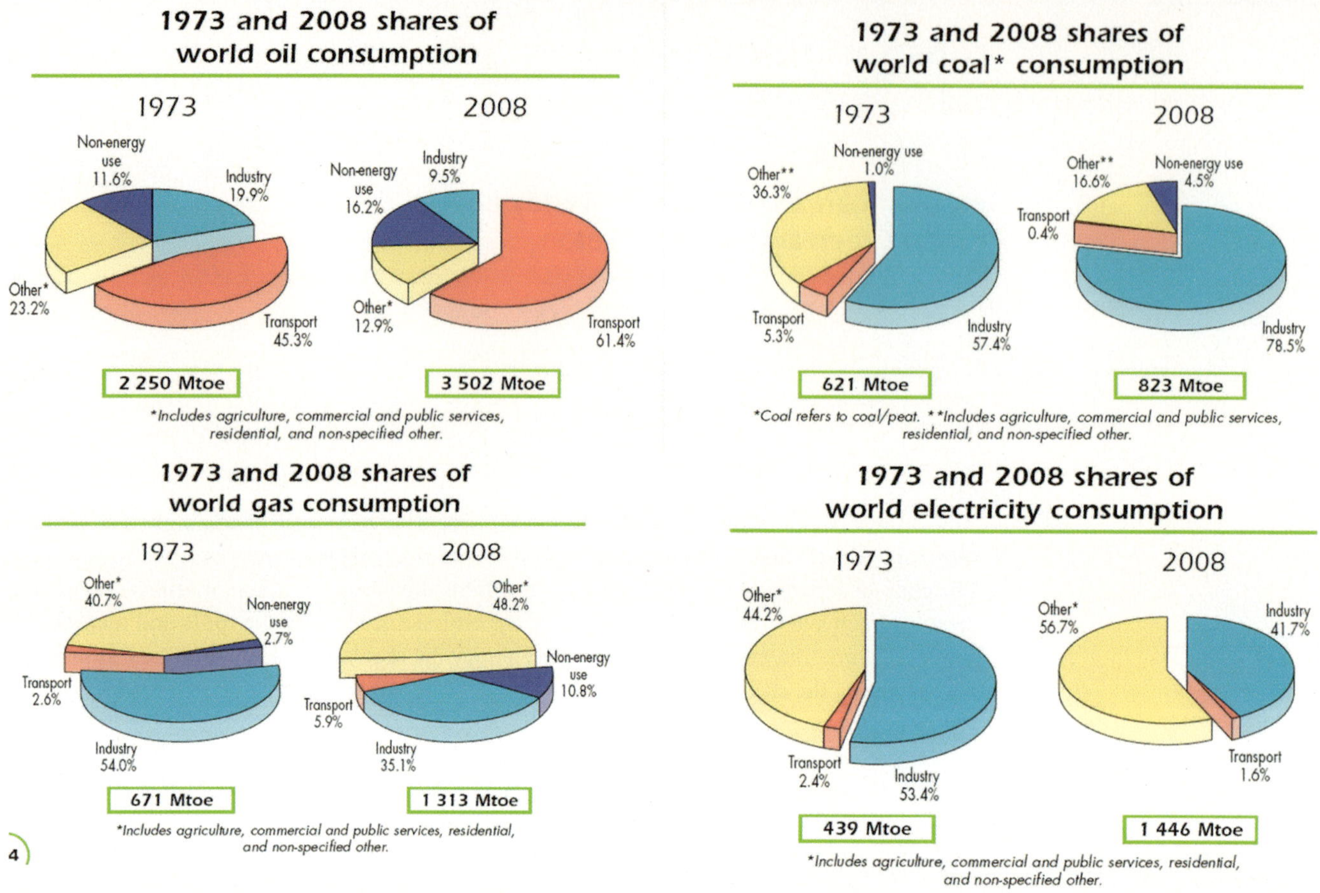

Fig. 10.1 Transportation and Global Primary Energy. Transportation is the largest consumer of oil, and uses oil virtually as the only source of primary energy. (Key World Energy Statistics 2010 © OECD/IEA 2010)

Fig. 10.2 Canadian GHG Emissions ($CO_{2\text{equivalent}}$) by Sector. Similar to global statistics on primary energy use Transportation represents the largest sectoral emissions due to its dependence on oil. (Reprinted by permission from Canada's Energy Outlook: The Reference Case 2006. NRCan 2006)

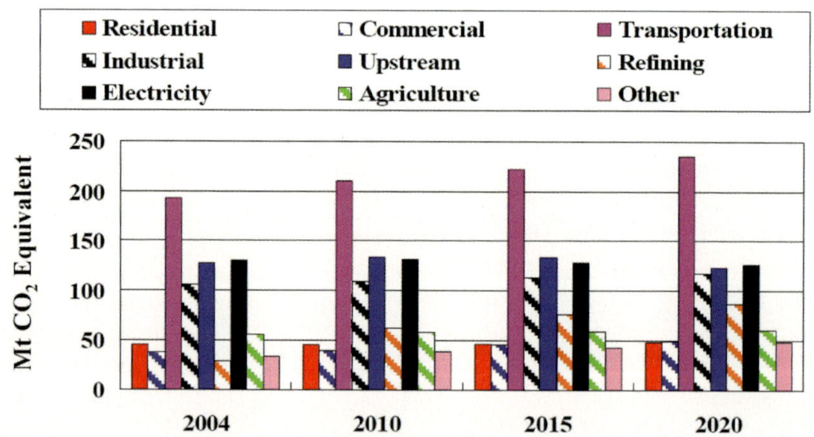

The Transportation sector as analyzed in this chapter comprises On-road vehicles, Off-road vehicles, Marine, Rail, and Aviation. On-road vehicles can be broadly divided into the Light Duty and Heavy Duty vehicle categories, although for regulatory and emissions inventory purposes up to 28 vehicle classes may be used. The off-road category of vehicles and equipment includes a broad spectrum of engines, equipment, vehicles, and vessels that do not operate on highways. The literature, both technical and regulatory, may refer to this transportation sector as "nonroad" or "off-highway," and within this sector there are a number of categories that are used to further describe the various sources including:

- outdoor power equipment,
- recreational vehicles,

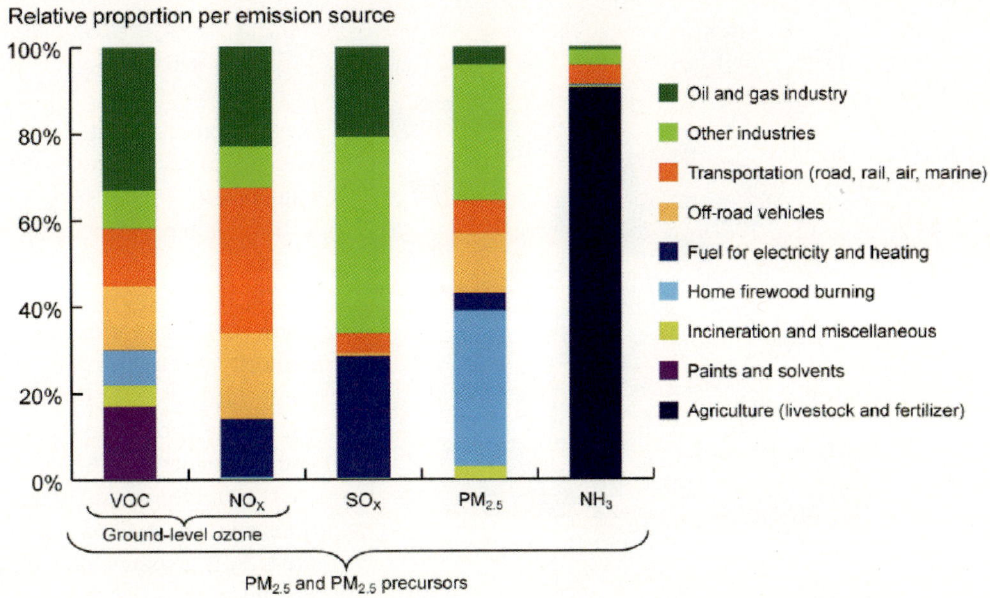

Fig. 10.3 Canadian Sources of CACs, 2007. (Adapted from CAMS 2010)

- farm and construction machinery,
- lawn and garden equipment,
- marine vessels,
- locomotives,
- aviation

Further breakdown of the transportation emissions in Figs. 10.4[1] and 10.5[2] show that road transportation, and in particular light duty vehicles dominate the totals for GHG and NO_x emissions, with the off-road sector playing an important in role in both.

With the exception of NH_3, emissions from mobile sources have shown a steadily, if slow, decreasing trend over the past 25 years (Fig. 10.6). These reductions have been achieved through a combination of fuel and vehicle emissions technologies that have been implemented in response to regulations that were primarily motivated by improving air quality. The trends in ambient concentrations of these pollutants and other air quality metrics have not necessarily followed the trends in mobile source emissions due to contributions from other sources and the complexities of atmospheric transport and transformations. Ambient air quality trends in Canada have also been in a positive direction generally. However, concern with the human health effects of air pollution even at low concentrations and the contributions from transportation sources to air pollution continue to motivate studies for a better understanding of these health effects. For example, in the often quoted 2005 study entitled *2005–2026 Health and Economic Damage Estimates* (OMA 2005), the Ontario Medical Association projected the *Illness Costs of Air Pollution* (ICAP) by linking regional air quality across Ontario with estimates of the premature deaths, hospital admissions, emergency visits, minor illness days, health care costs, and lost productivity costs due to the level of air pollution observations. The Canadian Medical Association followed this in 2008 by similar estimates at the national scale (CMA 2008).

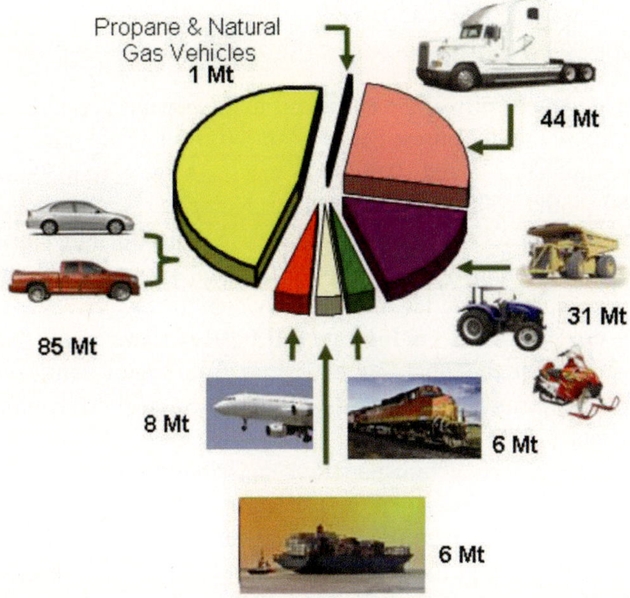

Fig. 10.4 Canada's 2005 GHG Emissions from Transportation. Within the Transportation sector On-road vehicles dominate GHG emissions, lead by Light Duty Vehicles. (Source: Environment Canada 2012–Part 3)

Emissions from on-road, and some off-road, vehicles are of particular concern from a human health perspective

[1] Figure courtesy of Andrew Giallonardo, Environment Canada.
[2] Figure courtesy of Andrew Giallonardo, Environment Canada.

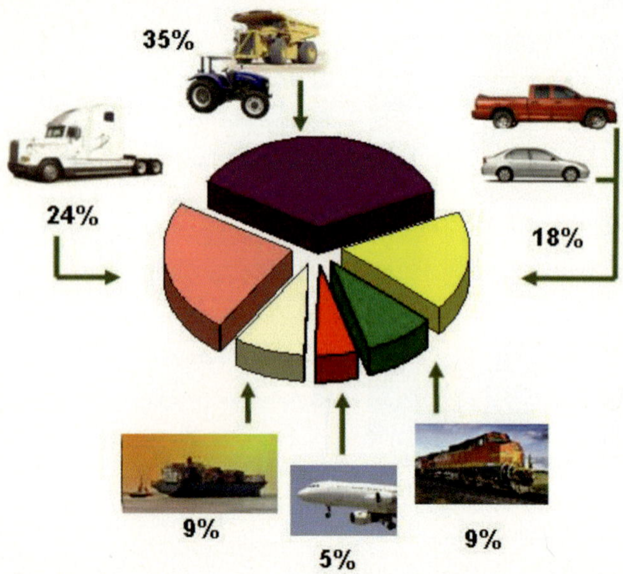

Fig. 10.5 Canada's 2005 NO_x Emissions from Transportation. Off-road vehicle and engines account for a significant fraction of total NO_x emissions from transportation. (Source: NPRI—Criteria Air Contaminants, National: 1985–2010)

due the proximity of the emissions to the general population living in urban centres. Evans et al. (2011) estimate that approximately 2, 4 and 10 million Canadians live within 50 m, 100 m and 250 m of a major road. Numerous studies have focused on correlations between various health indicators and the proximity of residential dwellings to major highways or busy arterial roadways. A comprehensive review by the Health Effects Institute (HEI 2010) summarizes the results of studies that link emissions from motor vehicles, direct exposure to these emissions, and the associated health effects.

The subject of transportation emissions in general and motor vehicle emissions in particular is thus an ongoing concern from a direct human health as well as general environmental impacts perspective. In the next sections of this chapter transportation emissions will be analyzed in terms of the particular regulatory and technical aspects of each subsector for estimating and minimizing greenhouse gas as well as air pollutant emissions.

10.2 On-Road Vehicle Emissions

The impressive reductions in Canadian mobile source emissions in Fig. 10.6 have been realized—despite the increased travel activity during the indicated period—mostly due to the continued reductions in new motor vehicle emission rates mandated by regulations which in turn have forced emission control technologies to meet the regulations.

10.2.1 Regulatory History and Harmonization with U.S.

Regulations aimed at controlling emissions from motor vehicles in Canada can be considered in two technically distinct categories:

- Emission regulations for new vehicles to be sold in Canada
- Regulations governing the emissions from the in-use fleet

These two categories also delineate the responsibility and authority for establishing regulations: the Federal Government fulfils the role for new vehicles via Environment Canada, while Provincial Governments establish the Inspection and Maintenance (I/M) programs for in-use vehicles.

The harmonization of new motor vehicle emission regulations and non-road emission standards between Canada and the U.S. is a somewhat unique feature since regulations that pertain to the other transportation categories are either less well established or are generally dealt with in an international forum. This section will therefore address specifically on-road vehicle emissions, reserving discussion on the other transportation sectors for their respective sections further below.

Since the Auto Pact agreement of 1965 between Canada and the United States the on-road vehicle manufacturing sector has been a practically integrated one between the two countries. The Auto Pact established a sectoral free trade agreement for vehicles and original equipment parts, requiring carmakers to build as many vehicles in Canada as they sold in Canada, being allowed to move their products across the Canada-U.S. border duty-free in exchange. Although the Auto Pact has now been technically replaced by the provisions of North American and international agreements, motor vehicle emissions regulations have evolved to ultimately become harmonized between the two countries. Figure 10.7[3] summarizes the milestones in this process while Table 10.1 summarizes the evolution of light duty gasoline vehicle emission standards and the emission control technologies that have become necessary to meet these standards in the U.S.

10.2.2 New Vehicle Emission Regulations and Testing

Regulations

Emission Regulations for new light duty vehicles are based on exhaust emission testing on a chassis dynamometer and evaporative emission testing in a Sealed Housing Evaporative Determination (SHED) test, described further below. Emissions from heavy duty vehicles are regulated based on

[3] Figure courtesy of Andrew Giallonardo, Environment Canada.

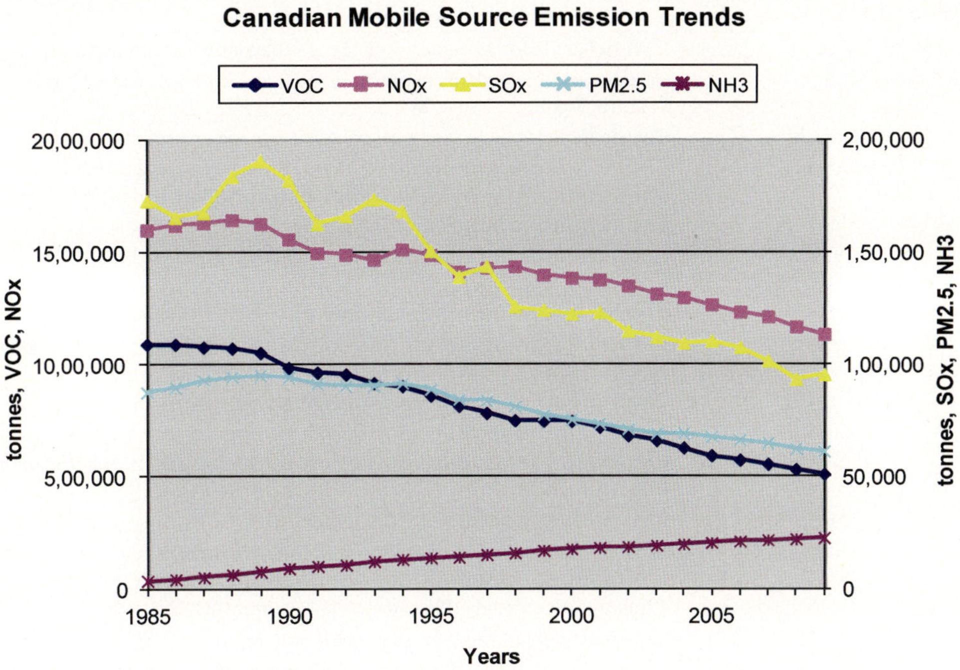

Fig. 10.6 Trends in Canadian Transportation Emissions of CACs. Emission Control Technology forcing vehicle and engine regulations as well as enabling regulations on fuels (for example reduced sulfur in transportation fuels) have resulted in significant reductions of criteria air contaminants from the Transportation sector. (Source: NPRI—Criteria Air Contaminants, National: 1985–2010)

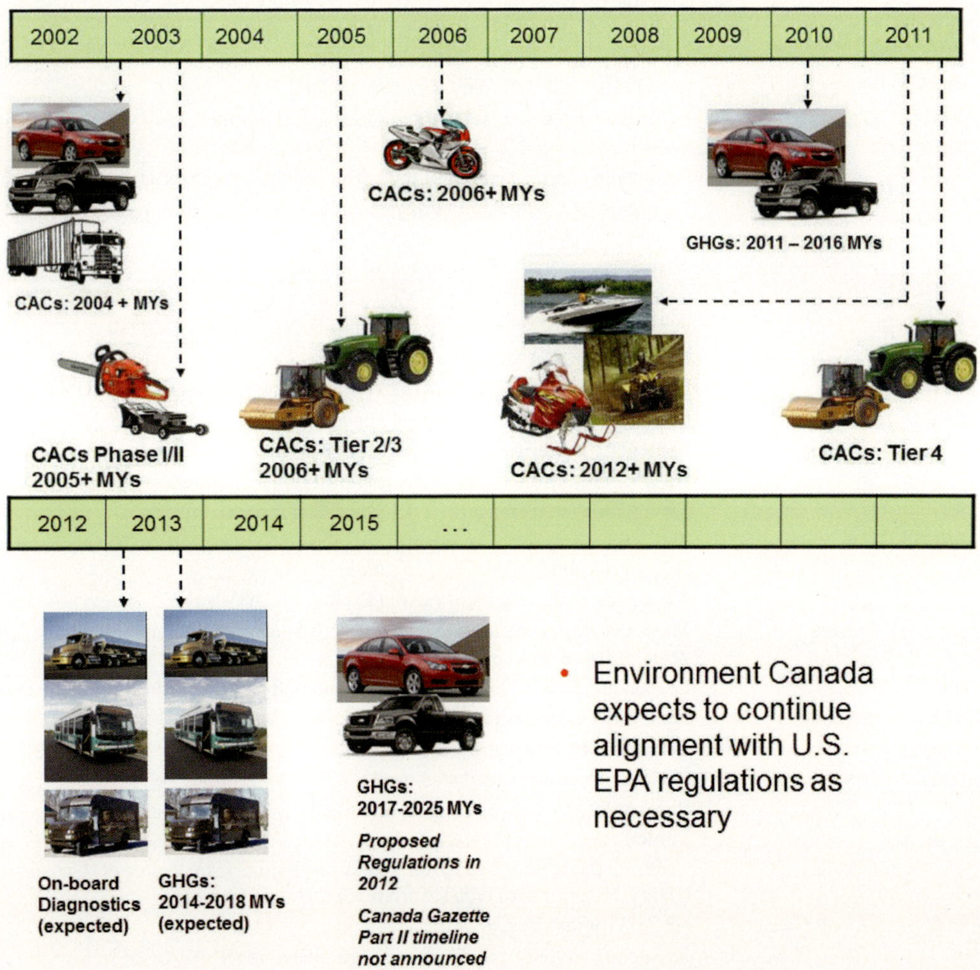

Fig. 10.7 Canada's regulatory history and alignment with U.S. (SOR/2003-2. On-Road Vehicle and Engine Emission Regulations, SOR/2010-201. Passenger Automobile and Light Truck Greenhouse Gas Emission Regulations)

Table 10.1 Evolution of Vehicle emission standards in the U.S. and Canada

1968	U.S.	First standards for 49-States, 7-mode test standards in volumetric units (ppm, %)
1970	U.S.	Introduction of standards in grams/mile HC=2.2 g/mile, CO=23/mile, still 7-mode cycle but with constant volume sampling
1971	Canada	Canada motor vehicle safety act enacted, 7-mode test standards in volumetric units
1972	U.S. and Canada	CVS-72 test procedure introduces LA-4 driving cycle and 2-Bag test, HC=3.4 g/mile, CO=39 g/mile, new driving cycle, more stringent, despite apparently more lenient standards
1973	U.S.	First EPA NO_x standard NO_x=3.0 g/mile
	Canada	Approximate date of introduction of unleaded gasoline to market. Price approx 5 cents per Gallon more than leaded fuel
1974	U.S.	13-mode test for heavy diesel engines, HC+NO_x=16 g/bhp-hr, no PM limit
1975	U.S.	Reduced HC, CO standards CVS-75 cycle adds 3rd Bag, hot start mode HC=1.5 g/mile CO=15 g/mile NO_x=3.1 g/mile Light-Duty Truck CO=20 g/mile
		Catalytic converters introduced (oxidizing type)
		Unleaded fuel required on vehicles with catalytic converters, creating a greater market share
	Canada	Reduced HC, CO standards CVS-75 cycle adds 3rd Bag, hot start mode HC=2.0 g/mile CO=25 g/mile NO_x=3.1 g/mile
1977	U.S.	Reduced NO_x standard NO_x=2.0 g/mile
1979	U.S.	Separate standards introduced for light-duty trucks, light-duty truck definition revised to include up to 8500 lbs. GVWR HC=1.7 g/mile CO=18.0 g/mile NO_x=2.3 g/mile
1981	U.S.	1.0 g/mile NO_x standard in effect (~70% control) CO=3.4 g/mile
		Waiver available from EPA administrator to allow up to 7.0 g/mile CO for 1981–1982
		Three-Way catalytic converters used on almost all passenger cars
		Closed-loop fuel control
1984	U.S.	More stringent standards for light-duty trucks, HC=0.80 g/mile, CO=10.0 g/mile, NO_x unchanged
		EPA transient test cycle optional for heavy-duty diesel engine certification, NO_x Limit 10.7 g/bhp-hr
1985	U.S.	Transient engine test mandatory for heavy duty diesels
1987	Canada	Harmonization of motor vehicle emissions standards between Canada and the US. Legislation takes effect on Sept 1, 1987 for light-duty cars and trucks
1988	U.S.	Light-duty truck NO_x standard reduced to 1.2 g/mile for GVWR <6000 lbs. and 1.7 g/mile for 6000–8500 lbs. GVWR
		Particulate matter standard of 0.6 g/bhp-hr takes effect for HDDV's
	Canada	1.0 g/mile NO_x standard in effect (~70% control) CO=3.4 g/mile, HC=0.41 g/mile, Light-Duty Truck NO_x Standard Reduced to 1.2 g/mile for GVWR <6000 lbs
		Three-way catalytic converters used on almost all passenger cars
		Closed-loop fuel control
		Particulate matter standard of 0.6 g/bhp-hr effective for HDDV
1990	U.S & Canada	HDDV NO_x limit reduced to 6.0 g/bhp-hr
1991	U.S.	Particulate matter standard of 0.25 g/bhp-hr for truck engines HDDV NO_x limit reduced to 5.0 g/bhp-hr
	Canada	Leaded gasoline banned from sale in Canada
		Particulate matter standard of 0.25 g/bhp-hr for truck engines
		HDDV NO_x limit reduced to 5.0 g/bhp-hr
1993	U.S. & Canada	Particulate matter standard of 0.1 g/bhp-hr for urban buses
1994	U.S.	Tier 1 standards phase-in begins (40%), cold CO standard at 20 deg. F, NO_x at 90% control LDGV NO_x=0.4 g/mile, NMHC=0.25 g/mile
		PM standard of 0.1 g/bhp-hr extended to all HDDV's
	Canada	MOU between Canada government and vehicle manufacturers allowing concurrent sales of Tier 1 and Tier 0 vehicles based on U.S. sales mix.
1995	U.S.	Leaded gasoline banned, phase I reformulated gasoline (RFG) for areas in non-attainment for ozone
1996	U.S.	Phase II RFG in California
		Urban Bus PM Standard =0.05 g/bhp-hr
1998	U.S.	HDDV NO_x standard reduced from 5.0 to 4.0 g/bhp/hr
		Consent Decree between engine manufacturers and EPA introduces supplemental testing and "Not to Exceed" Limits
	Canada	CEPA 99 transfers authority for vehicle emissions from Transport Canada to Environment Canada HDDV NO_x standard reduced from 5.0 to 4.0 g/bhp/hr
2000	U.S. & Canada	NLEV Program Begins
		Supplementary FTP requirements begin phase-in (SC03, US06)

Table 10.1 (continued)

2002	U.S.	Early compliance with NMHC + NO_x standard of 2.5 g/bhp-hr for 2003 MY Engines required by October as a result of the 1998 Consent Decree
2004	U.S. & Canada	Tier 2 emission standards begin phase-in (25 %) for passenger cars
		Start of 3-year phase-in for refueling vapour recovery for Light duty trucks (6000–8500 lbs. GVWR)
2005	U.S & Canada	Low sulfur gasoline required, 30 ppm average
2006	U.S. & Canada	Refiners required to produce ultra low sulphur diesel
		with 15 ppm S beginning June 1, 2006
2007–2010	U.S. & Canada	EPA heavy-duty engine standards PM = 0.01 g/bhp-hr, NO_x = 0.20 g/bhp-hr, NMHC = 0.14 g/bhp-hr, for 25 % of Sales in 2007, 100 % in 2010
		Standards in Canada come into effect in 2010 when the US are at 100 % fleet compliance
		Crankcase emissions must be controlled on turbocharged engines

The reductions in emissions per distance driven have been impressive by any standard but reductions in total emissions are tempered by the ever increasing distances driven by motor vehicles, mostly by light duty passenger vehicles

the engine emission performance on an engine dynamometer, discussed in Section 100.

The regulatory standards are primarily based on a complex matrix of weight categories (Table 10.2) but manufacturers have been provided certification flexibilities for exhaust emissions under the Tier 2 program through the choice of eight different "bins," each with their associated tailpipe emission requirements (Table 10.2). The emissions that are regulated on the basis of mass emission rates (e.g. g/km) include the criteria air contaminants CO, NO_x, HC (as NMHC non-methane hydrocarbons, or NMOG non-methane organic gases), PM, and formaldehyde (HCHO) (Table 10.3).

Emissions Testing

In the chassis dynamometer test the vehicle is essentially placed on a treadmill where the driver follows the speed vs. time trace of a prescribed driving cycle.

The Federal Test Procedure (FTP) driving cycle represented "average" urban driving in Los Angles in the late 1960s. The test procedure consists of three parts:

- Cold start ("Bag 1"): 3.6 miles; 25.6 mph
- Stabilized ("Bag 2"): 3.9 miles; 16.2 mph
- Hot Start ("Bag 3"): 3.6 miles; 25.6 mph

The FTP "composite" mass emission rate is found by a weighted average of the mass of pollutant emitted in the three phases:

$$\frac{[(0.43 \times Bag1) + (0.57 \times Bag3)] + Bag2}{7.5miles} = grams/mile$$

The weighting assumes that 43 % of all starts are cold and 57 % are hot starts.

Two new driving cycles were designed to capture the differences in driving behaviour from the 1970's by representing more aggressive driving (US06), and the effects of starts and speed fluctuations, as well as the effect of using air conditioning (SC03). The Supplemental Federal Test Procedure (SFTP), effective with 2001 + modelyear LDV/LDT, uses the weightings 35 % FTP, 37 % SC03, 28 % US06 to arrive at the average emission rate.

Evaporative Emission Standards and Tests

With the reductions in unburned hydrocarbons emitted from the vehicle tailpipes over the years (Table 10.1) evaporative emissions associated with the operation of a vehicle became a relatively significant contribution to total emissions of volatile compounds. Evaporative emissions are considered in the following effects/categories[4]:

- **Diurnal** breathing losses occur as the fuel tank heats up during the day.
- **Resting** losses result from vapor permeation and liquid leaks through various parts of the evaporative control system.
- **Hot Soak** losses occur after the vehicle has been turned off and result from evaporation of fuel in the engine and fuel delivery system.
- **Running** evaporative losses occur as the vehicle is being operated over the road.
- **Refueling** losses are a result of vapor space displacement and spillage.
- **Crankcase** losses are primarily the result of defective PCV[5] systems.

Initially, the test for evaporative emissions involved measuring the build-up of hydrocarbons in the sealed housing into which the vehicle had been placed immediately after the engine was turned off, or heating the fuel tank over the course of one hour. Recent testing methods simulate diurnal temperature variations in the sealed enclosure over 2 or 3 days. Table 10.4 lists the evaporative emission standards on the basis of emissions per test.

[4] MOBILE6 Training, http://www.epa.gov/otaq/m6.htm#m6train.

[5] The Positive Crankcase Valve (PCV) vents vapours from the crankcase and recirculates them to the cylinder in the engine. Previously these would have been vented directly into the ambient air.

Table 10.2 Vehicle weight classifications. (http://www.epa.gov/otaq/standards/weights.htm)

	Gross Vehicle Weight Rating (lbs)									
	6,000	8,500	10,000	14,000	16,000	19,500	26,000	33,000	60,000	
Federal	LDV		MDPV	-	-	-	-	-	-	-
	LDT		HDV / HDE							
	LLDT	HLDT	LHDDE				MHDDE		HHDDE / Urban Bus	
	LDT 1 & 2 [a]	LDT 3 & 4 [b]	HDV2b	HDV3	HDV4	HDV5	HDV6	HDV7	HDV8a	HDV8b

GVW gross vehicle weight, *GVWR* gross vehicle weight rating, *HDE* heavy-duty engine, *HDV* heavy-duty vehicle, *HHDDE* heavy heavy-duty diesel engine, *HLDT* heavy light-duty truck, *LDT* light-duty truck, *LDV* light-duty vehicle, *LHDDE* light heavy-duty diesel engine, *LLDT* light light-duty truck, *MDPV* medium-duty passenger vehicle, *MHDDE* medium heavy-duty diesel engine

[a] Light-duty truck (LDT) 1 if loaded vehicle weight (LVW)=3,750; LDT 2 if LVW>3,750

[b] LDT 3 if adjusted loaded vehicle weight (ALVW)=5,750; LDGT 4 if ALVW>5,750

10.2.3 Heavy Duty Engine Regulations and Testing

In principle, the exhaust emissions from a heavy duty vehicle can also be measured and regulated on the basis of a chassis dynamometer test, similar to those described in Section 10.2.2 for light duty vehicles. Two factors make it a more practical approach to measure and regulate the emissions from the engine rather than the vehicle as a whole:

- a heavy duty engine of a particular design is typically used for different purposes and in different vehicle platforms
- chassis dynamometer testing for heavy duty vehicles requires increasingly more demanding testing facilities as vehicles get bigger.

It is therefore the engine in heavy duty vehicles that gets certified to meet regulatory standards by operating the engine mounted on an engine dynamometer where the engine shaft is directly coupled to the power absorbing dynamometer. Since no distance traveling is involved, the emissions cannot be expressed in terms of g/mile but are rather expressed in terms of the mass of emission per energy delivered to the dynamometer, g/bhp-hr. These are presented in which also demonstrates the evolution of the standards in the U.S. over the years (Table 10.5).

For heavy duty engine testing *opacity,* i.e. the % blockage of light across the exhaust, is specified in addition to the mass emission rate of PM although the latter is generally the more stringent criteria to meet.

An alternative way of characterizing and regulating particulate matter emissions is on the basis of number of nanometer size particles emitted per unit distance driven. It is generally acknowledged that ultrafine particles in the nanometer size range represent a significant threat for human health due to their potential to penetrate deep into the lungs and deposit there. On the other hand, these particles account for only a small fraction of the mass of particles collected on a filter paper when characterizing ambient air PM concentrations in terms of mass concentration, $\mu g/m^3$, or of PM emissions in terms of g/km or g/bhp-hr. The monitoring and regulation of particulate matter in increasingly smaller size fractions (e.g. from total suspended particles TSP, to particles less than 10 µm aerodynamic diameter PM_{10}, and now to $PM_{2.5}$) has been motivated by this understanding. With the advances in particulate matter measuring technologies it is now possible to measure PM number concentrations in ambient air and emission sources. However there are also challenges in making such measurements, particularly for emission sources because the number concentrations of particles in the nanometer size range appear to depend strongly on the sampling conditions such as the time-temperature history of exhaust gases as they exit the source and the rate at which the exhaust gases are diluted with ambient air. Thus, while number concentration based PM regulations have been considered/implemented in Europe for both light duty and heavy duty vehicles in the Euro 5b and Euro 6 regulations[6], the PM regulations in the U.S. and Canada continue to be based on $PM_{2.5}$ mass.

[6] http://www.dieselnet.com/standards/eu/.

Table 10.3 Light-Duty Vehicle, Light-Duty Truck, and Medium-Duty Passenger Vehicle Tier 2 Exhaust Emission Standards. (MECA 2007; http://www.epa.gov/otaq/standards/light-duty/tier2stds.htm)

	Standard	Emission Limits at 50,000 miles					Emission Limits at Full Useful Life (120,000 miles)[2]				
		NO_x (g/ml)	NMOG (g/ml)	CO (g/ml)	PM (g/ml)	HCHO (g/ml)	NO_x (g/ml)	NMOG (g/ml)	CO (g/ml)	PM (g/ml)	HCHO (g/ml)
Federal	Bin 1	–	–	–	–	–	0	0	0	0	0
	Bin 2	–	–	–	–	–	0.02	0.01	2.1	0.01	0.004
	Bin 3	–	–	–	–	–	0.03	0.055	2.1	0.01	0.011
	Bin 4	–	–	–	–	–	0.04	0.07	2.1	0.01	0.011
	Bin 5	0.05	0.075	3.4	–	0.015	0.07	0.09	4.2	0.01	0.018
	Bin 6	0.08	0.075	3.4	–	0.015	0.1	0.09	4.2	0.01	0.018
	Bin 7	0.11	0.075	3.4	–	0.015	0.15	0.09	4.2	0.02	0.018
	Bin 8	0.14	0.100/0.125[c]	3.4	–	0.015	0.2	0.125/0.156	4.2	0.02	0.018
	Bin 9[b]	0.2	0.075/0.140	3.4	–	0.015	0.3	0.090/0.180	4.2	0.06	0.018
	Bin 10[b]	0.4	0.125/0.160	3.4/4.4	–	0.015/0.018	0.6	0.156/0.230	4.2/6.4	0.08	0.018/0.027
	Bin 11[b]	0.6	0.195	5	–	0.022	0.9	0.28	7.3	0.12	0.032

Manufacturers must meet a corporate average emission requirement for the entire fleet of vehicles sold in a given model year after certifying vehicles in one of the available categories. Reaching these regulations required advanced spark ignited engines, advanced engine control strategies, clean fuels, clean lubricants, and advanced emission control technologies

[a] In lieu of intermediate useful life standards (50,000 miles) or to gain additional nitrogen oxides credit, manufacturers may optionally certify to the Tier 2 exhaust emission standards with a useful life of 150,000 miles

[b] Bins 9–11 expire in 2006 for light-duty vehicles and light light-duty trucks and 2008 for heavy light-duty trucks and medium-duty passenger vehicles

[c] Pollutants with two numbers have a separate certification standard (1st number) and in-use standard (2nd number)

Table 10.4 Light-Duty Vehicle, Light-Duty Truck, and Medium-Duty Passenger Vehicle Tier 2 Evaporative Emission Standards. (http://www.epa.gov/otaq/standards/light-duty/tier2evap.htm)

	Vehicle Type	Model Year	3 day Diurnal + Hot Soak (g/test)	Supplemental 2 day Diurnal + Hot Soak (g/test)	Running Loss (g/ml)
Federal	LDV/LLDTs[a]	2004	0.95	1.20	0.05
	HLDTs[b]	2004	1.20	1.50	0.05
	MDPVs[a,b]	2004	1.40	1.75	0.05
	LDV[a]	2009	0.50	0.65	0.05
	LLDT[a]	2009	0.65	0.85	0.05
	HLDT[b]	2010	0.90	1.15	0.05
	MDPV[a,b]	2010	1.00	1.25	0.05

The emission test involves measuring the hydrocarbons emitted from the vehicle in an enclosed space over a period of time

[a] For liquefied petroleum gas-fueled light-duty vehicles (LDV), light-duty trucks (LDT), and medium-duty passenger vehicles (MDPV): 0.15 g hydrocarbon/gal (0.04 g/l) of fuel dispensed

[b] Refueling standards for heavy light-duty trucks (HLDT) are subject to phase-in requirements. MDPVs must also comply with the phase-in requirement and must be grouped with HLDTs to determine phase-in compliance

Table 10.5 Heavy-duty highway compression-ignition engines and urban buses exhaust emission standards. (http://www.epa.gov/otaq/standards/heavy-duty/hdci-exhaust.htm)

<table>
<tr><th></th><th>Year</th><th>HC (g/bhp-hr)</th><th>NMH C (g/bhp-hr)</th><th>NMHC + NOx (g/bhp-hr)</th><th>NOx (g/bhp-hr)</th><th>PM (g/bhp-hr)</th><th>CO (g/bhp-hr)</th><th>Idle CO (percent exhaust gas flow)</th><th>Smoke[a] (Percentage)</th><th>Useful Life (hours/years /miles)</th><th>Warranty Period (years/miles)</th></tr>
<tr><td rowspan="10">Federal[b]</td><td>1974-78</td><td>-</td><td>-</td><td>16</td><td>-</td><td>-</td><td>40</td><td>-</td><td>20 / 15 / 50</td><td>-</td><td>-</td></tr>
<tr><td>1979-84</td><td>1.5</td><td>-</td><td>10</td><td>-</td><td>-</td><td>25</td><td>-</td><td>20 / 15 / 50</td><td>-</td><td>-</td></tr>
<tr><td>1985-87</td><td>1.3</td><td>-</td><td>-</td><td>10.7</td><td>-</td><td>15.5</td><td>-</td><td>20 / 15 / 50</td><td>LHDDE: - / 8 / 110,000
MHDDE: - / 8 / 185,000
HHDDE: - / 8 / 290,000</td><td>-</td></tr>
<tr><td>1988-89</td><td>1.3[d]</td><td>-</td><td>-</td><td>10.7</td><td>0.6</td><td>15.5</td><td>0.5[c]</td><td>20 / 15 / 50</td><td rowspan="5">1990-97 and 1998+ for HC, CO, and PM:
LHDDE: - / 8 / 110,000
MHDDE: - / 8 / 185,000
HHDDE: - / 8 / 290,000
1994+ urban buses for PM only:
- / 10 / 290,000
1998+ for NOx:
LHDDE: - / 10 / 110,000
MHDDE: - / 10 / 185,000
HHDDE: - / 10 / 290,000</td><td rowspan="5">5 / 100,000 [q]</td></tr>
<tr><td>1990</td><td>1.3[d]</td><td>-</td><td>-</td><td>6.0</td><td>0.6</td><td>15.5</td><td>0.5[c]</td><td>20 / 15 / 50</td></tr>
<tr><td>1991-93</td><td>1.3</td><td>-</td><td>-</td><td>5.0 [ABT]</td><td>0.25 [ABT]
0.10[e]</td><td>15.5</td><td>0.5[c]</td><td>20 / 15 / 50</td></tr>
<tr><td>1994-97</td><td>1.3</td><td>-</td><td>-</td><td>5.0 [ABT]</td><td>0.1 [ABT]
0.07[f], 0.05[g]</td><td>15.5</td><td>0.5[c]</td><td>20 / 15 / 50</td></tr>
<tr><td>1998-2003</td><td>1.3</td><td>-</td><td>-</td><td>4.0 [ABT]</td><td>0.1 [ABT]
0.05[g]</td><td>15.5</td><td>0.5[c]</td><td>20 / 15 / 50</td></tr>
<tr><td>2004-2006[h]</td><td>-</td><td>-</td><td>2.4 (or 2.5 with a limit of 0.5 on NMHC)[o] [ABT[i,j]]</td><td>-</td><td>0.1
0.05[g]</td><td>15.5</td><td>0.5</td><td>20 / 15 / 50</td><td rowspan="2">For all pollutants: [p]
LHDDE: - / 10 / 110,000
MHDDE: - / 10 / 185,000
HHDDE: 22,000 / 10 / 435,000</td><td rowspan="2">LHDDE: 5 / 50,000
All other HDDE: 5 / 100,000 [q]</td></tr>
<tr><td>2007+[h,k,l,m,n]</td><td>-</td><td>0.14[o]</td><td>2.4 (or 2.5 with a limit of 0.5 on NMHC) [ABT]</td><td>0.2[o]</td><td>0.01</td><td>15.5</td><td>0.5</td><td>20 / 15 / 50</td></tr>
</table>

10.2.4 Greenhouse Gas Regulations

Greenhouse gas emissions regulations in both Canada and the U.S. are effectively based on a U.S. Supreme Court decision in November 2006 where the Court determined that greenhouse gases fit well within the Clean Air Act's definition of air pollutants. This in effect enabled the U.S. EPA and the National Highway Traffic Safety Administration to issue, on May 7th 2010, their *Light-Duty Vehicle Greenhouse Gas Emission Standards and Corporate Average Fuel Economy Standards; Final Rule*[7] establishing a new set of standards. The rule applies to light-duty vehicles, light-duty trucks, and medium-duty passenger vehicles for model years 2012 through 2016. There are a number of paths for vehicle manufacturers to comply with this new regulation[8]; however the

[7] http://edocket.access.gpo.gov/2010/2010-8159.htm.

[8] http://www.epa.gov/otaq/climate/regulations/420f10014.htm.

Table 10.6 GHG Regulations and Fuel Economy (http://www.epa.gov/otaq/climate/regulations/420f10014.htm)

Projected Fleet-Wide Emissions Compliance Levels under the Proposed Footprint-Based CO_2 Standards (g/mi) and Corresponding Fuel Economy (mpg)					
	2012	2013	2014	2015	2016
Passenger Cars (g/mi)	261	253	246	235	224
Light Trucks (g/mi)	352	341	332	317	302
Combined Cars & Trucks (g/mi)	295	286	276	263	250
Combined Cars & Trucks (mpg)	30.1	31.1	32.2	33.8	35.5

standards are designed so that they gradually become more stringent each year from 2012–2016 (Table 10.6). The regulation also includes specific standards for N_2O and methane, 0.010 and 0.030 g/mile respectively.

On October 1, 2010, Canada announced final regulations that establish progressively more stringent greenhouse gas emission standards for new passenger automobiles and light trucks for the 2011–2016 model years. These are aligned with regulations in the United States. Analysis by Environment Canada projected[9] that the average GHG emissions of new vehicles of the 2016 model year will be about 25 % lower than that of the vehicles that were sold in Canada in 2008. The cumulative GHG emission reduction expected from this regulation is estimated to be 92 Mt of carbon dioxide-equivalent and fuel savings of about 28 billion l over the lifetime of the operation of vehicles of the 2011–2016 model years sold in Canada.

The corresponding regulations for Medium- and Heavy-Duty Engines and Vehicles begins with the 2014 model year and increases in stringency through 2018, and breaks the diverse truck sector into 3 distinct categories:

- Line haul tractors
- Heavy-duty pickups and vans
- Vocational trucks (everything else, buses, refuse trucks, concrete mixers, ambulances, etc.)

The regulation[10] published in September 15, 2011 established separate standards for engines and vehicles, ensures improvements in both, and sets separate standards for fuel consumption, CO_2, N_2O, CH_4 and HFCs.

EPA's N_2O, CH_4 and Air Conditioning Leakage Standards In addition to the CO_2 standards described above, the U.S. EPA has adopted emission standards for N_2O and CH_4 emissions given that they are important GHGs that contribute to global warming but recognizing that they are a small fraction of the exhaust composition of diesel and gasoline engines. The agency has also adopted standards to limit the leakage of refrigerants.

In May 2010, the Government of Canada announced its intent to develop emission standards aligned with U.S. national standards and on August 9, 2011 released a detailed consultation document describing the key elements of the future Canadian GHG regulations for the medium and heavy duty vehicle sector. Canada's regulations, similar to the light duty vehicle GHG regulations, would fall under CEPA 1999 and would only apply to GHGs, and not to fuel consumption as per the U.S. joint rulemaking. It is anticipated that the proposed regulations will be published in the Canada Gazette, Part II in 2012, effective for the 2014 and later model years.

10.2.5 Road Transportation Fuel Regulations

The impressive reductions in regulatory standards for on-road vehicles reviewed earlier Table 10.1 became possible only with a combination of engine and emission control technologies, as well as transportation fuels that could support the performance expected of these technologies.

The phase-out of lead was probably the first major fuel based measure to affect vehicle emissions by removing emissions of lead that posed adverse health effects particularly for developing children and enabling the three-way catalytic converters that reduce CO, HC, and NO_x emissions from gasoline fuelled vehicles. A corollary can be found in the reduced sulfur levels currently mandated for transportation fuels. Although the direct emissions of sulfur dioxide from motor vehicles due to fuel sulfur is also a concern, the motivation for the sharp drop in diesel and gasoline sulfur levels (Table 10.7) has primarily been to enable the more aggressive emission control technologies required to meet exhaust emission regulations. Gasoline regulations limiting benzene levels is another example of a fuel regulation directly limiting the emissions of a toxic pollutant.

Environment Canada has compiled a *Fuels Regulations Compliance Promotion Package 2011*[11] which contains summaries and reporting requirements for the following nine CEPA 1999 fuels regulations:

- Fuels Information Regulations, No. 1;
- Benzene in Gasoline Regulations;
- Gasoline Regulations;

[9] http://www.ec.gc.ca/default.asp?lang=En&n=714D9AAE-1&news=0F384925-9836-4936-B20F-A551607EEC95.

[10] http://www.epa.gov/otaq/climate/regulations.htm#1-2.

[11] http://www.ec.gc.ca/energie-energy/default.asp?lang=En&n=48F8FEEC-1.

Table 10.7 Sulfur in diesel fuel regulations and effective dates. (EC 2010b)

	Sulphur Limit (mg/kg)	On Road Diesel Fuel	Off Road Diesel Fuel	Rail and Marine Diesel Fuel
500	Production or Import	Since 1998	June 1, 2007	June 1, 2007
22	Sales	Since 1998	October 1, 2007[b]	October 1, 2007[b]
	Sales	September 1,2006	N/A	N/A
15	Production or Import	June 1, 2006	June 1, 2010	June 1, 2012
	Sales	October 15, 2006[a]	October 1, 2010[c]	N/A

[a] September 1, 2007, in the Northern Supply Area, defined as the area corresponding to the following geographical areas: a) that part of the Yukon that is north of latitude 67° N; b) the Northwest Territories, except those areas within 1 km of the centre line of highways 2–7, and that portion of Hwy 1 south of Fort Simpson, and those areas within the municipalities of Yellowknife, Detah, Hay River, NWT, Fort Simpson, Rae, Edzo, Entreprise, Fort Resolution, Fort Smith and Fort Liard; c) Nunavut; d) those parts of Manitoba, Ontario and Quebec within 50 km from the coast of Hudson Bay or James Bay; e) that part of Quebec north of latitude 51° N if west of longitude 63.5° W, and north of latitude 50° N if east of longitude 63.5° W; and f) Newfoundland and Labrador, except the island of Newfoundland

[b] December 1, 2008, in the Northern Supply Area

[c] December 1, 2011, in the Northern Supply Area

- Sulphur in Diesel Fuel Regulations
- Sulphur in Gasoline Regulations
- Gasoline and Gasoline Blend Dispensing Flow Rate Regulations
- Contaminated Fuel Regulations
- Regulations Prescribing Circumstances for Granting Waivers Pursuant to Section 147 of the Act
- Renewable Fuels Regulations

The latest of these regulations, *Renewable Fuel Regulations,* is part of the Canadian Government's broader *Renewable Fuels Strategy*. The Strategy is aimed at reducing Canada's greenhouse gas emissions, with a target of 4 Mt in 2012, about the equivalent of taking one million vehicles off the road. Fig. 10.9[12]

Inspection and Maintenance Programs

Environment Canada's policy of alignment with US EPA standards for new vehicles has been detailed above. I/M programs through which in-use vehicles are expected to demonstrate their continued compliance with new vehicle emission regulations based on their model year are much more widespread in the U.S. than Canada. The only two provinces which, as of 2011, have Motor Vehicle Emission Inspection and Maintenance programs in Canada are British Columbia and Ontario, Quebec has considered but not yet implemented a program.

British Columbia's AirCare program

The B.C. AirCare program[13] starting in 1992 is an example of a *centralized* I/M program where special stations do only the inspection, with failing vehicles getting repairs elsewhere. The program applies to vehicle registered in Vancouver and the Lower Fraser River Valley where air quality concerns are significantly related to motor vehicle emissions. 1991 and older vehicles (except diesels) are tested using the **ASM/Idle** steady-state dynamometer test to measure emission levels for HC, CO, and NOx while the vehicle is driven at a steady speed and while the engine idles. 1992 and newer vehicles (except diesels) are tested using the **IM240** dynamometer test which follows the initial part of the regulatory FTP test Fig. 10.8[14]. Eligible 1998 and newer light-duty vehicles receive a scan of the vehicle's built-in **On-Board Diagnostics** monitoring system which stores codes in the vehicle's computer indicating potential malfunctions of the emission control system. Other tests include the **D147** for testing diesel vehicles (other than 1998 and newer light-duty vehicles) for exhaust smoke opacity, the **Gas Cap Pressure Test** to ensure that it does not allow fuel vapours to escape to the atmosphere, and the visual **Tampering Inspections** to ensure that the most important emission control devices have not been tampered with. There is also a roadside spot inspection program called AirCare ON-ROAD (ACOR)[15] for heavy duty vehicles. In June of 2012, the BC government announced that the AirCare program for cars and light-duty trucks would be phased out as of December 31, 2014. In the announcement, the government noted the significant reduction in emissions in new vehicles sold today as compared to 1992.

Ontario's Drive Clean

Starting in 1999, The Drive Clean program[16] in Ontario is an example of a *decentralized* program where existing repair shops are licensed to inspect, and carry out necessary repairs. The primary test procedure has been the **ASM/Idle** already mentioned for the AirCare program above. Changes in the Fall of 2011 will replace this with On-board diagnostic (**OBD**) testing for 1998 and newer light-duty vehicles, two-

[12] http://www.ec.gc.ca/energie-energy/default.asp?lang=En&n=0AA71ED2-1.

[13] http://www.aircare.ca/inspinfo-desc.php.

[14] http://www.epa.gov/otaq/standards/light-duty/ftp.htm.

[15] http://www.th.gov.bc.ca/ACOR/.

[16] http://www.ene.gov.on.ca/environment/en/category/drive_clean/STDPROD_075529.html.

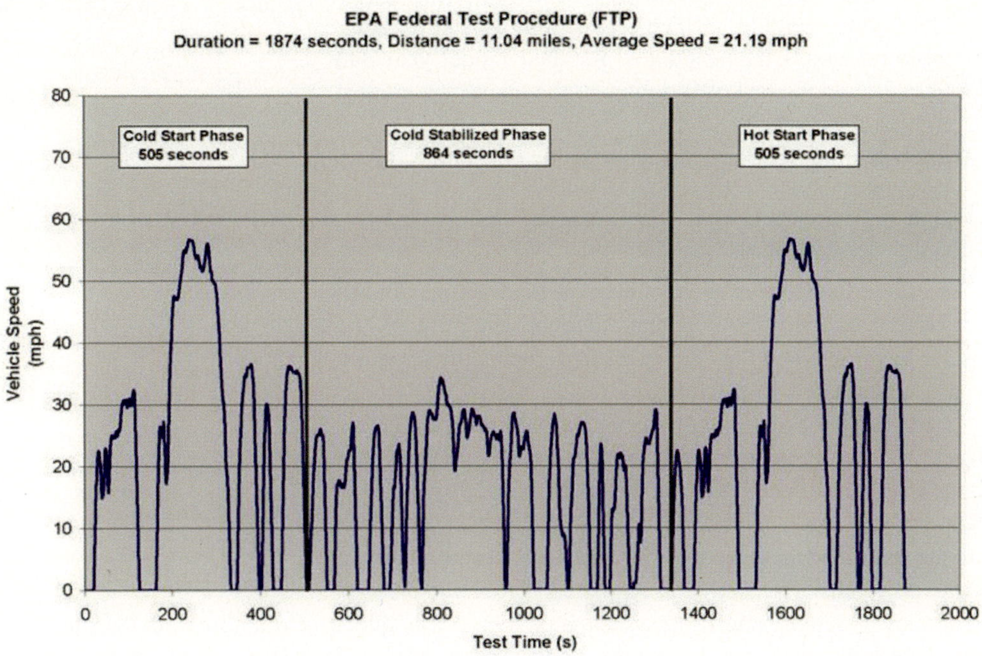

Fig. 10.8 Regulatory test drive cycle for Light Duty Vehicles. The U.S. Federal Test Procedure is also used in Canada and is now supplemented by two additional test cycles: US06 to incorporate the effects of aggressive driving and SC03 for air conditioning effects

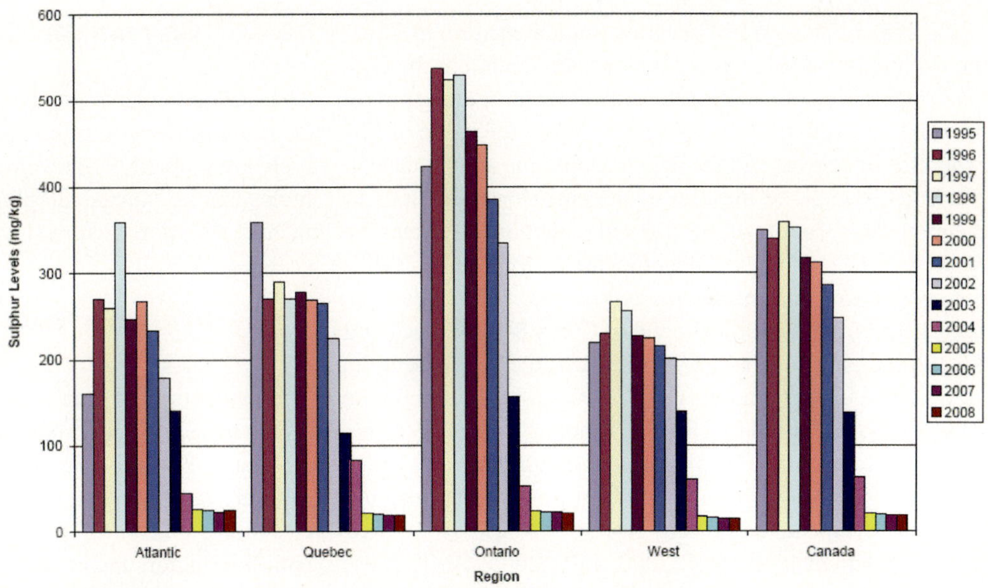

Fig. 10.9 Sulphur Levels in Gasoline in Canada, 1995–2008. Reduced sulfur levels enable the advanced emission control technologies that deliver reduced emissions. (Adapted from Environment Canada 2010b)

speed idle (**TSI**) tailpipe test on 1988–1997 vehicles, and the phase out the current dynamometer tailpipe test by December 31, 2012. Other changes relate to the frequency of testing and exemptions from testing under specific conditions (e.g. exemption form the testing requirement for transfers or license renewals if the vehicle had passed the test in the previous calendar year, or exemption for transfer of a vehicle among family members). There is also a heavy-duty component of the program based on exhaust opacity testing. Vehicles need a test for registration renewal beginning at seven years of age.

I/M Program Effectiveness

AirCare has one purpose—to reduce vehicle exhaust emissions The succinct statement above from British Columbia's *AirCare* program can be generalized in principle to all I/M programs, and it has indeed been the objective of increasingly stringent emission regulations implemented over the years for new vehicles. However, while the benefit from stringent new vehicle emission regulations can readily be estimated, the effectiveness of I/M programs for reducing emissions from the in-use vehicle fleet is a more challenging question to answer. I/M programs aim to achieve the objective by:

a. identifying vehicles which are emitting significantly more pollutants than regulated for their model year due to malfunctions or break-downs in the emissions control technology on the vehicle
b. implementing required repairs to reduce the emissions to acceptable levels
c. achieving a) and b) in a timely manner after the vehicle has started emitting more

Thus, the effectiveness is dependent on the fraction of high emitting vehicles in the fleet that are identified and repaired as well as how soon they are identified, factors that require some assumptions and estimations.

Criticism of I/M programs have focused on their effectiveness in reducing emissions from the motor vehicle fleet and/or the cost effectiveness of doing so. A study by the U.S. National Research Council specifically aimed at the issue of the quantifying the emission reductions expected from these programs. The study found that I/M programs provide much lower actual benefits than estimated by models, quoting MOBILE's estimate that a fully implemented enhanced I/M program would produce overall emissions reductions of 28 % for HC, 31 % for CO, and 9 % for NO_x.

Criticizing the first I/M program implementation in Canada, Coninx (1998) also raised the concerns found in the U.S. NRC report about the optimistic expectations of emission reductions. However, Coninx also questioned the value of I/M programs in general and for British Columbia and Ontario in particular, compared the cost of the emission reductions to the monetary benefits of the claimed reductions. It should be noted that the estimates of benefits quoted by Coninx (1998) for *averted damage to human health, materials, crops, etc. per tonne*[17] of less than $ 1/tonne of CO, $ 84/t of VOC, and $ 120/t NO_x are substantially lower than the health benefits alone found in later studies (OME 2005, CMA 2008) and the AQBAT model[18] used by Health Canada.

Evaluations of the effectiveness of the *AirCare* and *Drive Clean* programs have been undertaken from operational and technical perspectives, giving rise to changes from their initial implementation (dKC 2000, ERG 2005, OME 2001, OME 2007, OME 2010). *AirCare* publishes annual reports on the effectiveness of the program in terms of reductions in air pollutant emissions from motor vehicles in British Columbia (PVTT 2011). A sample of these evaluations from the most recent years available is presented in Table 10.8. Given the improving vehicle systems and fleet turnover, the effectiveness of I/M programs can be expected to diminish as time marches on. The general trend across North America is to phase them out and even strong programs (e.g. in British Columbia and Washington) are committed to phase themselves out sometime in the 2015–2020 timeframe.

Table 10.8 I/M Program Effectiveness—AirCare

Effect of Repairs on In-Use Light-Duty Vehicles Mass Emissions Inventory (tonne/year)						
	2009			2010		
	HC	CO	NO_x	HC	CO	NO_x
Base Inventory for year (tonnes)	5,105	77,926	6,712	4,490	69,652	6,200
Inventory after Repairs (tonnes)	4,617	70,591	6,412	4,099	63,887	5,922
Reduction from repairs	488	7,334	299	391	5,765	278
Reduction from Base (%)	10%	9%	4%	9%	8%	4%

Reprinted by permission from *AirCare—Results and Observations in 2009 and 2010*, Pacific Vehicle Testing Technologies Ltd, 2010

10.2.6 On-road Vehicle Emission Inventories

The on-road vehicle population in Canada is of the order of 20 million, travelling distances well above 300 million vehicle km annually[19]. The emission inventory for on-road vehicles can be viewed simply as the product of the emission factors emission factors (g/veh-km) and distances travelled (veh-km):

$$Emissions = (km\ travelled)(emissions/km)$$

The estimation of emissions by considering individual vehicles traveling on a road network is of course an impossible task. Emission factor models such as MOBILE aim to estimate *fleet average emission factors* (i.e. g/km, gram of emission per distance traveled) for classes of vehicles. These emission factors then need to be multiplied by the distances traveled by vehicles of that class to arrive at the quantity (grams) of emissions. Some of the emissions of concern are not directly proportional to the distances traveled, for example the evaporative emissions of hydrocarbons during refueling, or when parked after a trip. Furthermore, emissions associated with the initial period after the start of an engine may be quantified by the number of times that an engine is started and the period that the engine had been off since the last trip rather than the distance covered during a trip. Nevertheless, for the purpose of emission modelling all emissions

[17] attributed to ARA Consortium Sholtz & Associates (1995).

[18] http://www.bc.lung.ca/mediaroom/news_releases/documents/AQBATEstimatingHealthImpactsforChangesinCanadasAirQuality.pdf.

[19] The Natural Resources Canada web site quotes 18,608,297 vehicles, and 315,297.3 million vehicle-km for 2005. http://oee.nrcan.gc.ca/Publications/statistics/cvs05/chapter1.cfm?attr=0.

are ultimately brought to a common base of emissions per distance traveled by considering the necessary factors such as number of starts per day and average distances traveled per day.

The emissions from the on-road fleet made up of different size and age of vehicles is arrived at by considering the groups of vehicles whose emission performance can be expected to be similar by virtue of their model year and class. The vehicle classes in the commonly used MOBILE6 emission factor model are listed in Table 10.9 and correspond to the regulatory classification earlier presented in Table 10.2. Thus, to arrive at the fleet average emission rate, the emission factors for 28 different vehicle classes are weighted by the distances traveled by those vehicle classes:

$$[Fleet - ave\ emission\ rate] = \sum_{Veh=1}^{28} [VMT\ mix]_{vehicle\ class} \times [Fleet - ave\ emission\ rate]_{vehicle\ class}$$

The travel fraction of on-road vehicles is typically the focus of traffic and transportation demand models such as EMME. However, in contrast to emission factor models such as MOBILE6, travel demand models such as EMME and the vehicle traffic counts which are used in the calibration of such models typically characterize the vehicle fleet using only a handful of vehicle classes (e.g. *Light Duty Vehicles, Light Duty Trucks, Medium Duty Vehicles, Heavy Duty Vehicles, Buses*) Thus, emission factors for EMME vehicle classes need to be established from the fleet average emission factors estimated for 28 vehicle classes by MOBILE6. For this task, the registration records for the vehicle fleet can be used, including information on the age distribution of vehicles and annual distances traveled by vehicle classes.

The fleet average emission rate for vehicle classes modelled in MOBILE6 are based on thousands of FTP tests (Section 0) and Inspection Maintenance records (Section 100) to take into account many factors such as:

- Emission rate for the new vehicles by model year and the deterioration of emission control technology with the age (distances travelled) of the vehicle
- The average speed by different road type (e.g. highway, arterial, local roads)
- Characteristics of fuel used (e.g. volatility, oxygenate content, sulphur content)
- Meteorological parameters (average temperature, temperature by hour of day)

The MOBILE emission factor model which has been used extensively in the U.S. and Canada has been the subject of much research and development over the years. The last version, MOBILE6, was the result of efforts in the 1990s to improve emission factor estimates over the previous version, MOBILE5. Complete documentation on this R&D effort is available through technical reports available in a U.S. EPA

Table 10.9 On-road vehicle classes used in the emission factor model MOBILE6. (U.S. EPA 2003)

Number	Abbreviation	Description
1	LDGV	Light-Duty Gasoline Vehicles(Passenger Cars)
2	LDGT1	Light-Duty Gasoline Trucks 1
3	LDGT2	Light-Duty Gasoline Trucks 2
4	LDGT3	Light-Duty Gasoline Trucks 3
5	LDGT4	Light-Duty Gasoline Trucks 4
6	HDGV2B	Class 2b Heavy-Duty Gasoline Vehicles (8501-10,000 lbs. GVWR)
7	HDGV3	Class 2b Heavy-Duty Gasoline Vehicles (10,001-14,000 lbs. GVWR)
8	HDGV4	Class 2b Heavy-Duty Gasoline Vehicles (14,001-16,000 lbs. GVWR)
9	HDGV5	Class 2b Heavy-Duty Gasoline Vehicles (16,001-19,500 lbs. GVWR)
10	HDGV6	Class 2b Heavy-Duty Gasoline Vehicles (19,001-26,000 lbs. GVWR)
11	HDGV7	Class 2b Heavy-Duty Gasoline Vehicles (26,001-33,000 lbs. GVWR)
12	HDGV8A	Class 2b Heavy-Duty Gasoline Vehicles (33,001-60,000 lbs. GVWR)
13	HDGV8B	Class 2b Heavy-Duty Gasoline Vehicles (>60,000 lbs. GVWR)
14	LDDV	Light-Duty Diesel Vehicles (Passenger Cars)
15	LDDT12	Light-Duty Diesel Trucks 1 and 2(0-6,000 lbs. GVWR)
16	HDDV2B	Class 2b Heavy-Duty Diesel Vehicles (8501-10,000 lbs. GVWR)
17	HDDV3	Class 3 Heavy-Duty Diesel Vehicles (10,001-14,000 lbs. GVWR)
18	HDDV4	Class 4 Heavy-Duty Diesel Vehicles (14,001-16,000 lbs. GVWR)
19	HDDV5	Class 5 Heavy-Duty Diesel Vehicles (16,001-19,500 lbs. GVWR)
20	HDDV6	Class 6 Heavy-Duty Diesel Vehicles (19,001-26,000 lbs. GVWR)
21	HDDV7	Class 7 Heavy-Duty Diesel Vehicles (26,001-33,000 lbs. GVWR)
22	HDDV8A	Class 8a Heavy-Duty Diesel Vehicles (33,001-60,000 lbs. GVWR)
23	HDDV8B	Class 8b Heavy-Duty Diesel Vehicles (>60,000 lbs. GVWR)
24	MC	Motorcycles (Gasoline)
25	HDGB	Gasoline Buses (School, Transit and Urban)
26	HDDBT	Diesel Transit and Urban Buses
27	HDDBS	Diesel School Buses
28	LDDT34	Light-Duty Diesel Trucks 3 and 4(6,001-8,500 lbs. GVWR)

The vehicle classes are differentiated by weight: GVWR is the maximum design loaded weight, *VCW* vehicle curb weight or net weight, *LVW* loaded vehicle weight: VCW+136 kg, *ALVW* average loaded vehicle weight, (VCW+GVWR)/2

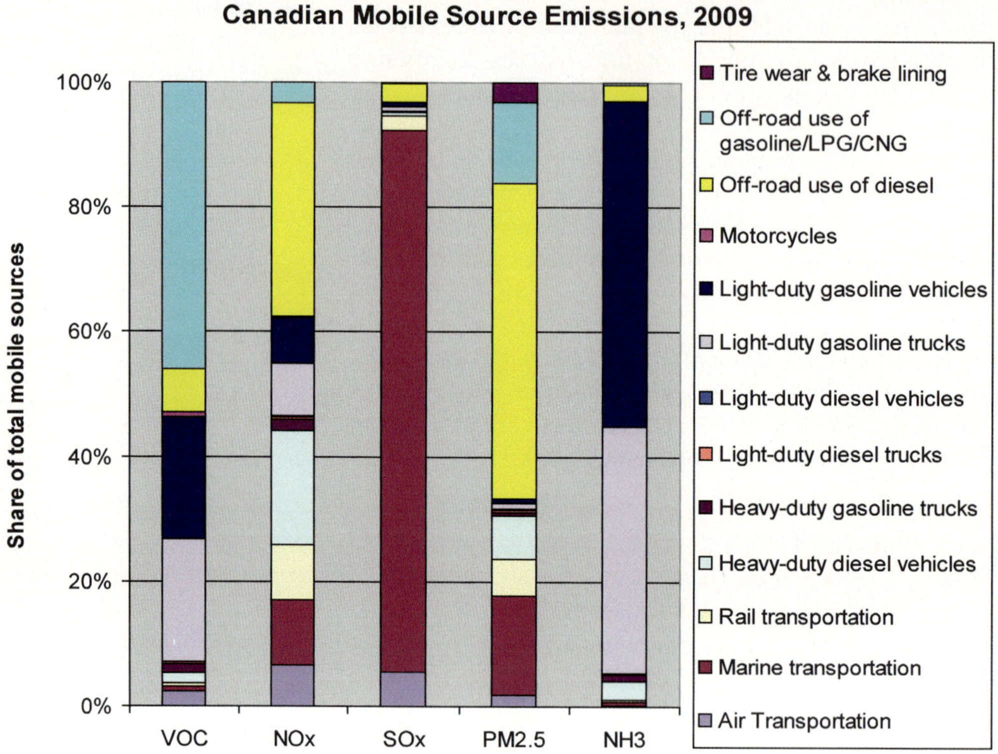

Fig. 10.10 Canadian Mobile Source Emissions of CACs, 2009. The estimations of detailed results for on-road vehicle classes depends on MOBILE6 modelling, the emissions from off-road, aviation, marine and rail sectors are estimated separately. (Source: NPRI—Criteria Air Contaminants, National: 1985–2010)

web site dedicated to the model[20]. A Canadian version of the model has existed (MOBILE5C and MOBILE6C), primarily to account for differences in the emission performance of the U.S. and Canadian fleets in the 1980s, but also to account for other Canadian specific conditions and default values that may be used in the model.

The U.S. EPA released MOVES[21] (Motor Vehicle Emission Simulator) in 2010 which has become the emissions model for state implementation plan (SIP) development, transportation conformity determinations, and other purposes outside of California (U.S. EPA 2012). Plans are underway in Canada as well for using MOVES for emission inventory development (Taylor 2012) Fig. 10.10.

10.3 Off-Road Vehicles and Equipment Emissions

From the North American perspective, the emissions from the engines and equipment that comprise the Off-road category were largely uncontrolled until the mid-1990s. This is in stark contrast to highway vehicles that had been the subject of emission regulations since the 1960's and where there had been significant progress in lowering the emissions from these mobile sources through cleaner fuels, and advances in engine and emission control technology.

The nonroad sector began to come under the attention of regulators in 1991 when the U.S. EPA conducted an analysis[22] of nonroad engines to examine their potential contribution to urban air pollution. The Agency found that the engines and equipment in this sector were emitting relatively large amounts of NO_x, hydrocarbons, carbon monoxide, and particulates. In some instances the emissions were nearly equivalent to the highway sector, and in the case of PM, the nonroad emissions were even higher than their highway counterparts. In response the EPA initiated a regulatory development process to address this sector.

In Canada, this sector is also an important contributor to the overall emissions attributed to the transportation sector. Figure 10.4 and Fig. 10.5 are good indicators of their contribution highlighting the total NO_x emissions from the various transportation categories as well as the total GHG emissions.

The regulatory approach to controlling the emissions from the aviation, marine, and rail sectors are discussed under separate sections of the Chapter. For the remaining nonroad categories Canada has used the authority legislated under the Canadian Environmental Protection Act, Part 7 Division 5 to introduce regulations that align with those established by Unites States Environmental Protection Agency. The following sections provide additional information specific to each of the regulations.

[20] http://www.epa.gov/otaq/m6.htm.

[21] http://www.epa.gov/otaq/models/moves/index.htm.

[22] Nonroad Engine and Vehicle Emission Study Report (EPA-21A-2001 or EPA460/3-91-02, November 1991).

Table 10.10 Small Spark-Ignition Engine Classes and Exhaust Emission Standards. (SOR/2003-355. Off-Road Small Spark-Ignition Engine Emission Regulations)

Engine class	Engine type	Engine displacement (cm^3)	Effective date (model year)	$HC+NO_X$ (g/kW-hr)	$NMHC+NO_X$[c] (g/kW-hr)	CO (g/kW-hr)
I-A	Non-handheld	<66	2005 and later	50[a]	–	610[a]
I-B	Non-handheld	<100 and ≥66	2005 and later	40[a]	37[a]	610[a]
I	Non-handheld	≥225 and ≥100	2005 and later[d]	16.1[b]	–	519[b]
			2005 and later[e]	16.1[a]	14.8[a]	610[a]
			2007 and later[f]	16.1[a]	14.8[a]	610[a]
II	Non-handheld	≥225	2005 and later	12.1[a]	11.3[a]	610[a]
III	Handheld	<20	2005 and later	50[a]	–	805[a]
IV	Handheld	<50 and ≥20	2005 and later	50[a]	–	805[a]
V	Handheld	≥50	2005	119[a]	–	603[a]
			2006	96[a]	–	603[a]
			2007 and later	72[a]	–	603[a]

[a] Standards apply throughout the engine useful life
[b] Standards apply only when the engine is new
[c] Some engine classes include a combined non-methane hydrocarbons (NMHC) and NO_X standard that applies only when the engine is fuelled by natural gas
[d] For models already in production at coming into force of the proposed Regulations
[e] For models initially produced after coming into force of the proposed Regulations
[f] For all models

10.3.1 Off-Road Small Spark-Ignition Engine Emission Regulations

In terms of technology, the small spark-ignition engines are either two or four stroke engines, with the 4-stroke versions offering lower emissions and greater fuel efficiency while the 2-stroke versions offer the weight reductions needed for many of the handheld applications.

First proposed by Environment Canada in 2003 the emission regulations addressing this segment of the nonroad sector came into effect with the 2005 model year engines. The Regulatory Impact Analysis Statement[23] for these regulations estimated that in the year 2000, engines in this category accounted for 20.7, 0.9, and 27.7% of the VOC, NO_X, and CO emissions attributed to the nonroad sector. The Regulations apply to off-road engines of model year 2005 and later that use sparkplugs and develop no more than 19 kW (25 hp) of power. The emissions standards are divided into seven classes based on engine displacement and usage in either a handheld or non-handheld application as shown in Table 10.10.

At the time of engine certification, a manufacturer can select one of three specified useful life periods, which range from 50 to 1000 h depending on the engine class. For example, for a class I engine, the useful life can be 125, 250 or 500 h. The selection of useful life duration must be supported by technical information. Longer useful lives, which entail a higher manufacturing cost, are typically found in commercial equipment while home consumer products are often designed for shorter useful lives[24].

10.3.2 Off-Road Compression-Ignition Engine Emission Regulations

This regulation was proposed by Environment Canada in 2004, with the regulations coming into effect with the 2006 and later model year engines and equipment, and encompassed the U.S. EPA Tier 2 and Tier 3 standards. The regulation is applicable to diesel engines found in construction, mining, farming, and forestry machines, such as tractors, excavators, and log skidders.

As can be seen from Table 10.11 the engines covered through this regulation represent a significant fraction of the off-road source contribution to air pollution. Similar to the small spark-ignited engine market, the engines in this sector are imported into Canada while some off-road diesel machines are manufactured in Canada. In either case, the engines sold in Canada must comply with the respective standard.

The *Off-Road Compression-Ignition Engine Emission Regulations* (SOR/2005-32) in Table 10.12 align the engine certification values with those of the US EPA Tier 2 and Tier 3 values, however there are some differences that are worth noting. The regulations specifically exempt engines:

- designed exclusively for competition
- regulated by the On-Road Vehicle and Engine Emission Regulations;

[23] http://canadagazette.gc.ca/archives/p1/2003/2003-03-29/html/reg1-eng.html.

[24] http://www.dieselnet.com.

Table 10.11 Off-road Diesel Engine Emissions in Canada in 2000. (SOR/2005-32 Off-Road Compression-Ignition Engine Emission Regulations—REGULATORY IMPACT ANALYSIS STATEMENT)

	Emissions (kilo-tonnes)	Percentage contribution to national inventory	Percentage contribution to Off-road sources of emissions
VOC[a]	43.4	1.7%	14.4%
NO_x	356	13.3%	87.1%
CO	207	2.5%	8.7%
PM_{10}	38.9	5.9%	84.5%
SO_x	15.3	0.6%	92.5%

[a] Non-methane hydrocarbon (NMHC) emissions have a mass of 41.2 kt

Table 10.12 Emission Standards Under the Off-Road Diesel Engine Regulations. (SOR/2005-32 Off-Road Compression-Ignition Engine Emission Regulations)

Engine power	Tier	Effective date (model year)	NMHC + NO_x (g/kWh)	CO (g/kWh)	PM (g/kWh)
kW<8	Tier 2	2006 and later	7.5	8.0	0.80
8≤kW<19	Tier 2	2006 and later	7.5	6.6	0.80
19≤kW<37	Tier 2	2006 and later	7.5	5.5	0.60
37≤kW<75	Tier 2	2006, 2007	7.5	5.0	0.40
	Tier 3	2008 and later	4.7	5.0	0.40
75≤kW<130	Tier 2	2006	6.6	5.0	0.30
	Tier 3	2007 and later	4.0	5.0	0.30
130≤kW<225	Tier 3	2006 and later	4.0	5.0	0.20
225≤kW<450	Tier 3	2006 and later	4.0	3.5	0.20
450≤kW≤560	Tier 3	2006 and later	4.0	3.5	0.20
kW>560	Tier 2	2006 and later	6.4	3.5	0.20

- designed to be used exclusively in underground mines;
- with a per-cylinder displacement of less than 50 cm^3;
- for military machines used in combat or combat support;
- being exported and not sold or used in Canada;
- designed to be used in a vessel and for which the fuel, cooling and exhaust systems are integral parts of the vessel.
- While not specifically exempted by the regulations, Environment Canada does not have legislative authority to regulate emissions from railway locomotive engines.

10.3.3 Regulations Amending the Off-Road Compression-Ignition Engine Emission Regulations

In 2011, Environment Canada proposed an amendment to the off-road compression- ignition engine emission regulations to further reduce emissions from this category by establishing more stringent Canadian off-road diesel emission standards and test procedures for those engines produced in 2012 and later. These new standards align the Canadian emission standards with those of the Tier 4 emission standards implemented by the U.S. EPA. Specifically the amendments apply to engines used in machines found in construction, farming, forestry, and some mining machines.

The proposed amendments would incorporate the U.S. EPA's steady state and transient emission standards for exhaust, crankcase, and evaporative emissions for 2012 and later model years. The allowable emission levels from individual engines would be significantly reduced from current standards, including reductions of 37% of non-methane hydrocarbons and 50–95% reductions of PM.

The U.S. EPA Tier 4 emission standards—to be phased-in from 2008 to 2015—introduce substantial reductions of NO_x (for engines above 56 kW) and PM (above 19 kW), as well as more stringent HC limits. CO emission limits remain unchanged from the Tier 2–3 stage Table 10.13.

10.3.4 Marine Spark-Ignition Engine and Off-Road Recreational Vehicle Emission Regulations

The U.S. introduced emission standards for marine outboard and personal watercraft engines in 1998. These standards required manufacturers to meet increasingly stringent HC levels over a nine-year phase-in period. Historically the marine outboard market was dominated by 2-stroke engines, however since the introduction of these emission standards there has been a growth 4-stroke engines which have much lower emission and as well there are an increasing number of engines that are now utilizing catalytic converters to achieve these emission levels.

The recreational vehicle category includes off-highway motorcycles, all-terrain vehicles, and snowmobiles that operate on gasoline. In 2002 the U.S. EPA adopted emission standards for new recreational vehicles that were phased in starting in 2006. These standards reduced HC emissions from these vehicles by 67% and reduced CO emissions by 28%[25].

In Canada, Environment Canada has decided to combine engine emission standard regulations for both the marine and recreational vehicles[26] into a single regulatory package. The approach will be modeled on current regulations outlined under Division 5 of CEPA 1999, and continue the policy of harmonization with the U.S. standards. These standards will be effective for engines and vehicles in the 2012 model year and beyond.

There is only one domestic manufacturer in this sector, with the majority of marine engines and recreational vehicles being imported. Statistics Canada data show that importers accounted for 98% of the outboard engines, personal watercraft, snowmobiles, and all-terrain vehicles imported in 2003.

[25] EPA420-F-03-011 Program Update 2003.

[26] http://www.ec.gc.ca/lcpe-cepa/default.asp?lang=Enxml=82A7A188-8719-7589-5755-14C854A97DCE.

Table 10.13 Amended Regulations for Off-Road Compression-Ignition Engine Emissions (Tier 4). (SOR/2005-32 Off-Road Compression-Ignition Engine Emission Regulations)

Engine power	Tier	Effective date (model year)	NMHC + NO_x (g/kWh)	NMHC (g/kWh)	NO_x (g/kWh)	CO (g/kWh)	PM (g/kWh)
kW<8	Tier 4	2008	7.5			8.0	0.4
8≤kW<19	Tier 4	2008	7.5			6.6	0.4
19≤kW<37	Tier 4	2008	7.5			5.5	0.3
19≤kW<37	Tier 4	2013	4.7			5.5	0.03
37≤kW<56	Tier 4	2008	4.7			5.0	0.3
37≤kW<56	Tier 4	2013	4.7			5.0	0.03
56≤kW<130	Tier 4	2012–2014		0.19	0.40	5.0	0.02
130≤kW<560	Tier 4	2011–2014		0.19	0.40	3.5	0.02

Table 10.14 U.S. EPA Regulations for Marine Spark-Ignition Engines

Model year	P<4.3 kW HC + NO_x (g/kW-hr)	P>4.3 kW HC+NO_x (g/kW-hr)
1998	278.00	$(0.917\times(151+557/P^{0.9}))+2.44$
1999	253.00	$(0.833\times(151+557/P^{0.9}))+2.89$
2000	228.00	$(0.750\times(151+557/P^{0.9}))+3.33$
2001	204.00	$(0.667\times(151+557/P^{0.9}))+3.78$
2002	179.00	$(0.583\times(151+557/P^{0.9}))+4.22$
2003	155.00	$(0.500\times(151+557/P^{0.9}))+4.67$
2004	130.00	$(0.417\times(151+557/P^{0.9}))+5.11$
2005	105.00	$(0.333\times(151+557/P^{0.9}))+5.56$
2006 and later	81.00	$(0.250\times(151+557/P\ 0.9))+6.00$

P=the average power of an engine family in kW (sales weighted). The power of each configuration is the rated output in kilowatts as determined by SAE J1228

Table 10.15 U.S. EPA Regulations for Snowmobiles

Model year	Phase-in %	HC g/kW-hr	HC + NO_x g/kW-hr	CO g/kW-hr
2006	50	100		275
2007–2009	100	100		275
2010 & 2011	100	75		275
2012 and later	100	75	(1)	(1)

Table 10.16 U.S. EPA Regulations for Off-Highway Motorcycles[33]

Model year	Phase-in %	HC + NO_x g/km	CO g/km
2006	50	2.0	25
2007 and later	100	2.0	25

Table 10.17 U.S. EPA Regulations for All Terrain Vehicles[33]

Model year	Phase-in %	HC + NO_x g/km	CO g/km
2006	50	1.5	35
2007 and later	100	1.5	35

The following Tables outline the U.S. EPA exhaust emission standards for marine spark-ignition engines and those of the off-road recreational vehicles. The Canadian standards will align with the lowest applicable standard for each category and will include averaging provisions which are not included on the following Tables (Table 10.14, 10.15, 10.16 and 10.17[27, 28]).

10.3.5 Evaporative Emissions

There are evaporative emission standards for small spark ignited gasoline engines, recreational vehicles, as well as engines used in marine vessel applications. These standards are expected to reduce the venting of evaporated gasoline from fuel tanks and reduce the permeation of fuel through the walls of fuel tanks and hoses. For recreational vehicles and engines the U.S. EPA stipulates that new vehicles must meet these evaporative emission standards for their useful life. Commencing with the 2008 model year permeation emissions from the fuel tank may not exceed 1.5 g/square-meter/day when measured with the test procedures for tank permeation outlined in the Code of Federal Regulations. As well, the permeation emissions from the fuel lines may not exceed 15 g/square-meter/day. Full details on the evaporative emission standards as well as the respective section on the Code of Federal Regulations are available on the U.S. EPA web site[29].

The emission regulations planned for Canada will align with those of the US for marine spark-ignition engines and off-road recreational vehicles [30]. For small spark ignited engines will rely on the likelihood that those engines will meet evaporative standards imposed by the U.S.

10.3.6 Off-Road Emission Inventories

The realization of the potential scale of criteria air contaminant emissions from the off-road sector led the U.S. EPA

[27] U.S. 40 CFR Ch.I Part 91_03.

[28] U.S. 40 CFR Ch.I Part 1051.103(a)(2) (7–1–03 Edition).

[29] http://www.epa.gov/oms/standards/nonroad/nonroadsi-evap.htm.

[30] http://www.ec.gc.ca/lcpe-cepa/default.asp?lang=En&n=15195F9A-1&offset=4#s4_2.

to develop the NONROAD[31] model to estimate emissions for six exhaust pollutants: hydrocarbons (HC), NO_X, carbon monoxide (CO), carbon dioxide (CO_2), sulfur oxides (SO_X), and PM. The model predicts emissions for all nonroad equipment listed earlier with the exception of commercial marine, locomotive, and aircraft emissions. The model includes more than 80 basic and 260 specific types of nonroad equipment, and further stratifies equipment types by horsepower rating. Fuel types include gasoline, diesel, compressed natural gas (CNG), and liquefied petroleum gas (LPG). The model estimates emissions for each specific type of nonroad equipment by multiplying the following input data estimates:

- Equipment population for base year (or base year population grown to a future year), distributed by age, power, fuel type, and application;
- Average load factor expressed as average fraction of available power;
- Available power in horsepower;
- Activity in hours of use per year; and
- Emission factor with deterioration and/or new standards.

One of the utilities of the model for estimating off-road emissions in the United States is the database of nonroad equipment population characteristics by county and state which are part of the model. Thus in using the model, national total emissions, emissions by state, or for one or more counties within a state can be estimated.

The emission factors in the NONROAD model have also been used for estimating off-road emission inventories in Canada[32] although the absence of a Canadian database of engine populations and activity levels comparable to the U.S. database brings higher uncertainty when estimating national level inventories.

10.4 Marine Vessel Emissions

A wide variety of marine vessels ply Canadian waters, from transoceanic ships over 75 m long, to small recreational sailboats. The marine vessels that emit the most air pollutants relevant to other emissions sources are transoceanic cargo and passenger ("cruise") ships. Transoceanic ships emit large quantities of nitrogen oxides (NO_x), sulphur oxides (SO_x) and fine particulate matter ($PM_{2.5}$), for three main reasons: (1) they have very large engines (megawatts to tens of megawatts rated power), (2) they have no air pollution control equipment, unlike most land-based vehicles, and (3) most burn so-called "heavy fuels," which contain substantially more sulphur and other impurities than fuels used by most land-based vehicles (up to 3,000 times more in the case of sulphur). Some bulk cargo "lakers" serving the Great Lakes share these features, although they have smaller engines. Other large non-transoceanic commercial marine vessels, such as ferries, tugboats, towboats and fishing vessels, can also emit significant quantities of nitrogen oxides and particulate matter, for the same reasons as transoceanic ships. Because most of these vessels burn distillate "diesel" fuels containing much less sulphur than heavy fuels, they emit much smaller amounts of sulphur oxides than transoceanic ships emit. Recreational vessels are more likely to be powered by small spark-ignition engines burning gasoline. As a result, they emit more pollutants associated with incomplete gasoline combustion, such as carbon monoxide and hydrocarbons (volatile organic compounds).

Table 10.18 Emissions from Commercial Marine Vessels in Canada. (SENES Consultants Limited 2008 and Metro Vancouver 2007)

Geography	Atlantic coast, Great lakes and St. Lawrence river	Pacific coast
Vessel classes included	Commercial marine vessels	Oceangoing ships
Time period (12 months)	2002/2003	2005/2006
Units	tonnes per year	tonnes per year
Emissions		
Nitrogen oxides (NO_x)	73,974	26,500
Sulphur oxides (SO_x)	54,959	18,413
Fine particulate matter ($PM_{2.5}$)	6,296	1,438
Carbon monoxide (CO)	6,165	2,236
Hydrocarbons (HC)	2,520	934
Carbon dioxide (CO_2)	3,620,092	1,278,084

Includes ships carrying containers, dry bulk, liquid bulk, motor vehicles, general cargo, and passengers (cruise), as well as coastal ferries, warships and ocean tugs
ships carrying containers, dry bulk, liquid bulk, motor vehicles, general cargo, and passengers (cruise)

10.4.1 Marine Emissions Inventory

Table 10.18 shows the results of the most recent detailed inventories of emissions from commercial marine vessels in Canada.

Table 10.19[33] shows how total marine transportation emissions in Canada (based on the original inventories in Table 10.18 but extrapolated to 2009) compared to emissions from other sources. Marine transportation makes a substantial contribution to total national emissions of nitrogen oxides and sulphur oxides, despite being heavily present only along the Pacific and Atlantic coasts and in the Great Lakes/

[31] http://www.epa.gov/otaq/nonrdmdl.htm#docs.

[32] See for example, Levelton 2004, RWDI 2007.

[33] Does not include recreational marine vessels.

Table 10.19 Emissions from Marine Transportation vs Other Sources in Canada, 2009. (Environment Canada 2011b)

	NO_X	SO_X	$PM_{2.5}$	CO	VOC
	All values in tonnes per year				
Marine transportation	119,368	82,766	9,773	9,982	3,941
Other mobile sources	1,012,711	12,589	51,378	6,595,717	505,634
Industrial, power generation, space heating & incineration	891,778	1,383,930	186,444	2,194,839	801,770
Open, natural & miscellaneous sources	151,492	1,262	932,658	896,077	23,852,786
Total all sources	2,175,349	1,480,547	1,180,253	9,696,616	25,164,130
Marine transportation per cent of total	5%	6%	0.8%	0.1%	0.02%

Table 10.20 Emissions from Marine Vessels Compared to Emissions from Other Sources in British Columbia's Lower Fraser Valley, 2005. (Metro Vancouver 2007 and Metro Vancouver 2010)

	NO_X	SO_X	$PM_{2.5}$	CO	VOC	CO_2
	All values in tonnes per year					
Oceangoing vessels	3,825	3,545	311	418	156	169,559
Harbour craft	2,159	161	93	253	71	116,304
Ferries	1,330	122	25	233	50	69,302
Fishing vessels	20	1	0	3	1	1,107
Recreational vessels	513	9	49	9,041	3,033	71,076
Total marine vessels	7,847	3,838	479	9,947	3,312	427,348
Other mobile sources	31,497	570	1,111	346,393	24,880	7,449,292
Point sources	5,208	957	1,167	4,281	3,942	3,323,010
Area sources	6,290	179	2,842	14,687	41,785	5,250,837
Total all sources	50,842	5,544	5,599	375,309	73,918	16,450,488
Marine vessels per cent of total	15%	69%	9%	3%	4%	3%

This number includes only emissions occurring during voyages between Canadian ports, as per International Panel on Climate Change reporting rules. The total estimated CO_2 emissions from all oceangoing activities were 248,919 t

St. Lawrence. Table 10.20[34] shows how emissions from the different types of marine vessels compare to each other, and how they compare to emissions from other sources, in British Columbia's Lower Fraser Valley, an area which includes metropolitan Vancouver and Canada's largest port. Levels of activity of the different emissions sources vary by region. For example the Atlantic coast has greater fishing activity, while the Great Lakes and St. Lawrence region have greater industrial activity. In small coastal communities, marine vessel emissions may dominate local emissions totals.

10.4.2 Emissions Estimation Methods

Marine vessel emissions are estimated using coarser methods than for land-based emissions sources. Generally speaking, the practice is to obtain vessel traffic or call information and use that to estimate the time each vessel spent within the study area in various activity modes (e.g. underway, at dock). This information is combined with vessels' propulsion and auxiliary engines' rated power and estimated load factors for each activity mode, and then multiplied by accepted "power-based emission factors" expressed in terms of grams of pollutant emitted per unit of energy output (kilowatt-hour). For transoceanic ships, sometimes a particulate matter emission factor that varies with fuel sulphur content is used. In some cases "fuel-based" emission factors (grams emitted per tonne of fuel consumed) are used, for example for vessel boilers. The accepted emission factors are based on a comparatively small dataset of a few dozen actual marine vessel engine emissions tests, as a result of the challenges associated with testing emissions from large engines that can usually only be found in the confined spaces of an actual marine vessel, operating in the uncontrolled (sometimes pitching) at-sea environment (SENES Consultants Limited and Air Improvement Resource, Inc. 2004).

10.4.3 Effects on Ambient Air Quality

Emissions from marine vessels can have substantial effects on the air quality in port and coastal communities along Canada's Pacific and Atlantic coasts, and to a lesser extent in the Great Lakes. Figure 10.11 shows the percent contribution to 2020 ambient fine particle concentrations that Environment Canada projected would result from emissions from transoceanic ships, prior to the recent adoption of more stringent

[34] In some cases the values in this table were calculated by addition or subtraction of values in these two reports.

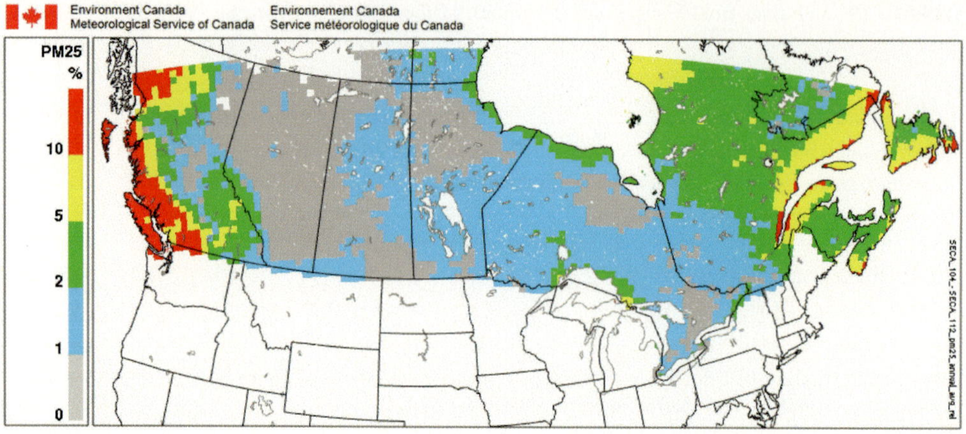

Fig. 10.11 Projection of Ships' Contribution to 2020 Ambient Fine Particle Concentrations in Canada -Absent 2008 Annex VI Revisions and 2010 North American ECA (United States and Canada 2009). (Reproduced with the permission of Environment Canada, November 23, 2012)

emissions requirements. The figure was generated by comparing national-scale photochemical grid modeling of ambient air quality with and without including transoceanic ship emissions as model input.

Particulate matter from diesel cycle combustion sources, such as marine engines, contributes the bulk of lifetime cancer risk attributable to outdoor toxic air pollutants in urban areas, for example in British Columbia's Lower Fraser Valley (Levelton Consultants Ltd. 2007).

Over the next several years a comprehensive suite of regulatory requirements is expected to dramatically reduce emissions and air quality impacts from commercial marine vessels.

10.4.4 Regulation of Emissions from International Ships

Ships that travel between nations are potentially subject to the laws of several nations. International law ensures global consistency. Air pollution from international ships is governed by Annex VI to the *International Convention for the Prevention of Pollution from Ships* (the MARPOL Convention). Annex VI is titled *Regulations for the Prevention of Air Pollution from Ships*. Almost every coastal nation in the world is bound by Annex VI by virtue of being a member of the International Maritime Organization, a United Nations agency. Nations must incorporate the requirements of Annex VI into their national laws governing shipping. Canada is represented in international deliberations related to Annex VI by Transport Canada, with participation from Environment Canada and other stakeholders.

In 2008, International Maritime Organization members made major revisions to Annex VI to substantially strengthen nitrogen oxides, sulphur oxides and particulate matter emissions limits. In 2011 they again revised Annex VI to add energy efficiency standards. The key elements of Annex VI are:

- **Emission Control Areas (ECAs)**. Annex VI restricts emissions of nitrogen oxides, sulphur oxides and particulate matter from international ships within two types of geographic area. One set of standards applies generally throughout the world, and a more stringent set of standards applies within so-called "Emission Control Areas" (ECAs). Effective August 2012, waters along Canada's Pacific and Atlantic coasts (south of 60 degrees North latitude), out to a distance of approximately 200 nautical miles, will be part of the North American ECA shown in Fig. 10.12.
- **Fuel sulphur content limits (or equivalent)**. Since sulphur oxides emissions are a direct result of sulphur in fuel, Annex VI limits the sulphur content of fuels used by international ships, although it does allow the use of alternate methods that achieve equivalent sulphur oxides emissions, for example pollution control aftertreatment. Until August 2012, the sulphur content of fuels used by ships in waters under Canadian jurisdiction is limited to 3.5 per cent. At that time, the North American ECA becomes enforceable, and sulphur content will be limited to 1.0 per cent. In January 2015, sulphur content will be further limited to 0.1 per cent[35]. This last limit will effectively require the use of either distillate "diesel" fuels or aftertreatment (a "scrubber"). Annex VI assumes that particulate matter emissions are also limited by fuel sulphur limits (or any alternate methods). There are no separate particulate matter limits. Note that the primary reason for limiting sulphur in transoceanic ship fuels is to directly reduce the otherwise very large emissions of sulphur oxides and sulphates. Enabling the use of aftertreatment technology is not a primary motivation, unlike the situation for land-based vehicles and smaller marine vessels.
- **Nitrogen oxides limits for new or converted engines**. Annex VI limits emissions of nitrogen oxides from a ship's diesel engines, based on the date of construction of the ship, or the date an engine underwent a major conversion. The limit for an engine is expressed as mass of NO_2

[35] The few remaining steamships operating in the North American ECA are exempt from the fuel sulphur content requirements.

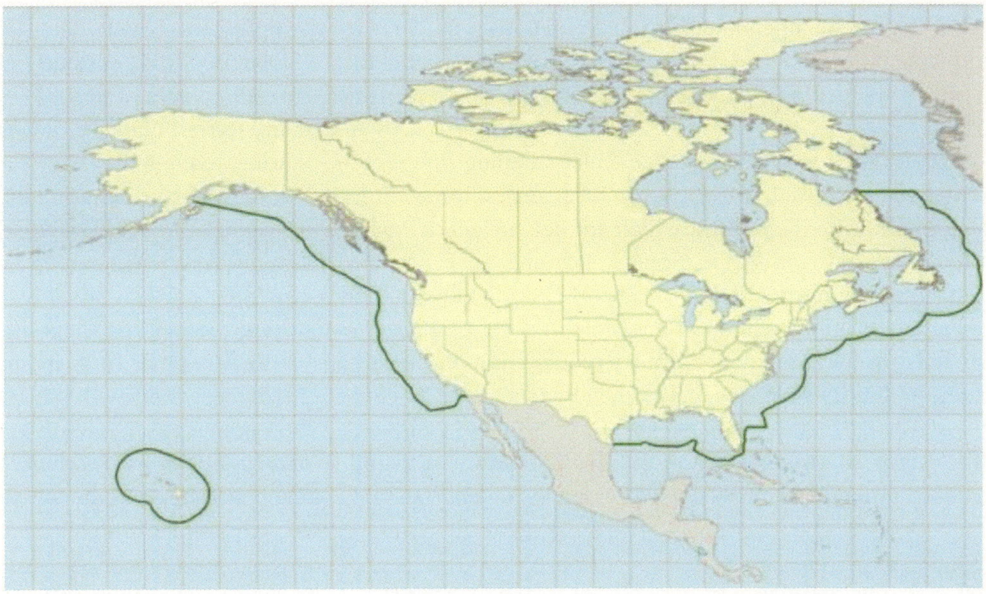

Fig. 10.12 The North American Emission Control Area (ECA). (Reproduced with the permission of Environment Canada, November 23, 2012)

emitted per unit energy produced by the engine, and is a function of that engine's rated engine speed. So-called "Tier I" limits applied to ships built from 2000 through 2010. Tier II limits are about 15 per cent more stringent than Tier I, and apply to ships built in 2011 or later. Tier III limits will apply only to operation within Emission Control Areas, to ships built in 2016[36] or later. Tier III limits are about 80 per cent more stringent than Tier I. Tier I limits also apply to some existing engines on ships built from 1990–1999, where a certified upgrade method is cost-effective.

- **Energy efficiency (carbon dioxide) standards.** Annex VI now requires that most cargo ships built in 2015 or later meet energy efficiency standards. A ship's "Energy Efficiency Design Index" will be expressed in terms of grams of carbon dioxide emitted per tonne of cargo moved a nautical mile, and calculated using several additional factors that account for different classes of vessels designed for different types of service. For large[37] ships built in 2015–2019, the standard will be a 10 % improvement compared to a baseline based on similar ships built between 1999 and 2009. The standard will tighten to a 20 % improvement for most large ships built in 2020–2024, and a 30 % improvement for large ships built after that. Member states may delay implementation by up to four years without penalty, so the schedule for ships actually achieving the standards may lag these official dates. Annex VI also now requires that all operating ships have a "Ship Energy Efficiency Management Plan," a management mechanism that does not prescribe performance in itself but may lead operators to discover fuel savings and consequent carbon dioxide emissions reductions.
- **Shipboard incineration conditions**. Annex VI sets conditions for shipboard incineration, including prohibiting incineration of certain hazardous substances. It also sets requirements for shipboard incinerators.
- **Volatile organic compounds collection**. Annex VI sets conditions for vapour collection systems to control volatile organic compound emissions from tanker cargoes.
- **Ozone depleting substances prohibitions**. Annex VI prohibits deliberate emissions of ozone depleting substances. It prohibits installations containing these substances (except hydrochlorofluorocarbons) on ships built since May 2005, and it will prohibit installations containing hydrochlorofluorocarbons on ships built in 2020 or later.
- **Record-keeping, ship surveys, certification and administration.**

As mentioned, Canada must incorporate the requirements of Annex VI into its laws governing shipping. The regulation of marine vessels in Canadian waters, and Canadian-registered marine vessels worldwide, is federal jurisdiction. The federal Minister of Transport is authorized, by the *Canada Shipping Act* 2001, to regulate "the control and prevention of air pollution by marine vessels". The *Regulations for the Prevention of Pollution from Ships and for Dangerous Chemicals* (SOR/2007-86) set out the requirements.

These *Regulations* were amended in 2007 to closely follow the requirements of Annex VI as it then stood, prior to the 2008 amendments to Annex VI. Transport Canada ex-

[36] This timetable will be reviewed by the end of 2013 to ensure it is technically feasible.

[37] There are different size thresholds for different vessel types. A ship smaller than the relevant threshold must still meet a target, but the target is less stringent.

pects to propose amendments to the *Regulations* in 2012, to implement the 2008 and 2011 revisions to Annex VI, and the North American ECA adopted in 2010 (Transport Canada 2011).

Great Lakes and St. Lawrence River shipping line representatives have said that applying ECA fuel sulphur limits and timelines to their ships, which are not subject to Annex VI if they stay within the Great Lakes and St. Lawrence, would be counterproductive. Transport Canada may include an alternate implementation schedule for these ships, intended to achieve greater emissions reductions over the long term, within its proposed 2012 amendments to the *Regulations*. The current *Regulations* also restrict smoke emissions from ships according to a "smoke chart" measure of visible smoke, although smoke is not addressed by Annex VI.

10.4.5 Regulation of Emissions from Engines Aboard Domestic Commercial Marine Vessels

The United States limits emissions from new and remanufactured marine engines, rated over 600 kW but with cylinders smaller than 30 l each, installed aboard U.S. registered vessels including domestic vessels not subject to Annex VI. Nitrogen oxides, particulate matter, carbon monoxide and hydrocarbons are limited. The limits on nitrogen oxides and particulate matter were strengthened in 2008, by adding "Tier 3" limits beginning in 2009 and "Tier 4" limits based on high-efficiency catalytic aftertreatment technology beginning in 2014. The 2008 changes also established emissions standards for remanufactured engines in addition to new engines (United States Environmental Protection Agency 2008). Emissions from similar engines installed aboard Canadian domestic vessels are not currently regulated, although in some cases these engines might meet the U.S. limits anyway, for market convenience. Transport Canada will likely include limits matching the U.S. limits within its 2012 proposed amendments to the *Regulations* (Transport Canada 2011). The energy efficiency and carbon dioxide emissions of domestic marine vessels are not regulated. Transport Canada's Marine Safety inspectors verify that international and domestic marine vessels meet a wide range of requirements under the *Canada Shipping Act* 2001, including emissions-related requirements. They will likely continue that role for new requirements as the *Regulations* are amended.

10.4.6 Regulation of Marine Fuel Manufacture, Import and Sale

The *Canadian Environmental Protection Act* 1999 authorizes the federal Minister of the Environment to regulate the level of sulphur in marine fuels manufactured or sold in, or imported to, Canada. The *Sulphur in Diesel Fuel Regulations* (SOR/2002-254) currently use this authority to regulate the sulphur content of marine *diesel* fuels (such as those used by many ferries, tugboats, towboats and fishing vessels), but not marine *heavy* fuels (which are mostly used by transoceanic ships). The concentration of sulphur in *diesel* fuel produced or imported for use in vessel engines is limited to 500 mg/kg until May 2012 and 15 mg/kg after May 2012. These requirements are enforced by Environment Canada. Once the sulphur content of fuels used by transoceanic ships in the North American ECA is limited to 0.1 per cent sulphur in 2015, these fuels will also qualify as marine diesel fuels, and their manufacture, import and sale will be subject to the *Sulphur in Diesel Fuel Regulations*. In 2011 Environment Canada proposed amendments to these Regulations to allow the concentration of sulphur in diesel fuel produced, imported or sold for use in the large marine vessel engines normally used in transoceanic ships to be as high as 1,000 mg/kg (which is 0.1 per cent), effective in June 2012.

10.4.7 Regulation of Small Recreational Boats and Their Gasoline Fuel

Under the *Canadian Environmental Protection Act* 1999, the Minister of the Environment regulates the level of air pollutants emitted by marine engines with brake horsepower under 37 kW, which are manufactured or sold in, or imported to, Canada. These types of engines are found on small recreational boats, discussed in more detail in Section 0. Gasoline is typically used in these engines. It must meet the same requirements as gasoline used for most other land-based purposes, such as in motor vehicles.

10.4.8 Alignment with the United States

As with other vehicle, engine and fuel requirements, most Canadian emissions-related requirements for marine vessels, engines and fuels are—or will soon be—aligned with those of the United States, even beyond the alignment that derives from both nations being subject to Annex VI.

10.5 Rail Emissions

10.5.1 Introduction

The powering of freight and passenger trains in Canada is, except for one electrified commuter line serving Montreal, exclusively by diesel engines. Despite the fact no regulations are in force in Canada concerning railway-generated diesel emissions, voluntary arrangements have existed for the railway sector to steadily reduce diesel emissions. A

Memorandum of Understanding (MOU) was signed in 1995 and renewed in 2007 between the Railway Association of Canada, Environment Canada and Transport Canada concerning emissions reduction targets for greenhouse gases (GHG) and criteria air contaminants (CAC) from locomotives operating in Canada. Progress to achieve the targets is reported in annual *Locomotive Emissions Monitoring* (LEM) reports which are available on the publications websites of the Railway Association of Canada, Environment Canada and Transport Canada. Regulations to control CAC emissions from diesel locomotives are being developed by Transport Canada under the *Railway Safety Act*. When they will come into force, the Canadian Locomotive Emissions Regulations are expected to be aligned with those of the U.S. Environmental Protection Agency (EPA) currently in place in the U.S.A.

Table 10.21 Railway Activity and Emissions at a Glance, 2009

48,000 km of track, transporting ▶	307.9 billion revenue tonne-kilometres (RTK) of freight (93.8% by the Class 1 railways, CNR and CPR)
	65.96 million commuter passengers
	4.54 million intercity passengers
	323 thousand excursion passengers
2,727 in-service diesel locomotives consuming 1.87 billion l of diesel fuel and producing ▶	5,627 kt of $CO_{2\ equivalent}$
	96.41 kt of NO_x
	13.22 kt of CO
	4.68 kt of HC
	2.46 kt of PM
	0.34 kt of SO_2 (based on a diesel fuel sulphur content of 110 parts per million (ppm)

10.5.2 Emissions from the Canadian Railway System

The parameters characterizing the railway sector's activity in Canada in 2009 and its cumulative constituent emissions are displayed in Table 10.21 below.

The data in Table 10.21 above and for all tables and figures throughout Section 10.5 were reproduced with permission from the Railway Association of Canada as extracted from its publications, *Locomotive Emissions Monitoring Program 2009* and *Rail Trends 2010*.

The cumulative emissions reported in the annual LEM reports are calculated from data collected from the Railway Association of Canada (RAC) member railways. The data include traffic volumes, diesel fuel consumption and locomotive fleet inventories for freight train, yard switching, work train and passenger operations. Freight data are differentiated between Class 1[38], Regional and Short Line operations. Passenger data are differentiated between Intercity, Commuter, and Tourist and Excursion operations. The GHG emissions of concern are CO_2, CH_4 and N_2O which are aggregated as $CO_{2\ equivalent}$. The CAC emissions of concern include NO_x (expressed as NO_2), CO, HC, PM and SOx (expressed as SO_2)

[38] Class 1 Railway: This is a class of railway within the legislative authority of the Parliament of Canada that realized gross revenues that exceed a threshold indexed to a base of $250 million annually in 1991 dollars for the provision of Canadian railway services. The two freight railways in Canada in this category are Canadian National and Canadian Pacific. A similar nomenclature exists for Class 1 in the U.S.A. There are seven Class 1 freight railways registered in the U.S.A. (two of which are Canadian National and Canadian Pacific because of their extensive operations in the U.S.A.

10.5.3 Railway GHG Emissions Reduction Record

Compared to the target levels set out in the MOU for 2010, analysis of railway data contained in the RAC 2009 *Locomotive Emissions Monitoring* report shows that aggregate GHG emissions intensity levels (expressed as $CO_{2\ equivalent}$ per productivity unit) by category of railway line-haul operation for 2006–2009 were as shown in Table 10.22. The productivity unit metric for freight railways is Revenue Tonne Kilometres (RTK).

The GHG emissions reported are calculated according to the amount of diesel fuel consumed using the emission factor value for $CO_{2\ equivalent}$ of 3.00715 kg/l. In meeting the GHG commitments under the MOU, the Canadian railways have focused primarily on acquiring new more fuel-efficient high-horsepower line-haul locomotives. These new units are also contributing to reducing CAC emissions as, being produced by American original equipment manufacturers (OEMs), they are designed to meet the latest U.S. EPA regulations for NO_x, CO, HC and PM emissions. A broad range of fuel conservation tactics are also being introduced such as anti-idling protocols, switching locomotives with multiple-engines that can be selectively shut off, lighter-weight railcars and a variety of new rolling stock equipment designs, train handling techniques and infrastructure improvements that increase operational fluidity, all resulting in reduced fuel consumption and, hence, reduced emissions. Government funding support has facilitated the take-up of some of the technological advancements.

10.5.4 Railway CAC Emissions Reduction Record

CAC emissions, namely NO_x, CO, HC and PM, are calculated based on the amount of diesel fuel consumed, the

Table 10.22 Railway GHG Emissions Intensity and Reduction

Railway operation	GHG emissions intensity	2006	2007	2008	2009	2010 (target)
Class 1 Freight	kg CO_{2eq}/1000 RTK	17.40	17.32	17.61	16.94	16.98
Regional and short lines	kg CO_{2eq}/1000 RTK	14.77	15.22	15.80	14.20	15.38
Intercity passenger	kg CO_{2eq}/passenger-km	0.13	0.13	0.12	0.13	0.12
Commuter rail	kg CO_{2eq}/passenger	1.70	1.71	1.74	1.95	1.46

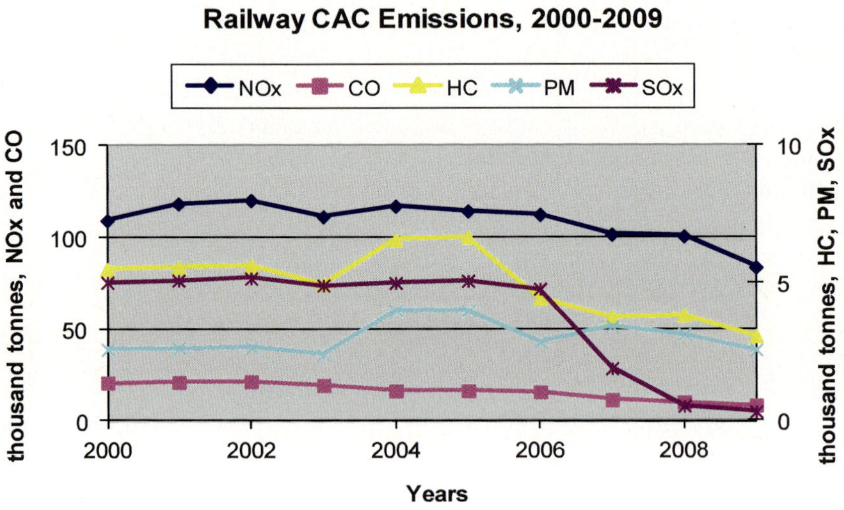

Fig. 10.13 CAC Emissions from Total Railway Operations. Most CAC emissions have shown a gradual decrease over the past decade

respective CAC emission factor and duty cycles that reflect a locomotive's operational service. Prior to the U.S. EPA setting locomotive-specific not-to-exceed limits, CAC emissions factors in grams per brake-horsepower hour (g/bhp-hr) were obtained from test cell measurements for each CAC emission at each locomotive power notch setting. Duty cycles are determined by evaluating the time spent at each power notch level for a statistically significant sample of locomotives. SOx emitted varies mostly according to the sulphur content of the diesel fuel that, effective June 2012, is limited in Canada to 15 ppm in compliance with the Environment Canada *Sulphur in Diesel Fuel* regulations. Emissions metrics are expressed in terms of absolute mass as well as intensity, that is, a ratio relating emissions to productivity. The CAC emissions from all railway operations between 2000 and 2009 are shown in Fig. 10.13. The NO_x reduction trend for total freight operations between 2000 and 2009 are shown in Fig. 10.14. More details can be viewed on the Railway Association of Canada website[39].

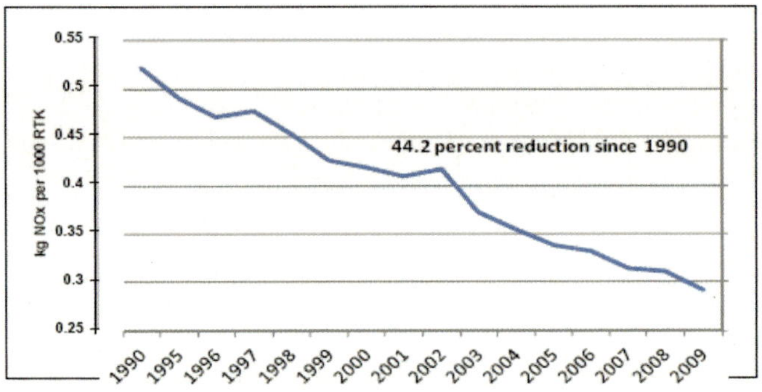

Fig. 10.14 Railway Freight NO_x Emission Intensity Trend. Railway Freight NOx Emissions Intensity (kg of NO_x per 1000 revenue tonne-kilometres (RTK)) has decreased ~44.2 % since 1990

10.5.5 Outlook for Further Reductions of Locomotive Emissions

It is expected that Canadian railway emissions intensity reductions will continue with the systematic purchase of new locomotives meeting the U.S. EPA emissions standards corresponding to the year of manufacture, the introduction of regulations in Canada, remanufacture with new low-emissions engines, installation of emissions reduction kits in existing locomotives and further operational and infrastructure improvements that reduce emissions.

10.5.6 Developing Locomotive Emissions Regulations in Canada

Transport Canada is developing new regulations regarding diesel locomotive emissions under the *Railway Safety Act* in two phases. Firstly, regulations aligned with those of the U.S. EPA are being developed to limit the release of CAC emis-

[39] www.railcan.ca/publications/trends.

sions from the railway sector; and secondly, regulations to limit the release of GHG emissions will be developed in step with the U.S.A. More information pertaining to the proposed regulations is contained on the Transport Canada website[40].

10.5.7 Locomotive Emissions Regulations in the United States

The U.S. EPA rulemaking promulgated in 1998 contains three levels of locomotive-specific emissions limits corresponding to the date of a locomotive's original manufacture—Tier 0, Tier 1 and Tier 2 (as listed in Table 10.23). The significance of the U.S. EPA regulations for Canadian railways is that the new locomotives they traditionally acquire from American OEMs are manufactured to meet the latest EPA emissions limits. Hence, emissions in Canada are reduced as these new locomotives are acquired. Information about the EPA compliance limits can be viewed on a U.S. EPA website[41].

Referencing the above-listed limits for locomotives operating in the U.S.A., the EPA in 2008 put into force revisions, which tighten the existing Tier 0 to Tier 2 standards. The revisions are now referred to as Tier 0+, Tier 1+ and Tier 2+ standards. As indicated in Table 10.24 and Table 10.25, the revised standards take into account the year of original manufacture of the locomotive. Also, two new, more stringent standards levels were introduced, designated Tier 3 and Tier 4. The revised and new standards are to be phased-in between 2010 and 2015 for locomotives as they become new (new in this case includes both when locomotives are originally manufactured and when remanufactured). It is envisaged that to meet the Tier 4 standards, locomotives manufactured starting in 2015 will require additional exhaust gas treatment technologies to be installed and be dependent upon diesel fuel having a sulphur content capped at 15 ppm.

Table 10.23 Compliance Schedule for U.S. EPA Locomotive-Specific Emissions Limits (g/bhp-hr)

Duty cycle	HC	CO	NO_x	PM
Tier 0 (1973–2001)				
Line-haul	1.0	5.0	9.5	0.60
Switching	2.1	8.0	14.0	0.72
Tier 1 (2002–2004)				
Line-haul	0.55	2.2	7.4	0.45
Switching	1.2	2.5	11.0	0.54
Tier 2 (2005 and later)				
Line-haul	0.3	1.5	5.5	0.20
Switching	0.6	2.4	8.1	0.24
Estimated Pre-Regulation (1997) Locomotive Emissions Rates				
Line-haul	0.5	1.5	13.5	0.34
Switching	1.1	2.4	19.8	0.41

Table 10.24 Line-Haul Locomotive Emission Standards g/bhp·hr

Tier	[a]MY	Date	HC	CO	NO_x	PM
Tier 0+[b]	1973–1992	2010[d]	1.00	5.0	8.0	0.22
Tier 1+[b]	1993–2004[c]	2010[d]	0.55	2.2	7.4	0.22
Tier 2+[b]	2005–2011	2013[d]	0.30	1.5	5.5	0.10[e]
Tier 3[f]	2012–2014	2012	0.30	1.5	5.5	0.10
Tier 4	2015 or later	2015	0.14[g]	1.5	1.3[g]	0.03

[a] MY—Year of original manufacture
[b] Tier 0+ to Tier 2+ line-haul locomotives must also meet switch standards of the same Tier
[c] 1993–2001 locomotives that were not equipped with an intake air coolant system are subject to Tier 0+ rather than Tier 1+ standards
[d] As early as 2008 if approved engine upgrade kits become available
[e] 0.20 g/bhp-hr until January 1, 2013 (with some exceptions)
[f] Tier 3 line-haul locomotives must also meet Tier 2+ switching standards
[g] Manufacturers may elect to meet a combined NO_x + HC standard of 1.4 g/bhp-hr

Table 10.25 Switching Locomotive Emission Standards g/bhp-hr

Tier	[a]MY	Date	HC	CO	NO_x	PM
Tier 0+	1973–2001	2010[c]	2.10	8.0	11.8	0.26
Tier 1+[b]	2002–2004	2010[c]	1.20	2.5	11.0	0.26
Tier 2+[b]	2005–2010	2013[c]	0.60	2.4	8.1	0.13[d]
Tier 3	2011–2014	2011	0.60	2.4	5.0	0.10
Tier 4	2015 or later	2015	0.14[e]	2.4	1.3[e]	0.03

[a] MY—Year of original manufacture
[b] Tier 1+ and Tier 2+ switching locomotives must also meet line-haul standards of the same Tier
[c] As early as 2008 if approved engine upgrade kits become available
[d] 0.24 g/bhp-hr until January 1, 2013 (with some exceptions)
[e] Manufacturers may elect to meet a combined NO_x + HC standard of 1.3 g/bhp-hr

10.6 Aviation Emissions

Air transportation is of immense importance to Canada. Based on land mass, Canada is the second largest country in the world, with its population of 35 million scattered across 9 million km^2. Therefore, necessity often dictates that air transportation is used to support Canada's domestic and international trade, as well as to connect Canadians within the country and to the rest of the world. Canada's air industry also serves remote communities where it is often the only mode of transportation for people, service providers, and basic commodities, resulting in a low volume, high cost service. Average distances flown domestically per passenger are considerably higher in Canada than in countries with smaller landmass, providing a unique and challenging context for Canada to oversee safety, security, efficiency and sustainability. Aviation also plays a key role in the Canadian economy. It bears noting that Canada is a leader in the manu-

[40] www.tc.gc.ca/locomotive-emissions.

[41] www.epa.gov/otaq/locomotives.htm.

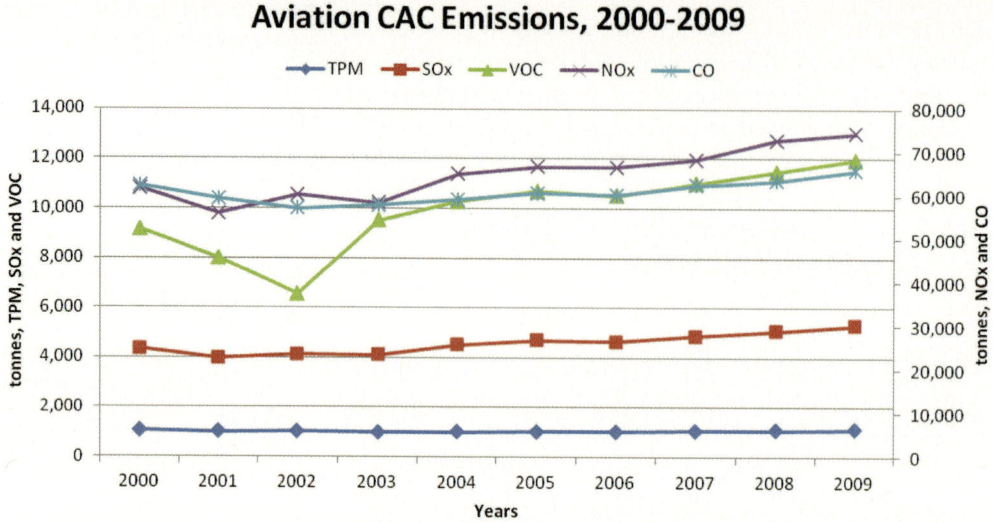

Fig. 10.15 Trends in Aviation Emissions of CACs. In spite of cleaner engines and more efficient operations, CAC emissions from aviation sources are increasing due to growth in air traffic. (Source: NPRI)

facture of regional aircraft, avionics, business jets, commercial helicopters, aircraft engines and flight simulators.

Aircraft emissions occur in and around airports during the landing and take-off phases (LTO) of flight, which may affect local air quality, as well as in the air during climb, cruise and descent phases. Air pollutant emissions from aircraft engines include NO_x, CO, SOx, unburned hydrocarbons (HC), also referred to as volatile organic compounds (VOC), and particulate matter (Fig. 10.15).

In 2010, aviation emissions accounted for 3.2 % of domestic greenhouse gas emissions from transportation, and 0.9 % of total Canadian emissions. The aviation sector globally contributes 2 % of the GHG emissions.[42]

Transport Canada forecasts that domestic air traffic will grow at an average rate of 2.8 % until 2020, while international air traffic will grow at a rate of 4.4 %.

10.6.1 Regulation of Emissions

In Canada, aviation emissions are regulated by the federal government, with Transport Canada as the lead department. Due to the global nature of the industry, Canada has a policy of harmonization with the international standards and recommended practices developed and adopted by the International Civil Aviation Organization (ICAO). Canada adopts by reference these ICAO standards for aircraft and engine emissions into its Airworthiness Manual Chapter 516, made pursuant to the Canadian Aviation Regulations and under the *Aeronautics Act*.

ICAO emissions standards are found in *Annex 16* to the *Convention on International Civil Aviation, Volume II Aircraft Engine Emissions.* These standards are developed by ICAO through its Committee on Aviation Environmental Protection (CAEP). Canada is a member of CAEP and actively participates on the CAEP technical working groups, which develop environmental standards according to the underlying principles of technical feasibility, economic reasonableness and environmental benefit, while taking into account the environmental interrelationships and tradeoffs. (An example of a trade-off that has proved challenging for aviation is where efforts to reduce NO_x emissions through lower temperatures in the combustion chamber decrease thermodynamic efficiency and lead to higher fuel burn and thus increased CO_2 emissions.) Current ICAO emissions standards limit emissions from engines of NO_x, unburned hydrocarbons, carbon monoxide and soot (measured as a smoke number (SN)). Historically, the focus of international efforts has been on the reduction of NO_x emissions, with ICAO standards having been reduced four times since 1981. The latest ICAO standard for NO_x emissions for new aircraft engines takes effect in 2014.

CAEP is currently working on several additional new standards. A certification requirement is under development in support of a non-volatile particulate matter standard applicable to aircraft engines, to be ready for 2016. A standard for volatile particulate matter will subsequently be developed. CAEP's current priority however is the development of a CO_2 standard for fixed wing aircraft (airplanes). While emissions standards for air pollutants typically apply to the aircraft engine, CO_2 emissions are affected by much more than engine performance, such that the standard will need to apply to the airplane. The CO_2 standard is thus complex to develop. The new CO_2 standard is expected to be ready by 2016. Once completed and adopted by ICAO, Canada

[42] Environment Canada 2012 National Inventory Report (NIR), http://unfccc.int/national_reports/annex_i_ghg_inventories/national_inventories_submissions/items/6598.php.

will adopt the new particulate matter and CO_2 standards domestically.

A regulatory approach for the management of greenhouse gas emissions from aviation is not currently envisaged. Canada is participating in global discussions at ICAO to identify emission reduction goals and to develop a global framework to guide the management of CO_2 from international aviation. Domestically, the federal government continues to work with industry stakeholders on Canada's Action Plan to reduce greenhouse gas emissions (see voluntary actions below).

10.6.2 Compliance and Enforcement

The compliance and enforcement of emissions standards is unusual compared to other transportation modes. For safety reasons aircraft and aircraft engines types are designed and certified to strict standards, including emissions standards. As the Canadian certificating authority, Transport Canada ensures during the certification process that the aircraft/aircraft engines meet the emissions standards: demonstration of compliance is one step in the process. Compliance is thus a certification issue. All aircraft and engines flying in Canada must be in compliance with the emissions standards. No in-service compliance verification is necessary, because major design changes carried out during the life cycle of the aircraft/aircraft engine type design have to be re-certified.

10.6.3 Emission Inventories, Modelling and Reporting

In compliance with the requirements of Annex 16, measurements of exhaust emissions are performed at the manufacturer's test facilities as part of the certification process. The data are published in an ICAO exhaust emissions databank. Engine emissions are given for the standardized landing and take-off (LTO) cycle represented by an engine power setting for taxiing, take-off, climb and approach, and given times in each mode. Together with fuel flow and maximum SN, emission indices of HC, CO, and NO_x in grams per kilogram of fuel burned are reported.

Transport Canada works in collaboration with Canadian airport authorities to minimize aviation-related emissions impacts at the airports. Airport ambient air quality studies over the past 40 years have provided real-time information on airport air quality and have formed the basis for the establishment of greenhouse gas inventories at the major Canadian airports. Working in collaboration with the airports, Transport Canada developed a methodology and tool to conduct GHG inventories at the airports: inventories have been conducted for all of Canada's major airports. At the request of the international association for the world's airports, Transport Canada is expanding this tool so that it can be used worldwide. These efforts assist Canadian airports, as well as ICAO and international airports, in understanding emissions impacts.

Environment Canada models aviation GHG, CAC and PM emissions using their Aviation GHG Emission Model (AGEM), which models the emission profile of both the LTO portion (below 3000') and the cruise portion (above 3000') for every departure occurring in Canada (with the exception of piston aircraft). Flight specific data is obtained annually from the national air navigation service provider, NAV CANADA and identifies airplane type and destination for each flight. Radar tracks are used to develop a true flight path length, and known airframe/engine performance characteristics identify operating parameters for each airplane type. This GHG information is submitted by Canada to the UN to meet its international reporting obligations on GHG emissions under the United Nations Framework Convention on Climate Change.[43]

10.6.4 Voluntary Measures

Despite the high and increasing level of demand for airline services in Canada, significant progress has been made to mitigate the carbon footprint of the sector, through voluntary measures undertaken by industry, the Government of Canada and NAV CANADA.

In June 2005, Canadian air carriers and Transport Canada signed the world's first voluntary agreement to address GHG emissions from both domestic and international aviation operations. The agreement established an average goal of 1.1 % per annum improvement in fuel efficiency from 1990–2012, a cumulative improvement of 24 %. Both the Air Transport Association of Canada (ATAC) and the National Airlines Council of Canada (NACC) fulfill their commitments under the voluntary agreement, annually reporting fuel burn and activity measurements, such as available seat-kilometres. As of 2010, Canada's aviation industry had achieved a 1.9 % average annual fuel efficiency improvement since 1990 or a 31 % cumulative improvement, surpassing the agreed-upon goal in the voluntary agreement. While absolute domestic and international emissions have grown at an average annual rate of 1.4 % between 1990 and 2010, this rate would have been significantly higher without these fuel efficiency improvements (Table 10.26).

In March 2010, Transport Canada joined with Canadian aviation stakeholders from the aerospace industry, air carriers, airports, the air navigation service provider, and the business aviation sector to form the Working Group on Aviation Emissions, with the objective of improving envi-

[43] Environment Canada—Pollutant Inventories and Reporting Division.

Table 10.26 Aviation activity and GHG emissions at a glance 1990, 2005–2010. (Source: 2010 Canadian Aviation Industry Report on Greenhouse Gas Emissions Reductions, March 2012)

	1990	2005	2006	2007	2008	2009	2010
Fuel use (million litres)	4,616	4,887	5,186	5,543	5,575	5,077	5,659
GHG emissions (millions of tonnes of CO_2-eq)	11.801	12.495	13.258	14.171	14.254	12.980	14.467
Traffic (billions)							
Revenue passenger-kilometres (RPK)	66.37	105.22	112.98	124.15	125.55	117.62	128.77
Passenger revenue-tonne-kilometres (pass. RTK)	6.64	10.52	11.30	12.42	12.55	11.76	12.88
Cargo revenue-tonne-kilometres (cargo RTK)	1.72	1.56	1.52	1.81	1.56	1.37	1.93
Total revenue-tonne-kilometres (RTK)	8.36	12.08	12.81	14.22	14.12	13.14	14.81
Emission rates							
CO_2e grams/RPK	177.81	118.75	117.35	114.14	113.53	110.36	108.65
CO_2e grams/Total RTK	1,412	1,034	1,035	996	1,010	988	969

ronmental performance. The Working Group provides a collaborative forum for information sharing and discussion, and has developed *Canada's Action Plan to Reduce Greenhouse Gas Emissions from Aviation,* which was released in June 2012.[44] Setting an aspirational goal to improve fuel efficiency by an average annual rate of 2% per year from the 2005 baseline until 2020, the Action Plan outlines a strategy to reduce GHG emissions from aviation. Contributing measures include fleet renewal and upgrades, increased efficiency in air operations and in airport aircraft ground operations, and improved capabilities in air traffic management. The Action Plan builds on and supersedes the 2005 voluntary agreement.

NAV CANADA, Canada's private civil air navigation service provider, plays a key and active role in assisting pollution prevention programs in the aviation industry, by focusing on performance-based navigation, en-route surveillance and communication, airspace utilization and airport operations. NAV CANADA estimates that between 1997 and 2009, newly deployed technologies and procedures have helped air operators save over 2 billion l of jet fuel (representing about 5.4 million metric tonnes of CO_2 emissions). By 2016 NAV CANADA forecasts additional reductions of over 3 billion l of jet fuel (representing about 8 million metric tonnes of CO_2 emissions).[45]

10.6.5 Fuel Impacts and Inter-Relation with their Regulation

Fuel Standards Aviation fuel standards are set by the Canadian General Standards Board, with the participation of users, producers, and regulators. Aviation fuels used in Canada are Jet A/ Jet A-1 for gas turbine engines in commercial aviation, and aviation gas for piston engines used in general aviation. CAN/CGSB-3.23 is the Canadian standard for Jet A/Jet A-1, while CAN/CGSB-3.25 is the standard for aviation gas (Avgas). The use of a particular aviation fuel, as specified by the manufacturer, is a condition for the certification of the aircraft.

To limit emissions of PM and SO_x from combustion, fuel standards have maximum limits on the sulphur content of fuels. The current global average jet fuel sulphur content is between 400–800 ppm[46], which is well below the specification maximum of 3,000 ppm. While other modes of transportation have moved to reduce allowable levels of sulphur in fuel, there are no plans at the present time to lower the allowable sulphur content in jet fuel. Studies show that lower sulphur levels would produce health benefits; however, there would be a trade-off in that the production of ultra low sulphur fuel requires more energy and would increase CO_2 emissions by 2% compared to the existing fuel refining process, resulting in negative climate impacts.

Aviation gasoline (100LL), or Avgas, is an aviation fuel used to power piston-engine aircraft. Avgas is the only transportation fuel in widespread use containing lead; the antiknock additive tetra-ethyl lead is put in gasoline to help maintain the rated performance of high compression piston engines. Despite mounting pressure and effort to find alternatives to leaded aviation fuels, no replacement fuel has yet been identified that can serve the broad range of aircraft currently certified to use leaded aviation gasoline. However, a new grade of Avgas containing 19% less lead (100VLL) has recently been approved by the United States Federal Aviation Administration. In Canada in 2010, emissions of lead from piston-engine aircraft using leaded Avgas accounted for 19% of the national inventory of lead emitted to air.[47]

[44] http://www.tc.gc.ca/aviation-emissions/.

[45] NAV CANADA, 2011. "NAV CANADA and the Environment" http://www.navcanada.ca/NavCanada.asp?Language=en&Content=ContentDefinitionFiles\AboutUs\Environment\CIFER\default.xml.

[46] http://web.mit.edu/aeroastro/partner/projects/project27.html.

[47] Environment Canada, 2010, "National Pollutant Release Inventory—Air Pollutant Emissions Summaries and Trends, 2010 Air Pollutant Emissions for Canada" http://ec.gc.ca/pdb/websol/emissions/ap/ap_result_e.cfm?year=2010&substance=all&location=CA&sector=&submit=Search.

Alternative Fuels The potential to use alternative fuels is being examined as an important part of the global strategy to reduce aviation emissions of CO_2. Air quality benefits will also be realized, as feedstocks for sustainable biofuels do not contain sulphur. The use of biofuels therefore will result in the virtual elimination of SO_x emissions as well as substantial reductions in particulate matter emissions.

Canadian aviation fuel standards CAN/CGSB 3.23 (Jet A and Jet A-1) were amended in 2011 to allow up to a 50 % blend of synthetic source material in Jet fuel. Due to the difficulty in meeting certain properties such as volatility and vapour pressure however, there is no intent to blend synthetic fuels into Avgas.

Canadian researchers and stakeholders are conducting biofuel research to assess the viability of alternative aviation fuels, and to evaluate how engine operations will be affected. Biomass feedstocks currently showing promise in Canada are algae, carinata, and camelina. The Government of Canada, through the *SD Tech Fund* of *Sustainable Development Technology Canada*[48], has allocated $ 10 million to two alternative jet fuel projects.

10.6.6 Regulatory/Management Challenges

Engine testing to develop and verify sampling and measurement methodologies and certification requirements for new standards, such as the new particulate matter standard, is very expensive, costing millions of dollars. In challenging financial times, certificating authorities, research organizations and industry need to work together to leverage resources to accomplish their goals.

With the issue of climate change emerging as the main environmental priority, and a strong push to show results in the aviation sector, the present challenge is for states to work collaboratively to reduce GHG emissions and achieve a goal of carbon neutral growth by 2020. Canada will continue to support this process.

10.7 Closure: Towards Sustainable Transport?

Transportation today depends nearly totally on oil with an ever increasing demand as the primary source of energy. This is clearly an unsustainable trend in terms of the depletion of a non-renewable resource and the climate change effects of the associated GHG emissions. Despite the significant achievements in terms of reductions in air contaminant emissions, better air quality also remains a target and is possible with alternative energy sources, fuels, technologies, and transportation systems. As we aim to move towards what might approximate "sustainable" transportation a few developments merit mention.

The developments in alternative vehicles and fuels are motivated primarily by GHG and climate change concerns, as well as the diminishing availability of conventional oil. However, in many cases there are criteria air contaminant emission benefits that are concurrently realized. A number of comprehensive studies are available on the life-cycle emissions of both GHGs and CACs from various combinations of technologies and fuels (Tiax 2007, CONCAWE 2007, Zah et al. 2007, Argonne 2010). Natural Resources Canada have supported the development and application of *GHGenius*[49] as a specialized life-cycle analysis model for transportation related emissions. Most of these published studies have focused on quantifying GHG and CAC emissions over the fuel and vehicle life-cycles.

A study (Zah et al. 2007) by the Swiss Federal Laboratories for Material Science and Technology (EMPA) stands out in terms of the broader scope in which they have attempted to include interactions of transportation biofuels with the environment. They examine a number of factors grouped in three areas: *Damage to Human Health*, *Damage to Ecosystems*, and *Depletion of Non-renewable resources.* Two overall indicators are ultimately arrived at: *Environmental Impact Points, UBP 06*, which evaluates the difference between environmental impacts and legal limits, and *European Eco-indicator 99* method which quantifies the damage done to human health and ecosystems. Despite challenges associated with quantifying and weighting some of the factors (e.g. "respiratory diseases" vs. "land occupation and transformation") the methodology is a refreshing one in its approach to dealing with life-cycle analysis in the context of sustainable transportation.

The Canadian Government's *Renewable Fuels Strategy* mentioned briefly in Section 0 aims at an average renewable fuel content of 5 % in gasoline and of 2 % in diesel. The renewable fuels used to meet these requirements are essentially first generation biofuels, in large part ethanol from corn and grain, and biodiesel from canola. Although "renewable" these alternative fuels are unlikely to represent a significant part of the primary energy demand for the on-road vehicle fleet in Canada even if they maintain a niche in the overall mix (Karman et al. 2008).

Natural gas has long been used in Canada as a lower emission, lower cost fuel particularly in commercial vehicle fleets. New opportunities for natural gas in transportation were identified recently by the Natural Gas Use in Transportation Roundtable with the *Natural Gas Use in the Canadian Transportation Sector—Deployment Roadmap* (CNGVA 2010). Westport Innovations of Vancouver supplies Canadian clean transportation technology in the form of heavy

[48] http://www.sdtc.ca/index.php?page=home&hl=en_CA.

[49] http://www.ghgenius.ca/.

duty engines burning natural gas or their innovative dual fuel engines burning natural gas and diesel to customers around the world.

Electricity and hydrogen are the energy carriers that have the potential to include other sources of primary energy in the mix for transportation. Gasoline-electric hybrid vehicles, plug-in hybrid vehicles and all electric vehicles are already in the on-road vehicle fleet in North America and are expected to represent a gradually increasing fraction of the fleet. The *Electric Vehicle Technology Roadmap for Canada* (Electric Mobility Canada 2010) estimates:

> By 2018, there will be at least 500 000 highway-capable plug-in electric-drive vehicles on Canadian roads, as well as what may be a larger number of hybrid-electric vehicles.

The Research and Development agenda of the Federal Government in terms of alternative fuels and technologies for sustainable transport are coordinated primarily by Natural Resources Canada's Office of Energy Research and Development (OERD)[50] and the National Research Council of Canada's Institute for Chemical Process and Environmental Technology (ICPET)[51]. Research units from Environment Canada and Health Canada are active participants in various areas of the R&D programs. A recent report submitted to OERD presents the history, current status and possible future directions for Clean Transportation R&D (Reilly-Roe 2011).

The *Program on Energy R&D* (PERD), under the Clean Transportation Systems (CTS) portfolio of NRCan coordinates federally funded research in five areas:

- Particles and Related Emissions (P&E)
- Advanced Fuels and Technologies for Emissions Reduction (AFTER)
- Advanced Materials for Transportation (AMT)
- Hydrogen and Fuel Cells (H2FC)
- Electric Mobility (EM)

NRC's ICPET identifies the following four areas with relevance to alternative fuels and technologies

- Chemical Technologies for Biomass Upgrading
- Clean Energy Technologies
- Environment Monitoring Technologies
- Technologies for Sustainable Oil Sands Development

Technology forcing regulations for air contaminant emissions from motor vehicles have achieved significant results over the 40 years of history reviewed at the beginning of this chapter. Comparable success in addressing the greenhouse gas emissions from the transportation sector will be required with technologies and regulations over the next 40 years to reach what we might start describing as "Sustainable Transport".

[50] http://www.nrcan.gc.ca/eneene/science/resres-eng.php.

[51] http://www.nrc-cnrc.gc.ca/eng/programs/icpet/clean-energy.html.

References

Aeronautics Act (R.S.C., 1985, c. A-2) Department of Justice Canada, Consolidated Acts. http://laws.justice.gc.ca/eng/acts/A-A2/

Argonne National Laboratory (2010) Well-to-wheels analysis of energy use and greenhouse gas emissions of plug-in hybrid electric vehicles, energy systems division, ANL/ESD/10-1, 2010. http://www.transportation.anl.gov/pdfs/TA/629.PDF

Canada Shipping Act (2001) (S.C. 2001, c. 26). Department of Justice Canada, Consolidated Acts. http://laws.justice.gc.ca/eng/acts/C-10.15/

CAMS (2010) Comprehensive air management system: a proposed framework to improve air quality management, CAMS Steering Committee, Canadian Council of Ministers of the Environment.

Canadian Environmental Protection Act (1999) (S.C. 1999, c. 33). Department of Justice Canada, Consolidated Acts. http://laws-lois.justice.gc.ca/eng/acts/C-15.31/

The Chamber of Shipping (2007) 2005–2006 BC Ocean-Going Vessel Emissions Inventory. http://www.pyr.ec.gc.ca/airshed/documents/ChamberofShippingEmissionInventory.pdf

CMA (2008) No breathing room: national illness costs of air pollution, Canadian Medical Association.

CNGVA (2010) Natural gas use in the canadian transportation sector—deployment roadmap, natural gas use in transportation roundtable, Canadian Natural Gas Vehicle Alliance.

CONCAWE (2007) Well-to-wheels analysis of future automotive fuels and powertrains in the European context, conservation for clean air and water in Europe (CONCAWE), European Council for Automotive R&D (EUCAR), and European Commission Joint Research Centre (JRC). ies.jrc.ec.europa.eu/media/scripts/getfile.php?file = fileadmin/H04/Well_to_Wheels/WTW/WTW_Report_010307.pdf

Coninx P (1998) Vehicle emissions testing—aircare, drive clean, and the potential of inspection and maintenance programs in Canada, The Fraser Institute

dKC (2000) Review of the British Columbia aircare program, prepared by: Rob Klausmeier de la Torre Klausmeier Consulting, Inc., Prepared for: British Columbia Ministry of Environment, Lands and Parks, Air Resources Branch,.September 15, 2000

Electric Mobility Canada (2010) Electric Vehicle Technology Roadmap for Canada- A strategic vision for highway-capable battery-electric, plug-in and other hybrid-electric vehicles

Environment Canada (2010) Renewable Fuels Regulations, Canada Gazette, Part I, Vol. 144, No. 15—April 10, 2010

Environment Canada (2010b) Sulphur in Liquid Fuels 2008, Oil, Gas, and Alternative Energy Division, Environment Canada, August 2010

Environment Canada (2011a) Discussion Paper—Proposed Amendment's to Canada's Sulphur in Diesel Fuel Regulations. http://www.ec.gc.ca/lcpe-cepa/694C8126-0652-41C5-B54D-08C1E15F1398/discussion_paper-eng.pdf

Environment Canada (2011b) National Emission Trends for Key Air Pollutants. http://www.ec.gc.ca/inrp-npri/0EC58C98-7D33-43BD-B7BA-3E9F1BBFB0AB/CAC_Trends_March2011_En.xls

Environment Canada (2012) National Inventory Report 1990–2010: Greenhouse Gas Sources and Sinks in Canada—The Canadian Government's Submission to the UN Framework Convention on Climate Change. http://unfccc.int/national_reports/annex_i_ghg_inventories/national_inventories_submissions/items/6598.php

ERG (2005) Evaluation of Ontario drive clean program FINAL REPORT, Eastern Research Group Inc., Prepared for: Ontario Ministry of the Environment. July 18, 2005

Evans Greg J, Cheol-Heon J, Kelly S, Parnian J, Stephanie A, Hugo L, Dennis H (2011) Design of a Near-Road Monitoring Strategy for Canada, report submitted to Environment Canada, Southern Ontario Centre for Atmospheric Aerosol Research, SOCAAR REPORT # CR-WB-2011-01–01

HEI (2010) Traffic-related air pollution: a critical review of the literature on emissions, exposure, and health effects, Health Effects Institute Special Report 17, January 2010
IEA (2010) Key World Energy Statistics, International Energy Agency. http://www.iea.org/textbase/nppdf/free/2010/key_stats_2010.pdf
MARPOL 73/78 Annex VI—Regulations for the Prevention of Air Pollution from Ships, International Maritime Organization.
Karman D, Smith M, Rowlands D, Cottes J, Nadar D, Said D, Tanha T (2008) Technical and policy implications of transportation biofuel regulatory approaches, submitted to Natural Resources Canada
Levelton Consultants Ltd (2004) Analysis of best management practices and emission inventory of agricultural sources in the LOWER Fraser River Valley Levelton Consultants Ltd. In association with Golder Associates Ltd., prepared for Fraser Valley Regional District and Environment Canada, December 2004. http://www.pyr.ec.gc.ca/airshed/Agr_BMP_EI_exec_summary_en.pdf
Levelton Consultants Ltd (2007) Air toxics emission inventory and health risk assessment—summary report. http://www.metrovancouver.org/about/publications/Publications/Air_Toxics_Emission.pdf
MECA (2007) Tier 2/LEV II Emission Control Technologies for Light-Duty Gasoline Vehicles, Manufacturers of Emission Controls Association
Metro Vancouver (2007) 2005 Lower Fraser Valley Air Emissions Inventory & Forecast and Backcast. http://www.metrovancouver.org/about/publications/Publications/2005_LFV_Emissions.pdf
Metro Vancouver (2010) 2005 Lower Fraser Valley Air Emissions Inventory & Forecast and Backcast, Detailed Listing of Results and Methodology. http://www.metrovancouver.org/about/publications/Publications/2005-LFV-Air_Emissions_Inventory-Forecast-Backcast.pdf
NPRI. National Pollutant Release Inventory, Air Pollutant Emissions (All Sources). http://www.ec.gc.ca/pdb/websol/emissions/ap/ap_query_e.cfm
NRCan (2006) Canada's Energy Outlook: The Reference Case 2006. Natural Resources Canada. http://www.oee.nrcan.gc.ca/Publications/statistics/trends09/chapter6.cfm?attr=0
OMA (2005) 2005–2026 Health and economic damage estimates, Ontario Medical Association. http://www.oma.org/Resources/Documents/e2005HealthAndEconomicDamageEstimates.pdf
OME (2007) Drive clean program emissions benefit analysis and reporting—light duty vehicles and non-diesel heavy duty vehicles—1999 to 2005, Ontario Ministry of the Environment, June 2007
OME (2010) Drive clean guide—emission standards, emission test methods and technical information relating to Ontario regulation 361/98 as Amended, Drive Clean Office, Ontario Ministry of the Environment
PVTT (2011) AirCare—results and observations in 2009 and 2010, Pacific Vehicle Testing Technologies Ltd
RAC (2011) Locomotive emissions monitoring report 2009, Railway Association of Canada, Ottawa
Reilly-Roe & Associates (2011a) Strategic priorities for transportation energy R&D—a guide for r&d program leads, prepared for the Office of Energy Research and Development, Natural Resources Canada, March 2011
Reilly-Roe & Associates (2011b) Environmental scan of the electric mobility area, prepared for the National Research Council of Canada, Institute for Chemical Process and Environmental Technology, March 2011
RWDI (2007) 2005 Nonroad Engine Fleet Characterization in the Canadian Lower Fraser Valley, RWDI Air Inc., submitted to Metro Vancouver. http://www.metrovancouver.org/about/publications/Publications/2005_Nonroad_Report.pdf
SAE J1228-2002. Small Craft-Marine Propulsion Engine and Systems-Power Measurements and Declarations, Marine Technical Steering Committee, Society of Automotive Engineers.
SENES Consultants Limited (2008) Marine Emission Factors, Engine Profiles and SECA Forecast
SENES Consultants Limited and Air Improvement Resource, Inc. (2004) Review of methods used in calculating marine vessel emission inventories. http://www.westcoastcollaborative.org/files/sector-marine/Review%20of%20Methods%20Used%20in%20Calculating%20Marine%20Vessel%20Emission.pdf
Statutory Orders and Regulations, SOR/2002-254 Sulphur in Diesel Fuel Regulations, Canada Gazette, Part II, Vol. 136, No. 16—July 31, 2002.
Statutory Orders and Regulations, SOR/2003-2. On-Road Vehicle and Engine Emission Regulations, Canada Gazette Part II, Vol. 137, No. 1—January 1, 2003
Statutory Orders and Regulations, SOR/2003-355. Off-Road Small Spark-Ignition Engine Emission Regulations, Canada Gazette Part II, Vol. 137, No. 24—November 19, 2003
Statutory Orders and Regulations, SOR/2005-32 Off-Road Compression-Ignition Engine Emission Regulations, Canada Gazette Part II, Vol. 139, No. 4—February 23, 2005
Statutory Orders and Regulations, SOR/2010-201. Passenger Automobile and Light Truck Greenhouse Gas Emission Regulations, Canada Gazette Part II, Vol. 144, No. 21—October 13, 2010.
Statutory Orders and Regulations, SOR/2007-86. Regulations for the Prevention of Pollution from Ships and for Dangerous Chemicals, Canada Gazette Part II, Vol. 141, No. 10
Statutory Orders and Regulations, SOR/2011-10 Marine Spark-Ignition Engine, Vessel and Off-Road Recreational Vehicle Emission Regulations, Canada Gazette Part II, Vol. 145, No. 4—February 16, 2011
Taylor B (2012) Inventory trends and modelling, environment Canada. Personal communication, 4 May 2012
Tiax LLC (2007) Full fuel cycle assessment: well-to-wheels energy inputs, emissions, and water impacts, prepared for California Energy Commission. http://www.energy.ca.gov/2007publications/CEC-600-2007-004/CEC-600-2007-004-D.PDF
Transport Canada (2009) Canadian Aviation Industry Report on Emissions Reductions. http://www.tc.gc.ca/media/documents/programs/2009atac.pdf
Transport Canada (2011) Vessel pollution and dangerous chemicals regulations—Briefing for Canadian Marine Advisory Council, Spring 2010
TC & RAC (2007) Memorandum of understanding between environment Canada, transport Canada and the Railway Association of Canada (Appendix A, Locomotive Emissions Monitoring Program 2008). http://www.tc.gc.ca/eng/programs/environment-ecofreight-about-voluntary-voluntaryagreementsrail-1844.htm)
U.S. National Research Council (2001) Evaluating vehicle emissions inspection and maintenance programs, National Academy Press, 2001
U.S. EPA (2003) User's Guide to MOBILE6.1 and MOBILE6.2, Mobile Source Emission Factor Model, Office of Transportation and Air Quality, EPA420-R-03–010
U.S. EPA (2008) EPA Finalizes More Stringent Emissions Standards for Locomotives and Marine Compression-Ignition Engines. http://www.epa.gov/otaq/regs/nonroad/420f08004.pdf
EPA US (2012) Policy guidance on the use of MOVES2010 and subsequent minor revisions for state implementation plan development, transportation conformity, and other purposes, EPA-420-B-12–010
U.S. and Canada (2009) Proposal to Designate an Emission Control Area for Nitrogen Oxides, Sulphur Oxides and Particulate Matter. http://www.epa.gov/otaq/regs/nonroad/marine/ci/mepc-59-eca-proposal.pdf
United States Code of Federal Regulations, U.S. 40 CFR Ch. I Part 1051—control of emissions from recreational engines and vehicles
United States Code of Federal Regulations, U.S. 40 CFR Ch. I Part 91—Control of emissions from marine spark-ignition engines
Zah R, Boni H, Gauch M, Hischier R, Lehmann M, Wager P (2007) Life cycle assessment of energy products: environmental assessment of biofuels, report by the Swiss Federal Institute of Materials Science and Technology Research (EMPA), Switzerland. www.bioenergywiki.net/images/8/80/Empa_Bioenergie_ExecSumm_engl.pdf

Effects of Agriculture on Air Quality in Canada

11

S. Bittman, D. I. Massé, E. Pattey, M. Cournoyer, G. Qiu, A. Narjoux, S. C. Sheppard and A. Van der Zaag

Abstract

The modernization of animal production as well as demographic changes in rural areas has resulted in difficult cohabitation problems in some regions of Canada. This chapter focuses on three predominant air quality issues associated with modern farming: odour, particulate matter and ammonia. Odour can adversely affect quality of life, lead to socio-emotional nuisances and cause genuine physical symptoms. Odours are mainly associated with confined animal farming operations, especially exhaust from animal housing, storage of animal wastes, application of manure to land. Particulate matter (PM) is now well known to be potentially toxic to humans. Agricultural PM is emitted from soil erosion tillage and livestock buildings. Ammonia contributes somewhat to odour but more importantly reacts in the atmosphere to form fine secondary particulates, also a potential health concern. Ammonia is emitted from livestock wastes (especially urine and uric acid) and from urea or ammonia containing fertilizers. Ammonia in the particulate form travels long distances and can damage certain plant communities and ecosystems. Good farming practices often reduce impact on air quality: examples include precise feeding of protein which reduced excretion of urea, injection of manure and fertilizer, and reduced tillage and summerfallow in crop production which reduces PM. In some cases, to reduce complaints, it may be necessary to regulate separation in space or time between the public and farming operations.

S. Bittman (✉)
Pacific Agrifood Research Centre, Agassiz, British Columbia, Canada
e-mail: shabtai.bittman@agr.gc.ca

D. I. Massé
Dairy and Swine Research and Development Centre, Quebec, Canada
e-mail: daniel.masse@agr.gc.ca

E. Pattey · A. Van der Zaag
Eastern Cereal and Oilseed Research Centre, Ottawa, Canada
e-mail: elizabeth.pattey@agr.gc.ca

M. Cournoyer
Groupe Conseil UDA Inc., St. Jean, Quebec, Canada

G. Qiu
Ontario Ministry of Environment, Ontario, Canada

A. Narjoux
Odotech Inc., Montreal, Canada

S. C. Sheppard
ECOMatters, Inc., Pinawa, Manitoba, Canada
e-mail: sheppards@ecomatters.com

A. Van der Zaag
e-mail: andrew.vanderzaag@agr.gc.ca

E. Taylor, A. McMillan (eds.), *Air Quality Management,* DOI 10.1007/978-94-007-7557-2_11,

Keywords
Odour · Particulate matter · PM · Ammonia · Nuisance · Environmental impact · Best management practices

11.1 Introduction—The Interconnection Among Air Pollutants from Agriculture

Probably due to the intensification of agriculture with large individual operations juxtaposed against expanding cities and towns, there has been a rapid shift in public perception away from agriculture as a bucolic healthful vocation to agriculture as an egregious industrial polluter.

Complaints by the general public against the impact of agriculture on air quality unusually concern odours; ranging from occasional if acute incidents to continual exposure, and from natural odiferous compounds to man-made chemicals usually pesticides. Also impacting human comfort and health is the emission of fine particles from fields exposed to wind erosion, and from crop and animal production activities. For example, particles are emitted during soil tillage and crop harvesting and from livestock raised in barns or feedlot. It is worthy of mention that fine particles are an important odour vector; particles influence the transport and perception of odour nuisance.

Beyond the long-range impact on the general public, agricultural activities can expose farm workers and livestock to acute concentrations of toxic or allergenic air pollutants. The air in animal facilities may contain toxic or even fatal levels of ammonia, hydrogen sulphide and dust. Similarly, carbon monoxide in enclosed places like silos is responsible for mortalities on farms. Particles can also carry microorganisms, some possibly pathogenic endotoxins, fungal spores and antibiotics that can affect the health of people and farm animals

An agricultural pollutant long overlooked, that has a relation with both odour and particulates, is ammonia. In typical environmental concentrations, it is not perceived as either a direct threat to health or to comfort. However, ammonia reacts with common acid gases in the atmosphere to form very fine particulates, hence referred to as secondary particulates. Fine particulates, though not necessarily those containing ammonia, are strongly implicated in a number of health related issues. Furthermore, due to the sheer volumes of emissions, the deposition of ammonia and ammonium containing PM, can harm sensitive plant communities by direct toxicity, nutrient enrichment and acidification. The emission of ammonia from farms is a tangible economic loss to farmers, and has a significant implication for energy consumption and carbon foot print of agriculture. This is because of the intensive use of energy to manufacture the fertilizer to replace the nitrogen lost from the farms as ammonia.

Volatile organic carbon compounds (VOCs), not only odiferous ones, can lead to formation of ground level ozone which is a major component (with fine PM) in smog. Many crops are sensitive to ozone so this type of smog can reduce farm productivity. Smog also harms human health and even aesthetics by reducing visibility. Farm VOCs consist of very unpleasant chemicals associated with animal wastes and certain pesticides, but also less offensive odours released by fermented plant materials notably silage, and even pleasant odours such as freshly cut grass and flowers.

Finally, agriculture contributes to the net release of greenhouse gases- notably nitrous oxide from land application of nitrogen fertilizer or manure, and methane (an odourless VOC) from enteric fermentation by ruminants and liquid manure storage.

The current chapter focuses on the three most prominent air quality impacts associated with farming, odour, fine particulate matter and ammonia.

11.2 Odour Emission Sources: How Are They Managed?

11.2.1 Introduction

The modernization of Canadian agriculture has resulted in the expansion and specialization of farms as well as more intensive use of land. This new agriculture is resulting in greater impact on the public leading to conflicts over issues such as nuisance odours. Odour can adversely affect quality of life in aspects of comfort, social behaviour and health (Sucker et al. 2009). Accordingly, a nuisance odour adversely affects the quality of life, leads to socio-emotional losses and, finally, may cause somatic and systemic vegetative symptoms. Although distinct from toxic-type reactions and occur at much lower odour levels, the symptoms cannot be occurring from the concept of health as defined by the World Health Organization (WHO). According to the WHO, odour nuisance may indirectly affect human health, since it defines the term as "a state of complete physical, mental and social well-being and not merely the absence of disease or infirmity" (Frechen 2000).

Exposure to odours, which is often associated with pollution, can cause serious disturbance among communities close to the emission source. Siting of new facilities or expansion of existing facilities for animal production is often opposed by neighbours due to concerns about prevalence of unpleasant odours, reduction in property values, and potential for adverse impact on their health and well being. Expansion of the livestock industry will rely on implementation of measures and policies designed to mitigate those impacts (Jacobson 2002). Nuisance odours are generally local problems but can become a major economic issue. Increasingly strict regulations are emerging regarding the siting of production facilities, separation distance from residences and regulations on the management and on the timing of manure, pesticide and fertilizer application.

11.2.2 Determinants of an Odour Nuisance

Odorous compounds are produced primarily from anaerobic decomposition of manure in animal guts and in manure storages. More than 150 organic and inorganic volatile compounds have been identified to contribute to the malodors (ASAE 2007; Filipy and Rumburg 2006). An odour becomes a nuisance when it causes major and regular disturbances of the normal olfactory landscape in a specific environment. The impact and the likelihood of complaints result from five interacting odour factors, known as FIDOL factors: Frequency, Intensity, Duration, Offensiveness, and Location (Furberg and Preston 2005). Intensity (I) refers to the perception of the odour strength or concentration; duration (D) refers to the time that an odour is experienced by an individual; offensiveness (O) is the subjective unpleasantness of an odour; and location (L) suggests proximity to and size of the exposed population, and the type of activity they are engaged in. The intensity and emission rate of odours are strongly dependent on manure composition, and climatic condition such as temperature, wind direction, atmospheric stability, and air movement above the odorous surface (ASAE 2007). The assessment of the odour strength and nuisance level is therefore site specific.

11.2.3 Farm Activities that Contribute to the Emission of Odorous Gas

In recent years, several literature reviews were conducted on odour emissions associated with livestock operations (Sweeten et al. 2006; Huang et al. 2005; AFOTW 2004; Ortech 2004, 2005). Some reviews identify the main sources and provide some emission factors or rates (Huang et al. 2005; AFOTW 2004; Ortech 2004; Lyngbye et al. 2006). Very often the emission factors or rates are expressed on a different basis or established with different methodologies. Several of those reviews also cover other emissions affecting air quality inside and outside livestock housing, specifically toxic gases (such as hydrogen sulphide and ammonia), greenhouse gases, particulate matter that includes pathogenic microorganisms. In North America, most odour conflicts are between agricultural producers and neighbouring populations and concern confined animal farming operations (Huang et al. 2005; NRC 2000; GEIS 1999). Other agricultural odours arise from production of compost, mushroom growing operations, and application of other organic wastes (Noble et al. 2001; Groeneveld and Hébert 2002). Although the latter sources have local importance, they are less widespread than livestock production activities. The following sections present the main odour sources and mitigation methods associated with agricultural activities, including certain social and geographical considerations.

11.2.4 Livestock Operations

Historically, livestock operations were small and scattered so no single farm presented an odour problem for a large number of people (Sweeten et al. 2006). In contrast, large modern animal feeding operations may impact many people. Impact is greater when there is a concentration of many animal operations in a region. This regional concentration exists because there is an economic incentive, both for the industry and for consumers, to bring together producers, feed suppliers and agri-food processors, labour force and consumers (Sweeten et al. 2006; Mwansa 2004). Rapid expansion of livestock industries, together with inadequate regulations pertaining to the location and size of livestock production operations, has resulted in increasing odour nuisance complaints (Sweeten et al. 2006; Chapin et al. 1998). In Canada, the primary regions where odour emissions are significant and abatement is becoming a priority are: (1) central (swine sector) and southern Alberta (beef sector), (2) southern Manitoba (beef and swine sectors), (3) southern Ontario (poultry, swine, dairy and beef sectors), (4) southern Quebec (swine, dairy and poultry sectors), (5) the lower Fraser River area in southern British Columbia (dairy and poultry sectors), (6) Prince Edward Island (beef and dairy sectors) and (7) the Annapolis region of Nova Scotia (beef and dairy sectors).

Odour emissions from livestock operations are caused primarily by: (1) exhaust from livestock housing, (2) manure storages, (3) field application of manure and (4) handling of dead animals (Huang et al. 2005). In the case of pig production operations in the United States, it is estimated that 40 % of odours and gases (ammonia and nitrous oxide) originate from manure application, 35 % from livestock housing, 20 % from manure storage and treatment, and 5 % from disposal of dead animals (Chapin et al. 1998). In Quebec, 70 % of odour complaints against pig farms are associated with manure application, 20 % with manure storage sites, and 10 % livestock housing (AAFC 1998). In the areas of livestock concentration the impact of odorous emissions has greatly increased, resulting in imposition of moratoria on pig production in several areas of the United States since the early 1990s (Chapin et al. 1998). In Denmark farmers are required to reduce odour from housing by more than 90 % before expansion of pig barns is allowed in regions with high pig density (Lyngbye et al. 2006). Such regulations are economically difficult for farmers since they cannot transfer the costs to the consumers either within or outside the countreis (Lyngbye et al. 2006). Therefore countries with high pig populations and tight regulations must achieve greater productivity, meat quality, animal health and food safety standards to be competitive. If these objectives cannot be achieved, pig production will move to countries with lower human population densities and lower environmental standards (Lyngbye et al. 2006).

Over the past decade most of the expansion of the swine and cattle sector has occurred western Canada, driven mainly by abundance of low cost feed and wish to diversify farm production, (Statistics Canada 2002; Laborde 2004). Between 1996 and 2001 cattle numbers in Alberta rose by 673,000 head (Statistics Canada 2002). From 1981 to 2001, the proportion of pigs in Manitoba, Saskatchewan and Alberta rose from 30 to 41 % of the Canadian herd, while the proportion in Quebec and Ontario fell from 67 to 55 % (Laborde 2004). But even in this thinly-populated region there have been conflicts within communities. The westward movement of livestock production in the USA is apparently due to higher standards and more constraints in eastern states (Chapin et al. 1998).

11.2.5 Measurement of Odour Emission Rates

Odour measurements are performed on air samples collected from a source according to international standards which allows the emission rate to be estimated. Three types of sources are generally identified: point, volume and area sources. Point sources are the easiest to characterize, as representative concentrations and flow rates can easily be determined in the duct where the emission is confined. Volume sources are typically buildings with several openings. The emission points are less well defined and controlled. The estimation of outside flow rates may be performed by ventilation calculations. Area sources are composting piles, manure spreading, cattle feedlots, manure storages, sewage treatment facilities, etc. Measurements are used to develop emission factors which attribute a certain amount of odour to a particular activity and the factors can be used to track changes in odour over time or place (AFOTW 2004; Huang et al. 2005). However, the limitations of emission factors are that they are too general and do not represent variations in emissions due to climate, geography, seasonal phenomena and agricultural practices. The two major limitations generally identified are accuracy of emission factors and availability of data (NRC 2000). Obtaining applicable mean values requires a small sample of operations representative of all operations in each defined category for various types of production and in a variety of geographic situations. In addition, a large number of direct measurements need to be performed to estimate the emissions, and the number of measurements is still inadequate at present to obtain reliable results.

It is recognized in North America that there are gaps in protocols used to measure the emissions arising from animal production facilities (NRC 2000). Inappropriate characterization of operations is also a cause of variability in results. Furthermore, ventilation rate is difficult to quantify especially in passively ventilated buildings. A standard protocol is required to reduce variability in emission rates. It may include factors associated with atmospheric conditions, ventilation and air quality in animal housing and relevant parameters in animal production practices. The standard protocol is required to establish typical profiles and dynamics of odour emissions based on type of facility and will allow studies on the relationship between odour impact and effects on health.

11.2.6 Best Management Practices to Mitigate Agricultural Odours

The acceptable size of a livestock production site or the required level of odour mitigation will depend on the site location compared with the effective odour impact distance. Proper siting of new livestock production facilities is therefore considered a simple and economic method of controlling odours and forestalling legal actions for nuisance odour (NCPP 2006; Sweeten et al. 2006). In certain regions, consideration may be given to locating new livestock facilities in the centre of an area required for manure or excreta application, depending on crop needs (Sweeten 1998). However, for existing facilities without adequate separation distance, the only alternatives to downsizing or relocation are methods for odour mitigation at source or improvement of odour dispersion. (Sweeten et al. 2006). Odour-control methods associated with livestock production facilities can be grouped under four areas of action (Sweeten et al. 2006): (1) changes in feed regimen to reduce nutrient content of manure to reduce gaseous emissions during storage and handling; (2) improved manure management: chemical or biological manure treatment, bedding type additives and cleaning practices, manure injection and incorporation during land application; (3) capture and treatment of air exhausted usually from actively ventilated housing or covered manure storages; and (4) improved atmospheric dispersion of odours with physical barriers or stack design. The above measures that will significantly influence odour mitigation in the vicinity of livestock production or application sites are not clearly defined. A measure applied to the interior of livestock housing may be effective with respect to its interior air quality but fail to significantly affect odour emissions in the surrounding environment. Therefore, to achieve satisfactory results at a specific distance, several odour mitigation measures probably need to be applied at various emission points of a livestock production site. A single control measure at a single point of intervention, or implementation of an incomplete measure, may not result in adequate mitigation of odours and could represent pointless expense for the operation, if the only purpose is odour control. Odour management programs therefore need to take these factors into account.

Of the three principal odour sources associated with livestock production, manure application probably provides the greatest potential and benefit for odour emissions abatement. First, spreading usually leads to the most complaints. Second, it is a periodic activity with affordable measures such as rapid incorporation into the soil or direct injection (Sweeten et al. 2006). Full injection where the manure is completely covered can reduce odour emission by 50 to 60 % (AFOTW 2004). In Manitoba and Denmark, for example, odours from pig manure application are less of a concern since more and more farmers must inject the manure or deposit it directly on the soil (trailing hoses or trailing shoes) (Zhang et al. 2005). Manure tank covers made of impermeable or permeable materials reduce odour emissions by 50 to 80 % (AFOTW 2004). Emissions from animal housing can be attenuated by up to 25 % via diet modification, by up to 40 to 50 % via exhaust air filtration, by up to 20 % via enhance odour dispersion with wind break and finally by 40 to 50 % with oil sprinkling for dust controlled inside the building (AFOTW 2004). Odours are also substantially reduced via manure treatments. Solids separation can reduce manure odour by 25 %, aerobic treatment by 40 to 60 %, composting by 30 % and anaerobic digestion by 50 to 80 % (AFOTW 2004). The planning and timing of land application and the communication with the neighbourhood are proactive initiatives that substantially improve the cohabitation.

11.2.7 Conclusion

The modernization of animal production as well as demographic changes in rural areas has resulted in difficult cohabitation problems in some regions of Canada. Odour can adversely affect quality of life, lead to socio-emotional nuisances and cause genuine physical symptoms. The main sources of odours are essentially associated with confined animal farming operations. They are associated with the ventilation of livestock housing, storage of animal wastes, application of manure to land, and the disposal of dead-animals.

New livestock operations have become impossible to implement in certain geographical areas because of increasingly strict regulations regarding where production sites can be located, the distances that certain agricultural enterprises must be from residences and restrictions on the timing of manure, pesticide and fertilizer application. Odour nuisance has therefore become an economic issue for the agriculture industry.

Proper location of new livestock production facilities is therefore considered a simple and economic method of controlling odours. For existing facilities without adequate distance separation, methods for odour mitigation at source or improvement of odour dispersion will need to be applied. Generally speaking, odour control methods associated with livestock production facilities can be grouped under five areas of action: (1) changes in feed regime, (2) improved management and treatment of excreta, (3) capture and treatment of gaseous emissions, (4) improved atmospheric dispersion (Sweeten et al. 2006), and (5) relocation of the facilities. The expansion of the livestock industry will rely on a large scale adoption and implementation of these measures to mitigate odorous emissions from several farm operations within a region.

11.3 Impact of Management Practices on Primary Particulate Matter Emissions

11.3.1 Introduction

Particulate matter (PM) refers to a mixture of solid particles and liquid droplets (aerosols) suspended in the air. PM is classified by its aerodynamic diameter into three categories. Particles with a diameter of 100 μm or less are called total particulate matter (TPM). They include PM_{10}, particles with a diameter of 10 μm or less also called inhalable particles, $PM_{2.5}$, particles with a diameter of 2.5 μm or less referred as respirable PM, and PM_1 with a diameter of 1 μm of less. Intact particles released into the air are described as primary PM. The main sources of primary PM in agriculture include: soil dust from fields, plant debris during harvest and grain handling, dust from animal feedlots and houses, including animal, waste, dander and feather fragments, agrochemical dust or droplets, pollen, spores, bacteria, mould, etc. If particles are generated through chemical reaction of gases in the atmosphere they are called secondary PM. The main source of secondary PM is the reaction of ammonia from agricultural sources with acid gases from non agricultural sectors (see Section on ammonia below).

11.3.2 Impact Of PM on Environment and Health

Particulate matter has been recognized as an air pollutant due to its adverse environmental and health impacts. Ambient PM contributes to visibility impairment, smog, acid rain, stratospheric ozone depletion and also influences climate by changing surface energy balance. Of particular concern is the increasing concentration of PM, especially $PM_{2.5}$, which is associated with adverse health effects such as increased respiratory diseases and premature death.

Particulate matter is an occupational hazard to agricultural workers either in or near confined environments (American Thoracic Society 1998). Respiratory issues for workers include the well-known farmer's lung, which is asthma symptoms caused by exposure to grain dust and various allergens (Warren 1977; Canadian Centre for Occupational Health and Safety 2008). The main toxic components in PM are inert materials like silica and organic materials such as bacteria, endotoxins and spores. Inert particles may cause inflammation of the air sacs or alveoli. Depending on the antigenicity of the material and host susceptibility, organic substances may affect the airways, and cause asthma, asthma-like symptoms, or chronic obstructive pulmonary disease (American Thoracic Society 1998). Inhalation of fungal spores leads primarily to parenchymal dysfunction such as hypersensitivity pneumonitis. Endotoxins found in moldy hay, grain, and similar organic materials may create a complex of constitutional symptoms known as the organic dust toxic syndrome. In Canada, a limit on permissible exposure level of 5 mg m^{-3} of grain dust over an 8-hour average period in the workplace to control short-term effects on workers was approved by the Standards Committee of the Canadian Thoracic Society (Becklake et al. 1996; Becklake 2007), while no limits were set on long-term exposure.

Workers and animals are also exposed to high PM concentrations in animal housing, especially in winter. The particles are made of feed grain dust, dried faeces, bits of hair, animal dander, pollen, insect parts, fungal spores, bacteria, peptidoglycans and/or endotoxins. Among livestock confinement operations, swine buildings (Cormier 2007) and poultry buildings (Guillam et al. 2007) present the highest health hazards. Although farm workers may adapt to the animal building environment, they still have a high occurrence of chronic bronchitis (Cormier 2007). The Canadian Centre for Occupational Health and Safety establishes limits to dust exposure of 5 mg m^{-3} for the inhalable fraction and 10 mg m^{-3} for all other unclassified dusts. Choinière and Munro (1997, 2001) indicate that these values are commonly exceeded inside pig and poultry housing.

11.3.3 Ambient PM Concentrations

The effect of PM on air quality may be reported as ambient concentrations or in emission inventories (Brown and Palacios 2005). The ambient concentrations indicate the mass of PM in a given volume of air, usually expressed in micrograms per cubic metre (μg m^{-3}), which can be directly compared to the ambient air quality standards and objectives. In 2006, the U.S. Environmental Protection Agency (EPA) revised the national ambient air quality standard in recognition of the adverse health and welfare impacts of ambient PM, and announced more stringent standards for $PM_{2.5}$ (US EPA 2006). The $PM_{2.5}$ standard for mean 24-hour concentration of the 98th percentile (3-year average) was lowered from 65 to 35 μg m^{-3} whereas the standard weighted annual mean concentration was left at 15 μg m^{-3}. The Canada-wide standard (CWS) for $PM_{2.5}$ (CCME 1999) was implemented by all provincial and territorial governments in Canada in 2010. The 24-hour standard was set at 30 μg m^{-3} based on the 3-year mean of 98th percentile ambient concentrations. The ambient PM concentrations are reported either from actual measurements or from modelling estimates.

Ambient PM concentrations have been monitored in Canada through The National Air Pollution Surveillance network since 1969. In general, ambient PM concentrations from all sources show a decreasing trend over time, for both TPM and $PM_{2.5}$ (Canada- United States Air Quality Committee 2004). One-year 98th percentile $PM_{2.5}$ concentrations in 2001 were greater than 30 μg m^{-3} at 17 out of 69 sites. The sites in Quebec and Ontario have higher $PM_{2.5}$ concentrations than those in western or Atlantic Provinces (Canada-United States Air Quality Committee 2004). Generally ambient PM concentrations are higher in urban than rural sites, but only 8 of 69 monitoring sites are rural.

11.3.4 PM Emission Inventories

Emissions inventories for a region are calculated by adding emissions from individual sources over a given time period. Particulate matter emission inventories help to indicate air quality trends. There is no direct relationship between emission inventories and ambient air quality because other processes such as long range transport, generation of secondary PM and deposition influence ambient PM concentration.

A complete estimate of the contribution of agriculture to PM should include secondary PM although agriculture contributes only the basic component of each particle while industry contributes the acidic component. This is a complex computation performed by meteorological models, which include atmospheric chemistry such as AURAMS (A Unified Regional Air-Quality Modelling System; Gong et al. 2006) and require information on the distribution of the PM precursor gases and pre-existing particles (National Agri-Environmental Standards Initiative (Anon 2010). The current version of AURAMS requires addition of coarse particle chemistry and improved estimates of emissions of soil dust. The 2008 Canadian Atmospheric Assessment of Agricultural Ammonia (Makar et al. 2010) reports that the impact of reducing ammonia emissions on the formation of secondary PM varies over the seasons and regionally, and that ammonia reduction might have more effect at high levels of $PM_{2.5}$ than at the median or average concentrations (see Ammonia section below).

Primary PM emissions occur from both natural and anthropogenic sources. The natural sources include natural fires (e.g., forest fires), volcanic ash and sea spray, etc. while the major anthropogenic sources of PM are combustion of fossil fuel by industry, power plants and automobiles, as well as wind blown dust from construction sites and agricultural land. Although urban sources have been extensively assessed, this is not the case for agricultural sources.

Agriculture operations have been recognized as a significant source of PM emitted into the atmosphere (Saxton 1996). In 2006, the Canadian Criteria Air Contaminants (CAC) emissions inventory estimated that agriculture accounted for 12 % of TPM, 18 % of PM_{10} and 4 % of $PM_{2.5}$ emissions in Canada (Environment Canada 2011). The CAC PM Emission inventory only accounted for emissions from animal operations, tillage and wind erosion. Thus, several other potentially significant sources, such as PM emitted during crop harvest, were ignored. Although PM emissions from agricultural sources appear to be of somewhat minor importance on a national perspective, these emissions are more prominent in regions with more agricultural activity than industrial activity. Agricultural PM emissions are often intense but short term and constitute an emerging air quality issue, especially at the boundaries between rural and urban environments.

11.3.5 The Main Agricultural Sources of Primary Particulate Matter

Primary fugitive PM from the agricultural sector is made of dust from soil and biogenic material from crops and livestock, droplets and particles from agrochemicals, bacteria, endotoxins and spores. The main causes and sources of agricultural emissions include wind erosion, land preparation, crop harvest, fertilizer application, grain handling, crop residue burning, animal feeding operations, animal cremation, chemical application and pollen and spore dispersion. Chemical application and pollen emissions contribute only to TPM, while crop residue burning contributes mainly to $PM_{2.5}$. The Agricultural Particulate Matter Emissions Indicator (APMEI) reported that wind erosion, crop harvest and land preparation were the main sources of all three PM size classes (TPM, PM_{10} and $PM_{2.5}$) in Canada (Pattey et al. 2010).

More than half of the total mass of agricultural PM emissions in Canada was generated by wind erosion. Wind erosion commonly occurs as wind blows across exposed dry surfaces of agricultural land, lifting entrained particles into the air (Fig. 11.1). Even on wet soils, the wind may dry a thin layer leaving it susceptible to erosion. While wind erosion is an issue in many areas it is most prevalent on sandy soils because of typically poor aggregation of particles. The most intense cases of wind erosion are reported in the semi-arid prairies of AB & SK where the growing season is short and cover crops relatively uncommon (Fig. 11.2). There are several well established practices for reducing soil erosion that are classified as Beneficial Management Practices (BMPs). The risk of wind erosion can be reduced by retaining crop residues because they physically shield the soil surface from wind and keep the soil surface moist. Similarly, reducing summerfallow has lessened soil erosion in western regions of Canada. Summerfallow is a practice of leaving fields unplanted every second or third year to increase soil water and nitrogen reserves and provide a hedge against dry years. Controlling weeds with tillage during the fallow year leaves the soil highly subject to erosion whereas controlling weeds with chemicals helps to reduce soil erosion. On sandy soils that are especially subject to erosion, perennial forages can be grown. Where sandy soils are used for production of potatoes or sugar beets, winter cover crops can be planted after harvest (Lobb et al. 2010). Other BMPs consist in planting and preserving windbreaks and shelterbelts to protect the soil surface, to reduce the soil evaporation and to retain the snow cover in fields. They may also protect the crops from lodging. Since 1901, the Prairie Shelterbelt Program (http://www4.agr.gc.ca/AAFC-AAC/display-afficher.do?id=1286907236630&lang=eng) actively contributed to land conservation by providing both technical services and adapted tree and shrub seedlings to farmers in the Prairie Provinces (MB, SK, AB) and part of BC.

Tillage is a common practice across Canada both to prepare the land for seeding and to control weeds. Particulate matter emissions are generated from airborne soil particles during tillage due to the mechanical disturbance of the soil surface (Fig. 11.3). The intensity of PM emissions depends on the soil, machinery type (e.g., plough, disk, and cultivator), and number of tillage operation per year. In order to reduce soil degradation and preserve soil organic matter and moisture, no-till and reduced tillage practices were successfully introduced to the Canadian Prairies. Since the benefits of minimum tillage are less obvious in Eastern Canada and the Lower Fraser Valley (BC), traditional tillage practices have remained an important source of PM emission in these regions (Fig. 11.2). Higher PM emissions occur when ambient relative humidity is low, and when tilled soil layers have less aggregates and moisture and more silt (Holmen et al. 2001). Canada-wide, land preparation contributes to 28 % of TPM and $PM_{2.5}$ emissions and 14 % of PM_{10} emissions (Pattey et al. 2010).

Crop harvesting emits 15–20 % of agricultural PM emissions (Figs. 11.3 and 11.4). Combine harvesters are used to harvest corn, soybean, flax, canola and small grain crops such as wheat. After cutting the plants, combines separate

Fig. 11.1 Wind erosion episode on November 13 2007 near Swift Current (SK). Visibility was substantially reduced for several hours while winds were blowing at more than 70 km/h (top pictures). Wind erosion was intense over a doubly chemfallowed lentil field (*left* bottom picture) while it was non existent over a nearby tilled fallow field (*right* bottom picture). Pictures taken by Brian McConkey, used with permission

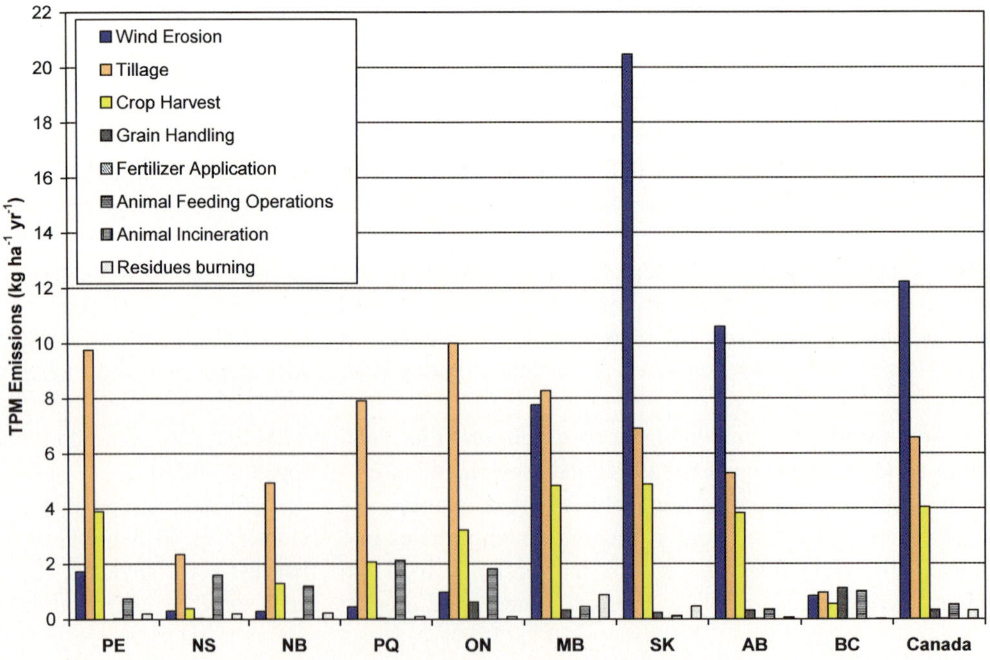

Fig. 11.2 Annual emissions of total particulate matter (TPM) from the various agriculture sources in Canada by province (NFD not reported)

Fig. 11.3 Dust generated during tillage (top *left*) and harvest (top *right*) of a corn field in Ontario (Pictures taken by Dr. E. Pattey's research team, used with permission) and during harvest of canola fields in Saskatchewan (Pictures taken by Brian McConkey, used with permission)

the grains or seeds from the rest of the plant, clean and store the grains, and return the stems and leaves with other debris to the field surface. These operations generate PM composed of grain and other plant material, but they might also emit moulds, pollen, spores, silica, bacteria, fungi, insects and pesticide residues. Not much is known about reducing PM emissions during harvest. Good maintenance and optimum operation of combine harvesters, and keeping fields free of pests, especially fungal diseases, helps to mitigate PM emissions. Also, avoiding harvesting during windy and dry conditions helps reduce the dispersion of grain debris and soil particles. Modern combine harvesters are equipped with filtered and air conditioned cabins to protect the operator.

Burning crop residues has been, historically, the most common practice for reducing the impact of leftover plant materials from agricultural land in Canada and worldwide. Residues can interfere with plant emergence, host disease organisms and keep the soil cool and moist hindering early plant growth. However, not only does burning crop residues eliminate protective soil cover but it also emits substantial amounts of PM and other pollutants into the atmosphere. Of agricultural emission in Canada, crop residue burning contributes 13% of $PM_{2.5}$ emissions and less than 3% of coarse particle emissions. Crop residue burning is no longer recommended in Canada as other residue management BMPs exist, such as chopping, removing or deeply cultivating plant materials. Marginal use of this practice continues across Canada (<2%) with slightly higher rates in SK (4%) and MB (8%).

Grain handling, pollen distributed by wind (mostly from corn or certain trees), animal feeding operations, livestock incineration, chemical fertilizer and other agrochemical applications, each contributes a small percentage (<2%) of Canadian PM emissions from agricultural operations (Pattey et al 2010; Pattey and Qiu 2012). However, they can locally affect mammals and unmanaged environment. Provincial regulations regarding distance between operations and population, conditions of usage and application exist to minimize impact on the public.

Fig. 11.4 PM emissions during crop harvest in the Prairies. Pictures provided by Doug McKell, P.Ag., used with permission

11.3.6 Trends in Canadian Agricultural Emissions of Primary PM

In 2006, estimates of PM emissions from agricultural operations in Canada were 1.64 Mt for TPM, 0.65 Mt for PM_{10} and 0.16 Mt for $PM_{2.5}$ (Pattey et al. 2010). Most agricultural PM (>90 %) was emitted in the Prairies (Fig. 11.5). A decrease in PM emissions between 1981 and 2006 was estimated for all provinces except Quebec which already had low emissions (Fig. 11.5). An overall reduction of 40–48 % was achieved over the 25-yr period despite increases in animal populations, fertilizer application and cropped land area over the period. Adoption of soil conservation practices in the Prairies in response to cropland degradation was the major cause of the reduction in PM emissions (Pattey and Qiu 2012). Indeed, producers adopted conservation tillage/no-till practices, increased usage of cover crops and reduced summerfallow, in order to increase the content of organic matter and moisture of their fields. Introducing and maintaining windbreaks and shelterbelts also reduce wind erosion. These practices produced a major co-benefit for air quality: lower CO_2 emissions from soils.

11.3.7 Other Beneficial Management Practices to Reduce the Nuisance of PM Emissions

Increasing amount and duration of soil cover with living plants or crop residues reduces exposure of soil to wind hence reducing the risk of wind erosion and emission of PM from fields. Field practices that increase soil cover include reduced tillage (or no-tillage), continuous cropping (no summerfallow), planting perennial forages, in permanent stands or in crop rotations, and planting fall cover crops. Strip cropping and contour cultivation are other BMPs contributing to reduce wind erosion. Using herbicides instead of tillage to control weeds before planting and in fallowed fields reduce PM emissions although herbicide droplets may drift during application.

PM emissions during crop harvest are generated from combines and vehicles traveling over the fields. PM emissions from harvesting vary with the crop type. While there are few specific practices that reduce PM emissions from crop harvesting operations, harvesting during high relative humidity and low wind speed conditions help to reduce PM emissions. Some practices used for controlling wind erosion will also limit the transport of harvested crop fragments in the wind. Reducing tillage and maintaining crop residues will decrease PM emissions from vehicles traveling in the fields.

Particulate matter is emitted from fertilizer application through suspension by the wind or soil disturbance by land preparation and wind erosion. Combining injection of fertilizer with seeding reduces field operations and obviates the need for incorporation of the fertilizer, which help to reduce PM emissions. Delaying fertilizer application into growing crops is more commonly done and this should minimize emissions due to the crop cover.

The application of pesticides to croplands is widely practiced in Canada although the number of applications is much lower than in some countries. Despite improving productivity, the drift of these chemicals due to operator error and non-optimal environmental conditions may contribute to PM emissions. Although these emission estimates are currently very low compared to other agricultural sectors, chemical drift may be further reduced by restricting chemical application to calm conditions and by using nozzles with larger droplets.

Primary PM from animal feeding operations is generated through animal activities in barns or feedlots. PM emissions can be reduced by changing the production environment, such as decreasing the confinement time (or increasing the grazing period), collecting litter and manure more frequently, installing dust extraction or ventilation systems and using mist or oil sprinkling to reduce dust. Windbreaks

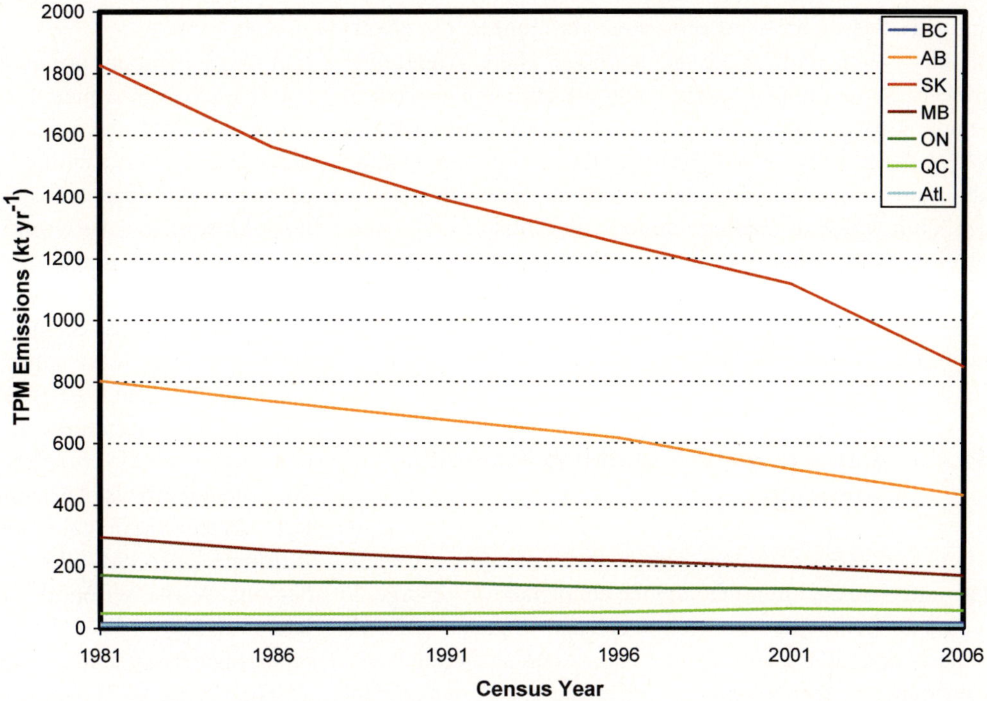

Fig. 11.5 Annual emissions of total particulate matter (TPM) from agriculture sources in Canada by province (NB, NS and PE are grouped as Atlantic Provinces Atl.)

around facilities may help in deposition and to reduce dispersion of dust.

11.4 Atmospheric Ammonia in Canada: Sources, Effects and Mitigation

11.4.1 Ammonia Emission: A Public Concern?

Ammonia concentrations in confined spaces such as animal housing and enclosed composting facilities can reach levels of concern to both comfort and health of livestock and humans. Ventilation, both active (fans) and passive, is used in animal housing to improve air quality, and lowering ammonia levels in housing is one of the key objectives of ventilation. Poor air quality in animal housing reduces the performance of the animals suggesting also a reduction in well-being (see PM section).

In unconfined spaces, ammonia does not reach concentrations where it is generally a direct concern to humans, except near sources such as ventilation systems where ammonia is an important constituent of livestock odours. In recent years there has been a growing understanding of the more subtle effects of atmospheric ammonia and this has caused considerable concern in many countries. The concern relates to the possible impact on both human and environmental health.

Effect on human health relates to the formation of $PM_{2.5}$ when ammonia, a basic gas, reacts with one of several acid gases in the atmosphere, notably NO_x and SO_x forming the salts NH_4NO_3 and $(NH_4)_2SO_4$. Such fine particulates can be deleterious to health by entering into the alveoli of lungs potentially causing inflammation which can result in cardiovascular symptoms. At present it is not known if ammonium-containing particulates contribute to particulate toxicity in humans (see Chap. 12).

As a basic gas, ammonia deposited on trees increases bark pH and thus shifting populations of epiphytic lichens on trees from those that are acidophilic to those that are nitrophillic (Sutton et al. 2009a). The impact of ammonia on lichens, can be seen at the local, regional and even the national scale in the UK, and lichens are now used widely as indicators of atmospheric ammonia (Sutton et al. 2009b). Recent work on lichens and sensitive higher species has led to the proposal that the critical atmospheric level for long term exposure be set at 2–4 $\mu g\ m^{-3}$ with the lower level for the lichens (Cape et al. 2009). Once deposited on soils, ammonia converts to ammonium except under high pH conditions and enriches the soil thereby favouring plants that respond well to nitrogen, such as grasses, over plants that are adapted to low fertility. Low nutrient (oligotrophic) ecosystems such as bogs are very vulnerable to loadings of atmospheric ammonia. Furthermore, in the soil ammonium oxidizes (a process called nitrification) to nitrate, releasing hydrogen ions which acidify the soil. In Europe, ammonia is currently considered an increasingly important acidifier compared to sulphur dioxide or other atmospheric pollutant (Sutton et al. 2009). Likewise ammonia is considered a prime source of nitrogen enrichment, often exceeding NO_x, though its relative importance varies regionally, depending on the importance of agriculture compared to transportation and manufacturing activities. In terms of ecosystem enrichment, the effects of ammonium and nitrate are additive. There is certainly concern about anthropogenic ammonia enrichment even in remote areas of North America such as the alpine tundra of WA, CA and CO (Jovan 2008) but there is little information about the effects of ammonia on Canadian ecosystems. The economic value of abating ammonia in terms of ecological services such as biodiversity is difficult to estimate but the benefits of abatement in terms of impact on climate change were recently estimated in the UK as equivalent to benefits for health (Smart et al. 2011).

11.4.2 Reporting and Regulating Ammonia Emissions

There are several initiatives worldwide to monitor and regulate emissions of ammonia into the atmosphere. The Organization for Economic Cooperation and Development (OECD) collects and reports emission inventories on an annual basis. Canadian emission inventories are prepared by Environment Canada (2011). In recent years, AAFC has assisted Environment Canada to develop and improved the inventory from agriculture which was previously in use (Anon. 2010). Many European countries have lowered their ammonia emissions while emissions in Canada and US have risen between 1990 and 2001 (Table 11.1)

In terms of acting to reduce ammonia emissions, the EU member states have taken the initiative. They have mandatory emission ceilings for several pollutants including ammonia under the National Emission Ceilings (NEC) directive (2001/81/EC) adopted in 2001. The binding emission ceilings were to be achieved by each member state by 2010 and not to be exceeded thereafter. In parallel to the NEC Directive, the European Economic Commission for Europe (UNECE)[1] has negotiated the Gothenburg Protocol under the Convention on Long-Range Transboundary Air Pollution in 1999. The emission reductions in the protocol were often less ambitious than those in the NEC Directive but encompassed more countries. The Gothenburg protocol is presently under revision with the intention that an amended version will be ready at the end of 2011. (The revision will consider intercontinental transport of air pollutants, but these will not include ammonia.) Ammonia emission is an important factor in soil N balances which Canada reports to the OECD; ammonia emissions are considered outputs and improve balances. In the EU, soil N balances are covered by the Directive on Integrated Pollution Prevention and Control which does not recognize ammonia emission as an output.

As a member of UNECE, Canada has signed but not ratified the Gothenburg protocol. The protocol exempts Canada and the US from Annex 9 on ammonia abatement practices based on the argument that while ammonium can travel long distances, ammonia emitted in North America is unlikely to cause harm in Europe. Instead, ammonia is dealt with in bilateral discussions between US and Canada. An interdepartmental study (National Ago-Environmental Standards Initiative) between Environment Canada and Agriculture and Agri-Food Canada recently produced a report called "The 2008 Canadian Atmospheric Assessment of Agricultural Ammonia" (Anon. 2010). This report represents the state-of-the-art on agricultural ammonia in Canada and its role as a precursor to PM formation while generally ignoring its environmental effects. Agriculture and Agri-Food Canada provides an environmental report card on agriculturally sourced ammonia, (also PM, odour and greenhouse gases) under the National Agri-Environmental Health Indicator Program (Eilers et al. 2009). This chapter makes extensive use of these reports and associated publications.

The USEPA (Environmental Protection Agency) has shown concern about ammonia emission from point sources, defined as animal housing and manure stores on Concentrated Animal Feeding Operations (CAFOs). The National Air Emissions Monitoring Study (NAEMS) begun in 2005 resulted from an agreement between the EPA and the dairy, pig and poultry sectors. The beef sector, small to medium farms, and diffuse emissions form from land applied manure or fertilizer were excluded although these are important sources. The intent of NAEMS was to ensure compliance with: Clean Air Act, Comprehensive Environmental Response, Compensation and Liability Act, and Environmental Planning and Community Right-to-Know Act. Importantly, the NAEMS program also promotes a national consensus on methodologies for estimating emissions from CAFOs.

In Canada, atmospheric concentrations of ammonia are monitored under the CAPMoN program which focuses mainly on southern Ontario with a few sites scattered across southern parts of the country. The US recently implemented a long term program for tracking ambient ammonia under the long-running program for tracking wet deposition of pollutants including ammonium (NADP). The goal of the US program is to assess the long-term trends in ambient ammonia concentrations and deposition of reduced nitrogen species, to validate atmospheric models; to better estimate total nitrogen inputs to ecosystems; to assess changes in atmospheric chemistry due to SO_2 and NO_x reductions; and to assess compliance with $PM_{2.5}$ standards.

[1] 56 members includes EU Member States, Switzerland and Norway, along with Caucasus, Central Asia, and Eastern European countries (EECCA), the United States and Canada.

Table 11.1 Estimated ammonia emissions from agriculture for selected countries. (OECD 2008)

Country	Average annual emission from agriculture, kt of NH_3		Agricultural emission, % of total
	1990–1992	2001–2003	2001–03
Canada	468	482‡	80
United States	3,421	3,945†	88
France	744	742	97
Germany	645	580	95
Italy	454	411	94
Spain	317	383	93
Poland	407	317	97
Japan	–	289	–
United Kingdom	302	277	89
Korea	143	181	–
Netherlands	236	123	90
Australia	–	61	–

† Anon (2010) reported emissions for the year 2002 from Canadian agriculture at 432 kt with 26% arising from fertilizer use
‡ USEPA (2011a) estimates emissions for the year 2008 from United States agriculture at 3,260 kt with 35% arising from fertilizer use

11.4.3 Ammonia Emissions

Unlike other major atmospheric pollutants, the primary source of atmospheric ammonia is agriculture. Worldwide annual emissions from agriculture were estimated at 18,000 ktonnes, with most emissions originating from application of fertilizers in tropical and subtropical countries (Bouwman et al. 2002). Of the 505 ktonnes of ammonia emitted annually into the atmosphere in Canada, only 14% is from non-agricultural sources which include industry, transportation and combustion (Fig. 11.6). The total agricultural emissions are 432 ktonnes per annum (359 ktonnes of nitrogen). This is 21% of the 1,680 ktonnes of commercial nitrogen fertilizer used annually, indicating that conserving ammonia on farms could significantly lower fertilizer consumption in Canada. Agricultural emissions are comprised of mainly livestock (64%) and fertilizer application sources (22%). Pesticides which are included in the fertilizer category are a negligible source of ammonia. Fertilizer is a more important source of emissions in Canada than in Western European countries (typically less than 15%) because of the relative importance of the crop sector in Canada. However, worldwide 59% of emissions come from fertilizer application and about 73% of this is emitted in tropical or subtropical countries (Bouwman et al. 2002). Ammonia loss during manufacture and transportation of fertilizers is included with non-agricultural sources. Emissions from other agricultural sources, such as those from crop surfaces and silage are not included in the Canadian inventory. Other anthropogenic sources such as pets, waste treatment plants etc. are also not included, nor are non-anthropogenic sources such as natural vegetation and wild animals. In most of these cases there is a lack of data and the sources are difficult or impossible to control. Nonetheless, from the standpoint of modelling the fate of emitted ammonia, these sources may need to be considered in future. Uncertainty in national emission inventories is under considerable attention; for inventories having local emission factors the emission estimate is thought to have a certainty of+/- 20% (Webb et al. 2009). While uncertainty of the Canadian emission model has not been determined, estimates are very sensitive to values relating to N excretion rates from livestock (Sheppard et al. 2007a).

Ammonia sources from agriculture in Canada, in order of magnitude, are beef cattle (37%) (Sheppard and Bittman 2012), fertilizer (26%) (Sheppard et al. 2010b), swine (17%) (Sheppard et al. 2010a), dairy (12%) (Sheppard et al. 2011) and poultry (6%) (Sheppard et al. 2009) (Fig. 11.6). The current estimate of cattle emission is less certain than those of other sectors at present due to scarcity of Canadian emission measurements (Sheppard and Bittman 2013). Highest emissions naturally occur in regions with highest concentrations of livestock, namely the lower Fraser Valley of BC, southwest and central Alberta, southeast Manitoba, southern Ontario, and southern Quebec (Fig. 11.7). Emissions from fertilizers are more diffuse but take place from a larger area, mainly the crop growing regions of the prairies, and Saskatchewan has the greatest area of cropland (40% of Canada's cropland).

Ammonia emission varies seasonally across Canada with highest rates occurring in May. Land application of fertilizer and manure is greatest in May and the warmer temperature increases emissions from manure storages and unheated or open housing for animals (Fig. 11.8). Emissions are very low from November to March when there is little land application and many emitting surfaces are very cold or frozen. Emissions are curtailed during periods of rainfall when wet soils constraint field activities (Sheppard et al. 2007b).

11.4.4 Farm Sources of Ammonia

The source of most ammonia from livestock operations is the urea contained in excreted urine (uric acid in poultry). Urea on soil and other surfaces breaks down rapidly into ammonia and carbon dioxide because the urease enzyme is almost ubiquitous. The hydrolysis may occur even in low temperatures making ammonium the predominant soluble form of nitrogen in farm manures. Poultry litter may contain some uric acid because of slower hydrolysis. The proportion of volatile ammonia (NH_3) to dissolved ammonium (NH_4^+) increases with pH, so maintaining a low pH will reduce ammonia volatilization.

To calculate emissions and identify possible mitigation methods it is important to understand the loss pathways on farms (Fig. 11.9). Volatilization of ammonia takes place in livestock housing, feedlots, pastures, manure storages and

Fig. 11.6 Sources of ammonia emissions in Canada in 200; top is all sources and bottom is agricultural sources. (From Ayres et al. 2010)

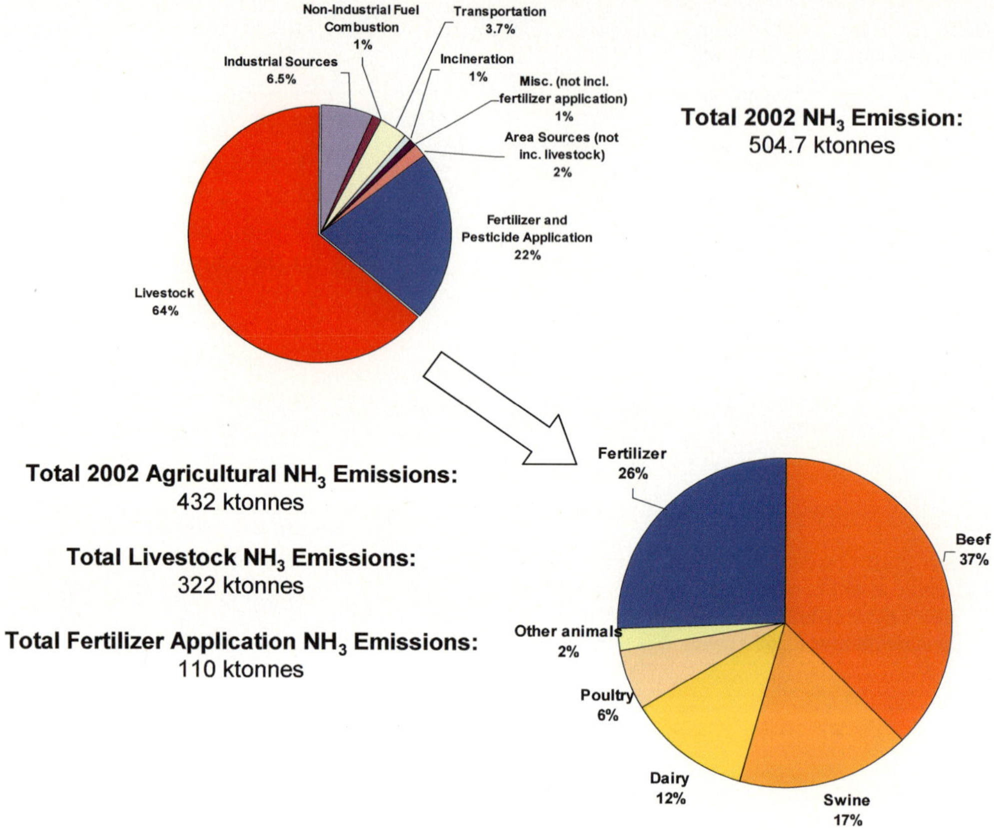

after land application of manure. The amount of urea excreted varies with the animal and type of feed; in all animals high excretion rates result from overfeeding protein. Advanced methods for avoiding overfeeding protein include phase feeding and supplementing amino acids. Feed mills can analyze feed ingredients quickly and cheaply allowing precise formulations and reducing incidents of excessive protein.

There are many types of animal housing, and their most important attributes with respect to emissions and emission control is type of manure system (solid or slurry) and type of ventilation (active or passive). Emissions occur from soiled surfaces, and so any practices that increase the surface area of excreta will increase ammonia emissions. Thus, the lower animal housing densities (e.g. free stall dairy barns) introduced for economic or animal welfare reasons may in fact inadvertently increase ammonia emissions. In Canada, most pigs (growers and sows) are raised on liquid manure systems and most table birds (broiler chickens and turkeys) and confined beef cattle are raised on solid manure, also called farmyard manure, systems which use a lot of straw or sawdust bedding. Dairy cattle systems are divided, with the newer and larger facilities almost exclusively handling manure as slurry. There are more tools for mitigating ammonia emission from liquid systems. Liquid systems tend to have lower emissions from storage and during land application than solid systems. It is generally difficult and costly to control emissions from housing, although there are now designs for new pig barns that reduce emissions by 20–30 % with minimal additional cost. Such innovations cannot be retrofitted to existing barns and are not yet available for dairy cattle. However, type and amount of bedding may affect emissions for housing (Misselbrook and Powell 2005). For table birds certain materials that adsorb ammonia or acidify the bedding can be added to reduce emissions but this is not yet commonly done due to costs. A new process for acidification of swine slurry with sulphuric acid, commercialized recently in Denmark, effectively reduces emissions from houses and storages, and land application. Lowering bird densities in barns for animal welfare (e.g. cages to aviaries for laying hens), as has been done in Europe, increases ammonia losses if no counter measures are taken. It is technically possible to install ammonia collecting filters on actively ventilated buildings, which are used for most pigs and poultry, but such filters are seldom installed in Canada because they tend to reduce fan efficiency and require substantial maintenance. Such filters are not possible in passively ventilated housing typically used for both dairy and beef cattle. Beef cattle in the finishing stages are often raised in outdoor feedlots in Western Canada; the animals are well adapted to cold, their health outdoors is better than indoors, and the feedlots are cost effective. Emissions from outdoor sources are most difficult to control. Very high emission rates (>50 % of feed protein) have been measured from beef feedlots (McGinn et al. 2007; van Haarlem et al. 2008).

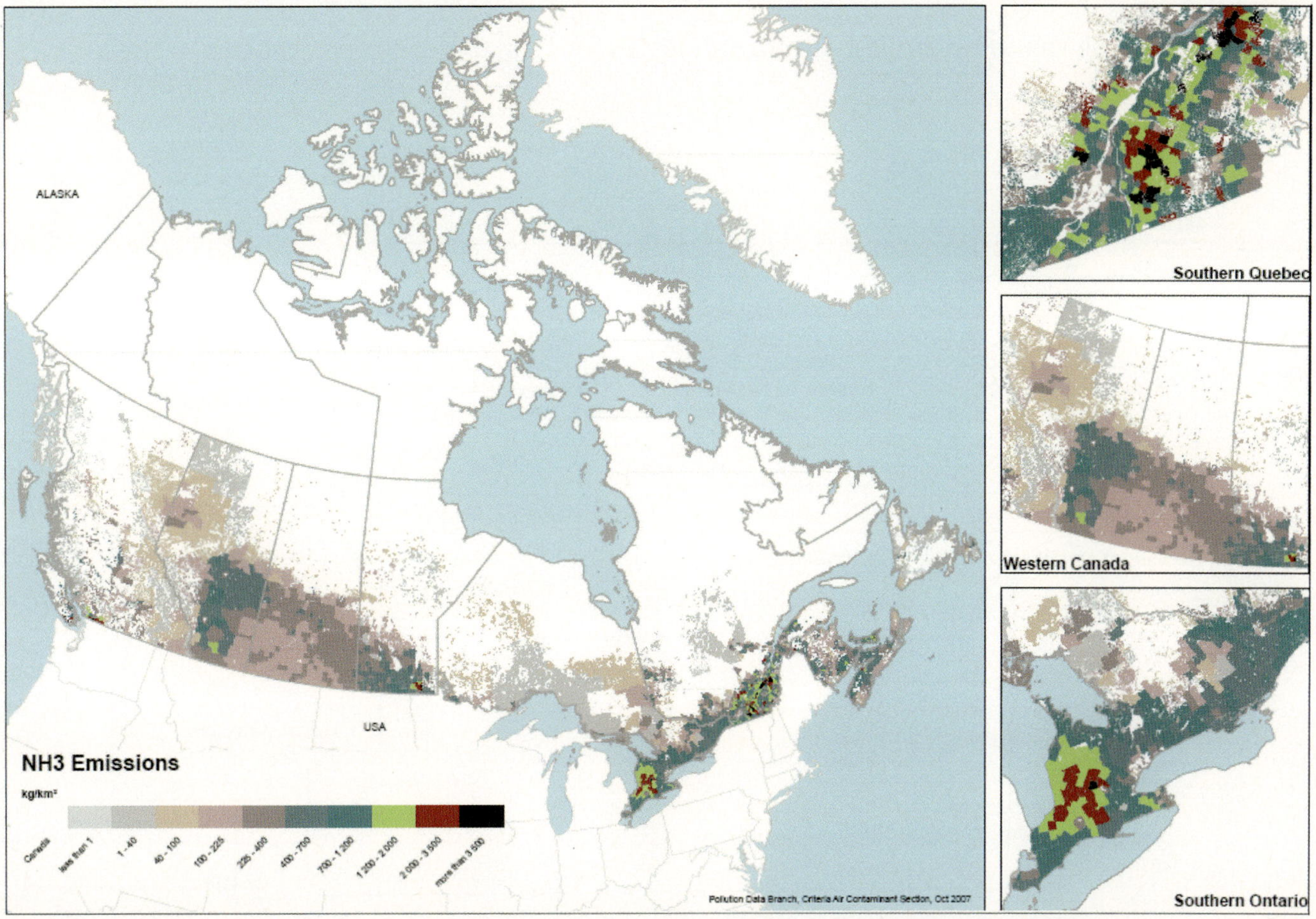

Fig. 11.7 Annual ammonia emission (kg NH_3 km^{-2}) on 42 km grid from all agricultural sources in Canada. (From Ayres et al. 2010)

Fig. 11.8 Monthly emissions of ammonia (tonnes NH_3) from all agricultural sources in Canada. (From Ayres et al. 2010)

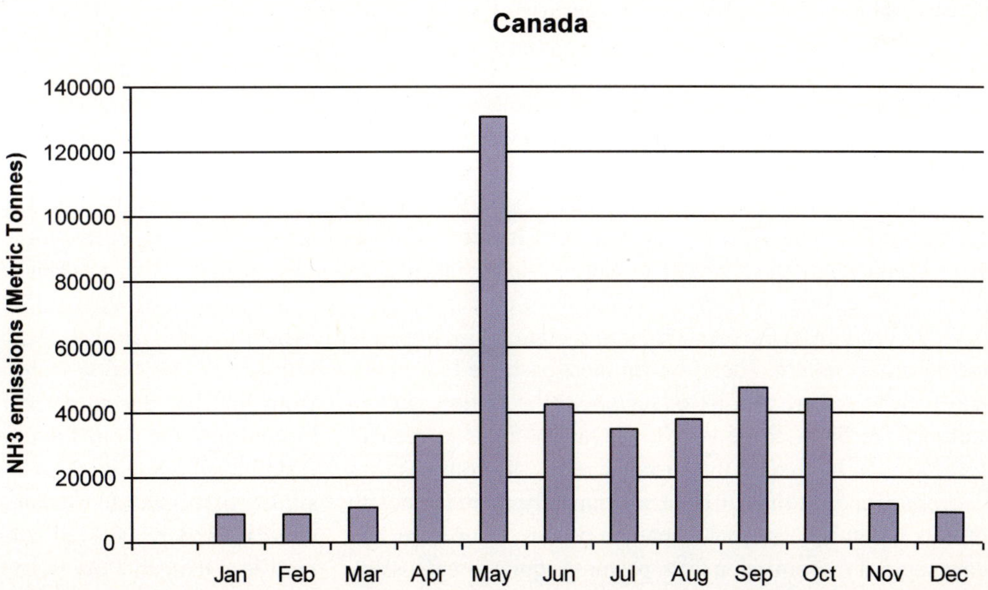

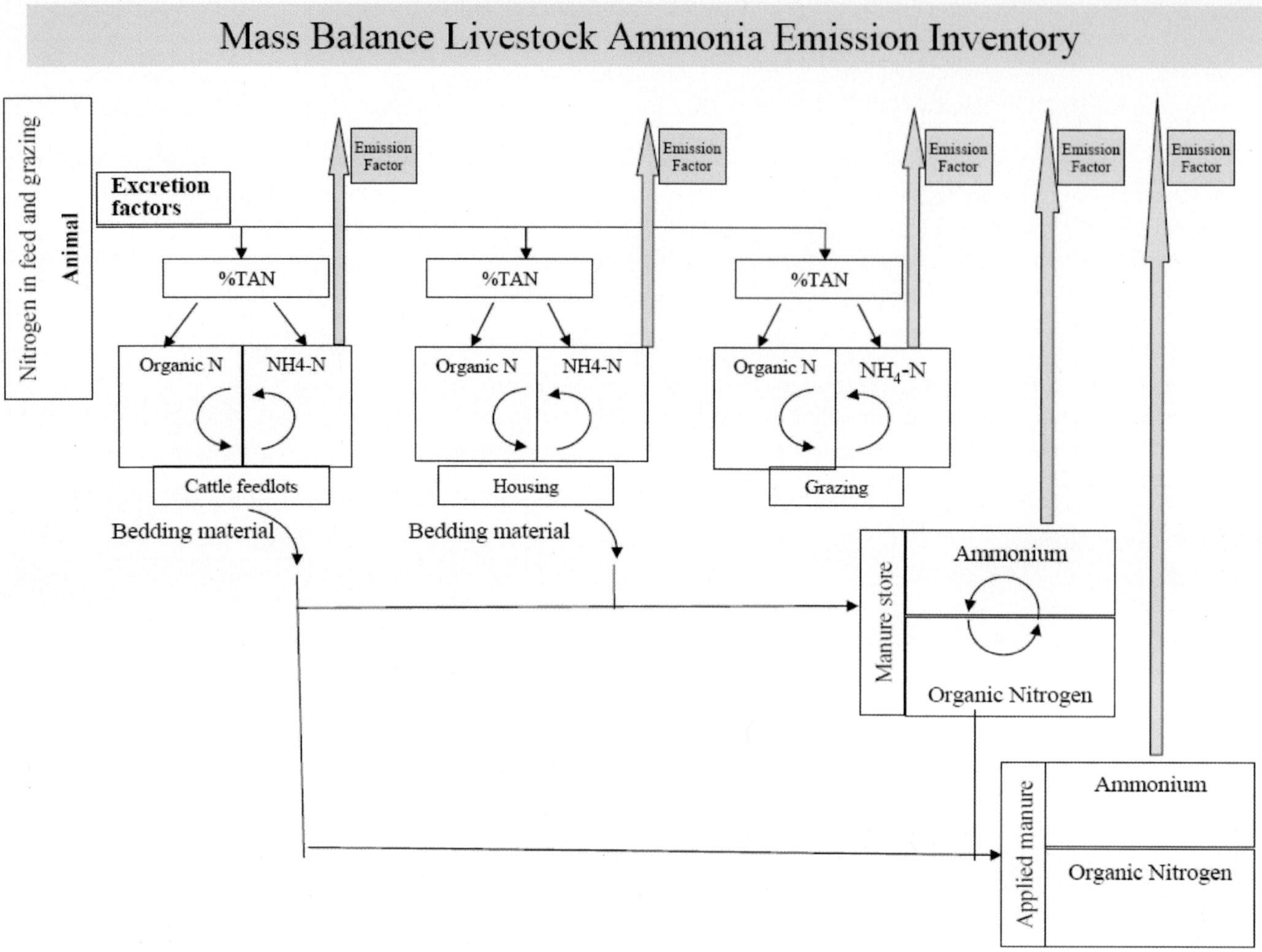

Fig. 11.9 Emissions sources of ammonia showing the structure for calculating farm emissions for inventories; TAN refers to total ammoniacal nitrogen which is the total of ammonia and ammonium. (From Bittman et al. 2010)

Cows kept for production of calves typically spend much of their time on pasture and depend on available pasture herbage for their complete diet. Because of the spring flush of vegetation growth, there are high urea excretion rates in spring when herbage is very rich in protein. In much of western Canada where the greatest amount of grazing takes place there is often very little growth after the first flush due to summer drought so the plants mature accumulating fibre and losing crude protein. Therefore for most pastures in Canada, except where there is a lot of legume, N excretion rates in summer are likely to be very low. Quality tends to recover somewhat in fall before the cows are removed from pastures (Sheppard and Bittman 2011). Unfortunately, there is a dearth of data on emissions from Canadian pastures, but European data suggest that emission from pasture is quite low (Bussink 1994). This is because excreted urine rapidly infiltrates into soil and ammonium released within soil is quickly adsorbed to negatively charged sites on clay organic matter. In Canada, extensive pastures have compacted or wet areas around corrals and water troughs and these areas may have lower rates of infiltration hence higher emissions than reported in Europe, but there are few data (Sheppard and Bittman 2011).

In Canada most farmers have at least 6 months of manure storage allowing spreading to take place in early spring before planting and, in drier regions, also in fall. Wintertime application on frozen soil leads to manure runoff and summer-time application is on growing crops which is usually impractical. Emission from storage is generally lower than from housing due to a smaller exposed surface area. Ammonia emission from storage can be substantially reduced by installing flexible or rigid covers and there is a greater possibility of mitigation for storage by installing flexible or rigid covers (vanderZaag et al. 2008). Deep storages such as tanks emit less than shallow storages and are easier to cover than lagoons, but few storages of any kind are covered in Canada. In fact, crusts that form naturally in slurry storages efficiently impede emission for at least part of the year (vanderZaag et al. 2008). Covers for solid manure are not well studied and are rarely used in Canada except to reduce runoff of nutrients from poultry litter during the rainy winters in coastal BC. Also, emissions from composting manures may be high if piles are left uncovered. Bio-digesters increasingly being used to produce methane from manure (often slurry) leave an ammonium-rich residue with high pH that needs to be handed carefully to avoid emissions.

The amount of ammonia loss during land application of manure will depend first on how much of the excreted urea in the manure remains after losses in housing and storage. Cool, overcast and rainy weather is associated with low emissions rates from manure applied on land. During warm and sunny weather, field losses can be controlled by injection (liquid manure) or rapid incorporation (both solid and liquid manure), although incorporation is not possible in forage and in no-till cropping systems. Injection is especially favourable in the arid Prairie regions where injection can be done in the fall after harvest, and the nutrients are largely preserved over winter because of nil leaching and freezing temperatures. With injection not an option, solid manure applied to forages or no-till land may be have high losses unless the ammonia was lost from the manure in storage.

Finally, ammonia is lost when urea or ammonia containing fertilizers are applied on soils. While rare in many European countries, urea containing fertilizers are very common in Canada, the USA and many other parts of the world because of their low cost and safety (Sheppard et al. 2011). As with manure, fertilizer urea hydrolyzes quickly with urease enzyme which is abundant in soils, but hydrolysis of urea granules cause a sharp increase in pH near granules which may result in particularly large emissions, such as when there is a lot of crop residue present (Rochette et al. 2009). However, when applied in bands a few cm below the soil surface, emissions from are greatly reduced. This is practiced widely for grain crops in Canada, but is not possible in most forage fields, which are therefore prone to high losses. Technologies designed to slow urea hydrolysis, such as urease inhibitors and coatings, lessen the pH spike around granules and may help to reduce emissions from surface applied urea. In parts of western Canada, gaseous ammonia is an important and effective fertilizer. Typically, the ammonia is injected of into soils in fall when temperature and moisture help to preserve the ammonia in the soil.

Fertilized crops may emit ammonia through open stomata when internal partial pressure exceeds that in the ambient atmosphere. Emissions from crops are highest after fertilizer application, after plants are cut for forage, or when the leaves senesce.

11.4.5 Fate of Ammonia in the Atmosphere

Ammonia has several unique attributes among the trace gases that are atmospheric pollutants. Typically emitted close to or at ground level, ammonia is likely to contact surrounding surfaces as it is carried in eddies. Ammonia will adsorb strongly to negatively charged sites in soils and plants and to any wet surface. Adsorption of gaseous ammonia to surfaces is referred to as dry deposition. Ammonia molecules that enter into substomatal cavities in leaves through open stomata may be incorporated into plant proteins. This occurs when ambient concentrations are high, especially near point sources, and internal leaf (apoplastic) concentrations are low (low plant N content). As mentioned above, transport is away from plants when the gradient is reversed.

There have been few Canadian studies on dry deposition of ammonia. The methodology is relatively complex and the deposition rates are variable because they depend on the nature of local surfaces; smooth surfaces such as bare soils have low rates than rough surfaces such as trees because of effects on airflow (Loubet et al. 2009). Previously, Asman and van Jaarsveld (1992) modelled deposition rates of over 40 % for typical European land cover. In contrast, deposition around feed lots in southern AB has recently been estimated at only 5–10 % of emission and the difference was attributed to drier surfaces in AB than in Western Europe (Staebler et al. 2010). A simple technique using soil sorption collectors over several months has produced deposition estimates downwind of feedlots in southern AB of about 19 % (Hao et al. 2006). Using a similar technique we have recently estimated depositions hotspots with over 20 kg NH_3-N ha^{-1} over a 24 hour period near principle fans on broiler barns in south-western BC but the deposition declines sharply with distance from the fans (Bittman unpublished data). Deposition hotspots near housing may lead to local high rates of emissions of N_2O at about 0.2 to 15 % of deposited N (Skiba et al. 1998). Research is underway in coastal BC to test if deposition hotspots are a pathway for nitrate into ground water. Note that when emission rates are determined from measurements taken more than 100 meters downwind of source, the emissions are net of local deposition, whereas measurements taken at the source are pre deposition (Staebler et al. 2010). Local deposition will affect nitrogen balances and may lead to local hotspots for N_2O emissions or even leaching. Although the quantities often are relatively low, some ammonia may be locally deposited in rain (referred to as wet deposition).

Atmospheric concentrations of ammonia are a determined by rates of emission, dry deposition, chemical reaction and atmospheric dispersion and transport. As expected, ammonia concentrations in Canada are highest in regions of highest livestock emissions, notably the lower Fraser Valley of BC, central and southern AB, and southern MB (Vett et al. 2010). In these regions median concentrations (generally 3–7 day averages) ranged from 2.4 to 19.3 μg m^{-3}. Somewhat lower concentrations (1.2 to 2.6 μg m^{-3}) were found in southern ON and southern QC, whereas moderate levels (~1.0 μg m^{-3}) were reported for southern SK, south central

ON and some parts of the Maritimes. Some of Canada's cities (Toronto, Montreal, Windsor, and Burnaby) located near farm sources may have elevated levels of ammonia. Very low concentrations ($<0.5\ \mu g\ m^{-3}$) have been measured in rural areas with little agricultural activity. Ammonia in such remote areas may arise from natural vegetation or may be transported into the area, although long range transport of gaseous ammonia is limited. The importance of emissions from farm animals was well demonstrated in a study conducted near Abbotsford, BC in 2004 when poultry barns were emptied due to an outbreak of Avian Influenza. During the cull period concentrations of gaseous ammonia averaged 11 $\mu g\ m^{-3}$ compared to 30 $\mu g\ m^{-3}$ at the same site and same time period of the following year when the barns were fully stocked (Staebler et al. 2010; Vett et al. 2010). Similar results showing the importance of source strength were reported for the foot and mouth outbreak in the UK in 2002 (Bleeker et al. 2009).

For understanding the fate of ammonia in the atmosphere it is important to note that emission and concentrations vary over time, not only seasonally but also hourly (Vett et al. 2010). Emissions of ammonia are higher in spring and summer compared to winter because manure application and emissions from unheated facilities are greater at these times. Because land spreading of fertilizer and manure are the major source of ammonia, emissions often peak when there is spike in land spreading activity. In locales where spring rain and soil conditions restrict the number of spreading days, the emissions may be particularly intense during favourable weather (Sheppard et al. 2007b). Atmospheric concentrations also have a diurnal cycle that follows livestock and indeed human activity on farms, with highest concentrations generally at midday associated with animal feeding or manure handling and lowest during night-time. The more stable and cool night-time air supports less emission especially from field sources and air from heated buildings will rise more rapidly at night; however, as air at night is stable there is less dispersion. Thus, the lower ambient concentrations reported at night-time are due to a combination of factors.

As the only important basic gas in the atmospheric, ammonia reacts readily with the dominant acids H_2SO_4 and HNO_3 which are often abundant (Loubet et al. 2009). When ammonia concentrations are low, most ammonia will react with H_2SO_4 forming $(NH_4)_2SO_4$. Under high ammonia concentrations, there is generally enough NH_3 remaining to react with HNO_3, which originates from NO_x, to form NH_4NO_3. Thus, the ratio of ammonia to the acid precursors affects the reaction rate, hence, concentrations of atmospheric gaseous ammonia. Relatively low levels of acid precursors contribute to the high ammonia concentrations in the LFV, southern AB and southern MB. This contrasts with heavily industrialized southern ON where there is an abundance of the SO_x precursor leading to rapid reaction of atmospheric ammonia into the fine particulates. There are similar (median) concentrations of $PM_{2.5}$ in the LFV (4.5–7.2 $\mu g\ m^{-3}$), southern AB (2.6–7.8 $\mu g\ m^{-3}$), southern MB (4–6.8 $\mu g\ m^{-3}$) and the Maritimes (2.5–8.3 $\mu g\ m^{-3}$) but these are lower than concentrations in southern ON (4–12.2 $\mu g\ m^{-3}$) and southern QC (5–12.7 $\mu g\ m^{-3}$) (Anon 2010). The 98th percentile for concentrations, which is of potential importance for acute human toxicity has a range of 14.8–19.9 $\mu g\ m^{-3}$ for the LFV compared to 27.4–50.1 $\mu g\ m^{-3}$ for southern ON and 20.5–69.3 $\mu g\ m^{-3}$ for southern QC.

The ratio of ammonia to acid gases in the atmosphere has other important implications. Particulates containing more ammonium are less acidic and this is characteristic of the LFV relative to southern ON (Makar et al. 2010). Also, there are implications for the expected effects of ammonia reductions on PM formation. The 2008 Canadian Assessment of Agricultural Ammonia revealed that particulate formation is relatively insensitive to reductions in ammonia across most of Canada, except in southern ON where the sensitivity was judged as medium to high with a fairly high (Anon 2010). In ON, a 30 % reduction of ammonia is expected to reduce PM by 7.5 % compared to less than 4 % in other parts of Canada. In contrast, emission reductions of 30 % were likely to reduce ambient ammonia concentrations more in the LFV than anywhere else in Canada. This is due to the surplus of ammonia relative to acid gases in the LFV.

Despite this overall trend, short-term episodic events of high PM even in the LFV may be amenable to ammonia reductions (Makar et al. 2010). Therefore, episodes of low visibility in the LFV attributed to $PM_{2.5}$ (Fig. 11.10) can be averted by abating ammonia. For example, it may be possible to encourage farmers to minimize emissions during periods of high PM probability, typically in late August and Sept. This could be accomplished by using low emission applicators (Fig. 11.11), deferring manure application, or by adding emission suppressing materials (like alum) to poultry bedding. It is interesting that PM formation in ON is more sensitive to ammonia concentration in winter than in summer, probably because ammonia is more limiting in winter (Li et al. 2010), but the opportunities for abatement are also more limited because emissions are already low and mainly from buildings. Abatement of emissions from pig housing may offer the greatest possibility for PM reduction in winter in ON.

11.4.6 Long Range Atmospheric Transport

It is in the particulate form, with its very low deposition rate, that ammonium can travel very long distances. The 2008 Canadian Ammonia Assessment showed that reductions of ammonia emissions have substantial impact on PM and deposition at significant distances from sources (Anon 2010).

Fig. 11.10 Poor visibility in the Lower Fraser Valley in late summer (*left*) due in part to fine secondary particulates containing ammonia emitted mainly from the poultry (Poultry house on lower *left*) and dairy agricultural sectors. Good winter visibility form the same viewpoint is shown on *right* photo

Fig. 11.11 A new low emission applicator for dairy slurry in the Lower Fraser Valley

The effect of long range transport of ammonia on PM and its trans-national implications can be seen in Fig. 11.12. The data which are derived from the AURAMS model, show Canadian sourced PM moving into the Midwest regions of the USA, and the USA sourced PM travelling well north into ON, QC and the Maritimes (Makar et al. 2010).

The long range transport of ammonium means the impact of emissions occur far from sources. Ammonium transported form agricultural regions in the western States is blamed for causing the destruction of vegetation, especially lichens, in alpine regions of the coastal mountains and the Rockies (Jovan 2008). Alpine vegetation is susceptible to ammonium loadings because these ecosystems are typically oligotrophic. Bogs are also typically oligotrophic and the bryophytes are susceptible to both atmospheric ammonia and to loading of ammonium onto the ground (L. Sheppard et al. 2009). Ammonia deposition may cause plant succession from nitrophobic to nitrophillic species like grasses. The relationship between critical atmospheric concentrations (called levels) and critical amounts of deposition (called loading) is explained by Sutton et al. (2009b).

The influence of wet deposition of ammonium in Canadian ecosystems is complicated by both atmospheric sulphur and NO_x. Oxidation of ammonium acidifies soils and water bodies. Acidification is primarily caused by SO_x, but as regulations have reduced SO_x emissions, ammonium has become more important for soil acidification. It is well known for example that forests in eastern Canada are susceptible to soil acidification and hence ammonia deposition (May et al. 2005). In its role as a soil nutrient, effects of ammonia are augmented by deposition of nitrate although the latter is not an acidifier. In AB and SK ammonia/ammonium is 50–65 % of total N in rain water but in most of eastern Canada, ammonia/ammonium is 35–50 % of total N, except in the lower St Lawrence where there is more ammonium (Vett et al. 2010). On the prairies, Canadian agricultural emissions are responsible for most of the ammonium in precipitation. However in Eastern Canada the pattern of deposition is influenced by long range transport of ammonium within Canada and from the USA. The high concentrations of ammonium in the air and in precipitation in the Canadian Shield of eastern Canada is due to long range transport from agricultural regions in southern QC, ON and Midwestern USA. And the large contribution of ammonium from the USA into ON means that ammonia abatement measures in ON will have greater benefit downwind than locally and that emission reductions in ON will require binational agreements such as the multinational Gothenburg Protocol.

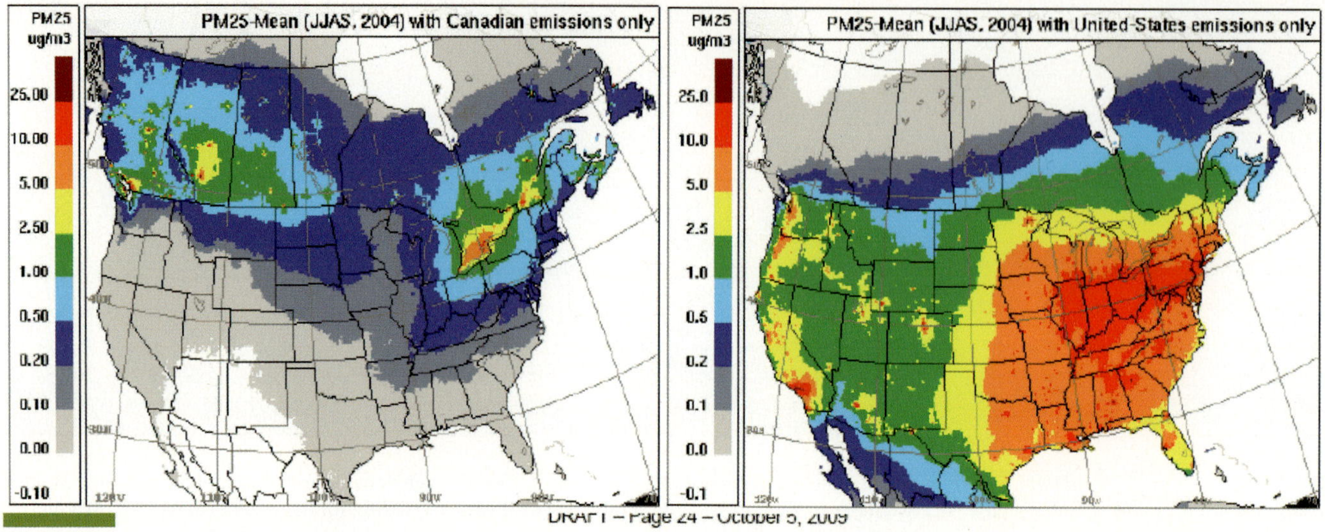

Fig. 11.12 Predicted contributions to mean ambient $PM_{2.5}$ concentrations (ug m^{-3}) in summer of 2004 due to Canadian (*left*) and USA (*right*) emission sources (from Makar et al. 2010). Courtesy Environment Canada

11.4.7 Conclusion: Abating Agricultural Emissions in Canada

In Canada most regulations relating to farming practices are under provincial jurisdiction, so developing national standards or codes of practices for curtailing emissions is challenging. While examples exist of concerted initiatives among the provinces (see PM section), with varying regional socioeconomic perspectives and vastly different climatic, and soil conditions, setting national standards for air quality will always be difficult. Nevertheless, a number of farming practices common in Canada are very beneficial towards air quality, even if the farmers have other reasons for employing them.

Canada has greatly reduced the land area devoted to summerfallow over the past decades, and farmers have aggressively adopted reduced tillage practices. These changes, intended to benefit the soil, have improved air quality by reducing particulate matter and emissions of CO_2. Many farmers employ in the crop rotation legumes such as red clover, field peas, lentils, soybeans, chickpeas and alfalfa. Crops grown after legumes are given much less N fertilizer, reducing the potential for ammonia emissions. Also, many farmers especially in western Canada make judicious use of soil nitrogen tests prior to planting their crops in order to optimize the rate of fertilizer. Research on in-crop testing promises to improve the accuracy of fertilizer applications and this technology would be aided greatly by improved long-term weather forecasts. And while urea containing fertilizers are widely used, emissions are mitigated because much of N fertilizer is placed in bands under the soil surface.

Likewise there are few practices on livestock farms designed specifically to reduce emissions. The widespread use of phase feeding in the poultry and pig sectors allows farmers to closely match feed protein with needs of the growing animal which reduces excretion of surplus protein as urea. Many poultry and pig farmers also supplement feed with amino acids which allows lower total protein concentrations in feeds and lower N excretion rates. Protein consumption by cattle is not easy to control on pastures, but unlike western European countries, excess protein consumption is partly avoided because typically little or no fertilizer is applied to pastures which often contain some legumes such as alfalfa and white clover. This is especially helpful since fertilizer cannot easily be applied below the soil surface on perennial forage crops. Extensive use of pastures by the beef cow-calf sector in Canada may be viewed as an ammonia abatement practice since emissions from pastures are generally low (Bussink 1994). Many cow-calf operators in western Canada, to save costs, have shifted their over winter feeding from feedlots to open fields where urine infiltrates into the soil. There is evidence of better N recovery in these systems and this is attributed to less ammonia loss, hence a trend towards ammonia abatement (Jungnitsch et al. 2011). In contrast, emissions from beef feedlots are high and may be increasing due to the feeding of distiller's grain which is a protein rich by-product form biogas plants (van Haarlem et al. 2008).

While frequent cleaning helps to reduce odour and ammonia emissions from dairy barn floors and crust formation in slurry storages similarly improves air quality, a lot more can be done by Canadian farmers to curtail emissions from these facilities. It helps that animal production facilities are surrounded or partly surrounded by buffer trees in certain parts of the country as these trees help to mitigate dust, odour and ammonia. Depending on their position and type, trees may help with local deposition of air contaminants or dispersion of pollutants thereby reducing the local nuisance effect. Some farms are also placed adjacent to woodlots which provide even greater benefits for air quality. Injection of manure below the soil surface and rapid incorporation of manure after broad-spreading, both very effective in reducing emissions of ammonia and odour, is becoming more commonplace. Where these practices are not possible, such as stony land or forages or untilled land, low disturbance methods such as surface banding or shallow injection are occasionally practiced. It is generally thought that reducing emissions from field applied manure is relatively cost effective (Webb et al. 2009), but in Canada the principal motivation to avoid odour complaints.

Even with beneficial practices employed in Canada, impact on air quality remains a significant concern and new technologies are needed to reduce emission of the various pollutants. As most Canadian farms are strongly subject to international competition, new technology should not reduce farm profitability or viability and should in general not place excessive demands on the producers. At the same time, many countries are tightening their environmental regulations and may expect their Canadian competitors to do the same.

References

AAFC (1998) Research strategy for hog manure management in Canada. Research branch agriculture and agri-food Canada. Department of Supply and Services Canada, Cat. No. A42–77/1998E

AFOTW (2004) Air emissions characterization, dispersion modeling, and best management pactices. Animal feeding operations technical workgroup report. Iowa department of natural resources. December 15, 2004

American Thoracic Society (1998) Respiratory health hazards in agriculture. Am J Respir Crit Care Med 158:S1–S76

Anon (2010) The 2008 Canadian atmospheric assessment of agricultural ammonia. Environment Canada, Gatineau, QC. ISBN 978-1-100-12420-9

ASAE—American Society of Agricultural and Biological Engineers (2007) Management of manure odors. EP379.4

Asman WAH, Jaarsveld HA van (1992) A variable-resolution transport model applied for NH_3 in Europe. Atmos Environ Part A General Topics 26:445–464

Ayres J, Bittman S, Girdhar S, Sheppard S, Niemi D, Ratte D, Smith P (2010) Sources of ammonia pp 77–91 in Anon. The 2008 Canadian Atmospheric Assessment of Agricultural Ammonia. Environment Canada, Gatineau, QC. ISBN 978-1-100-12420–9

Becklake M (2007) Grain dust and lung health: not just a nuisance dust. Can Respir J 14:423–442

Becklake M, Broder I, Chan-Yeung M, Dosman JAH, Ernst P, Mex Herbert F, Kennedy SM (1996) Recommendations for reducing the effect of grain dust on the lungs. Can Med Assoc J 155:1399–1403

Bittman S, Ayres J, Sheppard S, Girdhar S (2010) Emission inventory development 2010 pp 57–75 in Anon (2010). The 2008 Canadian Atmospheric Assessment of Agricultural Ammonia. Environment Canada, Gatineau, QC. ISBN 978-1-100-12420–9

Bleeker A, Sutton MA, Acherman B, Alebic-Juretic A, Aneja VP, Ellermann T, Erisman JW, Fowler D, Fagerli H, Gauger T, Harlen KS, Hole LR, Horvath L, Mittosinkpova M, Smith RI, Tang YS, Pul A van (2009) Linking ammonia emissions trends to measured concentrations and deposition of reduced nitrogen at different scales. In: Sutton MA, Reis S, Baker S (eds) Atmospheric ammonia. Springer Science and Business Media B.V, pp 123–180

Bouwman AF, Boumans LJM, Batjes NH (2002) Estimation of global NH3 volatilization loss from synthetic fertilizers and animal manure applied to arable lands and grasslands. Global Biogeochem Cycles 16:1024–1037

Brown J, Palacios M (2005) The state of urban air in Canada. http://www.policy.ca/policy-directory/Detailed/The-State-of-Urban-Air-in-Canada-1095.html#details

Bussink DW (1994) Relationship between ammonia volatilisation and nitrogen fertiliser application rate, intake and excretion of herbage N by cattle on grazed swards. Fertil Res 38:111–121

Canada-United States Air Quality Committee (2004) Canada—United States transboundary particulate matter science assessment. Catalogue No. En56–203/2004E. pp 129. http://www.ec.gc.ca/Publications/default.asp?lang=En&xml=3BC8288E-E427-4E0E-A624-F67334D31BB0. Accessed 19 Oct 2011

Canadian Centre for Occupational Health and Safety (2008) OSH answers: Farmer's lung. http://www.ccohs.ca/oshanswers/diseases/farmers_lung.html. Accessed 19 Oct 2011

Cape JN, van der Eerden LJ, Sheppard LJ, Leith ID, Sutton MA (2009) Reassessment of critical levels for atmospheric ammonia. In: Sutton MA, Reis S, Baker S (eds) Atmospheric ammonia. Springer Science and Business Media B.V, pp 15–40

CCME, Canadian Council of Ministers of the Environment (1999) Canadian national ambient air quality objectives: process and status. In: Canadian environmental quality guidelines. http://ceqg-rcqe.ccme.ca/. Accessed 19 Oct 2011

Chapin A, Boulind C, Moore A (1998) Controlling odor and gaseous emission problems from industrial swine facilities. A handbook for all interested parties. Yale Environmental Protection Clinic.

Choinière Y, Munroe JA (1997) Air quality inside livestock barns. OMAFRA Factsheet 93-001. http://www.omafra.gov.on.ca/english/livestock/swine/facts/93-001.htm. Accessed 19 Oct 2011

Choinière Y, Munroe JA (2001) Farm workers health problems related to air quality in live stock barns. Canada Plan Service, Canada Plan Service Report M-9708

Cormier Y (2007) Respiratory health and farming: an essay. Can Respir J 14(7):419–422

Eilers W, Mackay R, Palmer L, Lefebvre A (eds) (2009) Environmental sustainability of Canadian agriculture: Agri-environmental indicator report series—Report #3. Ottawa, Ont., Agriculture and Agri-Food Canada

Environment Canada (2011) Criteria air contaminants (National: 1985–2009). http://www.ec.gc.ca/inrp-npri/default.asp?lang=En&n=0EC58C98-#Emission_Summaries. Accessed 19 Oct 2011

Filipy J, Rumburg B et al (2006) Identification and quantification of volatile organic compounds from a dairy. Atmos Environ 40:1480–1494

Freschen FB (2000) Odour measurements and odour policy in Germany. Water Sci Technol 41:17–24

Furberg M, Preston K (2005) Odour management in British Columbia: Review and recommendations—Final Report, RWDI Air Inc. p 277

GEIS (1999) Generic environmental impact statement on animal agriculture: A summary of the literature related to air quality and odor (H). Prepared for the Environmental Quality Board of Minnesota. U Minn

Gong W, Dastoor AP, Bouchet VS, Gong S, Makar PA, Moran MD, Pabla B, Meynard S, Crevier L-P, Cousineau S, Venkatesh S (2006) Cloud processing of gases and aerosols in a regional air quality model (AURAMS). Atmos Res 82:248–275

Groeneveld E, Hébert M (2002) Perceptions d'odeur des matières résiduelles fertilisantes en comparaison avec les engrais de ferme. Vecteur environnment 35:3

Guillam MT, Claude C, Dewitte JD, Michel V, Ségala C (2007) Aérocontaminants et morbidité chez les éleveurs de volailles. Arch Mal Prof Environ 68:161–168

Hao X, Chang C, Janzen HH, Clayton G, Hill BR (2006) Sorption of atmospheric ammonia by soil and perennial grass downwind from two large cattle feedlots. J Environ Qual 35:1960–1965

Holmen BA, James TA, Ashbaugh LL, Flocchini RG (2001) Lidar-assisted measurement of PM_{10} emissions from agricultural tilling in California's San Joaquin Valley—Part II: emission factors. Atmos Environ 35:3265–3277

Huang G, Auld H, Mills B, Lin QG, Li HL, Cai YP, Zhang XD, He L, Zou Y (2005) Modeling tools for supporting agricultural odor management in Canada. Center for Studies in Energy and Environment, University of Regina. Report submitted to Meteorological Services Canada, Environment Canada. November 2005

Jacobson LD (2002) Danish environmental and pig housing trends. Minnesota/Wisconsin Engineering Notes. University of Minnesota, Biosystems and Agricultural Engineering Web site: www.bae.umn.edu

Jovan S (2008) Lichen bioindication of biodiversity, air quality, and climate: baseline results from monitoring in Washington, Oregon, and California. Gen. Tech. Rep. PNW-GTR-737. Portland, OR: U.S. Department of Agriculture, Forest Service, Pacific Northwest Research Station. p 115

Jungnitsch PF, Schoenau JJ, Lardner HA, Jefferson P (2011) Winter feeding cattle on the Canadian Prairies: Impacts on soil nutrient cycling and forage growth. Agric Ecosyst Environ 141:143–152

Laborde L (2004) Canadian agriculture at a glance—pig poduction is getting bigger and more specialized. Statistics Canada, no. 96-325-XPB in the catalogue

Li S-M, Vet R, Liggio J, Makar P, Hayden K, Staebler R, Chan E, Shaw M (2010) Sensitivity of particulate matter to NH3 in major agricultural region of Canada. pp 149–183 in Anon (2010). The 2008 Canadian Atmospheric Assessment of Agricultural Ammonia. Environment Canada, Gatineau, QC. ISBN 978-1-100-12420-9

Lobb DA, Li S, McConkey BG (2010) Soil erosion pp 46–53 in: Eilers W, MacKay R, Graham L, Lefebvre A (eds) Environmental sustainability of Canadian agriculture: Agri-Environmental Indicator Report Series—Report #3. Agriculture and Agri-Food Canada, Ottawa, ON. ISBN 978-1-100-15576-0

Loubet B, Asman WAH, Theobald MR, Hertel O, Tang YS, Robin P, Hassouna M, Daemmgen U, Genermont S, Cellier P, Sutton MA (2009) Ammonia deposition near hotspots: processes, models and monitoring methods. In: Sutton MA, Reis S, Baker S (eds) Atmospheric ammonia. Springer Science and Business Media B.V, pp 205–268

Lyngbye M, Hansen MJ, Riis AL, Jensen TL, Sørensen G (2006) 1000 Olfactometry analyses and 100 TD-GC/MS analyses to evaluate methods for reducing odour from finishing units in Denmark. The National Committee for Pig Production, Danish Bacon & Meat Council, Copenhagen, Denmark. Proceedings of the Agricultural Air Quality Workshop, Washington, DC, USA. June 5–8, 2006

Makar P, Moran M, Zheng Q, Cousineau S, Sassi M, Duhamel A, Besner M, Crevier L-P, di Cenzo C, Vingarzan R, Nissen R, Bouchet V (2010) Modelling the effects of ammonia on regional air quality pp 185–232 in Anon (2010) The 2008 Canadian Atmospheric Assessment of Agricultural Ammonia. Environment Canada, Gatineau, QC. ISBN 978-1-100-12420–9

May JD, Burdette SB, Frank S, Gilliam FS, Adams MB (2005) Interspecific divergence in foliar nutrient dynamics and stem growth in a temperate forest in response to chronic nitrogen inputs. J For Res 35:1023–1030

McGinn SM, Flesch TK, Crenna BP, Beauchernin KA, Coates T (2007) Quantifying ammonia emissions from acattle feedlot using a dispersion model. J Environ Qual 36:1585–1590

Misslebrook T, Powell JM (2005) Influence of bedding material on ammonia emissions from cattle excreta. J Dairy Sci 88:4304–4312

Mwansa PB (2004) Canadian agriculture at a glance—growing dominance of a few large poultry farms—a continuing legacy. Statistics Canada, no. 96-325-XPB in the catalogue

NCPP (2006) Facts on environmental impact and odour. National Committee for Pig Production, Denmark

Noble R, Hobbs PJ, Dobrovin-Pennington A, Misselbrook TH, Mead A (2001) Olfactometry response to mushroom composting emissions as a function of chemical concentration of atmospheric pollutants and trace gases. J Environ Qual 30:760–767

NRC (2000) Air emission from animal feeding operations: Current knowledge, future needs. Final Report. National Research Council. Prepublication copy. National Academy Press, Washington D.C., U.S., December 2002

OECD (2008) Environmental performance of agriculture in OECD countries since 1990. ISBN 978-92-64-04092-2. Organization for Economic Cooperation and Development, Paris

Ortech (2004) General characteristics of agricultural livestock waste odor sources. Canadian Ortech Environmental Inc. Report submitted to Meteorological Services of Canada

Ortech (2005) Preliminary compilation of agricultural livestock waste odour control technologies. Canadian Ortech Environmental Inc. Report submitted to Meteorological Services of Canada

Pattey E, Qiu G (2012) Trend in primary particulate matter emissions from Canadian agriculture. J Air Waste Manag Assoc 62:737–747 . doi:10.1080/10962247.2012.672058

Pattey E, Qiu G, van Haarlem R (2010) Particulate matter pp 126–132 in Eilers W, MacKay R, Graham L, Lefebvre A (eds) Environmental sustainability of Canadian agriculture: Agri-Environmental Indicator Report Series—Report #3. Agriculture and Agri-Food Canada, Ottawa, ON. ISBN 978-1-100-15576-0

Rochette P, Angers DA, Chantigny MH, MacDonald JD, Bissonnette N, Bertrand N (2009) Ammonia volatilization following surface application of urea to tilled and no-till soils: A laboratory comparison. Soil Tillage Res 103:310–315

Saxton KE (1996) Agricultural wind erosion and air quality impacts: a comprehensive research program. Am J Altern Agric 11:64–70

Sheppard LJ, Leith ID, Crossley A, Dijk N van, Cape JN, Fowler D, Sutton MA (2009) Long term cumulative exposure exacerbates the effects of atmospheric ammonia on an ombrotrophic bog: implication s fort critical levels. In: Sutton MA, Reis S, Baker S (eds) Atmospheric ammonia. Springer Science and Business Media B.V, pp 49–58

Sheppard SC, Bittman S, Swift ML, Tait J, Sommer SG, Webb J (2007a) Sensitivity analysis of alternative model structures for an indicator of ammonia emissions from agriculture. Can J Soil Sci 87:129–139

Sheppard SC, De Jong R, Sheppard MI, Bittman S, Beaulieu MS (2007b) Estimation of ammonia emission episodes for a national inventory using a farmer survey and probable number of field working days. Can J Soil Sci 87:301–313

Sheppard SC, Bittman S, Tait J (2009) Monthly NH_3 emissions from poultry in 12 Ecoregions of Canada. Can. J. Animal Science 89:21–35

Sheppard SC, Bittman S, Bruulsema TW (2010a) Monthly ammonia emissions from fertilizers in 12 Canadian Ecoregions. Can J Soil Sci 90:113–127

Sheppard SC, Bittman S, Swift ML, Tait J (2010b) Farm practices survey and modelling to estimate monthly NH_3 emissions from swine production in 12 ecoregions of Canada. Can J Animal Sci 90:145–158

Sheppard SC, Bittman S (2011) Farm survey used to guide estimates of N intake and NH_3 emissions for beef cattle, including early season grazing and piosphere effects. Anim Feed Sci Technol 166–167:688–698

Sheppard SC, Bittman S, Swift ML, Tait J (2011) Modelling monthly nh3 emissions from dairy in 12 ecoregions of Canada. Can J Animal Sci 91:1–13

Sheppard SC, Bittman S (2012) Farm practices as they affect NH3 emissions from beef cattle. Can J Animal Sci 92: 525–543

Sheppard SC, Bittman S (2013) Estimated net application of ammoniacal and organic N from manure,and potential for mitigating losses of ammonia in Canada. Agric Ecosys Environ 171:90–102

Skiba U, Sheppard L, Pitcairn CER, Leith I, Crossley A, Dijka S van, Kennedy VH, Fowler D (1998) Soil nitrous oxide and nitric oxide emissions as indicators of elevated atmospheric N deposition rates in seminatural ecosystems. Environ Poll 102:457–461

Smart JCR, Hicks K, Morrissey T, Heinemeyera A, Sutton MA, Ashmore M (2011) Applying the ecosystem service concept to air quality management in the UK: a case study for ammonia. Environmetrics 22:649–661

Staebler R, Pattey E, McGinn S, Jones K, Vet R, Li S-M (2010) Concentrations and flux measurements of NH_3 pp 33–56 in Anon. 2010. The 2008 Canadian Atmospheric Assessment of Agricultural Ammonia. Environment Canada, Gatineau, QC. ISBN 978-1-100-12420-9

Statistics Canada (2002) Census of agriculture: Canadian farm operations in the 21st century. The Daily, Statistics Canada, Wednesday. May 15, 2002

Sucker K, Both R, Winneke G (2009) Review of adverse health effects of odours in field studies. Water Sci Technol 59:1281–1289

Sutton MA, Sheppard LJ, Fowler D (2009a) Potential for the further development and application of critical levels to assess the environmental impacts of ammonia. In: Sutton MA, Reis S, Baker S (eds) Atmospheric ammonia. Springer Science and Business Media B.V, pp 41–48

Sutton MA, Wolsley PA, Leith ID, Dyke N van, Sim Tang Y, James PW, Theobald MR, Whitfield C (2009b) Estimation of the ammonia critical loading level for epiphytic lichens based on observations at farm, landscape and national scales. In: Sutton MA, Reis S, Baker S (eds) Atmospheric ammonia. Springer Science and Business Media B.V, pp 71–86

Sweeten JM (1998) Separation distances for swine odor control in relation to manure nutrient balances. Texas A&M University Agricultural Research and Extension Center. Manuscript #SE-1669

Sweeten JM, Jacobson LD, Heber AJ, Schmidt DR, Lorimor JC, Westerman PW, Miner JR, Zhang RH, Williams CM, Auvermann BW (2006) Odor mitigation for concentrated animal feeding operations: white Paper and Recommendations. In Animal Agriculture and the Environment: National Center for Manure and Animal Waste Management White Papers. ASABE Pub. Number 913C0306.

US EPA (2006) PM standards revision–2006. http://www.epa.gov/particles/naaqsrev2006.html. Accessed 19 Oct 2011

van Haarlem RP, Desjardins RL, Gao Z, Flesch TK, Li X (2008) Methane and ammonia emissions from a beef feedlot in western Canada for a twelve-day period in the fall. Can J Anim Sci 88:641–649

VanderZaag AC, Gordon RJ, Glass V, Jamieson RC (2008) Floating covers to reduce gas emissions from liquid manure storages: a Review. Appl Eng Agric 24:657–671

Vett R, Li S-M, Beaney G, Friesen K, Vingarzan R, Jones K, Belzer W, Chan E, Dann T, Hayden K, Hou A, Iqbal S, Leithead A, Liggio J, Makar P, Narayan J, Qui W, Ro-C-U, Shaw M, Sukloff B (2010) Characterization of ambient ammonia, PM and regional deposition across Canada pp 93–147 in Anon (2010). The 2008 Canadian Atmospheric Assessment of Agricultural Ammonia. Environment Canada, Gatineau, QC. ISBN 978-1-100-12420-9

Warren CPW (1977) Lung disease in farmers. Can Med Assoc J 116:391–394

Webb J, Hutchings NJ, Bittman S, Baker SMH, Reidy B, Raes C, Smith, Ayres J, Misselbrook T (2009) Reliability of emission estimates and Abatement efficiencies. In: Sutton MA, Reis S, Baker S (eds) Atmospheric ammonia. Springer Science and Business Media B.V, pp 423–431

Zhang Z, Zhu J (2005) Effectiveness of short-term aeration in treating swine finishing manure to reduce odour generation potential. Agric Ecosys Environ 105:115–125

Air Quality, Health Effects and Management of Ammonia Emissions from Fertilizers

12

S. Bittman, Jeffrey R. Brook, Albert Bleeker and T. W. Bruulsema

Abstract

Emissions of ammonia have been demonstrated to play a role in the formation of airborne fine particulate matter, by reacting with gaseous emissions of sulphur dioxide and oxides of nitrogen. Agriculture emits ammonia to the atmosphere, and atmospheric processes can transport the particulate form of ammonium long distances. Fertilizer use-related emissions have been estimated to comprise 10–35 % of all agricultural emissions. The potential contributions of ammonia emissions from agriculture to the formation of particulate matter, and to its harmfulness, are questions being addressed by current research. Recent expert reviews have concluded that exposure to airborne particulate matter air pollution contributes to human mortality. Epidemiological evidence suggests that exposures as short as a few hours to a few weeks are associated with small but significant increases in cardiovascular disease-related mortality, and that the size of the effect increases with longer-term exposures. Plausible biological mechanisms for inhaled particulate matter triggering physiological responses in the cardiovascular system have been described, though much remains uncertain. The specific constituents of particulate matter causing the responses have not been identified with certainty, and thus a direct link between emissions from fertilizer use and human health risks has not been demonstrated. Ammonia emissions from fertilizer can be controlled through the choice of source, rate, timing and placement. Key practices already in use to minimize emissions include placement in soil of any fertilizer with potential to release ammonia, and, for urea specifically, use of urease inhibitors, coatings or formulations to slow release. Even though the reduction in risk to human health is not currently well-

S. Bittman (✉)
Pacific Agrifood Research Centre, Agassiz,
British Columbia, Canada
e-mail: shabtai.bittman@agr.gc.ca

J. R. Brook
Environment Canada, Toronto, Canada
e-mail: Jeff.Brook@ec.gc.ca

A. Bleeker
Energy Research Centre of the Netherlands,
Heerhugowaard, Netherlands
e-mail: a.bleeker@ecn.nl

T. W. Bruulsema
Northeast Region, North American Program,
International Plant Nutrition Institute, Ontario, Canada
e-mail: tom.bruulsema@ipni.net

E. Taylor, A. McMillan (eds.), *Air Quality Management*, DOI 10.1007/978-94-007-7557-2_12,

known enough to be predictable, the evidence is sufficient to warrant recommendations for continued increase in the use of practices designed to reduce ammonia emissions, and for increased research efforts to develop such practices further.

Keywords

Ammonia · Emission · Fertilizer · Particulate matter · Aerosol · Exposure · Health risk · Practices · Cardiovascular · Cardiopulmonary · PM · NH_3

12.1 Introduction

Environmental agencies have drawn attention to the potential role of ammonia emissions as one of the contributors to the formation of fine particulate matter, $PM_{2.5}$, a substance which has been linked to detriments to human health in epidemiological studies (Anon 2010; Brook et al. 2010). A large quantity of ammonia (NH_3) is emitted to the atmosphere from agricultural sources, including livestock feeding facilities, manure stores, pastures, and land application of manure and fertilizer. The objective of this chapter is to describe emission of NH_3 from fertilizer application into the atmosphere and the possibility of its impact on formation of secondary $PM_{2.5}$, and to consider whether these emissions have potential to harm human health.

Global emission of NH_3 from natural and anthropogenic sources for the year 1990 was estimated at 45–75 Tg NH_3-N (Bouwman and Van der Hoek 1997). The portion of the emission arising from agricultural applications of fertilizer and manure on both arable land and grassland was more recently estimated at 18 Tg of N per year, with 59 % from mineral fertilizers and 41 % from manures (Bouwman et al. 2002). Most of the emissions from mineral fertilizer (~73 %) occurred in warm or tropical countries. This can be attributed to widespread use of urea and other NH_3-containing fertilizers, warm temperatures during application and extensive application on the soil surface (Bouwman and van der Hoek 1997).

In many developed countries, the ammonia emission attributed to agriculture comprises a large proportion of the total anthropogenic emissions (Table 12.1). The proportion attributed to fertilizer use, however, is generally smaller than that attributed to livestock production and manure management. As indicated in the footnotes to Table 12.1, the published US and Canadian environmental agency estimates of the amount and proportion of emissions attributed to fertilizer are larger than those reported in OECD (2008). Reasons are unknown.

While ammonia losses from agriculture can potentially result in large economic costs and numerous environmental impacts through the "nitrogen cascade" (Galloway et al. 2003), the focus of this chapter is on its potential effect on human health through the formation of $PM_{2.5}$ in the atmosphere (Sutton et al. 2009; Dämmgen and Erisman 2005). There is little concern about the direct toxicity of NH_3 on human health except in enclosed spaces where concentrations are several orders of magnitude greater than would normally be found in open spaces. The latter situation may be of concern in and around animal feeding operations and will not be considered in this chapter.

The contribution of NH_3 emissions to $PM_{2.5}$ formation is variable and difficult to estimate, requiring detailed air quality models and emission inventories. Agricultural sources are listed as contributing only 0.03 % of the 4.5 Mt of reported $PM_{2.5}$ emissions for the year 2005 in the USA (USEPA 2011b), but these estimates appear to report direct emission only, and do not appear to account for the role of NH_3 in secondary particle formation.

Across the USA between 2000 and 2009, annual average $PM_{2.5}$ is reported to have declined from 14 to 10 μg m^{-3} (USEPA 2011c), based on 724 monitoring sites. Similarly in southern Ontario (the most industrialized region of Canada) annual mean $PM_{2.5}$ declined from 8.5 to 6.5 μg m^{-3} between 2003 and 2009 (OME 2011). The same report indicates declines from 1990 to 2009 in annual mean levels of ambient nitrogen dioxide (NO_2) declining from 22 to 12 μg m^{-3} and of ambient sulfur dioxide (SO_2) declining from 6.6 to 2.6 μg m^{-3}.

Scientific findings of the European CAFE (Clean Air For Europe) showed that ammonia emissions significantly contribute to the formation of secondary particulate matter in the atmosphere (~20 % by mass). The main source of this ammonia is considered to be agriculture (in the form of emissions from cattle, pig and poultry farming and use of N-fertilizers).

There is widespread recognition that PM has a detrimental impact on human health. The EU estimates that anthropogenic PM reduces life expectancy, averaged across EU countries, by a mean of 9 months, although with considerable geographical variation (European Environmental Agency 2007). Secondary PM cannot be directly controlled so the strategy must be to control its precursors. In Canada, NH_3 was designated as a Criteria Air Contaminant (under the Canadian Environmental Protection Act) due to its ability to enter into chemical reactions in the atmosphere with SO_2 and NO_x [nitric oxide (NO) and NO_2] to form $PM_{2.5}$. Therefore, one of the pertinent questions is: can the concentration of $PM_{2.5}$ in the atmosphere be reduced by reducing NH_3 emissions arising from fertilizer use?

Table 12.1 Estimated ammonia emissions from agriculture for selected countries. (OECD 2008)

Country	Average annual emission from agriculture, kt of NH_3		Agricultural emission, % of total	Emissions from fertilizer use, % of agricultural
	1990–1992	2001–2003	2001–03	Mid-1990s
United States†	3,421	3,945†	88	11†
France	744	742	97	17
Germany	645	580	95	10
Canada‡	468	482‡	80	19‡
Poland	407	317	97	36
Japan	–	289	–	21
United Kingdom	302	277	89	14
Korea	143	181	–	27
Netherlands	236	123	90	5

†USEPA (2011a) estimates emissions for the year 2008 from United States agriculture at 3,260 kt with 35 % arising from fertilizer use
‡Anon (2010) reported emissions for the year 2002 from Canadian agriculture at 432 kt with 26 % arising from fertilizer use

12.2 Emissions from Application of Mineral Fertilizers

Emissions of NH_3 from the use of N fertilizers result either directly as volatilization from the soil of the applied fertilizer, or indirectly from the surfaces of N-enriched plants (Sutton et al. 1993a, b).

12.2.1 Emissions from Soils

Factors affecting direct NH_3 emissions from N fertilizers applied to soil have been reviewed most recently by Sommer et al. (2004) and Harrison and Webb (2001). Also, Misselbrook et al. (2004) proposed a process-based model to predict emissions from fertilizer using standard emission factors for fertilizer formulations and applying modifiers based on soil (pH), weather (temperature and rainfall), fertilizer practices (crop height at application and method of application), and rate of fertilizer application (because rate may affect soil pH).

When applied to the surface of the soil, fertilizers containing reduced forms of N may emit NH_3 to the atmosphere. Urea is more prone to such NH_3 loss than other forms of N fertilizers (Harrison and Webb 2001; Sommer et al. 2004). The reason is that when urea is hydrolyzed there is a rapid increase in soil pH near the fertilizer granule. This can happen very rapidly on most soils with a wide range of moisture because the urease enzyme is practically ubiquitous. The transient rise in pH occurs near the fertilizer granules and hence is difficult to measure and predict, but micro-sites around granules are thought to rise to pH 9. At this pH, ammonium dissociates and NH_3 may be released into the atmosphere if it is on or near the soil surface. The amount of NH_3 volatilized is affected by soil parameters such as texture, cation exchange capacity and moisture, and can be controlled by placement and use of technologies that prevent sharp increases in pH, such as urease inhibitors or slow-release forms.

Emissions may be much higher when urea is surface applied to untilled than to tilled soil because of: (1) higher abundance of urease enzyme on no-till due to presence of crop residue, and (2) less fertilizer-to-soil contact on no-till soil (Rochette et al. 2009). Citing mainly European studies on grasslands and winter cereals to which urea was applied as a topdress, Harrison and Webb (2001) noted measured emissions of NH_3 ranging from 6 to 41 % and average emissions of 10–20 % of the N applied. Emissions from other sources such as ammonium nitrate were usually much smaller. Lower emission rates are associated with factors like low soil pH, low temperature, and rainfall at or just after fertilizer application. Fortunately, fertilizers are frequently applied early in the growing season when temperatures are typically cool and often moist.

A recent study on dryland wheat soils in Montana, measuring ammonia losses following application of urea by surface broadcast (at dates ranging from October to March) found losses of 3–44 % of the fertilizer N, with a mean of 22 % (Jones and Engel 2010). This loss was reduced to 8 % with the use of a urease inhibitor. While this study is as yet unpublished, it indicates that cool temperatures may not sufficiently control losses when urea is surface-applied to very dry soil. The same rates of loss may not occur on wetter soils, particularly soils frozen at water contents close to saturation.

Table 12.2 shows published NH_3 emission factors for various N fertilizer formulations. These and similar emission factors are used in national NH_3 emission inventories. It is evident that urea has the greatest emission factor. Emission from the urea-containing fluid fertilizer UAN (urea ammonium nitrate) has been found to be intermediate between urea and ammonium nitrate (Harrison and Webb 2001). Ammonium nitrate is frequently given an emission factor of 1–10 % but usually < 4 % (Harrison and Webb 2001) and it is known to have more consistent agronomic results when surface-applied than urea or NH_3-containing fertilizers. In trials conducted with wind tunnels on silty loam soil in

Table 12.2 Mean emission factors (%) for fertilizer forms

Reference	Country	Urea	AN	UAN	AA	CAN	Other
Asman 1990	Europe	15	2	2.5	1		4
Whitehead and Raistrick 1990	UK	16.5	2.5	–	–	–	2–5
Goebes et al. 2003	USA	15	2	–	1	2	2–8
Hutchings et al. 2001	Denmark	15	2	–	–	–	–
Sheppard et al. 2010a	Canada	11	–	6	4	–	4
Bouwman et al. 2002	Global	21	6	5	2	3	1–25
Bouwman et al. 2002	Global range	18–26	5–9	2–11	1–3	2–4	1–25

AN ammonium nitrate, *UAN* urea ammonium nitrate, *AA* anhydrous ammonia, *CAN* calcium ammonium nitrate

Table 12.3 Global Fertilizer Consumption, kt N per annum, by region and form; mean of 2003–2008. (IFA 2011)

Region	Urea	AN	UAN	AA	CAN	AS	Other	Total
Africa	1,364	610	4	5	114	84	558	2,739
East Asia	23,058	107			49	1,064	12,627	36,905
Eastern Europe & Central Asia	221	901	55		18	26	1,859	3,080
Latin America & Caribbean	3,519	436	111	62	43	757	1,219	6,147
North America	3,010	413	3,121	3,549	19	336	2,664	13,112
Oceania	781	10	63	54	6	78	263	1,255
South Asia	14,404				126	116	2,634	17,280
West Asia	1711	356	16	0	280	94	556	3,013
Western & Central Europe	1976	2,227	1,173	7	2,477	314	2,882	11,056
World	50,044	5,060	4,543	3,677	3,132	2,869	25,262	94,587

AN ammonium nitrate, *UAN* urea ammonium nitrate, *AA* anhydrous ammonia, *CAN* calcium ammonium nitrate, *AS* ammonium sulfate, *other* mainly composite fertilizers

British Columbia, Canada, emissions from ammonium nitrate were 10–12% of the N applied, which was much lower than broadcast urea or ammonium sulfate but much higher than banded or incorporated urea (Bittman, unpublished data). Anhydrous ammonia has a low emission coefficient. Because of worker safety concerns and the obvious potential rapid loss if released to the air, handling is very carefully managed to minimize NH_3 loss (Souther et al. 2000; Shutske 2005).

The potential loss of NH_3 is determined by the product and the soil, but the actual loss is also much affected by the weather. One centimeter of rainfall soon after fertilizer application, especially on soils with high cation exchange capacity, greatly reduces emissions from all fertilizers including urea. However, if only a small amount of rain falls on a dry soil, enabling urea hydrolysis but not soil infiltration, very high losses can occur. Emissions can be reduced by incorporating or injecting the fertilizer into the soil, although incorporated bands may have high emissions on some soils (Rochette et al. 2009). If there is a delay in incorporation, some fertilizers have faster initial losses (Sommer et al. 2004) so delays will be more detrimental.

In situations where incorporation or injection are not possible (e.g. spring application to winter wheat or application on forages) use of urea products that delay release (polymer coatings and urease enzyme inhibitors) will reduce emissions by dampening the rise in pH (Misselbrook et al. 2004). Also, applying beneath a dense canopy reduces air flow and creates a higher atmospheric concentration, which facilitates some direct adsorption into leaves (Sommer et al. 1997). In wetland rice systems, losses of NH_3 from broadcast urea can be considerable, but timing the application a few days before panicle initiation leads to lower losses than earlier broadcast applications, owing to a larger canopy and lower pH in the floodwater (Bouwman et al. 2002).

For calculating regional emissions of NH_3, most inventories assume a single emission factor for each fertilizer. This has been criticised because countries and regions have different average temperatures (Klimont 2005). Therefore, in Europe, emission factors have been adjusted for three different geographical regions. Emission factors for low-emitting fertilizers like ammonium nitrate range by just 1% across regions whereas emission factors for urea range by more than 8% (i.e. 11.7–20%). Curiously, in Canada, despite considerable differences in climate across the country, average temperatures during months of peak fertilizer application are quite similar (Sheppard et al. 2010a).

Urea comprises a substantial proportion of world fertilizer N use, especially in developing countries (Table 12.3). Use of urea and UAN is increasing as a result of practical advantages in transportation, handling, and safety as compared to other sources. Bouwman et al. (2002) made estimates of global NH_3 emissions from fertilizers and showed that the highest losses occurred in South and East

Table 12.4 Area, use of synthetic fertilizers and NH_3 volatilization loss for fertilized grasslands, upland crops and wetland rice for different world regions for 1995. (From Bouwman et al. 2002)

	Fertilized Grasslands			Upland Crops			Wetland Rice		
Region[a]	Area, Mha	N use[b], kt	NH_3-N loss, kt	Area, Mha	N use[b], kt	NH_3-N loss, kt	Area, Mha	N use[b], kt	NH_3-N loss, kt
1 Canada	0	0	0	46	1576	140	0	0	0
2 U.S.A.	0	0	0	189	10982	788	1	168	15
3 Central America	1	25	3	40	1392	215	0	32	5
4 South America	1	12	1	109	2049	348	3	234	43
5 North Africa	0	0	0	21	1126	214	1	78	16
6 Western Africa	0	0	0	73	130	19	1	26	4
7 Eastern Africa	0	0	0	40	109	17	1	1	0
8 Southern Africa	3	31	3	42	477	51	0	3	0
9 OECD Europe	37	3074	156	90	6384	448	0	32	3
10 Eastern Europe	3	210	13	48	1834	123	0	1	0
11 Former U.S.S.R.	33	760	59	229	1856	157	1	14	1
12 Middle East	4	17	3	57	2305	422	1	71	18
13 South Asia	0	0	0	162	8295	1828	44	4646	1031
14 East Asia	0	0	0	69	19855	3318	26	4490	829
15 Southeast Asia	0	0	0	52	2405	421	35	1811	335
16 Oceania	20	175	23	49	639	108	0	12	2
17 Japan	0	27	4	2	265	36	2	171	24
Total	103	4331	265	1319	61678	8654	117	11788	2328

[a] Regions designated for global change research, as defined by Kreileman et al. (1998) and as indicated in Figs. 12.1 and 12.2

[b] Somewhat lower here than total presented by IFA (1999), because of scaling errors

Asia (Table 12.4). The relation between atmospheric NH_3 and $PM_{2.5}$ has been studied much less in Asia than in Europe and North America, but recent studies have pointed to a role for NH_3 in Shanghai (Ye et al. 2011) and Beijing (Meng et al. 2011).

12.2.2 Emissions from Plants

There have been relatively few studies on emissions from fertilized crops and those estimates tend to be fairly low (1–15 kg NH_3-N ha^{-1}) with most estimates under 6 kg NH_3-N ha^{-1} (Sommer et al. 2004). Higher emissions (7–34 kg N ha^{-1}) have been occasionally reported, such as from post-anthesis corn, and losses of this magnitude can affect calculations of N balances in field studies (Francis et al. 1993).

Many national inventories of NH_3 emission do not include estimates of emissions from crops for several reasons, among them the shortage of reliable data and the complexity of this emission source. In part, the complexity arises from the difficulty in separating natural vegetation emission (considered background) from crop emissions. Also, emission rate (and direction) from plant surfaces is transient since it is a function of the concentration gradient between the NH_3 emitted into sub-stomatal spaces from the plant apoplast (mesophyll cell walls) and the NH_3 concentration in the atmosphere outside the leaves or leaf boundary layer.

The compensation point for NH_3 transfer (the atmospheric concentrations at which there is no net movement of NH_3 into or out of plants) varies with rate of fertilization and is particularly high soon after fertilizer application (Sommer et al. 2004). Well-fertilized growing crops with high protein concentrations will have high levels of apoplastic ammonium and generally emit more NH_3 than they absorb, but plants that do not receive much fertilizer may be sinks for atmospheric NH_3 (Sutton et al. 1993a, b). Rate of transfer from live plants is also mediated by the stomatal resistance to gas transfer which, in combination with the resistance of the cuticle, is referred to as leaf resistance. There is little leaf NH_3 transfer when stomata are closed (resistance is high) such as at night or during periods of plant water deficit. Little emission occurs through the cuticle although, if moist, it can serve as a temporary reservoir of NH_3 which may be re-emitted to the atmosphere or washed onto the soil by irrigation or rainwater.

Ammonia is also lost from plants as they ripen in the field or after they are harvested, during the curing process (Sommer et al. 2004; Whitehead et al. 1988). While a portion of the N released from the lysed protein of senescent leaves is translocated to other tissues within the plant, some of the lysed N will be lost to the atmosphere via the apoplast. The amount of NH_3 loss after harvest will vary with the concentration of N in the leaves at harvest, and with rate of leaf drying, since there is little loss of NH_3 from dry leaves.

Legumes and pulse crops tend to have inherently high N concentrations without addition of N fertilizers, so losses during desiccation and senescence may be naturally high. Janzen and McGinn (1991) reported NH_3 emission rates as high as 14 % of the N from lentil crop residue over 14 days.

In considering the effects of the plants on emissions from applied fertilizer, two other factors should be mentioned. First, the plant canopy acts to reduce airflow over fertilizer applied on the soil surface to growing crops, effectively mitigating diffusion of NH_3 into the atmosphere. Second, the leaves may directly absorb NH_3 released from the fertilizer to the atmosphere (Sommer et al 1997; Whitehead and Lockyer 1987). This is most likely to occur if the fertilizer is applied under a full crop canopy so that the concentration in the sheltered air will be high and there is more likely to be a gradient favourable for absorption. Also, stomata must be open for leaf absorption to take place, and it probably helps if many stomata are found on the lower leaf surfaces that are more likely to be exposed to the atmospheric NH_3. This may be important for fertilizer applied in spring to fall-seeded crops such as winter wheat or winter canola (rape), especially if the plant-N concentration is low at time of application. It may also be important where fertilizer application is split or delayed particularly for long-season crops like corn or some perennial crops.

While the plant source of N into the atmosphere is very diffuse, this loss pathway needs some attention and further research, especially in countries and regions that have vast areas of fertilizer-enriched crops.

12.2.3 Emissions Estimates as Input to Air Quality Models

To understand the fate of NH_3 in the atmosphere as it relates to air quality and health issues, emission estimates are used as inputs to atmospheric models (see 2.2 and 2.3 below). The models (e.g. CMAQ and AURAMS used in Canada) have very short time steps (several minutes) in comparison to the annual NH_3 emission values obtained from national emission inventories (Anon 2010). The models also require fine spatial resolution (AURAMS currently uses 42 km square grids but model resolution is improving and future versions are planned with smaller grids). To provide the fine temporal information, typically models have algorithms that partition emissions temporally by such factors as time of day and day of week. For example, Chinkin et al. (2003) assumed that diurnal emissions follow air temperature and are 4 times higher near mid-day than at night. Also, there are GIS tools such as demarcating low-emitting land uses to help resolve spatial variation within census districts.

The spatial and temporal variation in emissions from fertilizers was demonstrated in a recent US study (Goebes et al. 2003). This study reported fertilizer emissions on a monthly basis by adjusting emission factors for temperature based on date of application which was inferred from average planting date in a region. No adjustment was made for soil properties or for application method. Across the US the average emission from fertilizer application was 0.063 Tg per month but emissions ranged from 0.143 Tg in April to under 0.001 Tg

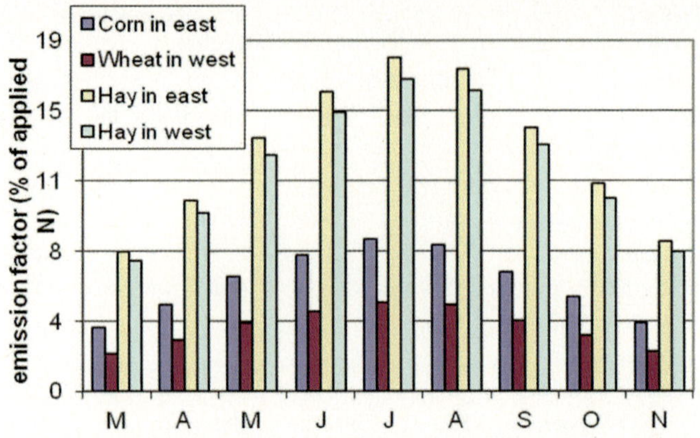

Fig. 12.1 Monthly NH_3 emission factors (% of applied N) for N fertilizer applied to various crops in eastern (Lake Erie Lowlands of Ontario) and western (*Dark Brown* Soil zone of Manitoba, Saskatchewan and Alberta) in Canada. The variation in emission factors is due to differences in fertilizer type, application method, temperature and soil properties. (from Sheppard et al. 2010a, with permission from Canadian Journal Soil Science)

in January. Seasonal patterns varied by state with greatest emission occurring in March in Kansas, in April in Iowa and in May-June in California. In Iowa the month with the second highest emission was November.

A more detailed inventory of emission from fertilizer was recently reported in Canada (Sheppard et al. 2010a). Emissions calculations were based on information on fertilizer forms, obtained from the fertilizer industry, and application methods, rates and timing, obtained from a targeted farm survey, and crop areas based on census data. In Canada, the most important N fertilizer forms are urea, urea-ammonium nitrate(UAN) and anhydrous NH_3, while ammonium nitrate and ammonium sulfate are important regionally. Mono- and di-ammonium phosphate (MAP and DAP) are used at low rates as P sources and are usually placed below the soil surface. The survey revealed large regional differences in method of fertilizer application with 86% full injection (with the seeder) practiced on the prairies compared to 25% in central Canada so emission factors were varied to reflect the contrasting application methods. [Less injection in central Canada was due to a greater proportion of fertilization of perennial forages, growing winter wheat (in spring) and corn (side-dressing), and more fertilizer incorporation in tilled fields.] These fertilizer emission values were adjusted further for soil and temperature factors using the equation of Bouwman et al. (1997). In both western and central regions of Canada urea and UAN were the primary fertilizer formulations.

This information was combined to calculate combined emission factors for various crops grown in each region of the country; hence, the variation in emission factors was due to differences in fertilizer form, method of application and soil properties (Fig. 12.1). In accordance with fertilization activity, most emissions from fertilizer were concentrated in

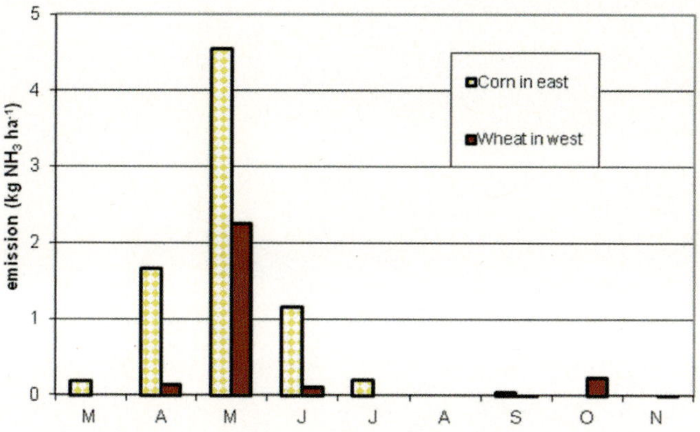

Fig. 12.2 Ammonia emissions per hectare in each month from all N fertilizers used on corn grain in the east (Lake Erie Lowlands Ecoregion, Ontario) and wheat in the west (*Dark Brown* Soil Ecoregion, Prairies) of Canada. (from Sheppard et al. 2010a, with permission from Canadian Journal Soil Science)

the month of May across Canada with smaller amounts emitted in April and June (Fig. 12.2). Clearly the concentration of emissions into a short period is important for modeling and has implications for formation of secondary particulates. Also, this coincides with the period of maximum emissions from land application of manure (Sheppard et al. 2009 and 2010b). Hence the effect of fertilizer application on emissions of NH_3 is likely to be more pronounced in regions within Canada where there is a lot of livestock and fertilizer use, such as parts of the Windsor-Quebec City corridor, the lower Fraser Valley and perhaps southern Manitoba and Alberta (Fig. 12.3). The proportion of total agricultural emissions attributable to fertilizer application varies greatly (<10 to >45 %) across Canada (Fig. 12.4).

12.3 Transport of Atmospheric Ammonia

The most important processes related to the overall transport through the air of NH_3 after emission from agricultural sources are: dispersion, conversion and eventually deposition.

After emission into the atmosphere, NH_3 does not remain in its gaseous form for long. Some of the NH_3 is directly adsorbed to surfaces such as soil and vegetation, usually within 500 m of the source, referred to as dry deposition (Asman et al. 1998). Dry deposition is important near source because on most farms NH_3 is emitted at or near the soil surface. The percentage of NH_3 that is adsorbed varies greatly with atmospheric conditions and with the attributes of the contacting surfaces, which can range from strong receptors (e.g. soil surfaces) to surfaces that are themselves NH_3 sources, such as N-rich vegetation. Local deposition may be large enough to significantly affect soil nitrogen balances downwind of major sources such as feedlots (Hao et al. 2006).

Limited data on dry deposition affects the accuracy of estimates of net emissions from agricultural sources to the atmosphere and may contribute to the gap between modelled and measured atmospheric NH_3 concentrations (Erisman et al. 2008). Ammonia that is not adsorbed may react with acid gases (nitrate, sulfate, chloride and organic acids) to form fine secondary particulates and aerosols with aerodynamic diameters of less than 2.5 μm ($PM_{2.5}$). The reaction rate is very rapid provided that reactants (often pollutants from industry or transportation) are available. These particles may stay aloft for a long time and can travel long distances; they deposit in rainfall (called wet deposition) enriching the soil or water. The impact of the particles on air quality and ecosystems is usually at some distance downwind (~1 km to more than 1000 km) of the source and is therefore potentially a transnational issue.

12.3.1 Dispersion

The degree to which NH_3 will eventually be dispersed depends on the source and the atmospheric conditions. Ammonia being released from fertilizer spread on to the field will disperse much less in the mixing layer, compared to NH_3 from mechanically ventilated animal houses. This is because the NH_3 is released higher up in the atmosphere in the latter situation. Therefore, for the dispersion of NH_3 it is important to make a distinction between the sources according to increasing height: surface sources (grazing and land spreading of fertilizer and manure), animal houses, and industrial sources.

Once released into the air, NH_3 is caught by the wind and turbulent air movements, resulting in a quick dispersion in the atmosphere. After NH_3 is volatilized from manure or fertilizer it becomes part of the atmosphere. The atmosphere is the layer of air around the earth upon which all life is dependent. When going up, the atmosphere starts with the boundary (or mixing) layer, being the first 1.5 to 3 km above the earth's surface. On top of that there is the troposphere, reaching up 10 to 15 km. Since NH_3 is a very reactive and soluble gas (in clouds and rain), only a limited amount of NH_3 will leave the boundary layer.

12.3.2 Conversion

There are four types of reactions possible when NH_3 travels through the atmosphere:

i. Reactions in the water phase, leading primarily to the formation of ammonium-sulfate and ammonium-nitrate aerosols. (NH_3 can also react with soot particles and other molecules (organic acids), forming carboxyl- and hydroxyl-ammonia complexes).
ii. Thermal reactions of NH_3 with sulfuric acid, sulfur dioxide and ozone.
iii. Photochemical reactions forming the amino radical NH_2.
iv. Reactions with sulfuric acid and ammonium sulfate particles that neutralize acidity but add to $PM_{2.5}$ mass.

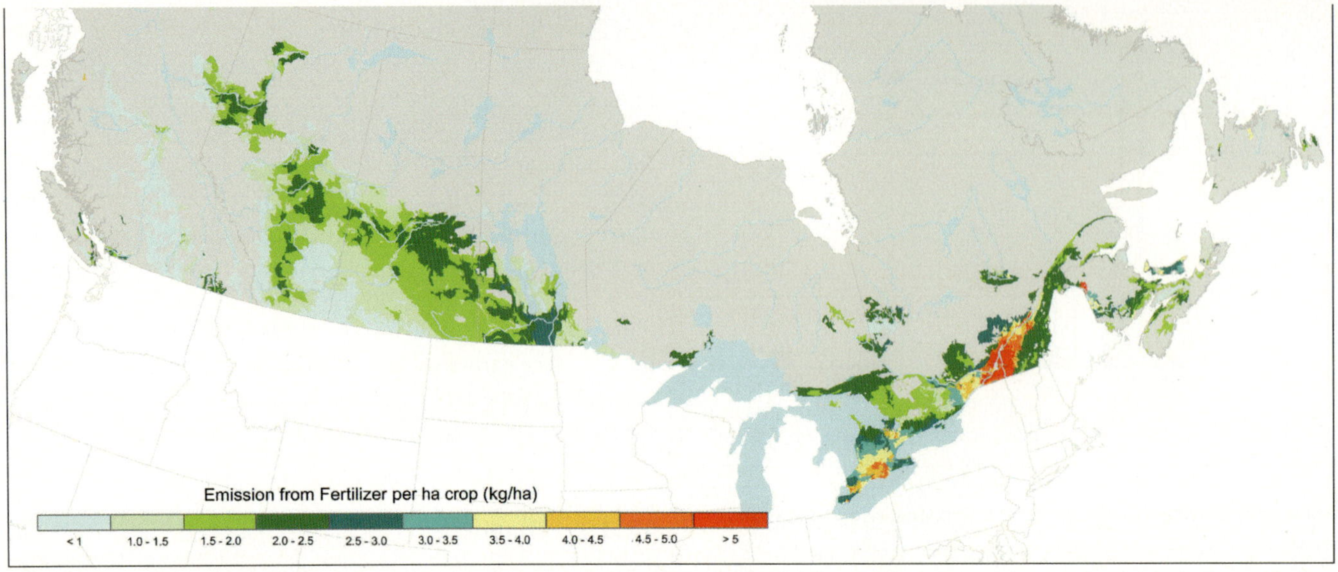

Fig. 12.3 Annual NH_3 emissions (kg NH_3 ha^{-1} crop land) from fertilizer in 2006 in Canada. (from Sheppard et al. 2010a, with permission from Canadian Journal Soil Science)

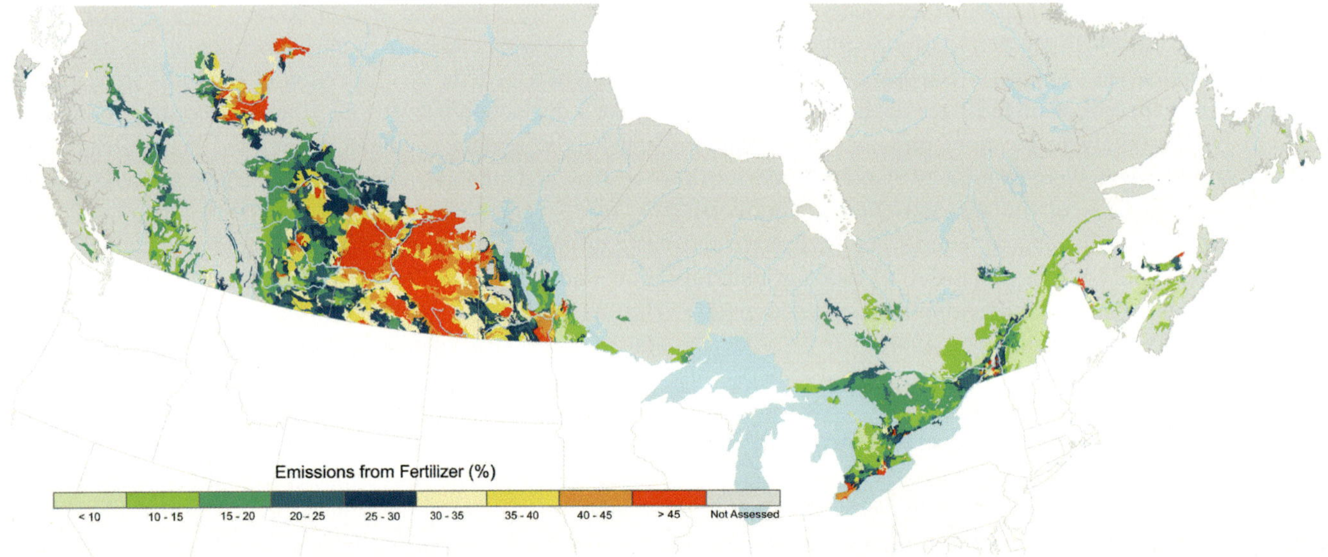

Fig. 12.4 Fraction (%) of total emitted nh_3 from all agricultural sources that was directly attributed to fertilizers in Canada. (Eilers et al. 2009)

The formation of sulfate aerosol has been studied for a long time, mostly due to the role of sulfur in acidification. The formation of sulfate aerosols is principally dependent on the SO_2 concentration in the atmosphere. In addition, relative humidity, temperature, pH and the presence of metal ions as catalysts are important. Ammonia enhances the overall reaction rate by increasing pH (Junge and Ryan 1958) and its participation with water and sulfate is one of four proposed possible mechanisms for nucleation of atmospheric nanoparticles (Kulmala 2003). In the presence of NH_3, ammonium sulfate will be rapidly formed. The experiments of McKay (1971) showed that 50% of the NH_3 in the atmosphere can be converted to ammonium sulfate in about 35 min, assuming typical atmospheric concentrations for that era: 20 μg m^{-3} SO_2 and 2.7 μg m^{-3} NH_3.

The thermal reactions include the reactions with SO_2 forming ammonium sulfate and the reaction forming ammonium nitrate. Ammonium nitrate can be formed through reactions of NH_3 with nitric acid (HNO_3), nitrous acid (HNO_2) and ozone (O_3). The reaction with ozone is as follows:

$$2NH_3 + 4O_3 \rightarrow 4O_2 + H_2O + NH_4NO_3$$

Ammonium nitrate (NH_4NO_3) can again dissociate into NH_3 and HNO_3, according the following equilibrium that is dependent on ambient relative humidity and temperature and light:

$$NH_4NO_3 \leftrightarrow NH_3 + HNO_3$$

Such an equilibrium also exists with HCl (Pio and Harrison 1987).

$$NH_4Cl \leftrightarrow NH_3 + HCl$$

There is no photochemical reaction known that leads to the formation of NH_3 in the atmosphere. However, two photochemical reactions can lead to the destruction of NH_3: (i) photolytic dissociation of NH_3 with formation of amino radicals, and (ii) photochemical reactions with ozone, atomic oxygen and the hydroxyl radical (OH). The reaction with OH is the most important photochemical destruction reaction for NH_3, and causes NH_3 to have a half-life of about 16 days in the free troposphere (McConnell 1973). This is much greater than the half-life of, for example, the formation of ammonium sulfate. So, the photochemical reactions are negligible in comparison with the reaction with acids (Chameides 1987). If the reaction with O_3 would have been important, little NH_3 would be present in the atmosphere, since there is 50 times more ozone than NH_3 available.

Under European conditions a major part of the gaseous NH_3 released from low-level agricultural sources reacts with sulfuric acid (H_2SO_4) to yield an ammonuim-containing aerosol. These are one-way reactions so the aerosol will not evaporate again (Asman and Janssen 1987). A minor part of the NH_3 will reversibly react either with gaseous HNO_3 forming NH_4NO_3 or with HCl forming NH_4Cl (Pio and Harrison 1987). For Europe as a whole enough acid precursors are released to neutralize (in time) all emitted NH_3. A constant rate of 8×10^{-5} s^{-1} (29% h^{-1}) was reported by Asman and Van Jaarsveld (1992). This value was consistent with empirical measurements (Erisman et al. 1988) and, when used in transport models, the value resulted in favorable comparisons with observed ammonium levels (Asman and Janssen 1987). However, since atmospheric concentrations of sulfur dioxide in Europe have decreased by a factor of 5 over the past 10 years while NH_3 concentrations have remained unchanged, the conversion rate should now be more like 10% h^{-1}. Further, on a local scale or during periods of intensive spreading of manure and fertilizer, there may be insufficient acid present to convert all NH_3. Therefore, conversion rates are likely to vary in both space and time.

12.3.3 Transboundary Atmospheric Transport

The above-mentioned processes are all important for the overall transport distances of reduced nitrogen. While NH_3 can only be transported over relatively short distances (0.5–50 km), once converted into an ammonium aerosol, the transport distances can exceed 1000 km. Thus, NH_3 emitted in a specific country can contribute to nitrogen-related effects across national boundaries. The distance over which NH_3 and/or ammonium is transported through the air thus depends strongly on the conversion of NH_3 to ammonium aerosol. The decline in atmospheric sulfate concentrations has resulted in longer transport distances for gaseous NH_3.

Fagerli and Aas (2008) analyzed trends in atmospheric nitrogen compounds in Europe using long-term measurements and a chemical transport model. They found statistically significant declines at the majority of sites for NH_x (sum of ammonia and ammonium) in air and for nitrate and ammonium in precipitation. They concluded, "Model simulations indicate that the simultaneous decrease of SO_2 and NH_3 emissions has led to larger reductions in NH_x concentrations in air than what we would have seen if SO_2 emissions had remained constant. Ammonium wet deposition is somewhat less affected by the reductions in SO_2 emissions as the change towards relatively more NH_3 and less ammonium aerosols is compensated by a somewhat more efficient scavenging of NH_3 than of ammonium."

The often large spatial dislocation between source and effect for ammonia is important from a policy perspective. In Canada, for example, ammonia emitted in the agricultural areas of southern Ontario has more effect on air quality many kilometres downwind in eastern Ontario and southern Quebec. The PM in southern Ontario attributable to agricultural ammonia would quite possibly be related more to emissions in the Midwest of the USA than to local emissions. Thus it is evident that such air quality concerns are transboundary in nature.

12.3.4 Deposition

After transport and possible conversion of NH_3 in the atmosphere, it will eventually return to the earth's surface by means of wet, dry and/or cloud deposition. Wet deposition is the process where atmospheric pollutants are transported to the surface via rain, fog, hail or snow. Dry deposition is the process that transports gases and particles onto vegetation, soil or other surfaces directly, without the intervention of some sort of precipitation.

Wet deposition Wet deposition takes place by uptake of pollutants in precipitation (rain, fog, hail, snow) as well as in cloud droplets. These two processes are below-cloud and in-cloud scavenging, respectively. For most pollutants, in-cloud scavenging is a more efficient removal process than below-cloud scavenging. However, wet deposition is an important removal process for NH_3 due to its high solubility in water. The below-cloud scavenging of NH_3 may be of importance in source regions. It has been shown from experimental results that the wet deposition of NH_4^+ is correlated to the local NH_3 emission density (Hertel et al. 2006). When considering the contribution from a single farm, the wet deposition of NH_3 will, however, be very limited. This is due to the short periods with precipitation compared with the dry periods, and to the short residence time of the pollutants near the farm.

Dry deposition Dry deposition is the most important removal process of NH_3 from the atmosphere. The dry deposition process is a strong function of the transport rate of the NH_3 in the air to the surface and the physical, chemical and biological characteristics of the surface. NH_3 is able to stick to almost any surface, and the dry deposition is therefore often limited by the transportation rate to the surface. One of the important pathways for dry deposition of NH_3 is uptake through the stomata of plants. Two other major pathways for transport of NH_3 to plants are absorption of NH_3 to dew on the plants or to the thin water film on the leaf epidermis. In transport-chemistry models, the dry deposition of gases and particles is often described with a deposition velocity and the concentration of the substance at a reference height. In turn the deposition velocity is often parameterized with a so-called resistance model in which the transport to the surface and the surface uptake is described with resistances. The dry deposition velocity (V_d) for a gaseous compound is expressed as the reciprocal value of the total resistance to transport down to and removal on to the surface:

$$V_d = \frac{1}{R_t} = \frac{1}{R_a + R_b + R_c}$$

where R_t is the total resistance, R_a is the aerodynamic resistance, R_b is the quasi-laminar sub-layer resistance and R_c is the surface resistance. In principle the resistances that describe the physical or meteorological part of the transport are well known under the assumption that the roughness characteristics are known. Using this resistance scheme it is assumed that the surface concentration of the air pollutant is zero. However, for NH_3 this is not the case and the deposition process is in principle the net result of an exchange process which is bi-directional. The surface concentration of NH_3 is often referred to as the compensation point, being the concentration where the exchange of NH_3 changes from deposition to (re-) emission or vice versa. Experimental studies have shown that bi-directional dry deposition transport is possible also over marine waters. Net flux depends on the meteorological conditions, the pH and content of NH_4^+ in the upper surface waters, and the NH_3 concentrations in ambient air just above the water surface on the other side.

After the reaction of NH_3 with the different acids in the atmosphere (see previous sections), small particles (aerosols) can be formed, during which the acids are partly or completely neutralized. This conversion affects transport distance since the dry deposition velocity of NH_3 is greater than that of ammonium particles. The dry deposition velocity is a measure of the speed with which the gases or particles are removed from the atmosphere. The transport distance of NH_3 is always less than 100 km, while it can be more than 1000 km for ammonium particles.

The division of the total deposition into dry, wet and fog deposition largely depends on the climatological conditions. Orography is an important factor for the amount of wet and/or fog deposition since in hilly or mountainous regions, fog/cloud deposition can play an important role in terms of its contribution to the overall deposition of reduced nitrogen.

12.4 Particulate Matter and Human Health

An expert group tasked by the American Heart Association with assessing the state of knowledge regarding the impact of $PM_{2.5}$ on cardiovascular (CV) health concluded that exposure to $PM_{2.5}$ air pollution contributes to cardiovascular morbidity and mortality (Brook et al. 2010; Fig. 12.5). They pointed out that in the previous 5–10 years strong evidence has accumulated that exposure to $PM_{2.5}$ over a few hours-to-weeks can trigger CV disease-related mortality and non-fatal events and that longer-term exposure (e.g. few years) further increases the risk for CV mortality. The overall opinion of the expert group was that $PM_{2.5}$ exposure is a modifiable factor that contributes to CV morbidity and mortality.

12.4.1 Epidemiological Evidence

The adverse human health effects of $PM_{2.5}$ have been a priority scientific and policy issue for the past 15 years. Two seminal American studies in the early 1990's can be credited with bringing the health effects of $PM_{2.5}$ to the forefront. The Harvard Six Cities Cohort study (Dockery et al. 1993) and the American Cancer Society (ACS) Cancer Prevention Cohort (Pope et al. 1995) study linked exposure to $PM_{2.5}$ or $PM_{2.5}$-related pollutants to reduced life expectancy. Several subsequent studies worldwide have uncovered associations between PM exposure and acute mortality and cardiac and respiratory morbidity (e.g., Burnett et al. 1995; Dominici et al. 2003; HEI 2004; Katsouyanni et al. 2003; Samet et al. 2000). Subsequent re-analysis of both the Six Cities and ACS Cohorts continued to show robust relationships between mortality and $PM_{2.5}$ (Krewski et al. 2000, 2004; Pope et al. 2002, 2006).

An extended follow-up of the Six Cities Study demonstrated that long-term exposure to particulate air pollution was positively associated with total mortality due to cardiovascular disease and lung cancer (Laden et al. 2006). This study further demonstrated that a reduction in annual

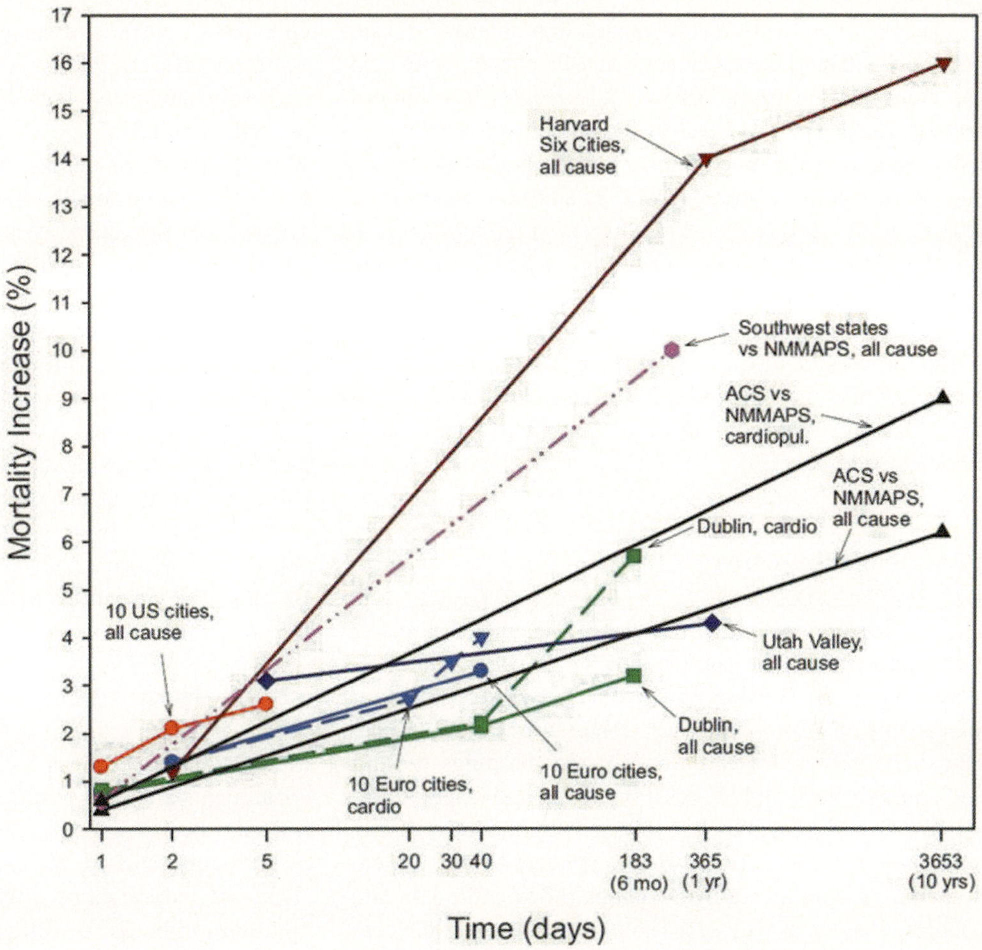

Fig. 12.5 Comparison of estimates of percent change in mortality risk associated with an increment of 10 μg m^{-3} in $PM_{2.5}$ or 20 μg m^{-3} of PM_{10} for different time scales of exposure (log scale of approximate number of days). Euro indicates European; cardio, cardiovascular disease; and cardiopul, cardiopulmonary disease. (from Brook et al. 2010)

average $PM_{2.5}$ concentrations from 18 μg m^{-3} to 14.8 μg m^{-3} resulted in a statistically significant reduction in mortality risk from cardiovascular and respiratory causes, though not from lung cancer, indicating for the first time that moderate reduction in fine PM pollution may yield health benefits. A re-analysis of the ACS data, focusing on neighborhood-to-neighborhood differences in urban air pollution in Los Angeles and employing more precise exposure assessment methods, revealed that PM pollution-related death rates from all causes including cardiopulmonary diseases were at least twice as high as reported previously in the ACS cohort study (Jerrett et al. 2006).

While the epidemiological work in the mid-1990s centered on mortality and administrative-type data (e.g., hospital admissions), a number of new health endpoint indicators for PM exposure were later introduced: (1) development of atherosclerosis with long-term PM exposure; (2) changes in cardiac rhythm and blood pressure; (3) increased risk of rupture of atherosclerotic plaques leading to clotting and eventually heart attacks or strokes; (4) health effects on developing children and infants; (5) markers of inflammation such as exhaled nitrogen oxide (NO); and (6) effects on non-cardiopulmonary organ systems (USEPA 2006). For example, there is now substantial evidence that PM_{10} exposure in children is associated with adverse effects on lung function including aggravation of asthma and increased incidence of coughs and bronchitis (WHO-Europe 2009). Moreover, there is evidence of increased risk of post-neonatal respiratory mortality as concentrations of PM_{10} increase. Pathological mechanisms for the effects of $PM_{2.5}$, such as inflammatory reaction in the alveoli, have been elucidated in recent years lending more credibility to epidemiological observations (Brook et al. 2010).

Current science does not support a safe threshold for $PM_{2.5}$. The relationship between $PM_{2.5}$ and cardiovascular risk appears to extend below 15 μg m^{-3} (the USEPA 2006 annual National Ambient Air Quality Standards level), and an increase above present-day levels by 10 μg m^{-3} "reduces life expectancy within a population probably by several months to a few years" (Brook et al. 2010). Note that all the studies shown in Fig. 12.5 were done in the United States of America and Europe. Many cities in the developing world have higher levels of $PM_{2.5}$ but studies on the human health effects are fewer (Meng et al. 2011; Ye et al. 2011).

Because many people are exposed to ambient PM, even small risks should be taken seriously. However, with relative risks ranging from 1.003 to 1.16 (Fig. 12.5), the evidence falls far short of the 3 to 4 required to establish causality, according to most noted epidemiologists, including Sir Richard Doll, (Taubes 1995). In addition, Koop et al. (2010), applying Bayesian Model Averaging, demonstrated that the statistical approaches used in studies of this nature could yield apparent pollution-health correlations that are not robust to reasonable variations in estimation methods. While the weight of the evidence supports a link between cardiovascular mortality and PM, the estimates of the increase in mortality and the reduction in life expectancy range considerably.

12.4.2 Mechanisms

The surface area of human lungs (70–100 m^2) is about 50 times greater than the surface area of the skin, and adults inhale about 18–23 m^3 or 15 kg of air per day, so the lungs are capable of substantial absorption of airborne substances. Transport of particles and aerosols into the lungs depends upon particle size, oral vs. nasal breathing and a person's respiratory health (e.g., chronic obstructive pulmonary disease alters particle deposition patterns). The fate of the particles in the lungs is mitigated by the cough reflex and by the mucociliary system, as well as the solubility of the particle compounds in lung fluids. Once in the lungs, the particles will induce a protective immune response and other endogenous processes (e.g., oxidative stress) for which chemical composition is believed to have a significant influence.

Particles smaller than 10 μm (PM_{10}) are considered inhalable. Those larger than 2.5 μm have a higher potential of being deposited in the upper airways and may be associated with bronchitis and asthma. Particles smaller than 2.5 μm ($PM_{2.5}$) more frequently penetrate deeply into the terminal bronchial tubes and alveoli of the lungs where they may produce more systemic effects, and the particles themselves can get into the circulation system and hence have direct impacts elsewhere in the body.

One criticism of the epidemiological studies is that they contrast with the results of clinical toxicological studies. Green and Armstrong (2003) surveyed laboratory-based studies and concluded "Toxicologic data on typical forms of pollution-derived PM strongly suggest that current ambient concentrations in the U.S. are too small to cause significant disease or death." Also, Mauderly (2006) notes that there are few clinical studies that involve exposures comparable to ambient air levels. Heuss and Wolff (2006), commenting on Pope and Dockery (2006), noted a considerable number of uncertainties in the conclusions from the multicity studies, one of which was that the range of effects among individual cities was an order of magnitude larger than the reported summary effect of PM on relative risks to human health.

Newly elucidated pathological mechanisms help to explain the epidemiological observations (Brook et al. 2010). These mechanisms include three pathways by which inhalation of PM can instigate effects beyond the lungs on the CV system. The first pathway is the release of proinflammatory mediators such as cytokines, activated immune cells, platelets, or vasculoactive molecules from lung cells. The second pathway is perturbation of the autonomic nervous system by particle interactions with lung receptors or nerves, increasing blood pressure, platelet aggregation and heart rate. The third pathway is translocation of the particles (only the smallest of them) or their constituents into the blood circulation. All three pathways can result in systemic oxidative stress and inflammation with effects on vasculature, metabolism and blood. While these putative pathways have not been proved operative, they add plausibility to the epidemiological observations.

12.4.3 Harmful Constituents

Much attention has been focused on determining which PM constituents are most harmful so that policy can be targeted towards specific sources. The major constituents of $PM_{2.5}$, by mass, are sulfate, nitrate and organic compounds consisting of thousands of different organic species. The ammonium ion bound to sulfate, nitrate and some organics is also a ubiquitous component, albeit less important on a mass basis due to its relatively small molecular weight. However, less ammonium is required by weight to form a particle, so for impact on particle number ammonia is relatively potent. Also present in $PM_{2.5}$, but in smaller proportions, are elemental carbon and trace metals associated with soil or various anthropogenic activities.

Given the chemical complexity of $PM_{2.5}$ and the differences in toxicity or carcinogenicity of the different constituent compounds it is generally accepted that some particles are likely to impact health more than others (Schlesinger et al. 2006). However, uncertainties in accurately characterizing exposure and identifying the source(s) of $PM_{2.5}$ make it difficult to elucidate which constituents are most related to the observed health effects. The various species tend to co-vary in time and space and the precision with which exposures can be assigned is not the same for all of them.

The potential of using multivariate receptor models to identify sources has been reviewed by Thurston et al. (2005). Based upon independent analyses of one dataset, they concluded that effect estimates generally have overlapping confidence bands, so large numbers of observations will be required in each of the domains (cities) studied to have enough power to significantly differentiate the various source impacts. The different health endpoints associated with PM or air pollution exposure is a further complication since it is reasonable to expect that not all of these endpoints (which include cardiovascular and respiratory mortality, elevated blood pressure, adverse birth outcomes and incidents of asthma and cancer) are related to the same components in the air pollutant mixture.

A further complicating factor is that the gases that co-exist with PM (eg, ozone, O_3, nitrogen dioxide, NO_2, carbon monoxide, CO) also have health effects, and these act independently, synergistically or antagonistically with $PM_{2.5}$ in ways that are poorly understood and are in need of systematic research (Mauderly and Samet 2009). Finally, there is uncertainty in estimates of source apportionment again suggesting the need for longer data records. Failing to identify specific sources and misidentifying commingled source factors have led to inconsistent or unreasonable conclusions (Grahame and Hidy 2007).

Schlesinger et al. (2006) reviewed both toxicological and epidemiological evidence to assess which constituents were most responsible for the adverse affects of $PM_{2.5}$. While several studies have examined the effects of sulfate and acidity, few have focused on ammonium nitrate. They concluded that in the past work the respiratory responses were tied to the acidity of the exposure atmosphere, namely, concentration of sulfuric acid, ammonium bisulfate and, under some unusual atmospheric conditions, nitric acid. As indicated above, ammonia neutralizes atmospheric acidity. Sulfate, in particular has been related to epidemiological health outcomes from ambient PM in many studies. However, the toxicological evidence does not support a role for these secondary inorganic aerosols in adverse health outcomes noted in epidemiological studies (Schlesinger and Cassee 2003).

Although sulfate has been implicated in many epidemiological studies, there are many reasons that this apparent effect may not reflect its inherent toxicity (Grahame and Schlesinger 2005), but controversy remains (Schwartz 2007 and see below). Schlesinger et al. (2006) reported that there was toxicological and epidemiological evidence for elemental or black carbon (EC), trace metals and organic carbon (OC) being the most important $PM_{2.5}$ constituents. Research published since that review has generally supported this finding. Peng et al. (2009) studied the associations between emergency room admissions and multiple $PM_{2.5}$ constituents in 199 counties across the U.S. They concluded that EC and OC were the most strongly linked to admissions of all constituents examined including ammonium, sulfate and nitrate. Interestingly, they attribute the association of ammonium with negative health outcomes to other correlated pollutants, based on the conclusions by Schlesinger et al. (2006).

In the last few years, epidemiological and toxicological studies have reported strong associations between adverse health outcomes of $PM_{2.5}$ and constituent trace metals such as Ni and V (Lippman et al. 2006; Bell et al. 2007), Zn (Hirshon et al. 2008) and As, Al, Si (Franklin et al. 2008; Ostro et al. 2010). In these studies, sulfate-containing PM was also often significantly associated with adverse health outcomes, even in statistical models with multi-pollutants (e.g., Franklin et al. 2008; Ostro et al. 2010).

Nitrate, on the other hand, was generally not as strongly implicated, especially in multi-pollutant models. Nitrate-containing PM may have less adverse health outcomes than those with sulfate due to its semi-volatile nature, which means concentrations are generally higher in the cooler than in the warmer months, particularly in the eastern part of North America (Dabek-Zlotorzynska et al. 2011). Also, when ammonium nitrate particles infiltrate into buildings they evaporate resulting in considerably lower PM concentrations (Lunden et al. 2003; Meng et al. 2007).

In California, where ambient atmospheric ammonium nitrate concentrations reach some of the highest levels in the world, they were associated with mortality (largely due to ischemic heart disease) in both single- and multi-pollutant (with sulfate) models, although the evidence in the latter was weaker (Ostro et al. 2010). Furthermore, in a two-pollutant model with organic carbon, the effect of ammonium nitrate was not significant. Thus, in this high nitrate environment, organic carbon species and elemental carbon showed a stronger association with health effects than nitrate, which is consistent with toxicological evidence. The relatively consistent epidemiological evidence for the importance of sulfate may reflect that it is a surrogate for photochemically-processed fine particulate matter and ozone (Franklin and Schwartz 2008), rather than that the ammonium sulfate comprising the PM is necessarily toxic.

The true toxic subcomponents of polluted airmasses, which may include ammonium and nitrate, remain to be fully elucidated. The goal of separating out the $PM_{2.5}$ species in epidemiological studies in order to fully elucidate their differential effects and target policy towards particular sources continues to be fraught with challenges. While the most consistent finding is that $PM_{2.5}$ containing elemental carbon, trace metals and other traffic-related pollutants are most deleterious, $PM_{2.5}$ mass itself remains one of the most consistent correlates with adverse health outcomes.

A better knowledge of PM chemistry and sources will help improve understanding of its biological effects and suggest more precise strategies for its management (Nel 2005). More toxicological and epidemiological research will be required to determine, with greater confidence, the extent to which reduction in NH_3 emissions and the subsequent reductions in ammonium nitrate and ammonium sulfate forms of particulate matter will lead to significant human health benefits and whether such emission reductions should be prioritized from a health perspective in comparison to other options (e.g., motor vehicles, power plants).

12.4.4 Can Harmful $PM_{2.5}$ be Mitigated by Reducing NH_3 Emission from Fertilizer?

What are the implications of mitigating agricultural ammonia emissions on air quality and hence human health? This question was in part tackled in Canada in the "The 2008 Canadian Atmospheric Assessment of Agricultural Ammonia" by considering the sensitivity of $PM_{2.5}$ to changes in concentrations of atmospheric ammonia (Anon 2010). The study reported that high concentrations of atmospheric ammonia in Canada indeed occurred in regions shown to have high rates of emissions from the agricultural sector. It also noted, however, that "areas of high NH_3 emissions and ambient NH_3 concentrations do not necessarily always correspond to the areas of highest PM concentrations."

Several studies using contrasting approaches were used to assess sensitivity of $PM_{2.5}$ concentration to changes in ammonia concentrations (Anon 2010). The approaches included assessment of long term atmospheric measurements, intensive field studies and atmospheric modelling. Although with varying levels of uncertainty, all of the studies suggested that there was a low sensitivity of PM concentrations to changes in NH_3 concentrations in most regions of Canada. This is due in part to the abundance of NH_3 relative to acid gases. Where NH_3 is most abundant and exceeds the acid gas concentrations, there is the least expected impact from a reduction in atmospheric NH_3. In these situations it would seem to be more effective to mitigate emissions of the acid gases.

In southern Ontario, where there is an abundance of SO_x and NO_x due to vehicles and manufacturing activities, a reduction in NH_3 would more directly impact PM formation than in other regions. It was estimated that a 30 % reduction in ammonia would reduce PM by up to 7 % (Anon 2010). Interestingly, the effect is stronger in winter than in the warmer weather of spring and summer (likely owing to atmospheric chemistry and the relatively lower concentration of NH_3 in winter). However there was evidence from these models that changes in ammonia concentration may cause substantial (as high as 33 % for Egbert, Ontario in November 2005) short term changes in $PM_{2.5}$ concentrations.

Also, the models showed that reductions in NH_3 may have greater impact on $PM_{2.5}$ some distance downwind of major agricultural sources. This is because of the low deposition rates of fine particulates and their potential long range transport. Deposition of ammonium particles mainly in rainfall was shown to contribute to N enrichment of natural environments far away from sources both in Eastern and Western Canada. The relative importance of ammonia deposition to acidification is likely to increase with declining levels of SO_2 and NO_3 in the atmosphere. The models also showed that in the Lower Fraser Valley, reducing ammonia may appreciably improve visibility, and this has aesthetic and possibly also health and economic advantages.

Particulate matter is more likely to harm human health when people spend more time outside or leave windows open. This usually occurs during moderate temperatures and potentially at times of significant farming activities including fertilizer application. The possibility that the effects of $PM_{2.5}$ combine with the health effects of other air pollutants or perhaps other health stressors such as high temperatures and humidity is an area of ongoing research. However, as noted in this review, the conditions required for NH_3 to induce PM formation are complex, and the PM impacts may be hundreds of kilometres removed from the source of the emissions.

The benefit to human health through emissions reduction that would be attained by increased use of fertilizing practices that cut ammonia loss is not yet predictable. Hence it remains difficult to determine how much to invest in such reductions. Ongoing efforts by agricultural producers to increase their N use efficiency through practices that reduce losses may entail some benefits to human health through reduced $PM_{2.5}$, but the size, duration and location of such benefits remain uncertain.

Ammonia emissions from fertilizer can be controlled through the choice of source, rate, timing and placement. Key practices already in use to minimize emissions include placement in soil of any fertilizer with potential to release ammonia, and, for urea specifically, use of urease inhibitors, coatings or formulations to slow release. Where surface application without incorporation cannot be avoided (e.g. when topdressing small grains), sources other than urea should be preferred and are often chosen since ammonia losses can be large enough to reduce N use efficiency and the profitability of N fertilizer application. To the extent possible, timing of application should coincide with an actively growing crop able to scavenge ammonia losses from the air.

Even though the reduction in risk to human health is not currently well-known enough to be predictable, the evidence regarding the role of ammonia in the formation of $PM_{2.5}$ is sufficient to warrant recommendations for continued increase in the use of practices designed to reduce ammonia emissions, and for increased research efforts to develop such practices further.

References

Anon (2010) The 2008 Canadian atmospheric assessment of agricultural ammonia. Environment Canada, Gatineau. ISBN 978-1-100-12420-9

Asman WAH (1990) A detailed ammonia emission inventory for Denmark. DMU A133. National Environmental Research Institute, Roskilde

Asman WAH, Janssen AJ (1987) A long-range transport model for ammonia and ammonium for Europe. Atmos Environ 21(10):2099–2119

Asman WAH, van Jaarsveld HA (1992) A variable-resolution transport model applied for NH_3 in Europe. Atmos Environ, Part A. General topics 26(3):445–464

Asman WAH, Sutton MA, Schjoerring JK (1998) Ammonia: emission, atmospheric transport and deposition. New Phytol 139:27–48

Bell ML, Dominici F, Ebisu K, Zeger SL, Samet JM (2007) Spatial and temporal variation in $PM_{2.5}$ chemical composition in the United States for health effects studies. Health Perspect 115:989–995

Bouwman AF, van Der Hoek KW (1997) Scenarios of animal waste production and fertilizer use and associates ammonia emissions for the developing countries. Atmos Environ 31:4095–4102

Bouwman AF, Lee DS, Asman WAH, Dentener FJ, Van der Hoek KW, Olivier JGJ (1997) A global high resolution emission inventory for ammonia. Global Biogeochem Cycles 11:561–587

Bouwman AF, Boumans LJM, Batjes NH (2002) Estimation of global NH_3 volatilization loss from synthetic fertilizers and animal manure applied to arable lands and grasslands. Global Biogeochemical Cycles 16:1024–1037

Brook RD, Rajagopalan S, Pope CA III, Brook JR, Bhatnagar A, Diez-Roux AV, Holguin F, Hong Y, Luepker RV, Mittleman MA, Peters A, Siscovick D, Smith SC Jr, Whitsel L, Kaufman JD (2010) Particulate matter air pollution and cardiovascular disease: an update to the scientific statement from the American Heart Association. Circulation 121:2331–2378

Burnett RT, Dales RE, Krewski D, Vincent R, Dann T, Brook JR (1995) Associations between ambient particulate sulphate and admissions to Ontario hospitals for cardiac and respiratory diseases. Am J Epidemiology 142:15–22

Chameides WL (1987) Acid dew and the role of chemistry in the dry deposition of reactive gases to wetted surfaces. J Geophysical Research-Atmospheres 92(D10):11895–11908

Chinkin LR, Ryan PA, Coe DL (2003) Recommended improvements to the CMU ammonia emission inventory model for use by LADCO. Prepared Lake Michigan Air Directors Consortium. Revised final report 902350-2249-FR2 http://www.vistas-sesarm.org/tech/NH3_Final_Report4.pdf. Accessed 3 March 2011

Dabek-Zlotorzynska E, Dann TF, Martinelango PK, Celo V, Mathieu D, Ding L, Austin C, Brook JR (2011) Canadian National Air Pollution Surveillance (NAPS) $PM_{2.5}$ Speciation program: methodology and $PM_{2.5}$ chemical composition for the years 2003 to 2008. Atmos Environ 45:673–686

Dämmgen U, Erisman JW (2005) Emission, transmission, deposition and environmental effect of ammonia from agricultural sources. In Kuczynski T, Dämmgen U, Webb J, Myczko A. Emissions from European Agriculture. Wageningen Academic Publishers, The Netherlands, pp 97–112

Dockery DW, Pope CA III, Xu X, Spengler JD, Ware JH, Fay ME, Ferris BG Jr, Speizer FE (1993) An association between air pollution and mortality in six U.S. cities. N Engl J Med 329:1753–1759

Dominici F, McDermott A, Daniels M, Zeger SL, Samet JM (2003) Mortality among residents of 90 cities, in revised analyses of time-series studies of air pollution and health. Special report. Health effects institute, Boston, pp 9–24. http://www.healtheffects.org/Pubs/Time-Series.pdf. Accessed 12 May 2004

Eilers W, Mackay R, Palmer L, Lefebvre A (eds) (2009) Environmental sustainability of Canadian agriculture: agri-environmental indicator report series—report #3. Agriculture and Agri-Food Canada, Ottawa (Ont)

Erisman JW, Vermetten AWM, Asman WAH, Waijers-Ijpelaan A, Slanina J (1988) Vertical distribution of gases and aerosols: the behaviour of ammonia and related components in the lower atmosphere. Atmos Environ (1967) 22(6):1153–1160

Erisman JW, Bleeker A, Hensen A, Vermeulen A (2008) Agricultural air quality in Europe and future perspectives. Atmos Environ 42:3209–3217

European Environmental Agency (2007) Air pollution in Europe 1990–2004. EEA Report No 2/2007. Copenhagen, DK. http://www.eea.europa.eu/publications/eea_report_2007_2. Accessed 10 Feb 2011

Fagerli H, Aas W (2008) Trends of nitrogen in air and precipitation: model results and observations at EMEP sites in Europe, 1980–2003. Environ Pollut 154(3):448–461

Francis DD, Schepers JS, Vigil MF (1993) Post anthesis nitrogen loss from corn. Agronomy J 85:659–663

Franklin M, Schwartz J (2008) The impact of secondary particles on the association between ambient ozone and mortality. Environ Health Perspect 116:453–458

Franklin M, Koutrakis P, Schwartz J (2008) The role of particle composition on the association between $PM_{2.5}$ and mortality. Epidemiology 19:680–689

Galloway JN et al (2003) The nitrogen cascade. Bioscience 53:341–356

Goebes MD, Strader R, Davidson C (2003) An emission inventory for fertilizer application in the United States. Atmos Environ 37:2539–2550

Grahame TJ, Schlesinger RB (2005) Evaluating the health risk from secondary sulphates in Eastern North American regional ambient air particulate matter. Inhal Toxicol 17:5–27

Grahame T, Hidy GM (2007) Pinnacles and pitfalls for source apportionment of potential health effects from airborne particle exposure. Inhal Toxicol 19:727–744

Green L, Armstrong S (2003) Particulate matter in ambient air and mortality: toxicologic perspectives. Regul Toxicol Pharmacol 38:326–335

Hao X, Chang C, Janzen HH, Clayton G, Hill BR (2006) Sorption of atmospheric ammonia by soil and perennial grass downwind from two large cattle feedlots. J Environ Qual 35:1960–1965

Harrison R, Webb J (2001) A review of the effect of N fertilizer type on gaseous emissions. Adv Agron 73:65–108

HEI (2004) Health effects of outdoor air pollution in developing countries of Asia: a literature review. Special Report 15. Health Effects Institute, Boston

Hertel O et al (2006) Modelling nitrogen deposition on a local scale: a review of the current state of the art. Environ Chem 3(5):317–337

Heuss JM, Wolff GT (2006) Comments on health effects of fine particulate air pollution: lines that connect. J Air Waste Manag Assoc 56:1369–1371

Hirshon JM, Shardell M, Alles S, Powell JL, Squibb K, Ondov J, Blaisdell CJ (2008) Elevated ambient air zinc increases pediatric asthma morbidity. Health Perspect 116:826–831

Hutchings NJ, Sommer SG, Andersen JM, Asman WAH (2001) A detailed ammonia emission inventory for Denmark. Atmos Environ 35:1959–1968

IFA (1999) Nitrogen-Phosphate-Potash, IFADATA statistics from 1973/74– 1973 to 1997/98– 1997 including separately world fertilizer consumption statistics, International Fertilizer Industry Association, Paris, 1999

IFA (2011) IFADATA Statistics. International Fertilizer Industry Association, Paris. http://www.fertilizer.org/ifa/ifadata/search. Accessed 31 Jan 2011

Janzen HH, McGinn SM (1991) Volatile loss of nitrogen during decomposition of legume green manure. Soil Biol Biochem 23:291–297

Jerrett M, Burnett RT, Ma R, Pope CA, Krewski D, Newbold KB, Thurston G, Shi Y, Finkelstein N, Calle EE, Thun MJ (2006) Spatial analysis of air pollution and mortality in Los Angeles. Epidemiology 17:727–736

Jones C, Engel R (2010) Ammonia volatilization from surface-applied urea and NBPT coated urea under cold temperatures. ASA, CSSA, SSSA Annual Meeting, Oct 31-Nov 3. Long Beach. Abstract 271–9

Junge CE, Ryan TG (1958) Study of the SO_2 oxidation in solution and its role in atmospheric chemistry. Q J R Meteorol Soc 84:46–55

Katsouyanni K et al (2003) Sensitivity analysis of various models of short-term effects of ambient particles on total mortality in 29 cities in APHEA2, in revised analyses of time-series studies of air pollution and health. Special report. Health Effects Institute, Boston, pp 157–164. http://www.healtheffects.org/news.htm. Accessed 16 May 2003

Klimont Z (2005) Projections of agricultural emissions of ammonia in the European Union. In: Kuczynski T, Dammgen U, Webb J, Myczko A (eds) Emissions from European Agriculture. Wageningen Academic Publishers Wageningen, The Netherlands, pp 231–250

Koop G, McKitrick R, Tole L (2010) Air pollution, economic activity and respiratory illness: evidence from Canadian cities, 1974–1994. Environmental Modelling & Software 25:873–885
Kreileman E, Van Woerden J, Bakkes, J(1998) RIVM Environmental Research, CIM Rep. M025/98. National Institute for Public Health and the Environment, Bilthoven
Krewski D, Burnett RT, Goldberg MS, Hoover K, Siemiatycki J, Jerrett M, Abrahamowicz M, White WH (2000) Reanalysis of the Harvard Six Cities Study and the American Cancer Society Study of particulate air pollution and mortality. Special report. Health Effects Institute, Cambridge
Krewski D, Burnett RT, Goldberg MS, Hoover K, Siemiatycki J, Abrahamowicz M, White WH (2004) Validation of the Harvard Six Cities Study of particulate air pollution and mortality. N Engl J Med 350(2):198–199
Kulmala M (2003) How particles nucleate and grow. Science 302(5647):1000–1001
Laden F, Schwartz J, Speizer FE, Dockery DW (2006) Reduction in fine particulate air pollution and mortality: extended follow-up of the Harvard Six Cities Study. Am J Respir Crit Care Med 73:667–672
Lippmann M, Ito K, Hwang JS, Maciejczyk P, Chen LC (2006)Cardiovascular effects of nickel in ambient air. Environ Health Persp 114(11):1662
Lunden MM et al (2003) The transformation of outdoor ammonium nitrate aerosols in the indoor environment. Atmos Environ 37:5633–5644
Mauderly J (2006) Comments on health effects of fine particulate air pollution: lines that connect. J Air Waste Manag Assoc 56:1375–1378
Mauderly JM, Samet JL (2009) Is there evidence for synergy among air pollutants in causing health effects? Environ Health Perspect 117:1–6
McConnell JC (1973) Atmospheric Ammonia. J Geophys Res 78(33):7812–7821
McKay HAC (1971) The atmospheric oxidation of sulphur dioxide in water droplets in presence of ammonia. Atmos Environ 5:7–14
Meng QY et al (2007) How does infiltration behavior modify the composition of ambient $PM_{2.5}$ in indoor spaces? Analysis of RIOPA data. Envir Sci Technol 41:7315–7321
Meng ZY, Lin WL, Jiang XM, Yan P, Wang Y, Zhang YM, Jia XF, Yu XL (2011). Characteristics of atmospheric ammonia over Beijing, China. Atmos Chem Phys 11:6139–6151. doi:10.5194/acp-11-6139-2011
Misselbrook TH, Sutton MA, Scholefield D (2004) A simple process-based model for estimating ammonia emissions from agricultural land after fertilizer applications. Soil Use Manag 20:365–372
Nel A (2005) Air pollution-related illness: effects of particles. Science 308:804–806
OECD (2008) Environmental performance of agriculture in OECD countries since 1990. ISBN 978-92-64-04092-2. Organization for economic cooperation and development, Paris
OME (2011) Air quality in Ontario 2009 report. http://www.ene.gov.on.ca/stdprodconsume/groups/lr/@ene/@resources/documents/resource/stdprod_081228.pdf. Accessed 29 Aug 2011
Ostro B et al (2010) Long-term exposure to constituents of fine particulate air pollution and mortality: results from the California teachers study. Environ Health Perspect 118:363–369
Peng RD, Bell ML, Geyh AS, McDermott A, Zeger SL, Samet JM, Dominici F (2009) Emergency admissions for cardiovascular and respiratory diseases and the chemical composition of fine particle air pollution. Environ Health Perspect 117:957–963
Pio CA, Harrison RM (1987) The equilibrium of ammonium chloride aerosol with gaseous hydrochloric acid and ammonia under tropospheric conditions. Atmos Environ 21:1243–1246
Pope CA 3rd, Dockery DW (2006) Health effects of fine particulate air pollution: lines that connect. J Air Waste Manag Assoc 56(6):709–742
Pope CA 3rd, Thun MJ, Namboodiri MM, Dockery DW, Evans JS, Speizer FE, Heath CW Jr (1995) Particulate air pollution as a predictor of mortality in a prospective study of U.S. adults. Am J Respir Crit Care Med 151(3 Pt 1):669–674
Pope CA, Burnett RT, Thun MJ et al (2002) Lung cancer, cardiopulmonary mortality, and long-term exposure to fine particulate air pollution. JAMA 287:1132–1141
Rochette P, Angers DA, Chantigny MH, MacDonald JD, Bissonnette N, Bertrand N (2009) Ammonia volatilization following surface application of urea to tilled land and no-till soils: a laboratory comparison. Soil Till Res 103:310–315
Samet JM, Dominici F, Curriero FC et al (2000) Fine particulate air pollution and mortality in 20 U.S. cities, 1987–1994. N Engl J Med 343:1742–1749
Schlesinger RB, Cassee F (2003) Atmospheric secondary inorganic aerosols: the toxicological perspective as a basis for health effects risk assessment. Inhal Toxicol 15:197–235
Schlesinger RB, Kunzli N, Hidy GM, Gotschi T, Jerrett M (2006) The health relevance of ambient particulate matter characteristics: coherence of toxicological and epidemiological inferences. Inhal Toxicol 18:95–125
Schwartz J (2007) Secondary sulphate effects; Schwartz responds. Environ Health Perspect 115:A532–A533
Sheppard SC, Bittman S, Tait J (2009) Monthly NH_3 emissions from poultry in 12 Ecoregions of Canada. Can J Animal Sci 89:21–35
Sheppard SC, Bittman S, Bruulsema TW (2010a) Monthly ammonia emissions from fertilizers in 12 Canadian Ecoregions. Can J Soil Sci 90:113–127
Sheppard SC, Bittman S, Swift ML, Tait J (2010b) Farm practices survey and modeling to estimate monthly NH_3 emissions from swine production in 12 Ecoregions of Canada. Can J Animal Sci 90:145–158
Shutske JM (2005) Using anhydrous ammonia safely on the farm. University of Minnesota extension FO-022326. [Online] Available: http://www.extension.umn.edu/distribution/cropsystems/DC2326.html
Sommer SG, Friis E, Bach A, Schjørring JK (1997) Ammonia volatilization from pig slurry applied with trail hoses or broadspread to winter wheat: effects of crop developmental stage, microclimate and leaf ammonia absorption. J Environ Qual 26:1153–1160
Sommer SG, Schjoerring JK, Denmead OT (2004) Ammonia emission for mineral fertilizers and fertilized crops. Adv in Agronomy 82:557–622
Souther L, Small-Johnson J, Messing RB (2000) A description of agricultural releases of anhydrous ammonia in Minnesota. Chem Health and Saf 7(6):16–22 doi:10.1016/S1074-9098(00)00142-8
Sutton MA, Fowler D, Moncrieff JB (1993a) The exchange of atmospheric ammonia with vegetated surfaces. I. Unfertilized vegetation. Q J Royal Meteorol Soc 119:1023–1045
Sutton MA, Fowler D, Moncrieff JB (1993b) The exchange of atmospheric ammonia with vegetated surfaces. II. Fertilized vegetation. Q J Royal Meteorol Soc 119:1047–1070
Sutton MA, Reis S, Baker SMH (2009) Atmospheric ammonia. Springer Science and Business Media B.V. pp 464
Taubes G (1995) Epidemiology faces its limits. Science 269:164–169
Thurston GD et al (2005) Workgroup report: workshop on source apportionment of particulate matter health effects—intercomparison of results and implications. Environ Health Perspect 113:1768–1774
USEPA (2006) Provisional assessment of recent studies on health effects of particulate matter exposure. National center for environmental assessment. Office of research and development. Research triangle park, NC. EPA/600/R-06/063. http://epa.gov/pm/pdfs/ord_report_20060720.pdf
USEPA (2011a) The national emissions inventory. Version 1.5. http://www.epa.gov/ttn/chief/net/2008inventory.html. Accessed 29 Aug 2011
USEPA (2011b) National summary of particulate matter emissions. http://www.epa.gov/air/emissions/pm.htm#pmnat. Accessed 18 Feb 2011
USEPA (2011c) National trends in particulate matter levels. http://www.epa.gov/airtrends/pm.html#pmnat. Accessed 18 Feb 2011
Whitehead DC, Lockyer DR (1987) The influence of the concentration of gaseous ammonia on its uptake by the leaves of Italian ryegrass, with and without an adequate supply of nitrogen to the roots. J Exp Bot 38:818–827

Whitehead DC, Raistrick N (1990) Ammonia volatilization from five nitrogen compounds used as fertilizers following surface application to soils. J Soil Sci 41:387–394

Whitehead DC, Lockyer DR, Raistrick N (1988) The volatilization of ammonia from perennial ryegrass during decomposition, drying and induced senescence. Ann Bot 61:567–571

WHO-Europe (2009) Exposure of children to air pollution (particulate matter) in outdoor air. World Health Organization—Europe. http://www.euro.who.int/__data/assets/pdf_file/0018/97002/enhis_factsheet09_3_3.pdf. Accessed 18 Feb 2011

Ye X, Ma Z, Zhang J, Du H, Chen J, Chen H, Yang X, Gao W, Geng F (2011) Important role of ammonia on haze formation in Shanghai. Environ Res Lett 6 024019 (5 pp) doi:10.1088/1748-9326/6/2/024019

Inventories of Air Pollutant Emissions

13

Warren McCormick

Abstract

Compiling and generating inventories of air pollutants is an important first step in any air quality assessment model or airshed management plan. One must know how much is being emitted into the atmosphere before the impact of those emissions can be determined or managed. Inventories can range from the facility scale to the national scale and must include all the pollutants of interest, possible precursors (A compound that participates in the chemical reaction that produces another compound) and all the associated source sectors. Source sectors are used to group similar types of sources, such as Point sources (large industrial facilities), Area sources (small, numerous sources like residential stoves) and Mobile sources (on and off road vehicles). Each sector requires different and specific methods to estimate emissions.

Keywords

Emission inventory · Air pollutants · Source sectors · NPRI · Emission estimation methods · Inventory trends · Air quality assessment · AQMS · NAFTA

13.1 Introduction

Air pollution degrades air quality and contributes to many current and potential environmental problems such as acidification, climate change, damage and soiling of buildings and other structures and stratospheric ozone depletion. It can also increase exposure to hazardous substances, thus elevating health risk, particularly among sensitive populations as described in the health impacts chapter by Dave Stieb in this book. An important tool in reducing this risk to the environment and human health is the ongoing development of accurate inventories of the emissions of pollutants, or their precursors.

Emission inventories provide consolidated information to quantify air pollutant emission rates associated with specific sources and time periods. Inventories are critical resources for air quality modelling, local and regional regulatory planning and international cooperation in the ongoing mission to reduce health risk. They can be used to inform policy makers and the public on emissions, identify the sources that are responsible for air pollutant problems and assess the potential impacts and implications of different strategies and plans (NARSTO 2005; European Environmental Agency 2011).

A common use of emission inventories is during the assessment of potential impacts on air quality when new emission sources are planned for a specific area or community. There are several components to these air quality assessments:

a. An emission inventory, or catalogue, of the types, amounts, rate and timing of pollutants that are emitted from all sources and the locations and characteristics of the emitting sources.
b. An estimate of the state of the atmosphere during the time period that the air quality assessment is needed. This includes the surface and upper level winds, turbulence, temperature, precipitation, solar radiation and other meteorological information that is fundamental to understanding how pollutants will be transported and changed over time.

W. McCormick
BC Ministry of Environment, Victoria, BC, Canada
e-mail: warren.mccormick@gov.bc.ca

E. Taylor, A. McMillan (eds.), *Air Quality Management,* DOI 10.1007/978-94-007-7557-2_13,

c. An air quality dispersion model that that uses the emissions and atmospheric information to estimate how pollutants will be transported, diluted and chemically and physically transformed over time and distance.

All three components are equally important in order to generate a realistic estimate of the ambient air quality at specific locations at specific times. However, using inaccurate or coarse resolution emission data in a highly refined air quality dispersion model will produce inaccurate or coarse results. Without some certainty in these air quality assessments, decision makers or political leaders cannot be expected to provide sound decisions about air quality management. It is therefore vitally important to develop a suitable inventory of all the pollutant emissions that could conceivably be affecting a community, region, country or continent of interest.

This chapter describes how Canada develops inventories of anthropogenic and natural emissions.

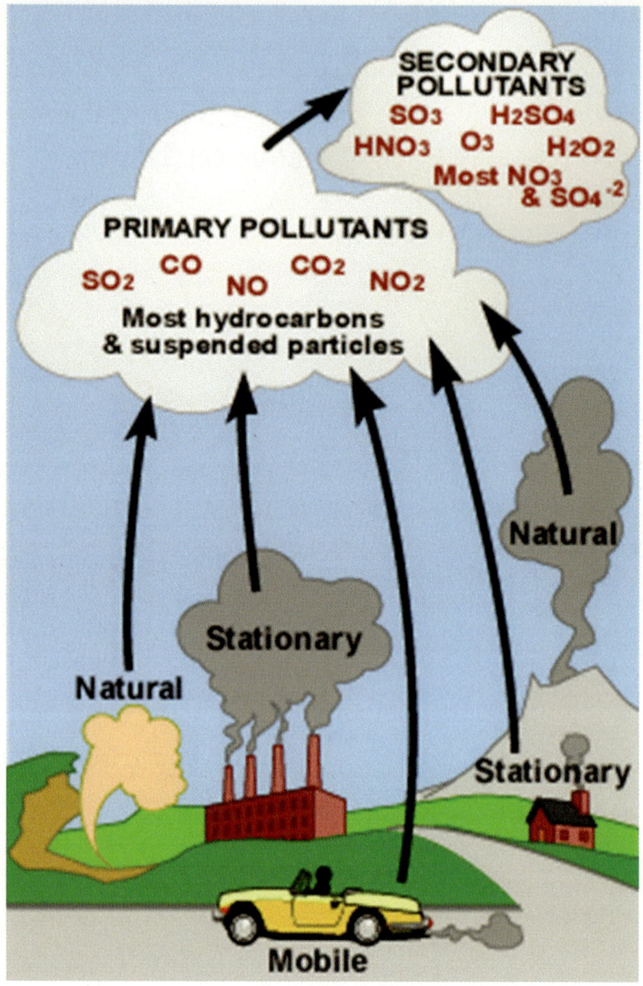

Fig. 13.1 Types and sources of air pollutants. (EPA 2010a)

13.2 Measuring Emissions

There are numerous types of emission sources and types of air pollutants (Fig. 13.1). There are also various methods to estimate emissions rates from these sources:

a. Direct sampling and monitoring is the most accurate technique to quantify emission rates from anthropogenic sources. This directly measures emissions. This is usually done through source or stack tests, which consists of taking quantitative air samples from exhaust stacks and analyzing these samples in a laboratory to determine pollutant concentrations. Certain sources may have Continuous Emission Monitoring (CEM) or Emission Predictive systems installed. This technique is usually limited to larger sources with gaseous pollutants like SO_2 (sulfur dioxide) and NO_x (nitrogen oxide and/or nitrogen dioxide). An example of sources that employ CEM is large power plants. Caution must be exercised when using source testing results for inventory purposes as many jurisdictions only require source tests for permit compliance purposes. Typically permit compliance procedures require the activity to be run at a high level to attempt to sample at maximum emission concentrations, as per the permit and this may not represent the average or actual emissions.
b. The Material or Mass Balance method is useful in situations where a substance flowing into a process is completely transformed into a new substance. Sulphur dioxide (SO_2) emission from fuel combustion is a good example. If one assumes that all the sulphur in the fuel is transformed into SO_2, then a measure of the sulphur in the fuel and the mass of fuel burned will provide the total SO_2 emissions.
c. Emission Factors: The above techniques are often impossible or prohibitively expensive to use on small industrial or mobile sources. Emission rates are therefore often estimated using a relationship linking our knowledge of the amount of pollution that is typically produced and the specific processes used. This relationship is typically developed from targeted studies and projects carried out by regulatory jurisdictions. This relationship, known as the Emission Factor, is used in following equation to estimate the pollutant emission rate:

emission rate = emission factor × activity factor × (1-control factor)

Where:

- **emission rate** is the amount of emissions per unit time (e.g. kilograms of nitrogen dioxide per month)
- **emission factor** is the amount of pollutant emitted by an activity associated with the source (e.g. kilograms of nitrogen dioxide per unit of fuel burned)
- **activity factor** is a measure of the activity that produces emission (e.g. kilograms of fuel burned per month)
- **control factor** is the fractional emission reduction in that source that is achieved by an add-on control device

(e.g. installation of a particulate collection device on a combustion source). For example, if the add-on device reduces emissions by 5 %, the control factor would be 0.05. The control factor is an optional term since in many cases it is included in the emission factor. (NARSTO 2005).

13.3 Generating Emission Inventories

Broadly speaking, an emission inventory is a database of air emissions. The details depend on the importance of the pollutants, the area of interest and the timeframe. Generally the types of pollutants inventoried in Canada are the common (or criteria) air contaminants (CAC): carbon monoxide (CO), nitrogen oxides (NO_x, which is NO and NO_2), sulphur oxides (SO_x which are compounds containing only sulfur and oxygen), volatile organic compounds (VOCs) and particulate matter (PM). However, specialized inventories could cover air toxics, hazardous air pollutants (HAPs) or persistent organic pollutants (POPs). Greenhouse Gases (GHG) can also be included, though inventory methods for GHGs differ in many respects.

Spatially, emission inventories can span scales from an individual industrial facility to the entire globe. Routine reporting of emissions are for a given timeframe, usually a year. Shorter timeframe inventories can be developed for analysis of individual air quality episodes.

Emission inventories in Canada are generally divided into many sectors and sub-sectors. These sub-sectors tend to be types of sources where similar estimation methods are used to generate the inventory. The first division is between Natural Sources and Anthropogenic Sources.

13.3.1 The Natural Sources Sector Inventory

This inventory (Fig. 13.2)accounts for air emissions related to natural events. While these events and emissions are generally considered uncontrollable in the true sense of air quality management, they can affect air quality management decisions in certain situations. For example, vegetation will emit certain volatile organic compounds (VOC) during the growing season, especially coniferous forests which emit isoprene-type organic compounds. Isoprene compounds are low on the reactivity scale in ground level ozone chemistry but in certain areas their impact can be significant if extensive forests are near the communities of concern. Typical Natural Sources Sector inventories in Canada include:

Vegetation Growth During the active growing season many plants emit certain types of VOCs that can play a part in the ozone chemistry or other local air quality. An example would be the Great Smoky Mountains in the eastern US. Another is the Lower Fraser Valley of British Columbia that is surrounded by vast areas of coniferous forests. In these areas this sub-sector is important. Generally, a computer model is used to apply emission factors for forests and other vegetative matter over different spatial scales and for different time periods. The current US Environmental Protection Agency (EPA) model is the Biogenic Emissions Inventory System (BEIS). It is part of a family of models which estimates emissions of VOCs that are the result of biological activity from land-based vegetative species and nitric oxide emissions that are the result of microbial activity from certain soil types (Environmental Protection Agency 2008).

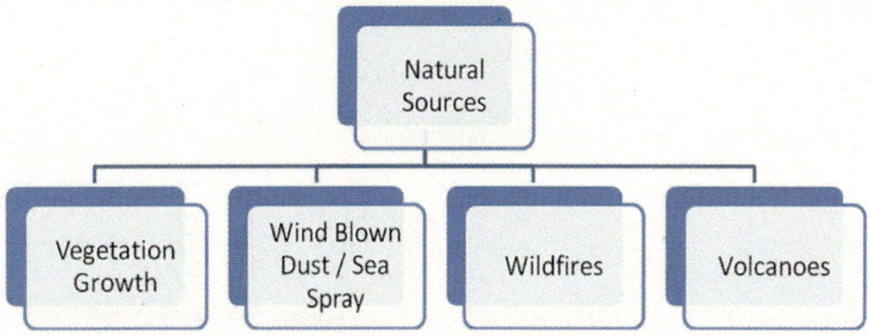

Fig. 13.2 The important natural sources of emissions

Wind-Blown Dust and Sea Spray Wind-blown dust and emissions of sulphur and chlorine from sea water can affect local air quality chemistry. Information about their relative source strength is sometimes necessary. Meteorological-based emission factors are used to estimate emissions of wind-blown dust and sea spray. Wind speed is the most important factor affecting emissions from these sources. In the case of wind-blown dust, soils dryness or days since last rain may also be used in the calculations. A secondary issue in coastal areas is the emission of sulphur compounds due to biological activity in sea water.

Wildfire/Forest Fire Smoke Smoke from wildfires can affect air quality hundreds or thousands of kilometers downwind. When analyzing for compliance of air quality objectives, emissions from uncontrollable extreme events such as wildfires generally need to be removed. To quantify these emissions, numerous forest fire consumption models have been expanded to calculate emissions for inventory purposes. One such model is Consume (US Forest Service 2009).

Volcanoes While even a small eruption can cause significant impacts, in most jurisdictions the events are so rare that this sub-sector is usually not included. Exceptions might be in areas that have more frequent impacts or continuous sources such as Alaska or Hawaii. No generalized models exist and emissions would have to be estimated with the assistance of a volcanologist.

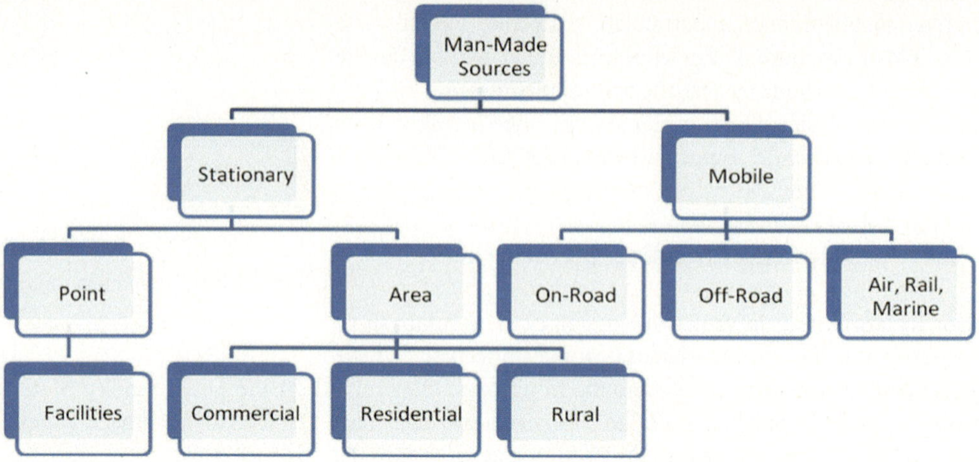

Fig. 13.3 The important man-made sources of pollutant emissions

13.3.2 Anthropogenic Source Sectors

Anthropogenic (man-made) emission sources are generally divided into *Stationary* sources and *Mobile* sources (Fig. 13.3).

Stationary Sources

In theory, *stationary sources* include all sources from the tallest industrial stack to the chimney on a house. However, in practice this sector is divided into two sub-sectors: **Point sources** and **Area sources**. Point sources tend to be larger industrial, institutional and commercial facilities that may be required to hold a permit or authorization from a government body in order to release air contaminants to the environment. All other smaller sources that do not require a permit are grouped into the Area Source sector.

Point sources Initially referring to the actual stack (chimney) at a facility, *point sources* now generally refer to all the emissions from a particular industrial facility. Facilities that are included in a Point source emission inventory are dependent on the permitting and/or reporting requirements of responsible jurisdiction. Provincial or local jurisdictions, such as Metro Vancouver in British Columbia, may have very extensive reporting requirements while other jurisdictions in Canada may not have any.

The need to have this reporting structure is usually driven by the requirement to provide results to managers and political leaders (as for airshed planning) or to satisfy inter-jurisdictional agreements (as in international agreements like the North American Free Trade Agreement—NAFTA).

Environment Canada operates an extensive reporting system called the National Pollutant Release Inventory (NPRI). The requirement to report to NPRI is based on the facility person-hours of work and/or the release of certain pollutants above set criteria. Authority to require this reporting is provided in the Canadian Environmental Protection Act.

Reporting facilities are required to follow Environment Canada's reporting procedure that describes what methods can be used to determine actual emissions. For sources or stacks that have actual stack test data or other continuous monitoring, they are required to use that data as the basis of their calculations. For smaller sources where the facility does not have actual monitoring data, the facility is required to use the best available emission factors and their activity data. (see http://www.ec.gc.ca/inrp-npri for more information).

Area sources All other stationary sources are grouped under the Area source category. The actual names and breakdown of sub-sectors under Area Sources can differ among jurisdictions due to local naming preferences and the detail that goes into their compilation of the Point Source list. For example, Metro Vancouver has a very detailed permitting list that includes sources that would not be on a point source list compiled outside of Metro Vancouver. Thus, direct comparisons of results between jurisdictions can be misleading without knowing the details of the Point Source list and the Area Sources.

The major area source sub-sectors are generally classified as Commercial, Residential and Rural.

- **Commercial (Sometimes Called Small Industrial) Area Sources** These are similar to the types of industries in the Point source list but are too small in size to require individual reporting. For example, while cement manufacturer facilities would likely operate under an air emissions permit that requires reporting and would thus be listed in the Point source sector, smaller concrete batch plants would be too small to require reporting and thus be grouped in the Area source sector. In some cases, this can lead to error or confusion between inventories. Also, a certain sub-sector could be counted in both the Point source sector and the Area source sector, leading to double counting of emissions. Also included in this sub-sector are businesses that deal in trade at a local level and that produce significant atmospheric emissions, particularly when emissions from a number of similar businesses are combined. Examples include dry cleaners, bakeries, printers, solvent and paint handling and use and restaurants.

- **Residential area sources** These include emission-generating activities such as heating of buildings (e.g. natural gas or wood heating), use of solvent-containing products (e.g. paints, glues, degreasers, etc.), structural fires, dust from construction activities and residential open burning, etc.
- **Rural Area Sources** These area sources include agricultural activities, logging and large scale open burning.
- **Estimation Methods of Area Sources** Methods to estimate the emissions from area sources vary widely depending on the accuracy needed in the output and objectives of the inventory. At a national or provincial scale many sub-sector estimates are based on fuel sales and other economic indicators or on general population-based emission factors. At an airshed level, estimates will be based on emission factors and actual activity data, where available, i.e. number of facilities in the airshed and their associated production data. A common place to find emission factors is the US EPA Emissions Factors and Policy Applications Center, Measurement Policy Group. This Group produces AP-42, the Compilation of Air Pollutant Emission Factors. AP-42 has been published since 1972 as the primary compilation of EPA's emission factor information. It contains emission factors and process information for more than 200 air pollution source categories. A source category is a specific industry sector or group of similar emitting sources (EPA 2011).

Mobile Sources

Generally, mobile sources refer to all types of vehicles, engines or equipment that can move under their own power (either for motion or work), usually with an internal combustion engine fueled by a hydrocarbon fuel. The Mobile sector has the following sub-sectors:

On Road Sources As a practical way of identifying vehicles, the On-Road Sources sector includes all vehicles that are licensed to travel on public roads, from everyday cars and trucks to heavy duty large transport trucks and buses. For the last several decades, the most commonly used estimation method was the MOBILE6.2 model by US EPA, and its Canadian counterpart MOBILE6.2C modified by Environment Canada to adjust for Canadian climate, operating environment and some emission criteria. In 2010, the US EPA replaced the MOBILE model with a new model called the Motor Vehicle Emission Simulator or MOVES (EPA 2012).

Non-Road Sources The Non-Road Sources sector encompasses all other vehicles and equipment that do not fall into the On Road sector. This includes:

- Construction and industrial equipment (forklifts, yard trucks, mining and forestry equipment, excavators, etc.)
- recreational equipment such as ATVs, snowmobiles, etc.
- hand-held equipment such as lawn and garden equipment (lawn mowers, weed wackers, blowers, etc.).

The US EPA model NONROAD is the primary emission estimation method, although there are current discussions indicating that NONROAD will be combined with the MOVES estimation method to provide one model for all land vehicles (EPA 2010b).

Aircraft Emission Sources Aircraft and airport operation equipment emissions can be estimated with the Emissions and Dispersion Modeling System, a complex microcomputer model designed to assess the air quality impacts of proposed airport development projects (FAA 2011). However, this model is designed for air quality assessments of proposed and existing airports and incorporates a dispersion model as well as the emissions model. With many airports being run by local agencies instead of the federal agencies (FAA or Transport Canada), the local agencies (e.g. Vancouver Airport Authority or Port of Seattle) will generally develop their own inventory system to accommodate their reporting requirements.

Rail Emission Sources There is no specific model to estimate rail emissions. Use of the NONROAD model methods or emission factors may provide some guidance for some rail activities.

Marine Emission Sources Port authorities in cooperation with government and regulatory agenciesmay develope their own inventory system for "at berth", harbor and land-side operations. Water side operations outside of the harbor (e.g. ocean going vessels) and other sub-sectors such as pleasure boats would be inventoried independently (some of these sub-sectors are included in the NONROAD model). On the west coast, Environment Canada and the BC Chamber of Shipping developed a very detailed emission inventory for 2005 (COS 2006).

Environment Canada and Transport Canada are doing further work on the Marine Emission Inventory Tool for 2010 and expanding the spatial coverage to all marine areas of Canada (Table 13.1).

Estimation Methods for Mobile Sources The models used in the mobile sector are largely empirical-style models where emission factors are developed from testing projects, or from engine manufacturer emission certification information and emission regulations. Data from Inspection and Maintenance

Table 13.1 Example of a Marine Emissions Inventory—the total emissions from vessels near the British Columbia coast, Juan de Fuca strait and Puget Sound in thousands of tonnes per year. (COS 2006)

Vessel class	NO_x	SO_x	CO_2	HC	PM_{10}	$PM_{2.5}$	CO	CH_4	N_2O	NH_3
Bulk vessel	6.04	4.51	272	0.212	0.365	0.326	0.532	0.029	0.007	0.006
Container ship	8.92	6.63	342	0.323	0.502	0.450	0.739	0.040	0.009	0.009
Cruise ship	6.10	3.49	415	0.203	0.411	0.370	0.489	0.032	0.015	0.009
General cargo	3.30	2.30	149	0.117	0.197	0.177	0.287	0.016	0.004	0.003
Misc	0.203	0.12	12.7	0.008	0.012	0.010	0.022	0.001	0	0
Motor vehicle carrier	0.878	0.56	35.2	0.032	0.051	0.046	0.073	0.004	0.001	0.001
Tanker	1.07	0.76	51.3	0.038	0.067	0.060	0.095	0.005	0.001	0.001
Total	*26.5*	*18.4*	*1,278*	*0.934*	*1.60*	*1.44*	*2.24*	*0.128*	*0.036*	*0.028*

programs (I&M) such as AirCare can be used in the mobile models. Activity level data must be gathered from users or assumed from other surrogates. The on-road models noted above (MOBILE and MOVES) have built-in climatological data and typical operating profiles for each class of vehicle that improve the accuracy of emission estimates. Model users can also modify built-in data, depending on the availability of improved local data.

These and other tools are described in the Canadian *Urban Transportation Emissions Calculator*, a user-friendly tool for estimating annual emissions from personal, commercial, and public transit vehicles. It estimates greenhouse gas (GHG) and criteria air contaminant emissions from the operation of vehicles. It also estimates upstream GHG emissions from the production, refining and transportation of transportation fuels, as well as from production of electricity used by electric vehicles (Transport Canada 2012; Fig. 13.4).

13.4 Airshed Inventory Trends (Forecasts and Backcasts[1])

In communities with long term airshed planning activities, questions arise as to the effect of past planning activities and regulations and/or the expected effect of proposed or recently implemented regulations on airshed emissions. In these communities, inventory reporting will also include forecasts and backcasts of the emission inventory at various points in the future and the past.

13.4.1 Forecasts

Forecasts of future emissions involve projecting the current inventory results into the future. Generally, economic or social indicators are used to assume a certain level of growth or shrinkage in sub-sectors of the inventory in order to estimate future emissions. Future regulations are also factored in when known with some certainty. This provides decision makers with some direction to address one sector over another in order to maintain or improve air quality in the airshed.

13.4.2 Backcasts

Backcasts involve projecting the current inventory results into the past, so that historical trends in emissions can be estimated. In this case, the economic indicators and trends are known and can be applied to the sub-sectors of the inventory. A backcast allows for past regulations to be "un-done", providing decision makers with an indication of the effectiveness of past regulations and policies.

Backcasts are not the previous inventories since inventory methods, emission factors and activity levels used to calculate the past inventory may be different from those used in the current inventory. Past inventories can be updated with current information but the use of past inventories as they were compiled could result in apparent changes in the inventory trend that were due to the changes in methods and thus not indicate a real trend.

13.5 Uses of Inventories

13.5.1 Generalized Public Reporting

Airshed management planning or general jurisdictional reporting out to the public (such as "State of the Air" or "State of the Environment" type reports) are examples of how inventory data is used. Emission inventories can inform airshed management plans and program or policy directions by providing information on the relative emissions from various emission sources, and the expected characteristics of or changes to those emissions. For example, regulatory intervention may be considered for a particular industrial sector if an emission inventory identified a large expected increase in emissions from that sector. The timeframe for this reporting and scope would be determined by the responsible

[1] An estimate of historical values, in this case emission inventories of a series of past years or decades.

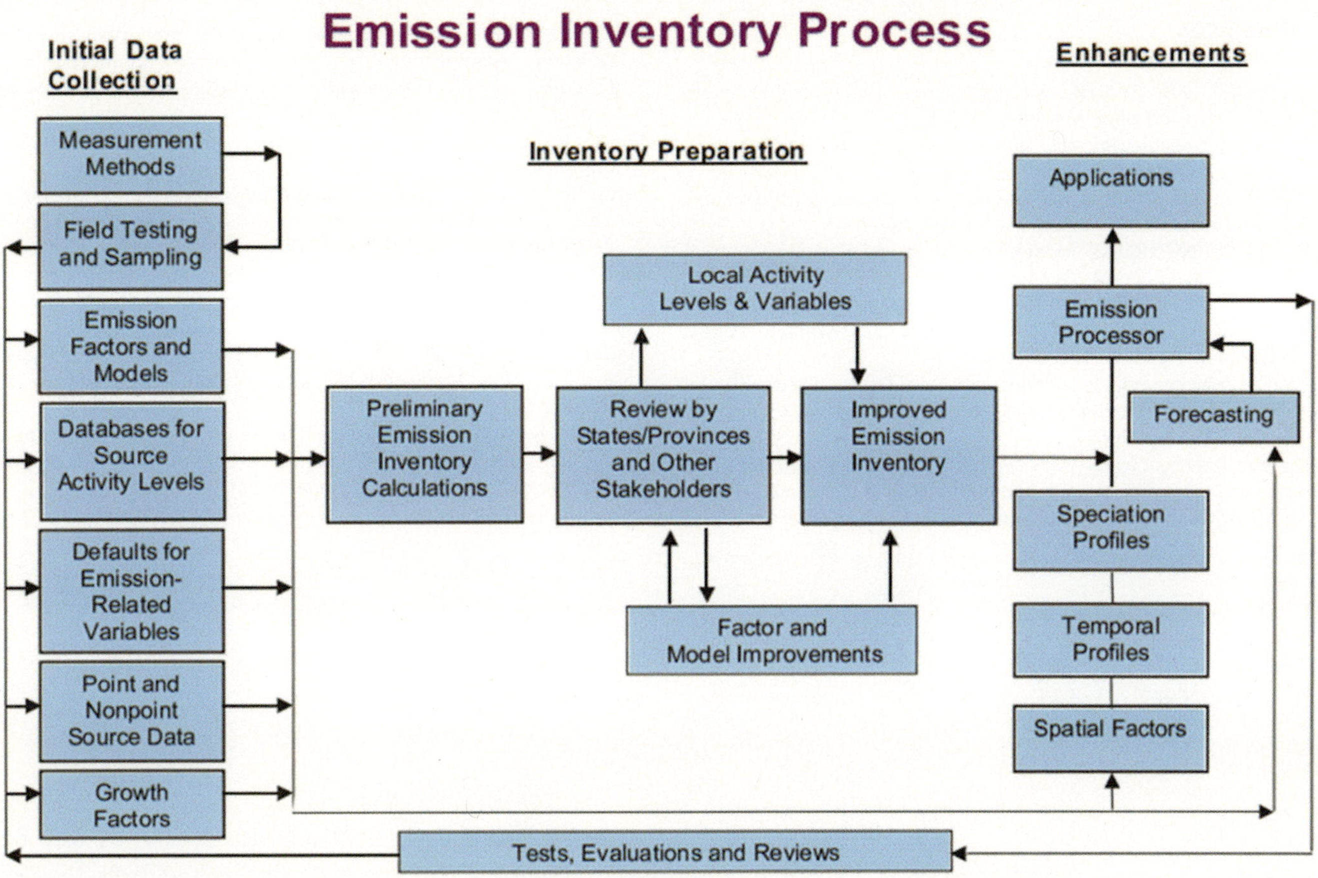

Fig. 13.4 A flowchart of the general process for compiling emission inventories in North America. (NARSTO 2005)

jurisdiction. A good example of this is the Emission Inventory reporting done by Metro Vancouver in support of their airshed plan (Metro Vancouver 2011).

13.5.2 Input for Comprehensive Air Quality Models

Regional comprehensive air quality models require input emissions inventory data to be spatially gridded to particular grid sizes. For example, on-road vehicle emissions would be geographically located on public roads and could be further allocated by the type of road (more emissions would be allocated to multi-lane highways than to residential streets). Advanced comprehensive air quality models may also require the emissions to be temporally allocated. This allocation could be done on timescales ranging from daily to annual depending on the emission source to be modelled, and a temporal profile would be developed that would apply specifically to the given emission source. For example, light duty on-road vehicles would have major peaks during weekday rush-hours and minimums in the overnight hours. Most emissions will be allocated to the lowest atmospheric level in the model, that is ground level grid cells. However some emissions from larger facilities with high stacks can be allocated to higher elevation grid cells. The US EPA endorses a program called SMOKE for this purpose, (MCNC 2012).

13.6 Future Directions

With the advent of a national air quality management system (currently called AQMS), with Environment Canada as the lead, more provincial jurisdictions will be developing interest and capacity in generating emission inventories. This proposed system will create the need for airshed driven, consistent inventories to be generated on a routine basis to meet the reporting requirements.

Since much of the activity data needed for many Area sub-sector calculations are available in GIS based data sources such as municipality zoning maps, many jurisdictions are looking to use GIS data in the calculations as well as developing GIS emission calculation layers. This new area of work holds great promise to greatly improve inventory estimates.

References

COS (2006) 2005–2006 BC Ocean-going vessel emissions inventory. Chamber of Shipping, Vancouver

EPA (2008) Biogenic emissions inventory system. http://www.epa.gov/asmdnerl/biogen.html. Accessed 2 Apr 2012

EPA (2010a) Air pollution control orientation course. http://www.epa.gov/apti/course422/ap3.html. Accessed 4 Apr 2012

EPA (2010b) NONROAD Model. http://www.epa.gov/oms/nonrdmdl.htm. Accessed 12 Apr 2012

EPA (2011) Basic emissions factors information. Emissions factors and policy applications center. http://www.epa.gov/ttn/chief/efpac/index.html. Accessed 12 Apr 2012

EPA (2012) Motor vehicle emission simulator. Office of transportation and air quality. http://www.epa.gov/otaq/models/moves/index.htm. Accessed 12 Apr 2012

European Environmental Agency (2011) Environmental terminology and discovery service. http://www.eea.europa.eu/publications/EMEPCORINAIR/page005.html. Accessed 4 Apr 2012

FAA (2011) Emissions and dispersion modeling system, Federal Aviation Administration. http://www.faa.gov/about/office_org/headquarters_offices/apl/research/models/edms_model/ Accessed 15 Apr 2012

MCNC (2012) SMOKE model. http://www.smoke-model.org/index.cfm. Accessed 17 Apr 2012

Metro Vancouver (2011) Air emissions & emission reduction programs. http://www.metrovancouver.org/services/air/emissions/Pages/default.aspx. Accessed 17 Apr 2012

NARSTO (2005) Improving emission inventories for effective air quality management across North America. NARSTO 05-001. Pasco. http://narsto.org/emission_inventory_1. Accessed 15 Apr 2011

Transport Canada (2012) Urban transportation emission calculator. http://wwwapps.tc.gc.ca/Prog/2/UTEC-CETU/Menu.aspx?lang=eng. Accessed 12 Apr 2012

US Forest Service (2009) Consume 3.0—A software tool for computing fuel consumption. US forest service. http://www.fs.fed.us/pnw/fera/research/smoke/consume/index.shtml. Accessed 2 Apr 2012

Part IV

Policy and Planning

Ambient Air Quality Objectives

14

Randolph P. Angle

Abstract

Ambient air quality objectives (AQOs), based on causal relationships between air pollution and environmental effects, are the cornerstone of air management. Nationally, Canada pioneered a three-tier framework in the 1970s, then experimented with a different approach, and is now moving to a structure echoing the original. A fully formulated ambient air quality objective has four main components—indicator, magnitude, averaging time, statistical form—and three supporting components—attainment date, endpoint, and measurement method. AQOs are developed following a three-stage risk paradigm consisting of prioritization, risk assessment and risk management. Reviewing scientific effects literature for a risk assessment is a challenging task in the selection of endpoints and the weighting of evidence. Provincial AQOs exist for a broad array of pollutants.

Keywords

Ambient air quality standard · Ambient air quality objective · Ambient air quality criteria · Air pollution effects · Air quality objective-setting · Air quality risk assessment · Air quality risk management

14.1 Introduction

As noted in the Introduction, an air quality management program is designed to achieve specific goals. Goals for the ambient environment are usually specified by ambient air quality objectives for different pollutants. An ambient air quality objective (AQO) is a numerical level of concentration or deposition that provides protection for human health and the environment. Two types of effects are considered—biological and physical. Biological effects include damage to human and animal health, damage to crops and damage to forests and native vegetation. Physical effects include damage to materials (metals, coatings, textiles, paper, leather, stone, and concrete), damage to structures (buildings, monuments, and art) and damage to atmospheric properties (visual range, colour, and clarity).

Ambient air quality objectives are based on scientific knowledge about the relationship between pollutant concentrations in the air and associated adverse effects. Such cause-effect relationships can be challenging to define. There are often wide ranges of response within a biological population. It can also be difficult to separate the effects caused by air pollution from those caused by the many other variables that influence biological systems. Ongoing research is needed to reduce uncertainties in the state of knowledge.

In the Canadian air quality community, ambient air quality objectives may appear under a variety of different, closely-related terms. Different writers and different jurisdictions will refer to *standards, objectives, benchmarks, guidelines, limits,* and *criteria* using specific definitions that stem from the particular legal or policy framework in which they are embedded. For example, the Canadian Council of Ministers of Environment defines *standard* as "a legally enforceable numerical limit or narrative statement, such as in a regulation, statute, contract, or other legally binding document, that has been adopted from a criterion or an objective" (CCME

R. P. Angle (✉)
R. Angle Consulting, Edmonton, Alberta, Canada
e-mail: rangle2009@gmail.com

E. Taylor, A. McMillan (eds.), *Air Quality Management,* DOI 10.1007/978-94-007-7557-2_14,

1999). The National Round Table (2008) used a similar definition. AQOs are used by government agencies for a variety of different purposes.

An AQO used in permitting (see the chapter on industrial emission management) is most likely to have a strong legal underpinning, and in this context might be regarded as a *limit*, "boundary that should not be crossed". When a permit application is being assessed, predicted concentrations using a dispersion model are compared with the AQO. Design modifications are required until the predicted values are below the AQO. After the facility is built and operating, monitored concentrations that exceed the AQO typically lead to some enforcement action by the regulator.

An AQO used for planning purposes serves either as an *objective*, "aim or goal" or as a *guideline*, "principle for determining a course of action". In long term planning, the objective specifies an end state that is desired at some future time. In short term planning, the objective provides guidance on how to proceed, for example, in managing an episode, or designing a monitoring study.

An AQO is also used as a communications tool for interpretation of measured or predicted concentrations, and as such might be regarded as a *criterion*, "test by which something can be judged". It is used to transform numbers into statements that will answer questions asked by politicians and the public, such as "is the air good or bad?" This usage is similar to the generic meaning of *standard*, "measure with which things are compared in order to determine their quality".

14.2 National Ambient Air Quality Objectives

The development of national ambient air quality objectives has been sporadic rather than ongoing. Legislative changes and the dynamics of federal-provincial relations have played an important role in the processes for establishing national AQOs.

14.2.1 The Canadian Clean Air Act

In 1969 the federal Department of National Health and Welfare formed an ad hoc Federal-Provincial Committee on Air Pollution. A subcommittee of senior officials was struck in 1970 to develop national ambient air quality objectives. The committee was formalized under the Clean Air Act of 1971. The committee articulated a three-level framework for ambient air quality objectives:

> The *maximum desirable level* is the long-term goal for air quality and provides a basis for an anti-degradation policy for unpolluted parts of the country and for the continuing development of pollution control technology. It provides guidance for land-use planners and technology developers. At lower levels, there is in essence "no effect" on any receptor. Persuasion and financial incentives would be the principal methods used to attain this objective (Federal-Provincial Advisory Committee 1976).

> The *maximum acceptable level* is intended to provide adequate protection against effects on soil, water, vegetation, materials, animals, visibility, and personal comfort and well-being. It represents the realistic objective today for all parts of Canada. When this level is exceeded, control action by a regulatory agency is indicated (Federal-Provincial Advisory Committee 1976).

> The *maximum tolerable level* denotes time-based concentrations of air contaminants beyond which, owing to a diminishing margin of safety, appropriate action is required without delay to protect the health of the general population (Working Group 1994).

The method for developing air quality objectives proceeded in three steps (e.g. Franson et al. 1982; Newill 1977): (1) scientific review: the relevant published literature was identified and then panels of experts systematically and critically reviewed the information to compile a report on what was known about the adverse effects of pollutants at various concentrations. The resulting documents were known as *air quality criteria*, or *guides*; (2) AQO selection: from the scientific knowledge summary, senior government officials developed the levels that would become the basis for air management; (3) implementation: regulators detailed the administrative steps necessary to achieve and maintain the AQOs.

Many of the criteria documents were compiled by the National Research Council's Associate Committee on Scientific Criteria for Environmental Quality (e.g. NRCC 1981, 1982). The Associate Committee evaluated available information on the probability of effects of contaminants on receptors in Canada together with the related fundamental principles and scientific knowledge. Members of subcommittees and panels were selected for individual competence and relevant experience with consideration for a balance among all sectors in Canada. Each report was reviewed according to a multi-stage procedure designed to preserve objectivity in the presentation of scientific knowledge. The scientific criteria provided a starting point for the Federal-Provincial Committee who was responsible for establishing the ambient objectives taking into account socioeconomic impacts and the state of technology.

By 1975 there were national ambient air quality objectives for sulphur dioxide, nitrogen dioxide, carbon monoxide, total suspended particulate and ozone. Table 14.1 shows the 2011 ambient air quality objectives for sulphur dioxide in all Canadian jurisdictions. The original national ambient air quality objectives have had an enduring influence on provincial AQOs. Manitoba adopted the same national three-level structure. British Columbia implemented a similar three-level structure: Level A (desirable levels), Level B (interim levels) and Level C (maximum levels).

Table 14.1 Canadian air quality objectives for sulphur dioxide in μg/m³

Jurisdiction	Averaging Time		
	Short (1-h except as noted)	Intermediate (24-h except as noted)	Long (annual)
Canada—federal			
Maximum desirable	450	150	30
Maximum acceptable	900	300	60
Maximum tolerable	–	800	–
Newfoundland and Labrador	900	300 600 as 3-h	60
Nova Scotia	900	300	60
New Brunswick	900	300	60
Prince Edward Island	90	300	60
Quebec	525 as 4-min	228	52
Ontario			
Ambient air criteria	690	275	55
Point-of-impingement	830 as ½-h		
Manitoba			
Maximum desirable	450	150	30
Maximum acceptable	900	300	60
Maximum tolerable	–	800	–
Saskatchewan	450	150	30
Alberta	450	125 30 as a 30-day	20
British Columbia			
Level A	450	160, 375 as 3-h	25
Level B	900	260, 665 as 3-h	50
Level C	900–1300	360	80
Yukon	450	150	30
Northwest Territories	450	150	30
Nunavut	450	150	30
Metro Vancouver	450	125	30
Communauté métropolitaine de Montréal	1300 500 as 10-min	260	52

14.2.2 The Canadian Environmental Protection Act

In 1988, the Canadian Environmental Protection Act (CEPA) subsumed the Clean Air Act into broader legislation aimed at the overall management of toxic substances. Risk assessments were internalized within Environment Canada and Health Canada. In 1992 the CEPA National Advisory Committee formed a new federal-provincial working group on air quality objectives and guidelines. During a review of human health and environmental effects literature, the working group found that many air pollutants had no thresholds for effects. It was unclear how to define three scientifically defensible levels; the originators of the three-level framework had not provided any procedural details. Consequently the working group proposed a new framework with two levels (CCME 1999):

> The *Reference Level* is a level above which there are no demonstrated effects on human health and/or the environment. Reference levels are defined for all receptors for which effects information is available (human health, animals, vegetation, materials, and aesthetic atmospheric parameters).
>
> The *Air Quality Objective* represents the air quality management goal for the protection of the general public and the environment in Canada. It is a level based upon consideration of scientific, social, economic and technological factors.

This was quite similar to the two-level system that Canada had been using for the management of acid rain since 1983 (Federal/Provincial/Territorial Ministers of Environment and Energy 1998; Nixon and Curran 1998). A *critical load* is the threshold above which pollutant deposition harms the environment. Ecosystems that can tolerate acidic pollution have high critical loads, while sensitive ecosystems have low critical loads. The critical load for aquatic ecosystems is defined as the amount of wet sulphate deposition that protects 95 % of lakes from acidifying to a pH level of less than 6. A *target load* is the amount of pollution that is deemed achievable and politically acceptable when other factors (such as ethics, scientific uncertainties, and social and economic effects) are balanced with environmental considerations. The aim of the Eastern Canada Acid Rain Program, established in 1983, was to reduce wet sulphate deposition to a target load of no more than 20 kilograms per hectare per year (kg/ha/yr). This load would protect moderately sensitive aquatic ecosystems from acidification.

European countries apply the concept of *critical load* for deposition amounts and *critical level* for pollutant concentrations (Bull 1991). *Critical load* is defined as "a quantitative estimate of exposure to one or more pollutants below which significant harmful effects on specified sensitive elements of the environment do not occur according to present knowledge". *Critical level* is defined as "concentrations of pollutants in the atmosphere above which direct adverse effects on receptors, such as human beings, plants, ecosystems or materials, may occur according to present knowledge" (UNECE 2012). The former refers to the quantity of pollutant deposited from air to the ground, whereas the latter is the gaseous concentration of a pollutant in the air. For comparison with the AQOs in Table 14.1, the sulphur dioxide critical levels are: for crops, an annual mean of 30 $\mu g/m^3$; for forests and natural vegetation, a winter mean (1 Oct to 31 Mar) of 20 $\mu g/m^3$; for forests and natural vegetation, an annual mean of 20 $\mu g/m^3$; for sensitive lichens, an annual mean of 10 $\mu g/m^3$ (APIS 2012). Critical levels have also been developed for ozone (Fuhrer et al. 1997).

The CEPA working group, operating with limited resources, had first reviewed the AQO for carbon monoxide using the three-level framework (Working Group 1994). After proposing the two-level framework, they recommended a protocol for the derivation of the reference level (Working Group 1996a) and a reference level for hydrogen fluoride (Working Group 1996b). The group also completed risk assessments for ozone (Working Group 1998) and particulate matter (Working Group 1999). These were used by the Canadian Council of Ministers of Environment in the development of the Canada-Wide Standards. A protocol for the derivation of the air quality objective and a risk assessment for total reduced sulphur compounds were in progress when federal resources were reallocated in 2000.

14.2.3 The Canadian Council of Ministers of the Environment

Under the auspices of the Canadian Council of Ministers of the Environment, in January 1998, the provincial and federal governments (with the exception of Quebec) signed the Canada-wide Accord on Environmental Harmonization. The Accord was designed for improved cooperation and better environmental protection across Canada. The standards development process included extensive stakeholder participation and several ancillary studies. In June 2000 the Canada-Wide Standards for Particulate Matter (PM) and Ozone was published (CCME 2000). These national ambient air quality objectives committed governments to reduce PM and ground-level ozone through jurisdiction-specific air quality management plans.

In April 2010 a new Comprehensive Air Management System for Canada was proposed (CCME 2010). Included in this system was the development of new Canadian Ambient Air Quality Standards (CAAQS) to drive air quality management actions in all jurisdictions. In areas where pollutant levels exceeded the standards, management efforts would focus on reducing emissions from all sources to move toward attainment of the standards. In areas where air quality met the standards, activities would be aimed at ensuring that pollutant levels did not rise above the standards. The standards would be developed, reviewed and strengthened over time through a process that included the consideration of practicality and achievability. Also included would be a set of action triggers at levels below the standards in order to keep them from being exceeded, and to prevent the standards from becoming "pollute-up-to" levels. Table 14.2 illustrates this approach. Such a tiered structure is reminiscent of the original three-level framework. A similar tiered response framework has been used in Alberta for managing acid deposition (CASA 1999) and for implementing the Canada-Wide Standards for Particulate Matter and Ozone (CASA 2003).

14.3 Expressing Ambient Air Quality Objectives

Our knowledge of air quality at a location is determined by various types of measurement data. For many pollutants automatic continuous air samplers draw air directly into robust pollutant-specific analytical equipment in the field, and can report the concentration immediately. Ideally an ambient air quality objective is related closely to the properties of the concentration data. There are four main components to an ambient air quality objective:

1. Indicator: generally the pollutant of concern in the air (e.g. sulphur dioxide), but it could also refer to rate of removal from the atmosphere (acidic deposition) or accumulation in a receptor.
2. Magnitude: the numerical value representing the concentration of the pollutant, usually a volumetric ratio (parts per million by volume, ppmv) or a gravimetric ratio (micrograms per cubic metre, $\mu g/m^3$) at a specified temperature and pressure (e.g. 25 °C and 1 atmosphere). For deposition, it would be expressed as mass per unit area per unit time (kilograms per hectare per year, kg/ha/y).
3. Averaging time: the interval of time over which continuously varying concentrations will be "averaged", that is, the sum of n measurements will be divided by n. The most

Table 14.2 Air quality levels and management actions under proposed new air management system for Canada

Level and description	Air management actions
Green: low pressure on air quality	Basic air quality surveillance Periodic reporting to the public on the state of air quality; public education Development planning based on principles of Keeping Clean Areas Clean and Continuous Improvement.
Yellow: air quality under pressure	Plan to reduce air quality deterioration Air quality monitoring to assess/identify relevant air quality issues Inventory and mapping of major emissions sources; modelling of emissions patterns Stakeholder involvement in air management efforts; public education and engagement. Economic and urban development policies to ensure air quality does not degrade
Red: Encroachment on Canadian Ambient Air Quality Standards	Rigorous action plan with key sources, contributions of stakeholders to reductions, milestones and timelines, periodic progress assessments, mapping and modelling Increased air monitoring and expansion of emissions inventory to all sources Public notification, education and engagement Sustainable economic and urban development policies to ensure improvements in air quality.
Black: non-attainment	Stronger action plan to achieve air quality improvements meeting the CAAQS Stepped-up air quality monitoring and source contribution assessment Stepped- up air zone management to meet CAAQS. Stepped-up coordination of actions at the regional airshed level

common averaging times are 1-h (1/2–h in Ontario), 24-h and annual (8760-h). Monitoring technology sometimes limits the available data for comparison and different types of models may be used to convert to a time period for which there is an objective (e.g. Alberta Environment 2011). Atmospheric dispersion models are also capable of producing estimates of concentrations averaged over virtually any time period, although for permitting shorter time periods are generally used. For deposition, the averaging time is usually one year.

4. Statistical Form (or metric): the calculations that must be performed on the basic data to determine whether an AQO is being met or exceeded. For example, the 30 μg/m^3 Canada-Wide Standard for Particulate Matter is based on the 98th percentile of 24-h averages over three consecutive years. For other pollutants, provincial AQOs sometimes specify an acceptable frequency of exceedance (e.g. Alberta Environment 2011). Most jurisdictions are silent on frequency although the stochastic nature of atmospheric turbulence is known to produce an exponential or log-normal concentration distribution (Barry 1977). The United States primary standards typically are not to be exceeded more than once per year.

Three additional supporting components are also need for full formulation of the AQO:

5. Date for Achievement: the time in the future when the objective is to be met. The Canada-Wide Standards for Particulate Matter and Ozone were to be achieved by 2010. Provincial AQOs generally apply from the time they are issued, after a fixed adjustment period (e.g. Alberta Environment 2011) or at a specified effective date (e.g. Alberta Environment 2010).
6. Endpoint: the specific type of outcome and measured effect that may result from exposure at or above the specified concentration. There is generally a lot of uncertainty in cause-effect data. Experts often disagree on the validity and interpretations of health effects data. Vegetation is subject to large species variability and distribution. Effects can be obvious or subtle. Odour may be an important consideration. Ultimately, some endpoint becomes

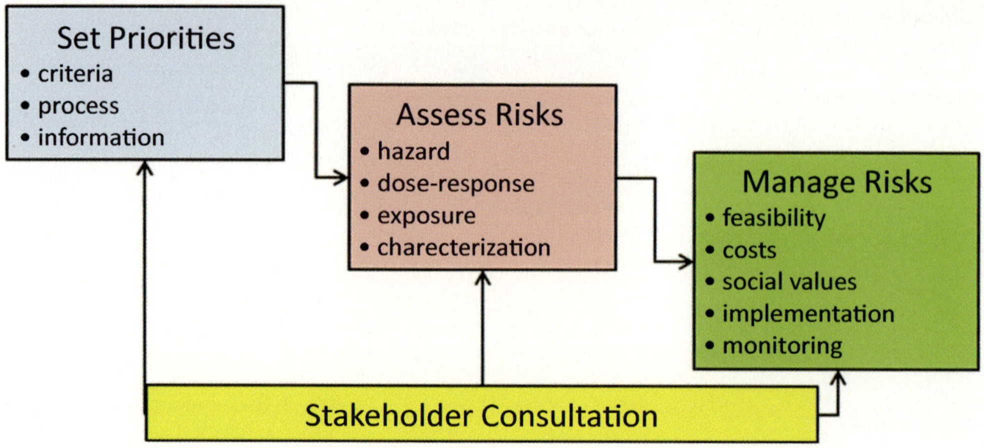

Fig. 14.1 Stages in setting ambient air quality objectives through a risk-based procedure

the determining factor in setting the AQO. Both Ontario and Alberta provide a general idea of the endpoints in their listings of AQOs (Ontario Ministry of Environment 2008; Alberta Environment 2010).

7. Methods for measurement: the techniques for determining concentrations or depositions. For consistency in making comparisons clearly defined methodologies are needed. This leads to supporting standards for measurement and testing (e.g. Saskatchewan Environment 2007) and the handling of data (e.g. CCME 2007).

Concentration in the air and deposition to ecosystems are not the only ways to represent air quality. Other important aspects of air quality are found in related atmospheric properties and in characteristics of the physical and biological receptors. Ambient air quality objectives can and have been set in terms of:

- Chemical content in vegetation, for example, fluoride in animal forage (Ontario Ministry of Environment 2008; Alberta Environment 2009)
- Sedimentation of particulate matter onto or into exposed receptacles, for example, dustfall (Ontario Ministry of Environment 2008; Alberta Environment 2010)
- Transmittance or reflectance of light through or from a spot of particulate matter collected on a filter, for example, coefficient of haze (Alberta Environment 2010)
- Rate of collection by molecular diffusion to a pollutant-specific absorbent material exposed in the field, for example, sulphation (Alberta Environment 2010)
- Odour perception, for example, in odour units (RWDI 2005)
- Visibility, for example, visual range or light extinction (RWDI Air 2008).

The statistical form of an AQO usually reflects its principal intended use. The Canada-Wide Standard for Particulate Matter and Ozone was developed to guide jurisdictional air quality planning. However, the statistical form for PM did not lend itself readily to comparisons with the output from simple dispersion model screening used in small source permitting or the short-term data from continuous monitoring. For these purposes Alberta developed an equivalent hourly AQO (Fu et al. 2000; Alberta Environment 2010).

14.4 Developing Ambient Air Quality Objectives

Most jurisdictions set their AQOs using some type of a risk-based approach to environmental safety. The fundamentals of risk assessment and risk management with some history are outlined in McColl et al. (2000). Adaptations for air quality are described by Bailey (1999) and Caton and Bates (2002). For most applications in air quality objective setting, there are three basic stages shown in Fig. 14.1. Risk communication is sometimes viewed as part of the process after managing the risks.

The procedure begins with issue prioritization, followed by risk assessment and consideration of risk management options (e.g. Caton and Bates 2002). The risk assessment stage looks at the nature of the hazard, the dose-response relationship, the population of receptors that are exposed, and the estimated impacts on the exposed receptor population. The risk management stage integrates information from the risk assessment with economic, technical, ethical, social, legal, ecological, and achievability factors. Stakeholders are generally engaged in some way throughout the process. This procedure is comprehensive and thus also very expensive and time-consuming. The ultimate decision is taken by the government agency or political leaders on behalf of its citizens. The Canada-Wide Standards for Particulate Matter and Ozone were developed in a lengthy process following this framework.

14.4.1 Setting Priorities

Priorities are generally set by air quality agencies either internally or in consultation with stakeholders. One or more

Table 14.3 Summary of some methods for setting priorities with stakeholders

Method	Brief Description	Advantages	Disadvantages
Expert Panels (Leiss 2008)	Independent specialists arrive at recommendations through consensus	Brings together best available expertise	Dominant scientist may over influence
		Yields high scientific credibility	Tendency to go beyond field of competence
Nominal Group Technique (Delbecq and de Ven 1971)	Participants give views, generate a list of ideas and concerns, vote or rank the list, have structured small group discussions, and repeat the ranking or voting	Minimizes conflict	Opinions may not converge
		Ensures relatively equal participation	Cross-fertilization of ideas may be constrained
		Provides sense of closure	Process may appear to be too mechanical
		Simpler multi-voting without discussion can be used with the internet	
Consensus Development Conference (Crowe 2009)	Meeting to debate summary statements and seek agreement on the most important	Accommodates a large number of participants	Requires extensive preparation
		Can be used with the internet	
Focus Groups (I-TECH 2008)	Structured conversation to obtain in-depth information about people's feelings, values, and ideas	Way to get feedback on specific proposals	Generally does not come to consensus or make a decision
Delphi Technique (Brown 1968)	Works through a number of cycles of anonymous written discussion and judgment, controlled by the process manager	Avoids groupthink and personality conflict	Slow and time consuming
		Gives time to think carefully, be rigorous and revise thinking	Intended primarily for experts
		Progressively refines content to convergence	
Criteria Weighting Method, or Multi-Criteria Analysis (DCLG 2009)	Participants establish a relevant set of criteria, assign a priority ranking, and then rank alternatives	Understandable, rational, structures the complexity	May be difficult to get agreement on weightings and criteria
			May be slow and protracted
Paired Comparisons (Brown and Peterson 2009)	Each member of the decision team compares each option with every other option	Quick and easy to set up simple binary choices	Works best with 6 to 12 alternatives
			May be tedious and time-consuming Mechanical

chemical screening tools may be used (Five Winds 2004; Caton et al. 1988). An air pollution agency typically sets its internal priorities based on considerations of: pending industrial developments, quantity of emissions, number of emitting sources, toxic substance lists; commitments to others (e.g. other governments, Council of Environment Ministers, environmental groups, consultative groups); scheduled reviews; new information related to environmental/ human health effects; and preliminary evaluation of risks.

If an agency chooses to involve its stakeholders, typically some sort of deliberative process is used to arrive at priorities. Table 14.3 summarizes some methods, many of which can also be used as ways of engaging stakeholders in both the risk assessment and risk management stages.

14.4.2 Assessing Risks

Risk assessment is the scientific evaluation of the likelihood of adverse health effects due to exposure of a human or an ecological component. Risk assessment is comprised of four major steps:

1. *Hazard identification* describes the type of adverse effect associated with the pollutant based on existing scientific literature.
2. *Dose-response assessment* determines the relationship between the amount of exposure and the probability of an adverse effect;
3. *Exposure assessment* determines the concentration, frequency, duration and continuity of exposure over time.
4. *Risk characterization* provides a summary of the risk assessment methods, sources of evidence, uncertainties and results for use by decision-makers and communicators.

Risk assessments rely heavily on the available scientific information. Aside from being very labor intensive and time consuming, they can be often be limited by the lack of good quality studies. There is a need for ongoing research programs to fill knowledge gaps and generate the scientific information needed to update AQOs.

Reviewing Scientific Literature

Hazard identification and dose-response (cause-effect) are assessments for air pollutants are generally based on existing scientific literature and are subject to all of the challenges of conducting a comprehensive literature review. One major problem is evaluating the quality of research and deciding whether or how to use lower quality information (Randolph 2009). Most reviewers do this implicitly and with their own personal biases; however explicit evaluation schemes do exist. Randolph (2009) provides one example of a scoring chart. Priestley et al. (2006) provided a summary of criteria to be used in selecting human and animal studies.

Davies and Haggerty (2002) used a rigorous scheme to rate the confidence in each study as high, medium or low. A set of eligibility criteria allowed for the systematic selection of studies. The technical quality of the eligible studies was judged against a pre-defined set of quality criteria derived from leading experts. A Confidence Rating was assigned to each study by weighing the strengths and weaknesses of the study, using a check-list for guidance. Each study was read and rated as "high", "moderate" or "low" independently by two or three knowledgeable professionals.

This systematic evaluation of studies is an example of an objective methodological approach to "weight of evidence". Priestley et al. (2006) concluded that weight-of-evidence is not a clearly defined approach. The term is often used metaphorically to imply that some evaluation has taken place, but generally there is no clear indication of how this was done. The term is also used theoretically to indicate simply a conceptual framework.

The other big challenge with the effects literature review is the selection of endpoints. An endpoint is a characteristic of a human or ecological component that may be affected by exposure to the pollutant. Endpoints can include any of the various levels of biological organization: molecules, organelles, cells, groups of cells (tissues, organs, and organ systems), organisms (individual living things), populations (groups of organisms of one type that live in the same area), communities (populations living together in the same area), and ecosystems (a community and its non-living surroundings (Odum 1993). Most studies of the effects of pollutants on living things focus on the organism and its groups of cells. Typical ecological endpoints include: visible injury, photosynthesis, stomatal conductance, dark respiration, biomass, crop yield, root growth, decomposition, nutrient uptake, mortality, gross anomalies, and fecundity. Medical endpoints can be concerned with the status of various organ systems (respiratory, cardiovascular, gastrointestinal, neurological, reproductive, ocular, immunological) in individuals or the distribution of such effects within an exposed population.

Health effects studies generally fall into three categories: animal toxicology, human experimental and human observational (see the chapter Air Quality Impacts on Health). Environmental effects studies can be categorized in much the same way. Analogous to animal toxicology there are laboratory studies of potted plants in growth chambers or greenhouses. The experiments are done with the plant species of interest, so there is no issue similar to extrapolation from animals to humans, but there are concerns about how well laboratory conditions reproduce actual outdoor field conditions. There is also a challenge in determining the appropriate plant species to consider, especially since there are wide variations in sensitivity to air pollutants. A jurisdiction may choose to ignore results for plants that do not grow in the area, or may focus on crops of economic value. Analogous to clinical trials, there are controlled field exposures, using open top chambers or free air enrichment fumigation systems. Analogous to human observational studies, there are uncontrolled field exposures which use "real-world" conditions, available data and additional measurements. Ecosystem techniques have been summarized by Chen and Goldstein (1986) and types of field exposures have been described by Eberhardt and Thomas (2008).

The risk analysis documentation is called a Science Assessment in the Canadian national process (e.g. Working Group 1999), a Rationale Document in Ontario (e.g. Ontario Ministry of Environment 2006), and an Assessment Report in Alberta (e.g. WBK & Associates 2004). While the content varies somewhat with the pollutant, typically these reports would contain:

a. General pollutant information (identification and uses; physical, chemical and biological properties; environmental fate and behaviour)
b. Emissions and concentrations (natural and anthropogenic sources; ambient concentrations)
c. Effects on humans and animals, Acute and Chronic (respiratory; cardiovascular; gastrointestinal; neurological; reproductive and developmental; genetic; cancer)
d. Effects on vegetation (visible injury; photosynthesis; growth and yield)
e. Effects on materials (electrochemical corrosion; chemical attack)
f. Ambient monitoring methods (continuous, integrated and passive)
g. Existing air quality objectives elsewhere.

In addition to narrative around each of the topics, such reports may contain tabular or graphical summaries of effects observed at various concentrations and exposure durations. Table 14.4 provides an example of a summary effects table.

Exposure assessment draws upon existing ambient air quality monitoring data, but since station density is rarely adequate for accurate categorization, interpolations and extrapolation need to be made using statistical or dispersion models. For humans who spend much of their time indoors and who move extensively through different regions of air quality, it can be challenging to estimate exposure accurately. For ecosystems and vegetation, relevant ambient air quality monitoring data tends to be even sparser.

If there insufficient effects literature, it may be necessary to initiate a specific research program to collect the relevant data for the pollutant and receptors of concern. For example, in the mid-1990s production and use of ethylene by the pet-

Table 14.4 Summary of effects on vegetation for ethylene exposures over 48 h. (Alberta Environment 2003)

Species	Concentration (μg/m³)	Duration of Exposure	Effect
Barley	57	3 days	41 % reduction in seed yield
Barley	34	14 days	63 % reduction in seed yield
Canola	57	31 days	20 % reduction in seed yield
Canola	40	87 days	Seed yield 63 % lower than in "background" treatment (12 μg/m³)
Field peas	288	16 days	50 % reduction in seed yield
Pea seedlings	12	60 h	Inhibition of epicotyl elongation
Bean seedlings	92	60 h	Inhibition of epicotyl and root elongation, radial expansion of hypocotyl
Begonia	173	10 days	Flower quality and number decreased significantly
Easter lily	58	77 days	Flowering and plan quality significantly decreased
Oat	40	100 days	Floret number 26 % lower than in "background" treatment (8 μg/m³)

rochemical industry in Alberta was increasing. Nearby agricultural producers were concerned that ethylene might affect plant growth and yield. The scientific literature yielded little about ethylene effects on crops commonly grown in Alberta. Government and industry cooperatively sponsored the Alberta Ethylene Crop Research Project from 1997 to 2001 (Alberta Research Council 2002). The results of the project (e.g. Archambault and Li 2001) provided key scientific information and were instrumental in finalizing Alberta's AQO for ethylene.

Risk Characterization

Risk characterization summarizes and integrates the information on hazard, dose-response, and exposure. It provides information about harmful effects for various exposures and the uncertainties or level of confidence in that information. The risk characterization provides a summary of the critical findings of the risk assessment process in language that can be understood by other scientists, regulators, stakeholders, and the general public (McColl et al. 2000). Effective risk characterization depends upon transparency, clarity, consistency and reasonableness (Science Policy Council 2000). The assessment documents produced by Ontario and Alberta may be regarded as "risk characterizations" although they do not use the term.

In the risk assessment documents emerging from Canadian federal process, risk characterization has appeared as a final chapter in the science assessment (e.g. Working Group 1999) or as a section in an Executive Summary (e.g. Working Group 1998). However, in these documents risk characterization is intended "to evaluate the weight of evidence presented in the Science Assessment Document to determine whether or not the findings support a causal association."

14.4.3 Managing Risks

The risk management stage integrates information from the risk assessment with economic and technical factors as well as ethical, social, legal, and achievability considerations. Information about existing background levels and trends in emissions is examined and various types of socio-economic studies may be conducted to inform the decision (see Introduction).

Background concentrations constitute one of the major uncertainties for some pollutants. Scientists often define "background" as the concentrations in pristine conditions unaffected by man-made disturbances, while air quality managers define background as the concentrations arriving at the upwind boundary of their jurisdiction. The air entering a jurisdiction will be a combination of natural and man-made pollution from upwind sources (McKendry 2006). Determining the actual concentrations for pollutants like ozone can be problematic. Reid (2007) summarizes the controversy about background ozone concentrations with reported values ranging from 15–60 ppb. High background concentrations due either to natural sources or transport from upwind source regions may severely limit the ability of a jurisdiction to meet the Canada-Wide Standard for Ozone. The achievement determination guidance document (CCME 2007) contains explicit procedures to account for transboundary flow, background levels and natural events.

The composition of the team making the risk management decision about the ambient air quality objective often determines the range of factors considered and the depth of analysis. Teams within a regulatory agency may be mindful of the means of implementation, the reaction of various stakeholder groups, and the disposition of senior decision-makers. Teams from multiple departments of a government may look more closely at impacts on affected parties, potential media attention, and consistency with the overall program of the government. There will be discussions about the significance of various effects, the importance of various species (indigenous and imported), public values, aesthetics and economics.

Teams drawn from the fourteen air quality jurisdictions in Canada bring different regional priorities, different economic concerns, different management structures and different philosophies to the table. Socio-economic impacts like facility closures and unemployment may need to be assessed and parts of the risk assessment may be questioned. The Canadian Council of Ministers of the Environment works with a consensus decision making model (CCME 2011a) and has developed a toolkit (CCME 2011b) to assist in forging agreements among governments with disparate interests. Teams that include stakeholders extend the perspectives even further, adding considerations such as industry competitiveness on the one hand and environmental justice on the other. The risk assessment and the socio-economic analysis may be questioned and external reviews may be sought (e.g. Royal Society 2001).

Individuals and groups are guided by their *worldview*, the comprehensive philosophy or conception of the world and human life that serves as a framework for organizing perceptions, shaping attitudes and interpreting reality. Inevitably there will be differences of opinion about the significance of effects, the relative importance of various endpoints, the weight of evidence, the cost of implementation, "fairness" and other factors that enter into risk management deliberations. A variety of tools are available to bring people together for constructive dialogue. Ultimately, the regulatory agency must balance a variety of perspectives and the AQO is a judgment on behalf of society as to what risks to health and ecosystems are acceptable.

14.4.4 Engaging Stakeholders

The Introduction provides an overview of the approaches and methods used to engage stakeholders. These can all be used in the process for setting ambient air quality objectives as can the techniques for group priority-setting outlined earlier in this chapter. All Canadian jurisdictions provide opportunities for stakeholder input at various stages of the objective-setting process. Ontario uses stakeholder meetings and the Environmental Bill of Rights registry website (Ontario Ministry of Environment 1999). Environment Canada uses the CEPA Registry website and other project-specific forms of consultation. Alberta involves stakeholder representatives in the entire objective-setting process, from priority-setting through risk assessment to risk management (Blair et al. 2007; Alberta Environment 2005).

CCME (1993) follows a set of principles in designing a stakeholder engagement process for its initiatives. The government of Canada (Privy Council Office 2000) has directed that a "citizen focus" be built into all federal government activities, programs and services. Health Canada (2000) developed a toolkit with a variety of methods for engaging the public in decision-making, Smith (2003) provided a practical guide, and the Health Council of Canada (Gauvin et al. 2005) explored the conditions for successful public involvement. The Canadian Institutes of Health Research have published both a framework (2009a) for engaging citizens in its research and a handbook (2009b) to introduce staff to the breadth of considerations in planning citizen engagement activities.

14.5 Provincial Ambient Air Quality Objectives

The risk assessment-risk management framework can be applied at various scales of effort. At one extreme, the development of Canada-Wide Standards for PM and Ozone required the pooling of talent and finances from the federal government, the ten provinces, and three territories. At a more modest level of effort many provinces, for example, Ontario and Alberta, have been able to develop AQOs for a wide variety of pollutants. There are also some shortcuts than can be used in some situations to streamline the process. Five useful approaches that have been used alone or in combination are:

1. **As Good As:** A community can decide it wants air quality "as good as" another community, "as good as" an earlier time in its own history, or "as good as" that experienced during particular time periods in the year (Newill 1977). No analysis of effects is required. The decision can be taken by community leaders or there can be extensive polling of residents. Non-degradation seeks to maintain current levels, that is, air quality in the future is to be "as good as" it is now.

   The Canada-Wide Acid Rain Strategy for Post 2000 (Federal/Provincial/Territorial Ministers of Environment and Energy 1998) included a policy called "keeping clean areas clean" that identified the need to manage emissions of SO_2 and NO_x to ensure deposition levels do not approach the critical loads. The Canada-Wide Standards for Particulate Matter and Ozone (CCME 2000) also contained provisions for "keeping clean areas clean" through the application of pollution prevention, continuous improvement and best management practices.

2. **Best of Class:** Air quality objectives from other jurisdictions are reviewed, and the most stringent from a relevant jurisdiction is selected. A relevant jurisdiction can be defined in terms of population, climate, types of industries, economic structure, legal framework, regulatory philosophy or other characteristic. This provides a quick and inexpensive way to put forward new objectives, relying on decisions taken elsewhere. However, there is the danger of adopting a number that was designed for an entirely different purpose (BC Ministry of Healthy Living and Sport 2010).

   Ontario (Bailey 1999) has reviewed its AQOs against other jurisdictions and Alberta (Alberta Environment 2010) has adopted some of its AQOs from other jurisdictions. Many provinces continue to use the numbers established under the old federal Clean Act in 1974.

3. **Effects-based:** At the end of risk assessment, the scientists conducting the assessment use their expert judgment

to select an air quality objective that protects against significant effects. There is no consideration of costs, achievability and other factors, all of which are deferred to implementation.

Ontario establishes Ambient Air Quality Criteria at levels below which adverse health and/or environmental effects are not expected. For non-carcinogens the criteria can be based on the no-observed-adverse-effect level (NOAEL) or lowest-observed- adverse-effect level (LOAEL) with or without "safety" or "uncertainty" factors applied. For carcinogens (cancer-causing pollutants) the Criteria are generally based on a probability of 1 in a million or 10 in a million over a lifetime. Odour, vegetation, soiling, visibility, corrosion or other effects are also considered and a pollutant may have multiple ambient air quality criteria. The criteria apply to general air quality independent of location and are used for environmental assessments, special studies using ambient air monitoring data, and the assessments of general air quality in a community (Ontario Ministry of Environment 2008).

Ontario also establishes *point of impingement* limits to review applications for permits and to assess compliance with Ontario regulations (Ontario Ministry of Environment 1999). Point of Impingement limits are used with dispersion models for permitting. They apply to a ½-h averaging time. The numerical values are generally derived from the criteria using a factor of 15 to convert from annual average concentrations and a factor of three to convert 24-hour averaging times (Bailey 1999).

4. **Exposure Reduction:** Applying the principle of continuous improvement, an AQO is expressed as a required percentage decrease in concentrations. Some analysis may be undertaken to determine the benefits of exposure reduction.

   Environment Canada's Clean Air Regulatory Agenda (Environment Canada 2007) set intended emission reductions targets ranging from 20 to 55% which, given the linear relationship between emissions and ambient air quality, implies similar reductions in ambient concentrations in the vicinity of the sources making the reductions.

   Under the proposed new Comprehensive Air Management System for Canada, new Canadian Ambient Air Quality Standards would be informed by using a new Population Exposure Improvement methodology (CCME 2010). This method links ambient concentrations to the size of the population exposed. The potential standards could then be set by specifying the desired reduction in exposed population.

5. **Worker Extrapolation:** Occupational health standards are modified to serve as objectives for ambient air outside the workplace (Cannon 1986; Bailey 1999). The adjustment factor typically ranges from 100 to 1000. An adjustment calculation could proceed as follows: start with the 8-hour Threshold Limit Value (TLV) of the American Conference of Government Industrial Hygienists; divide by 10 to take into account the greater susceptibility of the general population; multiply by 8/24 (the ratio of the number hours workers are exposed to the number of hours a member of the general public might be exposed); multiply by 5/7 (the ratio of the number of days in a week a worker is exposed to the number of days in a week that a member of the general public might be exposed); multiply by 30/70 (the ratio of the number of years a worker is exposed to the number of years a member of the general public might be exposed in a lifetime). Further adjustments might be made for uncertainty in knowledge (e.g. divide by another factor of 10) or some other difference depending on the pollutant. Occupational exposure limits are available for a wide range of chemicals and provide a relatively simple way to take advantage of the extensive risk assessment conducted by the occupational health organizations.

   The method is generally used when there is no literature for health effects at anticipated environmental exposure levels. It provides a quick way to create ambient objectives for use in permitting. Many US state agencies have applied this approach to manage toxic air pollutants, and thus Canadian jurisdictions adopting the same objective have indirectly also used this approach. It is not presently known whether an AQO derived in this way would approximate an AQO derived from an appropriate set of scientific health effects studies.

The "best of class", and "worker extrapolation" approaches rely upon risk assessments done by other jurisdictions. The "exposure reduction" and "as good as" approaches eliminate any explicit risk assessment and move immediately to risk management. In the effects-based approach the AQO emerges from the risk assessment, and risk management consists of planning for implementation.

Initially most provinces adopted the national maximum acceptable levels of 1974–75 while a few adopted the national maximum desirable levels. For many provinces, these levels are still in effect, as Table 14.1 illustrated for sulphur dioxide. More than three decades have passed since these AQOs were derived and much new information has emerged. Particulate matter and ozone have been the focus of recent national updates. For the other pollutants, a few provinces have independently undertaken reviews using their own processes.

Provinces have also found that AQOs are needed for a variety of other pollutants of concern within their jurisdictions. For example, Quebec, Ontario, Manitoba, Alberta and British Columbia have AQOs for formaldehyde. In 2011 Canadian air quality jurisdictions had the following numbers of AQOs as a reflection of their respective industrial bases and the concerns of citizens: Ontario (339), Quebec (79), Alberta (48), Manitoba (26), Newfoundland (23), British

Columbia (12), Saskatchewan (10), New Brunswick (6), Yukon (6), Northwest Territories (6), Nova Scotia (6), Prince Edward Island (5), Nunavut (2), Metro Vancouver (6), Communauté métropolitaine de Montréal (9).

14.6 Conclusion

Ambient air quality objectives have been an important part of air quality management in Canada since the 1970s. Under the Canadian Clean Air Act of 1971 a three-tier framework for objectives was developed and levels were established for sulphur dioxide, nitrogen dioxide, carbon monoxide, ozone and total suspended particulate. The Clean Air Act was subsumed into the Canadian Environmental Protection Act in 1988 and a different approach to AQOs was explored. In 2000, the Canadian Council of Ministers of the Environment produced updated air quality objectives for $PM_{2.5}$ and ozone. In 2010 the Canadian Council of Ministers of Environment proposed a new air management system for Canada with new national AQOs as a key component. AQO development has followed a three-stage risk paradigm in which priorities are set, risks are assessed and then risks are managed. Risk assessment has struggled with choosing appropriate endpoints and evaluating the quality of scientific studies. A variety of methods have been used to engage stakeholders. The provinces have produced AQOs for a large number of pollutants, sometimes applying shortcut methods.

References

Alberta Environment (2003) Assessment report on ethylene for developing ambient air quality objectives. Alberta Environment, Edmonton

Alberta Environment (2005) Alberta ambient air quality objectives work plan 2005–2008. Environmental Policy Branch, Edmonton

Alberta Environment (2009) Alberta ambient air quality objectives: hydrogen fluoride. Government of Alberta, Edmonton

Alberta Environment (2010) Alberta ambient air quality objectives and guidelines. Air Policy Branch, Edmonton

Alberta Environment (2011) Using ambient air quality objectives in industrial dispersion modelling and individual industrial site monitoring. Alberta Environment, Edmonton

Alberta Research Council (2002) Alberta's ethylene crop research project. Alberta Research Council. Edmonton

APIS (2012) Critical loads and critical levels—a guide to the data provided in APIS. Air pollution information system. www.apis.ac.uk. Accessed 28 Apr 2012

Archambault D, Li X (2001) Report III. Response of barley, field pea, canola and tree seedlings to ethylene exposure. Alberta Research Council, Edmonton

Bailey S (1999) Reviewing Ontario's air standards. Standards development branch. Ontario Ministry of Environment, Toronto

Barry PJ (1977) Stochastic properties of atmospheric diffusivity. In: Sulphur and its inorganic derivatives in the Canadian Environment, National Research Council, Ottawa

BC Ministry of Healthy Living and Sport (2010) Proposed provincial framework for the development of ambient air quality objectives. Victoria

Blair L, Fu L, Angle R (2007) Alberta's multi-stakeholder process for developing ambient air quality objectives. Extended abstract. Paper #164. Air and waste management association 100th annual conference & exhibition, Pittsburgh

Brown B (1968) Delphi process: a methodology used for the elicitation of opinions of experts. The Rand Corporation, Santa Monica

Brown TC, Peterson GL (2009) An enquiry into the method of paired comparison: reliability, scaling, and Thurstone's law of comparative judgment. United States Department of Agriculture, Ft. Collins

Bull KR (1991) The critical loads/levels approach to gaseous pollutant emission control. Environ Pollut 69:105–123

CASA (1999) Application of critical, target and monitoring loads for evaluation and management of acid deposition. Alberta Environment, Edmonton

CASA (2003) Particulate matter and ozone management framework. Clean Air Strategic Alliance, Edmonton

CCME (1993) Guidelines for consultations and partnerships involving stakeholders in CCME. Canadian Council of Ministers of the Environment, Winnipeg

CCME (1999) Canadian environmental quality guidelines. Canadian Council of Ministers of the Environment, Winnipeg

CCME (2000) Canada-Wide standards for particulate matter (PM) and ozone. Canadian Council of Ministers of the Environment, Winnipeg

CCME (2007) Guidance document on achievement determination, Canada-wide standards for particulate matter and ozone, revised. Canadian Council of Ministers of the Environment, Winnipeg

CCME (2010) Comprehensive air management system: a proposed framework to improve air quality management. Canadian Council of Ministers of the Environment, Winnipeg

CCME (2011a) Strengthening consensus: reinforcing the CCME Consensus Decision-Making (CDM) model. Canadian Council of Ministers of the Environment, Winnipeg

CCME (2011b) Consensus building: a table guide & toolkit. Canadian Council of Ministers of the Environment, Winnipeg

Cannon JA (1986) The regulation of toxic air pollutants: a critical review. JAPCA 36(5):562–573

Caton RB, Bates DV (2002) Updating BC provincial air quality objectives—an options discussion paper. BC Ministry of Water, Land & Air Protection, Victoria

Caton RB, Schroeder WH, Young JWS (1988) Priority-setting strategies for airborne toxic chemicals. Intern J Environmental Studies 31:111–127

Chen CW, Goldstein RA (1986) Techniques for assessing ecosystem impacts of air pollutants. In: Legge AH and Krupa SV (eds) Air pollutants and their effects on the terrestrial ecosystem. John Wiley & Sons, New York

Canadian Institutes of Health Research (2009a) CIHR's framework for citizen engagement. Canadian Institutes of Health Research, Ottawa

Canadian Institutes of Health Research (2009b) CIHR'S citizen engagement handbook. Canadian Institutes of Health Research, Ottawa

Crowe S (2009) Setting priorities for treatment uncertainties—a review of methods. James Lind Alliance, Oxford

DCLG (2009) Multi-criteria analysis: a manual. Department of Communities and Local Government, London

Davies DB, Haggerty SE (2002) Health effects associated with short-term exposure to low levels of hydrogen sulphide (H_2S)—a technical review. Alberta Health and Wellness, Edmonton

Delbecq A, Van de Ven AH (1971) A group process model for problem identification and program planning. J Appl Behav Science 7:466–492

Eberhardt LL, Thomas JM (2008) Designing environmental field studies. Ecological Monographs 61(1):53–73

Environment Canada (2007) Regulatory framework for air emissions. Government of Canada, Ottawa

Federal-Provincial Advisory Committee on Air Pollution (1976) Criteria for national air quality objectives: sulphur dioxide, suspended particulates, carbon monoxide, oxidants (ozone) and nitrogen dioxide. Fisheries and Environment Canada, Ottawa

Federal/Provincial/Territorial Ministers of Environment and Energy (1998) Canada-Wide acid rain strategy-post 2000: strategy and supporting document. Halifax, Nova Scotia. October 19, 1998. Minister of Public Works and Government Services Canada, Ottawa

Five Winds International (2004) Setting priorities for ambient air quality objectives. Alberta Environment, Edmonton

Franson MAH, Franson RT, Lucas AR (1982) Environmental standards: a comparative study of Canadian standards, standard setting processes and enforcement. Environment Council of Alberta, Edmonton

Fu L, Myrick B et al (2000) One-hour equivalent of a 24-hour average particulate matter standard and its potential application in the index of the quality of the air (IQUA). Presented at the air and waste management association conference, Banff, Alberta, 10–12 April 2000

Fuhrer J, Skirby L, Ashmore MR (1997) Critical levels for ozone effects on vegetation in Europe. Environ Pollut 97(1–2):91–106

Gauvin FP, Abelson J, MacKinnon MP, Watling J (2005) Primer on public involvement. Health Council Canada, Toronto

Health Canada (2000) Health Canada Policy toolkit for public involvement in decision making. Ministry of Public Works and Government Services Canada, Ottawa

I-TECH, the international training and education center for health (2008) Organizing and conducting focus groups. University of Washington, Seattle

Leiss W (2008) Development of the expert panel process in Canada, 1995–2005. Royal society of Canada, Ottawa

McColl S, Hicks J, Craig L, Shortreed J (2000) Environmental health risk management: a primer for Canadians. Institute for risk research, Waterloo

McKendry IG (2006) Background concentrations of $PM_{2.5}$ and ozone in British Columbia, Canada. British Columbia Ministry of Environment, Victoria

NRCC (1981) Hydrogen sulphide in the atmospheric environment: scientific criteria for assessing its effects on environmental quality. National research council Canada associate committee on scientific criteria for environmental quality, Ottawa

NRCC (1982) Effects of aerosols on atmospheric processes. National research council Canada associate committee on scientific criteria for environmental quality, Ottawa

National Round Table on the Environment and the Economy (2008) Developing ambient air quality objectives for Canada: advice to the minister of the environment. National round table on the environment and the economy, Ottawa

Newill VA (1977) Air quality standards. In: Air pollution, 3rd edn, Vol. V. Academic Press, New York

Nixon A, Curran T (1998) Acid rain. Parliamentary research branch, government of Canada, Ottawa

Odum EP (1993) Ecology and our endangered life-support systems, 2nd edn. Sinhauer, Sunderland

Ontario Ministry of Environment (1999) Setting environmental quality standards in Ontario: the ministry of the environment's standards plan. Ontario ministry of environment, Toronto

Ontario Ministry of Environment (2006) Rationale for the development of Ontario air standards for toluene. Ontario ministry of environment, Toronto

Ontario Ministry of Environment (2008) Ontario's ambient air quality criteria (sorted by chemical name). Ontario ministry of environment, Toronto

Priestly B, Drew R, McNeil J, Abramson M, Forbes A (2006) Ambient air quality standards setting: an approach to health-based hazard assessment. National Health and Medical Research Council and Environmental Health Committee, Canberra

Privy Council Office (2000) Policy statement and guidelines on consulting and engaging Canadians. Government of Canada, Ottawa

Randolph JJ (2009) A guide to writing the dissertation literature review. Practical Assessment, Research and Evaluation Volume 14, Number 13

Reid N (2007) A review of background ozone in the troposphere. Ontario ministry of environment, Toronto

Royal Society of Canada (2001) Report of an expert panel to review the socio-economic models and related components supporting the development of Canada-wide standards for particulate matter and ozone. Royal Society of Canada, Ottawa

RWDI Air (2005) Final report odour management in British Columbia: review and recommendations. BC Ministry of Water, Land and Air Protection, Victoria

RWDI Air (2008) Establishing a visibility goal for wilderness and urban areas in British Columbia and Canada. BC Ministry of Environment, Victoria

Saskatchewan Environment (2007) Air monitoring directive for Saskatchewan. Saskatchewan Environment, Regina

Science Policy Council (2000) Risk characterization handbook. United States Environmental Protection Agency, Washington

Smith BL (2003) Public policy and public participation: engaging citizens and community in the development of public policy. Atlantic Regional Office, Health Canada, Halifax

UNECE (2012) ICP modelling and mapping critical loads and levels approach. http://www.unece.org/env/lrtap/WorkingGroups/wge/definitions.html. Accessed 29 April 2012

WBK & Associates (2004) Assessment report on ammonia for developing ambient air quality objectives volume 1. Alberta Environment, Edmonton

Working group on air quality objectives and guidelines (1994) National ambient air quality objectives for carbon monoxide: desirable, acceptable & tolerable levels. Environment Canada and Health Canada, Toronto

Working group on air quality objectives and guidelines (1996a) A protocol for the development of national ambient objectives, Part 1, Science assessment document and derivation of the reference level(s). Environment Canada and Health Canada, Toronto

Working group on air quality objectives and guidelines (1996b) National ambient air quality objectives for hydrogen fluoride (HF) science assessment document. Environment Canada and Health Canada, Toronto

Working group on air quality objectives and guidelines (1998) National ambient air quality objectives for particulate matter. Executive Summary. Public Works and Government Services, Ottawa

Working group on air quality objectives and guidelines (1999) National ambient air quality objectives for ground-level ozone science assessment document. Health Canada and Environment Canada, Ottawa

15 Federal-Provincial Relations in Air Quality Management

Randolph P. Angle

Abstract

Canada is a federation where two orders of government—provincial and federal—have environmental jurisdiction based on the allocation of related powers in the Constitution. The concurrency of environmental authority over many similar matters necessitates federal-provincial diplomacy for effective air quality management. Cooperation has been the preferred mode of interaction, punctuated by occasional periods of competition. Coordination of air quality management takes place both under the Canadian Environmental Protection Act and at the Canadian Council of Ministers of the Environment. In addressing four major air quality issues in Canada, there was a complex interaction between the mechanisms reflecting the changing circumstances as issues evolved. The preconditions for effective governmental partnerships include both interpersonal and institutional factors.

Keywords

Federal-provincial diplomacy · Federal-provincial cooperation · Coordination mechanisms · Partnerships · Canadian Council of Ministers of the Environment · Cooperative federalism

15.1 Introduction

Canada is a federation, that is, a sovereign nation in which partially self-governing regions are united under a central government with respective powers entrenched in a constitution that cannot be altered by a unilateral decision of the central government. A federation has inherent tensions between the central government and the constituent regions, tensions that do not exist in a unitary state where regions are administrative divisions of the central government. As a consequence of a federal structure, air quality decisions in Canada are taken by the Canadian federal government, ten provincial governments, three territorial governments, hundreds of municipal governments and by the courts. In general, the federal government has an international focus and pushes for provincial action to support negotiations and commitments. The provinces have focused on establishing industrial air quality management systems. Municipal governments exercise control over local infrastructure and planning. Courts-of-law can also get involved if civil suits are filed because of air pollution by a nearby industrial source, but this has not been a significant influence on pollution control in Canada (McKitrick 2005).

The environment involves some of the most difficult policy and regulatory challenges for government. Environmental problems are scientifically complex and involve significant uncertainties concerning cause and effect linkages. Environmental problems rarely respect geographic or jurisdictional boundaries, and increasingly are linked to major questions of economic policy and international equity. While both federal and provincial governments have relevant powers, there is no sharp dividing line between them and jurisdiction frequently overlaps or is said to be "concurrent". In practice cooperation and partnerships are necessary for success in addressing such complex issues. The shared nature of environmental jurisdiction makes partnerships between the federal, provincial and territorial governments vital to the success of national environmental policies and objectives. Some problems re-

R. P. Angle (✉)
R. Angle Consulting, Edmonton, Alberta, Canada
e-mail: rangle2009@gmail.com

E. Taylor, A. McMillan (eds.), *Air Quality Management,* DOI 10.1007/978-94-007-7557-2_15,

quired coordinated global action. The federal government can negotiate international conventions, but much of the responsibility for implementation will rest with the provinces (OAG 1991).

15.2 The Constitutional Division of Powers

The Constitution Act of 1982, like the British North America Act of 1867 before it, does not specify jurisdiction for the environment or for air quality. As public concern over environmental issues has increased, the federal and provincial governments have scrambled to derive environmental authority from the powers that were enumerated in the Canadian Constitution. The courts have confirmed that the environment is not a single matter falling entirely within either federal or provincial jurisdiction (Simeon and Papillon 2006). Table 15.1 shows the constitutionally allocated powers from which the federal and provincial governments have derived their environmental authority.

With such varying sources of authority and the cross-cutting nature of environment issues in general and air quality issues in particular, it is not surprising that legislation and management activities would overlap and perhaps be duplicated by the two orders of government. While regulated industry often rankles at being governed by two sets of rules, court challenges have confirmed that jurisdiction is "concurrent" and that both sets of rules must be obeyed. In the rare instance that it is impossible to comply with one set of rules without contravening the other, the federal rules are paramount. In practice, no such conflicts have been reported. The potential problems that could arise from the overlap of environmental legislative ability have largely been mitigated by political diplomacy and pragmatic administrative working arrangements.

Three general modes of intergovernmental interaction have been identified (after MacKay 2004) herein called: cooperation, competition, conformity. In the cooperative approach agreements are reached between the provinces and the federal government on national environmental policies and the means of implementation. Each government is considered equal and consensus among all governments is sought. In the competitive approach each government aggressively pursues its own environmental objectives independently, although perhaps with knowledge of what the others are doing. In the conformist approach the federal government provides incentives to the provinces either for implementation of a federal initiative or for allowing control over fields of provincial legislative competence. The United States Clean Act uses Highway Sanctions (potential withdrawal of billions of dollars in federal transportation funding) to ensure that each state submits an acceptable State Implementation Plan (SIP). Canada's federal government has used the conformist approach in the field of health care (through funding transfers), but not in environmental protection.

In Canada, cooperative federalism has dominated the environmental field, with occasional bursts of competitive federalism when the federal government sought greater environmental legitimacy (Macdonald 2009). In the early 1970s, under the Trudeau Liberal government, Environment Canada used its powers under the newly enacted Clean Air Act to regulate directly pollutant emissions from several industrial sectors, most notably the pulp and paper industry. In the late 1980s, under the Mulroney Conservative government, the Eastern Canada Acid Rain Program was announced (1985), the Canadian *Environmental Protection Act* (CEPA) was passed (1988) and shortly thereafter the Canada-US Air Quality Agreement was signed (1991). In the late 2000s the Harper Conservative government announced the Clean Air Regulatory Agenda (2007) promising "mandatory and enforceable reduction targets for emissions of greenhouse gases and air pollutants from all major industrial sources" (Government of Canada 2007).

Each round of federal activism has prompted organized resistance by the provinces who view any federal unilateral actions to regulate industry as "intrusion" on provincial jurisdiction. Often the provinces will take action in order to occupy "the regulatory space" and avoid gaps that might offer the federal government a reason to step-in. The perceived threats associated with the extensive federal powers in the new Canadian Environmental Protection Act of 1988 spurred the provinces into strengthening the existing federal-provincial coordination mechanism and creating the Canadian Council of Ministers of the Environment. The potential for "competition" from the federal government has worked to unite the provinces in coordinated responses to air quality issues. It has also led to provincial actions that might not otherwise have occurred.

MacKay (2004) pointed to a natural reluctance on the part of provinces to harm the industries on which they rely for jobs, taxes, and political contributions. Many provinces have a heavy dependence on one or two resources, for example, forestry in British Columbia; oil and gas in Alberta; pulp and paper in Quebec. Industry generally prefers to be regulated by the provinces who are viewed as "having a better understanding" of their situation. The federal Clean Air Regulatory Agenda of 2007 prompted an unlikely coalition of industry and environmental non-government organizations to produce an alternative plan, in parallel with a similar effort by provincial and territorial governments under the Canadian Council of Ministers of the Environment. Eventually the three tracks converged in a proposal for a new Comprehensive Air Quality Management System for Canada (Steering Committee 2010).

The Canadian Council of Ministers of the Environment approved the new air quality management system in October

Table 15.1 The constitutional basis of authority for air quality management

Exclusive federal jurisdiction	Exclusive provincial jurisdiction
Navigation and shipping	Property and civil rights
Criminal law	Natural resources
Interprovincial and international transportation and communications	Municipal institutions
Works for the general advantage of Canada	Local works and undertakings
Any mode of taxation	Direct taxation
Census and statistics	All matters of a local or private nature
Federal lands, and lands reserved for Indians	Electrical energy
Peace, order and good government	
Any other powers not assigned to the provinces	

2010 and established milestones for progress. Officials were directed to develop the major elements of the system in 2011 using a collaborative process and engaging stakeholders. Implementation will begin in 2013. The new air quality management system includes more ambitious Canadian air quality standards for particulate matter and ozone as well as consistent industrial emissions standards across the country. Regionally coordinated airsheds and air zones within individual provinces and territories will be established. Collaborative work on transportation emissions will be undertaken and a new system of monitoring and reporting will be designed. A stakeholder advisory group has been convened to provide advice to governments on the finalization of the system (CCME 2012).

15.3 Federal Air Management Coordination

In 1969 the federal Department of National Health and Welfare established an Air Pollution Control Division and an ad hoc Federal-Provincial Committee on Air Pollution. The National Air Pollution Surveillance (NAPS) program was also established in 1969 to monitor and assess the quality of ambient (outdoor) air in the populated regions of Canada. A subcommittee of senior officials was struck in 1970 to develop national ambient air quality standards. With the creation of the Department of Environment in 1971, responsibilities for air pollution control were transferred to the Air Pollution Control Directorate of the new department and the Federal-Provincial Committee on Air Pollution was formally established under the Act. The Canada Clean Air Act of 1971 had three main objectives: to protect the health of the public of Canada from air pollution; to promote a uniform approach across Canada; and to make provisions for the mechanisms and institutions needed to ensure that all measures to control air pollution can be taken (Environment Canada 1979). The Canada Clean Air Act authorized the regulation of: emissions that constitute a significant danger to health, emissions that would violate an international agreement, or emissions from all federal works and undertakings, and the composition of fuels or fuel additives.

National emissions regulations were formulated, in consultation with the province, for three industrial operations: mercury from chlor-alkali plants, lead form secondary lead smelters, and asbestos from asbestos mining and milling. The first nation-wide emissions inventory for four hazardous pollutants—mercury, beryllium, asbestos and lead—was completed in 1973. Earlier, in 1970, a national inventory had been completed for the emissions for five common pollutants: sulphur oxides, particulates, carbon monoxide, nitrogen oxides, and hydrocarbons. Emission standards for new vehicles were also promulgated under the Motor Vehicle Safety Act. In 1973, the lead and phosphorous content of gasoline was regulated to accommodate the introduction of catalytic converters expected on some automobiles in 1975.

"Since the provinces have jurisdictional control of most air pollution sources, the Department…adopted the basic strategy of promoting and supporting viable provincial control agencies through federal-provincial cooperative efforts" (Environment Canada 1973). In addition to exchanging data, training air pollution technicians, providing monitoring equipment and related services, the federal government also coordinated the development of national emissions guidelines for a number of industry sectors (based on best practicable technology) to provide for uniform minimum controls across the country and prevent the pollution havens. The guidelines were to be adopted as regulations by the provinces, although it was recognized that more stringent controls might be required in some geographical locations with heavy industrialization. Since that time the Supreme Court of Canada has repeatedly affirmed a substantial role for federal environmental regulation, calling environmental protection a fundamental value of Canadian society and a major challenge requiring action by all levels of government (Wood et al. 2010).

In the mid-1970s the federal government signed agreements for the protection and enhancement of environmental quality with seven of the ten provinces. (No agreements were reached with Newfoundland, Quebec or British Columbia). The provinces agreed to establish and enforce environmental requirements at least as stringent as federal requirements. The federal government agreed, after consultation with the provinces, to establish national ambient air and water quality objectives and to develop national baseline effluent and emission requirements and guidelines for specific industrial groups and specific pollutants. Federal enforcement was to occur only at federal facilities (Douglas et al. 1997).

In 1988 the Clean Air and other environmental legislation was consolidated into the Canadian Environmental Protection Act (CEPA). CEPA provided a framework for the management and control of toxic substances at each stage of their life cycle, from development and manufacture or importation through to transportation, distribution, use, storage and ultimate disposal as waste. Specific to air quality, it authorized regulation of emissions that cause "acid rain", and restrictions on the import and export of ozone depleting substances such as chlorofluorocarbons (CFCs), and hazardous air pollutants such as polychlorinated bi-phenols (PCBs). CEPA also emphasized the necessity of national standards and the need to avoid conflict and duplication among the federal, provincial and territorial governments.

A National Advisory Committee was created to advise the Minister "on a cooperative, coordinated intergovernmental approach for the management of toxic substances" and "environmental matters that are of mutual interest". Environment Canada or Health Canada chaired the committee which comprises a representative of each of the provinces and territories as well as up to six representatives of Aboriginal governments (with Crown agreements empowering the enactment of environmental protection laws). Air issues then became only a small part of the broad toxics management business of the Advisory Committee. To many old-timers in air quality management, this represented a loss of focus on air and a weakening of federal coordination efforts.

There was also provision for federal-provincial agreements. Administrative agreements aim to establish work sharing partnerships relating to federal and provincial legislation and can cover a range of activities from research and monitoring to inspection and enforcement. Such agreements were signed with Quebec and British Columbia in 1994 (Douglas et al. 1997) but in 2011 there was only one active agreement with Saskatchewan. Equivalency agreements provide for federal regulations under CEPA to "stand-down" (not apply) in a province or territory where provincial regulations having equivalent effect are in place. The criteria for equivalency were: an equal level of control as sanctioned by law; comparable compliance measurement techniques; comparable penalties; comparable enforcement policies and procedures; and comparable rights for individuals to request investigation of a suspected offence and to receive a report of the findings. One equivalency agreement was signed with Alberta in 1994 (Douglas et al. 1997).

In 1999 the Canadian Environmental Protection Act was revised after an extensive review of the earlier legislation. Focused on pollution prevention and sustainable development, the renewed Act signaled a departure from pollution control as a priority. It set deadlines for evaluating and controlling toxic substances and included the power to require pollution prevention plans for toxic substances (Craig et al. 2008). It also envisioned greater citizen participation, and more effective cooperation and partnership with other governments and aboriginal peoples.

15.4 The Canadian Council of Ministers of the Environment (CCME)

Both federal and provincial governments established environmental departments in the early 1970s, leading to some competition for leadership and relevance to citizens. Both orders of government legislated and there seemed to be overlap and duplication in many areas. The Canadian Council of Resource Ministers, established in 1964 to coordinate resource management across Canada, was expanded in 1971 (Boardman 2009) to become the Canadian Council of Resource and Environment Ministers with a full time secretariat in Toronto, Ontario. However, as environment issues rose to the forefront, in 1988 the organization refocused as the Canadian Council of Ministers of the Environment, to serve as "a principal forum for members to develop national strategies, norms, and guidelines that each environment ministry across the country can use" (CCME 2011). Offices were moved to Winnipeg, Manitoba, near the geographical centre of Canada. Participation in CCME allowed provinces and territories to leverage limited resources, exchange ideas; build neighbour relationships, and showcase environmental successes.

Of the 12 task groups originally operated by CCME (CCME 1992) the Atmospheric Issues Task Group (AITG) was responsible for air quality and global warming. CCME developed its business plan around four themes: building partnerships, building a sustainable development vision, overcoming jurisdictional fragmentation, and adjusting to the international reality. AITG was charged with overseeing work on smog and ozone depleting substances (ODS) as well as developing a comprehensive national air quality management agreement that would integrate all air-related activities and include the cooperation of the departments of energy. In 1990 a Management Plan for Nitrogen Oxides and Volatile Organic Compounds was published. This "smog" plan contained a prevention program led by the federal government, a remedial program for non-attainment areas led by the respective provincial governments, and a series of studies and investigations to support future planning.

15.4.1 The Joint Ministers Mechanism

In 1993 the Canadian Council of Ministers of the Environment and the Council of Energy Ministers met for the first time and then regularly thereafter in what became known as the Joint Ministers Meetings (JMM). At the JMM, the ministers dealt with a variety of common air quality issues. In November 1993, the JMM approved a Comprehensive Air Quality Management Framework for Canada. This framework agreement established "a cooperative framework for the coordination of actions by governments on air quality issues of regional, national and international scope, especially those with transboundary or global effects" (Joint Ministers of Environment and Energy 1993). It also provided for the establishment of related goals and objectives in accordance with a defined set of principles and again anticipated the signing of future interjurisdictional agreements.

A National Air Issues Coordinating Mechanism (NAICM) was created to implement the framework agreement. The NAICM generally served as a forum to foster partnerships and co-operative discussions among environment and energy departments and to build consensus among jurisdictions on identifying and resolving multi-jurisdictional air quality issues. Its purpose was to co-ordinate progress domestically and to provide advice to the federal government on the development of international positions on such issues as climate change. The NAICM consisted of a steering committee and a coordinating committee. Reporting to JMM, the National Air Issues Steering Committee (NAISC) comprised federal, provincial and territorial deputy ministers of environment and energy, reporting to the Joint Ministers. Reporting to NAISC, the National Air Issues Coordinating Committee (NAICC), consisting of assistant deputy ministers or senior officials from the environment and energy departments, had the primary responsibility for implementation of the mechanism under the Framework. The NAICC was supported by a Stakeholder Advisory Committee. The NAICC established a Climate Change Working Group to act as a focal point for climate change issues and, in 1994, an Acid Rain Task Group to develop, with industry and other stakeholders, a national strategy for a long-term domestic acid rain program for post-2000.

The climate change issue soon dominated the NAICC to the point where Environment Ministries grew concerned that other air issues were not receiving enough attention. In April 1997, the Joint Ministers decided to operate the National Air Issues Co-ordination Mechanism with two coordinating committees. NAICC-A addressed priority air issues and monitored developments on other non-climate change air issues. NAICC-CC focused on climate change. NAICC-A and NAICC-CC acted in parallel, each charged with recognizing collateral effects of their activities on those of the other group. Both groups continued to report to the National Air Issues Steering Committee (NAISC). This structure is shown in Fig. 15.1. Working groups comprised government members only while Task Groups included representatives of industrial and environmental interests.

In 2001 NAICC-A added another initiative focussed on petroleum refineries. The Canadian Petroleum Products Institute (CPPI) had initially approached the federal government about establishing a performance-based regulation for refinery emissions. The intention was to "level the playing field" across Canada and perhaps gain a competitive edge over US refineries subject to complex prescriptive regulation. Federal officials, mindful of provincial sensitivities over major industries, suggested that a modified proposal be presented to the relevant provinces. All parties agreed to the development of a new way to regulate air emissions from Canadian petroleum refineries with a view to stimulating innovation while preserving or enhancing the competitiveness of the Canadian industry. Under the auspices of CCME, governments (federal, provincial and municipal), the Canadian Petroleum Products Institute, and non-government health and environmental organizations worked together to deliver the National Framework for Petroleum Refinery Emission Reductions (CCME 2005). The framework had four key elements: a methodology to assist jurisdictions with prioritizing and setting emission caps; a strategy to monitor and report on refinery emissions and reductions; a 10 year plan to keep the framework tools updated, measure performance and report on progress, and jurisdictional management of refineries.

15.4.2 Parallel Environmental Coordination

In 1990 the first ministers of the federal and provincial governments endorsed a Statement of Interjurisdictional Co-operation on Environmental Matters (CCME 1990). It listed principles for cooperation and identified objectives for harmonizing regulatory and environmental assessment programs, developing national environmental standards and strategies, enhancing the scientific database, and improving the linkages between international and domestic activities. The statement provided the overall framework for joint environmental action between the two levels of government and anticipated the signing of a number of interjurisdictional agreements on a variety of environmental topics.

Initially, CCME had focused on individual areas of environmental protection but in 1993, efforts to harmonize environmental programs and policies became a priority. In May 1994, CCME launched a "harmonization initiative", calling for the development of a new environmental management framework for Canada based on cooperation and an effective and efficient definition of roles and responsibilities (Croes

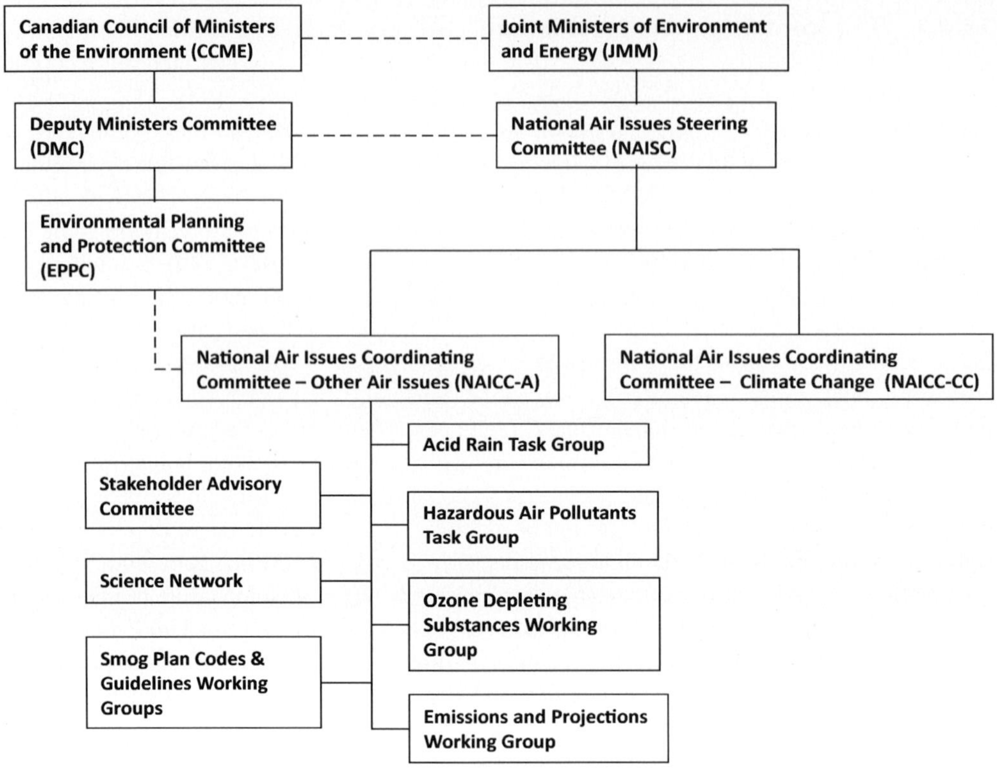

Fig. 15.1 The structure of air issue coordination under the Comprehensive Air Management Framework Agreement of 1993

et al. 2007). As initially proposed, the harmonization initiative would have consisted of an intergovernmental framework agreement and 11 schedules dealing with the following areas of environmental management: monitoring; compliance, licensing and approvals; environmental impact assessment; international agreements; research and development; guideline development; legislation, regulation and policy; communications and education; state of the environment reporting; and pollution prevention and emergency response (Douglas et al. 1997).

In December 1994, the Environmental Management Framework Agreement (EMFA) and four schedules (monitoring; compliance, licensing and approvals; environmental impact assessment and international agreements) were released in draft form for public comment. A modified main framework agreement was released to the public in October 1995, along with 10 of the 11 schedules (excluding the schedule on environmental impact assessment). Work on the harmonization initiative was subsequently suspended, due in part to the proposal's controversial nature (Douglas et al. 1997).

Negotiations resumed, however, following a meeting of the Council of Ministers, in May 1996 when it was decided to continue working toward harmonization, but with a new approach calling for a more gradual progression towards this goal.

A Canada-Wide Accord on Environmental Harmonization was signed in January 1998 by the federal government and all territories and provinces except Quebec. The Accord emphasized consensus-based decision making and single-window delivery of services by the government best situated to deliver the service. The objectives of harmonization were to: (1) enhance environmental protection; (2) promote sustainable development; and (3) achieve greater effectiveness, efficiency, accountability, predictability and clarity of environmental management for issues of Canada-wide interest. Harmonization was intended to encourage cooperation among the provincial governments in development of consistent environmental measures such as policies, standards, objectives, legislation, and regulation across federal and provincial jurisdictions.

A series of principles anchored the Accord, including: (a) polluter pays, (b) precautionary principle, (c) pollution prevention as a preferred approach, (d) performance, results, and science-based, (e) transparency, (f) cooperation with Aboriginal people, (g) flexible implementation, and (h) consensus-based decision-making. The individual governments retained legislative authority and were not prevented from legislating stricter standards than those determined under harmonization.

Three sub-agreements were also signed: Inspections and Enforcement Sub-agreement, Canada-Wide Environmental Standards Sub-agreement, and Environmental Assessment Sub-agreement. Canada Wide Standards (CWS) were intended to provide an alternative regulatory tool for the management of environmental issues of national interest. Eight of the eleven Canada-Wide Standards were related to air emissions: benzene, mercury (2), dioxins and furans (4), and

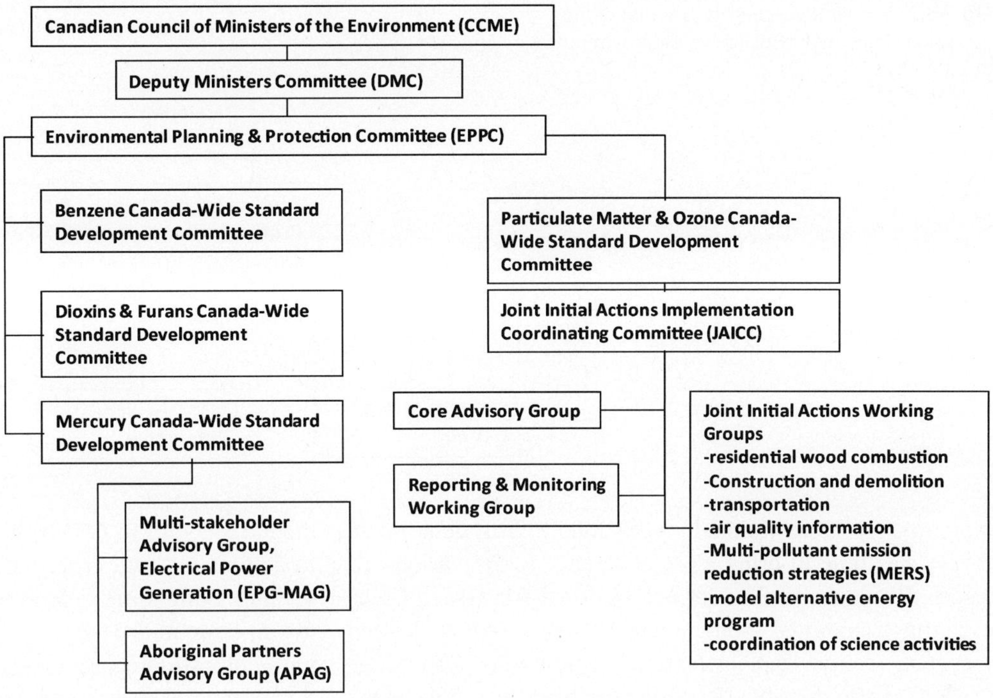

Fig. 15.2 Additional air management structure existing in 2004 as a result of the Canada-Wide Environmental Standards Sub-agreement

particulate matter and ground-level ozone. While specific responsibilities varied according to the specific standard, the following general functions emerged (Thomas 2008): Federal Government—provide scientific and technical support; implementation at international borders; implementation on federal lands; represent Canada internationally; promote actions at the international level to achieve reductions in Canada; and implement standards that require a product/substance approach. Provincial/Territorial Governments—provide technical and scientific support; and implementation requiring action from industrial, municipal, and other sectors (Thomas 2008). While the province of Quebec did not sign the Accord or the sub-agreements for broader political reasons, officials continued to participate is most initiatives and to follow the standards recommendations.

Part two of the Canada-Wide Standards for Particulate Matter and Ozone outlined the implementation actions to be taken by each jurisdiction (CCME 2000). This includes provisions for the development of "jurisdictional implementation plans" to achieve the target levels, the establishment and maintenance of monitoring networks to assess and track progress across Canada, implementation of activities to meet the principles for Continuous Improvement and Keeping Clean Areas Clean, and activities to reduce transboundary flow of particulate matter and ozone into Canada. A set of "joint initial actions" included activities related to emissions from transportation, residential wood burning appliances, and industry sectors, codes of practice for the construction and demolition sector, databases for ambient air quality, improving scientific knowledge, and the development of alternative energy models (CCME 2000). A Joint Action Implementation Coordinating Committee (JAICC) with representatives from each jurisdiction was established to carry out the work.

All of the work on Canada-Wide Standards took place outside of the National Air Issues Coordination Mechanism. By 2004 intergovernmental air management activities had evolved into a very complex structure adding the pieces in Fig. 15.2 to those already in place shown in the previous figure.

15.4.3 Consolidation

Health concerns are the basis of many air quality issues, yet Health Ministries in the provinces and territories had not generally been engaged. CCME Ministers had long aspired to create a stronger institutional connection. In 2004, the Committee on Health and the Environment was created as a Federal-Provincial-Territorial (FPT) forum for advice and joint-action on health and environment issues. The CHE reports to the Canadian Council of Ministers of the Environment (CCME) and is a liaising body of the Pan-Canadian Public Health Network (PHN) Council. This was on top of the already complex structure that was created by the parallel operation of the National Air Issues Coordination Mechanism and the Canada-Wide Environmental Standards Sub-agreement.

However, by this time energy departments had lost interest in the work of NAICC-A. Representation had dwindled to a single member from the federal government. Conditions

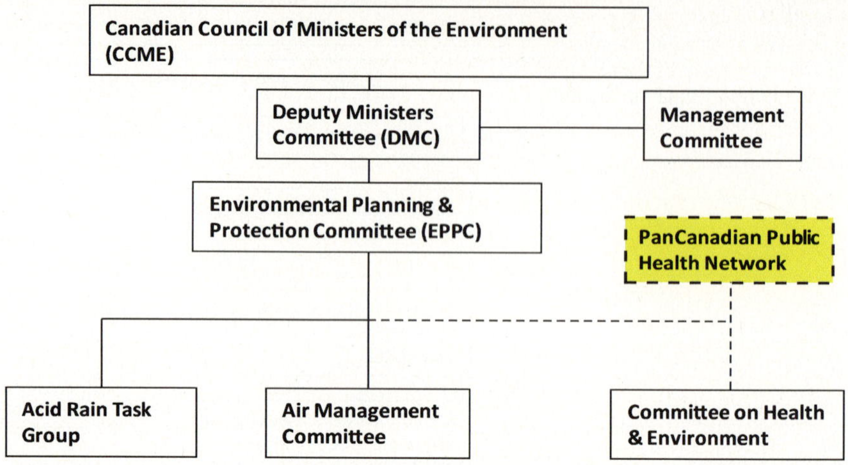

Fig. 15.3 The air management structure of the Canadian Council of Ministers of the Environment in 2011

were ripe for streamlining. In April 2005 CCME determined that it was time to bring more coherence to the various air issues mechanisms. The former National Air Issues Coordinating Committee Non-climate change (NAICC-A) and the Joint Action Implementation Coordinating Committee (JAICC) were dissolved and all responsibilities transferred to a new Air Management Committee (AMC). AMC is charged with recommending priorities for cooperative action, streamlining and managing the reviews for the Canada-Wide Standards, managing other air projects, and serving as a forum for air quality discussions. The 2011 structure of air quality management at CCME is shown in Fig. 15.3.

The new air quality management system for Canada proposed in 2010 (Steering Committee 2010) was from a committee that originally formed and operated independent of CCME. However, upon endorsement by the Canadian Council of Ministers of the Environment, implementation of that system became part of CCME's work.

15.5 Illustrative Air Issues

Canada's decentralized air management system makes it difficult sometimes to appreciate the extent of actions that have been taken in the country. There are emission limits for thousands of industrial operations and ambient air-quality standards for hundreds air pollutants. The environmental risks associated with air pollution have been substantially reduced over the past 30 years. However, there is no one large, omnibus piece of legislation like the US Clean Air Act, and that sometimes gives the erroneous impression that Canada does not impose emissions controls or manage air quality (McKitrick 2006).

When broad issues arise, there is a complex interplay between the federal government and the provinces and territories, either through direct negotiation or through the Canadian Council of Ministers of the Environment. Responses may start internationally with the federal government, then shift to CCME in whole or part, and finally lead to actions by both orders of government (federal and provincial). Sometimes federal coordination structures will handle an issue under the auspices of CCME as has been done with chlorofluorocarbons and the early smog management. Sometimes, both types of structures deal with an issue, as with toxic substances. Sometimes an issue weaves back and forth between federal coordination structures and CCME, as with acid rain. A brief description of the management of four major air issues serves to illustrate further.

15.5.1 Ozone Depleting Substances

Soon after Canada signed the Vienna Convention on the Protection of the Ozone Layer, CCME directed that a Federal-Provincial Working Group under CEPA coordinate the development of controls across all jurisdictions to avoid the creation of a regulatory patchwork quilt. The Working Group reported both to the National Advisory Committee under CEPA and to the AITG (later NAICC-A) under CCME. This crossover relationship has worked successfully in delivering Canada's program for protecting the ozone layer. Some of the key events are shown in Table 15.2.

Figure 15.4 shows the reductions in consumption of ozone depleting substances in Canada over the period 1986–1999 (CCME 2001).

15.5.2 Acid Rain

Some of the key events in the development of Canadian acid rain management are summarized in Table 15.3. A brief history of the acid rain issue is provided by Nixon and Curran (1998). The federal government played the major role in all aspects of managing this issue, brokering domestic deals that allowed international commitments to be made with some certainty that obligations could be met. At one point in the

Table 15.2 Some key events in Canadian management of ozone depleting substances

Year	Federal government of Canada	CCME
1985	*United Nations Environment Program (UNEP) Vienna Convention on the protection of the ozone layer*	
1987	*Montreal Protocol on substances that deplete the ozone layer*	
1989	*Helsinki Declaration*, accelerating the reduction of CFCs	Federal-provincial working group on controls harmonization (Ozone-depleting substances)
1990	*London Declaration*, phasing out the production of CFCs	
1992	*Copenhagen Resolution* on methyl bromide, to develop a general control scheme	*National Action Plan for recovery, recycling, and reclamation of chlorofluorocarbons (CFCs)*
1995	*Declaration on methyl bromide*, in Vienna, to limit consumption and adopt alternatives	National consultations leading to the report *Strengthening Canada's Ozone Layer Protection Program*
1998		*National Action Plan for the environmental control of ozone-depleting substances (ODS) and their halocarbon alternatives, addressing all ODS and alternatives*
2001	*Colombo Declaration on renewed commitment to the protection of the ozone layer to mark the forthcoming World Summit on sustainable development*, in 2002, the 15th anniversary of the Montreal protocol and the 10th anniversary of the establishment of the multilateral fund	*May 2001 Update National Action Plan for the environmental control of ozone-depleting substances (ODS) and their halocarbon alternatives, includes tasks to implement Canada's strategy to accelerate the phase-out of CFC and halon uses and to dispose of the surplus stocks*

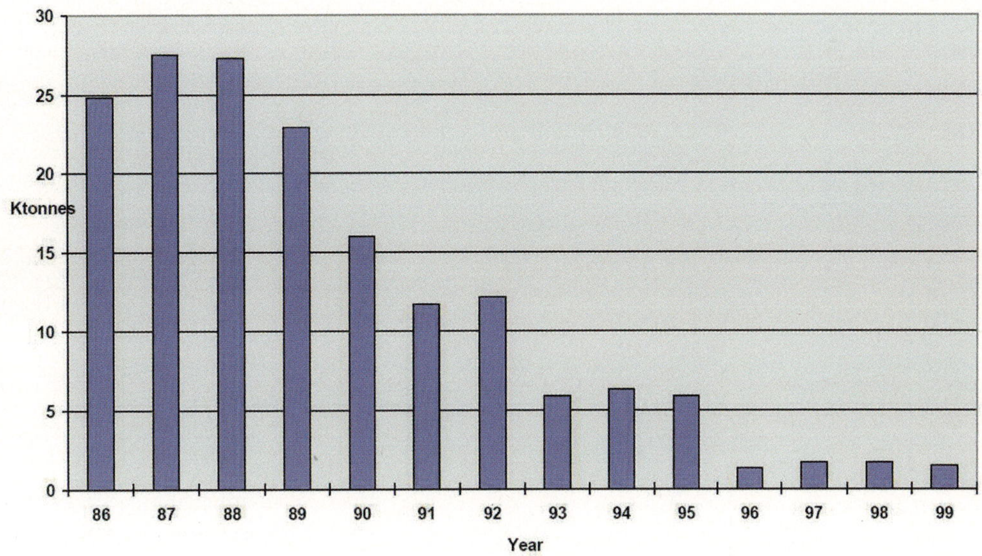

Fig. 15.4 Consumption of ozone depleting substances in Canada (CCME 2001)

negotiations with Ontario, Environment Canada threatened direct regulation of emissions, but at the end of the day, the 1985 federal-provincial program relied upon the provinces for that task (Macdonald 2009). The Canadian Council of Ministers of the Environment became involved only in the latter stages.

Figure 15.5 shows the reductions in Canadian emissions of sulphur oxides from 1985 to 2009 (Environment Canada 2011).

15.5.3 SMOG

Some of key events in the development of Canada's smog actions are summarized in Table 15.4. The Canadian Council of Ministers of the Environment provided an umbrella for the development of the NO_x/VOC Management Plan by the Federal-Provincial LRTAP Committee with federal officials playing a lead role. A majority of the 60 initiatives under the Plan were also led by the federal government. When agreement could not be reached on Phase 2, the federal government continued implementation of the components for which it was responsible. A few years later, CCME again addressed smog through the Canada-Wide Standards initiative. Thomas (2008) noted that "the biggest problem of the collaborative approach, both in terms of the development of the Standards and implementation activities, is an inequity among jurisdictions with respect to staff and financial resources. The federal government has played a larger role in the Canada-Wide Standards, mainly because they have more staff, technical expertise, and resources to contribute. .... Often, political commitments do not take into consideration the level of work and resources required

Table 15.3 Some key events in Canadian acid rain management

Year	Federal government	CCME
1950s	The acidification of lakes near Sudbury observed	
1976	The Canadian Network for Sampling Precipitation (CANSAP)	
1979	*UN Economic Commission for Europe Convention on Long-range Transboundary Air Pollution (LRTAP)*	
1980	*Memorandum of Intent Concerning Transboundary Air Pollution*	
	Western and Northern Canada LRTAP program	
1982	Northwest territories joins the Western and Northern Canada LRTAP program	
1985	*Eastern Canada Acid Rain program*	
	Helsinki Protocol, calling for a 30 % reduction in emissions	
1986	More stringent emission standards for cars and light-duty trucks	
1987	Signed agreements with Ontario, Prince Edward Island, Newfoundland, Quebec and New Brunswick	
1988	Signed agreement with Nova Scotia	
1991	Canada-United States Air Quality Agreement	
1993		Energy and environment ministers endorse Canada's signing of the UN-ECE Second Sulphur Protocol
1994	*Oslo Protocol on further reduction of sulphur emissions*	*Energy and environment ministers statement of intent* to develop a national strategy for a long-term domestic acid rain program for post-2000
1998		*The Canada-wide acid rain strategy for post 2000*
2007		The acid rain task group long-term strategic plan to implement the strategy *A National Acid Rain Science Plan*
2008		*A National picture of acid deposition critical loads for forest soils in Canada*

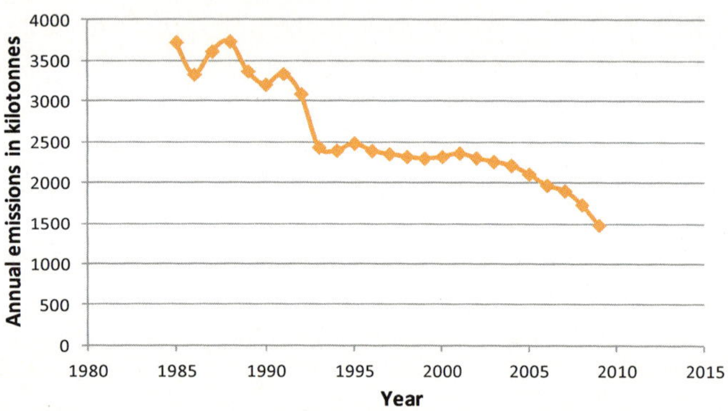

Fig. 15.5 Emissions of sulphur oxides in Canada

to meet the obligations and, thus, some regions may have a more difficult time meeting their commitments under the standards."

Figure 15.6 show the reductions in Canadian emissions of volatile organic compounds over the period 1985–2009 (Environment Canada 2011).

15.5.4 Hazardous Air Pollutants (HAPS)

Hazardous Air Pollutants are any potentially toxic substances that are not classified as criteria or common air contaminants. In the early 1990s CCME was concerned about dioxins and furans. Both the federal government and CCME developed policies for the management of toxic substances broadly, inclusive of those in the atmosphere. Canada-Wide Standards were established for dioxins and furans, mercury, and benzene. Some of the important events are shown in Table 15.5.

A subcategory of HAPS, Persistent Organic Pollutants (POPs), refers to substances which are stable, bio-accumulative and transported over great distances eventually accumulating in colder climates. Canada signed the Aarhus Protocol on POPS under the UNECE Convention on LRTAP in 1998. The National Air Issues Coordinating Committee established a Hazardous Air Pollutants Task Group to provide input to the federal government and develop implementation strategies. In the negotiations leading to the UNEP Stockholm Convention on Persistent Organic Pollutants, the federal government consulted regularly with provincial and territorial governments, aboriginal, environmental and health groups, and the chemical industry. Representatives of these governments and organizations were also included in Canada's negotiating team. The National Implementation Plan was similarly developed in close consultation with stakeholders. Many have regarded the federal government's handling of the POPs issue as the ideal model for cooperative environmental management.

Table 15.4 Some key events in Canadian Smog management

Year	Federal government	CCME
1988	*Sofia Protocol concerning the control of emissions of nitrogen oxides or their transboundary fluxes*	
1989		Endorsement of a federal regulation adopting the proposed California standards, and implementation of vehicle inspection and maintenance programs. provinces with ozone problems would
1990		*Management plan for nitrogen oxides (NO_x) and volatile organic compounds (VOCs): Phase I*
1995		Endorsement of *cleaner vehicles and fuels plan*
1997	*Phase 2 Federal Smog Management Plan*	Phase II of the National Smog Program put on "hold"
	Joint Plan of action for addressing Transboundary Air Pollution (with US)	
1998		*Canada-wide accord on environmental harmonization*
		Canada-wide standards sub-agreement
1999	*UNECE Gothenburg Protocol to abate acidification, eutrophication and ground-level ozone*	
2000	*Phase 3 Federal Smog Management Plan*	*Canada-wide standards for particulate matter (PM) and ozone*
	Canada-U.S. Air Quality Agreement—Ozone Annex	*Joint initial actions to reduce pollutant emissions that contribute to particulate matter and ground-level ozone*
2004	*Federal Agenda on the reduction of emissions of volatile organic compounds from consumer and commercial products*	
2005		*National framework for petroleum refinery emissions reductions*

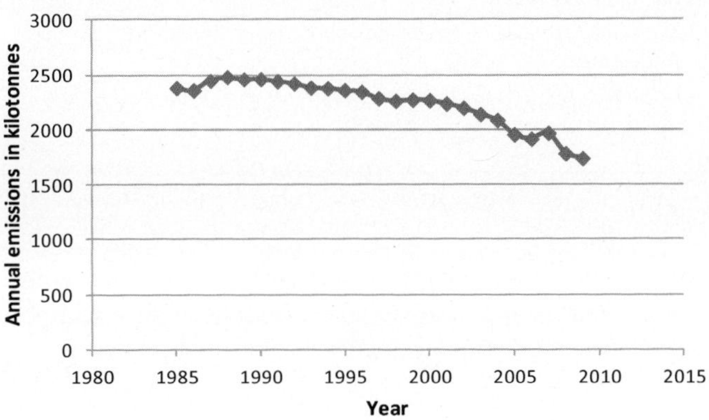

Fig. 15.6 Emissions of volatile organic compounds in Canada

15.6 The Challenges of Inter-jurisdictional Coordination

It is widely recognized that the concurrency of environmental jurisdiction is Canada makes partnerships between the federal, provincial and territorial governments vital to successful air quality management (e.g. OAG 1991). However, building and sustaining a successful partnership is not easy. There is an extensive literature on the subject (e.g. Alexander 1995) and numerous guides (e.g. World Economic Forum 2005). However, it is frequently acknowledged that "there is no simple checklist or blueprint for successful partnership building. It is often an intuitive and constantly evolving 'voyage of discovery' based on organizational and individual learning, trust and experimentation" (World Economic Forum 2005).

From a number of Canadian case studies, the Canadian Commissioner of Environment and Sustainable Development (OAG 2000) identified the five most important factors to a good working relationship as:

- clear and realistic objectives and expected results;
- shared or complementary goals;
- effective and committed individuals;
- clear benefits for the participating organizations; and
- senior management interest, support and commitment.

There is always tension between developing a good working relationship and effective accountability for the outputs and outcomes. Partnerships make accountabilities more complex and the partners must pay attention to:

- Clear objectives, expected performance, and operating conditions
- Clear roles, responsibilities and authorities of each partner
- Balance between expectations and capacities of each partner
- Defined performance measures and reporting
- Provision for review, program evaluation and audit.

Effective co-operation involves a mix of interpersonal and institutional factors. Attitudinal prerequisites for entering into any kind of relationship include: willingness to collaborate, belief in the desirability of working together, and perceived trustworthiness of the other parties. Institutional prerequisites for successful collaboration include: availability of resources, commitment of senior management, and support at the political level (OAG 2000). The success of the partnership will depend in large part on trust and respect, which are garnered largely through: (a) open and direct communication (straight talk); (b) listening for understanding; and (c) making and keeping commitments (reliability). Trust and

Table 15.5 Some key events in Canadian management of Hazardous Air Pollutants

Year	Federal government	CCME
1969	*Pest Control Products Act*	
1988	*Canadian Environmental Protection Act*	
1990	*Green Plan*	
		Comprehensive air quality management framework for Canada
1993		*Dioxins and Furans: a Canadian perspective*
1995	*Toxic substances management policy*	
1998	*Aarhus Protocol on persistent organic pollutants (POPs) pursuant to the 1979 United Nations Economic Commission for Europe (UNECE) Convention on Long-range Transboundary Air Pollution (LRTAP)*	*CCME policy statement for the management of toxic substances, and supporting rationale and implementation notes for CCME Policy for the management of toxic substances*
		CCME Canada-wide accord on environmental harmonization
		Canada-wide standards sub-agreement
2000		*Canada-wide standard for benzene Phase I*
		Canada-wide standards for mercury emissions
2001	*UNEP Stockholm Convention on persistent organic pollutants*	*Canada-wide standard for benzene Phase II*
		Canada-wide standards for dioxins and furans emissions from waste incinerators and coastal pulp and paper boilers
2002	*Pest Control Products Act* Revised	
2003		*Canada-wide standards for dioxins &furans: conical waste combustion of municipal waste*
		Canada-wide standard for dioxins and furans from iron sintering plants
		Canada-wide standard for dioxins and furans from steel manufacturing arc furnaces
2005		*National Framework for Petroleum Refinery Emissions Reductions*
2006	*Canada's National Implementation Plan under the Stockholm Convention on persistent organic pollutants*	*Canada-wide standards for mercury emissions from coal fired electric power generation plants*
	Chemicals Management Plan	

respect are earned slowly and lost quickly. Changes in personnel can easily undo hard-won agreements. Cooperation at the technical level can be over-ruled at the political level. Past experiences may prejudice future partnering. Provinces may perceive the federal government as an intruder while the federal government may view the provinces as captive to their major industries.

15.7 Conclusion

Air quality management in the Canadian federation faces many challenges ranging from the political stances of leaders to the administrative complexity of detailed bilateral and multi-lateral agreements. Cooperation, the main mode of federal-provincial interaction, requires that fourteen separate jurisdictions come together to find common ground. The struggle to coordinate the actions of two orders of government has led to the creation of quintessentially Canadian institutions for air quality management. The two main coordination mechanisms are fundamentally different in design. The federal-provincial committee under CEPA offers "advice" to the federal Ministers who retain all decision-making authority. At CCME federal and provincial governments have an equal voice and decisions are made by consensus. Each mechanism has advantages and disadvantages. Both have played important roles in Canadian air quality management.

References

Alexander ER (1995) How organizations act together: interorganizational coordination in theory and practice. Gordon and Breach, Amsterdam

Boardman R (2009) Environmental policy in the EU and Canada. Prepared for the 3rd annual research conference, EU centre of excellence. Dalhousie University, Halifax

CCME (1990) Statement of interjurisdictional cooperation on environmental matters. Canadian Council of Ministers of the Environment, Winnipeg

CCME (1992) 1992 Strategic overview. Canadian Council of Ministers of the Environment, Winnipeg

CCME (2000) Joint initial actions to reduce pollutant emissions that contribute to particulate matter and ground-level ozone. Canadian Council of Ministers of the Environment, Winnipeg

CCME (2001) National action plan the environmental control of ozone-depleting substances (ODS) and their halocarbon alternatives May 2001 update. Canadian Council of Ministers of the Environment, Winnipeg

CCME (2005) National framework for petroleum refinery emissions reductions. Canadian Council of Ministers of the Environment, Winnipeg

CCME (2011) What's the purpose of CCME? http://www.ccme.ca/about/index.html. Accessed 4 Dec 2011

CCME (2012) Air quality management system. http://www.ccme.ca/about/index.html. Accessed 18 May 2012

Craig L, Brook JR, Chiotti Q, Croes B, Gower S, Hedley A, Krewski D, Krupnick A, Krzyzanowski M, Moran MD, Pennell W, Samet JM, Schneider J, Shortreed J, Williams M (2008) Air pollution and public health: a guidance document for risk managers. J Toxicol Environ Health (Part A) 71:588–698

Croes B, Schneider J, Hedley A, Gower S, Craig L, Krupnick A (2007) Air quality management approaches and evidence of effectiveness. In: Air pollution and public health: a guidance document for risk managers. Institute for Risk Research, University of Waterloo, Waterloo

Douglas K, Johansen D, Hebert M (1997) Toxic substances: federal-provincial control. Depository services program. Government of Canada, Ottawa

Environment Canada (1973) The clean air act annual report 1972–1973. Environment Canada, Ottawa

Environment Canada (2011) Air and climate indicators. http://www.ec.gc.ca/indicateurs-indicators/default.asp?lang=En&n=03603FB3-1. Accessed 18 May 2012

Government of Canada (2007) Regulatory framework for air emissions. Canada Ministry of Environment, Ottawa

Joint Ministers of Environment and Energy (1993) A comprehensive air quality management framework for Canada. Canadian Council of Ministers of the Environment, Winnipeg

Macdonald D (2009) The government of Canada's search for environmental legitimacy: 1971–2008. Int J Can Stud 39/40:191–210

MacKay WR (2004) Canadian federalism and the environment: the literature. Geo Int'l Envtl L Rev 25:46–50

McKitrick R (2005) The politics of pollution: party regimes and air quality in Canada. Department of Economics, University of Guelph, Guelph

McKitrick R (2006) Air pollution policy in Canada: improving on success. The Fraser Institute, Toronto

Nixon A, Curran T (1998) Acid rain. Parliamentary Research Branch, Government of Canada, Ottawa

OAG (1991) Jurisdictional complexities. Chapter 11 In: 1991 Report of the auditor general of Canada. Office of the Auditor General, Ottawa

OAG (2000) Partnerships for sustainable development—overview. Chapter 5 and Co-operation between federal, provincial and territorial governments. Chapter 7. in: 2000 May Report of the commissioner of the environment and sustainable development. Office of the Auditor General, Ottawa

Simeon R, Papillon M (2006) The division of powers in Canada: an overview. In: A Global Dialogue on Federalism, Volume 2: Distribution of powers and responsibilities in federal countries. Akhtar Majeed, Ronald L. Watts and Douglas M. Brown, editors. McGill-Queens University Press

Steering Committee (2010) Comprehensive air management system: a proposed framework to improve air quality management. Canadian Council of Ministers of the Environment, Winnipeg

Thomas K (2008) Canada-wide standards for particulate matter and ground-level ozone: a shared approach to managing air quality in Canada. Institute of Intergovernmental Relations School of Policy Studies, Queen's University, Kingston

Wood S, Tanner G, Richardson BJ (2010) What ever happened to Canadian environmental law? Ecol Law Q 37:981–1040

World Economic Forum (2005) Partnering for success: business perspectives on multistakeholder partnerships. World Economic Forum Global Corporate Citizenship Initiative, Geneva

The Canada-US Air Quality Agreement and its Impact on Air Quality Management in Canada

16

Jean O. Melious

Abstract

Canada and the United States entered into the Canada-US Air Quality Agreement (Agreement) in 1991, focusing primarily on a single air pollution problem: acid rain in the eastern parts of both countries. The intent of the Agreement, however, was broad and ambitious. In addition to establishing specific objectives relating to emission limits and the reduction of air pollutants, the Agreement establishes an ongoing bilateral system of prior notification and impact assessment, coordinated scientific and economic research, ongoing progress reviews, and a dispute resolution procedure. This broader purpose has allowed the parties to expand the Agreement's scope to establish specific objectives for the constituents of ground-level ozone, a precursor of smog. It has also promoted cooperation and communication and the sponsorship of regional air quality initiatives. This chapter first discusses the reasons that the Canada and the United States found it necessary and beneficial to sign an international agreement addressing air quality issues. It then summarizes the Agreement, providing examples of some of the ways that its provisions have been used and discussing possible future initiatives. The last section assesses the effectiveness of this bilateral framework and its influence on Canada's air quality management.

Keywords

Air pollution · Ttransboundary treaty · Ozone Annex · Acid rain · Bilateral agreement · Dispute resolution

16.1 Introduction

The Prime Minister of Canada and the President of the United States signed the Canada-US Air Quality Agreement (Agreement) in 1991, focusing primarily on a single air pollution problem: acid rain in the eastern parts of both countries. The intent of the Agreement, however, was broad and ambitious. The two governments stated that their purpose was to "establish, by this Agreement, a practical and effective instrument to address shared concerns regarding transboundary air pollution" (Canada-US 1991). In addition to establishing specific objectives relating to emission limits and the reduction of air pollutants, the Agreement establishes an ongoing bilateral system of prior notification and impact assessment, coordinated scientific and economic research, ongoing progress reviews, and a dispute resolution procedure.

This broader purpose has allowed the parties to expand the Agreement's scope to establish specific objectives for the constituents of ground-level ozone, a precursor of smog. It has also promoted cooperation and communication and the sponsorship of regional air quality initiatives. While each country's implementation of the Agreement is under its own control, both countries have benefitted from the exchange of "best practices" and from the development of joint initiatives.

This chapter first discusses the reasons that the Canada and the United States found it necessary and beneficial to sign an international agreement addressing air quality issues. It then summarizes the Agreement, providing examples of some of the ways that its provisions have been used and discussing possible future initiatives. The last section assesses

J. O. Melious (✉)
Western Washington University, Bellingham, United States
e-mail: jean.melious@wwu.edu

E. Taylor, A. McMillan (eds.), *Air Quality Management*, DOI 10.1007/978-94-007-7557-2_16,

the effectiveness of this bilateral framework and its influence on Canada's air quality management.

16.2 Why Did Canada and the US Need a Transboundary Air Quality Agreement?

Canada and the United States both have sophisticated laws and bureaucracies dedicated to the regulation of air pollution. The air quality in both countries is among the best in the world. Why, then, should these countries dedicate time and resources to a bilateral approach to air quality management?

In light of the many domestic priorities that air quality managers on both sides of the border need to address, under conditions in which budgetary constraints are the norm, this is not a hypothetical question. Federal officials "facing busy agendas and limited resources would often prefer to avoid confrontations that could disrupt more important areas of their bilateral relationship" rather than seeking out involvement in environmental concerns (Springer 2007) Officials and air mangers at the provincial/state and local levels answer to local constituencies and may perceive cross-border cooperation as an unfunded mandate rather than a priority of government (Melious 2004).

Mere neighborliness, therefore, would not be sufficient to induce Canada and the United States to enter into a transboundary air quality agreement, much less to continue working together once the original terms were implemented. Rather, needs relating both to substance and structure motivated the parties to enter into the Agreement.

Pressing pollution issues of mutual concern, which could not be resolved without joint participation, are one basis of the Agreement. The most apparent need for joint problem-solving arises when pollution sources on one side of the border result in harm on the other side. Acid rain in Canada, which could not be solved without action by the US, is a classic example of this type of problem. Not surprisingly, it was the direct cause, and continues to be a primary focus, of the Agreement.

The Agreement additionally provides a long-term framework for cooperation and information sharing, establishing a forum that is both focused on air quality issues and broad enough to evolve to reflect the changing needs and priorities of the parties. This structural aspect of the Agreement is worth further consideration, because it demonstrates that even a binational relationship that spans a century may need to develop new procedures and forms to address the complexities of transboundary air pollution.

Canada and the United States have developed a working relationship over the course of many years that has helped to entrench interaction on the environment (Vannijnatten 2004). The US's Special Negotiator for the Agreement emphasized the importance of this long-standing "consultative relationship," noting that the two countries had developed a "toolchest of techniques" to prevent and manage transboundary disputes over the environment (Smith and Bizniaz 1991).

The Agreement itself commences by referring to this unusually well-stocked toolchest, acknowledging "the tradition of environmental cooperation" reflected in earlier agreements: the Boundary Waters Treaty of 1909, the Trail Smelter Arbitration of 1941, the Great Lakes Water Quality Agreement of 1978, the Memorandum of Intent Concerning Transboundary Air Pollution of 1980, the 1986 Joint Report of the Special Envoys on Acid Rain, and the Convention on Long-Range Transboundary Air Pollution of 1979 (US-Canada 1991). The tradition of consultation and cooperation established by these agreements paved the way for the creation of the Air Quality Agreement and established a foundation for consultation and dispute resolution provision in the Agreement, but it did not eliminate the need for a dedicated Agreement focusing on air pollution.

The 1909 United States-United Kingdom Boundary Waters Treaty, one of the earliest international agreements to address pollution, established the International Joint Commission (IJC). The IJC has several powers that are potentially relevant to transboundary air quality. It may advise the government on matters referred to it for consideration and has the authority to arbitrate issues referred to it by both governments (United States and United Kingdom 1909).

The IJC, sometimes referred to as the "watchdog of the Great Lakes," primarily is associated with water quality and quantity issues in eastern Canada and the U.S. It also has a formal role in transboundary air quality. In 1966, the governments of Canada and the United States asked the IJC to monitor air quality along the entire Canada-United States boundary and to draw air pollution problems to their attention. The IJC subsequently established the International Air Quality Advisory Board (IAQAB). The IAQAB is directed to "advise the Commission on matters of direct relevance to transboundary aspects of air quality in the United States and Canada" and to "be currently informed regarding any air quality or deposition problems involving both the United States and Canada, including transport across the international boundary" (International Joint Commission 1986). The IAQAB's role is entirely advisory, however, rather than regulatory. The Agreement fills a gap by providing a vehicle for the two nations to express binding commitments to be established through each country's laws.

The Trail Smelter Arbitration of 1941 between Canada and the United States resulted from a referral to the International Joint Commission. It resolved damage claims in the US based on emissions from a smelter in Canada, and is often cited as the source of the customary law principle of states' responsibility to ensure that activities within their jurisdiction or control do not cause damage to the environment of other states (Trail Smelter Arbitration 1941). This principle is somewhat inchoate, however, requiring proof of damage before liability provisions are triggered, and it does not implement an ongoing regime to monitor or implement states' responsibility. The Agreement was needed to overcome these shortcomings.

The Great Lakes Water Quality Agreement of 1978 (GLWQA) commits Canada and the U.S. to efforts to "restore and maintain the chemical, physical, and biological integrity of the waters of the Great Lakes Basin Ecosystem" (Canada and United States 1978). The GLWQA involves the International Joint Commission in this transboundary effort and reaffirms the Boundary Waters Treaty of 1909. It established a model for cooperation to address a large-scale, difficult transboundary resource issue, but its focus on water quality meant that air quality was not addressed with the same specificity until the Agreement was approved.

The Convention on Long-Range Transboundary Air Pollution (LRTAP) commits Canada and the United States, as well as European signatories, to reduce and prevent air pollution and to use the best available technology that is economically feasible (Convention 1979). As with the principles derived from the Trail Smelter arbitration, these obligations require the development of a monitoring and implementation regime in order to be effective.

The Agreement built on the principles and relationships developed through this long history of consultation, cooperation, and dispute resolution and avoidance. It also recognized that air pollution was a long-term, continental problem that could only be addressed through the concentrated efforts of the parties. The Agreement provided the specific focus on the regulation of air pollution that was missing from previous joint agreements between Canada and the US. This provided a means to implement the general principles of transboundary responsibility recognized by both parties.

16.3 Canada-US Air Quality Agreement: Structure and Substance

16.3.1 Overview

The Agreement covers a wide variety of topics. Its sixteen Articles and three Annexes include general and specific objectives; assessment, notification, and mitigation requirements; the establishment of a bilateral Air Quality Committee; and procedures for consultations, referrals, and settlement of disputes.

Because the Agreement is quite short for such a complicated topic, its provisions can be briefly summarized. Major provisions of the Agreement and its Annexes are then discussed in greater detail, in the following sections.

Article I "Definitions," defines "air pollution" and "transboundary air pollution" as follows:

> "Air pollution" means "the introduction by man", directly or indirectly, of substances into the air resulting in deleterious effects of such a nature as to endanger human health, harm living resources and ecosystems and material property and impair or interfere with amenities and other legitimate uses of the environment, and "air pollutants" shall be construed accordingly;
>
> "Transboundary air pollution" means air pollution whose physical origin is situated wholly or in part with the area under the jurisdiction of one Party and which has adverse effects, other than effects of a global nature, in the area under the jurisdiction of the other Party (Canada-US 1991).

Article II establishes that the "Purpose" of the Agreement is to establish a "practical and effective instrument to address shared concerns regarding transboundary air pollution." This purpose is reinforced by *Article III*, which provides that the objective is to "control" transboundary air pollution through all of the measures contained in the remainder of the Agreement.

Article IV requires each country to establish specific objectives for emission limitations or reductions of the air pollutants that the countries agree to address. These specific objectives are to be included as annexes to the Agreement.

Article V addresses assessment, notification, and mitigation. Each country commits to the assessment of "actions, activities and projects" within its territory that would be likely to cause significant transboundary air pollution, and to the consideration of mitigation. The parties are required to notify each other of such activities, and each party may request consultation about projects or changes in laws that may affect transboundary air pollution. Consultation is mandatory if a party becomes aware of an air pollution problem of joint concern requiring an immediate response.

Article VI requires the parties to carry out scientific and technical activities and economic research as described in *Annex 2*.

Article VII imposes an obligation to exchange information on monitoring; emissions, technologies and mechanisms for controlling emissions, atmospheric processes, and effects of air pollutants through the Air Quality Committee established by *Article VIII*.

Article VIII establishes a bilateral Air Quality Committee to assist in the implementation of this Agreement. The Committee is responsible for preparing biennial progress reports.

Article IX empowers the International Joint Commission to invite comments and hold public hearings on each progress report prepared by the Air Quality Committee pursuant to Article VIII and to report on these views to the public.

Article X requires the parties to consider and report on the Air Quality Committee's progress reports, and to review the Agreement every five years to consider whether any modification is needed.

Articles XI, XII and XIII address dispute resolution, with each Article addressing an escalating level of dispute. *Article XI* reiterates the obligation of consultation at the request of either party. *Articles XII and XIII* provides for a referral to a third party if consultation does not resolve the issue. *Article XIII* states that disputes over the interpretation or implementation of the Agreement may be referred to the International Joint Commission.

Article XIV obligates the party to seek the appropriation of funds, the adoption of legislation, and the cooperation of state and provincial governments to implement the Agreement.

Article XV provides that the Agreement does not diminish the rights and obligations of the parties in other international agreements between them, and *Article XVI* provides that the parties may amend or terminate the agreement.

Annex 1 contains specific programs and objectives concerning sulfur dioxide and nitrogen oxides (NO_x), designed to address acidic deposition in both countries.

Annex 2 relates to scientific and technical activities and economic research.

Annex 3 contains specific objectives concerning ground-level ozone precursors NO_x (both NO and NO_2) and VOCs.

The major issues addressed by the Agreement are discussed in greater detail in the following sections.

16.3.2 Acid Rain and Visibility (Annex I)

Acid Rain: Measures to Reduce Sulphur Dioxide and NO_x Acid precipitation has been a long-standing concern along the eastern Canada-US border. Coal-burning electrical generating plants in the US Midwest, in particular, contribute substantially to acid rain in Canada. In the 1980s, it was estimated that US sources were emitting approximately 25.7 million tons of sulphur dioxide and 23 million tons of NO_x annually, while Canada's annual emissions totaled 5.2 and 2 million tons, respectively (Hsu and Parrish 2007).
Annex 1 contains Canada and the United States' commitments to implement specific emission limitations and reductions of sulphur dioxide (SO_2) and NO_x emissions on a specified timetable. With respect to sulphur dioxide, Canada agreed to achieve a "permanent" national emissions cap of 3.2 million tonnes per year by 2000; in 2008, Canada reported total sulphur dioxide emissions of 1.7 million tons, about 47 % below the national cap. The US committed to a permanent national emission cap of 8.95 million tons of sulphur dioxide per year for electric utilities by 2010; it achieved this goal by 2007, three years early (Canada-United States Air Quality Committee 2010).

Both countries also report that their Annex 1 obligations to reduce NO_x emissions have been met. Canada committed to reduce NO_x emissions from power plants, major combustion sources and metal smelting operations by 100,000 t below the below the year 2000 forecast level of 970,000 t; in 2008, these emissions totaled approximately 665,000 t. The US committed to reduce total annual NO_x emissions to 2 million tons below projected annual emission levels for 2000; its 2009 NO_x emissions exceeded that obligation by 3 million tons (Canada-United States Air Quality Committee 2010).

Overall, Canada reduced its total sulphur dioxide emissions by 47 % between 1990 and 2008, and the United States reduced its sulphur dioxide emissions from the sources covered by the Annex by 51 %. Both countries reduced NO_x emissions by around a third in the designated transboundary ozone region (Canada-United States Air Quality Committee 2010).

These reductions represent a significant achievement that has resulted in signs of recovery in some ecosystems affected by acid precipitation. However, as Canada noted in the most recent (2010) progress report the Annex 1 commitments under the Agreement, "[d]espite Canadian emission reduction efforts, the control of acidifying emissions has not occurred to the extent necessary to reduce acid deposition below critical loads and ensure the recovery of aquatic and terrestrial ecosystems" (Canada-United States Air Quality Committee 2010). Neither country can afford to rest on its laurels. In particular, Canada's recovery from acid rain will depend on the US's continued efforts to address the constituent pollutants.

Prevention of Deterioration and Visibility Annex I, Part 4 further commits the two governments to take actions preventing significant air quality deterioration and protecting visibility from sources that could cause significant transboundary effects (Canada-US 1991). Canada's obligation is stated in terms of comparability with the US obligations, which in turn refer to the US Clean Air Act provisions specifically requiring "prevention of significant deterioration" and the protection of visibility in national parks and wilderness areas (Canada-US 1991, US Clean Air Act).

Visibility is designated as a second-tier "welfare" effect of air pollution in the Clean Air Act because it is not directly related to health effects. Visibility is related to other air pollution effects, however, and is the most-cited indicator of air quality by the public (Jacques Whitford-ACYS 2007). The fact that visibility is "a primary and highly obvious indicator of general air quality" has been called a "universal truism": "If you can see the air, it is polluted; if you cannot see the air, it might still be polluted" (RWDI 2008).

Visibility is impaired by "plume blight," consisting of smoke and dust from an identifiable stack or group of sources, and by "regional haze," referring to widespread haze from a multitude of sources. These classifications reflect rules adopted under the US Clean Air Act—the 1980 "Plume Blight Rule" and the 1999 "Regional Haze Rule"—as well as the differences in sources and approaches to addressing visibility impairment.

Plume blight is, at least conceptually, the easier issue to tackle. It can be addressed by identifying sources and imposing pollution control requirements. This is not necessarily easy, especially when the source of the plume is a significant economic activity, but it is a more straightforward problem than addressing regional haze. Regional haze is a particularly difficult challenge because it is caused by multiple sources that may be tens, hundreds or even thousands of kilometres away (Jacques Whitford-AXYS 2007).

Canada has addressed both visibility and prevention of significant deterioration through its overall air pollution management system. The Canadian Environmental Assessment Act and the Canadian Environmental Protection Act both address air pollution and pollution prevention. The Canada-Wide Standards for Particulate Matter and Ozone, which address pollutants that are important contributors to the impairment of visibility, include "continuous improvement" (CI) and "keeping clean areas clean (KCAC)" principles.

Although provincial as well as federal laws are aimed at pollutants that affect visibility, Canadian air quality legislation does not contain specific visibility goals (RWDI 2008). At the time that the Agreement was signed, this was also true of the US. The adoption of the Regional Haze Rule in the US, however, resulted in a much more detailed visibility regime, involving monitoring and regional planning to meet visibility goals. The Canadian approach to visibility protection does not address parks and wilderness areas and does not yet involve the use of regional planning.

Especially in western Canada, where clear vistas and dramatic scenery are essential to tourism, the need to address visibility directly has spurred interest in developing a visibility management framework. In response to concerns that health-based Canada-Wide Standards for Particulate Matter may not adequately protect visibility, the province of British Columbia has convened a British Columbia Visibility Coordinating Committee (BCVCC), which brings air quality managers from different levels of government together to work towards a visibility goal in B.C. (RWDI 2008; Canada-United States Air Quality Committee 2010). One goal is to extend the approach of the U.S. monitoring network (IMPROVE, or the U.S. Interagency Monitoring of Protected Visual Environments) into Canada.

The issue of transboundary visibility demonstrates the advantages of the long term nature of the Agreement. The initial charge required Canada to demonstrate comparable results with the U.S., despite differences in regulatory approaches and governmental structures. This approach was not capable of immediate implementation, but the fact that the Agreement establishes a permanent framework creates opportunities for long-term collaboration. Work done in the U.S. can be studied and assimilated to create an approach that works in the Canadian regulatory and policy context.

16.3.3 Ozone (Annex 3)

Ozone is a respiratory irritant that leads to decreases in lung function, aggravation of respiratory disease, and increases in hospital admissions. It also damages ecosystems, including forests and agriculture. Canada and the US have both recognized these effects for years and have applied increasingly strict standards to reduce ground-level ozone, which is a major component of smog.

Particularly in eastern North America, these individual efforts were not sufficient to prevent large-scale regional episodes of high ozone levels. Not long after the Agreement was signed, modeling and analysis of the spatial impacts of ozone established that transport occurs over distances of many hundreds of kilometres in the eastern United States and Canada. Because ozone concentrations are influenced by transported ozone as well as background concentrations and locally generated ozone, the reduction of pollution in one country can have a significant effect on pollution in the other country (Canada-US Air Quality Comm 1999; Brankov et al. 2003).

This evidence of the significance of ozone transport led to a recognition that ground-level ozone would be an appropriate issue under the Agreement as early as 1994. In 1995, Air Quality Committee members from both countries met to outline a joint work plan for cooperative analyses. These analyses included integrated air quality data analyses of patterns and episodes, consideration of emissions patterns together with analyses of ozone as a function of changing meteorological conditions and transport climatology, and joint modeling of regional scale ozone transport and responses to control scenarios. Based on these data, the Air Quality Committee concluded:

> These technical analyses clearly demonstrate the connections between emissions, transport, and ozone occurrences on both sides of the border. These results strongly support the commonsense conclusion that coordination of planning and execution of control strategies for ozone precursors (NO_x and VOC) for all source categories would be more beneficial than individual initiatives (Canada-US Air Quality Comm 1999).

This strong technical foundation supported the addition of the "Ozone Annex" to the Agreement in 2000. The purpose of the Ozone Annex is to control and reduce the "anthropogenic emissions of NO_x and volatile organic compounds (VOC) that are precursors to the formation of ground-level ozone and that contribute to transboundary air pollution, thereby helping both countries attain their respective air quality goals over time to protect human health and the environment" (Canada-United States Air Quality Agreement 2000, Annex 3).

The Ozone Annex established a transboundary region, known as the Pollutant Emission Management Area (PEMA), which includes central and southern Ontario, southern Quebec, 18 U.S. states, and the District of Columbia. Within this region, the parties agreed to specific country-specific approaches to reducing VOCs and NO_x. In addition, the parties agreed to report every two years on all anthropogenic NO_x and all anthropogenic and biogenic VOC emissions within the PEMA; on ambient NO_x and VOC; and on ten-year trends in ambient levels. Ambient reporting is undertaken for all sites within 500 km of the Canada-U.S. border that meet data completeness requirements, not solely within the PEMA. All reported data are summarized in biannual progress reports on the Agreement.

16.3.4 Scientific and Technical Cooperation and Research (Annex 2)

Under Annex 2, the parties agreed to coordinate their air pollutant monitoring activities; to add monitoring stations, networks, or tasks where necessary; to use compatible data formats; and to exchange monitoring data. The parties also agreed to cooperate and exchange information regarding impacts on ecosystems and human health, as well as the development of atmospheric models.

The most recent progress report under the Agreement points to improvements in emission inventories and trends for the pollutants addressed under the Acid Rain and Ozone Annexes: SO_x, NO_x and VOC. Canada and the U.S. also reported on collaborating on real-time air quality reporting and mapping for ozone and fine particles ($PM_{2.5}$). Both countries reported on health and ecosystem research, including studies gauging the recovery of acidified lakes and streams (Canada-US Air Quality Comm 2010). The Air Quality Committee, discussed below, meets regularly to help further the two countries' cooperative scientific efforts.

16.3.5 Air Quality Committee (Article VIII)

Article VIII of the Air Quality Agreement establishes a bilateral Air Quality Committee (AQC). The AQC is charged with reviewing progress in the implementation of the Agreement and preparing progress reports for the two governments. Progress reports are also submitted to the IJC for the solicitation of public comments and preparation of a synthesis of views presented during comments.

The AQC is composed of nine representatives from each government and has two subcommittees to assist in carrying out its work. Subcommittee 1 addresses the Acid Rain Annex, overseeing the implementation of programs and objectives contained in the Agreement. Subcommittee 2, which addresses Annex 2 of the Agreement, provides a forum for discussion of the additional scientific and technical activities contemplated by Annex 2 of the Agreement.

The Air Quality Committee helps to provide a forum for the pursuit of the long-term goals of the Agreement, as well as overseeing the Agreement's implementation and oversight goals. It brings together personnel from both countries and provides a focus and information source for other stakeholders, ranging from private sector representatives to environmental nongovernmental organizations. Some observers of Canada-US environmental relations believe that the development of such "epistemic communities" is particularly important to the development of transboundary environmental policy. Epistemic communities, defined as networks of "officials, experts, and non-governmental actors at multiple levels" (VanNijnatten 2003), comprise an informed constituency that can provide the needed impetus for new policies at the international level. Such networks also build trust and communication, helping to overcome the difficulties created by air quality personnel's perception of "how different" the system is in the other country (Melious 2004).

The Air Quality Committee's significant role in developing the Ozone Annex was discussed above. The Committee continues to work with the governments and stakeholders to improve bilateral cooperation and to provide a scientific basis for further expansions of the Agreement. For example, as discussed below, the Committee has published a scientific assessment of transboundary particulate matter, which may form the basis of an annex to address this issue (Canada-US Air Quality Committee Subcommittee 2 2004).

16.3.6 Notification of Significant New or Modified Sources (Article V)

A more site-specific, intensive form of the problem of cross-border air pollution is the issue of facility siting. Borders not only establish regulatory authority, but they also create perverse incentives. A border creates the political and economic opportunity to build a facility on one side that will impose costs on the other side. In economic terms, air pollution is a classic externality that can reduce the costs for the polluter while imposing costs on receptors. Even if the intent is not to disproportionately burden the other country, the siting of power plants and other large facilities near a border may create the perception that this is the case, causing tensions across the border and upsetting political constituencies.

Recognizing the transboundary effects of large facilities, Article V of the Agreement requires each country to notify the other of significant new or modified sources of pollutants within 100 kilometers of the Canada-U.S. border. Since this procedure was initiated in 1994, Canada has informed the US of 55 sources and the US has notified Canada of 61 sources (Canada-US Air Quality Committee 2010).

Notification alone, however, has not always served effectively to preempt disputes. One such example, along the western Canada-US border, involved a power plant in the small town of Sumas, Washington that was planned for construction only half a mile from the border with British Columbia. The Sumas Energy 2 Generating Facility (SE2) project would have resulted in the construction of a 660 MW natural gas-fuelled power plant in the City of Sumas, Washington, approximately one-half mile south of the Canadian border.

Regulators in the US approved the project, citing its commitment to offset NO_x and particulate emissions through project offsets in, or payments to, British Columbia. (EFSEC Council Order No. 768 17). The project's downfall, however, was its need for a power transmission line. The National Energy Board (NEB) turned down the application, following the receipt of approximately 25,000 letters opposing the project.

With respect to health impacts, NEB accepted that, "although pollutants from the Power Plant would result in a change in ground-level concentrations of the pollutants in the Lower Fraser Valley, the change would be small and not likely to result in a measurable effect on human health." Overall, NEB concluded "that the maximum concentrations of pollutants from the Power Plant, in combination with background levels, would likely be below the most stringent Canadian requirements and, hence, meet both the B.C. and Canadian Air Quality Objectives and Standards for these pollutants" (*ibid.* at 62).

Nonetheless, air quality impacts appear to have been the primary basis for NEB denial of SE2's application. NEB reasoned that standards were based on "acceptable" risk, and that the existence of any risk at all constituted a burden on Canadians. It noted that the acceptability of emissions to those living in the region was "a relevant factor for the Board to consider." Finally, NEB emphasized that SE2's cumulative effects, in conjunction with other new facilities contemplated in Canada, would make air pollution reduction efforts in Canada more expensive and difficult (*ibid.* at 62–63, 96). NEB concluded that, overall, any benefits to local communities would be either "external to these communities" or "negligible in value," while the burdens were "many and real" (*ibid.* at 96).

The SE2 controversy demonstrates the extent to which facility siting can create tensions between two generally amicable neighbors, despite formal frameworks of notification. It also demonstrates the border's strong influence on risk perception and cost-benefit analyses. The federal Agreement has not eliminated national loyalties or altered the localized nature of such disputes. It has been proposed that such disputes may increasingly be the subject of litigation between Canada and the US, rather than bilateral dispute resolution (Hsu and Parrish 2007). Although they have not been used, the Agreement includes dispute resolution provisions that could prevent the need for such adversarial approaches, as discussed in the following section.

16.3.7 Dispute Resolution Provisions (Articles XI-XIII)

The dispute resolution provisions of the Agreement provide for consultations at the request of either party. If a dispute remains over the interpretation or implementation of the Agreement, the parties are obligated to seek to resolve the dispute through negotiations. If negotiations fail, the parties may either submit the dispute to the IJC or submit it to another agreed-upon forum for dispute resolution (Canada-US Air Quality Agreement 1991).

These provisions have not been used, perhaps because many other forums are available for dispute resolution. In the SE2 border siting facility dispute discussed above, for example, the Governor of the State of Washington invited Canadian officials to a meeting to hear their concerns about the project, and the American sponsor of the SE2 project participated fully in the Canadian NEB hearings on the transmission line (Melious 2004). It is also possible that the provisions have not been used because they bring to mind the arbitration employed to resolve the Trail Smelter dispute, which may be viewed as creating a process that it lengthier and more formal than the parties need to resolve disputes to date. Finally, the fact that the two countries meet regularly through the Air Quality Committee and in regional meetings may help to defuse disputes.

The availability of dispute resolution techniques nonetheless is a useful reminder that the two countries have made a commitment to resolve disputes through negotiation, not through unilateral actions or adversarial approaches.

16.3.8 Possible Next Steps

Depending on the priorities and resources of the two federal governments, the future of the Agreement could take several different paths. It could continue along its present course, which focuses on pollution problems in eastern Canada and the United States, particularly acid rain and ozone. The Agreement may also evolve and expand. Several options that have been discussed over the years include expanding the Agreement to address additional transboundary pollution problems and expanding its geographical scope to include the border between western Canada and the US.

Potential new Annex on particulate matter Particulate matter (PM) air pollution is an air-suspended mixture of solid and liquid particles that vary in size, composition, and origin. "Fine" particles, less than 2.5 μm in diameter ($PM_{2.5}$), may have the most significant impacts on human health. They may be more toxic, and they can be breathed more deeply into the lungs. Fine particles also remain suspended for longer periods of time, penetrate more readily into indoor environments, and are transported over much longer distances than larger particles (Pope and Dockery 2006).

For over a decade, the countries have been considering the adoption of a new Annex addressing transboundary particulate matter (Ontario Pub Health Asoc 2003). In the run-up to the adoption of the Ozone Annex, the Parties signed a 1997 "Commitment to Develop a Joint Plan of Action for Addressing Transboundary Air Pollution," contemplating that the parties would address PM as well as ozone. The Ozone Annex moved ahead while science assessment activities were underway to provide information on transboundary PM issues (US Dept. of State 2011).

The Air Quality Committee released a report documenting the "Transboundary PM Science Committee" at the end of 2004. The report found that the transboundary transport of PM and PM precursors was significant enough in some regions to compromise the attainment of national standards and concluded that this scientific foundation supported the development of a PM Annex (Canada-United States Air Quality Committee 2004).

Representatives of both countries have indicated that the development of a PM Annex remains an important goal (Jacobson 2011; Health Canada 2011). An important step towards this goal would be Canada's adoption of commitments (as opposed to standards and objectives) for the reduction of PM and its precursors. The lack of such an emissions reduction commitment has stalled PM Annex negotiations to date.

Expansion of geographical focus to western Canada-US The Agreement's specific pollution reduction obligations, as set forth in the Acid Rain and Ozone Annexes, only apply to eastern Canada and the US. This reflects the fact that the monitoring of target pollutants shows that pollution levels generally are higher along the more urbanized eastern border. As population increases along the western Canada-US border, however, levels of pollutants such as ozone are increasing, especially in the airshed surrounding Vancouver, B.C.

In 2004, it appeared that the Ozone Annex might be extended to include a west coast border area (Melious 2004), but this has not occurred. As comments on the 2008 progress report on the Agreement noted, "although there has been some discussion of developing an annex to address transboundary issues in the western and Pacific regions, 'the progress report is silent on the status of a Western Annex, and the path forward is unclear'" (Int Joint Comm 2007). One commentator stated:

> We recognize the good work undertaken by the Parties in eastern Canada and the northeast states to reduce emissions of ozone precursors and that it has proven to be effective in reducing ozone levels. In light of increasing ozone trends in the Fraser Valley Regional District [outside of Vancouver, B.C.), we request that the Air Quality Committee recommend to the Parties that this area be similarly designated for specific action to reverse this trend (Int Joint Comm 2007).

It has also been suggested that the Agreement's provisions for scientific cooperation should be extended to the western border. One commentator noted that "[n]itrogen deposition appears to be having significant ecosystem impacts in the Georgia Basin/Puget Sound area, and more attention to this issue is needed. We recommend that the critical load assessment work in eastern regions be extended to include this international airshed" (Int Joint Comm 2007).

Canada and the US have developed cooperative relationships through pilot projects along the western border, as discussed below. These pilot projects and other joint efforts will help to establish the need for expanded efforts under the Agreement.

16.4 Achievements and Effects on Canadian Air Quality Management

16.4.1 Pollution Reduction: Acid Rain and Ozone

The development, implementation, and expansion of the Agreement embody all "four significant contributions" that the two national governments can bring to transboundary issues: resources of funding and scientific knowledge, regulatory authority, the ability to create new institutional frameworks, and breadth of perspective (Springer 2007).

There is widespread agreement that the Agreement has led to significant reductions in the emissions targeted by the Acid Rain and Ozone Annexes (Chiotti 2008). In his remarks commemorating the twentieth anniversary of the Agreement, Canada's Environment Minister observed that "After twenty years of cooperation, emissions causing acid rain have been cut in half and emissions causing smog have been cut by one-third in the region covered under this agreement." The Administration of the US Environmental Protection Agency stated that "Our joint efforts to clean up the air we breathe have saved lives and protected American and Canadian families from asthma and other respiratory illness, removed acid from rain and smog from air, and set the foundation for continued work together on our shared challenges" (US EPA 2011).

It is impossible to tell, of course, how much of the pollution reduction is directly attributable to joint action and how much each country would have achieved if acting unilaterally. It would be difficult to dispute, however, that the existence of binding commitments, bolstered by the accountability created by joint monitoring obligations and biennial progress reports, has promoted more diligent and rapid cuts in air pollution than would have occurred without the Agreement.

16.4.2 Best Practices and Harmonization

One benefit of information-sharing and cooperation is that each country can learn from the experience of the other country. This sounds like a truism, but the day-to-day reality of boundaries and bureaucracies is not conducive to the longer-term process of comparing and contrasting with neighbors unless there is a strong, legislated incentive to do so.

Recent comments on the progress of the Agreement summed up this benefit:

> The Air Quality Agreement has documented the elements of a successful process, that being establishing goals, collecting data that identify and track activities under the Agreement, and to routinely and objectively publicly report on progress toward achieving the goals of the Agreement ... To report progress has required the parties to work cooperatively to harmonize and standardize scientific and engineering methods of data collection, analyses and reporting in order to present accurate and comparable data (Int Joint Comm 2007).

The Ozone Annex, in particular, has been described as "a highly-detailed and ambitious recipe for harmonizing standards on both sides of the U.S.-Canada border" (Hsu and Parrish 2007). For example, it includes harmonization goals relating to national vehicle and engine standards for emissions of smog-forming pollutants (Canada-United States Air Quality Committee 2010). Harmonization provisions not only help to advance pollution regulations, but they also may help to prevent transboundary disputes over the countries' respective performance.

16.4.3 Innovative Regional Initiatives and Pilot Projects

Although the pollution control commitments under the Agreement are focused on the eastern border, and on specific measures to address pollutants related to acid rain and ozone pollution, the Agreement has helped to support several innovative initiatives that go beyond these issues.

Under the Border Air Quality Strategy announced in January 2003, Canada and the United States committed to the development of new cooperative transboundary air quality projects. While the Strategy was not an amendment or addition to the Canada-U.S. Air Quality Agreement, it was characterized as building on joint activities under the Agreement.

Three pilot projects, all aimed at reducing smog and improving health, were undertaken between 2003 and 2005. The Great Lakes Basin Airshed Management Framework explored the feasibility of a coordinated airshed management approach in the Southwest Ontario/Southeast Michigan airshed. It assessed monitoring capabilities, data harmonization, policy needs, and health effects. The report on this pilot project concluded that coordinated airshed management is feasible and desirable, and should be conducted under the auspices of the Agreement (Canada-US Border Air Quality Agreement 2005 Great Lakes).

The second pilot project, entitled "Maintaining Air Quality in a Transboundary Air Basin: Georgia Basin-Puget Sound," addressed the airshed along the western portion of the border between British Columbia and Washington state. Although air quality in this area generally meets federal standards on both sides of the border, this initiative responded to concerns about human health and ecosystem effects that may occur at the air pollution levels reported in the airshed (Canada-US Border Air Quality Agreement 2005a Maintaining).

The third pilot project, the Emission Trading Feasibility Study, reviewed the key components of U.S. cap and trade programs under the Clean Air Act and developed Canada-U.S. economic and air quality modeling tools to be able to assess the potential economic and environmental impacts of NO_x and SO_2 cross-border trading. The report on this pilot project observed that, despite progress made under the Agreement:

> A significant portion of the population in the eastern border region of both Canada and the United States is exposed to harmful levels of air pollutants, levels that often exceed both countries' air quality standards designed to protect human health. Acidic deposition remains at levels that cause concern for sensitive ecosystems in both countries, and significant degradation of visibility in national parks persists. Further emission reductions would improve air quality and reduce acid deposition in both countries, and an emissions cap and trading program can maintain those improvements at costs lower than traditional regulatory approaches (Canada-US Border Air Quality Agreement 2005b).

These pilot projects are testimony to the flexibility of the Agreement and its ability to adapt to the needs of the parties in addressing transboundary air pollution issues. The pilot projects also demonstrate that transboundary air quality issues are far from resolved. The initial friction between the two countries caused by acid rain may have been reduced, but ongoing concerns about health and ecosystems will ensure that air quality across the border will remain a matter of joint concern for many years to come.

16.5 Conclusion

The Canada-US Air Quality Agreement affects Canadian air quality management in many ways. It has spurred pollution reduction efforts in Canada as well as in the US and has helped Canada to coordinate its monitoring system with the

US. The Agreement has spurred the development and sharing of scientific information, especially through the ongoing efforts of the Air Quality Committee. Harmonization provisions help to ensure that the two countries can adopt each other's best practices.

The Agreement has been successful on its own terms. Both countries are substantially in compliance with its requirements, and both countries have met specific reduction requirements that have drastically reduced the emission of targeted pollutants. Nonetheless, pollution from the US continues to be a major determinant of air quality in parts of eastern Canada (Chiotti 2008), supporting the need for an ongoing effort by both countries to ensure that their pollution sources are not harming their neighbor. Ongoing cooperation related to monitoring, health, and other scientific aspects of transboundary air pollution is also clearly warranted.

As climate change-induced higher temperatures exacerbate the effects of air pollution, both directly and through synergistic effects on human health (Chiotti 2008), the challenges that Canada and the United States face in protecting their air quality will only grow more difficult. The strength and resilience of the framework established by the Agreement will be tested over the coming years. With two decades of cooperative experience to build upon, the two nations are well positioned to meet these challenges by coordinating their efforts to increase their effectiveness.

References

Barton J (2008) Canada-wide standards and innovative transboundary air quality initiatives. J Toxicol & Environ Health Part A 71:74–80. doi:10.1080/15287390701558022

Brankov E, Henry R, Civerolo K, Hao W, Rao S, Misra P, Bloxam R, Reid N (2003) Assessing the effects of transboundary ozone pollution between Ontario, Canada and New York, USA. Env Pollut 123:403–11

Canada and United States (1978) Great lakes water quality agreement

Canada and United States (1991) Canada-United States Air Quality Agreement

Canada-United States Air Quality Agreement (2000) Annex 3, Specific objectives concerning ground-level ozone precursors. http://www.ec.gc.ca/air/default.asp?lang=En&n=9992B080-1. Accessed 29 Oct 2011

Canada-United States Air Quality Committee (1999) Ground-level ozone: occurrence and transport in eastern North America. http://www.ec.gc.ca/pdb/can_us/can_us_glozone/intro_e.cfm. Accessed 29 Oct 2011

Canada-United States Air Quality Committee (2004) Canada-United States Transboundary Particulate Matter Science Assessment. http://dsp-psd.pwgsc.gc.ca/Collection/En56-203-2004E.pdf. Accessed 29 Oct 2011

Canada-United States Air Quality Committee (2010) Canada-United States Air Quality Agreement progress report 2010. http://www.ec.gc.ca/Publications/4B98B185-7523-4CFF-90F2-5688EBA89E4A/CanadaUnitedStatesAirQualityAgreementProgressReport2010.pdf. Accessed 29 Oct 2011

Canada- United States Air Quality Committee Subcommittee 2: Scientific Cooperation (2004) Canada-United States transboundary particulate matter science assessment. http://dsp-psd.pwgsc.gc.ca/Collection/En56-203-2004E.pdf. Accessed 29 Oct 2011

Canada-United States Border Air Quality Strategy (2005) Maintaining air quality in a transboundary air basin: Georgia Basin-Puget Sound. http://www.ec.gc.ca/Publications/CC269DEF-C1E9-4498-BD7C-387EA3107298/MaintainingAirQuality_GB_PS.pdf. Accessed 29 Oct 2011

Canada-United States Border Air Quality Strategy (2005a) Great Lakes Basin airshed management framework pilot project. http://www.ec.gc.ca/Publications/59A7520B-5066-463A-87E5-505312E8B94B/GreatLakesBasinAirshedMgtFramework.pdf. Accessed 29 Oct 2011

Canada-United States Border Air Quality Strategy (2005b) United States–Canada emissions cap and trading feasibility study. http://www.epa.gov/airmarkt/progsregs/usca/docs/feasstudy.pdf. Accessed 29 Oct 2011

Chiotti Q (2008) Principles for air quality management: an environmental nongovernment organization (ENGO) perspective. J Toxicol & Environ Health Part A 71:43–46. doi:10.1080/15287390701557644

Convention on Long Range Transboundary Air Pollution (1979) http://www.unece.org/fileadmin/DAM/env/lrtap/full%20text/1979.CLRTAP.e.pdf. Accessed 29 Oct 2011

Health Canada (2011) Planning for a sustainable future: health Canada's 2011–2014 sustainable development strategy

Hsu S, Parrish A (2007) Litigating Canada-U.S. transboundary harm: international environmental lawmaking and the threat of extraterritorial reciprocity. Virginia J Int Law 48:1–63

Int Joint Comm (1986) International Air Quality Advisory Board Mandate. http://www.ijc.org/conseil_board/air_quality_board/en/iaqab_mandate_mandat.htm. Accessed 29 Oct 2011

Int Joint Comm (2007) Synthesis of public comment on the 2006 progress report under the Canada-United States Air Quality Agreement. http://www.ijc.org/php/publications/pdf/ID1606.pdf. Accessed 29 Oct 2011

Jacques Whitford–AXYS (2007) The view ahead: identifying options for a visibility management framework for British Columbia. http://www.bcairquality.ca/reports/pdfs/view_ahead.pdf. Accessed 29 Oct 2011

Jacobson D (2011) Blog post of US ambassador to Canada. http://blogs.ottawa.usembassy.gov/ambassador/index.php/2011/03/13/march-13-2011-ottawa/. Accessed 29 Oct 2011

Melious J (2004) Transboundary air quality management models: options for western Canada/United States. Report prepared for Environment Canada

Ontario Pub Health Assoc (2003) Letter to International Joint Commission re 2002 Progress Report on activities under the 1991 Canada-United States Air Quality Agreement. http://www.ijc.org/rel/pdf/airquality/21_JLee.pdf. Accessed 29 Oct 2011

Pope C, Dockery D (2006) Health effects of fine particulate air pollution: lines that connect. Air & Waste Manage Assoc 56:709–742

RWDI Air Inc. (2008) Establishing a visibility goal for wilderness and urban areas in British Columbia and Canada. http://www.bcairquality.ca/reports/pdfs/visibility_goal_report_final.pdf. Accessed 29 Oct 2011

Smith R, Bizniaz S (1991) Beyond dispute: an air quality agreement in the context of a consultative relationship. Can-US L J 17:421–429

Springer A (2007) From Trail Smelter to Devils Lake: the need for effective federal involvement. Canadian-American Environmental Disputes. Am Rev Can Stud 37:77–94

Trail Smelter Arbitration (United States v. Canada) (1941) 3 R Int Arb Awards 1938, reprinted in Am J Int Law 35:684

United States and United Kingdom (1909) Treaty between the United States and great Britain Relating to Boundary Waters between the United States and Canada, 36 Stat. 2448

United States Department of State (2011) U.S. Canada Air Quality Agreement (AQA). http://www.state.gov/g/oes/env/83011.htm. Accessed 29 Oct 2011

United States Environmental Protection Agency (2011) Press release, twenty year anniversary 2991–2011. http://www.epa.gov/airmarkets/progsregs/usca/docs/anniversary.pdf. Accessed 29 Oct 2011

United States Clean Air Act, Title1, Part C, 42 USC §§ 7470–7479, 7491–92

VanNijnatten (2003) Analyzing the Canada-U.S. environmental relationship: a multi-faceted approach. Am Rev Can Studies 33:93–120

VanNijnatten D (2004) Canadian–American environmental relations: interoperability and politics. Am Rev Can Studies 34:649–664

17 Airshed Management

Norm Zirnhelt, Randolph P. Angle, D. Laurie Bates-Frymel, Monique Gilbert, Sonia Melancon, Natalie Suzuki and Rebecca Freedman

Abstract

Airshed management planning is a collaborative approach to air quality management usually involving a variety of stakeholders that includes the public, industry and local governments. The airshed management planning approach recognizes that poor air quality can often be the result of the cumulative impact of a multitude of activities and emission sources (regulated and unregulated), and this is often exacerbated by topographical and meteorological conditions that do not allow dispersion of pollutants. Airshed management planning processes in BC, Alberta and other parts of Canada are described with case studies ranging from small communities with small numbers and types of emission sources, to larger metropolitan areas with a variety of emission sources and complex air quality issues which require unique approaches to air quality management. The term *airshed* is defined, including a discussion of how airshed boundaries are delineated.

Keywords

Airshed · Airshed Management · Airshed plan · Air stakeholder · Airshed framework · Air quality management system · Airshed concept · Woodstoves · Cleaner woodstoves · BC Woodstove Exchange · Levoglucosan · Benzene · Community air quality

N. Zirnhelt (✉)
Cariboo Environmental Quality Consulting Ltd.,
Williams Lake, Canada
e-mail: norm@environmentalquality.ca

R. P. Angle
R. Angle Consulting, Edmonton, Alberta, Canada
e-mail: rangle2009@gmail.com

D. L. Bates-Frymel
Metro Vancouver, Canada
e-mail: laurie.bates-frymel@metrovancouver.org

M. Gilbert · S. Melancon
Ville de Montréal, Canada
e-mail: mgilbert@ville.montreal.qc.ca

S. Melancon
e-mail: sonia.melancon@ville.montreal.qc.ca

N. Suzuki · R. Freedman
BC Ministry of Environment, Victoria BC, Canada
e-mail: natalie.suzuki@gov.bc.ca

R. Freedman
e-mail: rebecca.freedman@gov.bc.ca

17.1 Introduction to Airshed Planning

An important tool used by communities to combat air pollution is airshed planning, a process to manage and coordinate activities to improve and protect air quality in a defined area or airshed. This approach recognizes that local air quality is influenced by a multitude of economic activities, emission sources and overlapping regulatory jurisdictions and therefore requires the support of numerous stakeholders (BC Ministry of Environment 2007).

Airshed management planning as described in this chapter is a collaborative approach to air quality management in a single community or a relatively small airshed, usually involving the public, industry, and local governments. An *airshed management plan* (AMP) outlines goals for improving air quality and makes recommendations on how to achieve them. It is one form of *air quality management* as discussed in Chap. 1.

E. Taylor, A. McMillan (eds.), *Air Quality Management*, DOI 10.1007/978-94-007-7557-2_17,

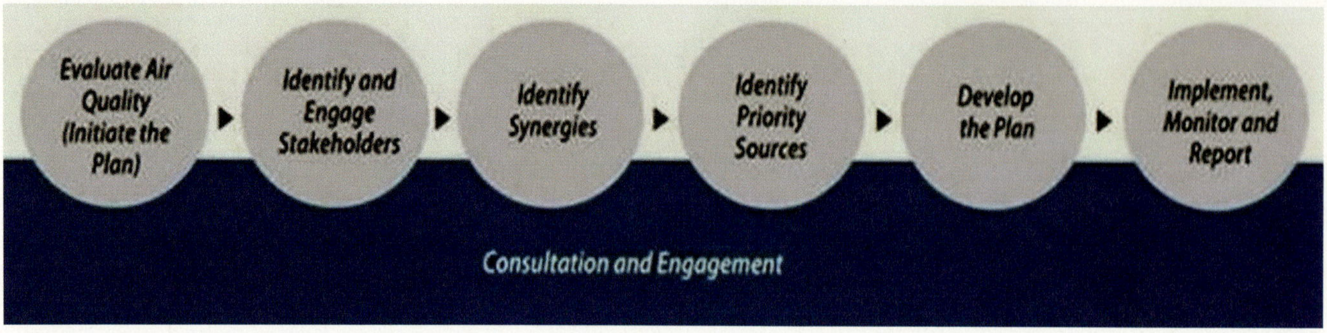

Fig. 17.1 One approach to the design and development of an airshed management plan. (Reproduced with permission from BC Ministry of Environment (2007))

This chapter will show how airshed management planning, primarily confined to western Canada, has evolved in some provinces. The chapter will also highlight how two large Canadian metropolitan areas, Metro Vancouver and the Montréal Metropolitan Community, manage air pollution within their boundaries as a result of their unique situation of having delegated air management authority from their provincial governments.

Airshed boundaries are defined differently for different situations.

- An airshed can be a limited geographic region such as a small valley community in which dispersion of air pollutants is limited by topographic constraints such as surrounding hills and water bodies. During stable, stagnant, light wind situations, these features can reduce the dispersion of pollutants emitted from local sources such as industry or wood stoves, leading to degraded air quality.
- An airshed can also be a large geographic area covering hundreds of square kilometers that, because of similar types of pollutant emissions, topography and meteorology, experience similar air quality issues.
- The boundaries of an airshed can also be based more on jurisdictional considerations such as municipal, regional or political borders. This occurs in heavily urbanized areas like Montréal or Metro Vancouver. This can also occur in vast, relatively level areas with no significant changes in land height, such as in the provinces of Alberta and Saskatchewan.

Air pollutants that impact human health and are therefore of interest in airshed management include:

- Particulate matter. This refers to extremely small solid and/or liquid particles, often including organic matter, which can be inhaled into the respiratory system. Several types of monitoring equipment can measure particulate matter less than 10 μm (millionths of a metre) in diameter (PM_{10}) and less than 2.5 μm in diameter ($PM_{2.5}$). Health professionals are becoming concerned about even smaller particulates that are less than 1 or 0.1 μm in diameter. The source of smaller particulates is usually combustion or industrial processes. Larger particulates are usually a component of dust from soils or industrial processes.
- Ozone. This is an irritating gas and a common constituent of smog.
- Nitrogen oxides. These pollutants aid in ozone production and are associated with vehicle traffic.
- Sulfur dioxide. This gas is associated with burning of wood, coal and other biomass as well as with oil and gas development.
- Hydrogen sulfide and total reduced sulfur compounds. These gases are associated with oil and gas development, petrochemical refining and other industrial activity.

Figure 17.1 illustrates in a general way how a typical airshed management planning process can be designed and implemented.

The need for airshed management planning has evolved in many instances because of the need to address sources of air pollution other than industrial sources. It is an important tool for communities and regional governments to coordinate activities affecting air pollution in an airshed. The airshed management planning approach recognizes that poor air quality can often be the result of the cumulative impact of a multitude of activities and emission sources (regulated and unregulated).

17.2 Airshed Planning Basics

Over the past twenty-five years, industrial permitting of point sources has been used as a means of regulating emissions in Canada. In British Columbia it has resulted in reductions in emissions from industrial point sources but the relative contributions from non-industrial sources that are largely unregulated have grown (Levelton 2009). Non-industrial emissions such as motor vehicle exhaust, domestic wood stoves, and all types of wood burning associated with human activity are significant sources of air pollution in BC (Province of BC 2008), hence the need for alternative approaches. Dust from roads and other commercial and industrial sources are also important air pollutants; however these sources primarily contribute to the coarse fraction of particulate matter which has a lower health impact.

The process of initiating an airshed management plan relies upon community support and/or awareness of the air quality problem. There also needs to be a local 'champion' to lead the process. Communities then determine whether an airshed management plan is needed, and if so, the scope of the plan. This will involve a preliminary air quality assessment to determine the kinds of emission sources and the concentrations of pollutants through monitoring.

If the factors contributing to an air quality problem are relatively few, then a simple plan can be developed in a relatively short time. For example, if the primary air quality issue is a contaminant from only one or two sources, permits from a regulatory authority requiring emission reductions may be implemented. If the contributing factors are complex, a more thorough air quality assessment involving detailed emission inventories, dispersion modelling, and identification of the prime sources of the most troublesome airborne contaminants will be needed for the development of an airshed plan. British Columbia's *Provincial Framework for Airshed Management Planning* (BC Ministry of Environment 2007) has provided guidance for this type of assessment.

17.3 British Columbia's Airsheds

There are approximately thirteen airshed management planning processes underway in British Columbia and these are at various stages of development or implementation. The airsheds represent diverse areas of BC (Fig. 17.2).

In the early 1990s, there was growing awareness in BC that regulation of point source industrial pollution often resulted in the unmasking of air quality problems resulting from *other* human activities previously considered minor contributors to air pollution (Levelton 2009). Mechanisms were needed to address these other air quality issues. In keeping with its commitment to implement Canada-wide Standards[1] for air quality and its provisions for Continuous Improvement and Keeping Clean Areas Clean, the B.C. Government developed a *Provincial Framework for Airshed Management Planning* (BC MOE 2007). This framework communicated a clear understanding of the BC government's expectations of an airshed management plan in terms of approach and content.

The Framework for Airshed Management Planning is founded on the following concepts:

- Shared stewardship through responsible planning and management of resources
- Sustainable development
- Integrated planning
- Continuous improvement and "keeping clean areas clean"
- Flexibility
- Principles of adaptive management

[1] Canada-wide Standards (CWS) are quantitative standards to reduce risks to human health that apply nation-wide—see the definitions for elaboration.

17.4 Case Study 1: Airshed Planning in Merritt, BC

Concentrations of particulate matter (PM_{10}) in the City of Merritt (population 7,000; Fig. 17.2) in the interior of British Columbia were once among the highest recorded in the province. Public meetings were held to identify the air quality problems and chart a course to resolve them through a stakeholder-run airshed committee. Stakeholders on the airshed committee included representatives of the forestry industry, government agencies and the community. An airshed management plan was subsequently developed and had early success toward measureable improvements in emissions. This may in part be due to the small size of the community and the relative ease of identifying emission sources, however strong airshed committee leadership, commitment of the public and all sectors of the local economy are also so very important (Levelton 2009, Taylor 2011).

A significant success of the airshed committee was the removal of beehive burners that burned wood chips and other waste, producing extensive smoke. Their removal was facilitated by the diversion of locally produced wood chips from a lumber mill to a power generating facility in a nearby city for incineration in a low emission burner. Other industry players recognized their contribution to a dust problem and paved large areas of their operations to reduce dust emissions. In order to reduce wind-blown dust, the municipal government reduced the amount of road sanding (used to improve vehicle traction) in the snowy winters, and increased spring street sweeping. Local news media were quick to commit to the plan and cooperated with other agencies to promote educational programs aimed at reducing residential emissions from domestic wood stoves and outdoor burning of yard waste.

Ambient pollutant concentrations naturally vary from year to year because of changing meteorological conditions. Therefore, while emissions have been reduced in Merritt, there is still insufficient data to determine whether ambient air quality has improved as a result of the airshed management plan (Taylor E 2011 Personal Communication).

17.5 Case Study 2: Airshed Planning in Quesnel, BC

Quesnel, a community that serves approximately 24,500 people, is located in a valley at the confluence of the Fraser and Quesnel Rivers 660 km north of Vancouver on the interior plateau of British Columbia (Fig. 17.2). At the north and south ends of the valley are plateaus which are at higher elevations than the downtown area. This creates a bowl that restricts atmospheric dispersion in Quesnel and leads to elevated pollutant concentrations. The Quesnel airshed and air quality monitoring stations are shown in Fig. 17.3.

Fig. 17.2 BC Airsheds. (Reproduced with permission from the BC Ministry of Environment)

Quesnel is primarily supported by forestry-based industries, mining, ranching and tourism. In 1998, air quality in downtown Quesnel was the worst in British Columbia out of 28 BC locations where air quality was monitored continuously. Studies determined that the Quesnel air quality problem was the result of local topography and the combined impact of a number of sources of air pollution ranging from small residential backyard burning to large industrial facilities. Local and provincial governments and local industry had taken measures to reduce air pollution for several years prior to the establishment of the Quesnel Air Quality Roundtable and the subsequent development of the Quesnel Airshed Management Plan (QAMP). This Roundtable included representatives from local industry, government, health authorities, environmental groups, and concerned citizens groups.

The air quality issue was complicated by multiple sources and types of emissions, so the Roundtable undertook a detailed air quality assessment. This consisted of four major components: air quality monitoring, an emission inventory, computer dispersion modeling, and pollutant source apportionment. The process can be visualized in Fig. 17.4.

Fig. 17.3 Quesnel airshed map showing air pollutant and meteorological monitoring sites. (Quesnel Air Quality Roundtable 2004, Reproduced with permission from the BC Ministry of Environment)

17.5.1 Quesnel Air Quality Monitoring

Air quality had been measured by a set of air quality monitors in Quesnel since the 1980s. The beginning of the air quality assessment process began with a comparison of these pollutant data to air quality objectives and guidelines that had been developed by the provincial and federal governments. Quesnel was in compliance with the provincial total reduced sulfur (TRS) objectives, as well as the Canada-Wide Standard for $PM_{2.5}$. However, the BC air quality objective for PM_{10} and both federal health reference levels were often exceeded in the community.[2]

The air quality monitoring system in Quesnel needed to be enhanced for the air quality assessment. This involved additional monitoring stations with state of the art equipment capable of monitoring fine particulates ($PM_{2.5}$). Locations of the Quesnel stations are shown in Fig. 17.3.

[2]At the time of development of the QAMP, BC did not have an air quality objective for $PM_{2.5}$, so the federal health reference levels were used as a rough guideline. Had the BC $PM_{2.5}$ objective been in place, Quesnel would have frequently exceeded it.

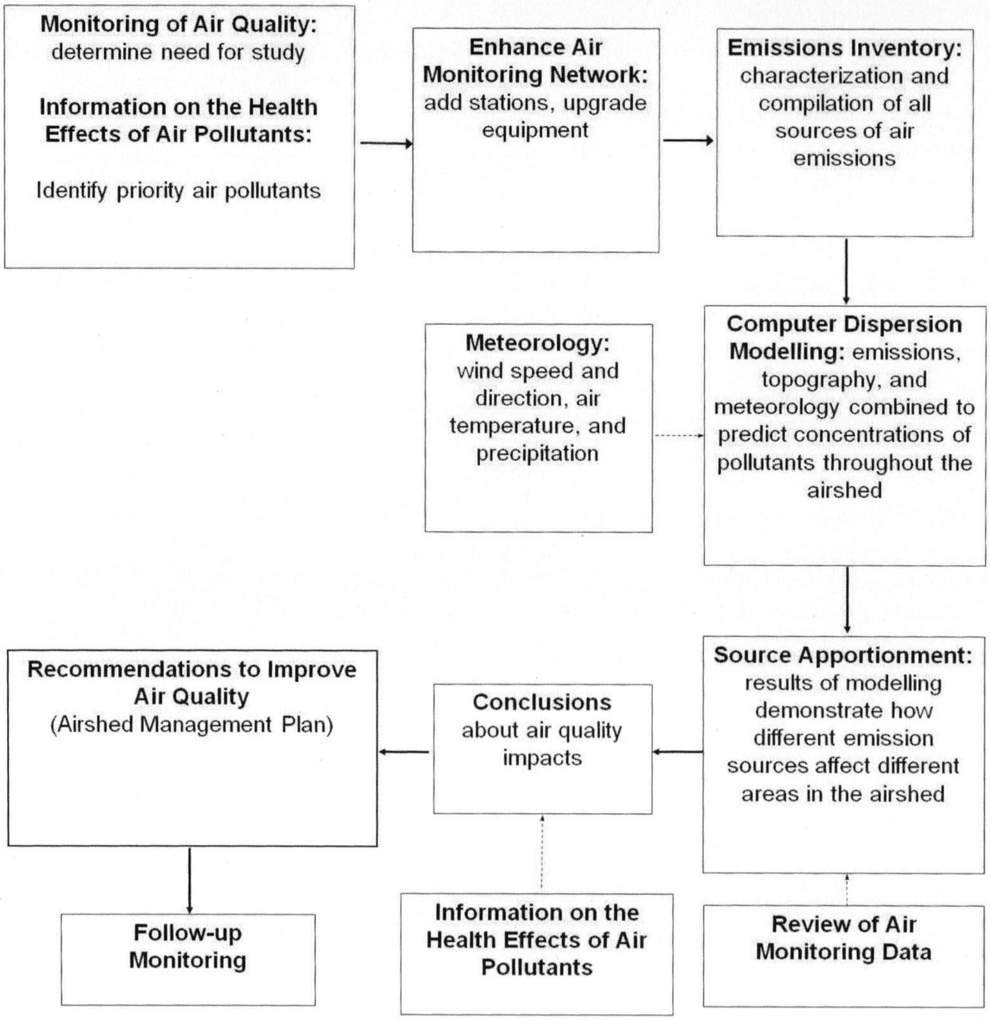

Fig. 17.4 Major components of the air quality assessment process in Quesnel, BC that formed the basis for the Quesnel Airshed Management Plan. (Quesnel Air Quality Roundtable 2004, Reproduced with permission from the BC Ministry of Environment)

17.5.2 Quesnel Emission Inventory

In order to determine the nature and quantity of substances being released into the atmosphere, emissions from all sources within the Quesnel airshed boundaries were inventoried for the year 2000. This included seven major source categories that emitted one or more of carbon monoxide (CO), nitrogen oxides (NO_x), sulfur oxides (SO_x), volatile organic compounds (VOC_s), total particulate matter (TPM), particulate matter less than 10 micrometres in diameter (PM_{10}), particulate matter less than 2.5 micrometres in diameter ($PM_{2.5}$) and Total Reduced Sulfur (TRS). The seven source categories included: industrial and commercial permitted sources, non—permitted commercial sources, mobile sources, residential sources, unpaved road dust, paved road dust, and natural sources.

Table 17.1 shows the total loading in tonnes per year of major pollutants from all emission sources in the Quesnel airshed (Quesnel Air Quality Roundtable, 2004). Table 17.1 also provides a breakdown for the PM_{10} and $PM_{2.5}$ for different emission sources (Quesnel Air Quality Roundtable 2004).

The emission inventory indicated that permitted industrial sources were responsible for the highest total loading of pollutants emitted in the airshed. Road dust was also a significant contributor to fine particulates. To estimate how these emissions affected pollutant concentrations throughout the community, dispersion modelling was initiated.

17.5.3 Quesnel Dispersion Modeling

When pollutants are discharged to the atmosphere their dispersion is complex. Concentrations at a given location at any hour will vary by pollutant type, the temperature of the emitted pollutants, time of day of the emission, and wind speed and direction, among other factors.

An air quality dispersion model (CALPUFF) was used to calculate the concentrations of pollutants listed in the previous section at various locations in the airshed. These estimates were then compared to pollutant concentrations measured at a number of ambient air quality monitors sited throughout the community to ensure that model estimates were within acceptable limits.

Table 17.1 Summary of air emissions by category for the Quesnel Airshed—2000. (Tonnes/year, Table reproduced with permission from the BC Ministry of Environment (Quesnel Air Quality Roundtable 2004))

	CO	NOx	SOx	TRS	VOC	TPT	PM_{10}	PM_{25}
Permitted sources	4488.39	1882.00	518.01	22.42	1039.43	3362.97	2027.76	1362.31
Commercial sources	469.52	21.90	2.38	0.00	249.49	192.78	44.05	31.61
Residential sources	1401.42	34.34	6.18	0.00	555.62	187.75	187.49	177.87
Natural sources	0.00	1.31	0.00	0.00	495.25	0.00	0.00	0.00
Mobile sources	4223.62	661.94	24.29	0.00	456.00	28.16	27.86	23.21
Paved road dust						3847.05	928.59	230.80
Unpaved road dust						1853.86	834.17	220.06
Total (Tonnes/year)	10582.95	2601.50	550.85	22.42	2795.80	9472.57	4049.93	2045.86

17.5.4 Allocating Pollutant Contributions from Each Source in Quesnel

The dispersion model was also used to estimate the individual contributions from each emission source (e.g. industrial, commercial, transportation, residential). The results of such modeling can be used to determine the most cost-effective method to improve air quality by calculating the effects of reducing emissions from each individual emission source. The modeling platform that was developed can also be used for future modeling work as new industrial projects are proposed for the airshed, or as the community emissions change. Other methods, such as receptor models, are available for emission source allocation but were not used in the Quesnel case.

The final air quality assessment determined that the primary pollutant driving poor air quality episodes in the Quesnel area was particulate matter (PM_{10} and $PM_{2.5}$). This includes particles from both natural sources (such as pollen, dust from soil erosion, forest fires, etc.) and from man-made sources (such as home heating devices, road dust, automobiles, open burning, and industrial processes).

Further details of the air quality impact assessment in Quesnel can be found in the Quesnel Airshed Management Plan (Quesnel Air Quality Roundtable 2004).

17.5.5 Recommendations for Improved Air Quality in Quesnel

This assessment found that no particular sector of the economy could solve the air quality problem on its own. Permitted industrial sources were generally in compliance with emission standards. However, the cumulative impact of both permitted and non-permitted emissions including transportation, dust sources and domestic wood stoves were degrading air quality in the community. As a result, a comprehensive set of recommendations was developed by the Roundtable, which became the essential elements of the airshed management plan. While the Merritt Airshed Management Plan was developed in 1 year, the Quesnel Airshed Management Plan took 5 years to complete.

The Roundtable recommended that the QAMP actions be implemented voluntarily by the parties concerned and implemented "as opportunities arise" over the 10-year implementation period. It was envisioned that there would be a declining trend over that period in concentrations of fine particulate. A key goal of the plan was to improve the air quality without job loss as a result of actions taken to implement the plan. Two fundamental reasons for the actions to be voluntary were that MOE permits were generally in compliance[3], and the community did not want sudden large emission reduction measures to negatively impact the local economy.

Prior to the development of the QAMP, there was considerable "polarization" on air quality issues in the City of Quesnel. This was particularly evident in the controversial permitting by the Ministry of Environment of a fiberboard plant in Quesnel in the mid 1990s (McMillan, D 2011 Personal Communication). Through the process of the formation of the Quesnel Air Quality Roundtable and the development of the QAMP, stakeholders began to work together and continue to do so with the implementation of the airshed plan. Airshed management planning, then, can be an effective tool for dealing with "polarization" over issues, by bringing stakeholders together to work collaboratively toward solving air quality issues.

After the completion of the three-year air quality assessment, twenty-eight recommendations were developed for improving air quality, specifically with regard to PM_{10} and $PM_{2.5}$. These recommendations pertained to reduction in emissions by all sectors: industry, municipal and regional governments, business owners, and local residents. Rather than relying on new legislation, many of the recommendations were based on public education and ranged from changes to backyard burning, home heating, improved dust control and reductions in emissions by industry. The recommendations are outlined in Sect. 7.2 of the Quesnel Airshed Management Plan (Quesnel Air Quality Roundtable 2004).

The Quesnel Airshed Management Plan was implemented in 2004 and air quality has shown an improving trend, although as of 2011 not all monitoring stations had achieved

[3] Compliance refers to a state of being in accordance with established conditions of an MOE air pollutant emissions permit.

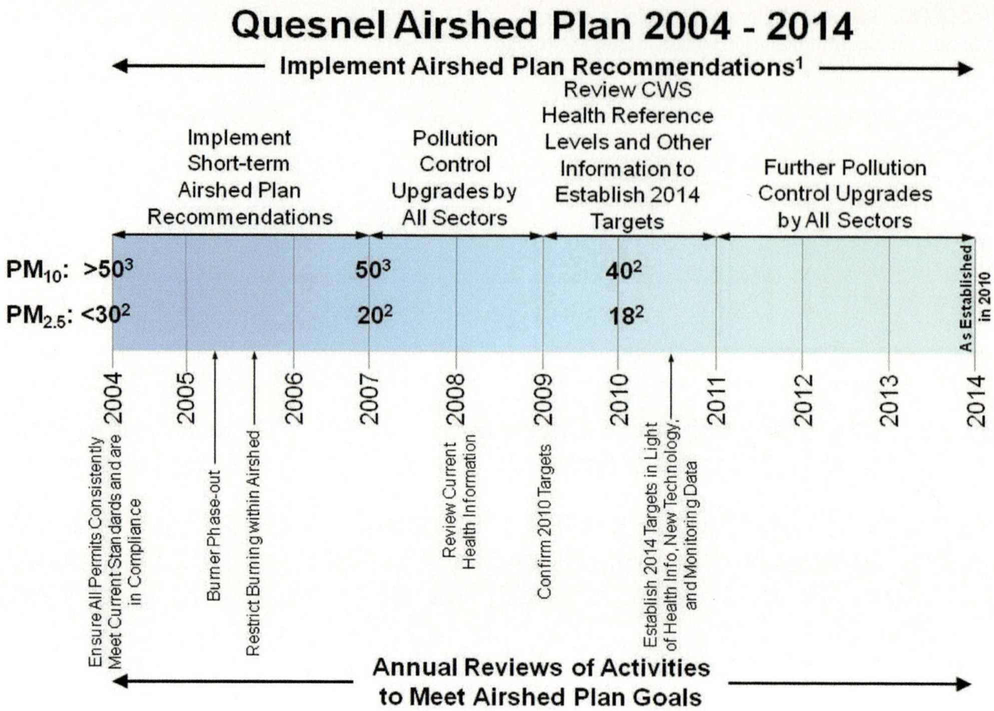

Fig. 17.5 Quesnel timetable to achieve clean air goals (Quesnel Air Quality Roundtable 2004, Reproduced with permission from the BC Ministry of Environment, [1]Incorporate into community and industry planning as opportunities arise. [2]$\mu g/m^3$: based on a 24-hour average, and achievement is based on the 98th percentile ambient measurement annually, averaged over three consecutive years (allows for seven days of exceedances each year). [3]$\mu g/m^3$: 24-hour average not to be exceeded)

the clean air goals of the airshed management plan (Zirnhelt 2011). This is illustrated in Fig. 17.5 and Fig.17.6.

17.6 Case Study 3: Airshed Planning in Metro Vancouver, BC

Vancouver and its surrounding municipalities differ from the previous two case studies in that Metro Vancouver[4] has been delegated the authority by the provincial government to manage air quality within its boundaries. This is similar to the City of Montréal, discussed in Sect. 17.9.

Metro Vancouver comprises twenty-two municipalities, one electoral area, and one treaty First Nation. The population of this region totals about 2.4 million people, more than half of the population of British Columbia. It lies in the western portion of the Lower Fraser Valley (Fig. 17.8) in the southwest corner of British Columbia. The Lower Fraser Valley is bordered on the north and east by mountains, in the south by the United States, and in the west by the ocean.

17.6.1 History of Air Quality in Metro Vancouver

Although Metro Vancouver generally enjoys good air quality, that wasn't always the case. In the 1930s and 40s there were few controls on industrial emissions and the waterfronts of False Creek and the Fraser River were lined with industrial stacks and beehive burners (Fig. 17.7). At times, air pollution in the area was so severe that sunlight was obscured. However, local air quality began to improve over the years as increasing pressure for residential development eventually pushed the heavy industry out of this area.

Metro Vancouver was formed in 1967 as the Greater Vancouver Regional District, one of 29 regional districts in the province of British Columbia. Four years later the Government of British Columbia issued the *Pollution Control Act* (subsequently replaced by the *Waste Management Act*, now the *Environmental Management Act)*. In 1972 the *Act* and the *GVRD Letters Patent* were amended, giving Metro Vancouver the authority to (a) administer certain provisions in the *Act*, (b) assume municipal pollution control activities, and (c) establish a regional ambient air monitoring program. Amendments to the *Waste Management Act* in 1992 confirmed Metro Vancouver's authority to manage air quality within its' jurisdiction, to pass Bylaws requiring air quality permits for stationary sources and to enforce emission limits that were more stringent than the provincial *Pollution Control Objectives*. In 1991, Canada and the United States entered into an agreement to address transboundary air pollution focused on nitrogen and sulphur oxides (later expanded to include ozone, see Ch. 16). This agreement was particularly relevant for the Lower Fraser Valley airshed, which includes Metro Vancouver, the Fraser Valley Regional District and Whatcom County in Washington State.

[4]Although legally still known as the Greater Vancouver Regional District, the Board changed the organization's name to "Metro Vancouver" in 2007.

Fig. 17.6 $PM_{2.5}$ levels in Quesnel relative to Airshed Management Plan goals for three $PM_{2.5}$ monitoring locations in Quesnel. See Fig. 17.3 for monitor locations. (Quesnel Air Quality Roundtable 2004, Reproduced with permission from the BC Ministry of Environment)

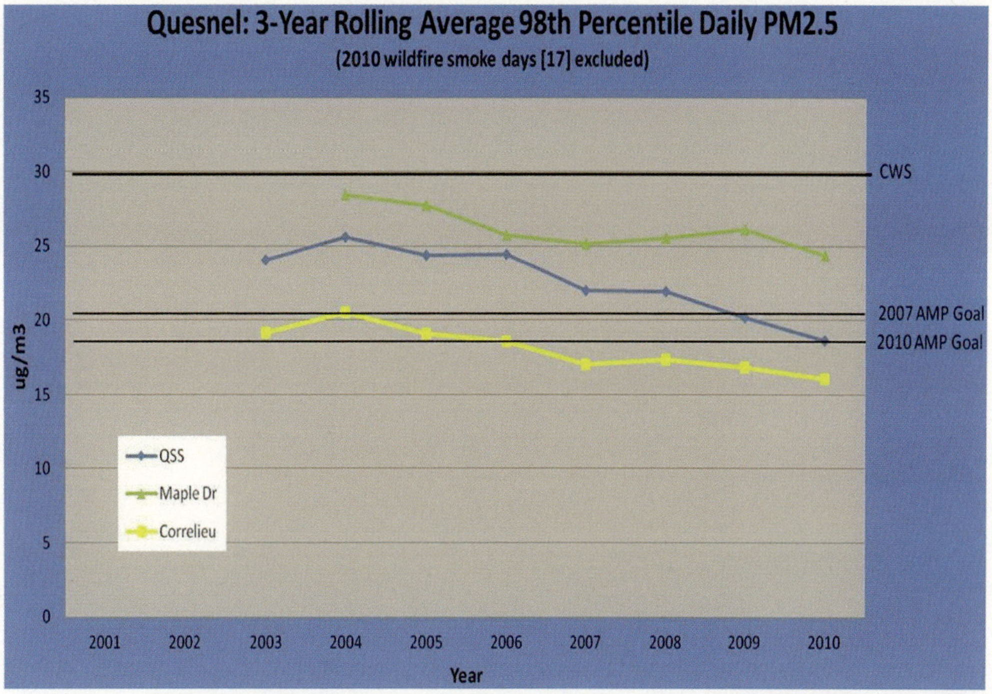

Fig. 17.7 Sawmill plumes in the Metro Vancouver Region circa 1940

In the late 80s and early 90s automobile use increased rapidly in Metro Vancouver and the region began experiencing frequent summertime ground-level ozone episodes during periods of hot, sunny, and stagnant weather. Metro Vancouver led a five-year planning process which involved the assessment of regional air quality trends, compilation of a detailed "bottom-up" emission inventory and forecast, as well as airshed modelling. These studies concluded that without concerted emissions curtailment, air quality in the Lower Fraser Valley airshed would become progressively worse. With this information in hand, Metro Vancouver consulted with stakeholders, collaboratively developed a set of emission reduction measures for various sources within the region, and assessed the costs and benefits of each measure. In 1994 Metro Vancouver became the first major urban centre in Canada to approve an Air Quality

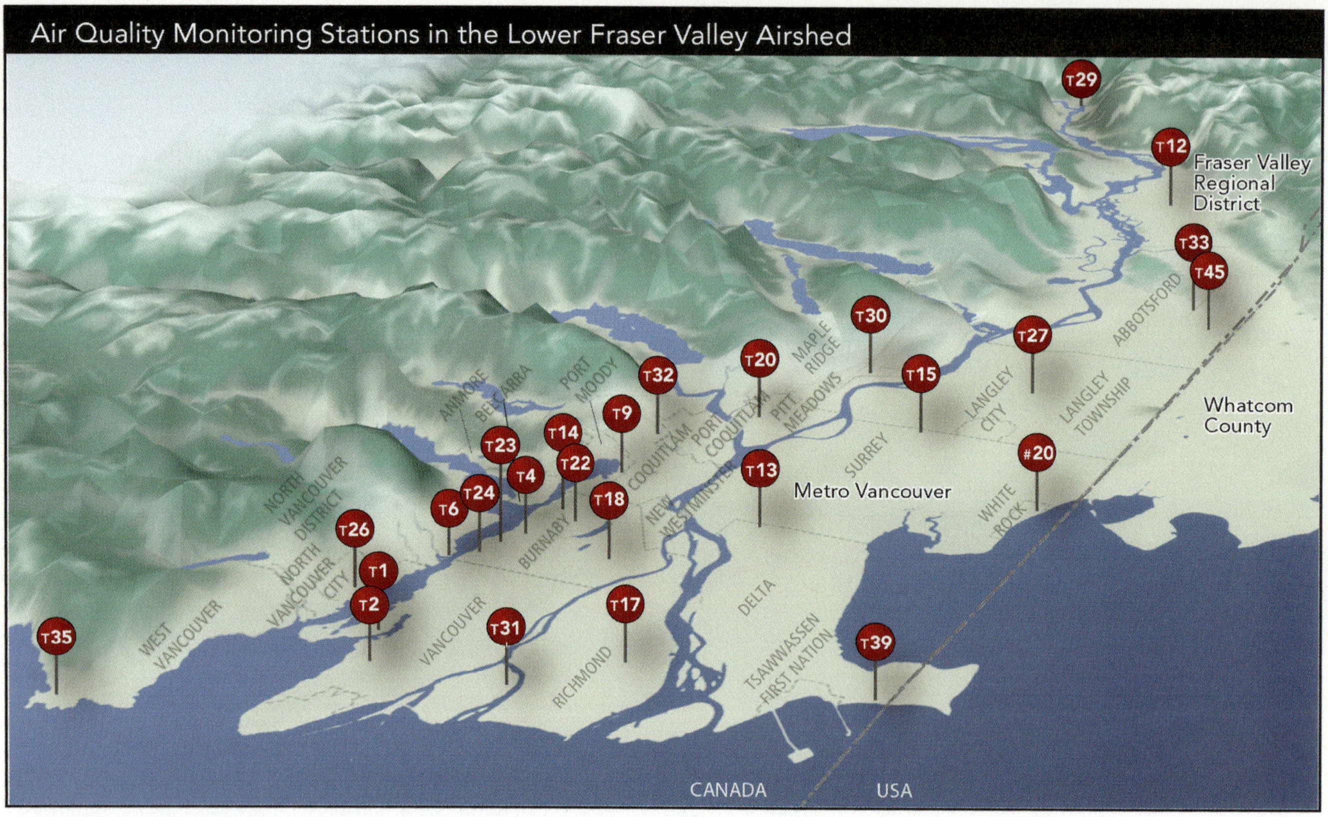

Fig. 17.8 The air quality monitoring network in Metro Vancouver and the Fraser Valley Regional District in 2012

Management Plan (AQMP), complete with fifty-four sector-specific emission reduction measures and a detailed implementation schedule. The 1994 AQMP achieved several key milestones, including:

- The first light-duty vehicle emission testing program in the world to impose a mandatory NO_x limit under dynamic conditions—the AirCare Program. The chapter on Transportation Emissions in this book contains more details on AirCare
- The first smoke testing program for heavy-duty vehicles in Canada—the AirCare On-Road Program.
- Some of the most stringent requirements for industry in Canada (at the time), including:
  - Stage 1 vapour recovery systems at all levels of the gasoline distribution system, including bulk petroleum storage tanks and loading, service stations and tanker trucks
  - Installation of selective catalytic reduction technology to meet stringent nitrogen oxides emission limits for natural gas-fired boilers at the Burrard Thermal Power Generation Plant, and
  - Installation of a comprehensive emission control program at the region's municipal solid waste incinerator.

In addition, Metro Vancouver also became one of the first agencies in Canada to implement the "polluter-pay principle" by requiring that permitted emission sources under its jurisdiction pay fees for designated pollutants.

The implementation of these and other actions achieved a region-wide reduction in total emissions of carbon monoxide, nitrogen oxides, volatile organic compounds, sulfur oxides and particulate matter emissions by 38% (57% per capita) between 1985 and 2000.

Metro Vancouver adopted its second AQMP in October 2005. Some key accomplishments within the term of the 2005 AQMP include:

- Progressive actions on marine vessels and port operations including new International Maritime Organization regulations that will see our coastal waters designated as an "Emission Control Area", the implementation of shore power at the Canada Place cruise ship terminal, an Integrated Northwest Ports Clean Air Strategy for the ports of Vancouver, Seattle, and Tacoma as well as improved collaboration with Port Metro Vancouver.
- Innovative actions to reduce harmful diesel particulate matter emissions including a new Non-Road Diesel Engine Emission Regulation.
- Initiation of a pilot program in the Lower Mainland to improve visual air quality.
- Enhancement of the tools to support measurement of performance and progress towards the region's air quality

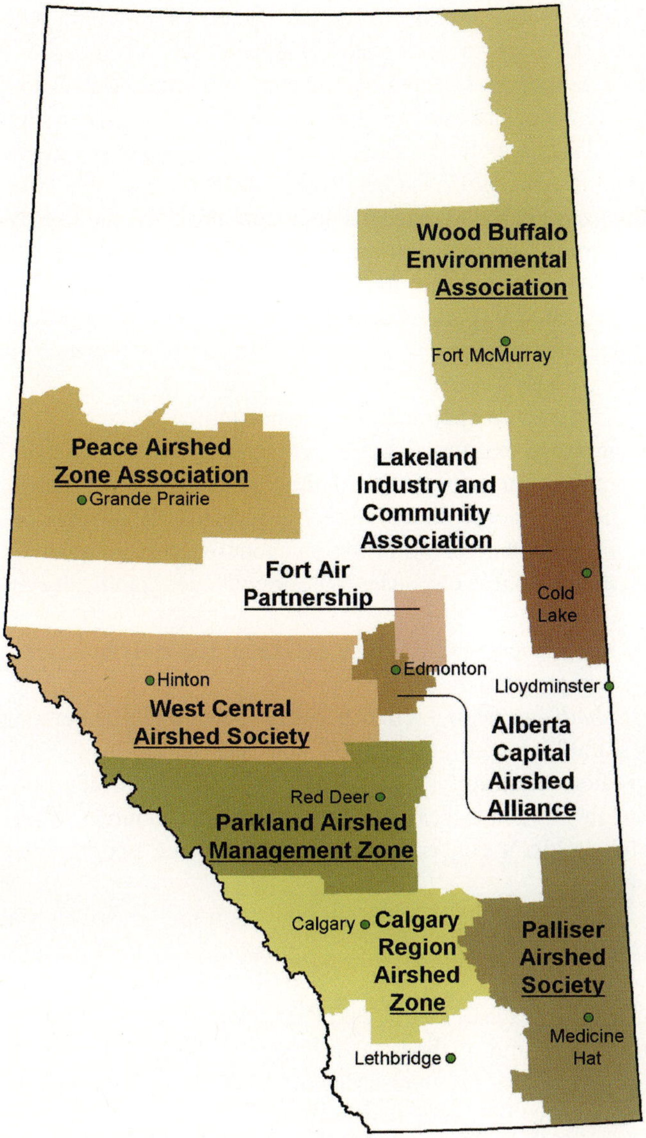

Fig. 17.9 Alberta's nine *airsheds* in 2009 (Reproduced with permission from Alberta Environment and Sustainable Resource Development)

management goals, including improvements to the ambient air quality monitoring network and the regional emissions inventory.

To the north of Vancouver, Whistler Village has developed an Integrated Energy, Air Quality and Greenhouse Gas Management Plan. This is the first planning document in Canada to integrate these three approaches to air quality management. In Whistler, more than 90% of air pollutants come from energy consumption, so that managing energy and GHG emissions can largely address air quality issues. The Whistler plan identifies a number of co-management opportunities as well as a streamlined approach to implementation (Municipality of Whistler 2004).

17.6.2 The Current Metro Vancouver Airshed Plan

Since the region's first plan was adopted in 1994, the population of Metro Vancouver has increased by over 30% to nearly 2.4 million residents. Although air quality in Metro Vancouver has improved since then, maintaining healthy air quality will be challenging with continued growth in population, trade and transportation.

In October 2011 Metro Vancouver adopted its third plan—*the Integrated Air Quality and Greenhouse Gas Management Plan* (Metro Vancouver 2011). This plan establishes three primary goals:

- Protect public health and the environment
- Improve visual air quality
- Minimize the region's contribution to global climate change

To achieve these goals, the 2011 plan identifies twelve strategies and eighty-one actions for Metro Vancouver and its partners. Key actions in this plan include implementing further diesel particulate emission reduction measures, developing and implementing a visual air quality improvement program, working with partners to address short-lived climate forcers, increasing public awareness of air quality and climate change issues, and exploring the establishment of a climate action fund to support regional greenhouse gas emission reduction efforts. The plan also establishes a set of performance measures for each goal and emission reduction targets for greenhouse gases and diesel particulate matter. Taking an adaptive management approach, Metro Vancouver will assess progress against these measures every two years and conduct a comprehensive review of the plan every five years, adapting programs as necessary.

To track performance, Metro Vancouver generates detailed emission inventories and forecasts every five years. In addition, Metro Vancouver operates the Lower Fraser Valley Air Quality Monitoring Network in partnership with Environment Canada, the Government of British Columbia, the Fraser Valley Regional District and several other partners. This extensive network consists of twenty six air quality monitoring stations from West Vancouver to Hope (Fig. 17.8).

Metro Vancouver's air quality regulatory program administers regional bylaws, including the issuance of permits, compliance promotion and enforcement of emission requirements for industrial, commercial and institutional stationary sources. Recognizing that Metro Vancouver does not have regulatory authority over some significant air emission sources in the region, such as marine vessels and on-road vehicles, Metro Vancouver works in close collaboration with other government agencies and coordinates these efforts through committees like the Lower Fraser Valley Air Quality Coordinating Committee, the Georgia-Basin Puget Sound International Airshed Strategy and the Regional Engineers Advisory Committee—Climate Protection Subcommittee.

17.7 Alberta Airshed Planning

Contributed by R. P. Angle

17.7.1 Introduction to Alberta Airsheds

The term "airshed" arises as a parallel with "watershed." A "watershed" is the geographic area of land that drains water to a shared destination. A watershed is bounded by heights of land that separate it from other watersheds. Unfortunately, air does not flow to a shared destination and there are no equivalents to "heights of land" in the atmosphere. In a generic sense an "airshed" has been defined in a number of different ways mainly to imply a geographic territory:

1. a region sharing a common flow of air, which may become uniformly polluted and stagnant;
2. an area, bounded by topographical features such as hills or water bodies, within which airborne contaminants can be retained for an extended period;
3. a part of the atmosphere that behaves in a coherent way with respect to the dispersion of emissions;
4. a geographic boundary for air quality standards;
5. the geographical area associated with a given air supply;
6. an area whose air quality is influenced by a set of emission sources;
7. the area responsible for emitting a significant portion of air pollution reaching a target region;
8. a specified volume of air with similarities in climate, weather and topography;
9. An area characterized by air with common qualities.

The well-defined boundaries of a "watershed" provide a convenient management unit for water. An "airshed", except in mountainous areas, does not have clearly defined natural boundaries. Air management units are therefore defined somewhat arbitrarily.

In Alberta the term "airshed" is used to denote an air quality management unit operated by a not-for-profit corporate entity. Hereafter in this section the italicized word *airshed* is used to denote an air management unit, and the word in normal font is used in reference to a geographical area.

17.7.2 Origins of Alberta Air Management Units: Airsheds

The 1991 Clean Air Strategy for Alberta (Advisory Group 1991) identified local air quality issues and problems as a priority and hence announced a goal to "develop and implement a zone approach to managing air quality within specific airsheds." Moving toward implementation of this recommendation a Working Group (1993) explored the establishment, function, operation and common needs of Zone Air Quality Management Systems. After the Clean Air Strategic Alliance (CASA)—a multi-stakeholder association composed of representatives selected by industry, government and non-government stakeholders—was formed in 1994, one of their first tasks was the production of zone management guidelines (CASA 1995a) to help stakeholders who wanted to set up a zone in their area. This document provided the foundation for the subsequent formation of *airsheds* in Alberta. The West Central Airshed, the first of the nine Alberta *airsheds*, was formed in parallel with the development of the CASA guidelines and there was considerable interplay between the two groups.

The *airshed* approach was intended to:

- improve existing air quality monitoring in the region and make local and regional monitoring systems more efficient,
- collect data to address specific regional air quality concerns, and obtain information about regional air quality,
- make flexible responses to issues with geographically targeted solutions,
- share responsibility, engage stakeholders and obtain more comprehensive and reliable data.

Airsheds were to facilitate open communication among stakeholders and create good will and credibility through greater community involvement. *Airsheds* were appropriate in circumstances of multiple emission sources where stakeholder's concerns were limited to one geographical area.

Although Alberta has a variety of terrain—relatively flat prairie along the eastern border, mountains along the southwestern border, and some deep river valleys—the topographical features do not lend themselves to a simple definition of natural airsheds. Consequently, *airshed* boundaries are determined by a consensus of stakeholders considering a complex set of geopolitical factors:

- Landforms, watersheds, climate, animal behavior patterns
- Wind, temperature stratification, turbulence, deposition patterns
- Boundaries of municipalities, national parks, and First Nations and aboriginal communities
- Effects of emissions on visibility, vegetation, animal and human health; chemical content of water, soil and plant/animal tissues
- Emission sources, volumes, types, and dispersion patterns
- Boundaries of existing zones.
- Land use patterns, and the type and number of industrial users and other stakeholder organizations in the area.

When the first *airshed* was formed, an important input to the boundary selection was a series of Alberta-wide concentration and deposition maps generated by a regional-scale atmospheric dispersion model (Cheng 1994). These maps and additional model runs also informed the boundary selections made by subsequent *airsheds*. In 2009, as shown in Fig. 17.9, there were nine *airsheds* operating in Alberta, eight collecting monitoring data, and one planning for air quality management in the Capital Region.

17.7.3 Role and Operation of Airsheds in Alberta

The purpose of Alberta *airsheds* is air quality management within a defined zone to address local air quality problems. The issues were addressed by developing and implementing an appropriate management plan, initially focused on collecting credible scientific information about air quality and potential effects. Establishing an *airshed* involved the assessment of a wide range of information about impacts, existing ambient air quality, current and projected emissions, modelled or predicted air quality, and ambient air quality trends. Alberta *airsheds* addressed such local issues as: soil acidification, pollution stress on vegetation, livestock health, visibility, odours, human health, water acidification, ozone, heavy metal deposition and accumulation, smoke and dust, cumulative effects on air quality, and air quality during industrial upset conditions.

Before the creation of *airsheds*, the operator of each emission source was required to monitor ambient air quality for one or two major pollutants at points of maximum predicted concentration. *Airsheds* allowed more comprehensive measurements to be made at locations of interest to the local community. *Airsheds* were seen as ways to improve the effectiveness and efficiency of air quality monitoring, and to raise the credibility of the results communicated. The strong emphasis on monitoring was reflected in the revised CASA *airshed* guidelines (CASA 2004). Table 17.2 provides summary information about the nine Alberta *airsheds*.

The Particulate Matter and Ozone Management Framework (CASA 2003), designed as Alberta's implementation mechanism for the Canada-Wide Standards for PM and Ozone, made *airsheds* responsible for three tiers of air management actions depending upon the magnitude of ambient pollution levels at monitoring stations within the zone. In 2006, Alberta Environment notified five *airsheds* that the planning trigger of the framework had been reached in their areas; consequently there was a need to develop air quality management plans. The three affected *airsheds* in northern Alberta formed a partnership to develop a management plan for the Edmonton and surrounding area (Capital Airshed Partnership 2008). The Parkland Air Management Zone developed a management plan for the Red Deer region (Stantec Consulting 2008) and the Calgary Regional Airshed Zone (CRAZ 2008) developed a plan for the Calgary region.

17.7.4 Alberta Airsheds: Membership and Operation

Alberta *airsheds* are grassroots organizations formed when a group of like-minded individuals and organizations comes together with a common interest in air quality and joint action. The following steps have generally been followed in setting up an Alberta *airshed*, but not always in this order (after CASA 2004):

1. Gather background information (air quality issues, impact studies, emissions inventory, trends in air quality, dispersion modelling results).
2. Incorporate as a society under the Alberta Societies Act (governance structure, members, bylaws, operating policies, conflict resolution).
3. Prepare a business plan (goals, objectives, targets, funding, planning processes).
4. Prepare a monitoring plan (objectives, locations, types of equipment, industry requirements, quality control/quality assurance).
5. Prepare a communications plan (purpose, issues, emissions, general geography, monitoring plans, public access to data).
6. Seek endorsement from the Board of Directors of the Clean Air Strategic Alliance.

Typical members of an *airshed* include: local municipalities, counties, and municipal districts; associations or companies representing major emission sources; local business groups; local agricultural groups; environmental organizations; academic or research-oriented organizations; community-based groups comprising citizens at large; regional health authorities; local aboriginal communities; Alberta Environment and other provincial government departments or agencies; federal government departments or agencies; public at large.

The Board of Directors of an *airshed* is generally selected from the broader membership which usually comprises representatives of the three stakeholder groups recognized by the Clean Air Strategic Alliance—industry, government, and non-government organizations (both human health and environmental). In some instances, each member of the Society is also entitled to name one Director to the Board, although large boards can become unwieldy.

The decision-making process for Alberta *airsheds* is based on the same collaborative and consensus process used by the Clean Air Strategic Alliance (CASA 2007). This process brings people together to address the interests or concerns that underlie all positions on an issue. The goal is to find solutions to the problems faced by each party so that all participants can agree to a set of recommendations or actions. Agreements reached by consensus are likely to be more innovative and long lasting than those reached through traditional negotiation processes.

Table 17.2 Overview of Alberta's nine airsheds

Name of airshed	Year of formation	Main issue	Area (km^2)	Population
Alberta capital airshed alliance	2006	Comprehensive air quality management in the region	4500	960,000
Calgary regional airshed zone	2005	Potential effects of projected population growth	29,900	1,124,300
Fort air partnership	1997	Cumulative effects and industrial upsets	4,500	76,800
Lakeland industry and community association	2000	Timely response to concerns about industrial development	18,000	30,000
Palliser airshed society	2003	Potential increase in NO_x from industrial expansion	40,000	100,000
Peace airshed zone association	1999	Relevant, scientifically credible information to stakeholders	38,500	85,000
Parkland airshed management association	1997	Concerns of area residents—human and livestock health, odour, ozone	42,000	260,000
West central airshed society	1995	Potential impact of air quality on soil acidity, crops and forests	46,000	113,000
Wood buffalo environmental association	1996	Potential vegetation and health effects related to oil sands extraction	68,500	104,300

In Alberta all permitted sources with substantial emissions to the atmosphere are required as a condition of their permit to conduct ambient monitoring, both continuous and passive. To date depending on the emissions, this compliance monitoring requirement has ranged from: (a) one to five continuous monitoring stations for one to three pollutants, and (b) two to forty passive monitors for one or two pollutants. In 1995 it was estimated that $ 12,900,000 was spent annually on ambient air quality monitoring (CASA 1995b). The Alberta Department of Environment supported the formation of *airsheds* by allowing the compliance monitoring to be replaced by the regional monitoring in an *airshed* under certain performance conditions (Sandhu 2000). Some *airsheds* also assumed responsibility for compliance monitoring on behalf of the permit holder. All monitoring is done in accordance with the provincial Air Monitoring Directive (Alberta Environment 2006).

Each *airshed* is responsible for obtaining and managing its own funds. The primary source of funding is the reallocation of monitoring resources and other environmental spending by the operators of emission sources in the area. *Airsheds* may develop a funding formula based on annual emission amounts, production levels, or other criteria. Pooling financial resources in this way allows for more comprehensive air quality monitoring and may include appropriate biomonitoring to address the issues in the airshed. Each *airshed* reports data on its own website in addition to transferring it to the central CASA Data Warehouse (www.casadata.org) on a monthly basis. *Airshed* websites differ in complexity, content and display techniques. One *airshed* does not operate a monitoring network and serves exclusively as a regional air quality planner. Table 17.3 summarizes several aspects of *airshed* operations in 2009.

Alberta's nine *airsheds* have come together to form The Alberta Airsheds Council (AAC 2009) to identify and advocate for common interests of Alberta *airsheds* and facilitate cooperation among *airsheds* and between *airsheds* and others, including government agencies. The Airsheds Council is also developing a closer working relationship with the Clean Air Strategic Alliance. Recent environmental management initiatives by both federal and provincial governments create both challenges and opportunities for Alberta *airsheds* (Angle 2010).

17.8 Saskatchewan Airshed Planning

The Province of Saskatchewan plans to establish a number of *airsheds* to manage air quality. The first *airshed,* the Southeast Saskatchewan Airshed Association (SESAA), was established in 2005. Key economic activities in this airshed include agriculture, oil and gas exploration and operation, power generation, mining, and transportation (Southeast Saskatchewan Airshed Association 2011). The airshed encompasses an area of approximately 36,800 sq km in the southeastern area of the province and includes 45 municipalities. The airshed boundaries were established based on common history, meteorology, and funding considerations. It was incorporated as a non-profit organization of public, industry, government, and non-government members (SESAA 2011).

The goal of the SESAA was to collect credible, scientifically defensible air quality data for the southeast Saskatchewan region, and to make this data freely available to all stakeholders. The *airshed* has brought together stakeholders from all backgrounds to identify local air quality issues and to develop innovative solutions for managing these issues.

A second Saskatchewan *airshed*, the Western Yellowhead Air Management Zone, has recently been established to meet similar goals to the SESAA. Several other Saskatchewan *airsheds* are planned.

Table 17.3 Summary of Alberta airshed operations

Airshed	Continuous monitoring stations	Passive stations	Biomonit-oring sites	PM and Ozone planning (Y/N)	Members	Expenditures (CAD dollars 2009)
Alberta capital airshed alliance	0	0	0	Y	18	134K[a]
Calgary regional airshed zone	3	0	0	Y	34	321K
Fort air partnership	8	57	0	Y	33	749K
Lakeland industry and community assn	4	25	1	N	12	344K
Palliser airshed society	2	20	0	N	61	203K
Peace airshed zone association	6	48	0	N	66	580K
Parkland airshed management assn	4	34	0	Y	60	746K[a]
West central airshed society	13	14	7	Y	64	840K
Wood buffalo environmental assn	15	40	20	N	27	8,389K

[a] 2008 amounts

17.9 Montréal, Quebec, Airshed Management

Contributed by Monique Gilbert and Sonia Melançon

Air quality regulations of the province of Quebec do not apply in Montréal. Montréal has the mandate to manage air quality within its boundaries through ambient air monitoring and By-Law enforcement. Industrial and commercial sectors are required to meet the emission and ambient air quality standards for various pollutants stipulated in applicable By-laws. There are air quality standards for a list of three hundred and seventy-six specific pollutants in Montréal. It is stipulated that any company emitting a polluting agent into the atmosphere must have a permit to do so and must comply with the bylaw requirements and standards. Random inspections are conducted by the Division of the "Contrôle des rejets industriels" and companies not meeting the regulatory requirements must make the necessary adjustments.

Montréal is unique, not only defined by its cultural diversity and urban livelihood but also by its air quality issues. It is a great challenge to maintain and improve air quality in Montréal because of its dense population, with residential areas interspersed with industrial, commercial and institutional activities; heavy traffic and precious natural areas (Fig. 17.10).

As the early hub for industrial and commercial activities in Canada, Montréal has been active in air quality control since 1872. In these earlier times the main pollutant of concern was coarse particulate emitted from the combustion of coal and bunker oil used for heating and steam engines, causing health concerns as well as soiling.

Montréal By-Laws evolved along with urban and industrial development, providing a structure enabling coexistence of all activities within one shared air resource. The By-Laws now have emission limits and ambient air standards for hundreds of pollutants and various durations: 1-hour(h), 8-h, 24-h, 1 year as well as provisions limiting emission of air toxics as well as nuisance smoke, odour and dust from a variety of industrial and commercial activities. These activities include petroleum, petrochemical, chemical, pharmaceutical, aeronautical, printing, textile, food and beverage industries, metal smelting, purifying, casting and electroplating, painting, wood transformation, quarries, cement and asphalt plants, incinerators and landfills for waste elimination, fur tanning, rendering plants, construction and maintenance of buildings and roads.

Close proximity of the human population to these industrial and commercial activities requires that all operations, regardless of their size, are subject to the same By-Law requirements. Similarly, with a few technological exceptions, there are no general grandfather clauses. That is, older industrial and commercial activities are subject to the same appropriate schedule of compliance as new activities.

When the Province of Québec's Ministry of the Environment was created in 1979, it recognized Montréal's involvement in air quality management within its jurisdiction by transferring some regulatory powers and responsibilities and exempting Montréal from Quebec's Air Quality Act. The provisions in the Montréal's By-laws are historically more stringent, by necessity, than those in the Provincial Act. Montréal's air quality management includes ambient air quality monitoring (since 1967) and By-Law enforcement activities. The latter includes permit issuance and inspections of emitting industries (including source sampling and analysis), corrective requirements and prosecuting as required. Air quality monitoring stations are located so as to assess all types of land usage and exposures (background, residential, heavy traffic, downtown, industrial areas, etc.). The case studies in this section provide examples of the successful coordination of ambient air monitoring and By-Law enforcement in Montréal.

The public is kept informed on air quality on a real time basis through Montréal's ambient air monitoring network's

Fig. 17.10 Montréal Islands are home to 2 million people living on a 500 km^2 territory, 76 % urbanized. (Photo courtesy Ville de Montréal)

website www.rsqa.qc.ca and on regulatory enforcement activities from www.ville.montreal.qc.ca/environnement. It is possible to lodge air pollution complaints through the Website or a hotline and inspectors report back after investigation.

As air quality, science and health concerns evolved, so did the monitoring network and By-Law provisions to address current priorities. Now more than ever participation of the entire society is essential to further reduce pollution from various stationary and mobile sources of emission.

A regulatory prohibition of installation of solid-fuel-burning equipment other than certified wood pellet appliances was instituted in 2009 in Ville de Montréal's 19 boroughs (see Sect. 17.9.2 for further discussion on woodstove impacts and regulation in Montréal).

The Montréal Community Sustainable Development Plan 2010–2015 has the Ville de Montréal, partner organizations and local administrations working together to collectively carry out and achieve specific objectives and actions, one of the five main orientations being the improvement of air quality and reduction of emissions (Ville de Montréal 2010).

Through constant personal awareness and commitment, citizens can make informed decisions to reduce the environmental impact of their actions to make a breathable difference!

17.9.1 Montréal Benzene Case Study

Montréal has a history of high concentrations of benzene in the east near an area of petroleum refineries and petrochemical plants. The air quality monitoring station located in this area has been known to have the greatest concentrations of benzene in Canada and it held that title for many years. Benzene is classified as carcinogenic to humans by the International Agency for Research on Cancer (IARC). Even exposure to low levels of benzene can cause a form of leukemia.

In 1995, Montréal asked refineries and petrochemical plants to implement measures in order to reduce their benzene emissions on a voluntary basis. A year later additional provisions to By-Law 90 were promulgated; By-Law 90–3 requires vapor recovery of organic compounds (OCs) emitted at gasoline service stations and storage tanks, and allows a limit of 35 mg of total OCs emitted per liter of gasoline transferred. Annual tests are mandatory to check for compliance.

Since July 1999, sale of gasoline containing benzene at a concentration exceeding 1.0 % by volume is prohibited in Canada.

By-Law 90 was further amended in 2001 (By-Law 90–6) with provisions for detection and repair of fugitive emissions from equipment in refineries, petrochemical and chemical industries. Improvement of floating roof tanks (double seal) was also part of this amendment as well as storage tanks vapor recovery for highly volatile compounds. Later on, refineries installed covers to enclose vapors from separators in their wastewater treatment systems.

All these corrective actions and regulatory updates resulted in an impressive 75 % reduction in benzene concentrations at the local air quality monitoring station from the 1997 annual 24-h average of 11.4 ug/m^3 to 3.0 ug/m^3 in 2002. Local ambient concentrations then increased for a couple of years due to establishment of a chemical plant in the neighborhood. Corrective actions were implemented and benzene levels resumed their downward trend, and appeared to stabilize at an annual 24-h average of 1.86 ug/m^3 by 2010 (Fig. 17.11)

Data from a continuous gas chromatograph installed in 2009 indicate that ambient benzene concentrations currently meet the By-Law 90 standards of 260 ug/m^3 (1-h average) and 150 ug/m^3 (8-h average). In 2010, the maximum 1-h concentration observed was 49.1 $\mu g/m^3$.

17.9.2 Montréal Woodstove burning: Source of comfort or pollution?

There are over 85,000 fireplaces and wood stoves on the island of Montréal.

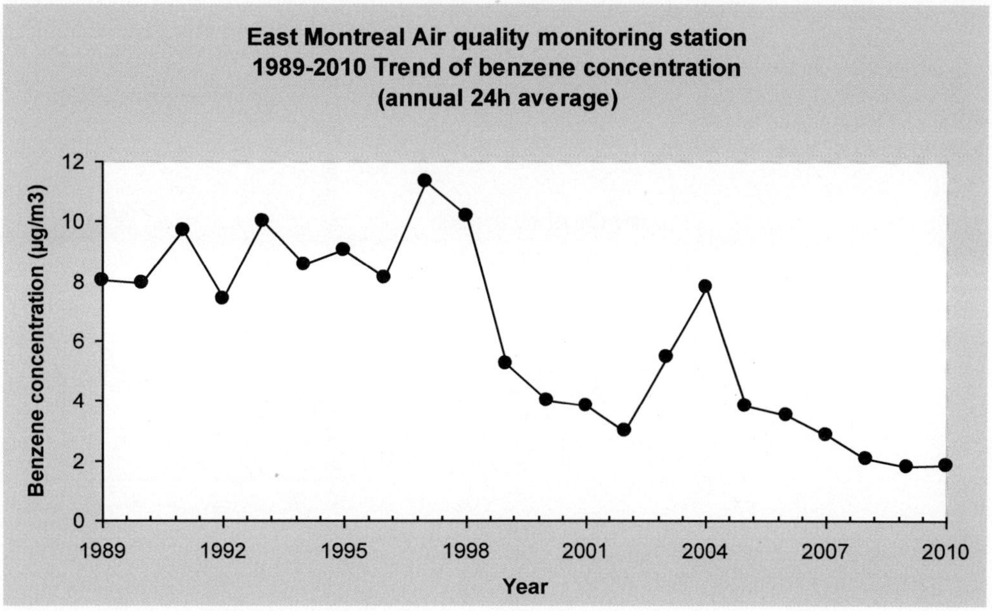

Fig. 17.11 Ambient benzene concentrations fell significantly in East Montréal as fugitive emissions of fossil fuels and volatile organic compounds were reduced and as the percentage of benzene in gasoline decreased. (Data from www.rsqa.qc.ca Montréal's Réseau de surveillance de la qualité de l'air)

The air quality monitoring station located in *Rivière-des-Prairies* borough is measuring the second highest density of domestic wood burning equipment on the island (11 %). When this monitoring station showed high $PM_{2.5}$ concentrations it became essential to refine the analysis in order to distinguish among various potential contributing factors.

Levoglucosan is a product of cellulose combustion and is found in $PM_{2.5}$. It has been recognized as a biomass tracer and is source-specific to the burning of any cellulose containing fuel (e.g., wood, paper, plants). Therefore in the winter of 2009–2010, *levoglucosan* analysis was used to assess the contribution of wood burning to the $PM_{2.5}$ concentrations in ambient air. It was determined that an average concentration of 188.8 µg/m^3 of this tracer over 24 hours was a reliable indication of « good » or « acceptable » air quality days.

On the evening of December 30th 2009, fine particulate concentrations were increasing at the *Rivière-des-Prairies* monitoring station (Fig. 17.12) and the tracer's concentration was later found to be about ten times the usual 24-hr average of 1813 µg/m^3. Air quality was « poor » locally because of wood stove burning while it was « good » at all other monitoring stations on the island, with average $PM_{2.5}$ concentration three-times lower than measured in *Rivière-des-Prairies*. Although not all of Montréal's area is covered by monitoring stations, the locally observed pattern is deemed representative of all similar residential areas.

The causal link between domestic wood burning activity and « poor » air quality has been demonstrated repeatedly; therefore a new regulation was introduced in 2009 in Ville de Montréal's 19 boroughs, prohibiting installation of solid-fuel-burning equipment other than certified wood pellet appliances (see articles 1 and 12 of By-Law 11-018 for the current prohibition). For existing combustion equipment, Québec Ministère du Développement durable de l'Environnement, de la Faune et des Parcs is providing funds for a replacement program called Feu Vert initiated in 2011. http://www.feuvert.org/accueil

Concurrently, at the municipal level, efforts are underway for a stricter and wider ban covering the whole Montréal agglomeration (composed of 19 boroughs and 15 reconstituted cities). Home heating from hydroelectricity is a possible choice everywhere in Montréal, although less romantic than a fireplace. Existing fireplaces can be retrofitted to use natural gas or wood pellets can be burned in certified appliances that have sufficiently low emissions of fine particulate matter.

When society and individuals lifestyle changes are required, mitigation may necessitate a mix of incentives and regulatory enforcement, as well as decades of effort and a strong commitment from officials to institute somewhat unpopular restrictions that are effective at improving air quality for the benefit of everyone.

17.10 British Columbia's Wood Stove Exchange Program—Communicating with the Public

Contributed by Rebecca Freedman

Woodstoves are a significant emission source (Fig. 17.13) that has been addressed in British Columbia through regulations and community airshed management planning. The Wood Stove Exchange Program is an example of a successful campaign to replace older, polluting wood stoves with cleaner burning stoves.

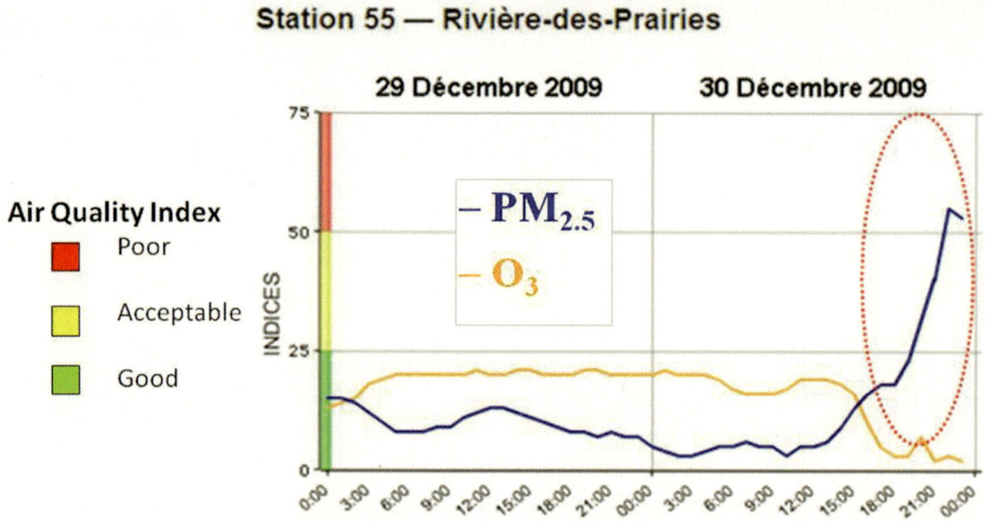

Fig. 17.12 *Rivière-des-Prairies* Air quality monitoring station. (Air quality index on December 29–30, 2009. Data from www.rsqa.qc.ca Montréal's Réseau de surveillance de la qualité de l'air)

People love wood fires and British Columbians along with many other Canadians have an abundance of firewood available to them. Changing opinions and behaviours is most difficult when people are attached to their current circumstances and lifestyles. It would be ineffective to try to shift people entirely away from wood burning in B.C., though there is growing pressure from health professionals and concerned British Columbians to do so because of evidence linking wood smoke to health impacts. For the time being, the provincial government has focused on helping British Columbians switch to cleaner burning technology and operate their wood stoves so as to minimize fine particulates.

Survey work and focus groups conducted in the mid 2000's have helped air quality managers to understand the challenges of communicating issues related to wood smoke and health to the public (NRG 2006, Xue and Wakelin 2006). Although many British Columbians are concerned about air quality in their communities, they are more concerned about big industry pollution and vehicle pollution, and tend to downplay or ignore the contribution that domestic heating plays. Even though 15 % of the province's fine particulate levels come from domestic home heating and exposure to this source in BC is high since it is emitted directly into residential neighbourhoods (BC MOE 2011), many people believe that their contributions to community air emission levels are insignificant. Information about the health hazards of wood smoke is also downplayed by many because it is easier to justify the use and enjoyment of fire if we believe that smoke is harmless and 'natural'. To change these perceptions, government and health care sectors must continue to educate about the human exposure to and health risks from, indoor and outdoor air pollution from domestic wood burning.

Over 120 000 old technology wood stoves (including stove inserts) are in operation in B.C.—these produce 70 % more fine particulate emissions than newer, cleaner burning models. However persuading people to switch over to new technology stoves is not easy since for many, the cost of replacing the stoves can be prohibitive. Therefore, providing financial or other incentives can help motivate people to make the replacement. Also, more communities are including standards in their legislation that address highly polluting, older wood stoves—for example, limiting or banning their use on poor air quality days, or requiring them to be removed upon sale or transfer of property, or in some cases, banning any stove in new neighbourhoods.

Changing the way people operate their stoves is another effective way of reducing the human exposure to fine particulate from wood burning. Using clean, dry firewood that has been properly stored and seasoned can substantially reduce emissions, as can the way the fire is built and maintained in the stove. The best indication on how well a fire is burning is the amount of smoke coming out the chimney—something that most people don't monitor. Many people feel confident in their ability to operate a wood stove and don't seek or want advice on how to change, so reaching them can be a real challenge. Connecting burning behaviour to the stove's operating efficiency and potential savings in fuel is actually a stronger motivator than reducing fine particulates for health reasons (NRG 2006). Most people want to be good neighbours and do not want their smoky chimney to be a problem for others. If people are made aware of their chimney smoke, they may be more receptive to learning about techniques to reduce it.

The BC Wood Stove Exchange Program was developed using principles of community-based social marketing to specifically address the challenges discussed above. Community–based social marketing uses strategies and techniques from psychology and marketing to promote behaviour change at the local level, recognizing that people are more motivated to change their behaviour in response to direct appeals from others (McKenzie-Mohr 2011). The Wood Stove Exchange Program targets a single behaviour change with

Fig. 17.13 Woodstove emissions polluting a small community in northwestern British Columbia—August 2003. (Reproduced with permission from BC Ministry of Environment)

incentives for people to replace their old technology wood stoves with new wood burning or alternate fuel appliances. It also targets the more difficult repetitive behaviour of wood stove operation with Burn-it-Smart education.

Customized, local wood stove exchange programs are run by municipalities or airshed management groups and involve a wide array of partners, including manufacturers and retailers of wood burning appliances. These individual programs have come up with some great ways of communicating to the public through advertising the incentives and delivering the educational component. Newspaper and radio ads, flyers and posters are traditional ways of reaching the public and can help get the word out about the incentives and raise awareness about the programs.

The most effective way of influencing beliefs and behaviours is through social diffusion—personalized conversations with respected, knowledgeable spokespeople—and the wood stove exchange programs try to maximize this strategy. Each program has a dedicated community coordinator who can promote the exchange in person at home shows and community events, as well as answer phone and email inquiries from the public. In-depth Burn-it-Smart workshops run by wood energy technicians provide technical information about optimal stove operation and offer the opportunity for participants to ask specific questions about their stoves. Appliance retailers also promote the program and are a good source of information about proper stove operation.

Several communities have conducted and publicized studies on where wood smoke hot spot neighbourhoods are—often determined by a combination of the topography in those neighbourhoods and the age of the home. Some programs have enlisted volunteers to conduct door-to-door campaigns and directly talk with neighbours about wood smoke and how to minimize it.

Since the pilot wood stove exchange program began in the Bulkley Valley region in 2007, B.C. has invested close to $ 1.9 million on rebates, education and support to 25 programs across B.C., ranging in size from small municipalities to large regional programs. By 2012, over 5000 old stoves had been changed out, preventing an estimated 310 t of $PM_{2.5}$ from being released into the atmosphere each year. Legislation has been enacted or strengthened in many communities to address installation and operation of wood burning appliances and in some cases, a complete phase out of old technology stoves. The BC regulation governing the manufacture and sale of wood burning appliances is also being tightened to ensure that high standards for particulate matter are met. Communities and jurisdictions can develop their own wood stove exchange programs using an online toolkit that includes survey and focus group reports, implementation and marketing guides and evaluation guides (Province of BC 2012).

References

AAC (2009) Alberta Airsheds Council. About Us. http://albertaairshedscouncil.ca. Accessed 5 July 2011

Advisory Group (1991) Clean air strategy for Alberta Report to ministers. Clean Air Strategic Alliance, Edmonton

Alberta Environment (2006) 2006 amendments to the air monitoring directive, 1989. Alberta Environment, Edmonton

Angle R (2010) Mechanisms for the alignment of air quality planning. Alberta Environment, Edmonton

BC MoE (2007) Provincial framework for airshed management planning. Victoria, BC. http://www.bcairquality.ca/reports/pdfs/airshedplan_provframework.pdf. Accessed Dec 2012

BC MOE (2011) A smoke management framework for British Columbia: A cross-government approach to reduce human exposure to smoke from biomass burning. BC Ministry of Environment (BC MoE), Victoria, BC
Capital Airshed Partnership (2008) Capital region ozone management plan. Alberta Capital Airshed Alliance, Edmonton
CASA (1995a) Zone air quality management guidelines. Clean Air Strategic Alliance, Edmonton
CASA (1995b) A strategic plan for air quality monitoring in Alberta. Clean Air Strategic Alliance, Edmonton
CASA (2003) Particulate matter and ozone management framework. Clean Air Strategic Alliance, Edmonton
CASA (2004) Airshed zones guidelines. Clean Air Strategic Alliance, Edmonton
CASA (2007) Beyond consultation: making consensus decisions. Clean Air Strategic Alliance, Edmonton
Cheng L (1994) Concentration and deposition of anthropogenic pollutants in Alberta. Alberta Department of Environmental Protection, Edmonton
CRAZ (2008) CRAZ particulate matter and ozone management plan. Calgary Regional Airshed Zone, Calgary. http://www.whistler.ca/residents/sustainability-environmental-protection/Whistler%20Policies%2C%20Initiatives. Accessed 11 March 2013
Levelton (2009) A review of airshed management planning in BC. Prepared for the Ministry of Healthy Living and Sport, Province of BC. Levelton Consultants Ltd. Richmond, BC
McKenzie-Mohr D (2011) Fostering sustainable behaviour: an introduction to community-based social marketing,3rd edn. New Society, Gabriola Island, B.C.
Metro Vancouver (2011) Integrated Air Quality and Greenhouse Gas Management Plan. http://www.metrovancouver.org/services/air/management/ReviewProcess/Pages/default.aspx. Accessed Jan. 2013
Municipality of Whistler (2004) Integrated energy, air quality and greenhouse gas management plan. http://www.whistler.ca/images/stories/sustainability_images/GHG%20information/whistler_int._energy_plan.pdf. Accessed Dec 2012
NRG (2006) Wood Stove Change-over: Focus Groups and Survey. BC Ministry of Environment (BC MoE), Victoria, B.C.
Province of BC (2008) BC Air Action Plan. Government of British Columbia, Victoria http://www.bcairsmart.ca/docs/bcairactionplan.pdf
Province of BC (2012) Provincial Wood Stove Exchange Program http://www.bcairquality.ca/topics/wood-stove-exchange-program/index.html. Accessed Dec 2012
Quesnel Air Quality Roundtable (2004) Quesnel airshed management plan (QAMP). http://quesnelairshed.files.wordpress.com/2010/12/airshed_management_plan.pdf. Accessed 19 March 2012.
Resort Municipality of Whistler (2004). Integrated energy, air quality and greenhouse gas management plan.
Sandhu HS (2000) Criteria for reducing compliance-based air quality monitoring requirements for zonal air quality management in Alberta. Alberta Environment, Edmonton
Southeast Saskatchewan Airshed Association (2011) http://www.sesaa.ca/. Accessed 1 Nov 2011
Stantec Consulting (2008) Parkland airshed management zone (PAMZ) ozone management plan. Parkland Airshed Management Zone, Sundre
Ville de Montréal (2010) Montréal community sustainable development plan. http://ville.montreal.qc.ca/pls/portal/docs/PAGE/d_durable_fr/media/documents/Action_Plan.pdf. Accessed Dec 2012
Whistler Municipality (2004) Integrated energy, air quality and greenhouse gas management plan. http://www.whistler.ca/images/stories/sustainability_images/GHG%20information/whistler_int._energy_plan.pdf. Accessed Dec 2012.
Working Group (1993) Zone air quality management: report of working group G-1, clean air strategy for Alberta. Clean Air Strategic Alliance, Edmonton
Xue H, Wakelin T (2006) Residential wood burning in British Columbia- public behaviour and opinion. BC Ministry of Environment (BC MOE), Victoria, B.C.
Zirnhelt N (2011) Quesnel airshed management plan review—2011 Prepared for the Quesnel Air Quality Roundtable. Cariboo Environmental Quality Consulting Ltd., Williams Lake, BC. http://quesnelairshed.files.wordpress.com/2010/12/qamp-review.pdf

Part V

Communicating Air Quality Information

Air Quality Management: Canadian Perspectives on a Global Issue

18

Eric Taylor

Abstract

Air pollution can cause both short and long term impacts on human health. A new tool, the Air Quality Health Index, has been developed in Canada to provide the public with a real time indication of short term health risks associated with air pollution concentrations. The Air Quality Health Index (AQHI) builds upon existing indices of air pollution but differs by being based on a rigorous statistical link between mortality and air quality. The development of the AQHI in Canada was aided by recent advancements in health effects research, pollutant monitoring and reporting, weather and air quality prediction systems, efficient dissemination processes and political support. More heavily urbanized areas generally experience higher AQHI values than smaller communities.

The AQHI is a function of three common air pollutants: ozone, nitrogen dioxide and $PM_{2.5}$ (Particulate matter of maximum diameters 10 and 2.5 microns (millionths of a metre) respectively). It normally varies from 1 to 10, though it can occasionally rise above 10 during severe pollution episodes, such as thick wildfire smoke. The AQHI and its predicted maximum values for the next 24–36 h are provided to the public through various websites, with specific health advice provided for each of four AQHI health risk categories. The vulnerable population receives more cautious health advice than the general public. Promotion campaigns are used to explain the AQHI and encourage its use.

Keywords

Air quality health index · Health risk of air pollution · Health effects · Health risk · Health communication · AQHI · Cote air santé · Québec

18.1 Introduction

It has been estimated that approximately 5900 premature deaths result from short- and long-term exposure to air pollution in Canadian cities annually (Judek et al. 2004). This has been compounded by a continuing demographic shift toward urban living, where high pollution concentrations are often more prevalent, thereby increasing exposure of the overall population. In addition, over three million Canadians have serious respiratory diseases, making them more vulnerable to air pollution events. The elderly are also at increased health risk (Pope et al. 1995), and since the fraction of the Canadian population 65 and older continues to climb—from 8% in 1971 to 15% today—more widespread adverse effects are expected in this vulnerable population.

Chapter 7 in this book provides ample evidence of the impacts of degraded air quality on human health. The significance of these impacts has been compounded by the political demand to reduce health-related air pollution costs. For example, air pollution impacts in Ontario have been estimated at one billion dollars annually by the Ontario Medical Association. For these reasons, Canada has developed the Air Quality Health Index (AQHI) to help the public reduce the short term health risk associated with air pollution. This

E. Taylor (✉)
British Columbia Ministry of Environment, Victoria, BC, Canada
e-mail: eric.taylor@gov.bc.ca

E. Taylor, A. McMillan (eds.), *Air Quality Management,* DOI 10.1007/978-94-007-7557-2_18,

chapter provides a description of the AQHI, including its development and how it is communicated to the public.

18.2 Background of Air Pollutant Indices

Indices of air pollution are among the longest running and most successful of public environmental risk communications (Goldberg 2002; Johnson 2003). They serve to simplify the interpretation of air pollution by transforming measured concentrations of pollutants into a single number and/or categorical description. These indices are usually accompanied by a clear descriptive term (e.g. *poor* air quality) along with understandable health advice and/or prescriptive messages that provide more information on the pollutants.

In 2001, the only daily health advice available to address air pollutant concerns in Canada was the Air Quality Index (AQI). This index had been developed in the 1970s to report the amount of air pollution relative to national and provincial air quality objectives. It focused only on the most predominant pollutant of several that are known to impact heath. It was also based on the assumption that health impacts to humans do not occur below a given threshold of pollutant concentration, an assumption that is now not generally shared among air quality health professionals. Below this pollutant threshold, concentrations generally equated with "good" or "very good" air quality, depending upon the jurisdiction. Above this threshold, the AQI was used to communicate pollutant concentrations that reflected arbitrary government air pollutant objectives or standards at different tiers of air pollutant concentrations (i.e., desirable, acceptable and tolerable) using a mixture of health and non-health endpoints incorporated into the objectives (e.g. damage to vegetation).

The algorithm used to calculate the AQI is not the same across all jurisdictions since each has different air quality objectives. They are therefore not directly comparable. They also can have different numbers of, and labels for, air quality descriptors. For example, the United States' version of the AQI currently uses six categories that imply different levels of health impacts of air quality:

1. Good
2. Moderate
3. Unhealthy for Sensitive Groups
4. Unhealthy
5. Very Unhealthy
6. Hazardous.

In contrast, the province of Ontario, Canada, defines only five categories of the AQI, with quite different labels:

1. very good
2. good
3. moderate
4. poor
5. very poor.

In the US, the UK and in some provinces of Canada, air quality alerts are issued when the AQI exceeds certain thresholds and enters the "poor" or "unhealthy" category, though the criteria for issuing these alerts varies with jurisdiction. These thresholds also use a number of different averaging periods ranging from 1 to 24 h, depending on the pollutant. Some Canadian provinces, including British Columbia, do not report the AQI explicitly but retain it as a basis for issuing air quality advisories or alerts.

While health impacts were likely the major consideration in developing the AQI, Canadian users such as health care organizations began to question the value of the AQI in the late 1990s since it had not incorporated evolving evidence of air pollutant impacts on health. The AQI also had not had the benefit of recent epidemiological evidence that showed airborne pollutants had cumulative health effects. It also suggested that there were concentrations where no health risk existed, while we now know that there are no such pollutant thresholds. Since the AQI was also being implemented independently and differently in each Canadian province, users felt that the AQI "system is not a very good indicator for the health impact of air quality." (Pengelly et al. 2001). At the same time, public opinion research indicated a strong demand to reduce the risk from air pollution.

18.3 The Origin of the Air Quality Health Index

In 2001, Dr. Sheela Basrur, the Toronto Medical Officer of Health recommended improving the Air Quality Index (AQI) and the smog warning system in the province of Ontario. She also recommended that the revised system be applicable to other jurisdictions in Canada. Her recommendations included continuous monitoring of PM_{10} and $PM_{2.5}$ in Toronto, using the latest health research to improve the descriptions of air quality in each index category and incorporating the cumulative health impacts of multiple pollutants.

These recommendations led to Environment Canada (EC) initiating the development of a daily prediction of short-term human health risks associated with poor air quality across Canada. It was assumed at the time that such a prediction could be used as a communication tool to express the cumulative risks to humans of atmospheric pollutants in order that those susceptible to poor air quality would make more informed choices to protect themselves and those in their care from short term health impacts.

The timing of EC's decision to proceed with tool development was as a result of a confluence of political, environmental, social, economic, and legal factors that had evolved during the previous decade. For example, various air quality stakeholders had expressed concern over the cumulative impacts of air pollution on Canadians and a variety of government agencies were working on various domestic and international fronts to improve air quality management regimes. Meanwhile, scientific and technical advances were being made in the form of increased meteorological data availability and computational processing speeds as well as advances in epidemiological methodologies that more accurately correlated human health impacts to air quality. These factors culminated politically in 2001 in a speech made at the 2001 Smog Summit in Toronto by the Minister for Environment Canada:

> We need an up-to-date national air quality index that…tells people how bad air pollution is in their community and that is supported by health and action messages. I am challenging scientists, health experts and communications specialists both inside and outside government to give us this tool within a year.

This challenge provided EC with the political capital needed to mobilize resources and marked its formal engagement in a process to innovate a new health communications tool that would later be called the Air Quality Health Index. Canada's Departments of Health and Environment, in collaboration with the provinces, territories, universities and other stakeholders, then began the development of the Index. It needed to be based on solid health science, take into account the multiple-pollutant impact on health and provide individuals with clear health advice that could help them reduce the health risk caused by degraded air quality. This advice would include guidance on changing behavior that could reduce exposure to pollutants. It would also increase awareness and response to early symptoms of disease that may be caused in part by air pollutants. It could also identify contributing risks such as excessive heat or strenuous exercise that should be avoided. Individuals, particularly the elderly, the very young and others at risk, were to use the Air Quality Health Index to help protect themselves from the negative effects of degraded air quality. This type of advice required convincing and informative communication tools that were able to update the public about air pollution in near real-time so that they can take actions to appropriately protect their health (Cantwell, personal communication 2012).

The stakeholder community was involved extensively in the development and implementation of the AQHI (Henderson et al. 2004). They provided advice on defining the index, identifying the audiences, and on testing the AQHI communication materials and processes. A testament to the success of this process was the declaration by the Auditor General of Canada that its development was a model of stakeholder engagement (Office of the Auditor General of Canada 2009).

18.4 The Development of the Air Quality Health Index

18.4.1 Foundation of the AQHI

Developers of the new Air Quality Health Index were able to build on several new scientific, engineering and political advances (Cantwell, M. Personal communication):

- **Health Effects Research**: As noted above, there had been significant advances in air health effects research since the old, single-pollutant AQI was introduced in the 1970s. The most significant finding is the linkages between adverse health effects and $PM_{2.5}$ (Pope et al. 1995). In addition, research has shown that adverse health effects have been found to occur at even low levels of air pollution, which were previously regarded as safe (WHO 2000), suggesting that all levels of air pollution are associated with at least some hazard. Also, since health is impacted by cumulative effects of all pollutants that are present in the air (WHO 2000), multiple pollutants were incorporated into the new Air Quality Health Index.
- **Monitoring and reporting**: Weather and air pollutant monitoring technology has improved significantly since the 1970s. This has resulted in more accurate measurement of pollutant concentrations, measurement of more types of pollutants and faster and more reliable transfer of air pollutant data from the monitoring equipment to data management centres. Also, the systems that organize, quality control and disseminate this air quality information to the public in real time have become more efficient and effective.
- **Prediction of air quality changes:** Variable weather conditions are an important predictor of hourly and daily changes to air quality and the AQHI in a community. Stagnant atmospheric conditions coupled with pollutant emissions are often a precursor to degraded air quality while increasing winds and atmospheric turbulence can dilute pollutants and quickly improve air quality. Prediction of air quality is generally aided by the use of both weather and air quality forecast models, which have advanced significantly since the 1970s. In addition, these models have benefitted from improved pollutant emissions inventories and data assimilation techniques as described in the chapters in this book on air quality modelling and forecasting.
- **Political commitment:** The national political environment played a significant role in the impetus to develop a new Air Quality Health Index. The most significant event was a commitment made in the spring of 2001 by the Canadian Minister of the Environment, the Honourable David Anderson, to pursue implementation of an improved "tool" for communicating air quality information to Canadians. In addition, bi-lateral advisory bodies like the International Joint Commission identified the availability of new air

quality information tools such as air quality indices and health based advisories as one of the six critical issues for Canada and the United States to address.

18.4.2 The Design of the Three-pollutant AQHI Formula

As evidenced by the series of deadly air pollution events in Europe and the US referred to in Dave Stieb's chapter in this book, increased air pollutant concentrations appear to coincide with increased mortality rates. However, even at lower concentrations experienced in Canada, air pollution is still associated with increased mortality (Stieb et al. 2008). The formulation of the new Air Quality Health Index was therefore based on a model linking daily deaths with daily concentrations of individual air pollutants. The modeling approach assumed a linear relationship between air pollution and mortality, consistent with the large body of evidence indicating that health effects are observed even at low levels of exposure.

The air pollutant and mortality associations were based on statistical links between daily 3-hour maximum concentrations of air pollution measurements and daily mortality rates in twelve Canadian cities over ten years. The AQHI was then adjusted to a 1–10 scale, and calculated hourly on the basis of trailing 3-hr average pollutant concentrations (Stieb et al. 2008). Though the AQHI was based on data that gave it a range of 1–10, in practice values well above 10 have subsequently been observed in extremely polluted air in Canada, mostly in smoke from severe wildfires.

Several different combinations of pollutants were considered for the AQHI in order to best reflect the mix of pollutants. The final formulation was based on mortality associations with only three pollutants: nitrogen dioxide, ozone, and $PM_{2.5}$. This combination of pollutants exhibited the strongest associations with mortality in large cities, and the addition of other pollutants did not add significantly to the overall health risk (Stieb et al. 2008). Because the AQHI was based on data from large cities, it should be noted that it may not be as applicable to small communities or rural settings that have a completely different pollutant mix than that used in the development of the index. This will be discussed in more detail later.

The current AQHI formula is:

$$\text{AQHI} = 10/10.4*\left(100*\left(e^{(0.000871*\text{NO}_2)} - 1 + e^{(0.000537*\text{O}_3)} - 1 + e^{(0.000487*\text{PM}_{2.5})} - 1\right)\right)$$

In this equation, NO_2, O_3 and $PM_{2.5}$ (nitrogen dioxide, ozone and fine particulate matter less than 2.5 microns in diameter, respectively) refer to concentrations averaged over three consecutive hours. Units are parts per billion for the gases and micrograms per cubic metre for $PM_{2.5}$.

18.4.3 Accounting for Other Pollutants

Though the three pollutants in the AQHI have direct health impacts, some, particularly nitrogen dioxide, are more likely surrogates for other pollutants originating from the same sources such as ultrafine particles, VOCs or other substances that impact health but that are not included in the AQHI formula (Hidy and Pennel 2010). These other pollutants may mimic the variations in concentrations of the three AQHI component pollutants, particularly when they are produced by the same processes.

For example, in large cities, vehicle exhaust is a prime source of NO_2. But in addition to NO_2, fuel combustion in vehicles produces emissions of hundreds of other products that may be included in the measured direct health effects of NO_2, even though only NO_2 is routinely monitored. It is unknown how much of the health impact assigned to NO_2 in the AQHI is caused by the other pollutants contained in vehicle exhaust. Future improvements in the AQHI formulation may be needed to improve its correlation with health impacts in those locations where vehicle exhaust is not the primary source of NO_2, such as in smaller communities in rural, forested or agricultural areas.

18.4.4 Mortality vs Health and the AQHI

Though the AQHI is intended to provide an estimate of *health* risk, its formula is based on the statistical relation between *mortality* and air pollution. However, the contention of health scientists is that there is coherence between mortality and health. That is, the association between air pollution and a severe outcome like mortality broadly reflects impacts on a variety of health outcomes (Bates et al. 2003). In addition, in the process of developing the AQHI, numerous sensitivity analyses were conducted, including substitution in the AQHI formula of risk estimates based on hospital admissions rather than mortality. It was found that the AQHI performed similarly—i.e. the distribution of AQHI values was similar and values based on the mortality and hospital admissions-based formulas were highly correlated. More recently, analyses have also found associations between the AQHI and other health outcomes including asthma hospital admissions in Ontario, cardiac and respiratory hospital admissions in 24 Canadian cities, stroke emergency visits in Edmonton and asthma symptoms in Windsor.

18.5 AQHI Values in Canadian Cities

The majority of hourly values of the AQHI across Canada fall within the Low Health Risk category (AQHI less than 3.5). Figure 18.1 shows distributions of the relative frequen-

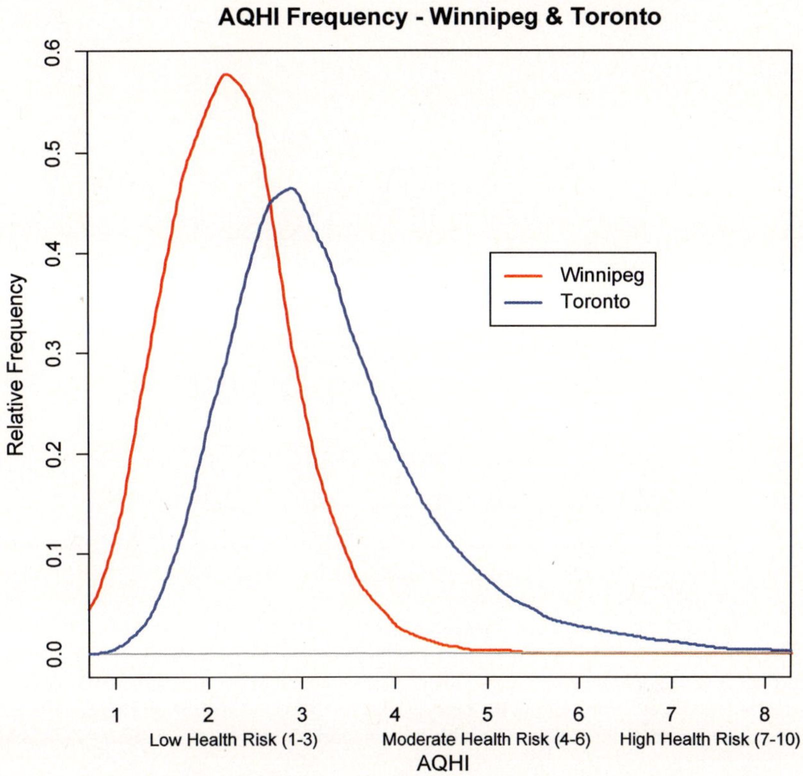

Fig. 18.1 A graph comparing the relative frequency of the AQHI at Winnipeg and Toronto. Toronto has the higher short-term health risk since the most frequent value of the AQHI is higher in that city, and the AQHI is also in the Moderate (3.5–6.5) and High (6.5–10.5) Health Risk categories more frequently. Areas under each curve are equal. Based on data from 2001–2009. (Data courtesy of Environment Canada)

cy of occurrence of the AQHI at two typical Canadian cities, Winnipeg (pop. 640,000) and Toronto (pop. 4,800,000). The shapes of the distributions are quite different, with the most frequent AQHI values for Winnipeg being 2.2, while Toronto is higher at 3.0. Figure 18.1 also shows that Toronto is in the Moderate Health Risk (AQHI between 3.5 and 6.5) and High Health Risk (6.5 to10.5) categories much more frequently than the less urbanized city of Winnipeg. This suggests that the short term health risk from air pollution is higher in Toronto than Winnipeg. Very high health risk (above 10.5) is rare in all locations in Canada, occurring much less than 0.1 % of the time (Hasselback and Taylor 2009), and usually being associated with infrequent events like wildfire smoke.

The AQHI at six Canadian cities is compared in Fig. 18.2, These figures suggest that Calgary, Montreal and Toronto experience a higher incidence of short term health risk due to degraded air quality than Winnipeg, Vancouver or Saint John.

The three constituent pollutants of the AQHI do not contribute equally to the AQHI. For example, in Kelowna, British Columbia, Fig. 18.3 shows that ozone and nitrogen dioxide account for the bulk of AQHI value, while $PM_{2.5}$ contributes only about 10 %. However, the relative contribution of $PM_{2.5}$ begins to climb as the AQHI exceeds 4, and it becomes the dominant pollutant at higher AQHI values. However, these high AQHI values are rare, usually only occurring during high smoke concentrations from forest fires. Similar results have been found in other communities in Canada.

18.6 AQHI Categories and Health Messaging

The relationship between the AQHI and its component pollutants is essentially linear for common ranges of these pollutants. Also, there is no health evidence that supports specific thresholds or step-changes for the AQHI as pollutant concentrations increase. However, stakeholder focus studies revealed that a key part of health messages would require separating AQHI values into separate health risk categories, each corresponding to a range of AQHI values. These categories were labelled *Low, Moderate, High* and *Very High* health risk. These categories and the corresponding health advice are described in Fig. 18.4.

A cornerstone in the development of the AQHI was the formulation of prescriptive health messages to communicate the risk posed by air quality and the specific actions which can be taken to reduce exposure and health effects. However, developing air quality messages that are both informative and focused towards fostering behavioral-change are complicated within the context of competing public health priorities (Henderson et al. 2004). An example is the recommendation for vulnerable people to stay indoors on days where the air quality is significantly degraded, while not discouraging the benefits of outdoor recreational activity. Therefore, an integrated program of mental models research and public opinion research was conducted to support the development of the health messages (Henderson et al. 2004)

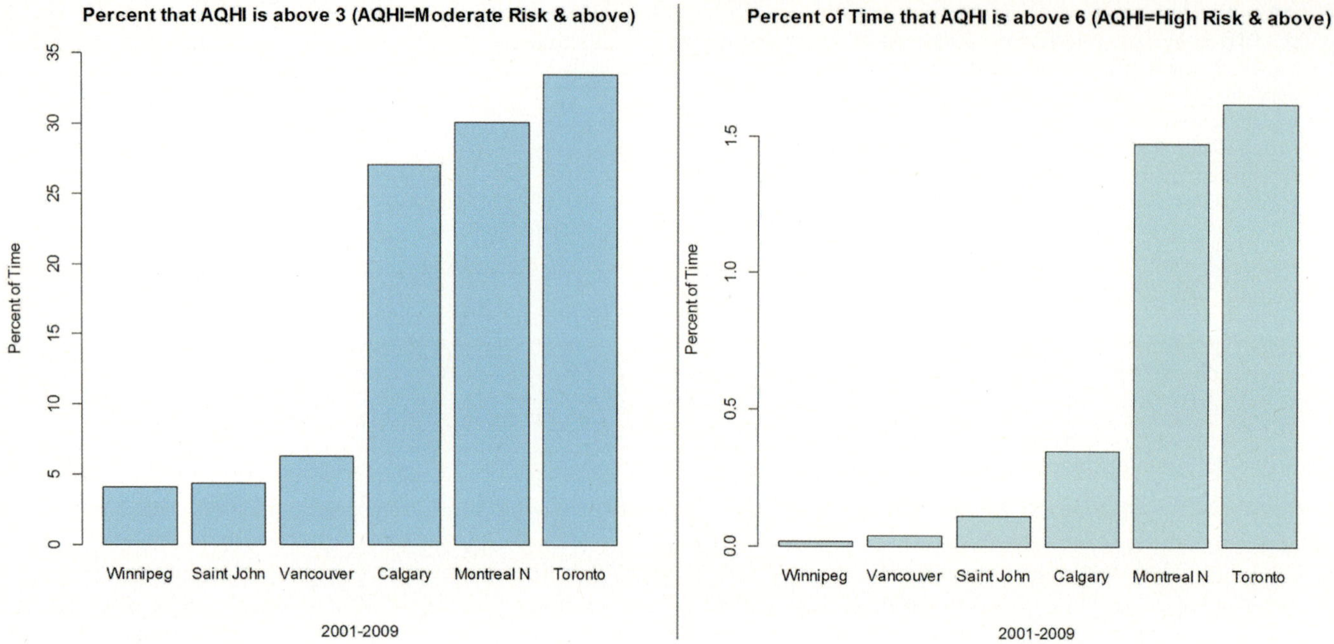

Fig. 18.2 The *left hand* graph shows the percent of time that the AQHI is in the Moderate (or above) Health Risk categories for six Canadian cities. The *right hand* graph shows the percent of time the AQHI is in High Health Risk or above category. These graphs indicate that Montreal and Toronto experience these elevated health risk categories much more frequently than Winnipeg, Vancouver, Saint John or Calgary

Fig. 18.3 The relative contribution of three pollutants, ozone, $PM_{2.5}$ and NO_2 to the AQHI value in Kelowna, BC. Starting with the graph in the *top left corner* and going clockwise, the graphs show the percent contribution of each pollutant by month, year, AQHI value and hour

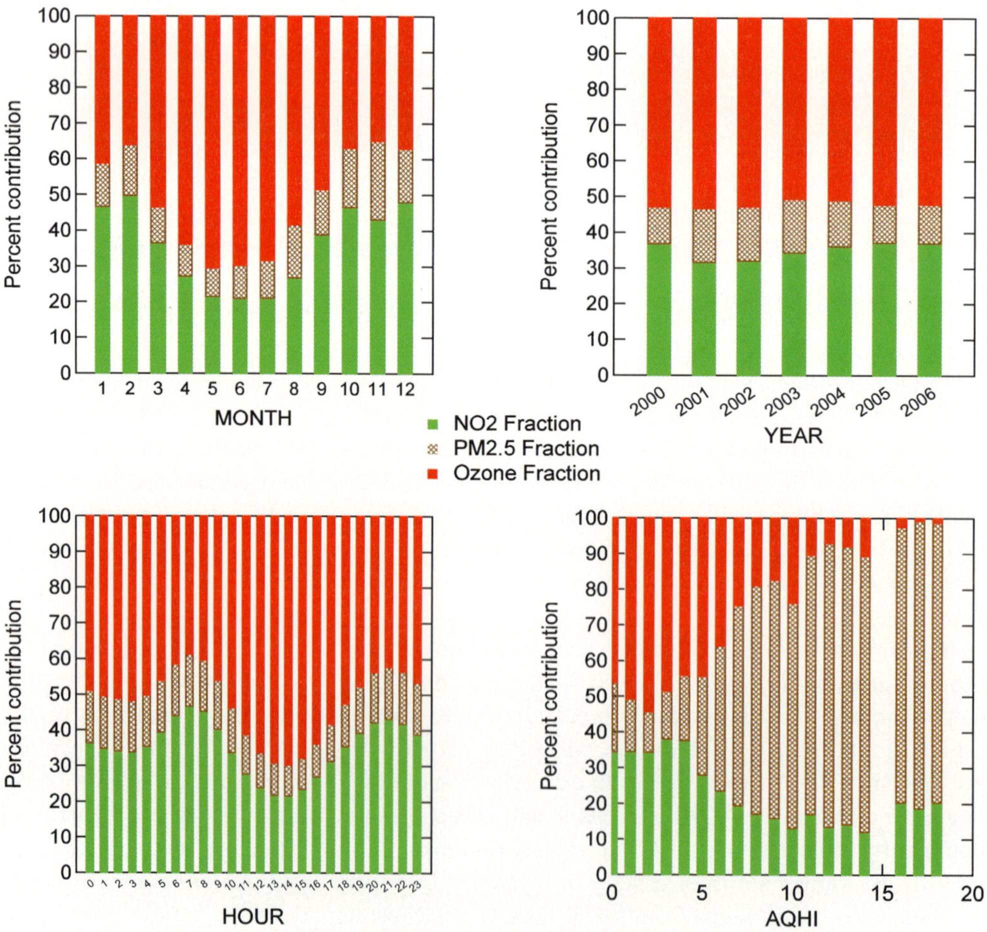

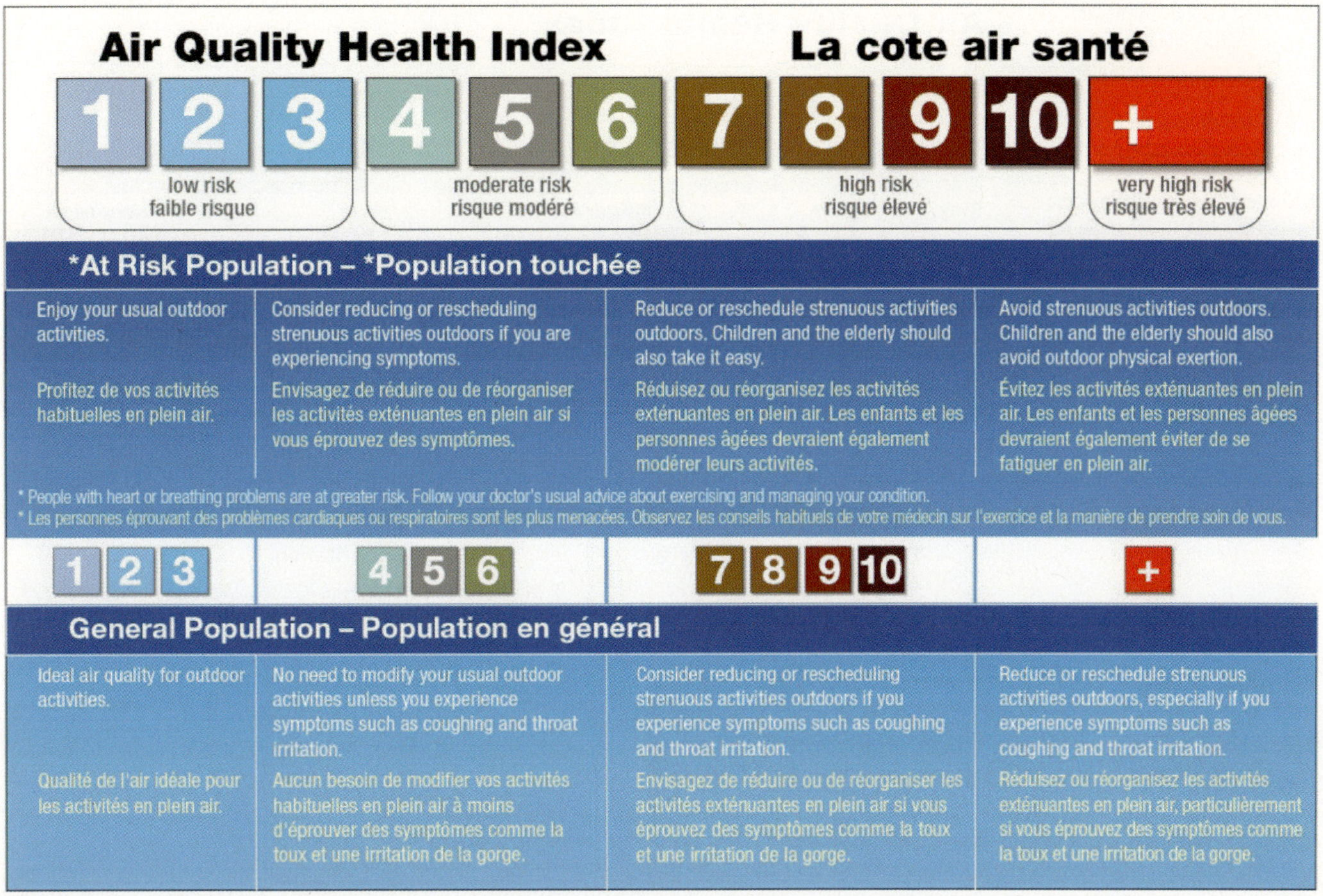

Fig. 18.4 The AQHI has four categories, corresponding to low, moderate, high and very high health risk. Specific health messaging is provided for the at-risk population and the general population. (Image courtesy of the Government of Canada)

Though all people have the potential to develop health symptoms as air quality deteriorates, health risk varies from person to person depending on their age, underlying health and sensitivity to air pollutants. The health advice provided by the AQHI system reflects this concept by providing health advice for the "at-risk" population that is more restrictive and cautious than for the general population. The "at-risk" population is defined as individuals with existing respiratory or cardiovascular conditions, young children, the elderly and those active outdoors.

Figure 18.4 associates the four AQHI categories with their corresponding AQHI values and health messages. For example, the health advice for an AQHI in the "moderate health risk" category for the "at-risk" population is:

Consider reducing or rescheduling strenuous activities outdoors if you are experiencing symptoms. People with heart or breathing problems are at greater risk. Follow your doctor's usual advice about exercising and managing your condition.

For the general population, the message is less restrictive:

- No need to modify your usual outdoor activities unless you experience symptoms such as coughing and throat irritation.

Figure 18.5 provides an example of the Environment Canada website display of the AQHI that provides the current and predicted AQHI values and categories, along with the specific health messages.

Individuals are also encouraged to "self-calibrate" by associating any pulmonary or cardiac symptoms to a specific AQHI value when air quality deteriorates. The theory behind this approach is that individuals can then take action in the future to reduce their exposure to air pollution when these AQHI values are either occurring or predicted. Anecdotal evidence suggests that self-calibration may offer a degree of protection, though the approach has yet to be tested rigorously.

Lacking from most discussions on air quality is what medical interventions can be offered for individuals who are at-risk for developing negative consequences from degraded air quality. Current health messaging is based on best available knowledge and expert opinion. Further exploration of the impacts of health messaging, and the acceptance by individuals to adopt recommended practices is required. For example:

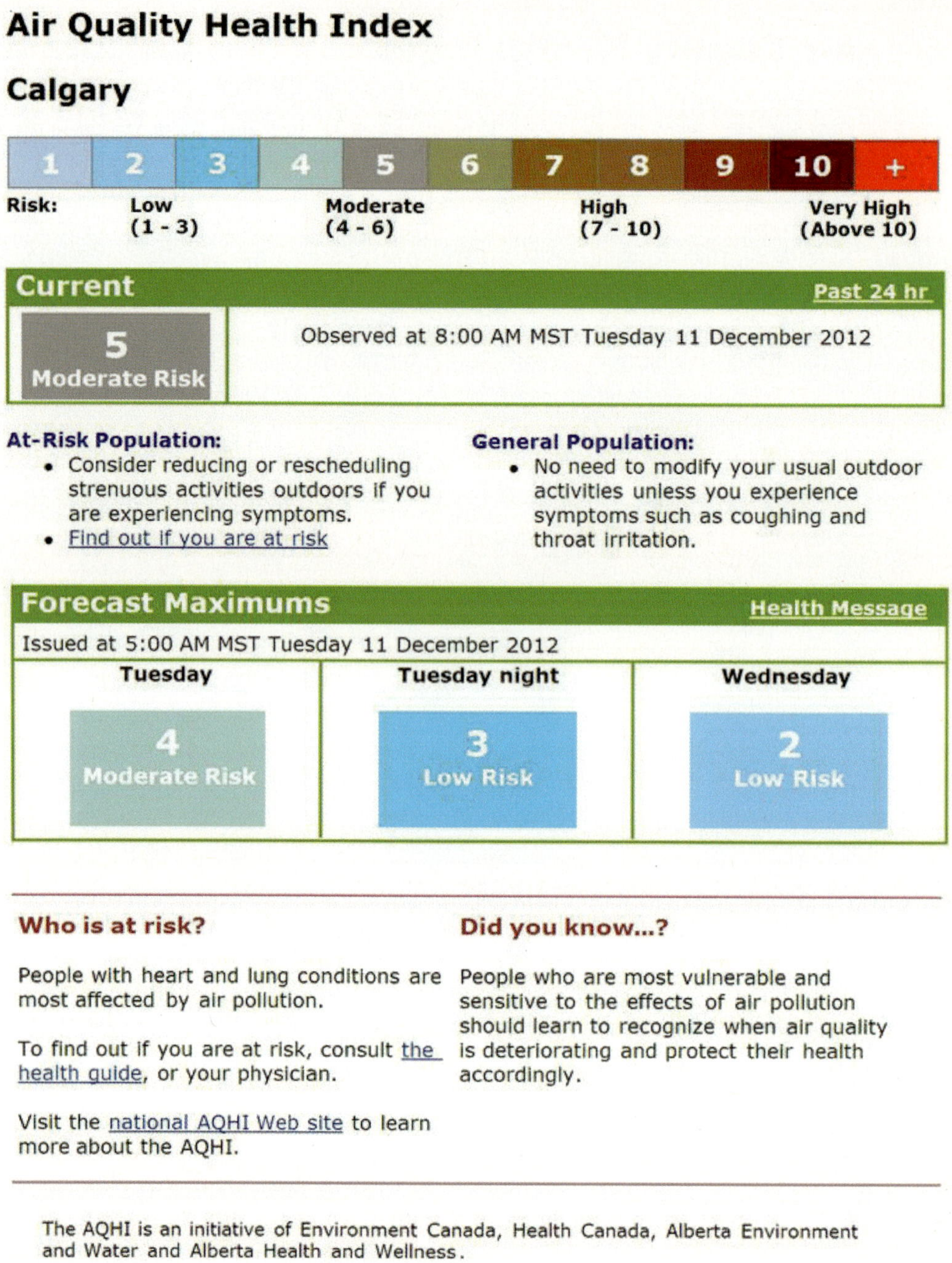

Fig. 18.5 This example of the Air Quality Health Index for Calgary on Environment Canada's website shows the current and predicted AQHI values as well as health advice. Clicking on the "forecast maximum" values displays the appropriate health message for that AQHI value

- What changes are required in personal asthma or COPD plans such as modification of medications or oxygen flow rates?
- Are there particular cardioprotective measures to be instituted in poorer air quality environments and days?
- Will some individuals have better health if relocated to different communities?

The usability of the AQHI will be improved as researchers, patients, and interested persons begin to explore how to employ the AQHI to both reduce negative health impacts and contribute to health improvement (Hasselback and Taylor 2009).

It is important to emphasize that the AQHI is not intended to address the health impact on individuals of long term (multi-year or multi-decadal) exposure to air pollutants. The AQHI was designed as a tool that can be used by the public to reduce their *short term* exposure to air pollution and plan, on a daily basis, to modify their behaviour and reduce their personal health risk. The AQHI therefore responds to the acute, or short-term, changing levels of risk associated with air pollution.

18.7 Communicating the AQHI and Health Messages

Environment Canada now broadcasts the hourly AQHI value for 74 communities across Canada through its website on a real-time basis. The air pollutant data needed for the calculation are provided in real time by provincial and territorial air quality agencies[1]. Environment Canada also routinely predicts the AQHI over the next 24–36 h for each community. More information on forecasting the AQHI can be found in Chap. 6 in this book. This information, the associated AQHI health risk category and the appropriate health messaging (Fig. 18.4) is communicated to the public through websites and other media. Several other Canadian jurisdictions also provide AQHI information in real-time through the internet.

18.7.1 Adaptation of the AQHI and Messaging for Specific Circumstances

The basic AQHI formula is the same for all communities in Canada. However, Environment Canada has agreed to accommodate some jurisdictions during extreme pollution events when there are conflicts between the AQHI and other measures of air quality. For example, in Alberta, additional pollutants are also considered when reporting the AQHI. These pollutants include sulphur dioxide, hydrogen sulfide, total reduced sulphur and carbon monoxide. Hourly concentrations of these pollutants are compared to Alberta Ambient Air Quality Objectives and on the rare occasion that objectives are exceeded, the AQHI value is increased to reflect a higher health risk. This modification of the index allows the AQHI to respond quickly to single-pollutant events on an hourly basis.

Special community-based messaging for odour or visibility is also a feature of the AQHI in Alberta, since the public identifies these factors with degraded air quality. This special messaging is provided when pollutant concentrations reach specific thresholds for odour and visibility in communities, for example during smog conditions, wildfire smoke episodes or as a result of nearby industrial activity. Despite this visibility or odour—related messaging, the AQHI values may not indicate that the air quality is significantly degraded

A parallel AQHI formula that replaces $PM_{2.5}$ with PM_{10} (particulate matter less than 10 microns in diameter) is used on rare occasions in some communities, particularly in British Columbia, when high dust levels are causing or expected to cause elevated PM_{10} concentrations.

[1] Canada is comprised of ten provinces (British Columbia, Alberta, Saskatchewan, Manitoba, Ontario, Québec, New Brunswick, Nova Scotia, Prince Edward Island and Newfoundland and Labrador) and three territories (Yukon, Northwest Territories, and Nunavut).

In Quebec, there are several differences in the way the AQHI is communicated to the public compared to other provinces and territories.

18.7.2 La cote air santé au Québec (English translation follows)

Depuis décembre 2008, le Québec étudie, par le biais d'un projet pilote, la possibilité de mettre en place la Cote air santé (CAS) de manière permanente sur son territoire. Trois localités urbaines ont été sélectionnées soit Québec, Gatineau-Ottawa et l'Île de Montréal.

Les aspects distinctifs du projet pilote québécois de la CAS Le projet pilote québécois de la CAS se distingue du programme national, principalement dans son aspect communicationnel. Le premier aspect distinctif du projet pilote est de cibler spécifiquement les personnes vulnérables aux effets à court terme de la pollution atmosphérique (personnes souffrant de problèmes respiratoires et cardiovasculaires) parce qu'elles sont plus susceptibles de ressentir les effets d'une élévation ponctuelle des concentrations ambiantes de pollution atmosphérique. Le second aspect distinctif est de communiquer aux personnes vulnérables *le bon message, par le bon moyen et au bon moment*. Afin de répondre à ces orientations distinctives, différentes adaptations ont été apportées à la CAS nationale afin de l'appliquer au Québec:

- un seuil d'alerte a été défini afin de communiquer les recommandations de santé aux personnes vulnérables *seulement* lorsque la CAS atteint ce seuil;
- les recommandations de santé de la CAS ont été révisées afin qu'elles soient mieux adaptées à la situation des personnes vulnérables;
- des mécanismes spécifiques de diffusion, notamment un service automatisé d'alertes téléphoniques, ont été développés afin de diffuser efficacement et au moment opportun les recommandations de santé aux personnes vulnérables;
- une stratégie de promotion de la CAS ciblant spécifiquement les personnes vulnérables a été mise sur pied.

Seuil d'alerte En l'absence de seuil, le risque à la santé et les recommandations associées sont communiqués continuellement, risquant ainsi de compromettre, avec le temps, l'intérêt des personnes vulnérables pour le message. C'est pour cette raison qu'un seuil d'alerte a été fixé dans le cadre du projet pilote pour une valeur de CAS supérieure ou égale à 6. De cette façon, les personnes vulnérables sont avisées de l'élévation du niveau de risque pour la santé seulement lorsque la CAS atteint ce seuil.

Recommandations de santé Les recommandations de santé du programme national ont été adaptées afin qu'elles soient claires sur les comportements préventifs à adopter et reflètent davantage l'expérience personnelle des personnes vulnérables. De surcroit, la variation des recommandations de santé selon le niveau de risque a été revue afin qu'elles suggèrent une modification de comportements seulement lorsque la valeur de la CAS atteint le seuil d'alerte (CAS ≥ 6). Enfin, les catégories de risque présentes sous l'échelle ont été supprimées et remplacées par une flèche afin de mieux illustrer que le risque pour la santé associé à la pollution atmosphérique est sans seuil.

Mécanismes de diffusion Les valeurs de la CAS sont diffusées au Québec par le biais de trois services:

- un site Internet dédié présentant la CAS (www.coteairsante.qc.ca), les populations vulnérables et les effets de la pollution atmosphérique sur la santé. Les niveaux observés actuels et prévus, l'évolution des niveaux atteints durant 48 dernières heures ainsi que les recommandations en matière de santé peuvent être consultés pour les 3 régions du projet pilote. De plus, le site propose gratuitement un service d'abonnement et de désabonnement aux alertes;
- une ligne téléphonique sans frais (1 866 688-3810) permettant à l'usager de s'informer de la valeur CAS actuelle, des recommandations santé et de s'abonner au service d'alertes téléphoniques;
- un service automatisé d'alertes par téléphone, message texte ou courriel permettant à l'abonné de recevoir des avis lorsque la CAS atteint le seuil d'alerte.

Promotion de la CAS et sensibilisation Les activités de communication visaient principalement la population vulnérable à la qualité de l'air, cible principale du projet pilote québécois. La stratégie adoptée consistait à relayer l'information grâce aux professionnels de la santé et certaines associations pertinentes. L'objectif de la promotion était de faire connaître la CAS aux personnes vulnérables, de les sensibiliser aux enjeux associés à la qualité de l'air et de susciter leur adhésion aux services de la CAS (service gratuit d'alertes téléphoniques, ligne téléphonique sans frais et site Internet). Les moyens de communication élaborés consistaient en un dépliant d'information et un aimant promotionnel. Ces outils ainsi que les moyens de communication mis en ligne (hyperliens, bandeaux) sur les sites Internet des associations partenaires pointaient systématiquement vers les trois services.

La CAS et les autres outils d'information liés à la qualité de l'air Il existe, au Québec, d'autres outils d'information liés à la qualité de l'air soit l'Indice de qualité de l'air (IQA) et le programme Info-Smog. L'IQA qualifie la qualité de l'air de bonne, acceptable ou mauvaise en fonction de la concentration de certains polluants. Pour sa part, le programme Info-smog fourni une prévision et des alertes en lien avec les épisodes de smog. Ces indices sont fondés sur le dépassement de normes et de critères ce qui les différencie de la CAS qui se base sur les risques à la santé engendrés par une combinaison de polluants.

Dans certaines conditions, lorsque la concentration de polluants, en particulier les $PM_{2,5}$, dépasse le critère utilisé par l'IQA, le seuil d'alerte de la CAS peut ne pas être atteint. Il en découle une contradiction entre les messages véhiculés par ces différents indices. Afin de remédier à cette situation, un système de bannière a été mis en place sur le site Internet de la CAS. Lorsque la qualité de l'air est jugée mauvaise selon l'IQA et que la CAS est inférieure à 6, une bannière informant de la mauvaise qualité de l'air et des recommandations de santé inhérentes est affichée. Si la bannière est affichée durant plus de trois heures consécutives, le service d'alertes envoie des alertes aux abonnés avec un message santé différencié de ceux reliés à la CAS.

Avenir de la CAS au Québec En mars 2012, les autorités de santé publique impliquées dans le projet se sont prononcées en faveur de la poursuite du projet pilote au Québec jusqu'en mars 2015 à condition de conserver l'IQA, le programme Info-Smog et de réviser l'importance des $PM_{2,5}$ dans la formulation de la CAS. Le projet se décline en trois étapes. La première consiste en l'ajout de stations de mesure dans la région de Québec et de Gatineau. Sous réserve d'approbation des différents partenaires et de la disponibilité des budgets requis, l'étape suivante verra le développement de la CAS dans les couronnes nord et sud de Montréal. Finalement, et sous les mêmes conditions, le service pourrait être étendu à 3 autres régions urbaines soit celles de Saguenay, Sherbrooke et Trois-Rivières. À la fin du projet pilote, les autorités ministérielles se verront proposer une recommandation favorable ou non à l'adoption définitive de la CAS au Québec.

The Air Quality Health Index in Québec In December 2008, Québec initiated a pilot project to assess the possibility of implementing the Air Quality Health Index (AQHI) throughout the province. Three urban centres were selected for this pilot: Québec City, Gatineau-Ottawa, and the Island of Montréal.

Distinctive aspects of the Québec AQHI pilot project The Québec AQHI pilot project differs from the national AQHI program, primarily in its approach to communication. Firstly, it specifically targets those who are vulnerable to the short-term effects of air pollution (people suffering from respiratory and cardiovascular problems) since they are more likely to suffer the effects of elevated concentrations of ambient air pollution. Secondly, it communicates to vulnerable people *the right message, in the right manner, and at the right time.* In order to meet these guidelines, the following adjustments have been made to the national AQHI program in Québec:

- An alert threshold has been defined to communicate health advice to vulnerable people only when the AQHI reaches this specific threshold;
- The AQHI health advice has been revised so that it is better tailored to vulnerable persons;
- Specific mechanisms of communication, most notably automated phone alerts, effectively disseminate health advice to vulnerable people in a timely fashion;
- A strategy to promote the AQHI by specifically targeting vulnerable people.

Alert Threshold In the absence of a defined AQHI threshold, the health risk and the associated health advice messages are communicated continuously to the public. This may cause vulnerable people to ignore this repetitive information because of so-called "message fatigue". This is the reason that an AQHI value greater than or equal to 6 has been established as a warning threshold. In this way, vulnerable persons are advised of the elevated level of health risk only when the AQHI reaches this threshold.

Health Advice The health advice provided by the national AQHI program has been modified to be clearer with respect to protective behaviours that users should adopt, and also to more accurately reflect the personal experience of those vulnerable to air pollution. In addition, the advice related to the level of health risk has been revised such that a change in behaviour is recommended only when the value of the AQHI reaches the alert threshold (AQHI greater than 5). Finally, the various risk categories presented immediately below the standard AQHI scale on the website have been replaced by a single, continuous arrow (extending from low to high AQHI values) to better demonstrate that the health risks associated with air pollution are not related to any specific AQHI threshold.

Dissemination mechanisms AQHI values are disseminated in Québec through three services:

- A website (http://www.coteairsante.qc.ca/default.aspx?lang=en) dedicated to the presentation of the AQHI. This describes those who are vulnerable to air pollution and summarizes the effects of air pollution on health. It also provides the current and forecasted AQHI values, the hourly values over the last 48 hours, as well as health advice for the three pilot project areas. In addition, the site offers a free subscription service for AQHI alerts.
- A toll-free line (1 866 688-3810). This provides the current AQHI observation and related health advice, and the opportunity to subscribe to the telephone alert service;
- An automated system of alerts by phone, text message, and e-mail. These permit subscribers to receive notifications when the AQHI reaches the prescribed alert threshold.

AQHI Promotion and Awareness Communication activities are primarily intended for those vulnerable to air pollution, the main target of the pilot project in Québec. The strategy is to relay information through health professionals and certain relevant associations. The objective of the promotion program is to inform vulnerable people about the AQHI, raise awareness of the issues associated with air quality, and encourage the use of the AQHI services (free alerts by phone, a toll- free phone line, and website). Communication materials consist of an information brochure and a promotional magnet. These tools, as well as those developed for the online environment (hyperlinks and banners on the websites of partner organizations) consistently point to these services.

AQHI and Other Information Tools Relating to Air Quality There are, in Québec, other information tools related to air quality. These are the Air Quality Index (AQI) and the Info-Smog program. The AQI describes the air quality as *good, fair,* or *poor* depending on the concentrations of certain pollutants. The Info-Smog program provides forecasts and alerts related to smog episodes. These indices are structured around the exceedance of standards or select criteria defined for single pollutants. This differentiates them from the AQHI which is based on the health risks caused by a combination of pollutants.

Under certain conditions, when the concentration of pollutants, in particular $PM_{2.5}$, exceeds the criterion used by the AQI, the AQHI may not exceed the defined threshold value for alerts. This leads to a disagreement between the messages conveyed by the AQHI and the AQI. To remedy this, a banner is used on the AQHI web site. If the air quality is judged to be degraded according to the AQI, but the AQHI remains below 6, a banner is posted informing the public of degraded air quality and provides appropriate health advice. If the banner is displayed for more than three consecutive hours, the alert service is activated to provide subscribers with a special health message, which differs from the AQHI health message.

The Future of the AQHI in Québec Following a decision made in March 2012, local public health authorities involved in the AQHI project supported the continuation of the AQHI pilot project in Québec until March 2015, provided that that the AQI and the Info-Smog program remain and that a review is undertaken of the importance of $PM_{2.5}$ in the AQHI formula. Expansion of this project has been divided into three stages. The first is the addition of monitoring stations in Québec City and Gatineau. Subject to the approval of the various partners and the availability of required budgets, the next step will be the implementation of the AQHI in the suburbs to the north and south of Montréal. Finally, and under the same conditions, the service could be extended to three additional urban areas: Saguenay, Sherbrooke, and Trois-Rivières. At the end of the pilot project, the Québec government will make a recommendation for or against the final adoption of the AQHI in Québec.

18.7.3 AQHI Applicability in Large and Small Communities

To establish robust epidemiological links between air pollution and mortality data, health studies need to rely on information from hundreds of thousands of individuals, typically in *large urban areas*, since only about 0.1 to 1 % of mortality is attributable to air pollution (British Columbia Provincial Health Officer 2004). We can therefore have the greatest confidence that the AQHI accurately reflects health risk in similar urban settings where the mix of air pollution is largely derived from local vehicle emissions together with regional "smog". In rural settings or small communities where sources are much different, the AQHI may predict health risks less accurately. Note that this difficulty pertains to any measure that links air quality and health, including provincial or federal air quality objectives and trigger levels for air quality advisories, not just the AQHI.

Similarly, both the AQHI as well as air quality objectives and standards may confer different levels of protection where air pollution sources and mixtures are different. For example, 25 μ/m^3 of $PM_{2.5}$ may differ in toxicity depending on whether it is derived from wildfire smoke, extremely fine crustal material, or urban smog.

18.8 Promoting the AQHI

Promotion of the AQHI to the public continues to be an essential component of the AQHI program (Fig. 18.4). A communication initiative was developed using a pan-Canadian process that engaged hundreds of stakeholders that included the public, private and non-profit/voluntary sector. This process was a partnership between Environment Canada, Health Canada and key municipal and provincial colleagues that assessed a wide range of views on how the AQHI information should be displayed, accessed and used. Issues resolved in this process included defining the objectives of the AQHI, identifying the audience for the index, assessing different audience needs and testing AQHI communication materials and approaches.

Health Canada and Environment Canada have supported a number of projects across Canada to raise awareness of the AQHI. One of these was in Toronto where a significant effort was made to promote the use of the AQHI (Fig. 18.5). Another was in British Columbia where an extensive, multi-year campaign was implemented to promote the recognition and use of the AQHI throughout the province, particularly with a focus on the "at risk" population and, during the summer of 2010, the health risks associated with wildfire smoke in the interior of the province. More information on communication and promoting the AQHI is included in Chap. 19 on communicating air quality information in this book (Fig. 18.6).

Since changing weather in a community is an important reason for hourly and daily changes to air quality and the AQHI, Environment Canada has incorporated the observed hourly AQHI values for each community into its website http://weather.gc.ca/. It also provides the 24–36 h AQHI forecast with the related health messaging on this site. A number of municipal, provincial and private websites also provide AQHI and other air quality information, including the following:

- Alberta: http://environment.alberta.ca/0977.html.
- British Columbia: http://www.bcairquality.ca/readings/index.html
- Manitoba: http://www.gov.mb.ca/conservation/pollution-prevention/airquality/aq-health/
- New Brunswick: http://www.gnb.ca/0053/public_health/air-e.asp
- Newfoundland and Labrador: http://www.env.gov.nl.ca/env/env_protection/science/aqhi.html
- Nova Scotia: http://www.gov.ns.ca/nse/aqhi/
- Prince Edward Island: http://www.gov.pe.ca/health/index.php3?number=1021134&lang=E
- Québec: http://www.coteairsante.qc.ca/Default.aspx?lang=en
- Toronto: http://www.toronto.ca/health/airquality/aqhi/faq.htm
- Weather Network: http://www.theweathernetwork.com/airquality/cancitiesaq_en/

18.9 The AQHI in the Future

Several areas of research are planned for the AQHI. Health-focussed research areas include self calibration of the index, exercise and exposure to air pollutants, applicability of the index to sensitive populations, and communication of the health risk. Other areas include the applicability of the AQHI to rural populations, indoor air quality vs. the outdoor AQHI, scientific formulation of the AQHI and the application of the index internationally (Health Canada 2011).

More methods will be developed to lower health risk. For example, a promising tool is being designed to give advice to asthmatics by integrating the AQHI into their asthma action plans. This system uses a combination of an electronic asthma action plan with the current and predicted AQHI to provide health advice to patients. This advice is based on clinical data on symptoms and other indicators that patients

Fig. 18.6 A poster promoting the relation between the AQHI and health risk. (Image courtesy Toronto Public Health)

routinely send to a server. The communication is based on a web-based smart phone application (Licskai et al. 2012).

18.10 Conclusion

The Air Quality Health Index promises to be a practical tool to provide Canadians with local air quality information on an hourly and daily basis that can be used to protect their health from the negative impacts of air pollution. It can also be used as a measure to compare the short term health risk due to air pollution in different communities, with some caveats.

The good news is that air quality tends to be improving through the collective efforts of public education, monitoring and enforcement, better engineering, and mitigation of known polluting sources. Air quality can be measured through existing air quality monitoring systems and can be synthesized into a simple communication tool in the AQHI. This Index can reach the majority of the Canadian population, although many communities still lack the monitoring systems required to generate the AQHI. Further experience with the AQHI will help to identify better how to intervene with individuals and populations to act to reduce our risk from poorer air quality (Hasselback and Taylor 2009).

References

Alberta Government (2012) What is the Air Quality Health Index. http://environment.alberta.ca/03603.html. Accessed November 2012

Bates D, Koenig J, Brauer M (2003) Health and air quality 2002—Phase 1, methods for estimating and applying relationships between air pollution and health effects. BC Lung Association. http://www.bc.lung.ca/pdf/health_and_air_quality_2002.pdf. Accessed 5 Feb 2012

British Columbia Provincial Health Officer (2004) Every Breath You Take…Provincial Health Officer's Annual Report 2003. BC Ministry of Health Services. http://www.health.gov.bc.ca/pho/pdf/phoannual2003.pdf. Accessed 5 Feb 2012

Goldberg E (2002) Aggregated environmental indices: review of aggregation methodologies in use. Organisation for Economic Co-operation and Development, Paris

Hasselback P, Taylor E (2009) Air Quality Health Index Variation across British Columbia, BC Ministry of Environment. www.bcairquality.ca/reports/pdfs/aqhi-variation-bc.pdf. Accessed 5 February 2012

Health Canada (2011) Air Quality Health Index Research Plan: 2011–2016-Draft

Henderson D, Stieb D, Economouc V (2004) A new approach to communicating ambient air quality information. Clean Air Society of Australia and New Zealand

Hidy G, Pennell W (2010) Multipollutant air quality management. J Air Waste Manag Assoc 60:645

Johnson BB (2003) Communicating air quality information: experimental evaluation of alternative formats. Risk Anal 23(1):91–103

Judek S, Jessiman B, Stieb D, Vet R (2004) Estimated number of excess deaths in Canada due to air pollution. Health Canada, Environment Canada http://www.metrovancouver.org/about/publications/Publications/AirPollutionDeaths.pdf. Accessed 19 Apr 2013

Licskai C, Sands T, Ferrone M (2012) Integrating the Air Quality Health Index into Asthma Action Plans. Am J Respir Crit Care Med 183:A4767 (2011)

Office of the Auditor General of Canada (2009) Status Report of the Commissioner of the Environment and Sustainable Development. Government of Canada

Pengelly D, Campbell M, Mcfarlane R, Li-Muller A (2001) Toronto Air Quality Index: Health Links Analysis. Toronto Public Health. http://www.toronto.ca/health/hphe/pdf/taq_index.pdf. Accessed 19 April 2013

Pope CA III, Dockery DW, Schwartz J (1995) Review of the epidemiological evidence of health effects of particulate pollution. Inhal Toxicol 7:1–18

Stieb DM, Burnett RT, Smith-Doiron M, Brion O, Shin HH, Economou V (2008) A new multipollutant, no-threshold air quality health index based on short-term associations observed in daily time-series analyses. J Air Waste Manag Assoc 58:435–450

World Health Organization (2000) Guidelines for Air Quality 2000. WHO. http://www.airimpacts.org/documents/local/AQGUIDE.pdf. Accessed 5 February 2012

Communicating About Air Quality—Lessons From Canada's AQHI

19

Sharon Stevens

Abstract

This chapter outlines the social marketing challenge of increasing personal and collective awareness, understanding and action related to air quality and its effects on health. It provides an overview of some of the key lessons learned through Air Quality Health Index outreach programs implemented in Canada. The power of outreach partnerships is explained, and illustrated, through a case study of one social marketing campaign implemented in western Canada. The chapter explores the role of social media within social marketing about air quality and concludes with some future gazing and samples of new opportunities to encourage people to care about the air.

Keywords

Health promotion · Social marketing · Outreach · Behavior change communication · Social media

19.1 Introduction

Driven by a shared interest in protecting Canadians from the health impacts of poor air quality, and to improve both population health and air quality itself, Canada's Air Quality Health Index (AQHI) is a pan-Canadian process that engaged stakeholders from the public, private, and non-profit/voluntary sector.

First presented to the public in 2004 through a pilot presentation in Western Canada, the AQHI has always been a partnership between two federal departments and a growing list of partners and players. Using a stakeholder engagement model, these federal agencies have provided funds, research results and social marketing resources about the AQHI to provincial and regional governments, health and environment organizations, NGO groups and media organizations. The goal was to raise awareness and engagement about air quality and health, and for people to use the AQHI to better manage their health.

All the while, these government agencies and their network of partners and stakeholders have collaborated on social marketing of the AQHI by sharing lessons learned as well as outreach tools and tactics.

19.2 Balancing the Science and the Social

Identifying pollutants of greatest harm and quantifying the cost of that harm is the work of science and economics. This work is intended to inform policy.

There is no doubt that the policies and programs, stemming from the acceptance of the science, are the subject of ongoing study and debate, within Canada and around the world.

The question that remains is how does ongoing scientific study and policy debate affect the approach toward social marketing related to air quality?

This is the answer—and the focus of this chapter. If we agree that people can have an effect on the quality of the air they share, and we accept that the benefits of our individual or collective attention and action outweigh the risks of inaction, then we must also accept that increased public interest and engagement about air quality should be encouraged, supported and celebrated, everywhere.

S. Stevens (✉)
Air Shift Group, Kamloops, Canada
e-mail: sharon@airshiftgroup.com

E. Taylor, A. McMillan (eds.), *Air Quality Management,* DOI 10.1007/978-94-007-7557-2_19,

Table 19.1 Air Quality Health Index values which characterize health risk on the left, are linked to appropriate health advice on the right

Health risk	Air Quality Health Index	Health messages	
		At risk population*	General population
Low risk	1–3	*Enjoy* your usual outdoor activities	*Ideal* air quality for outdoor activities
Moderate risk	4–6	*Consider* reducing or rescheduling strenuous activities outdoors if you are experiencing symptoms	*No need to modify* your usual outdoor activities unless you experience symptoms such as coughing and throat irritation
High risk	7–10	*Reduce or reschedule* strenuous activities outdoors. Children and the elderly should also take it easy	*Consider* reducing or rescheduling strenuous activities outdoors if you experience symptoms such as coughing and throat irritation
Very high risk	Above 10	*Avoid* strenuous activities outdoors. Children and the elderly should also avoid outdoor physical exertion	*Reduce or reschedule* strenuous activities outdoors, especially if you experience symptoms such as coughing and throat irritation

Another chapter in this book provides a comprehensive overview of the science and policy related to Canada's Air Quality Health Index. This chapter shifts the focus from science and policy to people's awareness and experience related to air quality. It is about the social marketing and outreach considerations related to encouraging and supporting people's increased awareness, engagement and action related to air quality and health.

19.3 From Those Who Have Gone Before

There are some simple truisms that should guide any communication and outreach efforts designed to ultimately influence someone's behaviour. No matter what an organization says or does, no matter which communication channels they use to spread their message, no matter who those messages are ultimately intended to inform or influence, there are two imperatives:

- Keep it simple
- Make it personal

This chapter provides food for thought in how to keep social marketing messages simple and make them personally relevant.

Case in point. Public opinion research in Canada found a significant difference between people's understanding of how air quality affects their health and their awareness of their own health risks related to air quality. About 96% of people studied believe air quality affects health to some degree, but very few said they would seek out air quality information on a regular basis. That's a big gap between awareness and action. Likewise, people reported they may consider the long term effects of poor air quality but few of them were able to identify the immediate impact on health, unless this connection was because of a personal experience related to air quality.

Helping people understand an issue and see how it is relevant to them personally is the precursor to action. What does that have to do with social marketing? When the previously mentioned study compared people with the highest awareness of air quality to those with the lowest awareness of air quality, they found a significant difference depending on if they lived in an area where a social marketing campaign on air quality, health and Canada's AQHI, had been implemented.

19.4 What's The Big Deal About Messaging?

The messages about how to protect your health are the foundation of Canada's AQHI.

The AQHI's health protective and environmental stewardship messages were crafted very carefully. Many health and environment organizations were consulted to ensure the AQHI's messages are as appropriate and applicable to as many people as possible (Table 19.1).

That said, these health messages are not the focus of this chapter. Instead, the focus is on the concept of social marketing *messages* used to attract and sustain attention and interest in air quality and health.

19.5 A Staged Approach

When it comes to prompting people toward increased awareness, engagement or action related to air quality and health, it is important to recognize that no two people are at the same starting point in this journey. And the journey is a process. Like any process, it can be chunked into steps or stages along the way, steps or stages integral to changing behaviour.

There are a great number of social marketing and outreach efforts underway by air quality policy, monitoring, regulatory and advocacy groups across North America and beyond. An array of social marketing tools, including everything from posters and brochures to websites, tweets, and smart phone apps, reveal a pattern related to social marketing messages. It's a concerning pattern worth noting for future social marketing programs.

The pattern is that not nearly enough of these social marketing efforts demonstrate an understanding of the importance

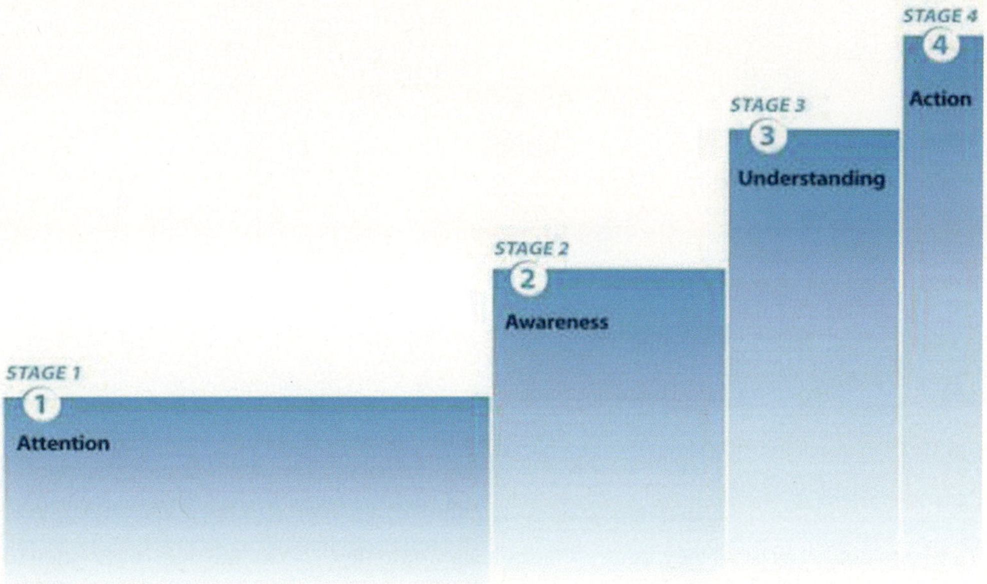

Fig. 19.1 A simplified version of McGuire's Theory, "Hierarchy for Constructing Persuasive Communication. (McGuire 1974)"

of each stage within behaviour change communication that people need to transition through before they will ever 'do anything' about the air quality's effect on their health.

Instead, far too many air quality outreach messages try to prompt a change in behaviour, starting at the end of the behaviour change communication process, without first trying to seek and sustain attention, build awareness or support a deeper understanding of air quality—all prerequisites to someone taking action or doing anything differently.

In any kind of marketing, it's tempting to focus communication on prompting a desired behaviour. But for social marketing about air quality and health, it's ill advised. Behaviour change communication is well studied. Just Google "behaviour change communication" or "integrated behaviour change model" and you will find a plethora of theories which explain how and why messages intended to prompt behaviours either fail or succeed. According to Owen and Lee's integrated model (1994), behaviour change is a cyclical process that involves five stages: (1) awareness of the problem and need to change; (2) motivation to make a change; (3) skill development to prepare for the change; (4) initial adoption of the new behaviour; (5) maintenance of the new activity and integration of it in to a person's lifestyle.

Social psychologist William J. McGuire, in his effort to further develop the Yale Attitude Change Approach Model (a method to making persuasive communications effective) created a 6-step persuasion process, which in summary suggests that before taking action, people must first:

- Tune into the message (attention)
- Attend to it (attention)
- Maintain interest in it (awareness)
- Understand it (understanding)
- Think about it (engagement)
- Agree with it (engagement)
- Store the message in memory (engagement)
- Decide to act on it (action or intention)

In the context of communicating about air quality, this means that air quality outreach messages should focus on supporting each of these responses within this continuum. Far too often, air quality health and advocacy messages skip straight to 'here is what you can do'.

> Behaviour change happens in stages. The marketing and communication designed to inspire and support behaviour change must address each stage to be effective.

Figure 19.1 shows a simplified version of McGuire's Theory, categorizing only four stages of communication, which has guided a good portion of the air quality outreach *message strategies* developed for Canada's AQHI.

Let's break these down and go a bit deeper into the goal of each stage and how to achieve it.

Stage 1—Get People's Attention Does this seem like a statement of the obvious? Perhaps, but this critical stage of social marketing is often overlooked. No one would argue that getting attention is a mandatory stage in all social marketing efforts. The problem is that it's often the most challenging stage to address, especially when the topic is as complex as air pollution, health and climate change. It's no coincidence that the graphic depicts stage 1 as shallow and wide. This is the stage where social marketing messages need to appeal to as wide an audience as possible through messages that are as simple and personally relevant as possible.

That means don't overload people with too many details too early in the process!

Or in other words, Beware the *curse of knowledge.* The curse of knowledge is what happens when air quality experts develop communication messages and tools and forget what

Fig. 19.2 An image from the Movember campaign to raise awareness of prostate cancer. Image Credit—Movember Canada

it's like to not know what they know. It lacks insight into the average person's level of interest, awareness and understanding. And it can have significant negative impact on the success of any social marketing program.

Here's the rub. It's highly likely that readers of this book already associate with people who are already aware of, and understand, complex issues related to air quality, health and environmental stewardship. These people are often already at, or beyond, **stage four** in the continuum of change communication. They get it, agree and act accordingly.

Yet it's not uncommon to see air quality social marketing messages and tools that have clearly overlooked the other 99.9 % of the intended audience, the very people social marketing efforts are designed to reach. The ones with little understanding of the details related to air pollution, or more specifically, its impact on health.

The irony is that the more complex the social marketing topic, the more imperative it is to make it simple.

It's an imperative that the social marketing strategists for the prostate cancer movement Movember understood well (Fig. 19.2). The Movember Movement set out to eliminate the social stigma related to prostate cancer by making this men's health issue an accepted dialogue between, and among, every man and every woman. In developing the Movember Movement, the creative minds behind the campaign suppressed the desire to lead with the risk factors or symptoms of prostate cancer, or the focus on an appeal for financial support, too early in the campaign. Instead, they led with a smart and simple message that would appeal to as wide an audience as possible—men (Mo Bros) and the women who love them (Mo Sisters). All people needed was a 'core idea' they could agree with—that men's health is worth thinking about, and talking about! By agreeing with this core message, people would be more empowered to become informed, engaged and motivated toward action related to this disease. That's when the mustache growing and fund raising challenges began—in homes and offices around the world.

In the 'get their attention' stage, those responsible for communicating with the public should do everything possible to ensure that social marketing messages about air quality and health do not suffer from *the curse of knowledge*.

They should also resist the urge to focus on the details of air quality indices, specific pollutant numbers, forecasts or suggested actions. Instead, it is important to recognize that before the average person is going to be interested or take action, they need to first say "Ah, ha… *this affects me and I should pay attention*".

How can social marketing messages make that happen? Chip and Dan Heath, the authors of the New York Times bestseller *Made to Stick: Why Some Ideas Survive and Others Die* state it with such simplicity—"by digging the hole before trying to fill it in". They explain that curiosity happens to people when they realize there is a gap in their knowledge. The goal is to open these so called 'gaps' in the first stages of communication—before trying to fill them in with too much detail.

This approach was used for the introduction of the Air Quality Health Index in Canada's most western province, British Columbia. The public introduction of the AQHI in BC illustrates how the four stages of behaviour change communication were applied.

Canada and other fortunate parts of the world, are blessed with relatively good air quality. There are, however regional and seasonal variations, as well as natural events such as forest fires, that significantly compromise local air quality in some areas. When this happens, poor air quality rapidly accelerates the speed through which people move through the behaviour change stages toward action. This is a key consideration in countries or communities routinely suffering from poor air quality where interest, awareness and engagement are high because of personal experiences.

The downside to being blessed with generally good air quality in many parts of Canada, is the challenge of getting people to pay attention to the quality of the air they share.

Fig. 19.3 A "simple and smart" water conservation message developed by Denver Water

Fig. 19.4 Twitter message promoting the Air Quality Health Index in Kamloops, British Columbia

Exactly the challenge that stage 1, Getting Attention, is intended to address. It recognizes that, for the majority of people, there is a gap in their knowledge related to air quality and how it affects their health. Let alone the indices that measure and report on it. It's the stage that focuses on this gap and sets the stage for filling it in.

Focus testing across Canada in 2008 confirmed that, for the majority of people, there was indeed a gap in awareness about the effects of air quality on health and the indices that report on it. These topics simply had not yet earned a place on most people's mental shelves.

Stage 2—Awareness The goal at this stage of behaviour change communication is to help the "unaware" by making them aware of their need for more information.

In British Columbia, social marketing tools at this stage of the Province's campaign proposed, first and foremost, that air was something people needed to get to know. *AIR QUALITY—GET TO KNOW IT!* The goal was to use the power of curiosity, consistent with Heath's theory of "digging the hole" before filling it in, by focusing on the "core of the idea" that maybe air was something people should get to know a bit better. The details related to the science, the policies or even the AQHI itself took backstage to the simple idea of 'knowing your air'.

The goal was to present air quality and health in a way that was simple, and smart. Figure 19.3 is an example of the simple and smart approach related to water conservation, developed by Denver Water in Colorado, USA. It gets attention, is easy to understand and is clever and memorable.

Stage 3—Understanding The goal at this stage in behaviour change communication is to support increased understanding. Even at this stage, messages should remain as simple and as unintimidating as possible. They are intended to start "filling in the hole" created when someone tunes into a topic and starts to want to know more about it.

Even in the confines of 140 characters on the social media platform Twitter, messages can still be created to address each stage of behaviour change communication (Fig. 19.4).

Stage 4—Engagement/Action The goal at this stage, now that a person has sustained their interest and increased their understanding, is to give them the information, and reassurance needed to act. Messages at this stage should be detailed, concrete and include clear calls to action.

At this stage in the AQHI social marketing campaign in British Columbia, the core message was "Know the Number—Protect Your Health". The outreach intended to address those people who were at stage 4 provided more detailed information about the Air Quality Health Index itself and reinforced the reasons why someone should check the Air Quality Health Index on a regular basis.

At the end of the day, social marketing about air quality, health and the environment is about motivating people to do something, for example: access an AQ index, turn off their idling car or reschedule their outdoor activities. Behaviour change communication is about people and their individual relationship with a topic. When sharing messages about air quality, communicators should be ever mindful of the fact that each person receiving these messages will be at a different stage in their relationship with air, how it affects them, and what they may do about it. Communicators should ask themselves:

- Are the majority of key messages aimed at the ACTION stage, with lots of detail, intended to prompt people to "do something"?
- Would equally as many fall into the AWARENESS and UNDERSTANDING stages with the appropriate amount of information required to sustain someone's interest and inform increased understanding?
- Do enough key messages focus on the ATTENTION stage, encouraging people to stop and say "*what is this about? Is this something I should know?*" *Does this affect me?* Are these messages simple and relevant enough to earn a place on someone's mental shelf that day? (Fig. 19.5)

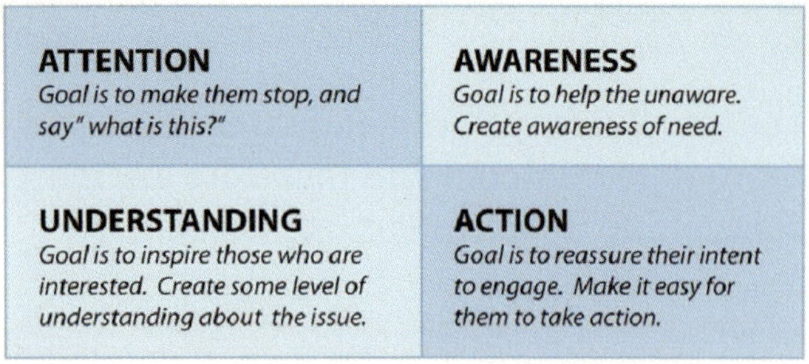

Fig. 19.5 Communication designed to support behaviour change is a continuum—and communicators can't assume what stage people are at when they are exposed to a message

19.6 The Power of Partnerships

When it comes to social marketing, there are few things as true as the power of partnerships. Just as "many hands make light work," the more organizations spreading a common message, the greater the results.

A key contributor to the success of the Air Quality Health Index outreach in Canada has been the program's willing and motivated partners. The potential of a collaborative model was recognized from the first introduction of the AQHI in Canada. The power of partnerships has been a guiding principle over the past 10 years.

As early as 2004, among municipalities, health and environment NGO's and related advocacy groups, Canada implemented a process of identifying the willingness and resources available to help spread the message about air quality and health and the AQHI.

Surveys and face-to-face, telephone and email outreach to these types of organizations across Canada, identified shared mandates, communication capacity and practices. More specifically, it provided key insights, such as:

1. The extent to which organizations, who regularly communicated news or information to employees, members of the public or the media, would share information about the AQHI,
2. Their willingness to include AQHI within existing tools,
3. The types and frequency of events they participate in that provide an opportunity to promote the AQHI.

Almost all organizations surveyed had practices and processes for communicating with the public. Most had formal processes for communicating regularly with partners. They used numerous communication and outreach tools, and 95 % of these organizations were willing to include information about the AQHI on their website. Just over half indicated they would be willing to create a dedicated AQHI page on their site, and many said they would promote the Index through their email distribution lists.

With such willing and engaged partners, the lead federal agencies in Canada guided the distribution of funding, shared tools, resources, and information about the AQHI through to provincial and regional governments, health organizations, NGO groups and media organizations. All the while, the government worked to improve information sharing processes through lessons learned and a variety of ongoing outreach recommendations, tools and tactics.

Outreach tools were created by the lead agencies, and equally as often by partnering organizations across Canada, to promote the AQHI to specific at-risk stakeholders and those that might influence them most. Key stakeholders included:

- **Dissemination Networks**—Organizations that could serve to disseminate tools, resources, and information about the AQHI, including media.
- **At-Risk Populations**—Those with respiratory or cardiovascular disease, children, the elderly, and other vulnerable populations, accessed through schools, nursing homes, rehabilitation centres, day care centres, and other community-based services.
- **Health Service Providers and Advocates**—Health professionals, hospitals, community health facilities and health information sources that would be most likely to influence people experiencing the health effects of air quality.

Overall, the AQHI outreach campaign garnered admirable levels of support and engagement by multiple partners across the country, in turn expanding the reach of air quality and health information and advocacy to levels not witnessed prior to the introduction of Canada's AQHI.

Equally important, it allowed for evaluation of social marketing implemented by a diverse group of government, non-government and non-profit organizations throughout the country. In December 2009, Health Canada commissioned "Consolidating AQHI Outreach Reports: Lessons Learned and Recommendations." This report was followed by another in 2010 which summarized the findings and recommendations from the AQHI pilot program. This evaluation identified a number of recurring lessons about social marketing of air quality including the continued importance of leveraging collaborative partnerships.

19.7 Case Study—One Province, Multiple Partners

Between 2004 and 2010, the Province of British Columbia played a leadership role in Canada's introduction of its Air Quality Health Index. Today, the AQHI is available in 18 communities in BC to more than 80 % of the province's population.

Over this 5 year period, provincial and federal government and NGO partners, applying social marketing strategies developed by the team at Air Shift Group, a division of

Communication Solutions Inc., implemented a comprehensive social marketing and outreach program designed to raise awareness, understanding and use of the Air Quality Health Index.

The focus of the program was to create and sustain multiple linkages with stakeholders in health and environment, reach at-risk populations, and encourage health professionals to share air quality and health information. Outreach was targeted to health care providers and at-risk populations including those vulnerable to respiratory conditions, participants in active living, as well as sport audiences and youth. The core mandate was to create and support promotion partnerships, opportunities and capacity among health and environment groups, government agencies and municipal governments in an effort to eventually sustain AQHI promotion without federal funding.

In BC, it all began with a key question: what do people think of when they think about 'air' and about 'health'? Are these two things connected in people's minds?

The planning included literature reviews, lessons from similar social marketing case studies, consultation with health experts (physicians, respiratory therapists, etc.) and health associations (BC Lung Association, etc.). It also included observation of focus testing. All of which confirmed that many British Columbians were not aware of the extent to which air quality affects their health.

"*Air Pollution: Information needs and the knowledge, attitudes and behaviour of Canadians*" (Environics Research Group 2002) reinforced the importance of creating an identity for this new health index, keeping it short, bold and easy to remember, and minimizing the scientific and technical jargon in how it is presented to the public.

In a part of the country where the ocean and mountains are synonymous with 'fresh air', air quality, for the majority of people, is not something that is top of mind. Nor was it commonly associated with personal health.

Therefore, the social marketing challenge was clear. If the majority of people don't realize the extent to which the air they breathe affects their health, they are unlikely to pay close attention to a new Air Quality Health Index.

Additional public opinion research explored awareness of air quality and health, introducing the Air Quality Health Index into the conversation for the first time. These kinds of comments were not uncommon:

> *This allowed me to understand that air pollution has a big impact on my health.*
>
> *It's something i've got to think about. How it fits in my life.*
>
> *I never really put much thought into environmental impacts on my respiratory system.*

So, before asking people to start using Canada's new health based air quality index, they were instead invited to "get to know" the air. The "tone" of the social marketing was both approachable and informative, and on occasion employed just the right amount of wit to attract the attention of those who, for the most part, did not give much thought to air—nor its effects on their health.

19.7.1 Getting Attention by Invoking Curiosity

Using the 'Gap Theory' described earlier, communication tools were developed to get people's attention by suggesting there was something they needed to know more about. This technique was designed to prompt people to recognize a gap in their knowledge—and a personal interest or desire to have that gap filled.

Numbers and visual synonyms are great tricks of the trade at this stage. People (and media) love numbers and visual shorthand helps them to understand by creating a picture of something simple in their mind.

These examples of social marketing messages used in British Columbia illustrate how the gap theory, numbers and visual short hands were used.

> Radio Advertisement Copy: "According to the latest research, 100 % of all humans require air for even the most basic life supporting functions. That includes breathing. So if you are among the 100 %, know more about your air—at airhealth.ca. For the other 0 %, uh, as you were. The Air Quality Health Index—at airhealth.ca—Get to know it."
>
> To listen to this radio advertisement, visit http://www.airhealthbc.ca/radio.htm
>
> Outreach Product Headline: On average we take 20,000 breaths a day. That's a lot of air—enough to fill 10,000 L. Find out how all that air affects your health by visiting www.airhealthbc.ca
>
> Print Advertisement Headline: You use it every day. You have used it since the moment you were born. You can't see it, touch it or hold it—And you cannot live without it. Isn't it time you got to know it?

Online outreach The initial web presentation of the AQHI in British Columbia was at www.airhealthbc.ca—a website introducing the relationship between air and health and providing additional "layers" of information to meet varying levels of interest, awareness and understanding about air quality (Fig. 19.6).

Branded web buttons (graphic or image provided with codes linking the image to www.airhealthbc.ca) were created and distributed to all partners and media. Short videos were provided to media as Public Service Announcements and posted on YouTube (Fig. 19.7).

Outreach and the Creation of Champions Third party champions (those with extended networks and credibility) were asked to extend the reach of the air quality messages to their membership or key stakeholders. These 'champions' included health professionals such as respiratory therapists

Fig. 19.6 Website used in British Columbia to present air quality and AQHI information to the public

Fig. 19.7 Image linking to a video presentation on how athletes can use the AQHI to reduce health risk

and public health agencies as well as municipal leaders, politicians, clean air advocates, athletes, people with chronic disease, seniors, children and their parents.

This approach saw, for example, outreach to at-risk and target audiences at flu clinics throughout the province. Imagine the increase in interest and understanding when people were introduced to the AQHI by the public health nurse administering their flu vaccine (Fig. 19.8).

Presentations were made to more than 100 Respiratory Therapy students and faculty at Thompson Rivers University, to the Respiratory Therapist professional association, and at healthy living fairs and events across the province. Similar outreach happened at outdoor athletic events, clean air and environmental stewardship events and healthy living events throughout the year.

Through repeated use of radio, transit bench and shelters, television and print advertising, and the distribution of promotional materials to partners and networks of provincial and municipal governments, not-for-profit organizations and environment and health advocates, the "Air Quality—Get to Know It" campaign reached millions of people in BC.

19.7.2 Early Integration of Social Media

In 2009 and again in 2010, when numerous social media channels were just beginning to gain popularity, social media was used to take advantage of "teachable moments" created by large forest fires affecting air quality through much of the province. It was an opportune time to share messages via social media channels to increase awareness and understanding of the AQHI, and most importantly, give people information on how to protect their health.

The result showed a 2,500 %, increase in daily visits to www.airhealthbc.ca, a significant increase in Twitter followers (Fig. 19.9) and re-tweets, increased reference to the Index by others, and an increase in media inquiries and reporting about air quality, health and the Air Quality Health Index.

Fig. 19.8 Face to face outreach is an important part of raising awareness of air quality and health

Fig. 19.9 A Twitter message promoting the use of the AQHI

The analytics in Fig. 19.10 were captured from the Google Analytics account for www.airhealthbc.ca on September 23, 2010. It shows that more than 75 % of visitors to British Columbia's air quality health index website were the result of referrals or links from other websites, primarily media and program partners. Once again, this illustrates the power of partners in social marketing programs.

Social media strategies and content were also used to build awareness and understanding of air quality and health and Canada's AQHI, during the 2010 Winter Olympics in Vancouver BC.

Air quality and AQHI information shared via @aqhivancouver and @aqhiwhistler Twitter accounts focused on Olympic, sport and tourist audience interest in sport and environment to draw attention to air quality at Olympic venues, and promote how people could check the AQHI when planning to attend events.

Below is a sample of tweets from these two Twitter @ aqhi accounts during the 2010 Vancouver Olympics, posted to encourage re-tweets and visits to the website

> Hitting Vancouver streets today to catch #Olympic excitement? Don't forget you can check the air quality anytime at http://bit.ly/ldyy0

> Did you know #air quality can affect athlete health & training. Learn more about AQ in Vancouver & Whistler http://bit.ly/9kwgwc #olympics

> Enjoying activities today in Whistler? The #Air quality Health Index is 2—ideal conditions for being outside #olympics http://bit.ly/dkuvbq

> #air quality Health Index for Richmond/Delta is currently 4 (MODERATE). Find out what that means to your #health http://bit.ly/b57oqu

> #Olympic ski jumping finals are happening at Whistler this morning where the #airquality is LOW RISK. http://bit.ly/cs2j2q

Throughout the multi-year social marketing effort in British Columbia, numerous goals were established and a number of methods used to measure success.

Here is a synopsis of a few of these goals and the results achieved:

Make the Connection 100 % of respondents to a web survey agreed the air affects their health. 42 % of them considered themselves at increased risk from air pollution. Most importantly, 62 % of respondents told others about the AQHI.

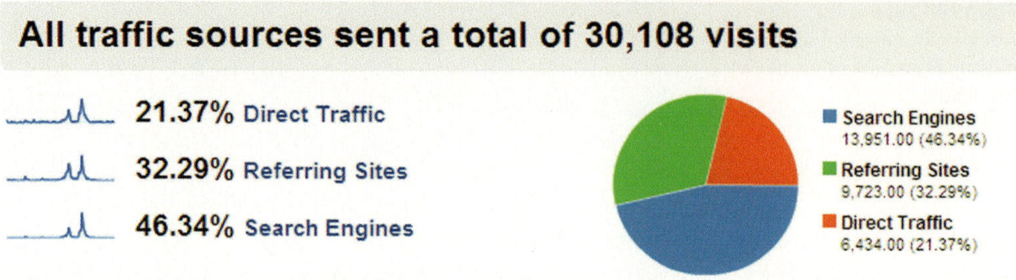

Fig. 19.10 Web activity is an important measurement in assessing online exposure to air quality communication tools like the AQHI

Maximize Partners 95% of media, healthcare, municipal and NGO partners who were asked to support the AQHI offered to host a link to the AQHI website. Two media partners become the top referring websites to the AQHI website. 100% of media partners offered in-kind support, doubling the exposure that would otherwise have been achieved. Several media partners included the AQHI with their regular weather report.

The quantity and quality of outreach partnerships, web visitation metric, online survey participation and results, and public opinion research, all point to the success of using a partnership model to inform and encourage the Air Quality Health Index as a personal health management tool.

19.8 Getting Social About Air Quality and Health

This section provides an overview of the social marketing channels and tools for consideration within an air quality and health social marketing program.

Before elaborating on marketing channels and tools, a point of clarification is necessary. Many people continue to confuse social media with social marketing or cause marketing. They are not the same.

Social Marketing Social marketing is marketing that builds awareness, understanding and engagement about a social issue. Any marketing designed to change, in the long term and in a measurable way, people's attitudes or behaviours *for the public good* is considered social marketing. Campaigns related to recycling, vaccinations, safe sex, prostate cancer and saving the whales, to name a few examples, all qualify as social marketing. Social marketing uses techniques developed and refined by more commercial marketers to "sell" healthy behaviours to children and adults around the world. It became a formal health communication discipline in the 1970s after Philip Kotler and Gerald Zaltman published an influential article in the Journal of Marketing titled "*Social Marketing: An Approach to Planned Social Change*."

By contrast, any marketing designed to influence people's vote, the car they drive, where they bank, or the mattress they purchase does not qualify as social marketing.

Cause Marketing Many for-profit organizations are now commonly building social causes into their marketing efforts to build business and increase sales. While raising awareness and money for a cause is commendable, Cause Marketing is, by virtue of its very purpose, to build business. The marketing associated with a cause may support increased awareness of that cause, and often raise funds, but this should not be confused with social marketing.

Social Media On the other hand, social media is online and emphasizes interactive conversations and information sharing between people, organizations and 'brands'.

Now let's focus on getting social about air quality and health, and explore why it's imperative for air quality, health and environmental stewardship marketing programs to include social media strategies and activities.

Social media represents a broad change in how people communicate with one another—using online platforms to share opinion, insights, experiences and information. The main difference between social media and traditional media is that the content and conversation, the dialogue, is shaped by the users (many to many) rather than a monologue controlled by an individual or organization (one to many).

When used strategically, it can lead to exponential reach, virally, by people sharing and contributing to a topic with their network(s). If the content is relevant and engaging, people will interact, inform, and share important messages. The content (stories, views, opinions and calls to action about air quality) lends itself to being archived, indexed by search engines, and shared by users in many ways.

Health and environment organizations are increasingly using social media platforms to personalize communication with large audiences, build trust and strengthen relationships, and engage key audiences in transparent dialogue, something that is difficult to do through traditional media channels. Here are some other reasons to include social media strategies in social marketing programs:

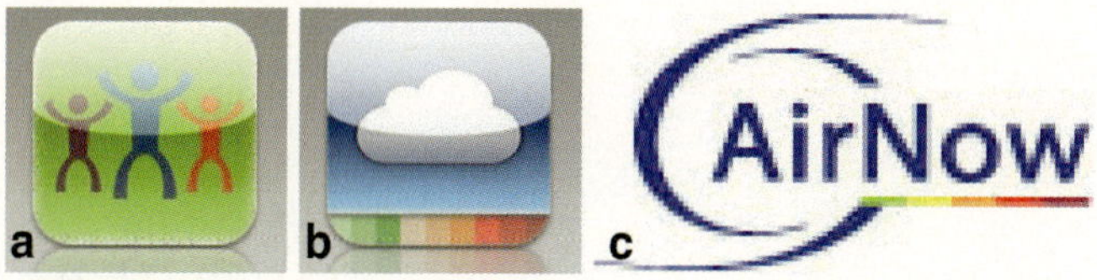

Fig. 19.11 A sample of just some of the apps offering easy access to local air quality

The Audience is Online People are spending less time watching television, reading newspapers or listening to the radio and more time on the Internet and on social media channels. Social networks and blogs now account for nearly 25 % of total time spent online and nearly 80 % of those people who are online are active on social networks and blogs. Likewise, social media is replacing traditional news outlets and journalists as a source of news.

Increasingly, accessing social media channels is a growing feature among mobile device users. An estimated 46 % of social media channel users are accessing these sites via their mobile phone, another 16 % via a tablet. In just 12 months from July 2011 to July 2012, the number of people accessing the web via their mobile device rose 82 % while the number of people using mobile apps grew 85 %, (Nielson, Social Media Report: December 2012).

And this trend includes both young and old. People aged 55 + are among the fastest growing category of mobile users accessing social networking sites.

With the explosion of Smart Phones has come an explosion of apps related to weather, health and local air quality. Of particular note, the 2nd most common category of app downloads is related to weather, (Neilson's Social Media Report, Q3 2011).

Figure 19.11 shows a few of the apps offering access to local air quality, in Canada and internationally.

They are Looking for Health Information People used to rely on only a few sources when looking for health information—including their physician, a friend or family member. Today, they are using Google, reading blogs and social media, listening to podcasts and posting comments. People are participating more actively in matters of their health and are turning to online networks for answers to health related information.

According to the *Pew Internet and American Life Project*, 80 % of internet users, which translates into 59 % of all adults, looked online for health news—the majority of them searching for health information on behalf of someone else. Their most recent report on *The Social Life of Health Information* found that "technology is not an end, but a means to accelerate the pace of discovery, widen social networks, and sharpen the questions someone might ask when they do get to talk to a health professional."

They Share What they Find Few people have not heard of something that has "gone viral" because the content was considered worthy of sharing. Social media has a unique power to introduce a topic or idea to a vast number of people quickly. When the content is engaging, users will converse about it on their own by republishing it (re-tweet, share, comment, blog or send to a friend). The result is greater exposure of, and trust in, the message through peer-to-peer sharing.

Other Media are Listening Social media conversations are not confined to only those active on social media channels. Social media channels are now one of the most accessed sources of news and information for traditional media. Newspapers, magazines, television and radio stations are increasingly using social media channels to access information, issues and opinion that increasingly makes news.

It's Accessible and Affordable With rare exception there is minimal cost to accessing social media channels. Most are easy to join and don't require software downloads or special system requirements.

The School of Public Health Services at George Washington University conducted a study called *New Media Cases in Public Health Communication & Marketing: The Promise and Potential* (George Washington University 2012).

The researchers reviewed several campaigns designed to change behaviour. They found that despite minimal budgets, every web and social media example they reviewed was found to be successful to expand the campaign message. "Whether measured with page views to websites, downloads of campaign videos, or numbers of text messages sent, all campaigns were able to generate significant exposure to their campaign materials with new media channels. Once traditional campaign materials are developed, the additional resources required for new media appear to be comparatively minimal, and those efforts may have long lasting effects."

If applied strategically, social media can enhance exposure of an issue and improve engagement around specific health and environment-related topics. It's become an essential part of any multi-dimensional social marketing effort—especially ones related to weather, health, environment and specifically air quality.

This chapter does not offer a social media boot camp on "how" to maximize the potential of social media to support air quality and health communication goals. A self-directed study can be accessed by downloading the U.S. Centre for Disease Control and Prevention's "*The Health Communicators Social Media Toolkit*".

19.9 The Future of Communicating About Air Quality

The only thing certain about the future of social marketing related to air quality and health is that the future is coming fast and the speed of change accelerates each year.

Accessibility to the internet, the rapid rise of social media and networking platforms, the explosion of personal mobile devices and the 'personalization' of information is already influencing how we reach out to people about air quality and health. A new reality of connectedness is creating new ways to encourage people to care about the air.

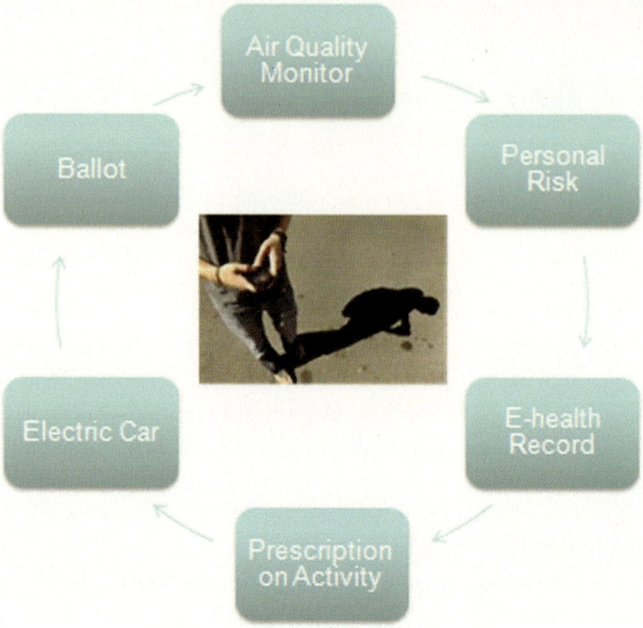

Fig. 19.12 Graphic illustration of how air quality monitoring could interact with electronic medical records (EMR) to prompt health protection and individual air stewardship actions in the future

19.9.1 From Imagination to Implementation

In 2011 at a US Environmental Protection Agency AIRNOW Conference on Air Quality (Stevens 2011), an audience of scientists, policy makers and communication professionals were asked to imagine a time in the future when:

- Everyone understands how air pollution affects both their short and long term health based on their personal health status and risk factors,
- Their mobile device is equipped with an air quality monitor,
- Air quality data is cross referenced to their cloud-hosted personal electronic health record to identify their air pollution health risk,
- A prescribed health management message, recommending personalized activity levels is instantly pushed to their mobile device, and
- This heightened personal relevance of air quality motivates people to better care for themselves, and for the environment.

All of which of course leads to increased political pressure by voters to elect leaders committed to regulating and rewarding improved air quality.

Ah, if it were only that simple.

Well, in less than 24 months since that presentation, this future gazing is starting to become a reality—and fast (Fig. 19.12).

The personalized and hyper-local air quality monitoring devices, and their integration with mobile phones, are well into development. At the University of California San Diego, a team representing the Jacobs School of Engineering, the School of Preventative Medicine and Calit2, has a prototype in field testing. It combines artificial intelligence, sensor technology, cell phones and the opportunities presented by crowd-sourcing to achieve hyper-local, almost personal, air quality monitoring (Fig. 19.13).

The Air Quality Egg is also combining technology and crowd sourcing of air quality. According to www.kickstarter.com, the Air Quality Egg is a community led sensor system designed to allow anyone to collect readings of NO2 and CO concentrations outside of their home, and then enable this data to be collected and aggregated online, to yield an international stream of air quality data. The development of the Egg was fully funded through www.kickstarter.com and Eggs have been shipped and are now collecting data around the world. Want to see where—just visit http://airqualityegg.com/ (Fig. 19.14).

The Government of Canada and the Ontario Lung Association funded a study that has concluded that a smart phone application is a feasible way to integrate Canada's AQHI into the action plan asthmatics use to control and self manage this chronic respiratory condition (Licskai et al. 2012).

Cloud-hosted electronic health records are now a reality. In Canada, as in many other countries, there is continuous progress toward a complete health system transfer to Electronic Medical Records (EMR). Patients are already happily reporting rapid e-access to diagnostic test results through e-health and telehealth programs in Canada.

This is just an introduction to the future. See Sect. 20.9 of Chap. 20 for more examples today of things that will become common place in air quality management of the future.

Fig. 19.13 Development of smart phone air quality monitor. Image Credit: CitiSense

Fig. 19.14 Air Quality Egg. Photo credit: Ed Borden

References

Environics Research Group (2002) Air pollution—information needs and the knowledge, attitudes and behaviour of Canadians—final report. Health Canada. http://www.hc-sc.gc.ca/ewh-semt/pubs/air/pollution/index-eng.php. Accessed 27 June 2012

Licskai C, Sands T, Ferrone M (2012) Integrating the air quality health index into asthma action plans. Am J Respir Crit Care Med 183(2011):A4767

McGuire W (1974) Psychological motives and communication gratification. In: Blumler J, Katz E (eds) The uses of mass communications: current perspectives on gratifications research. Sage, Newbury Park, pp 167–196

Stevens S (2011) Creating a social dialogue about air and health. In: Proceedings of the US EPA AIRNOW Conference on Air Quality, San Diego, California, March 2011. http://www.epa.gov/airnow/2011conference/communications/stevens.pdf. Accessed May 2013

The Future of Air Quality Management 20

Eric Taylor, Jeffrey R. Brook, Mike Moran, David M. Stieb, Randolph P. Angle, Deniz Karman, Judi Krzyzanowski, Ann McMillan, Sharon Stevens, James Young and Ed Piché

Abstract

This final chapter considers how air quality management may change in the future. Air quality monitoring will likely expand to measure more pollutants in denser networks in more communities, and may refocus on pollutant issues that negatively impact the vulnerable population. Satellite technology will be increasingly applied to meteorology and air pollution monitoring, especially to provide data in remote areas such as Canada's North. Monitoring equipment may become smaller, be less expensive to build and operate, and be less power demanding. Real-time air quality forecasting will undoubtedly become more accurate as meteorological modeling and air quality models improve and refocus on weather patterns conducive to degraded air quality. Emissions data from industry and transportation may become more widely available in near real time to enable these models to produce more accurate and useful predictions. Research into the impacts of ultrafine particles and pollutant mixtures on health, should pave the way for improved methods of reducing health risk. Increased awareness at both the domestic and international level of the health risks related to air pollution from industry, particularly resource-based industries, will likely lead to increased pressure to reduce industrial emissions. Remarkable reductions in on-road transportation emissions will likely continue through better post-combustion treatments and the inclusion of better pollutant control systems in off-road vehicles. Electric and hydrogen vehicles will eventually increase their share of the market, resulting in lower emissions within urban communities. National air quality management programs will increasingly mandate reductions in air pollutants and their health impacts for the whole country, including Northern regions. Canada's close binational relationship with the U.S. will continue and expand north. Air pollution management will make more effective use of the many new approaches to communications from new approaches to formulating air quality messages to new ways of linking air quality data to personal devices to enable individual actions in response to real time air quality.

Keywords

Future air quality management · Alternate fuels · Air quality monitors · Evolving technology · Pollutant trends · The North

20.1 Introduction

This chapter assesses what developments could occur in air quality management in the future based on information from authors of earlier chapters and from other health, air quality and meteorological experts.

As noted by Jeff Brook in Chap. 3 in this book on the State of Canada's air quality, current pollutant concentra-

E. Taylor (✉)
British Columbia Ministry of Environment, Victoria, BC, Canada
e-mail: eric.taylor@gov.bc.ca

J. R. Brook · M. Moran
Environment Canada, Toronto, Canada
e-mail: Jeff.Brook@ec.gc.ca

M. Moran
e-mail: Mike.Moran@ec.gc.ca

E. Taylor, A. McMillan (eds.), *Air Quality Management,* DOI 10.1007/978-94-007-7557-2_20,

tions in populated areas are considerably lower than they were when the Department of Environment was created in the 1970s, Though Canada has been successful in reducing its burden of air pollution, new emission sources will likely develop and existing sources will expand. The latter includes the rapid development of one of the world's great petroleum deposits—the oil sands and heavy oil resources in Alberta. This massive activity will continue to release large amounts of greenhouse gases, nitrogen oxides (NO_x) and sulphur dioxide (SO_2), the latter two contributing to acid precipitation and air pollution (Dyer et al. 2013).

In this milieu, it is not known whether we can accomplish further significant reductions in the impacts of pollutants on health. Will challenges still persist in areas in Canada where ozone (O_3), fine particulate ($PM_{2.5}$) and toxics continue to exceed standards? Will the public recognize significant improvements in visual air quality? Climate change may also play an important role in air quality issues as shifts in global patterns of temperature, precipitation, wind and atmospheric stagnation continue to occur.

20.2 Monitoring Air Quality in the Future

Since air pollution impacts health, and by extension the health care system, it will certainly remain a public policy issue of concern. An increasingly skeptical public needs assurance that pollutant concentrations adhere to standards and that any potential impacts are documented. It therefore follows that the routine monitoring and reporting of air pollution concentrations will likely continue to be important in the future, particularly in areas where large numbers of people are exposed to air quality that affects their health.

20.2.1 Improving Existing Monitoring

Canadian air quality monitoring networks that report data to the national air quality database now operate 318 air monitoring stations in 217 communities, using over 800 instruments. These include continuous analyzers for SO_2, CO, NO_2, O_3, and fine particulate matter. Toxic substances such as polycyclic aromatic hydrocarbons (PAH), dioxins and furans, and heavy metals such as arsenic, lead, and mercury are also measured at many of these stations (Environment Canada 2012). The techniques and systems to measure these ambient air pollutants have improved significantly in recent decades and will likely continue to do so in the future. The numbers of both monitors and monitored pollutants will also likely continue to rise as the population increases and as the natural resource sector expands into previously uncharted territory. The following recent moves by Environment Canada should help ensure that monitoring will provide ever-improving air quality data to the public in the future.

- A program has been implemented to ensure that monitoring instruments are properly maintained and replaced when an instrument has reached its end of life. This has reduced the average age of NAPS instruments from over 15 years old to 5 years old over the past decade.
- Existing continuous $PM_{2.5}$ instruments are being converted to U.S. Class III Federal Equivalent Method (FEM) instruments. (See Chap. 3 for information on FEM monitors.)
- Federal, provincial, and municipal data logging and data reporting systems are being modernized to allow timelier reporting of data, and to improve the quality of real time data.
- The existing $PM_{2.5}$ chemical speciation network is now using updated samplers and measurement programs for VOCs, a significant improvement over the previous network.
- The laboratories and analytical equipment used to carry out detailed chemical analyses, such as VOC and $PM_{2.5}$ speciation, have been expanded (Environment Canada 2012).

20.2.2 Emerging Monitoring Issues

Future air quality monitoring systems in Canada may refocus on a number of emerging issues.

- Air quality programs may increase their focus on long-range sources of pollution, specifically on sources from

D. M. Stieb
Environmental Health Science and Research Bureau, Health Canada, 3rd floor, 4595 Canada Way, Burnaby, BC, V5G 1J9, Canada
e-mail: dave.stieb@hc-sc.gc.ca

R. P. Angle
R. Angle Consulting, Edmonton, Alberta, Canada
e-mail: rangle@shaw.ca

D. Karman
Carleton University, Ottawa, Canada
e-mail: Deniz_Karman@carleton.ca

J. Krzyzanowski
Krzyzanowski Consulting, Stirling, Canada
e-mail: judi@krzyzanowski.ca

A. McMillan
Storm Consulting, Ottawa, Canada
e-mail: mcmillan@storm.ca

S. Stevens
Air Shift Group, Kamloops, Canada
e-mail: sharon@airshiftgroup.com

J. Young
SENES Consultants, Richmond Hill, Canada
e-mail: jwsyoung@rogers.com

E. Piché
Environmental Business Consultant, Toronto, Canada
e-mail: ed.piche@sympatico.ca

Asia. This may occur as emission rates increase in Asia, causing increased transport across the Pacific and ultimately higher ambient concentrations in North America. Concurrently, local emission reduction efforts may become more successful and relatively less important.

- Mobile monitoring could increase and become more focused on health impacts at "hot spots" near emission sources such as busy roads and in heavily industrialized areas. By being portable, these instruments need only remain in one location long enough to sufficiently characterize air quality before moving to another areas where other health risks exist. These mobile monitors would complement the information from fixed sites further away from sources.
- Satellite-based remote sensing of air pollutants may improve the amount and accuracy of air quality data collected, particularly if surface-based observations are routinely integrated with satellite-derived measurements. This would extend the spatial and temporal coverage of current monitoring networks.
- Improvements in technology should also be able to transfer air quality data from satellites and other sensors to systems more rapidly and efficiently, allowing air quality forecasting systems to better manipulate and make use of those data (see Sect. 20.9).
- As the climate changes, emissions, atmospheric chemical reaction rates, temperature and atmospheric patterns such as stability and precipitation regimes may also change. Measuring and documenting these changes in the future will provide valuable information for air quality managers.
- Monitoring for precursors to ozone, ultrafine particles and other particulate matter could increase in the future. This will provide better support for air quality forecasting models, improvements in emission control strategies and programs to reduce health risk.
- Future governments from around the world may pool resources to increase market incentives to support the commercialization of advanced methods of air quality monitoring.
- The boundary layer may be monitored more intensively and routinely in the future to ascertain its depth, dynamics, structure, pollutant concentration profiles, and temperature profile. (Committee on Environment and Natural Resources 2010).

20.2.3 Evolution of Air Quality Monitors

Emerging technology promises more widespread monitoring of air quality. For example, electrochemical sensors are beginning to offer the ability to monitor multiple pollutants and meteorological parameters in a single unit at a fraction of the cost of traditional 'active' monitoring. Nanotechnology is further increasing the reliability and precision of electrochemical sensors, which can now measure multiple pollutants at concentrations as low as one part per billion in a small, single unit.

Smart phone technology and wireless infrastructure are providing platforms for new air quality sensors that are becoming smaller, less costly and able to monitor other variables such as temperature. New monitoring units are also increasingly power stingy, loosening the logistical constraints of more traditional devices. Using these monitored data, mobile applications ("apps") are able to automatically handle two-way data transmission, data compilation and analysis, and share the resulting information with the public. By providing real-time data transmitted through cellular or satellite channels, communities and governments alike will increasingly be able to monitor air pollution hotspots, background sites or entire airsheds. The increased affordability and ease of use of these new technologies may revolutionize air quality monitoring as we know it (see Sect. 20.9). Deploying simple inexpensive and multifunctional sensors for routine monitoring and field studies can provide greater data density to better understand the spatial and temporal variations of pollutants and human exposure. In the future, widespread use of sensor networks for multiple pollutants may significantly improve our understanding of linkages between air pollutants and adverse health effects and provide insight on the cumulative impacts of multiple pollutants (White et al. 2012).

These new tools need to first prove themselves in Canada's challenging environments. As this is done, monitoring systems should be increasingly able to expand without significant increases in resources. This would be a boon to air quality management processes such as air quality assessment, reporting and pollution control.

20.3 Future Developments in Air Quality Forecasting

Currently, Environment Canada provides two-day forecasts of pollutant concentrations and the Air Quality Health Index to 74 communities across Canada, primarily through its Weatheroffice website (weather.gc.ca). However, as roles of governments change, some specialized air quality forecasts are now being produced at the provincial level. An example is the BlueSky wildfire smoke forecasting system now operating in British Columbia. In the future, additional specialized forecast products and related health advice could be developed at the local level using air quality models to focus on specific communities with unique air quality problems, such as smoke accumulation in deep valleys.

As discussed in Chap. 4, regional air quality can be impaired from the long-range transport of air pollutants.

However, as shown in Chap. 5, air quality is also often impaired from local emissions during stable, stagnant conditions, when meteorological dynamics are weak, allowing pollutants to accumulate near the surface. In these situations, local variations in cloud cover, solar intensity, temperature and wind can quickly lead to changes in small scale circulations, resulting in variations in local mixing depth and other local atmospheric conditions. Pollutant concentrations can vary quickly as a result of the development of these small-scale circulations, (Zhang et al. 2012), which in turn can be influenced by synoptic scale events. Unfortunately, current meteorological models are not designed to represent air pollution episodes that have very weak dynamical forcing (Baklanov et al. 2002). On top of this, errors in conventional numerical weather prediction models can propagate rapidly into air quality dispersion models and cause large errors in pollutant concentration forecasts (e.g. Hess et al. 2004).

Future air quality forecasting systems will therefore become more accurate only when models improve their predictions of small-scale meteorological events that are important for air quality. These include improved predictions of local circulations such as land-sea breezes and topographically-induced circulations, as well as physical processes such as turbulence, boundary layer depth, cloudiness, etc. This will require modelling of meteorology on fine scales (about 1 km grid resolution), particularly in heavily populated regions (Zhang et al. 2012). This will in turn require improved parameterization of elements such as stagnation, turbulence, deep convection, low clouds and nocturnal transport that are critical to determining pollutant concentrations (Zhang et al. 2012). Progress is already being made on this task. For example, during the 2010 Winter Olympic and Paralympic Games in Vancouver, Environment Canada produced meteorological forecasts at 1 km resolution over southwestern British Columbia (Mailhot et al. 2010).

Future air quality forecasting systems may also provide forecasts for longer periods, perhaps up to two weeks, to match some current atmospheric models such as the Global Forecast System, instead of the present one or two days. More and better pollutant data should also become available for input into air quality forecasting systems. This includes data from improved satellite measurements of near-surface concentrations and vertical profiles of many important chemical species (e.g., McLinden et al. 2012). Emission estimates should also improve when they account for variations in current weather conditions, since weather itself can affect emission rates (e.g., forest fire smoke, VOC emissions from agriculture and nitrogen oxide from motor vehicles). When current emissions from large industrial sources are available in real time and online for use by air quality models, air quality forecasts will also be more accurate and therefore useful (Zhang et al. 2012). In the future, the suite of pollutants being forecast may also expand to include selected toxic chemicals and allergens such as specific pollens.

20.4 Future Progress in Identifying Health Impacts

Future research should fill important gaps in our understanding of the impacts on health of air pollution, guiding policies to further reduce emissions, ambient concentrations and risks to health. As mentioned by Dave Stieb in Chap. 7 in this book, the health impact of mixtures of multiple pollutants is an important issue. Many sources, including traffic and other combustion processes, emit multiple pollutants that are comprised of hundreds of chemical species. As is the case with multiple drugs, health may be more dangerously affected by the synergistic action of multiple pollutants than the sum of their effects individually. Unfortunately, the public is exposed simultaneously to many pollutants whose concentrations are highly correlated with each other (Billionnet et al. 2012). This makes statistical linkages with health impacts difficult. Improved understanding of the relationships between pollutants and health will likely evolve to address this issue over time, particularly as our understanding of the biological and chemical mechanisms and the health impacts of these pollutant mixtures increases.

Future research is also likely to uncover more important information on the health risks of ultrafine particles (UFP), particles that are much smaller and more numerous than $PM_{2.5}$. Due to their ability to penetrate deep within the lung, UFPs may be a major concern for respiratory exposure and health. If warranted, there will likely be efforts to reduce exposure and related health risks. It may be more challenging, however, to reduce their concentrations, particularly indoors, since they appear to be ubiquitous and persistent. Health professionals could develop better warning systems that are more useful and targeted, particularly for those vulnerable to high concentrations of certain pollutants.

As Dave Stieb has noted in Chap. 7, studies of air pollution interventions have shown that reducing air pollution results in reduced health risks. This provides powerful support for the value of continuing to improve air quality. Future research will inevitably lead to further ways to reduce and even prevent pollution. This could also influence how we work, where we live and how we make personal decisions that affect our health. This will occur as research results find their way into policies to further reduce emissions from vehicles, industry and domestic heating. Health impacts will diminish as we plan to reduce traffic congestion and introduce wider setbacks from traffic for residential buildings, care facilities and schools. Health impacts will also moderate when individuals make more informed personal decisions to reduce their air pollution exposures and protect their health (see Sect. 20.9).

20.5 Industrial Emissions—Future Directions

Air quality management requires a good knowledge of current emissions from industrial sources, but acquiring these data is not always easy or prompt. In the future, various types of air quality management systems could obtain industrial emissions types and rates in real time. For large industrial emission sources, in-stack instruments can provide these data. For smaller sources, calculations could be made based on the real-time fuel consumed or the production rate. Perhaps improved methods for short-term forecasting of changing emissions will also be developed. Industrial emission inventories for any time period, facility or industrial area could also be calculated as needed. Future information and communications technology systems should be able to securely and rapidly transmit these data to where it is needed, including users such as air quality forecasters who could incorporate them into air quality models to produce more accurate and precise predictions. This would then allow health professionals to develop better warning systems that are more useful and targeted, particularly for those most vulnerable to high concentrations of certain pollutants.

The good news is that future emissions of unhealthy pollutants from industry are likely to diminish due to local and international pressure. If markets are threatened by public opinion, industry will be forced into greater action to restore consumer confidence and retain both a local and an international social license to operate. Emission reductions will likely be accomplished through cleaner industrial technology, including switching from burning coal or oil to cleaner fuels such as natural gas. A current example of this process is the international pressure on the perceived emissions from the industrial development of the Alberta oil sands.

In the future, emission *standards* should be able to be updated more frequently and reflected in more responsive and efficient permit systems. Permits should simultaneously become more effective and equitable while also being simple to implement. Environmental and health agencies will likely also push for more accurate and robust air quality models so that permits can be based on more reliable pollutant modelling scenarios. Real time air quality modelling could be also be part of the permit system in the future, allowing industrial emissions to be temporarily reduced in real-time if these models predict increased human exposure to unhealthy pollutants, particularly near schools and senior's homes where the population is more vulnerable to air pollution.

These and other approaches to future industrial emissions will help in building long term sustainability of businesses, particularly if shareholders and corporate executives take more of a multigenerational and global view.

20.6 The Future of Road Transportation and Air Quality Management

The subject of air emissions from transportation is an ongoing concern from an air quality and human health perspective as well as the significant contributions of the transportation sector to greenhouse gas emissions. Emissions from on-road vehicles are of particular concern from a human health perspective due the proximity of the emissions to the general population living in urban centres.

Impressive reductions in air pollutant emissions from on-road vehicle have been achieved in the past 25 years by the emission control technologies that have been mandated by increasingly strict regulations. While alternative fuels have played a role in the early reductions of air pollutant emissions, the more dramatic reductions currently being realized are due to aggressive after-treatment (i.e. post combustion) technologies. Even with the increasing travel demand, the future air pollution burden from on-road vehicles can be expected to continue decreasing, regardless of the fuel being used. Emission control technologies are also beginning to be implemented in the off-road, marine and rail sectors, albeit at a slower pace.

The overwhelming dependence of the transportation sector on petroleum as the source of primary energy is clearly an unsustainable trend in terms of the depletion of a non-renewable resource and the climate change effects. Electricity and hydrogen are the energy carriers that have the potential to include other sources of primary energy in the mix for transportation. Gasoline-electric hybrid vehicles, plug-in hybrid vehicles and all electric vehicles are already in the on-road vehicle fleet in North America and are poised to claim a gradually increasing share of the fleet. How quickly this change will occur is still up for debate.

The wide range of fuel economies that are achievable by on-road vehicles suggest that significant reductions in greenhouse gas emissions can be achieved by consumers' choices. However, even the relatively higher fuel prices in Canada do not appear to have a strong effect in this regard. Natural gas as a lower carbon intensity transportation fuel, or renewable biofuel, has the potential for modest reductions in the greenhouse gas emissions from the sector.

The developments in alternative vehicles and fuels are currently motivated primarily by greenhouse gas concerns, as well as the diminishing availability of conventional oil. However, in many cases there are co-benefits of lower emissions of criteria air contaminants that are concurrently realized since any pollutants emitted from internal combustion engines are emitted in much closer proximity to humans. Life-cycle analysis of alternative fuels and energy technologies should be employed in the future to quantify the associated upstream emissions, and regional air quality models should be used to analyze the effect of emissions associated with the production of the fuel. Only in this way can we expect to have an accurate picture of the effect of our multiple objectives and efforts on the air we breathe.

20.7 Canada's Developing North—Air Quality an Emerging Concern

Canada currently manages air quality across its southern border with the United States through the Canada/US Air Quality Agreement, which is described in Chap. 16. Of course we have a second border with the US, namely the Yukon/Alaska border. With the development of the oil sands as well as developments in Alaska and ultra-long range transport from Asia, we can expect that Northern people will identify air quality concerns, and, at present, there is no bi-national regulatory framework under which to discuss and manage these issues.

In 2008, the International Joint Commission's (IJC) International Air Quality Advisory Board (IAQAB) held a workshop in Anchorage to discuss transboundary air quality and in 2010 that workshop was followed up with one in Whitehorse. Reports from these workshops as well as the summary, "Report on Air Quality Issues related to the Northern Boundary Region between Canada and the United States" can be found on the IJC website (International Air Quality Advisory Board 2010, 2012).

Southern experience seems limited in applicability in the North because air quality concerns in the North are quite different from those in the South. The Inuit people have been exposed to a variety of chemicals (such as herbicides and pesticides), that were never used in the North, through long range transport from the South. It is expected that as Canada takes over the chair of the Arctic Council, for example, such issues will come forward into the international arena.

Alaska's Panhandle is volcanically active and one of the global Volcanic Ash Advisory Centres is located in Anchorage and run by the National Oceanographic and Atmospheric (NOAA). Volcanic activity with the gases released, can affect air traffic to the North, and, indeed, globally.

There are few roads in the North, but the few that are present can be the source of dust that may contain harmful particulates. Thus, air pollution sources in Alaska and probably the Yukon include road dust, resource development, diesel generators, and open burning of solid waste.

As the northern sea ice melts, ocean navigation is increasing rapidly. Eco-tourists are becoming more common, and with the collapse of winter ice roads, resupply of Northern communities is often done by sea. Marine air pollution is a possible emerging issue for some Northern communities. We should expect to see oil spill damage in the north as exploration and extraction increase. Conditions at northern spill locations will be much different than experienced in the Gulf of Mexico (for example).

To compound the difficulty of managing air quality in the North, there are few air quality monitors, especially on the Canadian side of the Alaskan border. Programs such as the Northern Contaminants Program (NCP) and the International Polar Year (IPY), both led by Aboriginal Affairs and Northern Development for Canada have laid a basis for understanding air quality issues in the North. Newer programs, emerging since devolution, such as the Nunavut General Monitoring Plan (NGMP 2013) and the Northwest Territories Cumulative Impacts Monitoring Program (CIMP) are beginning to establish the need for data, but it will be many years before scientists have the data required to be able to knowledgably advise jurisdictions on a science-based management approach. The Canadian High Arctic Research Station recently established in Cambridge Bay will evidently have environmental monitoring as one of its priorities when it starts up in 2017.

The advent of satellite programs such as RadarSat and the and the future Polar Communications and Weather (PCW) Program, can offer spatially dense data with rapid revisit rates. It will be important for the air quality management community to make use of these modern data to complement ground based measurements in remote Northern areas.

As we've seen throughout this book, monitoring and modeling together make up credible systems for supporting air quality management. We've seen that current models focus on transboundary transport of pollutants, examine urban and global scales and forecast weather. These forecasts predict the movement of volcanic ash, surface pollutants and upper level ozone and solar radiation and simulate movements of a number of pollutants from special sectors as well as smog, acid deposition and air toxics. At this time, there is not the same level of confidence in northern model results as there is in the south for such applications as weather. prediction and ecosystem management for a variety of reasons including lack of monitored data. This is an important future issue for Canada.

As development occurs in the North it will be expected that Canadian innovation will continue. Air quality management systems will be called on to extend their reach over new vast areas which are sparsely populated but expected to house major industrial facilities in some of the most sensitive ecologies on earth.

20.8 Communicating Air Quality Issues: From Imagination to Implementation

Raising the awareness of air quality issues among individuals will continue to be an important component of air quality management in the future. In 2011 Sharon Stevens, the author of Chap. 19, predicted it would not be long before personal health data intersected with air pollution data to give people personalized advice on air quality and their health. Soon after, in June 2012, the US Environmental Protection Agency and the National Institutes of Health created an air pollution sensor challenge called *My Air, My Health*—www.epa.gov/research/challenges. The challenge was created to encourage development of portable devices that would gather and integrate health and air quality data. Individuals and teams submitted designs for sensors that can be easily worn or carried and link airborne pollutants and a person's health measurements such as heart rate and breathing.

There are already a multitude of examples of clever, creative and collaborative efforts to advance awareness of air quality. The select few provided below offer a glimpse into the future of air quality outreach and communication. Many examples will have been imagined and put into practice by the time the ink on this page has dried—or is downloaded.

Flash Mobs, Die In's And School Strikes Children have gathered in Haifa, Israel, wearing charcoal-blackened ties and masks, to urge adult voters to think about air pollution issues when they vote. Environmentalists in Hong Kong have organized "Lights Out Hong Kong" as a way of sending a message to their government about the city's rapidly worsening air pollution. More than 15,000 students in southern Kuwait staged a "stay home from school strike" to protest against air pollution. And a London-based organization called Climate Rush held a mass 'die in' in Soho Square to raise awareness of the health impacts of air pollution in that urban city.

These grass roots events in today's world are an important part of a continuum of global outreach that is increasing awareness about the quality of air we share. What will tomorrow's world bring?

New Sectors in the Sandbox As more people stand up for the air we share, more governments, organizations and sectors of our economy will find ways to align with the issue of air quality. Even the fashion industry is contributing to awareness of, and potential solutions, related to air quality.

Dress for Success In an effort to show reduced environmental impact, the fashion industry has introduced a fabric they claim reverses the environmental impact of air pollution. The "Purifying Dress", developed by the Centre for Sustainable Fashion at the London College of Fashion and the University of Sheffield, claims to use a textile which can reverse the impact of air pollution. Through the power of a photo-catalyst, the fabric claims to break down air borne

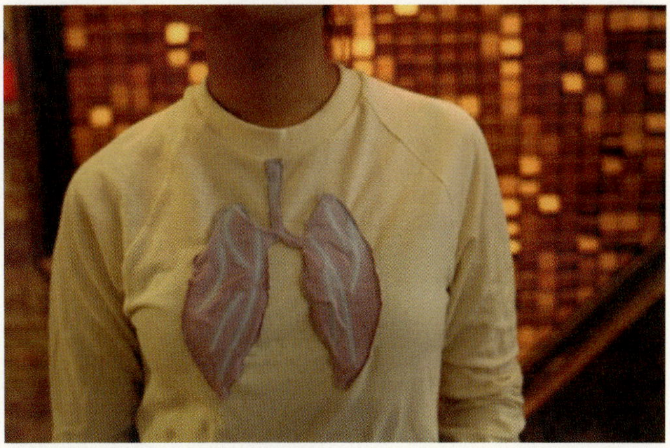

Fig. 20.1 NYU student's wearable technology T-shirt

Fig. 20.2 Non-Sign II. (Image by Ian Gill courtesy of Lead Pencil Studio)

pollutants (Storey 2012). A professor of physical chemistry at Sheffield University has teamed up with a popular fashion designer to develop a laundry additive they claim will stick to the surface fibers of cloths, and neutralize nitrogen oxide.

New York University grad students Sue Ngo and Nien Lam, as part of a class in Wearable Technologies, used thermo-chromic fabric to design a shirt featuring a graphic of lungs which change colours based on exposure to $CO2$ (Fig. 20.1). Using tiny $CO2$ detectors, the microcontroller sends electrical currents through the fabric and as the temperature of the fabric changes, so does the colour of the graphic on the colour of the shirt.

Building on Benefits What looks like a billboard, an outdoor marketing format most commonly used to market burgers, cars and beer, is making a silent statement about environment and air quality. This sculpture (Fig. 20.2) sits at the U.S.–Canada border crossing between British Columbia

Fig. 20.3 Semiamoo Sky Garden. (Image credit: Green Over Grey Living Walls and Design)

and Washington State. It was designed as a permanent billboard advertising, well, nothing. Not really. It is designed to promote the clean air of Blaine, Washington. Called Non-Sign II, a Seattle art and architecture firm, Lead Pencil Studio, was commissioned to produce it by the U.S. federal government.

Coca-Cola and the World Wildlife Fund unveiled a 60-by-60 foot billboard in the Philippines that is covered in Fukien tea plants to absorb CO_2 and alleviate air pollution within its proximate area.

In many major metropolitan areas, living walls are helping improve air quality. Designed and installed by Green Over Grey Living Walls and Design, the exterior face of the Semiamoo Sky Garden in Surrey, British Columbia was transformed into the largest outdoor vertical garden in North America. It has 10,000 individual plants representing 120 unique species (Fig. 20.3).

These are but two examples of the convergence of interest and awareness in air quality, urban architecture and outdoor art.

Maximizing Mobile to Promote Air Quality The use of Smart phone technology has exploded and will only continue to evolve at break neck speed. Marketing and communication innovators have been quick to combine web and mobile technology. Near Field Communication (NFC) is enabling mobile users to simply tap their devise against anything that has an interactive tag (from magazine ad to bus stop shelter) and go from awareness to action—either getting their local real time air quality report or sharing an air quality message with friends or family.

Learning by Doing The most effective way to attract attention and change behavior for the better is to give people a way to interact with your topic—and if possible, make it interesting and fun. Here are two examples of how this is being achieved:

Throughout the world's largest cities, interactive billboards offer increased opportunity to attract attention and engagement.

In Canada, the Agency of Health and Social Services of Quebec City (Agence de la santé et des services sociaux de la Capitale-Nationale), through their creative agency lg2, used interactive billboards to raise interest in careers in public health. A bus shelter poster features a person in a hospital bed with a "flat" electrocardiogram indicating no heartbeat. When passers-by push their hands on the image of the patient's heart, the electronic electrocardiogram beats, reinforcing the rewards of a career in public health (Fig. 20.4).

Referred to as a Fun Theory (www.thefuntheory.com), this Volkswagen initiative proposes that something as simple as fun is the easiest way to change people's behaviour for the better. The Piano Stairs Project demonstrates how turning public staircases into life-size piano keys, complete with

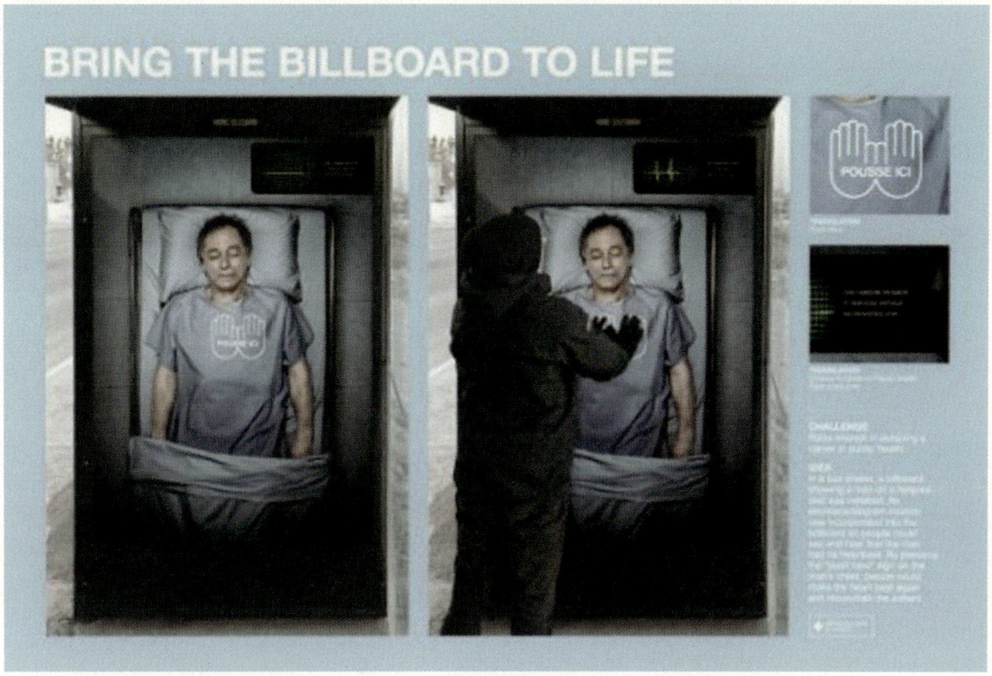

Fig. 20.4 Electrocardiogram interactive billboard

the appropriate audio reinforcement for each stair, or in this case key, encourages and rewards the choice to take the stairs instead of the escalator.

Consider the possibilities when technology and creativity come together to engage people on the topic of health and environment and prompt them to change their behavior for the better.

It's an exciting time to be creating social marketing programs about air quality and health. The more air quality becomes part of our social dialogue, around both the boardroom table and the kitchen table, and is integrated into our community planning, our expression of art and culture, and our economic growth, the more the collective consciousness will grow. And with that will come an increase in individual behaviors to protect oneself from the risks of air pollution, and to be better stewards of the air we share.

20.9 Canada's Air Quality Management System—Improving Future Air Quality

An Air Quality Management System (AQMS) has recently been developed to guide future air quality management in Canada. It is the product of collaboration among the federal, provincial and territorial governments and stakeholders and promises to be a comprehensive approach to improving and protecting air quality. The Canadian Council of Ministers of the Environment (CCME), the primary minister-led intergovernmental forum for collective action on environmental issues of national and international concern, has the responsibility for refining and implementing the AQMS (CCME 2012).

Federal, provincial and territorial governments all have roles and responsibilities in the implementation of the new Air Quality Management System as it is rolled out over the next several years. The AQMS has several key components which will have a significant impact on future air quality management in Canada. These are:

- New *Canadian Ambient Air Quality Standards* (CAAQS), benchmarks for outdoor air quality management across Canada
- Industrial emission requirements that set a base level of performance for major industries
- A framework for air zone air management within provinces and territories that enables action tailored to specific sources of air emissions in a given geographical area
- Regional airsheds that facilitate coordinated action where air pollution crosses a border
- Improved intergovernmental collaboration to reduce emissions from the transportation sector

The CAAQS will replace the existing Canada-wide Standards by strengthening standards for fine particulate matter ($PM_{2.5}$), ground-level ozone (O_3), nitrogen dioxide (NO_2) and sulphur dioxide (SO_2). Air quality management will be guided by an Air Zone Management Framework to ensure proactive measures are taken to protect and improve air quality in accordance with the principles of keeping clean areas clean and continuous improvement. Air zones will be a place-based approach to manage local air quality. They are to be both geographical regions as well as arenas for air quality management. The provinces and territories will delineate and manage air zones within their boundaries. The goal in all air zones will be to drive continuous improvements in air quality and to prevent the CAAQS from being exceeded. As in the past, the AQMS is expected to be based on the latest science. It will be challenging to reflect the things we now know (and the things we don't) especially about the impacts of pollutants on human health.

The AQMS has created six large, regional airsheds together covering all of Canada to coordinate efforts to reduce transboundary air pollution flows and report on regional air quality. However, it is unclear how the role of these AQMS airsheds will differ from the existing inter-jurisdictional air quality mechanisms. The AQMS airsheds should not be confused with the relatively small airsheds described in Chap. 17 on "Airshed Planning". The smaller airsheds were created in the past in British Columbia as an approach to managing local air quality in urban areas or narrow valleys, primarily to manage local emissions, the main contributor to poor air quality in these communities.

Base-level Industrial Emission Requirements (BLIERs) are quantitative or qualitative emissions requirements proposed under the AQMS for new and existing major industrial sectors and some equipment types. These requirements are based on what leading jurisdictions around the world are requiring of industry in "attainment areas," but adjusted for Canadian circumstances. The BLIERs are focused on nitrogen oxides (NO_x), sulphur dioxide (SO_2), volatile organic compounds (VOCs), and particulate matter (PM). BLIERs are intended to ensure that all significant industrial sources in Canada, regardless of where facilities are located, meet a required base-level of performance.

Emissions from mobile sources (transportation and small engines) are a major source of air pollution in many parts of the country, especially urban areas. However, there are multiple jurisdictions responsible for managing both these emissions and the resultant human exposure to harmful pollutants. Under the AQMS, governments at all levels will need to work together to ensure that emissions of pollutants, and the related human exposure, will continue to decrease in the future.

Monitoring and public reporting are critical to transparency, accountability and the effective implementation of the AQMS in the future. Provinces and territories, with assistance from the federal government, will be responsible for monitoring in the air zones and reporting to the general public in their jurisdictions on air quality and the measures taken to implement the AQMS (CCME 2012).

References

Baklanov A, Rasmussen A, Fay B, Berge E, Finardi S (2002) Potential and shortcomings of numerical weather prediction models in providing meteorological data for urban air pollution forecasting. Water, Air and Soil Poll Focus 2(5–6):43–60

Billionnet C, Sherrill D, Annesi-Maesano I (2012) Estimating the health effects of exposure to multi-pollutant mixture. Ann Epidemiol 22(2):126–141. doi:10.1016/j.annepidem.2011.11.004

CCME (2012) Air quality management system. http://www.ccme.ca/ourwork/air.html?category_id=146. Accessed 28 Apr 2013

Committee on Environment and Natural Resources (2010) Air quality observation systems in the United States. http://www.esrl.noaa.gov/csd/aqrs/reports/aqmonitoring.pdf. Accessed 13 Apr 2013

Dyer S, Grant J, Huot M, Lemphers N (2013) Beneath the surface—a review of key facts in the Oilsands debate. Pembina Institute http://www.pembina.org/pub/2404. Accessed April 2013

Environment Canada (2012) Canada-United States air quality agreement progress report 2012 http://www.ec.gc.ca/air/default.asp?lang=En-US&n=8ABC14B4-1&xml=8ABC14B4-ED53-4737-AD51-528F8DBA2B4C&offset=3&toc=show. Accessed 13 Apr 2013

Hess GD, Tory KJ, Cope ME, Lee S, Puri K, Manins PC, Young M (2004) The Australian air quality forecasting system. Part II: case study of Sydney 7-day photochemical smog event. J Appl Meteorol 43:663–679

International Air Quality Advisory Board (2010) Second expert consultation meeting—air quality issues related to the northern boundary region between the United States and Canada. http://ijc.org/php/publications/pdf/IAQAB_-_Whitehorse_Report_april-11.pdf. Accessed 30 Apr 2013

International Air Quality Advisory Board (2012) Report on air quality issues related to the northern boundary region between Canada and the United States. http://www.ijc.org/php/publications/pdf/IAQAB-Report-to-IJC-Northern-Air-Quality-April-2012.pdf. Accessed 30 Apr 2013

Mailhot J, Bélair S, Charron M, Doyle C, Joe P, Abrahamowicz M, Bernier NB, Denis B, Erfani A, Frenette R, Giguère A, Isaac GA, McLennan N, McTaggart-Cowan R, Milbrandt J, Tong L (2010) Environment Canada's experimental numerical weather prediction systems for the Vancouver 2010 Winter Olympic and Paralympic games (2010). Bull Am Meteorol Soc 91:1073–1085

McLinden CA, Fioletov V, Boersma KF, Krotkov N, Sioris CE, Veefkind JP, Yang K (2012) Air quality over the Canadian oil sands: a first assessment using satellite observations. Geophys Res Lett 39(4):L04804

Nunuvut General Monitoring Plan (2013) http://www.nunavutmining-symposium.ca/wp-content/uploads/2011/04/presentation-8-Nunavut-General-Monitoring-Plan.pdf. Accessed 30 Apr 2013

Storey H (2011) Catalytic clothing technology (2011) http://www.catalytic-clothing.org/home.html. Accessed 19 May 2013

White RM, Paprotny I, Doering F, Cascio WE, Solomon PA, Gundel LA (2012) Sensors and apps for community-based atmospheric monitoring. EM: air and waste management associations magazine for environmental managers. Air & Waste Management Association, Pittsburgh 5:36–40

Zhang Y, Bocquet M, Mallet V, Seifneur C, Baklanov A (2012) Real-time air quality forecasting, part II: state of the science, current research needs, and future prospects. Atmos Environ 60:656–676

Glossary and Abbreviations

µg/m³ micrograms (of contaminant) per cubic metre of air
14C carbon-14 or radiocarbon (radioactive isotope)
A&WMA Air and Waste Management Association
AAQC Ambient Air Quality Criteria (Ontario)
AAQO Ambient Air Quality Objectives
AAQS Analysis and Air Quality Section (Environment Canada)
Abatement The reduction or elimination of pollution.
Absorption The process by which incident radiant energy is retained by a substance by conversion to some other form of energy. In "visual air quality" it refers to visible light photons being "soaked up" by pollutants.
ACE Air Contaminant Emission (Model) or Aerosol Characterization Experiment
Acid rain Air pollution produced when acid chemicals are incorporated into rain, snow, fog or mist. The "acid" in acid rain comes from sulfur oxides and nitrogen oxides, products of burning coal and other fuels and from certain industrial processes. The sulfur oxides and nitrogen oxides are related to two strong acids: sulfuric acid and nitric acid. When sulfur dioxide and nitrogen oxides are released from power plants and other sources, winds blow them far from their source. If the acid chemicals in the air are blown into areas where the weather is wet, the acids can fall to Earth in the rain, snow, fog, or mist. In areas where the weather is dry, the acid chemicals may become incorporated into dusts or smokes. Acid rain can damage the environment, human health, and property.
Activity factor A measure of the activity that produces emission (e.g. kilograms of fuel burned per month)
Acute Exposure One or a series of short-term exposures generally lasting less than 24 h.
ADCL Acid-Deposition Critical Load
ADOM Acid Deposition and Oxidants Model
Aerosol Extremely small (less than 2–3 µm) particles of solid or liquid matter than can remain suspended in air from a few minutes to many months depending on the particle size and weight.
AES Atmospheric Environment Service
AGL Above Ground Level
Air Toxics Pollutants found in ambient air that is associated with adverse health effects. Generally excludes criteria air contaminants such as particulate matter and ozone. Used interchangeably with the term "Hazardous Air Pollutant".
Airshed In general terms, an airshed is defined as a geographic region in which dispersion of air emissions is limited by topographic constraints such as mountains and water bodies. When determining airshed boundaries, jurisdictional considerations are often taken into account as well e.g. municipal boundaries. See section on Alberta airsheds for more information.
ALNM AES Lagrangian Nitrogen Model
ALSM AES Lagrangian Sulphur Model
Alternative fuels Fuels that can replace ordinary gasoline or other common fossil fuels. Alternative fuels may have particularly desirable energy efficiency and pollution reduction features. Alternative fuels include compressed natural gas, alcohols, liquefied petroleum gas (LPG), and electricity. The 1990 US Clean Air Act encourages development and sale of alternative fuels.
Alveoli Small air sacs in the lungs of mammals, also called pulmonary alveoli. Alveoli are located at the ends of the air passageways in the lungs and have very thin (one cell thick), wet walls and are surrounded with a network of small blood vessels, or capillaries. This allows gases to diffuse across the surface of the alveolus.
AMAP Arctic Monitoring and Assessment Programme
AMP Airshed Management Plan. In British Columbia, this is one form of an Air Quality Management Plan which is the result of a collaborative multi stakeholder airshed planning process for a particular airshed. The result is a document that states goals for improving air quality and makes recommendations on how to achieve them.
AOT40 Accumulated dose of O_3 Over a Threshold of 40 ppb
AQ Air Quality
AQHI Air Quality Health Index. A Canadian index designed to help people understand the impact of air quality on health. The index value is calculated as the weighted sum of concentrations of NO_2, O_3 and $PM_{2.5}$, where the weights are concentration response functions relating each pollutant to mortality. This health protection tool can help people make personal decisions to reduce short-term exposure to air pollution by adjusting activity levels during increased levels of air pollution.

E. Taylor, A. McMillan (eds.), *Air Quality Management,* DOI 10.1007/978-94-007-7557-2,

AQI Air Quality Index. A numerical index used in the US for reporting severity of air pollution levels to the public. It replaces the formerly used Pollutant Standards Index (PSI). AQI incorporates five criteria pollutants—ozone, particulate matter, carbon monoxide, sulfur dioxide, and nitrogen dioxide—into a single index by comparing each pollutant to its standard, and basing the index value on the pollutant that is highest relative to its standard. AQI also incorporates the 8-hour ozone standard and the 24-hour $PM_{2.5}$ standard into the index calculation. AQI levels range from 0 (Good air quality) to 500 (Hazardous air quality). The higher the index, the higher the level of pollutants and the greater the likelihood of health effects.

AQMP Air Quality Management Plan. In British Columbia, this is a general term that details what needs to be done to improve air quality, the timelines, how it will be achieved, and who will do it. The AQMP comprises various methods that drive emissions reductions using regulatory instruments, economic incentives, and social marketing.

AQMS Air Quality Management System is a comprehensive approach for reducing air pollution in Canada now under development. It is the product of a collaboration by the federal, provincial and territorial governments and stakeholders. It will likely include new Canadian Ambient Air Quality Standards (CAAQS), Air zone air quality management and regional airshed coordination and Base-level industrial emissions requirements (BLIERs).

Area sources Those sources for which a methodology is used to estimate emissions. This can include area-wide, mobile and natural sources, and also groups of stationary sources (such as dry cleaners and gas stations). The federal air toxics program defines a source that emits less than 10 t per year of a single hazardous air pollutant (HAP) or 25 t per year of all HAPs as an area source.

ARM Approved Regional Methods (U.S.)

asl above mean sea level

atm atmosphere

Attainment area A geographic area in which levels of a criteria air pollutant meet the health-based primary standard (national ambient air quality standard, or NAAQS) for the pollutant. An area may have on acceptable level for one criteria air pollutant, but may have unacceptable levels for others. Thus, an area could be both attainment and non-attainment at the same time. Attainment areas are defined using federal pollutant limits set by EPA.

AURAMS "A Unified Regional Air-quality Modeling System" is an episodic, multi-pollutant, regional air-quality modelling system that predicts size-resolved and chemically-characterized particulate matter.

Back-Trajectory Analysis This is a useful tool for analyzing source regions for haze and other transport-related pollution phenomena. This approach involves using meteorological data to track the prior "path" of parcels of air arriving at a particular monitoring site over a period of hours or days.

BACT Best Available Control Technology, or for any specific source, the currently available technology producing the greatest reduction of air pollutant emissions, taking into account energy, environmental, economic and other costs.

BAM Beta Attenuation Monitor

BC British Columbia or black carbon

Bext The sum of the Rayleigh scattering coefficient (bRg) and the Mie scattering coefficient (bscat). Or, bext = bRg + bscat.

BLIERS Canadian "Base-level industrial emissions requirements"

BP Barometric Pressure: ambient atmospheric pressure.

Btu British thermal unit, a unit of energy: 1 Btu = 1060 J.

CAA The Clean Air Act (United States). Enacted in 1963 (Public Law 88–206) and subsequently amended (1967, 1970, 1977). A successor to the Air Pollution Control Act (1955).

CAAQS Canadian Ambient Air Quality Standards

CAC Criteria (or Common) Air Contaminants, including NO_x, CO, HC, PM and SO_x.

CAIR Clean Air Interstate Rule

CAPMoN Canadian Air and Precipitation Monitoring Network

Carbon dioxide (CO_2) A colorless, odorless gas that occurs naturally in the Earth's atmosphere. Significant quantities also are emitted into the air by fossil fuel combustion.

Carbon monoxide (CO) A colorless, odorless, poisonous gas, produced by incomplete burning of carbon-based fuels. including gasoline, oil, and wood. Carbon monoxide is also produced from incomplete combustion of many natural and synthetic products.

CARE Centre for Atmospheric Research, Egbert, Ontario, a rural site away from obvious urban influences

CAS registry number The Chemical Abstracts Service Registry Number (CAS) is a numeric designation assigned by the American Chemical Society's Chemical Abstract Service and uniquely identifies a specific compound. This entry allows one to conclusively identify a material regardless of the name or naming system used.

CCME Canadian Council of Ministers of the Environment

CDN Canadian

CEAA Canadian Environmental Assessment Act

CEM Continuous emission monitoring (at stationary sources)

CEP Carolina Environmental Program

CEPA Canadian Environmental Protection Act

CEPS Canadian Emissions Processing System

CFR Code of federal regulations (U.S.)

CH_4 Methane

Chlorofluorocarbons (CFCs) These chemicals and some related chemicals have been used in great quantities in industry, for refrigeration and air conditioning, and in consumer products. CFCs and their relatives, when released into the air, rise into the stratosphere, a layer of the atmosphere high above the Earth. In the stratosphere, CFCs and their relatives take part in chemical reactions which result in reduction of the stratospheric ozone layer, which protects the Earth's surface from harmful effects of radiation from the sun. The 1990 US Clean Air Act includes provisions for reducing releases (emissions) and eliminating production and use of these ozone-destroying chemicals.

Chronic exposure Long-term exposure, usually lasting one year to a lifetime

CHRONOS Canadian Hemispheric and Regional Ozone and NO_x System

CI/KCAC Continuous Improvement/Keeping Clean Areas Clean. A conservation concept expressed in the Canada-wide Standards.

Clean Air Act (CAA)—United States The original US Clean Air Act was passed in 1963, but our national air pollution control program is actually based on the 1970 version of the law. The 1990 US Clean Air Act Amendments are the most far-reaching revisions of the 1970 law. In this glossary, we refer to the 1990 amendments as the 1990 US Clean Air Act.

Clean fuels Low-pollution fuels that can replace ordinary gasoline. These are alternative fuels, including gasohol (gasoline-alcohol mixtures), natural gas, and LPG (liquefied petroleum gas)

CLRTP Convention for Long range Transport of air Pollutants

CMAQ "Community Multiscale Air Quality Modeling system" is capable of simulating regional through urban patterns of ozone and photochemical oxidants, fine and coarse particulate matter, visibility, and acid deposition. It was developed by the US EPA and is driven by the MM5 meteorological model and the Sparse Matrix Operator Kernel Emissions (SMOKE) model.

CO Carbon Monoxide

CO_2 Carbon Dioxide

CO_2 equivalent A quantity that describes, for a given mixture and amount of greenhouse gas, the amount of CO_2 that would have the same global warming potential (GWP), when measured over a specified timescale (generally, 100 years).

Code of federal regulations (CFR) The body of regulations written by federal government agencies, such as the EPA. Regulations relating to air pollution are found in Title 40 of the CFR, sections 50–99.

Combustion Burning. Many important pollutants, such as sulfur dioxide, nitrogen oxides, and particulates (PM-10) are combustion products, often products of the burning of fuels such as coal, oil, gas, and wood.

Compliance Advisory Panel (CAP) A seven member panel appointed by the governor state legislature and the Air permitting agency. The panel is composed of 4 small business owners, one permitting representative, and two people to represent the public at large. The CAP provides oversight to the Small Business Assistance Program (SBAP).

Continuous emission monitoring systems (CEMS) Machines which measure, on a continuous basis, pollutants released by a source. The 1990 US Clean Air Act requires continuous emission monitoring systems for certain large sources.

Continuous measurement Measurement of pollutants over an uninterrupted period of time

Control factor The fractional emission reduction in that source that is achieved by an add-on control device (e.g. installation of a particulate collection device on a combustion source). For example, if the add-on device reduces emissions by 5%, the control factor would be 0.05. The control factor is an optional term since in many cases it is included in the emission factor. (NARSTO 2005).

Control technology; control measures Equipment, processes, or actions used to reduce air pollution. The extent of pollution reduction varies among technologies and measures. In general, control technologies and measures that do the best job of reducing pollution will be required in the areas with the worst pollution. For example, the best available control technology/best available control measures (BACT, BACM) will be required in serious non-attainment areas for particulates, a criteria air pollutant. A similar high level of pollution reduction will be achieved with maximum achievable control technology (MACT) which will be required for sources releasing hazardous air pollutants.

Criteria air pollutants A group of common air pollutants regulated by EPA or other jurisdiction on the basis of criteria (information on health and/or environmental effects of pollution). Criteria air pollutants are widely distributed all over the country. A National Ambient Air Quality Standard exists for each criteria pollutant (particulate matter, sulfur dioxide, nitrogen dioxide, ozone, carbon dioxide, and lead).

Critical loads and levels A quantitative estimate of ecosystem exposure to one or more pollutants below which significant harmful effects on specified sensitive elements of the environment do not occur, according to present knowledge.

Crowd Sourcing A process whereby problems are broadcast to an unknown group of solvers in the form of an open call for solutions. Users—also known as 'the crowd'—submit solutions. Solutions are then owned by the entity that broadcast the problem in the first place—the crowdsourcer. The contributor of the solution is, in some cases, compensated either monetarily, with prizes, or with recognition. In other cases, the only rewards may be kudos or intellectual satisfaction. Crowdsourcing may produce solutions from amateurs or volunteers working in their spare time, or from experts or small businesses which were unknown to the initiating organization.

CTBT Comprehensive Nuclear Test Ban Treaty

CTG Control Techniques Guidelines issued by the EPA on ways to reduce pollutant emissions from stationary sources.

CTM chemical transport model

CWS Canada-wide Standards (e.g. for Ozone and $PM_{2.5}$)

d day

De Minimis In general, a level of emissions, etc., below which a particular process or activity is exempted or not regulated.

Deciview (Dv) A deciview (dv) is a visibility index, the scale of which is linear to humanly-perceived changes in visual air quality. A one dv change is approximately a 10 % change in the extinction coefficient, which is a small, but usually perceptible scenic change. Expressed in terms of extinction coefficient (bext) and visual range (vr) it is defined: $dv = 10 \ln(bext/0.01\ km^{-1}) = 10 \ln(391\ km/vr)$.

dichotomous samplers A dual-filter air sampler for the simultaneous collection of the fine $PM_{2.5}$ and the coarse $PM_{10-2.5}$ particles contained within PM_{10}.

Diesel engine A type of internal combustion engine that uses low-volatility petroleum fuel and fuel injectors and initiates combustion using compression ignition (as opposed to spark ignition that is used with gasoline engines).

Dose The amount of a pollutant that is absorbed. A level of exposure that is a function of a pollutant's concentration, the length of time a subject is exposed, and the amount of the pollutant that is absorbed. The concentration of the pollutant and the length of time that the subject is exposed to that pollutant determine dose.

DQOs Data Quality Objectives

Dry deposition The settling of particles on the earth's surface due to gravity or interception by objects.

Dust Solid particulate matter that can become airborne.

EC Environment Canada

EDXRF Energy Dispersive X-Ray Fluorescence

ELA Experimental Lakes Area

Emission Release of pollutants into the air from a source. We say sources emit pollutants. Continuous emission monitoring systems (CEMS) continuously measure pollutant releases.

Emission factor The amount of pollutant emitted by an activity associated with the source (e.g. kilograms of nitrogen dioxide per unit of fuel burned).

Emission inventory An estimate of the amount of pollutants emitted into the atmosphere from major mobile, stationary, area-wide, and natural source categories over a specific period of time such as a day or a year.

Emission rate The amount of emissions per unit time (e.g. kilograms of nitrogen dioxide per month)

Enforcement The legal methods used to make polluters obey the US Clean Air Act. Enforcement methods include citations of polluters for violations of the law (citations are much like traffic tickets), fines, and even jail terms. EPA and the state and local governments are responsible for enforcement of the US Clean Air Act, but if they don't enforce the law, members of the public can sue EPA or the states to get action.

EPA Environmental Protection Agency (U.S.)

Ethanol Ethyl-alcohol, a volatile alcohol containing two carbon groups (CH_3CH_2OH). For fuel use, ethanol is produced by fermentation of corn or other plant products.

ETSC Environmental Science and Technology Centre (Environment Canada)

Exceedance A measured level of an air pollutant higher than the national or state ambient air quality.

Exposure The concentration of the pollutant in the air multiplied by the population exposed to that concentration over a specified time period.

Extinction (Mm^{-1}) Atmospheric extinction, given in units of inverse megameters (Mm^{-1}). bext, p is an indicator of how much light is removed from a sight path by particle scattering and absorption. Higher values mean poorer visibility. Bext values should include additional clear air scattering of ~10 Mm^{-1}.

FDMS Filter Dynamic Measurement System

FEM Federal Equivalent Method (U.S.)

Flash Mob A group of people who assemble suddenly in a public place, perform an unusual and seemingly pointless act for a brief time, then disperse, often for the purposes of entertainment, satire, and artistic expression. Flash mobs are organized via telecommunications, social media, or viral emails.

Fly ash Air-borne solid particles that result from the burning of coal and other solid fuel.

Formaldehyde A chemical compound, the simplest aldehyde, chemical symbol CH_2O. Formaldehyde is a common pollutant, a VOC.

Fossil fuels Fuels such as coal, oil, and natural gas; so-called because they are the remains of ancient plant and animal life.

FR Federal Registrar (U.S.)

FRM Federal Reference Method (U.S.)

Fugitive dust Dust particles that are introduced into the air through certain activities such as soil cultivation, or vehicles operating on open fields or dirt roadways.

Fugitive emissions Emissions not caught by a capture system that are often due to equipment leaks, evaporative processes, and windblown disturbances.

g gram

g/s grams per second

GDAD Guidance Document on Achievement Determination (CWS)

GEM Global Multiscale Meteorological model

GEM-AQ Global Multiscale Meteorological model—Air Quality

GEM-MACH Global Multiscale Meteorological model—Modelling Air quality and Chemistry

Global warming An increase in the temperature of the Earth's troposphere. Global warming has occurred in the past as a result of natural influences, but the term is most often used to refer to the warming predicted by computer models to occur as a result of increased emissions of greenhouse gases.

Gothenburg Protocol The Protocol sets emission ceilings for 2010 for four pollutants: sulphur, NO_x, VOCs and ammonia. These ceilings were negotiated on the basis of scientific assessments of pollution effects and abatement options. Parties whose emissions have a more severe environmental or health impact and whose emissions are relatively inexpensive to reduce will have to make the biggest emission reductions. Once the Protocol is fully implemented, Europe's sulphur emissions should shrink by at least 63 %, its NO_x emissions by 41 %, its VOC emissions by 40 % and its ammonia emissions by 17 % compared to 1990.

GRAHM Global-Regional Atmospheric Heavy Metal model

Greenhouse effect The warming effect of the Earth's atmosphere. Energy from the sun that passes through the Earth's atmosphere is absorbed by the Earth's surface and re-radiated into the atmosphere as heat energy. The heat energy is then trapped by the atmosphere, creating a situation similar to that which occurs in a greenhouse.

Greenhouse gases Atmospheric gases such as carbon dioxide, methane, chlorofluorocarbons, nitrous oxide, ozone, and water vapor that absorb re-radiated heat through the Earth's atmosphere.

GWP Global Warming Potential. A relative measure of how much heat a greenhouse gas traps in the atmosphere.

h hour

H_2S hydrogen sulphide

HAMN Hamilton Air Monitoring Network

HAPs Hazardous Air Pollutant (HAP)—Pollutant found in ambient air that is associated with adverse health effects. Generally excludes criteria air contaminants such as particulate matter and ozone. Used interchangeably with the term "Air Toxic".

HCFC Hydrochlorofluorocarbon. A chemical compound that would be a hydrocarbon except that one or more hydrogen atoms in each molecule is replaced by a chlorine atom and one or more hydrogen atoms is replaced by a fluorine atom. Some HCFCs are implicated in the destruction of stratospheric ozone.

HDDV Heavy-Duty Diesel Vehicle

HNO_3 Nitric acid

HTAP Hemispheric Transport of Air Pollutants

Hydrocarbons Compounds containing various combinations of hydrogen and carbon atoms. They may be emitted into the air by natural sources (e.g., trees) and as a result of fossil and vegetative fuel combustion, fuel volatilization, and solvent use. Hydrocarbons are a major contributor to smog.

I/M Inspection and maintenance

IAQ Indoor Air Quality

IJC International Joint Commission

IMPROVE Interagency Monitoring of Protected Visual Environments (U.S.)

Indirect source control program Rules, regulations, local ordinances and land use controls, and other regulatory strategies of air pollution control districts or local governments used to control or reduce emissions associated with new and existing indirect sources. Indirect source control programs include regulatory strategies such as transportation control measures; parking charges; land use controls that reduce the need for vehicle travel and increase transit, bicycle, and pedestrian access; and source-specific regulations such as truck idling and travel schedule requirements.

Indoor air pollution Air pollutants that occur within buildings or other enclosed spaces, as opposed to those occurring in outdoor, or ambient air. Some examples of indoor air pollutants are nitrogen oxides, smoke, asbestos, formaldehyde, and carbon monoxide.

Inhalable particles All particles capable of entering the human respiratory tract.

Inspection and maintenance program (I/M program) Auto inspection programs are required for some polluted areas. These periodic inspections, usually done once a year or once every two years, check whether a car is being maintained to keep pollution down and whether emission control systems are working properly. Vehicles which do not pass inspection must be repaired. Under the 1990 US Clean Air Act, some especially polluted areas will have to have enhanced inspection and maintenance programs, using special machines that can check for such things as how much pollution a car produces during actual driving conditions.

Internal combustion engine An engine in which both the heat energy and the ensuing mechanical energy are produced inside the engine. Includes gas turbines, spark ignition gas, and compression ignition diesel engines.

Interstate air pollution In many areas, two or more states share the same air. We say these states are in the same air basin defined by geography and wind patterns. Often, air pollution moves out of the state in which it is produced into another state. Some pollutants, such as the power plant combustion products that cause acid rain, may travel over several states before affecting health, the environment, and property. The 1990 US Clean Air Act includes many provisions, such as interstate compacts, to help states work together to protect the air they share. Reducing interstate air pollution is very important since many Americans live and work in areas where more than one state is part of a single metropolitan area.

Inversion A layer of warm air in the atmosphere that prevents the rise of cooling air and traps pollutants beneath it.

Irritant A substance that causes irritation of the skin, eyes, or respiratory system. Effects may be acute from a single high-level exposure, or chronic from repeated low-level exposures to such compounds as chlorine, nitrogen dioxide, and nitric acid.

K Temperature in Kelvin

KCAC/CI Keeping Clean Areas Clean – Continuous Improvement (CWS)

km kilometre

km/hr kilometre per hour

Kpa kilopascal

kPa(g) Pressure in units of thousands of pascals above 101.3 kPa

Kyoto Protocol An international agreement adopted in December 1997 in Kyoto, Japan. The Protocol sets binding emission targets for developed countries that would reduce the emissions on average 5.2% below 1990 levels.

L Litre

Light extinction Light extinction (bext)

Local emergency planning committee (LEPC) Established under the Superfund Reauthorization Amendments, represent local governments, emergency response officials, environmental and citizen groups, industry, and other interested parties is established in each planning district.

Lowest achievable emission rate (LAER) Considered to be the lowest rate of emissions from a source category which is contained in the State Implementation Plan, or which is achieve in practice by such category of sources. This term is most often associated with a nonattainment area.

LRT long-range transport

LRTAP Long Range Transport of Air Pollutants

LST local standard time

m metre

m/sec metres/second

Major source Under the PSD regulations it is a facility, belonging to one or more of 28 source categories, having the potential to emit 100 t per year of a pollutant regulated under the federal US Clean Air Act (CAA). For categories other than the 28 sources, the potential emission level can not exceed 250 t per year. A major source for the purpose of Title V in the CAA is a stationary source that has the potential to emit 100 t per year a pollutant regulated under the CAA and/or a source that has the potential to emit 10 t per year for a single hazardous air pollutant or 25 t per year of a combination of all hazardous air pollutants.

Manual $PM_{2.5}$ measurement Manual $PM_{2.5}$ measurement

MARS Measurement and Analysis Research Section (Environment Canada)

Material safety data sheets (MSDSs) Product safety information sheets prepared by manufacturers and marketers of products containing toxic chemicals. These sheets can be obtained by requesting them from the manufacturer or marketer. Some stores, such as hardware stores, may have material safety data sheets on hand for products they sell.

Maximum achievable control technology (MACT) Federal emissions limitations based on the best demonstrated control technology or practices in similar sources to be applied to major sources emitting one or more federal hazardous air pollutants.

MeHg methyl-mercury

MeV million electron-volts

mg milligram

micro-(μ) The metric prefix for one millionth of the unit that follows.

microgram (μg) One millionth of a gram: $1\ \mu g = 10^{-6}\ g = 0.001\ mg$

micrograms per cubic meter (μg/m³) The mass in micrograms of a substance contained within a cubic meter of another substance or vacuum. This is the standard unit of measure for the mass density (concentration) of particles suspended in air; also sometimes used for the concentration of gases in air.

micrometer (μm), micron One millionth of a meter: $1\ \mu m = 10^{-6}\ m$

MM5 Fifth-Generation Penn State/NCAR Mesoscale Model

MnO_2 Manganese dioxide

Mobile sources Motor vehicles and other moving objects that release pollution; mobile sources include cars, trucks, buses, planes, trains, motorcycles and gasoline-powered lawn mowers. Mobile sources are divided into two groups: road vehicles, which includes cars, trucks, and buses, and non-road vehicles, which includes trains, planes, and lawn mowers.

MOE Ministry of Environment

Monitoring (monitor) Measurement of air pollution is referred to as monitoring. EPA, state, and local agencies measure the types and amounts of pollutants in community air. The 1990 US Clean Air Act requires states to monitor community air in polluted areas to determine if the areas are being cleaned up according to schedules set by law.

MRWG Monitoring and Reporting Working Group (CCME)

MSC Meteorological Service of Canada

Multimedia exposure Exposure to a toxic substance from multiple pathways such as air, water, soil, food, and breast milk.

N_2O Nitrous oxide

NAAQO National Ambient Air Quality Objectives (Canada)

NAAQS National Ambient Air Quality Standards (Untied States)

NAICS (North American Industry Classification System) A system developed jointly by the U.S., Canada, and Mexico, for classifying industries and other businesses. NAICS replaced SIC. They system assigns two-digit numbers to broad categories and adds more digits for increasingly finer classes within those categories.

NAPAP National Acid Precipitation Assessment Program (US)

NAPS National Air Pollution Surveillance

NAPS RM NAPS reference method

NARSTO North American Research Strategy on Tropospheric Ozone

National ambient air quality standards (NAAQS) Ambient standards developed by EPA that must be attained and maintained to protect public health. "Secondary" NAAQS are necessary to protect the public welfare. NAAQS exist for particular matter, sulfur dioxide, nitrogen dioxide, ozone, carbon dioxide, and lead.

National emission standards for hazardous air pollutants (NESHAPs) National emission standards for hazardous air pollutants.

National emissions inventory The new EPA database system for inventories of calculated emissions from point sources in the U.S. that incorporates data from the National Emissions Trends Inventory and National Toxics Inventory.

NAV CANADA A privately run, not-for-profit corporation that owns and operates Canada's civil air navigation system.

Near Field Communication (NFC) A set of standards for smartphones and similar devices to establish radio communication with each other by touching them together or bringing them into close proximity, usually no more than a few centimetres.

Near Field Communication (NFC) A set of standards for smartphones and similar devices to establish radio communication with each other by touching them together or bringing them into close proximity, usually no more than a few centimetres.

New source performance standards (NSPS) These are federal EPA emission standards for certain air pollutants that are emitted from new, modified, or reconstructed stationary emission sources which reflect the use of best available control technology.

New source review (NSR) New source review. NSR typically means any new source locating in a, e.g., ozone non-attainment area that will emit volatile organic compounds (VOCs) and/or oxide of nitrogen (NO) in certain amounts. These sources must: undergo a new source review that provides for offsetting emissions for any increases in the emissions of these two pollutants; use the lowest achievable emissions technology to control emissions; apply for a construction permit; and meet other state requirements before the new emission from the source can be permitted. Existing sources, located in the ozone nonattainment area, that emit these two pollutants and plan to change their operational methods that will cause an increase in the emissions of these two pollutants must apply for a modification permit and under go a review similar to a new source.

NH_3 Ammonia

NH_4NO_3 Ammonium nitrate

NIST National Institute of Standards and Technology (U.S.)

Nitrogen oxides (NO_x) A criteria air pollutant. Nitrogen oxides are produced from burning fuels, including gasoline and coal. Nitrogen oxides are smog formers, which react with volatile organic compounds to form smog. Nitrogen oxides are also major components of acid rain.

nm nanometre

NMHC NMHC

NO Nitric oxide or Nitrogen oxide

NO_2 Nitrogen Dioxide

NO_3– nitrate

Non-attainment area A geographic area in which a criteria air pollutant level is higher than allowed by the federal standards. A single geographic area may have an acceptable level for one criteria air pollutant, but have unacceptable levels of one or more other criteria air pollutants. Thus, an area can be both an attainment and non-attainment area at the same time. Sixty percent of Americans are estimated to live in non-attainment areas.

Non-point sources Diffuse pollution sources that are not recognized to have a single point of origin.

NO_X Oxides of nitrogen reported as NO_2

NOy Total reactive oxidized nitrogen species

NWP Numerical weather prediction

O_3 Ozone

°C Degrees Celsius

Octane number A numerical measure of the antiknock properties of gasoline used as a motor fuel. The higher the octane number, the greater the antiknock properties.

OECD Organisation for Economic Co-operation and Development

°F Degrees Fahrenheit

Offset A method used in the 1990 US Clean Air Act to give companies, which own or operate large (major) sources in non-attainment areas, flexibility in meeting overall pollution reduction requirements when changing production processes. If the owner or operator of the source wishes to increase releases of a criteria air pollutant, an offset (reduction of a somewhat greater amount of the same pollutant) must be obtained either at the same plant or by purchasing offsets from another company.

Opacity The amount of light obscured by particle pollution in the atmosphere. Opacity is used as an indicator of changes in performance of particulate control systems.

Organic compounds A large group of chemical compounds containing mainly carbon, hydrogen, nitrogen, and oxygen. All living organisms are made up of organic compounds.

Ozone A gas which is a variety of oxygen. The oxygen gas found in the air consists of two oxygen atoms stuck together; this is molecular oxygen. Ozone consists of three oxygen atoms stuck together into an ozone molecule. Ozone occurs in nature; it produces the sharp smell you notice near a lightning strike. High concentrations of ozone gas are found in a layer of the atmosphere—the stratosphere-high above the Earth. Stratospheric ozone shields the Earth against harmful rays from the sun, particularly ultraviolet B. Smog's main component is ozone; this ground-level ozone is a product of reactions among chemicals produced by burning coal, gasoline and other fuels, and chemicals found in products such as solvents, paints, and hair sprays.

Ozone depletion The reduction in the stratospheric ozone layer. Stratospheric ozone shields the Earth from ultraviolet radiation. The breakdown of certain chlorine and/or bromine-containing compounds that catalytically destroy ozone molecules in the stratosphere can cause a reduction in the ozone layer.

Pa Pascal

PAH Polycyclic Aromatic Hydrocarbon

PAMS Photochemical Assessment Monitoring Stations (U.S.)

PAN Peroxyacetyl Nitrate

Particulate matter (PM_{10}) Particulate matter is a criteria air pollutant and is a finely divided particle with an aerodynamic diameter of 10 μm or less. Particulate matter includes dust, soot and other tiny bits of solid materials that are released into and move around in the air. Particulates are produced by many sources, including burning of diesel fuels by trucks and buses, incineration of garbage, mixing and application of fertilizers and pesticides, road construction, industrial processes such as steel making, mining operations, agricultural burning (field and slash burning), and operation of fireplaces and wood stoves. Particulate pollution can cause eye, nose, and throat irritation and other more serious health problems.

Particulates: particulate matter ($PM_{2.5}$) Includes tiny particles with an aerodynamic diameter less than or equal to a nominal 2.5 μm. This fraction of particulate matter penetrates most deeply into the lungs.

Parts per billion (ppb)/parts per million (ppm) Units commonly used to express contamination ratios, as in establishing the maximum permissible amount of contaminant in water, land, or air.

PBMS Performance-Based Measurement System

PDF Adobe Acrobat Format of document

Peak levels A level of airborne pollutants that is much higher than average. They can occur over a short period of minutes or hours in response to sudden releases, or they can occur due to a longer term build-up over several days.

PEMA Pollution Emission Management Area (US)

Permit A document that resembles a license that is required by the US Clean Air Act for big (major) sources of air pollution, such as power plants, chemical factories and, in some cases, smaller polluters. Usually permits are issued by states, but if EPA has disapproved part or all of a state permit program, EPA will issue the permits in that state. The 1990 US Clean Air Act includes requirements for permit applications, including provisions for members of the public to participate in state and EPA reviews of permit applications. Permits contain information on all the regulated pollutants at a source. Permits include information on which pollutants are presently released, how much pollution the source is allowed to release, and the control measures necessary to meet pollutant release requirements. Permits are required both for the operation of plants (operating permits) and for the construction of new plants. The 1990 US Clean Air Act introduced a nationwide permit system for air pollution control.

Permit fees Fees paid by businesses required to have a permit. Permit fees are like the fees drivers pay to register their cars. Money from permit fees helps pay for state air pollution control activities.

pH The pH scale is used to measure acidity. The scale ranges from 0 to 14 where seven is the normal healthy stream pH, and normal precipitation is slightly more acidic (pH about 5.3). For reference, some common pH levels are battery acid at about 1, lemon juice at 2, and vinegar at 3. At the other end of the scale, lye has a pH of about 13 and ammonia about 12.

Plume A visible or measurable discharge of a contaminant from a given point of origin that can be measured according to the Ringelmann scale.

Plume Blight The US EPA defines plume blight as "smoke, dust, colored gas plumes, or layered haze emitted from stacks… relatable to a single source or small group of sources".

PM Particulate Matter

$PM_{1.0}$ Airborne particles with aerodynamic diameter smaller than 1 μm

PM_{10} Particulate Matter with an aerodynamic diameter less than or equal to 10 μm

$PM_{10–2.5}$ Particulate matter in the size range 2.5–10 μm in diameter (aka PM coarse fraction)

$PM_{2.5}$ Respirable particulate matter; airborne particles with aerodynamic diameter smaller than 2.5 μm (aka PM fine fraction).

Pollutants (pollution) Unwanted chemicals or other materials found in the air. Pollutants can harm health, the environment and property. Many air pollutants occur as gases or vapors, but some are very tiny solid particles: dust, smoke, or soot.

POP Persistent Organic Pollutant

ppb parts per billion

ppm parts per million

ppmv parts per million by volume

ppmvd parts per million by volume, dry

Primary standard A pollution limit based on health effects. Primary standards are set for criteria air pollutants.

PSD The Prevention of Significant Deterioration (PSD) permitting program is a U.S. Clean Air Act permitting program for new and modified major sources of air pollution such as power plants, manufacturing facilities, and other facilities that emit air pollution. PSD applies to all pollutants that do not exceed the National Ambient Air Quality Standards (NAAQS) in an area. The NAAQS establish maximum pollution concentration levels to protect public health and welfare from harmful levels of pollutants. Pollutants covered by the NAAQS are nitrogen oxides, volatile organic compounds (which are precursors to ground-level ozone), sulfur dioxide, fine particulate, carbon monoxide, and lead.

PTE (Potential to Emit) Capacity (usually measured in tons per year) of a pollution source to release a particular pollutant or class of pollutants.

QA/QC Quality assurance/quality control. A system of procedures, checks, audits, and corrective actions to ensure that all research design and performance, environmental monitoring and sampling, and other technical and reporting activities achieve the program's desired data quality objectives (DQOs).

Reasonably available control technology (RACT) It is usually an emission limit set by a state air program and is the basis for emission rates used in their SIP. It usually applies to sources in attainment areas and in most cases is less stringent than the NSPS level of control.

Reciprocating internal combustion engine (RICE) An engine in which air and fuel are introduced into cylinders, compressed by pistons, and ignited by a spark plug or by compression. Combustion in the cylinders pushes the pistons sequentially, transferring energy to the crankshaft, causing it to rotate.

Reflection The process by which light changes direction when it strikes and rebounds from a surface or the boundary between two media.

Reformulated gasoline Specially refined gasoline with low levels of smog-forming volatile organic compounds (VOCs) and low levels of hazardous air pollutants. The 1990 US Clean Air Act requires sale of reformulated gasoline in the nine smoggiest areas. Some reformulated gasolines were sold in several smoggy areas before passage of the 1990 US Clean Air Act.

Regional Haze The US EPA defines regional haze as "widespread, regionally homogeneous haze from a multitude of sources".

RELAD REgional Lagrangian Acid Deposition model

Respirable particle Airborne particles with aerodynamic diameter smaller than 2.5 μm (aka PM fine fraction).

RH Relative humidity

Ringelmann chart A series of charts, numbered 0–5, that simulate various smoke densities by presenting different percentages of black. A Ringelmann No. 1 is equivalent to 20 % black; a Ringelmann No. 5 is 100 % black. They are used for measuring the opacity or equivalent obscuration of smoke arising from stacks and other sources by matching the actual effluent with the various numbers, or densities, indicated by the charts.

Risk assessment An evaluation of risk that estimates the relationship between exposure to a harmful substance and the likelihood that harm will result from that exposure.

RM Reference Method

SBA Small Business Administration

SBDC Small Business Development Center

SBEAP Small Business Environmental Assistance Program

Scattering The process by which small particles in the atmosphere deflect radiation from its path into different directions. Rayleigh scattering (bRg) is the scattering and absorption of visible light by the component gases of the atmosphere (nitrogen and oxygen mainly). Mie scattering (bscat) is the scattering of light by very fine airborne particles.

SCC Sharp Cut Cyclone

Scrubber An air pollution control device that uses a high energy liquid spray to remove aerosol and gaseous pollutants from an air stream. The gases are removed either by absorption or chemical reaction.

Secondary particulate Particles that usually form over several hours or days and attain aerodynamic diameters between 0.1 and 2.5 μm. Several of these particles, particularly those containing ammonium nitrate, are volatile and transfer mass between the gas and particle phase to maintain a chemical equilibrium.

Secondary standard A pollution limit based on environmental effects such as damage to property, plants, or visibility. Secondary standards are set for criteria air pollutants.

SHARP Synchronized Hybrid Ambient Real-time Particulate

SIC (Standard Industrial Classification) An old federal system of classifying industries and other businesses. They system assigned two-digit numbers to broad categories and 4-digit numbers beginning with those 2-digit numbers to finer classes within those categories. SIC has been replaced by NAICS, but SIC numbers are still in common use.

SLEA Sarnia Lambton Environmental Association

Smog A mixture of pollutants, principally ground-level ozone, produced by chemical reactions in the air involving smog-forming chemicals. A major portion of smog-formers comes from burning petroleum-based fuels such as gasoline. Other smog-formers, volatile organic compounds, are found in products such as paints and solvents. Smog can harm health, damage the environment and cause poor visibility. Major smog occurrences are often linked to heavy motor vehicle traffic, sunshine, high temperatures and calm winds, or temperature inversion (weather condition in which warm air is trapped close to the ground instead of rising).

SMOKE Sparse Matrix Operator Kernel Emissions processing system

SO_2 Sulphur Dioxide

Solar radiation Electromagnetic energy emitted by the sun with wavelengths ranging from approximately 200 to 2,000 nm or nanometers (10–9 m).

SOP Standard Operating Procedure

Source Any place or object from which pollutants are released. A source can be a power plant, factory, dry cleaning business, gas station, or a farm. Cars, trucks, and other motor vehicles are sources. Consumer products and machines used in industry can also be sources.

SSI Size Selective Inlet

Stakeholder Stakeholder—an organization, department, business or individual who has a vested interest in the issue, the solution, and the outcome.

State implementation plan (SIP) A detailed description of the programs a state will use to carry out its responsibilities under the US Clean Air Act. This includes such things as rules and regulations, plans to control ozone, and ambient air standards used by a state to reduce air pollution. The US Clean Air Act requires that EPA approve each state implementation plan.

Stationary source A place or object from which pollutants are released which stays in place. Stationary sources include power plants, gas stations, incinerators, and houses.

STP Standard Temperature and Pressure

Sulfur dioxide A criteria air pollutant. Sulfur dioxide is a gas produced by burning coal, most notably in power plants. Some industrial processes, such as production of paper and smelting of metals, produce sulfur dioxide. Sulfur dioxide is closely related to sulfuric acid, a strong acid. Sulfur dioxide plays an important role in the production of acid rain.

SVOC Semi-Volatile Organic Compounds

SVPM Semi-Volatile Particulate Matter

Synthetic minor permit A permit with practically, enforceable conditions issued to a facility. These conditions limit the amount of regulated pollutant emissions so the permitted amount of actual emissions from the facility is below potential major source emission levels. These major source thresholds are usually 100 t per year under Title V, 50 t per year VOC/NO in the nonattainment area for ozone; 100/250 t per year for P.D; 10 t per year for a single hazardous air pollutant; 25 t per year for a combination of all hazardous air pollutants; and 50 t per year for a new source review.

TEOM™ Tapered Element Oscillating Microbalance

Terrestrial radiation Electromagnetic energy emitted with wavelengths ranging from approximately 8–14 nm or nanometers (10–9 m).

Title V permit A US federal operating permit program adopted and implemented by the state. The basic program elements typically specify that major sources will submit an operating application to the specified state environmental regulatory agency according to a schedule. EPA and the affected states will review the permit issuance. The pubic also has an opportunity to comment on the permit, which is renewable every five years. Minor changes to the permit can be made without opening the permit for public participation.

Total suspended particulates Particles of solid and liquid matter suspended in air. TSP is collected on filtration media and analyzed by weight only. Particle sizes represented by the method are up to 100 μm in aerodynamic diameter.

Toxic release inventory (TRI) Database of toxic releases in the U.S. compiled from SARA Title II Section 313 reports.

Transmission The fraction of radiant energy that passes through a substance.

TSP Total Suspended Particulate

$\mu g/m^3$ microgram per cubic metre

ULSD Ultra-Low Sulphur Diesel

Ultraviolet B (UVB) A type of sunlight. The ozone in the stratosphere filters out ultraviolet B rays and keeps them from reaching the Earth. Ultraviolet B exposure has been associated with skin cancer, eye cataracts, and damage to the environment. Thinning of the ozone layer in the stratosphere results in increased amounts of ultraviolet B reaching the Earth.

UNECE United Nations Economic Commission for Europe

USEPA United States Environmental Protection Agency

UTC Universal Time Coordinated

Vapor The gaseous phase of liquids or solids at atmospheric temperature and pressure.

Vapor recovery nozzles Special gas pump nozzles that reduce the release of gasoline vapor into the air gas is pumped into car tanks. There are several types of vapor recovery nozzles. Therefore, nozzles may not look the same at all gas stations. The 1990 US Clean Air Act requires the installation of vapor recovery nozzles at gas stations in smoggy areas.

Vapor recovery systems Mechanical systems that collect and recover chemical vapors resulting from transfer of gasoline from operations such as tank-to-truck systems at refineries, tanker-to-pipeline systems at offshore oil operations, and pump-to-vehicle systems at gasoline stations.

Variance Permission granted for a limited time (under stated conditions) for a person or company to operate outside the limits prescribed in a regulation.

Visibility See visual air quality

Visible light Electromagnetic energy with a wavelength between approximately 400 to 700 nm or nanometers (10^{-9} m). That which is visible to human eyes.

Visual air quality The degree with which visual range is degraded by atmospheric pollutants.

Visual Range The horizontal distance one can see through the atmosphere, the distance at which a given standard object can be seen with the unaided eye or a measurement of the ability to see and identify objects at different distances.

VOCs Volatile organic compounds. Organic chemicals all contain the element carbon (C). Organic chemicals are the basic chemicals found in living things and in products derived from living things, such as coal, petroleum, and refined petroleum products. Many of the organic chemicals we use do not occur in nature, but were synthesized by chemists in laboratories. Volatile chemicals readily produce vapors at room temperature and normal atmospheric pressure. Vapors escape easily from volatile liquid chemicals. Volatile organic chemicals include gasoline, industrial chemicals such as benzene, solvents such as toluene and xylene, and tetrachloroethylene (perchloroethylene, the principal dry cleaning solvent). Many volatile organic chemicals, such as benzene, are also hazardous air pollutants.

VSCC™ Very Sharp Cut Cyclone

Wet deposition The deposit on the earth's surface of airborne particles through their scavenging by rain, drizzle and snow.

WRF Weather Research and Forecasting model

Zero Air Pure air, used for calibrating air monitoring instruments. The EPA requires zero air to have less than 0.1 ppm of hydrocarbons.

Index

E. Taylor, A. McMillan (eds.), *Air Quality Management*, DOI 10.1007/978-94-007-7557-2,

Printed by Printforce, the Netherlands